FLUID MECHANICS

FLUID MECHANICS

DR. SADHU SINGH
Professor & Head, Mechanical Engineering Department (Former)
&
Dean, Faculty of Engineering & Technology (Former)
Govind Ballabh Path University of Agriculture & Technology,
Pantnagar – 263145 (Uttarakhand)
and
(Director Colleges), Punjab Technical University, Jalandhar (Former)

KHANNA BOOK PUBLISHING CO. (P) LTD.
Publisher of Engineering and Computer Books
4C/4344, Ansari Road, Darya Ganj, New Delhi-110002
Phone : 011-23244447-48 **Mobile:** +91-9910909320
E-mail : contact@khannabooks.com
Website : www.khannabooks.com

Dear Readers,

To prevent the piracy, this book is secured with HIGH SECURITY HOLOGRAM on the front title cover. In case you don't find the hologram on the front cover title, please write us to at contact@khannabooks.com or whatsapp us at +91-99109 09320 and avail special gift voucher for yourself.

Specimen of Hologram on front Cover title:

Moreover, there is a SPECIAL DISCOUNT COUPON for you with EVERY HOLOGRAM.

How to avail this SPECIAL DISCOUNT:

Step 1: Scratch the hologram
Step 2: Under the scratch area, your "coupon code" is available
Step 3: Logon to www.khannabooks.com
Step 4: Use your "coupon code" in the shopping cart and get your copy at a special discount
Step 5: Enjoy your reading!

Price: ₹ 495.00

ISBN: 978-93-81068-65-6

Fluid Mechanics
by Sadhu Singh

Edition: 2025

To view complete list of books, Please scan the QR Code:

Published by:
Khanna Book Publishing Co. (P) Ltd.
CIN: U22110DL1998PTC095547

Visit us at: **www.khannabooks.com**
Write us at: **contact@khannabooks.com**

Printed in Bharat.

PREFACE

Fluid Mechanics is a Text Book for B.E./B.Tech. students of all Indian Universities and Institutions. The book contains fourteen chapters comprising of Fluid Properties; Fluid Statics; Fluid Kinematics; Fluid Dynamics; Compressible Fluid Flow; Viscous Flow; Flow through Pipes; Boundary Layer Theory; Turbulent Flow; Fluid Measurements; Dimensional and Model Analysis; Notches and Weirs; Flow through Open Channels; and Potential Flow.

The book contains a large number of solved and unsolved problems. The special features of the book are : Summary, Short answer type questions and end of chapters Exercises. At the end of the book multi-choice questions from ESE/CSE and GATE have been given with explanatory notes. This shall be quite useful to the students.

The help provided by Mr. Mukul Seth, and Puneet Khanna during the preparation of the manuscript and otherwise is highly appreciated and acknowledged.

My grand children, Kanupreet Kaur and Amitoej Singh have been a source of great inspiration during the preparation of the manuscript. They kept my mood cheerful, which helped in improving the text. Mrs. Narinderpal Kaur, my daughter-in-law, took proper care during writing of the manuscript and otherwise. I have all appreciations for them.

It is hoped that the book in its present form shall be quite useful to the students and teachers. Further suggestions for the improvement of the book are invited by the publishers and the author.

July, 2012

Dr. Sadhu Singh

CONTENTS

1
FLUID PROPERTIES

1.1 BASIC DEFINITIONS

Hydraulics: It deals with the study of water at rest or in motion which is based on experimental observations.

Fluid: It is a substance which can flow. It can be sub-divided into a liquid and gas. The intermolecular bonds are very weak in a gas and weak in a liquid. The liquids take the shape of its container and a gas occupies whole of the space of the vessel containing it.

Fluid Mechanics: It deals with the study of behaviour of fluids at rest or in motion. The fluid mechanics may be further classified as: Hydrostatics, kinematics and dynamics.

Hydrostatics: It deals with the study of incompressible fluids at rest.

Kinematics: It deals with the translation, rotation and deformation of fluid elements without considering the forces and energy causing such motion.

Dynamics : It deals with the relationship between the velocities and accelerations of a fluid caused by forces or energy acting on it.

1.2 TYPES OF FLUIDS

A fluid may be defined as a substance which is capable of flowing, and deforms continuously under applied shearing force. The characteristic of a fluid are the following:

1. A fluid has no definite shape of its own and takes the shape of the vessel containing it.
2. The fluids undergo deformation when a shear force is applied on them. The deformation continues as long as the shear force remains.

The fluids may be classified as follows:

1. Based upon form:

   (i) Liquid (ii) Gas, and (iii) Vapour

2. Based upon fluid properties:
   (i) Ideal fluids, and (ii) Real fluids

Liquid: A liquid is a fluid which possesses a definite volume. The effect of temperature and pressure on volume is not significant. The liquids are incompressible for all practical purposes. They have bulk elastic modulus when subjected to compression. They also possess viscosity. Water, kerosene, petrol, milk, etc. are liquids.

Gas: A gas has no fixed volume for given mass. It is compressible. It readily responds to changes in temperature and pressure. Air, ammonia, carbon dioxide, etc. are gases.

Vapour: It is a gas whose temperature and pressure are very near the liquid state. Steam is an example of water vapour.

Ideal Fluids: An ideal fluid is one which has no viscosity (inviscid) and surface tension and is incompressible, i.e. bulk modulus is infinity. Ideal fluids do not exist in nature. However, some common fluids such as air and water have very low values of viscosity and can be treated as ideal fluids for all practical purposes.

Real Fluids: A real fluid possesses viscosity, surface tension, compressibility and density. The real fluids are actually available in nature.

Continuum: A continuum may be defined as a continuous and homogeneous fluid medium. From the continuum view point, the overall properties and behaviour of fluids can be studied without regard to their atomic and molecular structure.

1.3 FLUID PROPERTIES

The important properties of fluids are:

1. Density
2. Specific weight
3. Specific gravity
4. Viscosity
5. Vapour pressure
6. Cohesion
7. Adhesion
8. Surface tension
9. Capillarity, and
10. Compressibility

1.4 DENSITY

The density (or mass density) of a fluid is defined as the mass per unit volume at a standard temperature and pressure. It is denoted by the symbol ρ (Greek letter rho). It has the dimension $[ML^{-3}]$, kg/m^3.

$$\rho = \lim_{\Delta V \to 0} \frac{\Delta m}{\Delta V} = \frac{m}{V} = \frac{\text{mass}}{\text{volume}} \quad \text{... (1.1)}$$

where Δm = mass contained in a small volume ΔV.

Density of pure water is 1000 kg/m^3.

1.5 SPECIFIC WEIGHT

The specific weight (or weight density) is defined as the weight of a fluid per unit volume. It is denoted by the symbol γ (Greek letter gamma). Its units are N/m^3.

$$\gamma = \rho g \qquad \text{... (1.2)}$$

The specific weight of fresh water is 9.81×10^3 N/m^3.

$$\text{Specific volume} = \frac{V}{m} = \frac{1}{\rho} \qquad \text{... (1.3)}$$

1.6 SPECIFIC GRAVITY

It is defined as the ratio of the density of a fluid to the density of pure water at 4°C. It is denoted by the symbol.

$$s = \frac{\rho_{\text{fluid}}}{\rho_{\text{water}}} \qquad \text{... (1.4)}$$

1.7 VISCOSITY

The viscosity is the property by virtue of which a fluid offers resistance to deformation under the influence of a shear force. The fluid undergoes continuous deformation so long as the shear force remains in existence. The molecular friction or shear resistance within the fluid opposes such continuous deformation.

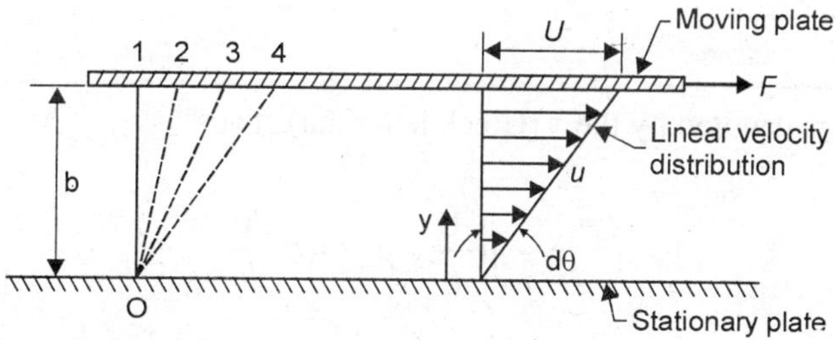

Fig. 1.1 Fluid motion between parallel plates

Consider a fluid contained in between two parallel plates as shown in Fig. 1.1 The bottom plate is being kept stationary and the top plate moves at a constant speed U under the action of applied shearing force F. As the top plate starts moving the fluid particles lying along the vertical line 0 –1 start moving with the same velocity *u* while those adhering to the bottom stationary plate are at rest. The velocity of the intermediate particles vary from 0 to U. For small gap between the plates, the velocity distribution will be linear. The fluid particles originally lying on line 0-1 after a short time interval Δt, will occupy the positions indicated by the line 0-2, and at time 2Δt by 0-3, and so on. The maximum deformation of fluid takes place at y = b, the gap between the plates.

$$\text{Time rate of deformation} = \frac{U}{b}$$

$$\text{Rate of angular deformation} = \frac{d\theta}{dt}$$

$$\therefore \qquad b\frac{d\theta}{dt} = U$$

or

$$y\frac{d\theta}{dt} = u$$

$$\therefore \qquad \frac{d\theta}{dt} = \frac{U}{b} = \frac{u}{y}$$

where u = velocity at a distance y from the stationary plate.

If A = area of the moving plate surface

F = force required to move the surface at a constant velocity

Then $\frac{F}{A} \propto \frac{d\theta}{dt}$, i.e. the shear stress $\frac{F}{A}$ is directly proportional to the time rate of deformation.

or

$$\frac{F}{A} = \mu\frac{d\theta}{dt} = \mu\frac{U}{b} = \mu\frac{u}{y}$$

where μ is the proportionality constant and is called the coefficient of viscosity or absolute or dynamic viscosity.

The shear stress $\frac{F}{A}$ is denoted by the τ (Greek letter tau). Thus

$$\tau = \mu\frac{U}{b} = \mu\frac{u}{y} = \mu\frac{d\theta}{dt} \qquad \text{... (1.5)}$$

Eq. (1.5) states that the rate of angular deformation is proportional to the shear stress.

Eq. (1.5) is differential form may be expressed as:

$$\tau = \mu\frac{du}{dy} \qquad \text{... (1.6)}$$

For a small gap between the plates, the velocity distribution is linear and $\frac{du}{dy}$ is a constant. But if the gap between the plates is large, the velocity distribution will be non-linear as shown in Fig. 1.2. The shear stress then will not be constant throughout the fluid.

In Eq. (1.6), $\frac{du}{dy}$ represents the rate of shear deformation or rate of shear strain and is often called the velocity gradient. It relates the shear stress with the viscosity. It is known as the Newton's law of viscosity or of fluid friction From Eq. (1.6), we have

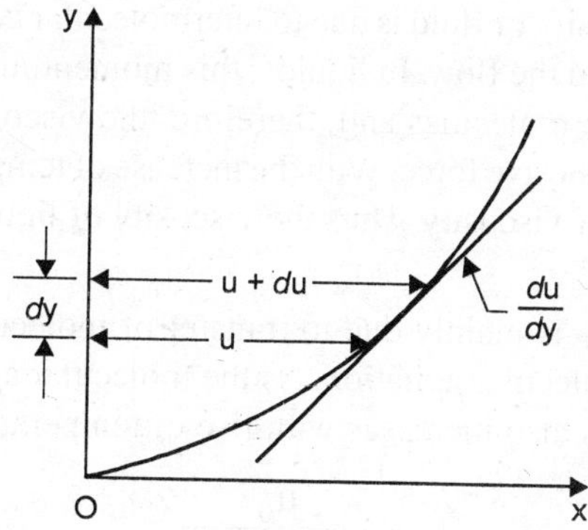

Fig. 1.2 Non-linear velocity distribution

$$\frac{du}{dy} = \frac{\tau}{\mu}$$

For a given shear stress,

$$\frac{du}{dy} \propto \frac{1}{\mu}$$

1.7.1 Units of Dynamic Viscosity (μ)

$$\mu = \frac{\tau}{\left(\frac{du}{dy}\right)} = \frac{\frac{\text{Shear force}}{\text{Area}}}{\frac{\text{Length}}{\text{Time}} \times \frac{1}{\text{Length}}} = \frac{\frac{\text{Shear force}}{(\text{Length})^2}}{\frac{1}{\text{Time}}} = \frac{\text{Shear force} \times \text{Time}}{(\text{Length})^2}$$

$$= \frac{\text{N} \times \text{s}}{\text{m}^2} = \frac{\text{N} \times \text{s}}{\text{m}^2} = \text{Pa.s}$$

1.7.2 Kinematic Viscosity (ν)

It is defined as the ratio of dynamic viscosity to the density of the fluid. It is denoted by ν (Greek letter nue).

$$\nu = \frac{\mu}{\rho} = \frac{\text{N.s}}{\text{m}^2} = \frac{\text{m}^3}{\text{kg}} = \frac{\text{kg} \times \text{m}}{\text{s}^2} = \frac{\text{s} \times \text{m}}{\text{kg}} = \frac{\text{m}^2}{\text{s}}$$

$$1 \text{ Stoke} = 1 \text{ cm}^2/\text{s} = 10^{-4} \text{ m}^2/\text{s}$$

1.7.4 Effect of Temperature on Viscosity of Fluids

The viscosity of fluid depends upon its intermolecular structure. On that account the viscosity of a gas is much different from that of a liquid. In gases, the molecules are widely spaced resulting in a negligible intermolecular cohesion, while in liquids the molecules being very close to each other, the cohesion is much larger. The viscosity of fluid is due to intermolecular cohesion and transfer of molecular momentum in a direction normal to the flow. In liquids, this momentum transfer is small as compared to the force of cohesion between the molecules and, therefore, the viscosity is primarily dependent upon the magnitude of intermolecular cohesive force. With the increase of temperature, cohesive force decreases rapidly resulting in the decrease of viscosity. Thus the viscosity of liquids decrease with the increase of temperature.

In case of gases, the viscosity is mainly due to transfer of molecular momentum in the transverse direction brought about by the molecular agitation. As the molecular agitation increases with the rise of temperature the viscosity of gases also increases with rise of temperature.

(i) For liquid,

$$\mu_t = \frac{\mu_0}{1 + \alpha t + \beta t^2} \qquad \text{... (1.7)}$$

where

μ_t = Viscosity of liquid at t°C

μ_0 = Viscosity of liquid at 0°C

α, β = constants for the liquid

For water,

μ_0 = 0.0179 Poise

α = 0.03368

β = 0.000 221

(ii) For gases,

$$\mu_t = \mu_0 + \alpha t - \beta t^2 \text{ Pa.s} \qquad \text{... (1.8)}$$

For air,

$\mu_0 = 1.7 \times 10^{-5}$ Pa.s

$\alpha = 0.56 \times 10^{-7}$

$\beta = 0.119 \times 10^{-9}$

1.7.5 Effect of Pressure on Viscosity of Liquid

The viscosity of liquids increases with increases in pressure. The relation between viscosity and pressure is given by,

$$\mu_p = \mu_0 \exp [A (p - p_0)] \qquad \text{... (1.9)}$$

where

μ_p = Viscosity of liquid at pressure p

μ_0 = Viscosity of liquid at initial pressure p_0

A = constant for a liquid.

1.8 VAPOUR PRESSURE

A change from liquid state to gaseous state is called evaporation. When evaporation of liquid takes place in a closed vessel the vapour molecules accumulate in the free space. These vapour molecules exert a partial pressure on the liquid surface. This partial pressure of vapour molecules is known as vapour pressure of liquid.

The vapour pressure increases with the increases in temperature as more liquid is evaporated at higher temperature, which will exert more pressure A liquid with a high vapour pressure evaporates readily and is known as a volatile liquid. The boiling of a liquid is closely related to its vapour pressure when the pressure impressed on the liquid surface is brought slightly below the vapour pressure limit, the liquid starts boiling. This is the reason that the liquids starts boiling at a lower temperature on the hills.

The low vapour pressure of mercury makes it very suitable for use in barometers and other pressure measuring devices. The problem of cavitation in hydromachinery such as turbines and pumps, is a result of local pressure being equivalent to or less than the vapour pressure of the liquid. When such a situation develops, vapour bubbles or cavities are formed in the flow. The unsteady nature of these bubbles and their ultimate collapse is responsible for the high pressure which often leads to vibration, noise, pitting and erosion of metal parts of machines.

1.9 COHESION AND ADHESION

The molecules of a liquid are held together by a force of attraction between them, which is known as *cohesion*. The force of attraction between the molecules of two different liquids or another body in contact is called adhesion.

When a liquid like mercury is spilled on a smooth horizontal surface, it gathers into droplets because the force of cohesion is more than the force of adhesion, as shown in Fig. 1.3 (a) Such liquids are called non-wetting liquids. But when water is poured on a horizontal surface, it spreads out and wet the horizontal surface because the force of adhesion is more than force of cohesion, as shown in Fig. 1.3 (b).

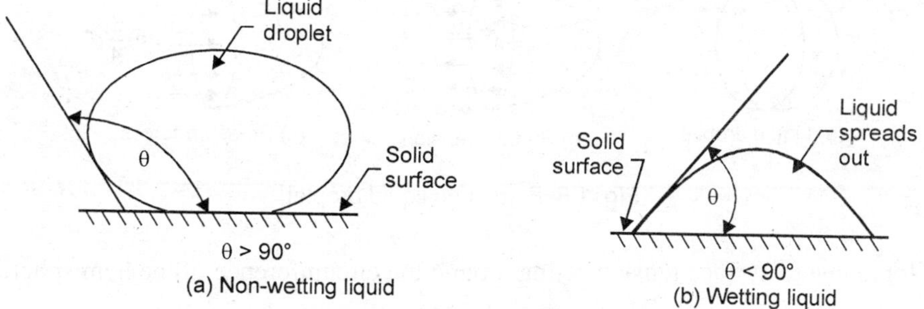

Fig. 1.3 Wetting and non-whiffing liquid

1.10 SURFACE TENSION

At the surface of contact between a liquid and a gas or between different immiscible liquids, molecular attraction introduces a force which causes the interface to behave like a membrane under tension. The magnitude of this force per unit length of the free surface is called surface tension. It is denoted by σ (Greek letter sigma) The surface tension is expressed in S.I. units as N/m.

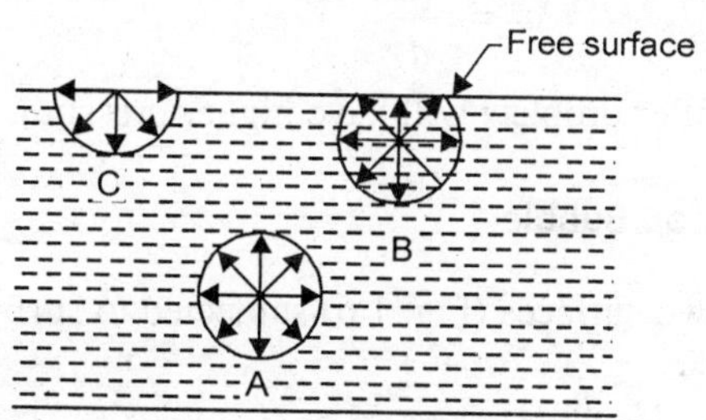

Fig. 1.4 Surface tension

The phenomenon of surface tension is explained in Fig. 1.4. Consider three molecules of a liquid in a container. The molecule A which is inside the liquid is attracted in all directions equally by the surrounding molecules. Thus the resultant force acting on the molecule A is zero. The molecule situated near the free surface is acted upon by a net resultant force in the downward direction. The molecule C, situated on the free surface of liquid is also subjected to a resultant downward force. Thus, all the molecules at the surface are subjected to net inward force of molecular attraction. Thus, the free surface of the liquid behaves like a very thin film under tension. The surface tension decreases with rise in temperature.

1.10.1 Surface Tension on Liquid Droplet

Consider a small spherical droplet of a liquid of diameter 'd' as shown in Fig. 1.5(a).

Let σ = surface tension of the liquid.

Δp = excess pressure in the droplet

Let the droplet be cut into two equal halves. The surface tension and pressure forces on the halves are shown in Fig. 1.5 (*b*) and (*c*) respectively.

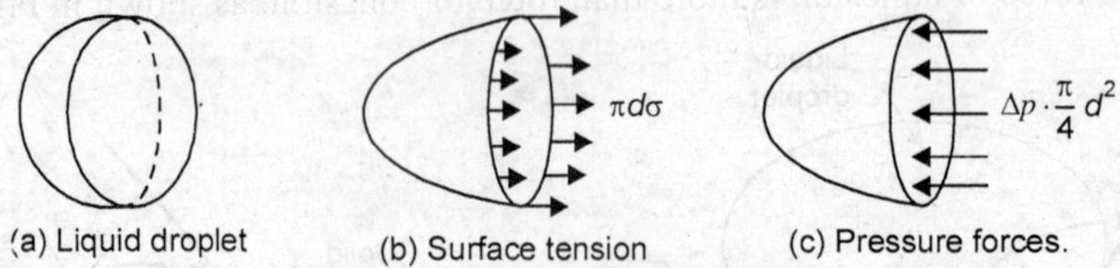

(a) Liquid droplet (b) Surface tension (c) Pressure forces.

Fig. 1.5 Forces on liquid droplet

Tensile force due to surface tension acting around the circumference of the hemisphere

$$= \pi d \times \sigma$$

$$\text{Pressure force} = \frac{\pi}{d} d^2 \times \Delta p$$

For equilibrium of the droplet,

$$\frac{\pi}{4} d^2 \times \Delta p = \pi d \times \sigma$$

$$\Delta p = \frac{4\sigma}{d} \qquad \text{... (1.10)}$$

Eq. (1.10) indicates that smaller droplets will have higher pressure.

1.10.2 Surface Tension on a Soap Bubble

In a soap bubble, there are two surfaces (Fig. 1.6) subjected to surface tension.

$$\therefore \quad \Delta p \times \frac{\pi}{4} d^2 = 2 \times \sigma \times \pi d$$

$$\Delta p = \frac{8\sigma}{d} \qquad \text{... (1.11)}$$

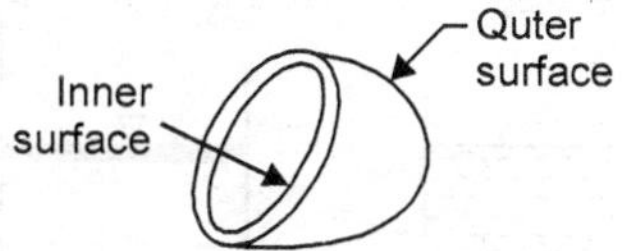

Fig. 4.6 Soap bubble

1.10.3 Surface Tension on a Liquid Jet

Consider a liquid jet of diameter d and length l as shown as shown in Fig. 1.7. Consider the equilibrium of semi-jet.

$$\text{Force due to pressure} = \Delta p \times ld$$

$$\text{Force due to surface tension} = \sigma \times 2l$$

For the equilibrium of jet,

$$\Delta p \times ld = \sigma \times 2l$$

$$\Delta p = \frac{2\sigma}{d} \qquad \text{... (1.12)}$$

(a) (b)

Fig. 1.7 Liquid jet

1.11 CAPILLARY ACTION

Capillary action or capillarity may be defined as a phenomenon of rise or fall of a liquid surface in a tube of small diameter when the tube is held vertically in the liquid. The rise of liquid surface is called capillary rise and the fall of liquid surface is called capillary depression.

The molecules of a solid surface attract liquid molecules with a greater force than that which exists between the liquid molecules (except mercury). Because of this, most of the liquids completely wet the surface. When the adhesion force is more than the cohesion force then if a glass tube is dipped vertically into water, the water rises in the tube, as shown in Fig. 1.8 (a). When the force of cohesion in more than the force of adhesion then if a glass tube is dipped in mercury, the surface of mercury inside the tube will be lower than the outside level, as shown in Fig. 1.8 (b)

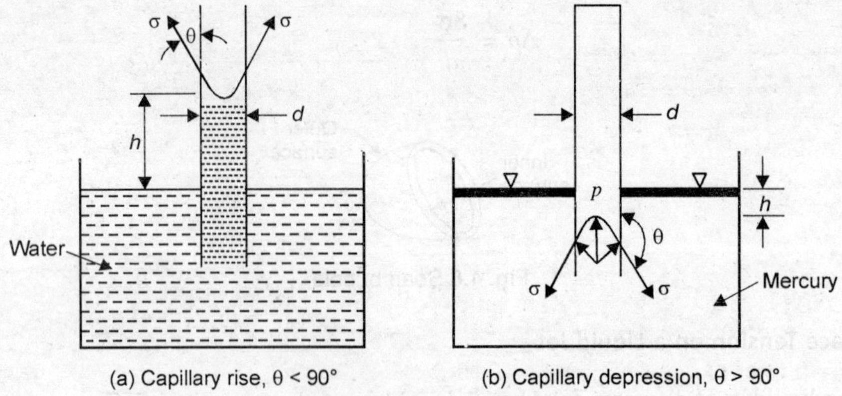

(a) Capillary rise, $\theta < 90°$ (b) Capillary depression, $\theta > 90°$

Fig. 1.8 Capillary action

Expression of capillary Rise

Let d = diameter of glass tube open at both ends.

h = height of the liquid in the tube

σ = surface tension of liquid

θ = angle of contact between liquid and glass tube.

Under the state of equilibrium, the weight of liquid of height h in the tube is balanced by the forces at the surface of the liquid in the tube due to surface tension.

Weight of liquid of height h in the tube = $\frac{\pi}{4} d^2 h\rho g$

where ρ = density of liquid.

Vertical component of the force due to surface tension = $\sigma \times \pi d \times \cos\theta$

$$\therefore \quad \frac{\pi}{d} d^2 h\rho g = \sigma \times \pi d \cos\theta$$

$$h = \frac{4\sigma\cos\theta}{\rho g d} = \frac{4\sigma\cos\theta}{\gamma d} \quad \text{... (1.13)}$$

$\theta = 0°$ between water and clean glass tube.

$$\therefore \quad h = \frac{4\sigma}{\gamma d} \quad \text{... (1.14)}$$

For water in contact with glass and air, $\theta = 25° - 32'$ and $\sigma = 0.0735$ N/m.

For h and d in mm,

$$h = \frac{27.07}{\alpha} \text{ mm} \quad \text{... (1.15)}$$

Expression for capillary Fall

When the glass tube is dipped in mercury, two forces act on the mercury inside the tube.

1. Force due to surface tension acting in the downward direction = $\sigma \times \pi d \cos\theta$

2. Force due to hydrostatic pressure acting upwards = $\frac{\pi}{4} d^2 p = \frac{\pi}{4} d^2 \times \gamma h$

For equilibrium of the two forces, we have

$$\frac{\pi}{4} d^2 \gamma h = \sigma \times \pi d \cos\theta$$

$$h = \frac{4\sigma\cos\theta}{\gamma d} \quad \text{... (1.16)}$$

which is the same equation as (1.13)

The value of θ for mercury and glass is 128°–52′, $\sigma = 0.52$ N/m and $s = 13.6$. For glass - air contact,

$$h = \frac{9.78}{d} \text{mm} \quad \text{... (1.17)}$$

1.12 COMPRESSIBILITY

The compressibility of a fluid is its ability to change its volume under pressure. Compressibility results in the change of fluid density due to change in volume by external pressure. This change in density is very small in liquids. For all practical purposes, liquids are considered incompressible.

Consider a cylinder fitted with a piston as shown in Fig. 1.9.

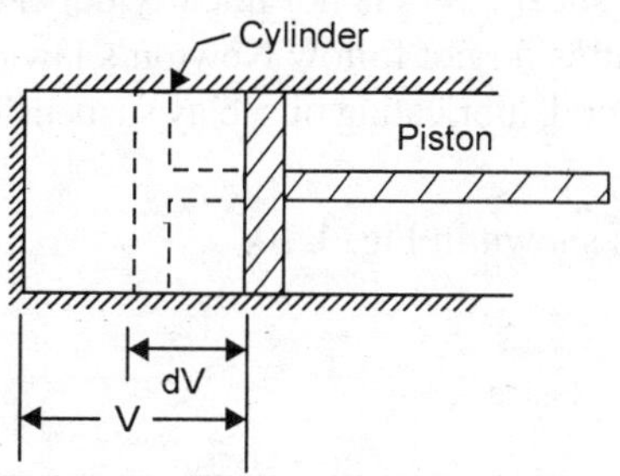

Fig. 1.9 Cylinder fitted with piston.

Let V = Volume of a gas enclosed in the cylinder.

p = pressure when volume is V.

Let the pressure be increased from p to $p + dp$, so that the volume decreases from V to V-dv.

$$\text{Increases in pressure} = dp$$

$$\text{Decrease in volume} = dV$$

$\therefore$ Volumetric strain, $\qquad \varepsilon_V = -\dfrac{dV}{V}$

Bulk modulus, $\qquad K = \dfrac{\text{Increase of pressure}}{\text{Volumetric strain}}$

$$= \frac{dp}{-\dfrac{dV}{V}} = -\frac{dp}{dV} \times V \qquad \text{... (1.18)}$$

Compressibility, $\qquad \beta = \dfrac{1}{K}$

For constant mass, $\qquad \dfrac{d\rho}{\rho} = -\dfrac{dV}{V}$

$$\therefore \qquad K = \rho \frac{dp}{d\rho} \qquad \text{... (1.19)}$$

1.13 NEWTONIAN AND NON-NEWTONIAN FLUIDS

A fluid which obeys Newton's law of viscosity $\left(\tau = \mu \dfrac{du}{dy}\right)$ is known as a Newtonian fluid. For a Newtonian fluid the viscosity is independent of the shear stress. Some common examples of Newtonian fluids are: air, water, light oils and gasoline.

For a non-Newtonian fluid the shear stress is not linearly dependant upon the velocity gradient. In other words, the non-Newtonian fluids do not follow Newton's law of viscosity. Common examples of non-Newtonian fluids are: human blood, lubricating oils, clay suspension in water, molten rubber, printer's ink, butter and sewage sludge.

The fluids may be classified as shown in Fig. 1.10.

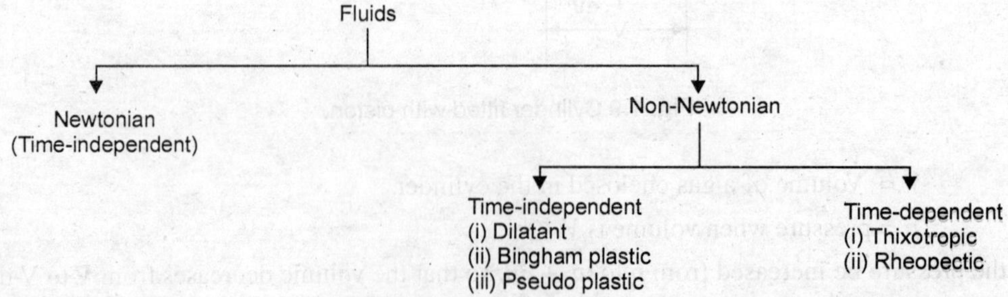

Fig. 1.10 Classification of fluids

A general relationship between shear stress and velocity gradient for a non-Newtonian fluid may be expressed as:

$$\tau = A\left(\frac{du}{dy}\right)^n + B \qquad \text{... (1.20)}$$

where A, B = constants which depend upon the type of fluid and conditions imposed on the flow. The fluids which obey Eq. (1.20) are called power-law fluids.

$B = 0$ for all fluids except Bingham plastic.

Based on the value of power index n in Eq. (1.20), the non-Newtonian fluids are classified as:

(i) Dilatant, if n > 1, e.g. quicksand, butter, printing inks.

(ii) Bingham (or ideal) plastic, if n = 1, e.g. sewage sludge, drilling muds.

(iii) Pseudoplastic, if n < 1, e.g. paper pulp, polymeric solutions such as rubber, suspension paints.

1. For a time-independent fluid the velocity gradient depends only upon the shear stress.
2. For a time-dependent fluid the rate of deformation and the viscosity depends upon both the shear stress and the duration of its application.

The shear or viscous characteristics of different fluids are illustrated in Fig. 1.11. Bingham plastic fluids require a certain minimum shear stress τ_y, called yield stress, before they start flowing.

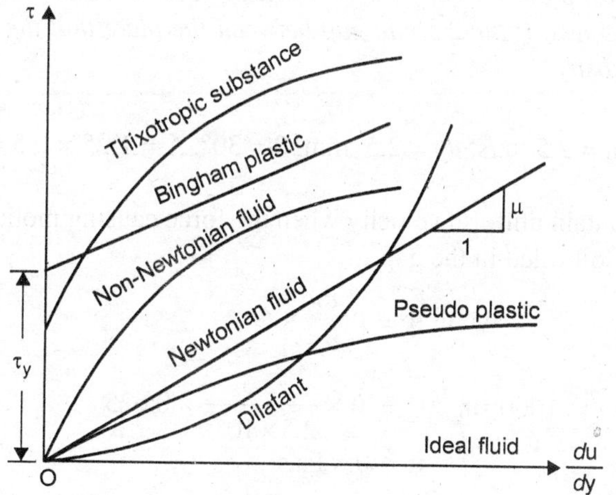

Fig. 1.11 τ V's $\frac{du}{dy}$.

Thixotropic fluids are those which show an increase in apparent viscosity with time. Lipstic and certain paints and enamels exhibit thixotropic behaviour when subjected to high shear by the brush during paint application, the apparent viscosity is reduced and brush marks disappear. Those fluids which show a decrease in the viscosity with time are called rheopectic. Gypsum suspension in water and bentonite solutions are examples of rheopectic fluids.

Example 1.1 *Two horizontal plates are placed 15 mm apart. The space between them is filled with an oil of viscosity 1.5 Pa. s. Calculate the shear stress in the oil if the upper plate moves with a velocity of 3m/*

Solution.

Given: dy = 15 mm, du = 3 m/s, μ = 1.5 Pa. s

Refer to Fig. 1.12.

According to Newton's law of viscosity,

$$\tau = \mu \frac{du}{dy}$$

$$= 1.5 \times \frac{3}{15 \times 10^{-3}} = 300 \text{ Pa}$$

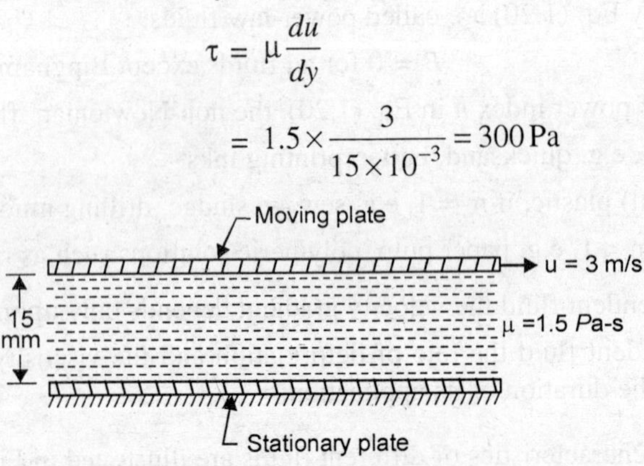

Fig. 1.12

Example 1.2 *A rectangular plate 1.25 m × 0.5 m, weighing 1000 N slides down a 30° inclined surface at a uniform velocity of 2.5 m/s. If the 2.5 mm gap between the plate and the inclined surface is filled with oil, determine its viscosity.*

Solution.

Given: W = 1000 N, u = 2.5 m/s, dy = 2.5 mm, θ = 30°, A = 1.25 × 0.5 = 0.625 m²

Refer to Fig. 1.13.

The sliding plate will attain uniform velocity when the force causing motion (W sin θ) balances the fluid resistance offered by oil filled in the gap.

$$W \sin\theta = \mu \frac{du}{dy} \times A$$

$$1000 \sin 30° = \mu \times \frac{2.5}{2.5 \times 10^{-3}} \times 0.625$$

$$\mu = 0.8 \text{ Pa.s}$$

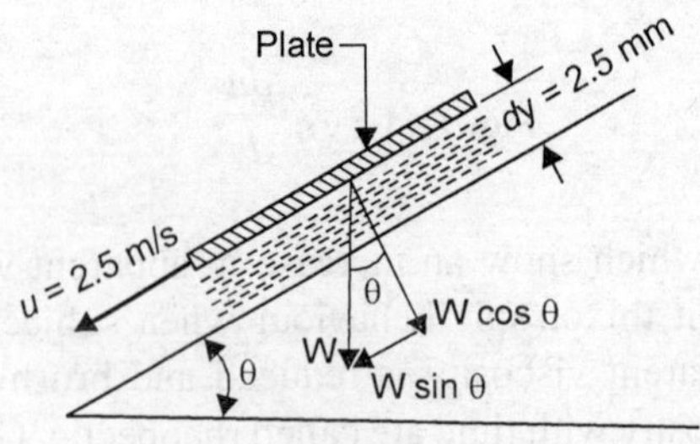

Fig. 1.13

Example 1.3 *A piston 746 mm diameter and 200 mm long works in a cylinder of 750 mm diameter. The annular space between the piston and cylinder is filled with an oil of viscosity 6 cP. Calculate the speed of descent of piston when the cylinder is in vertical position. Take weight of the piston and axial load as 10 N.*

Solution.

Given: d_p = 746mm, l = 200 mm, d_c = 750 mm,

μ = 6 cP = 6×10^{-3} Pa.s, W = 10 N

$$dy = \frac{1}{2}(d_c - d_p) = \frac{1}{2}(750 - 746) = 2\text{mm}$$

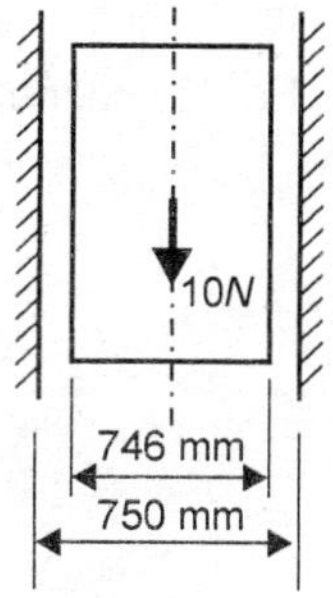

Refer to Fig. 1.14.

According to Newton's law of viscosity.

$$\tau = \mu\frac{du}{dy}$$

$$\tau = \frac{W}{A} = \frac{W}{\pi d_p l_p} = \frac{10}{\pi \times 0.746 \times 0.2} = 21.334\,\text{N/m}^2$$

Fig. 1.14

Let u = Velocity of descent of piston, m/s

Then
$$\frac{du}{dy} = \frac{u}{2 \times 10^{-3}}$$

$$21.334 = 6 \times 10^{-3} \times \frac{u}{2 \times 10^{-3}}$$

$$u = 7.11 \text{ m/s}$$

Example 1.4 *A fluid has an absolute viscosity of 0.048 Pa. s and a specific gravity of 0.913. For flow of such a fluid over a solid flat surface, the velocity at a point 75 mm from the surface is 1.125 m/s. Calculate the shear stress at the solid boundary and at a point 50 mm and 75 mm away from the surface in normal direction, if the velocity distribution across the surface is (a) linear, and (b) parabolic with vertex at the point 75 mm away from the surface.*

Solution.

Given: μ = 0.048 Pa.s, s = 0.918, du = 1.125 m/s, dy = 75 mm

(a) Linear velocity distribution.

Shear stress, $\tau = \mu\frac{du}{dy} = 0.048 \times \frac{1.25}{75 \times 10^{-3}} = 0.72\,\text{N/m}^2$

(b) Parabolic velocity distribution.

Let $u = Ay^2 + By + C$

At y = 0, u = 0, gives C = 0

At y = mm, u = 1.125 m/s, gives

$$1.25 = (0.075)^2 + 0.075\text{ B}$$
$$= 5.625 \times 10^{-3} \times \text{A} + 0.075\text{ B} \quad \text{... (1)}$$

At $\quad y = 75 \text{ mm}, \dfrac{du}{dy} = 0$

$$\frac{du}{dy} = 2Ay + B$$

$$0 = 2 \times 0.075 \times A + B$$

or $\quad B = -0.15\,A \qquad \text{... (2)}$

Eq. (1) becomes,

$$1.125 = 5.625 \times 10^{-3}\,A - 0.075 \times 0.15 \times A = -5.625 \times 10^{-3}\,A$$

$$A = -200$$

$$B = -0.15 \times (-200) = 30$$

$$u = -200y^2 + 30\,y$$

$$\frac{du}{dy} = -400\,y + 30$$

At $\quad \tau = 50 \text{ mm},$

$$\tau = 0.048\,(-400 \times 0.050 + 30) = 0 = 0.48 \text{ N/m}^2$$

At $\quad \tau = 75 \text{ mm},$

$$\tau = 0.048\,(-400 \times 0.075 + 30) = 0$$

At $\quad y = 0,\ \tau = 0.048\,(0 + 30) = 1.44 \text{ N/m}^2$

Example 1.5 *A shaft of diameter 200 mm is rotating at 30 rpm in a bearing of length 100 mm. The thickness of oil film is 2 mm and absolute viscosity of oil is 0.6 Pa.s. Calculate (a) torque required to overcome bearing friction, and (b) power lost in overcoming viscous resistance.*

Assume linear velocity profile.

Solution.

Given: $d = 200$ mm, $l = 100$ mm, N = 300 rpm, $\mu = 0.6$ Pa.s, $c = 2$ mm

Refer to Fig. 1.15.

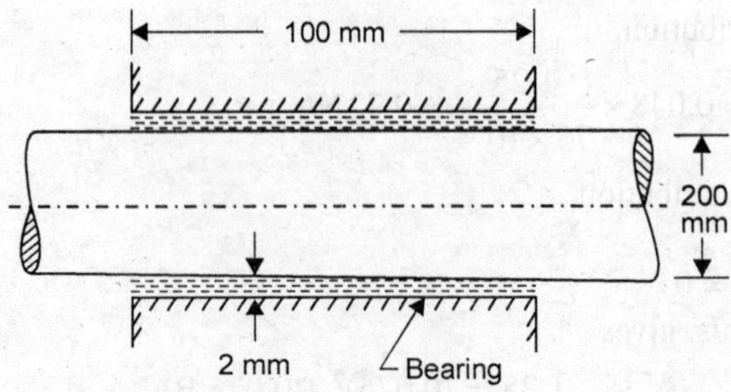

Fig. 1.15

Peripheral velocity of shaft, $u = \frac{\pi dN}{10^3 \times 60}$

$$= \frac{\pi \times 200 \times 300}{10^3 \times 60} = 3.14\,\text{m/s}$$

(a) $\tau = \mu \frac{du}{dy}$

$$du = u - 0 = 3.14 \text{ m/s}$$

$$dy = c = 2 \text{ mm}$$

$$\tau = 0.6 \times \frac{3.14}{2 \times 10^{-3}} = 942.5 \text{ N/m}^2$$

Shear force, $F = \tau \times \pi\, dl = 942.5 \times \pi \times 200 \times 100 \times 10^{-6} = 59.2 \text{ N}$

Viscous torque, $T = F \times \frac{d}{2} = 59.2 \times \frac{200}{2} \times 10^{-3} = 5.92 \text{ N.m}$

(b) Power lost, $P = \frac{2\pi NT}{60} = \frac{2\pi \times 300 \times 5.92}{60} = 186 \text{ Watts}$

Example 1.6 *A cylinder of 100 mm diameter rotates concentrically inside a fixed cylinder of 110 mm diameter. The length of both cylinders is 200 mm. Determine the viscosity of the liquid filling the space between the cylinders if a torque of 0.8 N.m is required to maintain an angular speed of 50 rpm.*

Solution.

Given: $dy = \frac{1}{2}(110 - 100) = 5$ mm, d = 100 mm, T = 0.8 Nm, N = 50 rpm, $l = 200$ mm

Refer to Fig. 1.16.

$$u = \frac{\pi dN}{60 \times 10^3} = \frac{\pi \times 100 \times 50}{60 \times 10^3} = 0.2618 \text{ m/s}$$

$$du = 0.2618 - 0 = 0.2618 \text{ m/s}$$

$$\tau = \mu \frac{du}{dy} = \mu \times \frac{0.2618}{5 \times 10^{-3}} = 52.36 \times \mu \text{ N/m}^2$$

$$T = \tau \times A \times \frac{l}{2} = \tau \times \pi dl \times \frac{d}{2}$$

$$0.8 = 52.36 \times \mu \times \pi \times 100 \times 200 \times 10^{-6} \times \frac{100}{2} \times 10^{-3}$$

$$\mu = 4.86 \text{ N.s/m}^2$$

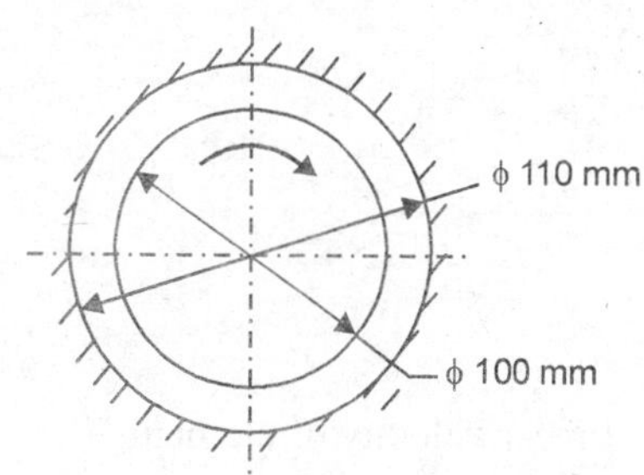

Fig. 1.16

Example 1.7 *A shaft of 150 mm diameter rotates at 1500 rpm in a 200 mm long journal bearing having 155 mm internal diameter. The gap between the shaft and the bearing is filled with an oil of dynamic viscosity 0.8 Poise. Calculate the power dissipated as heat.*

Solution.

Given: d_s = 150 mm, d_b = 155 mm, μ = 0.8 Poise = 0.08 Pa.s, N = 1500 rpm, l = 200 mm

Radial thickness of oil, $$\mathrm{dy} = \frac{1}{2}(155 - 150) = 2.5\,\mathrm{mm}$$

Tangential velocity of shaft, $$u = \frac{\pi d_s N}{60 \times 10^3} = \frac{\pi \times 150 \times 1500}{60 \times 10^3} = 11.78\,\mathrm{m/s}$$

$$du = u - 0 = 11.78\ \mathrm{m/s}$$

shear stress, $$\tau = \frac{\mu\, du}{dy} = 0.08 \times \frac{11.78}{2.5 \times 10^{-3}} = 376.96\,\mathrm{N/m^2}$$

$$\text{Power dissipated as heat} = \tau \times \pi d_s l \times \mathrm{u}$$
$$= 376.96 \times \pi \times 0.15 \times 0.2 \times 11.78$$
$$= 418.5\ \text{Watts}$$

Example 1.8 A circular disc of diameter d is rotated slowly in a liquid of viscosity μ at a small distance h from a fixed surface. Derive an expression for torque T required to maintain an angular speed ω.

Solution. Consider an element of disc at a radius r and of width dr as shown in Fig. 1.17.

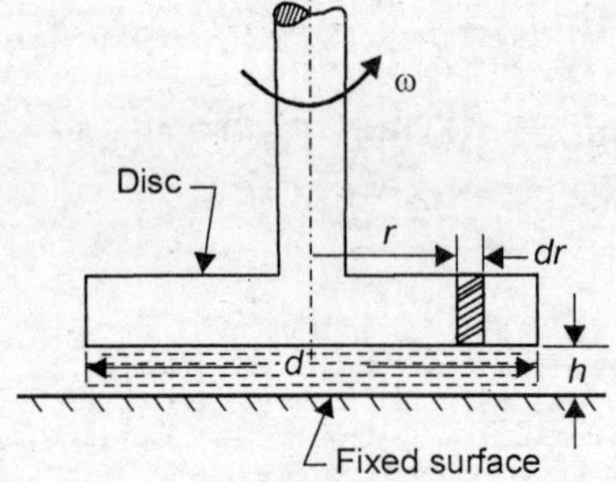

Fig. 1.17

Linear velocity of element, $$u = r\omega$$

Shear stress in the liquid, $$\tau = \mu \frac{du}{dy}$$

where $$du = r\omega - 0 = r\omega,$$
$$dy = h$$

Torque, $$dT = \tau \times (2\pi r \times dr) \times r$$

$$= \mu \times \frac{r\omega}{h} \times 2\pi r^2 \times dr$$

$$= \frac{2\pi\,\mu\omega}{h} \times r^3 dr$$

Total torque,
$$T = \int_0^{d/2} \frac{2\pi\mu\omega}{h} r^3 dr$$

$$= \frac{2\pi\mu\,\omega}{h}\left|\frac{r^4}{4}\right|_0^{d/2}$$

$$= \frac{\pi\mu\omega d^4}{32h} \qquad \text{... (1.21)}$$

Example 1.9 *A solid cone* of maximum radius R and vertex angle 2θ is to be rotated at angular speed ω. An oil of viscosity μ and thickness *h* fills the gap between the cone and the housing. Derive an expression for the torque required and the rate of heat dissipation in the oil.

Solution. Consider an elementary area dA of the cone of radius *r*, as shown in Fig. 1.18 (*b*).

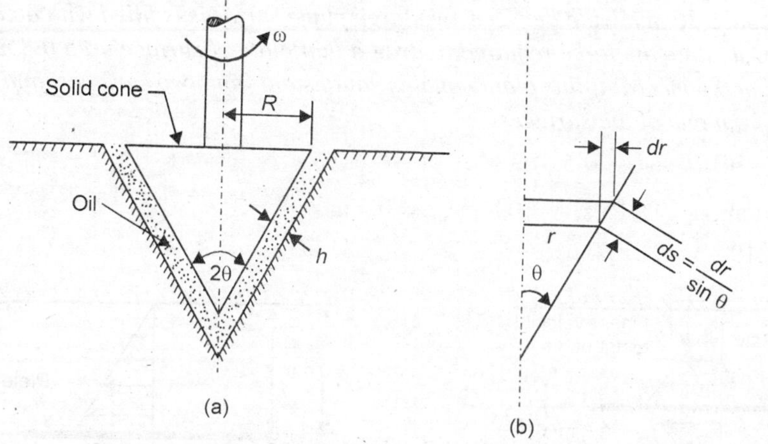

Fig. 1.18

$$dA = 2\pi r \times ds = 2\pi r \times \frac{dr}{\sin\theta}$$

Shear stress,
$$\tau = \mu\frac{du}{dy} = \frac{\mu u}{h}$$

Shear force, $$dF = \tau \times dA = \frac{\mu u}{h} \times 2\pi r \times \frac{dr}{\sin\theta}$$

$$= \frac{2\pi \mu u r}{h} \times \frac{dr}{\sin\theta}$$

Viscous torque on the element, $$dT = dF \times r$$

$$= \frac{2\pi \mu u}{h \sin\theta} \times r^2 dr$$

$$= \frac{2\pi\mu\omega}{h\sin\theta} \times r^3 dr \qquad (\because\ u = r\omega)$$

Total torque, $$T = \frac{2\pi\mu\omega}{h\sin\theta}\int_0^R r^3\, dr$$

$$= \frac{\pi\mu\omega}{2h\sin\theta}\left| r^4 \right|_0^R = \frac{\pi\mu\omega R^4}{2h\sin\theta} \qquad \text{... (1.22)}$$

Rate of heat dissipation in the oil, $$H_d = T\omega = \frac{\pi\mu\omega^2 R^4}{2h\sin\theta} \qquad \text{... (1.23)}$$

Example 1.10 *A space 30 mm wide between two large plane surfaces is filled with an oil of dynamic viscosity 0.8 Pa.s. Calculate the force required to drag a thin plate of surface area 0.8 m² between the surfaces at a speed of 0.5 m/s (a) if this plate remains equidistant from two surfaces, and (b) if it is at a distance of 10 mm from one of the surfaces.*

Solution.

Given: b = 30 mm, μ =0.8 Pa.s, A = 0.8 m², u = 0.5 m/s

Refer to Fig. 1.19.

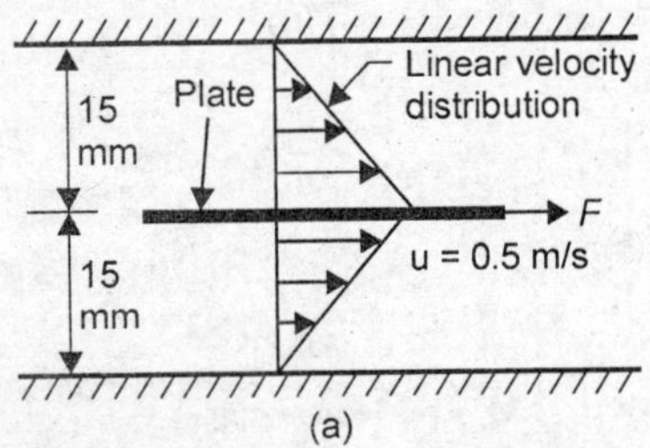

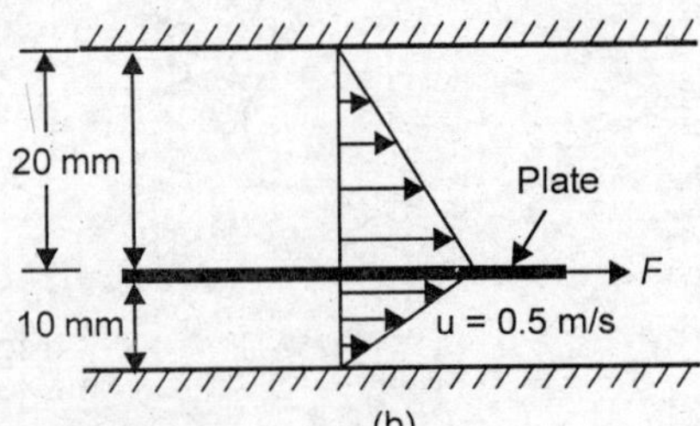

Fig. 1.19

(a) Since the space between the surface is small. The velocity distribution may be assumed linear.

$$F = \tau \times 2A = \frac{\mu\, du}{dy} \times 2A$$

$$= 0.8 \times \frac{0.5}{15 \times 10^{-3}} \times 2 \times 0.8 = 42.67\,\text{N}$$

(b) $F = F_1 + F_2$

$$= (\tau_1 + \tau_2) A$$

$$= \mu \left[\frac{du}{dy_1} + \frac{du}{dy_2} \right] A$$

$$= 0.8 \left[\frac{0.5}{20 \times 10^{-3}} + \frac{0.5}{10 \times 10^{-3}} \right] \times 0.8$$

$$= 48\ \text{N}$$

Example 1.11 *Two oils of dynamic viscosities 0.15 Pa.s and 0.45 Pa.s are placed between two fixed parallel surfaces 20 mm apart. A central plate of area 5m² is being pulled with a force of 150 N. Determine the velocity at which the central plate will move.*

Solution.

Given: $\mu_1 = 0.15$ Pa.s, $\mu_2 = 0.45$ Pa.s

$dy_1 = dy_2 = 10$ mm, $F = 150$ N, $A = 5\ \text{m}^2$

Refer to Fig. 1.20

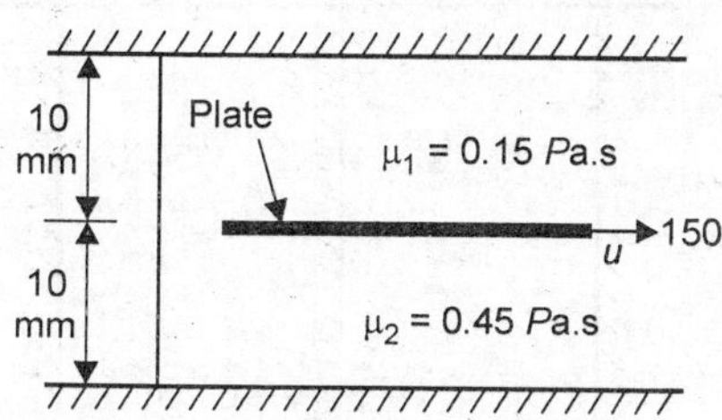

Fig. 1.20.

Let u = velocity of plate, m/s

$$\tau_1 = \mu_1 \frac{du}{dy_1}$$

$$\tau_2 = \mu_2 \frac{du}{dy_2}$$

$$F = F_1 + F_2 = (\tau_1 + \tau_2) A$$

$$= \left(\mu_1 \frac{du}{dy_1} + \mu_2 \frac{du}{dy_2} \right) A$$

$$150 = (0.15+0.45)\times\frac{du}{10\times10^{-3}}\times5$$

$$du = 0.5 \text{ m/s}$$

$$= u - 0$$

$$\therefore \quad u = 0.5 \text{ m/s}$$

Example 1.12: *A steel plate 1.5 m × 1.5 m × 6 mm thick and weighing 125 N is placed midway in the 30 mm gap between two vertical plane surfaces as shown in Fig. 1.21. The gap is filled with an oil of specific gravity 0.86 and dynamic viscosity 3 Pa. s. Determine the force required to lift the plate with a constant velocity of 0.2 m/s.*

Solution.

Given : $W = 125$ N, $s = 0.86$, $\mu = 3$ Pa. s, $u = 0.2$ m/s, $dy_1 = dy_2 = \frac{1}{2}(30-6) = 12\text{mm}$,

A = 1.5 × 1.5 = 2.25 m^2

Refer to Fig. 1.21.

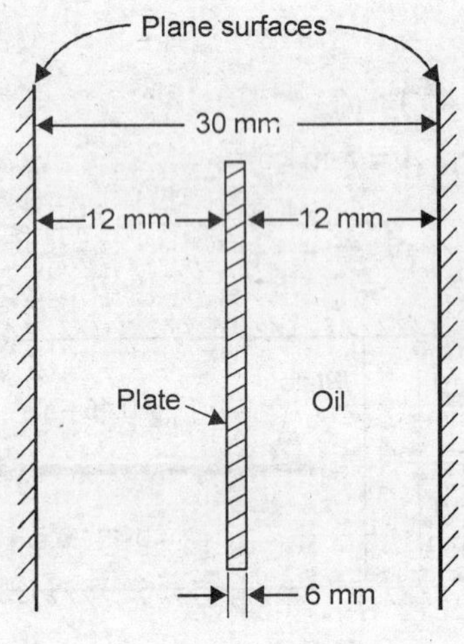

Fig. 1. 21

Drag (or viscous resistance) force on the plate,

$$F = (\tau_1 + \tau_2)A = 2\tau A$$

$$= 2\times\mu\frac{du}{dy}\times A = 2\times3\times\frac{0.2}{12\times10^{-3}}\times2.25 = 225\text{N}$$

Buoyant force on plate = specific weight × volume of oil displaced

$= 0.86 \times 9.81 \times 10^3 \times 1.5 \times 6 \times 10^{-3}$

$= 113.9$ N

Effective weight of plate = 125 – 113.9 = 11.1 N

Total force required to lift the plate = 225 + 11.1 = 236.1 N

Example 1.13 *A cylinder of 0.5 m diameter and 1.5 m long rotates coaxially inside a fixed cylinder. After the constant velocity is attained, determine: (a) the velocity gradient at the cylinder walls, (b) the resulting rpm, and (c) the power dissipated by fluid resistance Neglect end effects.*

Solution.

Given: $d_1 = 0.5$ m, $d_2 = 0.6$ m, $l_1 = l_2 = 1.5$ m , $\mu = 0.05$ Pa. s, T = 5 N.m

Since the gap between the cylinders is large, therefore the velocity profile through the fluid will be non-linear. The torque of 5 N.m is transmitted from inner cylinder to the outer cylinder through fluid friction.

Refer to Fig. 1.22.

Let r = radial distance of any fluid layer.

Then $$T = \tau\,(2\pi\, rl\ ) \times r$$

$$= \tau \times 2\pi r^2 \times 1.5 = 3\pi r^2 \times \mu \frac{du}{dy}$$

$$5 = 3\pi r^2 \times 0.05 \times \frac{du}{dy}$$

$$\frac{du}{dy} = \frac{10.61}{r^2}$$

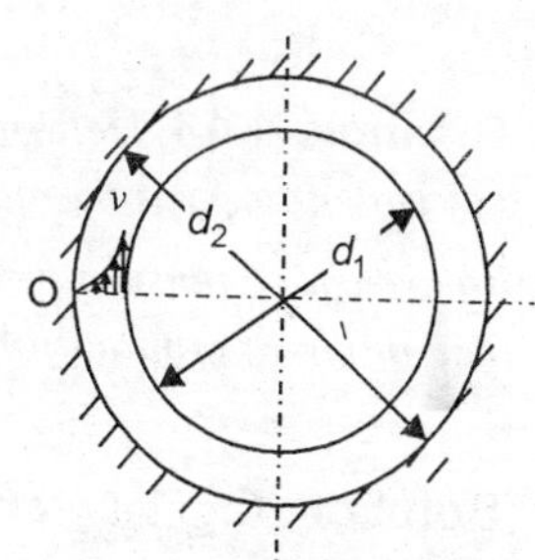

Fig. 1.22

$$\left(\frac{du}{dy}\right)_1 = \frac{10.61}{(0.25)^2} = 169.\ 76\ s^{-1}$$

$$\left(\frac{du}{dy}\right)_2 = \frac{10.61}{(0.3)^2} = 117.89\, s^{-1}$$

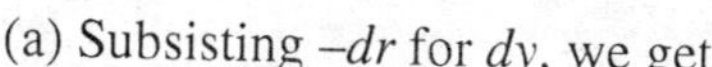

(a) Subsisting $-dr$ for dy, we get

$$\int_0^v du = -10.61 \int_{0.30}^{0.25} \frac{dr}{r^2}$$

Velocity of inner cylinder,

$$v = 10.61 \left| \frac{1}{r} \right|_{0.30}^{0.25} = 10.61\left(\frac{1}{0.25} - \frac{1}{0.30}\right) = 7.07 \text{ m/s}$$

(b) Angular speed of inner cylinder,

$$\omega = \frac{v}{r_1} = \frac{7.07}{0.25} = 28.28 \text{ rad/s}$$

$$N = \frac{60 \times 28.28}{2\pi} = 270 \text{ rpm}$$

(c) Power dissipated in fluid friction $= T\omega = 5 \times 28.28 = 141.4$ Watts

Example 1.14 *The surface tension of water in contact with air is 0.075 N/m. Calculate the capillary rise in tube of 2.5 mm diameter. The angle of contact for water-air-glass contact is 25°–32′.*

Solution.

Given: $\sigma = 0.75$ N/m, $d = 2.5$ mm, $\theta = 25.53°$

Capillary rise, $$h = \frac{(4\cos\theta)\sigma}{\gamma d}$$

$$= \frac{4\cos 25.53° \times 0.075}{9.81\times 10^{3}\times 2.5\times 10^{-3}} = 0.0123\,\text{m or } 12.23\,\text{mm}$$

Example 1.15 *Determine the capillary effect in a glass tube of 3 mm diameter, when immersed in (a) water, and (b) mercury.*

Take surface tension of water and mercury in contact with air as 0.075 N/m and 0.510 N/m respectively. The angle of contact for water is 25.53° and that for mercury 130°.

Solution. Capillary effect, $$h = \frac{(4\cos\theta)\sigma}{\gamma d}$$

For water, $$h = \frac{4\cos 25.53° \times 0.075}{9.81\times 10^{3}\times 3\times 10^{-3}} = 9.19\,\text{mm (rise)}$$

For mercury, $$h = \frac{4\cos 130° \times 0.510}{13.6\times 9.81\times 10^{3}\times 3\times 10^{-3}} = -3.27\,\text{mm (depression)}$$

Example 1.16 *The surface tension at water-air interface is 0.070 N/m. What is the pressure difference between inside and outside of an air bubble of diameter 0.01 mm?*

Solution.

Given: $\sigma = 0.070$ N/m, $d = 0.01$ mm $$p = \frac{4\sigma}{d} = \frac{4\times 0.070}{0.01\times 10^{-3}} = 28\,\text{kPa}$$

Example 1.17 *The surface tension at the soap bubble-air interface is 0.09 N/m. Calculate the internal pressure in a soap bubble of 30 mm diameter.*

Solution.

Given: $\sigma = 0.09$ N/m, $d = 30$ mm

$$p = \frac{8\sigma}{d} = \frac{8\times 0.09}{3\times 10^{-3}} = 24\,\text{Pa (above atmospheric pressure)}$$

Example 1.18 *Calculate the absolute pressure and the gauge pressure that would exist within: (a) a spherical droplet of water 6 mm in diameter, and (b) a jet of water 6 mm diameter. Take surface tension of water 0.075 N/m and the barometer reading 750 mm of mercury. The specific gravity of mercury is 13.6 and the specific weight of water 9810 N/m^3.*

Solution.

Given: $d = 6$ mm, $\sigma = 0.75$ N/m, $h_b = 750$ mm Hg, $s_{Hg} = 13.6$, $\gamma = 9810$ N/m^3.

(a) Spherical droplet:

$$\Delta p = \frac{4\sigma}{d} = \frac{4 \times 0.075}{6 \times 10^{-3}} = 50\,\text{N/m}^2$$

$$p_{atm} = s \times h_b \times \gamma = 13.6 \times 750 \times 10^{-3} \times 9810 = 110.062 \text{ kPa}$$

Absolute pressure inside the droplet $= \Delta p + p_{atm} = 50 \times 10^{-3} + 100.0.62 = 100.112$ kPa

(b) Liquid jet:

$$\Delta p = \frac{2\sigma}{d} = \frac{02 \times 0.075}{6 \times 10^{-3}} = 25\,\text{Pa}$$

Absolute pressure inside the liquid jet = $25 \times 10^{-3} + 100.062 = 100.\ 087$ kPa

Example 1.19 *A U-tube is made up of two capillaries of bores 1.5 mm and 3 mm respectively. The tube is held vertical and partially filled with a liquid of surface tension 0.065 N/m and zero contact angle. If the difference in the level of two menisci is 15 mm, determine the density of the liquid.*

Solution.

Given: $d_1 = 1.5$ mm, $d_2 = 3$ mm, $\sigma = 0.065$ N/m, $\theta = 0°$, $h_1 - h_2 = 15$ mm

$$h_1 - h_2 = \frac{4\sigma}{\gamma}\left[\frac{1}{d_1} - \frac{1}{d_2}\right]$$

$$15 \times 10^{-3} = \frac{4 \times 0.065}{9.81 \times \rho}\left[\frac{1}{1.5} - \frac{1}{3}\right] \times 10^3$$

$$\rho = 588.97 \text{ kg/m}^3$$

Example 1.20 *Two vertical parallel plates are placed at a distance b apart. They are dipped in a liquid having surface tension σ and contact angle θ. Determine and expression for the capillary rise.*

Solution. Consider two vertical parallel plates immersed in liquid as shown in Fig. 1.23. For the equilibrium of the system, the weight of liquid of height h is balanced by force at the surface of liquid between the plates.

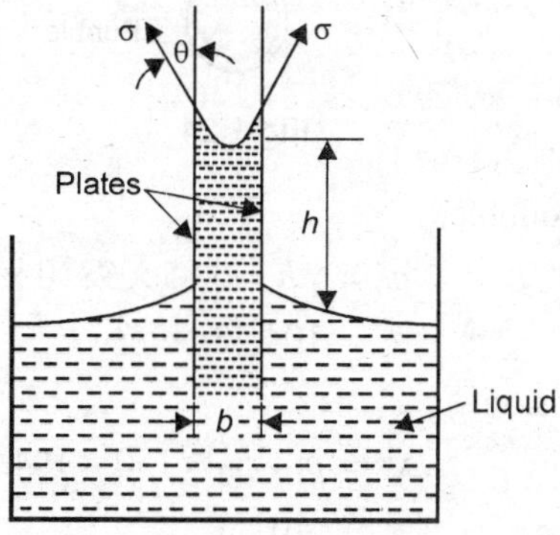

Fig. 1.23

Let L = length of each plate

γ = weight density of the liquid

Weight of the liquid of height h between the plates $= b \times L \times h \times \gamma$

Vertical component of surface tension force $= \sigma \times 2L \times \cos\theta$

For equilibrium, we have

$$bLh\,\gamma = 2\sigma L \cos\theta$$

$$h = \frac{2\sigma\cos\theta}{b\gamma} \quad \text{... (1.24)}$$

Example 1.21 *In measuring the unit energy of mineral oil (specific gravity = 0.85) by the bubble method, a tube having an internal diameter of 1.5 mm is immersed to a depth of 12.5 mm in oil. Air is forced through the tube forming a bubble at the lower end. What magnitude of the unit surface energy will be indicated by a maximum bubble pressure intensity of 150 N/m².*

Solution.

Given: $s = 0.85$ $d = 1.5$ mm, $h = 12.5$ mm, $p_i = 150$ N/m² (gauge)

Refer to Fig. 1.24.

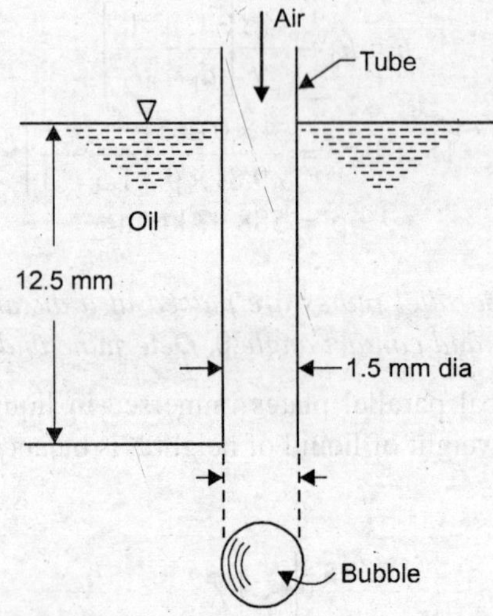

Fig 1.24

Gauge pressure outside the bubble,

$$p_0 = \gamma h = 0.85 \times 9810 \times 12.5 \times 10^{-3}$$

$$= 104.23 \text{ N/m}^2$$

Net pressure for surface tension,

$$\Delta p = p_i - p_0 = 150 - 104.23 = 45.77 \text{ N/m}^2$$

Also $$\Delta p = \frac{4\sigma}{d}$$

$$45.77 = \frac{4\sigma}{1.5 \times 10^{-3}}$$

$$\sigma = 0.0172 \text{ N/m}$$

Example 1.22 *When a pressure of 21 MPa is applied to 100 litres of a liquid its volume decreases by 1 litre. Find the bulk modulus of the liquid.*

Solution.

Given: dp = 21 MPa, V = 100 litres, dV= – 1 litre

Bulk modulus,
$$K = \frac{dp}{-\left(\frac{dV}{V}\right)} = \frac{21}{\frac{1}{100}} = 2100 \text{ MPa or } 2.1 \text{GPa}$$

Example 1.23 *A thin plate of very large area is placed in a gap of height h with oils of viscosities μ_1 and μ_2 on the two sides of the plate. The plate is pulled at a constant velocity v. Calculate the position of plate so that:*

(a) The shear force on the two sides of the plate is equal and

(b) The force required to drag the plate is minimum.

Assume viscous flow and neglect all end effects.

Solution. Refer to Fig. 1.25.

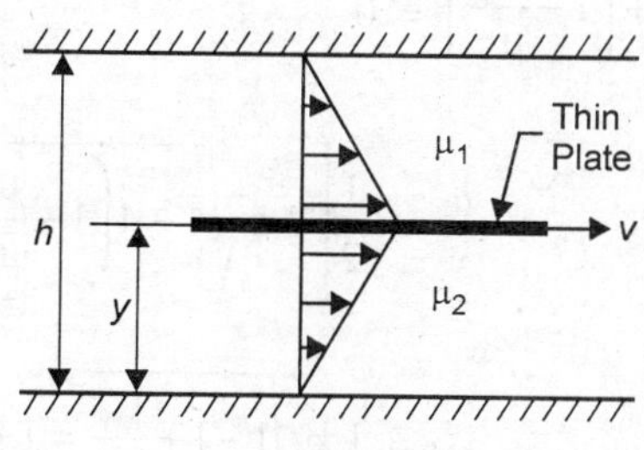

Fig. 1.25

(a) Let y = distance of the thin plate from the bottom surface

Shear force on the upper side of plate,

$$F_u = \mu \frac{du}{dy} \times A = \mu_1 \left(\frac{v}{h-y}\right) A$$

Shear force on the lower side of plate,

$$F_l = \mu_2 \left(\frac{v}{y}\right) A$$

For $F_u = F_l$, we have

$$\mu_1\left(\frac{v}{h-y}\right)A = \mu_2\left(\frac{v}{y}\right)A$$

$$\mu_1 y = \mu_2(h-y)$$

$$y = \frac{\mu_2 h}{\mu_1+\mu_2}$$

(b) Total drag force, $F = F_u + F_l = \mu_1\left(\dfrac{v}{h-y}\right)A + \mu_2\left(\dfrac{v}{y}\right)A$

For F to be minimum,

$$\frac{dF}{dy} = 0$$

$$\frac{\mu_1 vA}{(h-y)^2} - \frac{\mu_2 vA}{y^2} = 0$$

$$\frac{\mu_1}{\mu_2} = \left(\frac{h-y}{y}\right)^2 = \frac{h^2}{y^2} + 1 - \frac{2h}{y}$$

$$\frac{h^2}{y^2} - \frac{2h}{y} + \left(1-\frac{\mu_1}{\mu_2}\right) = 0$$

$$\frac{h}{y} = \frac{1}{2}\left[2 \pm \sqrt{4-4\left(1-\frac{\mu_1}{\mu_2}\right)}\right]$$

$$= 1 \pm \sqrt{1-1+\frac{\mu_1}{\mu_2}} = 1 + \sqrt{\frac{\mu_1}{\mu_2}}$$

$$y = \frac{h}{1+\sqrt{\dfrac{\mu_1}{\mu_2}}}$$

Example 1.24 *Two coaxial glass tubes forming an annulus with small gap are immersed in water in a trough. The inner and outer radii of the annulus are r_i and r_o respectively. What is the capillary rise if σ is the surface tension of water in contact with air?*

Solution. Refer to Fig. 1.26.

The water in the annulus will continue to rise until the vertical component of the surface tension force which acts on the wetted length (outer curve of the inner tube and inner curve of the outer tube) equals the height of the water column.

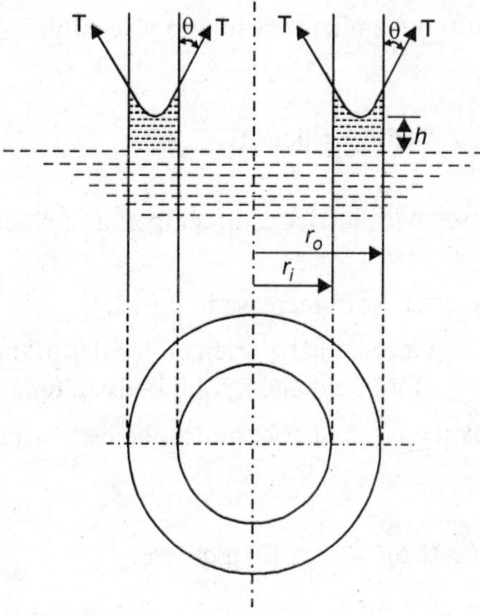

Fig. 12.6

$$\therefore \qquad T\cos\theta = \pi(r_0^2 - r_i^2)h\rho g$$

where

$$T = \pi(r_0 + r_i)\sigma$$

$$\therefore \qquad \pi(r_0 + r_i)\sigma\cos\theta = \pi(r_0^2 - r_i^2)h\rho g$$

$$h = \frac{\sigma\cos\theta}{(r_0 - r_i)\rho g}$$

For

$$\theta = 25° - 32', \; h = \frac{0.902\sigma}{(r_0 - r_i)\gamma}$$

SUMMARY

1. Fluid mechanics deals with the study of behaviour of fluids at rest or in motion.
2. Density is the mass per unit volume.
3. Specific weight is the weight of the fluid per unit volume.
4. Specific gravity is the ratio of density of a fluid to the density of a standard fluid.
5. The viscosity is the property of a fluid by virtue of which it offers resistance to shearing stresses or deformation.

6. The Newton's law of viscosity states that the shear stress on a fluid element layer is directly proportional to the rate of shear strain.

$$\tau = \mu \frac{du}{dy}, \text{ where } \mu = \text{dynamic (or absolute) viscosity.}$$

7. Kinematic viscosity, $\nu = \frac{\mu}{\rho}$, where ρ = density.

8. Viscosity of liquids decreases with increases in temperature whereas those of gases increases with rise in temperature.
9. The viscosity of liquids increases with increase in pressure.
10. Cohesion is the inter molecular attraction between molecules of the same liquid Adhesion is the attraction between molecules of a liquid and a solid body in contact.
11. Surface tension is caused by the force of cohesion at the free surface of a liquid. It is expressed as the force per unit length.

$$\text{Inside pressure excess, } \Delta p = \frac{4\sigma}{d} \text{ for a droplet}$$

$$= \frac{8\sigma}{d} \text{ for a soap bubble}$$

$$= \frac{2\sigma}{d} \text{ for a liquid jet.}$$

12. Capillarity is the phenomenon of rise or fall of a liquid surface in a thin tube when held vertically in the liquid.

$$h = \frac{(4\cos\theta)\sigma}{\gamma d}, \text{ where } \theta = \text{angle of contact}$$

13. Compressibility is the property due to which the fluid undergoes a change in volume under the action of external pressure.

$$\text{Bulk modulus, } K = \frac{dp}{\left(\frac{dV}{V}\right)}$$

$$\text{Compressibility, } \beta = \frac{1}{K}$$

Short Answer Type Questions

1. Define a fluid. How do you classify fluids?
2. Differentiate between an ideal and a real fluid.

3. Define viscosity. How viscosity is affected by temperature and pressure?
4. State Newton's law of viscosity.
5. Define dynamic viscosity and give its units.
6. Define kinematic viscosity and give its units.
7. Differentiate between Newtonian and non-Newtonian fluids.
8. Explain the meaning of cohesion and adhesion.
9. Explain the mechanism of surface tension.
10. What is capillarity? Explain wetting and non-wetting type liquids.
11. Define bulk modulus and compressibility.
12. Explain how certain insects are able to walk on the surface of water?
13. Explain how the surface tension accounts for (a) formation of droplet, and (b) rise of oil in a kerosene lamp.
14. Why does oil spread when it is poured on water surface?
15. What will happen to the capillary rise in a tube when (a) the tube is of insufficient length, and (b) the top of the tube is closed.
16. Why in a capillary tube, the meniscus of water is concave while that of mercury is convex?
17. Why soap bubbles rise up in air?
18. List some manifestations of surface tension.
19. What are time-dependent and time independent fluids?
20. Explain various non-Newtonian fluids.

EXERCISES

1. The velocity distribution for flow over a flat plate is given by:

$$u = \frac{3}{2}y - y^{3/2}$$

where u = velocity of flow at a distance y metre above the plate in m/s.
If dynamic viscosity is 0.75 Pa. s, determine the shear stress at $y = 0.1$ m **[Ans.** 0.769 N/m^2]

2. A shaft 10 cm diameter rotates at 200 rpm in a journal bearing of 10.2 cm diameter. If the viscosity of oil is 0.15 Pa.s, calculate the shear stress. **[Ans.** 157 N/m^2]

3. Two plates are placed at a distance of 0.2 mm apart. The lower plate is fixed while the upper plate having surface area of 1.25 m^2 is pulled at 0.25 m/s. Determine the force and power required to maintain this speed. The fluid viscosity is 0.15 Pa.s. **[Ans.** 234.375 N, 58.6Watts]

4. A square plate of size 1m × 1m and weighing 350 N slides down an inclined plane with a uniform velocity of 1.5 m/s. The inclined plane is laid on a slope of 5 vertical to 12 horizontal and has an oil film of 1mm thickness. Calculate the dynamic viscosity of oil. **[Ans.** 0.0897 Pa.s]

5. The velocity distribution over a plate is given by: $u = \frac{2}{3}y - y^2$. Determine the shear stress as $y = 0.1$ and 0.2 m. Take $\mu = 6$ Poise. **[Ans.** 0.028 N/m^2, 0.16 N/m^2]

6. Calculate the surface tension in a soap bubble of 30 mm diameter when the inside pressure is 2 Pa above atmosphere. **[Ans.** 0.0075 N/m]

7. The surface tension of water in contact with air is 0.0725 N/m. The pressure outside the droplet of water of diameter 0.02 mm is atmospheric at 1.01325 bar. Calculate the pressure within the droplet of water. **[Ans.** 1.15825 bar]

8. Calculate the capillary rise in a glass tube of 3mm diameter when immersed vertically in (a) water of surface tension 0.0725 N/m, and (b) mercury of surface tension 0.52 N/m, in contact with air. Take specific gravity for mercury as 13.6 **[Ans.** 8.89 mm, 3.26 mm (depression)]

9. A 150 mm diameter vertical cylinder rotates concentrically inside another cylinder of diameter 151 mm. Both the cylinders are of 250 mm height. The space between the cylinders is filled with a liquid of viscosity 10 Poise. Determine the torque required to rotate the inner cylinder at 100 rpm. **[Ans.** 13.88 N.m]

10. A flat plate of weight 0.48 kN has a surface area of 0.1 m^2. It slides down an inclined plane at 30° to the horizontal at a constant speed of 3m/s. If the inclined plane is lubricated with an oil of viscosity 0.1 Pa. s, find the thickness of oil film. **[Ans.** 0.125 mm]

11. A square metal plate 2 m side and 2 mm thick weighing 120 N is to be lifted through vertical gap of 25 mm of infinite extent. The oil in gap has a specific gravity of 0.96 and viscosity of 2.5 Pa.s. If the metal plate is to be lifted at a constant speed of 0.1 m/s, find the force and power required. **[Ans.** 218.57N, 21.83 W]

12. Two large fixed parallel planes are 24 m apart. The space between the surfaces is filled with an oil of viscosity 0.8 Pa.s A flat thin plate 0.5 m^2 area moves through the oil at a velocity of 0.6 m/s. Calculate the drag force when (a) the plate is equidistant from both the planes, and (b) when the thin plate is at a distance of 80 mm from one of the plane surfaces. **[Ans.** 40N, 45N]

13. A cylinder of 100 mm diameter and 300 mm length rotates about a vertical axis inside a fixed cylindrical tube of 105 mm diameter and 300 mm length. If the space between the tube and cylinder is filled with a liquid of dynamic viscosity of 0.125 Pa.s, determine the speed of rotation of the cylinder which will be obtained if an external torque of 1 N.m is applied to it **[Ans.** 81 rmp]

14. Air is introduced through a nozzle into a tank of water at 20°C to form a stream of bubbles. The process requires 2.5 mm diameter bubbles to be formed. Calculate by how much the air pressure at the nozzle must exceed that of surrounding water. Take surface tension of water at 20°C = 0.0735 N/m **[Ans.** 117.6 N/m^2]

15. Determine of bulk modulus of elasticity of a liquid, if the pressure of the fluid is increased from 6 MPa to 12 MPa and the liquid volume decreases by 0.15%. **[Ans.** 4GPa]

16. A 20 mm/wide gap between two vertical plane surface is filled with an oil of specific gravity 0.86 and dynamic viscosity 2.5 Pa.s. A metal plate 2 m × 2m × 4 mm thick and weighing 180 N is placed midway in the gap. Determine the force required to lift the plate with a constant velocity of 0.2 m/s. [Ans. 545 N]

17. Calculate the capillary effect in a glass tube of 4mm diameter when immersed in (a) water, and (b) mercury. Take surface tension of water and mercury in contact with air a is 0.07357 N/m and 0.49 N/m respectively; The contact angle for water is 0° and for mercury 130°–24′.

[**Ans.** 7.5 mm, 2.47 mm]

18. Two large parallel flat plates are placed 12.5 mm apart. A 2.5 mm thick plate of 0.2 m^2 area is being towed in glycerine filled between the above plates with a constant force of 9.81 N. Calculate the towing speed of the plate when it is held equidistant from the two parallel plates. Take $\mu = 0\ 0.096$ Pa.s

[**Ans.** 1.277 m/s]

2

FLUID STATICS

2.1 INTRODUCTION

The Study of fluids at rest or undergoing uniform acceleration in a closed space is called fluid statics. A fluid at rest has no relative motion between adjacent or neighbouring fluid layers. Under such conditions there is no velocity gradient which results in zero shear. Thus the viscosity of the fluid has no effect on fluids at rest. Therefore, the behaviour of the fluid is exactly similar to ideal and real fluids

There are two types of forces to be considered on a fluid element or mass at rest. These are the body forces and surfaces forces. Body forces arise due to the mass of the fluid such as gravitational field. Surface forces represent the action of the surrounding fluid on the element under consideration through direct contact such as static pressure.

2.2 VARIATION OF STATIC PRESSURE

Consider a fluid element of size $dx \times dz$ and unit length as shown Fig. 2.1 Let

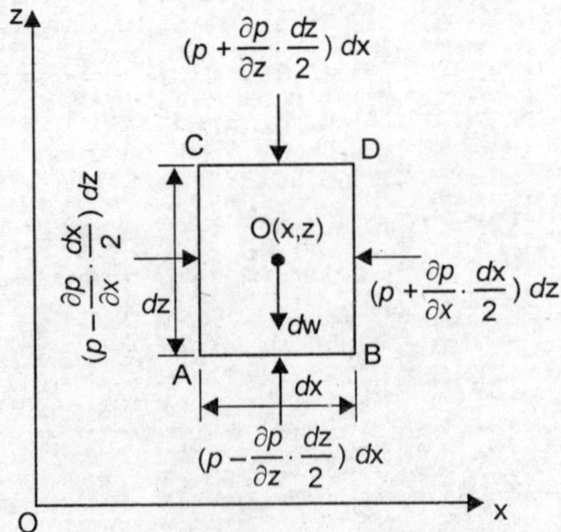

Fig. 2.1 Forces on a static fluid element.

p = static pressure at the centre of fluid element at O.

The fluid element $ABCD$ is in equilibrium under the action of following forces:

1. The static pressure force on the left face AC of the fluid element = $\left(p-\frac{\partial p}{\partial x}\cdot\frac{dx}{2}\right)dz$

2. The static pressure force on the right face BD of the fluid element = $\left(p+\frac{\partial p}{\partial x}\cdot\frac{dx}{2}\right)dz$

3. The static pressure force on the bottom face AB of the fluid element = $\left(p-\frac{\partial p}{\partial z}\cdot\frac{dz}{2}\right)dx$

4. The static pressure force on the top face CD of the fluid element = $\left(p+\frac{\partial p}{\partial z}\cdot\frac{dz}{2}\right)dx$

5. The weight of the fluid element, $dw = dx{\cdot}dz.1\times\gamma$
   where γ = specific weight of fluid element.
   For static equilibrium, $F_x = 0$ and $F_z = 0$.

$$\therefore \quad \left(p-\frac{\partial p}{\partial x}\cdot\frac{dx}{2}\right)dz-\left(p+\frac{\partial p}{\partial x}\cdot\frac{dx}{2}\right)dz = 0$$

$$\therefore \quad \frac{dp}{dx} = 0 \qquad \text{... (2.1)}$$

$$\left(p-\frac{\partial p}{\partial z}\cdot\frac{dz}{2}\right)dx-\left(p+\frac{\partial p}{\partial z}\cdot\frac{dz}{2}\right)dx - dx\cdot dz.\gamma = 0$$

$$-\frac{dp}{dz} = \gamma \qquad \text{... (2.2)}$$

Integrating Eq. (2.2) for incompressible fluids (γ = const.) We have

$$\int_{p_1}^{p_2} dp = \int_{z_1}^{z_2} \gamma.dz$$

$$p_2 - p_1 = \gamma(z_2 - z_1) \qquad \text{... (2.3)}$$

or

$$\frac{p_2}{\gamma}+z_2 = \frac{p_1}{\gamma}+z_1 \qquad \text{... (2.4)}$$

The expression $\left(\frac{p}{\gamma}+z\right)$ is called the *piezometric head*, and is constant throughout a static fluid.

The result of integration of Eq. (2.2) may also be written as:

$$p = -\gamma z + A \qquad \text{... (2.5)}$$

where A = a constant of integration

Let H = depth of fluid in a container.

Then at $z = H$, gauge pressure, $p = 0$, so that $A = \gamma H$, and

$$p = \gamma (H - z) = \gamma h \qquad \text{... (2.6)}$$

where h = depth below the free surface of fluid.

Eq. (2.6) describes the variation of static pressure, and is termed as *hydrostatic law of pressure variation.*

Pressure head,
$$h = \frac{p}{\gamma} \qquad \text{... (2.7)}$$

2.3 ABSOLUTE AND GAUGE PRESSURES

Absolute pressure, p_{abs}: The pressure measured above the absolute zero or complete vacuum is called the absolute pressure. It is measured by the Aneroid barometer

Gauge Pressure, p_g: The pressure measured and expressed as the difference between the absolute value and the local atmospheric pressure is called the gauge pressure. It is measured by the Bourdon gauge.

Vacuum pressure, p_{vac}: The pressure below the local atmospheric pressure is known as the vacuum or negative or suction pressure. The relationship between the various pressure is:

$$p_g = p_{abs} - p_a \qquad \text{... (2.8 a)}$$

$$p_{vac} = p_a - p_{abs} \qquad \text{... (2.8b)}$$

where p_a = local atmospheric pressure

There relations are schematically shown in Fig. 2.2.

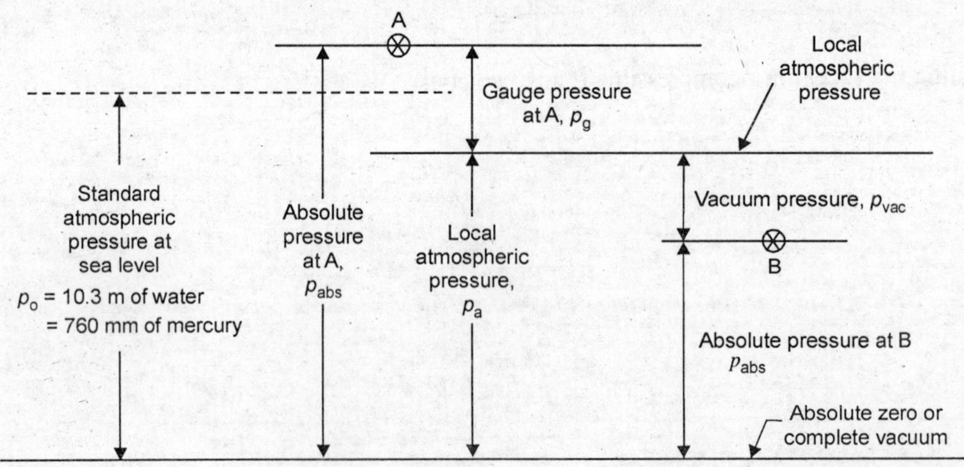

Fig. 2.2 Representation of various pressures.

The local atmospheric pressure is measured by means of a mercury barometer shown in Fig. 2.3. A mercury barometer consists of a glass tube closed at one end and is filled with mercury. It is inverted so that the open end is submerged in mercury. The space above the mercury contains mercury vapour exerting a negligible pressure, which may be ignored. The pressure at A is the local atmospheric pressure, p_a. If p_v is the vapour pressure of mercury, then

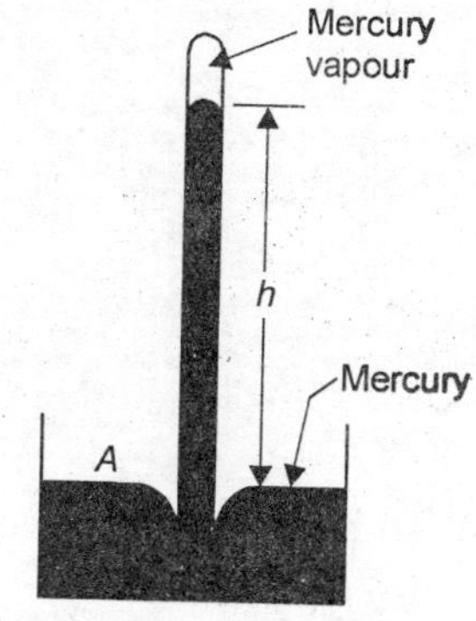

Fig. 2.3 Mercury barometer

$$p_a - p_v = \gamma h$$

or

$$\frac{p_a}{\gamma} = -\frac{p_v}{\gamma} = h \quad \text{... (2.9a)}$$

Neglecting p_v, being small,

$$\left.\begin{aligned}\frac{p_a}{\gamma} &= h \text{ metre of mercury column}\\ &= 13.6\,h \text{ metre of water column}\end{aligned}\right\} \quad \text{...(2.9}b\text{)}$$

2.4 STATIC AND TOTAL PRESSURE

The static pressure may be defined as the force per unit area. It is measured by inserting a tube into a pipe in which the fluid is flowing as shown in Fig. 2.4. The total or stagnation pressure is defined as the pressure that would be obtained if the fluid stream is brought to rest by inserting a bent tube at the centre of the pipe. The difference between the total and stagnation pressures is called the dynamic pressure.

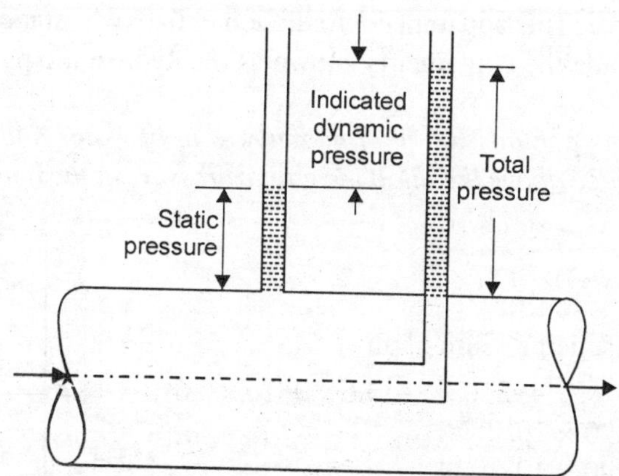

Fig. 2.4 Static and total pressure

$$\text{Total pressure, } p_t = \text{Static pressure, } p_s + \text{Dynamic pressure, } \frac{v^2}{2g} \quad \text{... (2.10)}$$

2.5 HYDROSTATIC PARADOX

Consider three containers of different shapes having the same base area A, which are filled by the same fluid of specific weight γ upto the same height h, as shown in Fig. 2.5.

The pressure at the base of each container is,

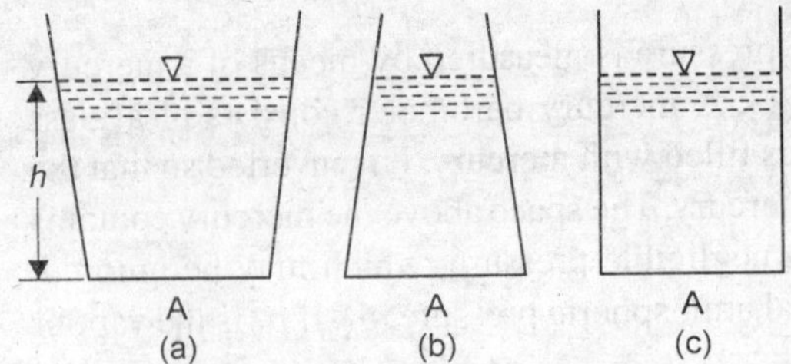

Fig. 2.5 The hydrostatic paradox

$$p = p_a + \gamma_h$$

or $$p - p_a = \gamma h \quad \text{... (2.11a)}$$

$$= p_g$$

where p = absolute hydrostatic pressure

p_a = local atmospheric pressure acting over the liquid surface.

Hydrostatic force exerted on the base of each container,

$$P = (p - p_a) \times A = \gamma h A \quad \text{... (2.11 b)}$$

Since A, h and γ are same for all the containers, the hydrostatic force on the bottom is same irrespective of their shapes.

Hydrostatic force,

p = Weight of the liquid prism of base area A and height h.

In container (a) the weight of liquid is more than the hydrostatic force, whereas in container (b), it is less than the hydrostatic force. This apparent contradiction in the hydrostatic force on the container base and the weight of liquid inside the container is known as the hydrostatic paradox.

Example 2.1 *A closed tank contails 0.6 m of mercury, 2 m of water and 2.5 m of an oil of specific gravity 0.75. There is air space above the oil. If the gauge pressure at the bottom of tank is 220 kPa, what is the pressure of air?*

Solution. Refer to Fig. 2.6

Let p_g = gauge pressure of air

Then, $p_g + 0.75 \times 9810 \times 2.5 + 9810 \times 2 + 13.6 \times 9810 \times 0.6$

$= 220 \times 10^3$

$p_g + 178.493 \times 10^3 = 220 \times 103$

$p_g = 41.507$ kPa

Fig. 2.6

Example 2.2 *The pressure at the base and top of the mountain are 750 mm and 630 mm of mercury respectively. If specific weight of air is 12 N/m³, find the height of the mountain.*

Solution.

$$p_{base} - p_{top} = \gamma h$$

$$(750 - 630) \times 10^{-3} \times 13.6 \times 9810 = 12 \times h$$

$$h = 1334.16 \text{ m}$$

Example 2.3 *A cylindrical tank of cross-sectional area 500 mm² and 2.5 m height is filled with water upto a height of 1.5 m and remaining 1 m with oil of specific gravity 0.75. The tank is open to atmospheric pressure. Calculate:*

(a) *Pressure at the interface,* (b) *Absolute and gauge pressure at the base of the tank in terms of water head, and* (c) *the net force at the base of the tank.*

Assume atmospheric pressure = 1.01325 bar.

Solution.

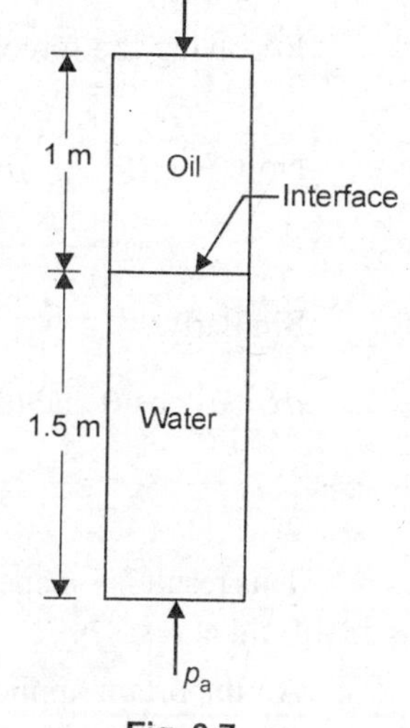

Fig. 2.7

Given: $A = 500 \text{ mm}^2$, $h_\omega = 1.5 \text{ m}$, $h_O = 1 \text{m}$, $s_O = 0.75$, $p_a = 1.01325 \text{ bar}$

(a) $p_{inteface} = s_O \gamma h_O = 0.75 \times 9810 \times 1 = 7357.5 \text{ N/m}^2$

(b) $p_{base} = p_{inteface} + \gamma h_\omega = 7357.5 + 9810 \times 1.5 = 22,072.5 \text{ N/m}^2 \text{ (gauge)}$

$$= \frac{22,072.5}{9810} = 2.25 \text{m of water}$$

$$p_a = \frac{1.01325 \times 10^5}{9810} = 10.329 \text{m of water}$$

$$(p_{base})_{abs} = p_a + (p_{base})_{gauge} = 10.329 + 2.25$$

$$= 12.579 \text{ m of water}$$

(c) Net force at the base, $P = (p_{base}) \times A = 22,072.5 \times 500 \times 10^{-6}$

$$= 11.036 \text{ N}$$

2.6 PASCAL'S LAW

The Pascal's law states that the pressure at any point in a fluid at rest is same in all directions.

Proof: Consider an infinitesimal wedge shaped fluid element at rest having unit depth perpendicular to the plane of the paper, as shown in Fig. 2.8.

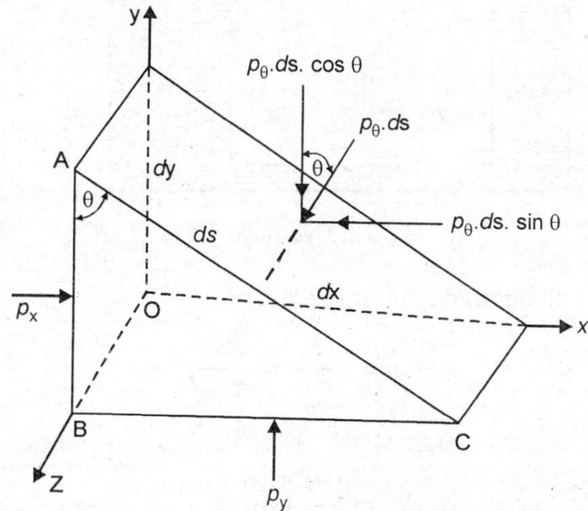

Fig. 2.8 Pressures on wedge shaped fluid element

Let p_x, p_y, p_θ = pressures on faces AB, BC and AC respectively

Then force on face AB $= p_x \times (d_y \times 1) = p_x \cdot dy$

Force on face BC $= p_y \cdot dx$

Force on face AC $= p_\theta \cdot ds$

$$\text{Weight of fluid element} = \frac{1}{2} dx \times dy \times 1 \times \gamma$$

$$= \frac{\gamma}{2} \times dx \cdot dy$$

Resolving the forces in x – and y-directions for the equilibrium of the fluid element, we have

$$p_x \cdot d_y = p_\theta, ds \sin\theta$$

Now $$dy = ds \cdot \sin\theta$$

∴ $$p_x = p_\theta$$

Similarly, $$p_y \cdot dx = p_\theta \cdot ds \cdot \cos\theta + \frac{\gamma}{2} dx \cdot dy$$

$dx = ds \cos\theta$ and $(dx \cdot \mathbf{dy}) << ds$, which can be neglected.

∴ $$p_y = p_\theta$$

or $$p_x = p_y = p_\theta$$

This result is independent of angle θ. Therefore, it follows that pressure acts equally in all directions in a fluid at rest.

An important application of the Pascal's law is the hydraulic press shown in Fig. 2.9.

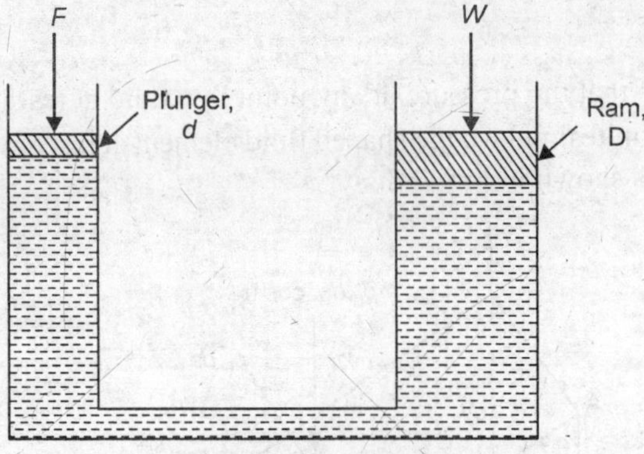

Fig. 2.9 Hydraulic press

Let a_1 D = diameter of plunger and ram respectively .

Then $$p = \frac{F}{\frac{\pi}{4} d^2} = \frac{W}{\frac{\pi}{4} D^2}$$

∴ $$W = \left(\frac{D}{d}\right)^2 F \qquad \text{... (2.12)}$$

Example 2.4 *For the hydraulic jack shown in Fig. 2.10, find the load lifted by the ram when a force of 500 N is applied on the plunger. The specific gravity of the oil in the jack is 0.8.*

Solution.

Given: d = 40 mm, D = 100 mm, h = 0.4 m

$$A_1 = \frac{\pi}{4}d^2 = \frac{\pi}{4}(0.04)^2 = 1.257 \times 10^{-3}\,\text{m}^2$$

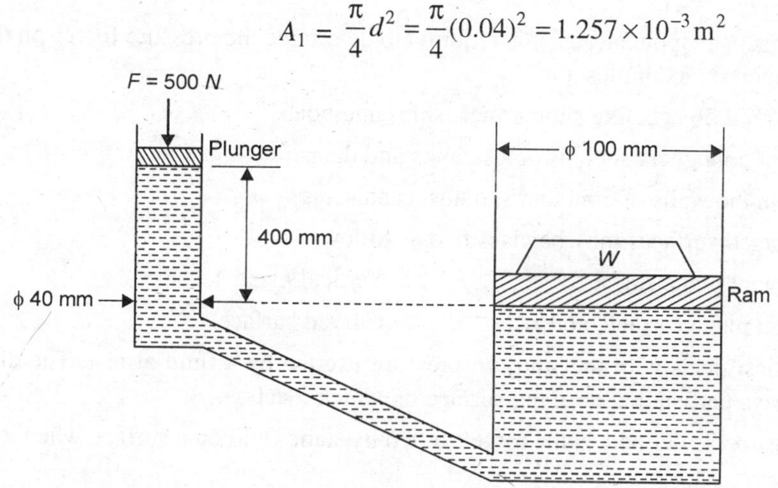

Fig. 2.10 Hydraulic jack

$$A_2 = \frac{\pi}{4}D^2 = \frac{\pi}{4}(0.1)^2 = 7.854 \times 10^{-3}\,\text{m}^2$$

Pressure on plunger, $p_1 = \dfrac{F}{A_1} = \dfrac{500}{1.257 \times 10^{-3}} = 3.978 \times 10^5\,\text{N/m}^2$

Pressure on ram, $p_2 = p_1 + \gamma h = 3.978 \times 10^5 + 0.8 \times 9810 \times 0.4 = 4.009 \times 10^5\,\text{N/m}^2$

Force on the ram, $W = p_2 A_2 = 4.009 \times 10^5 \times 7.854 \times 10^{-3} = 3148.7\,\text{N}$

Example 2.5 Determine pressure at a point 5 m below the free surface in a fluid that has a variable density given by the expression:

$$\rho = (300 + Ay)\ \text{kg/m}^3$$

where A = 10 kg/m^4 and y is the distance in m measured from the free surface.

Solution. According to hydrostatic law,

$$dp = -\gamma dy, \text{ when } y \text{ is measured upward.}$$
$$= \gamma dy, \text{ when } y \text{ is measured downward.}$$
$$dp = g(300 + Ay)\,dy$$

Integrating both sides, we get

$$p = g\left(300y + \frac{Ay^2}{2}\right)$$

For $y = 5$ m and $A = 10$ kg/m^4, we get

$$p = 9.81\left(300 \times 5 + 10 \times \frac{25}{2}\right)$$

$$= 15{,}941.25 \text{ N/m}^2$$

2.7 HYDROSTATIC FORCES ON SUBMERGED SURFACES

In many practical applications, it is required to determine the pressure forces on the entire surface. Some of the important examples are:

1. Submerged objects like submarines, ships and boats.
2. Walls of containers such as pipes, tanks and dams.
3. Gates in the walls of containers, dams, canals, etc.

The submerged surfaces may be classified as follows:

1. Horizontal plane surface
2. Vertical plane surface
3. Inclined plane surface, and
4. Curved surface.

Hydrostatics: deals with the study of pressure exerted by a fluid at rest. The direction of such a pressure is always perpendicular to the surface on which it acts.

Total pressure: is defined as the force exerted by static fluid on a surface when it comes in contact with the surface.

Centre of pressure: is defined as the point of application of the total pressure on the surface.

2.7.1 Horizontal Submerged Plane Surface

Consider a plane horizontal surface submerged in a static fluid shown in Fig. 2.11. As every point of the surface is at the same depth from the free surface of the liquid, the pressure will be same on the entire surface and equal to, $p = \gamma h$, where h = depth of the surface. In this case, the centre of gravity G of the surface coincides with the centre of pressure C.

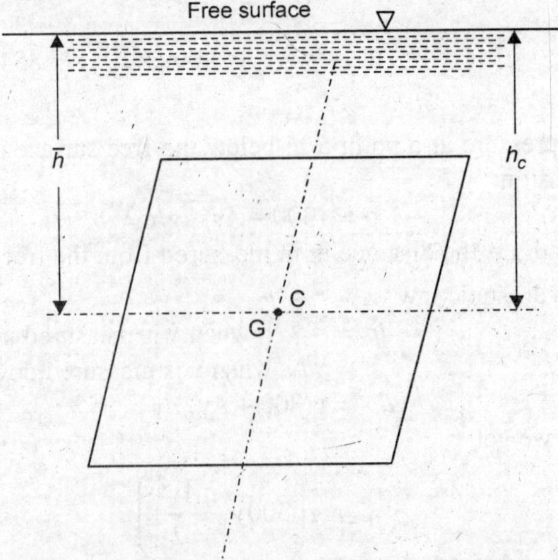

Fig. 2.11 Horizontal submerged plane surface

Let A = total area of the surface.

Then total force on the surface, F = $p \times A = \gamma h A = \gamma \bar{h} \times A$ ... (2.13)

where $\bar{h}$ = depth of C.G. from the free surface of liquid = h

Also h_c = depth of centre of pressure from the free surface of liquid = h

2.7.2 Vertical plane submerged surface

Consider a plane vertical surface of arbitrary shape submerged in a liquid as shown in Fig. 2.12

Let A = total area of surface

G, C = C.G and centre of pressure of plane surface respectively from the free surface of liquid.

h_c = depth of centre of pressure from the free surface of liquid.

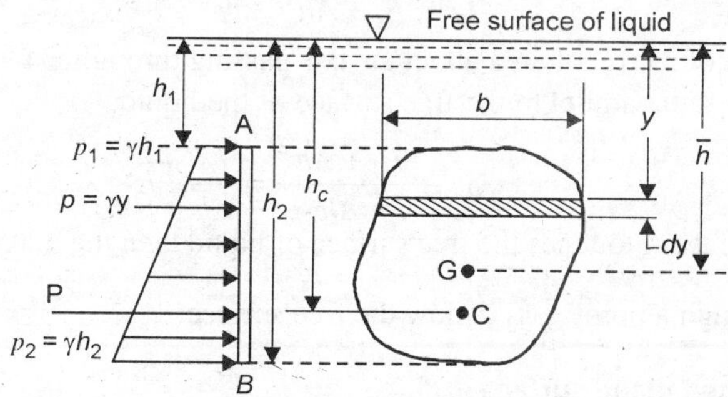

Fig. 2.12 Vertical submerged plane surface

Total pressure, P

Consider a strip of thickness dy and width b of the plane area at a depth y from the free surface of the liquid.

pressure on the strip, $p = \gamma y$

Area of strip, $dA = b \times dy$

Total force on the strip, $dp = \gamma y \times b \times dy$

Total force on the whole surface, $P = \int dp = \int \gamma y \times b \cdot dy = \gamma \int by \cdot dy$

Now $\int by \cdot dy = \int y \cdot dA$ = moment of surface area about the free surface of liquid.

= area of surface × depth of C.G. from the free surface

= $A\bar{h}$

$\therefore \quad P = \gamma A \bar{h}$ ... (2.14)

Centre of pressure, h_c

Moment of force dP about the free surface of liquid,

$$dM = dP \times y = \gamma y b \cdot dy \times y = \gamma b \cdot dy \cdot y^2$$

Total moment about the free surface of liquid,

$$M = \gamma \int b \cdot dy \cdot y^2 = \gamma \int dA \cdot y^2$$

Now $\int dA \cdot y^2$ = moment of inertia of the surface about free surface of liquid

$$= I_0$$

$$\therefore \quad P \times h_c = \gamma I_0$$

$$\gamma A \bar{h} \times h_c = \gamma I_0$$

$$h_c = \frac{I_0}{A\bar{h}} \quad \text{... (2.15)}$$

Using parallel axis theorem, we have

$$I_0 = I_G + A\bar{h}^2$$

When I_G = moment of inertia of area about an axis passing through C.G. of the area and parallel to the area and parallel to the free surface of the liquid.

$$\therefore \quad h_c = \frac{I_G + A\bar{h}^2}{A\bar{h}} = \frac{I_G}{A\bar{h}} + \bar{h} \quad \text{... (2.16)}$$

If the top of vertical surface A touches the free surface of liquid then the magnitude of hydrostatic force $= \frac{1}{2}\gamma h_2 A$, and acts through a point $\frac{2}{3}h_2$ below the free surface.

2.7.3 Inclined submerged plane surface

Consider a plane surface of arbitrary shape submerged in such a way that the plane of the surface makes an angle θ with the liquid free surface as shown in Fig. 2.13

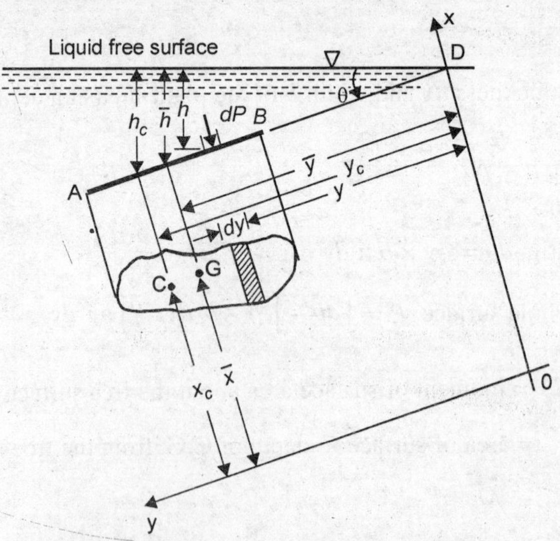

Fig. 2.13 Inclined submerged plane surface.

(a) Total pressure, p

Let A = area of inclined surface

$\bar{h}, h_c$ = depth of C.G and centre of pressure of inclined surface from the liquid free surface respectively.

Let the plane of the inclined surface, when produced meet the liquid free surface at D.

At D draw rectangular x-axis with origin at O and Oy as y-axis.

Let $\bar{y}, y_c$ = distance of C.G. and centre of pressure C from x-axis respectively.

$\bar{h}, h_c$ = depth of G and C from liquid free surface

Consider an elementary strip of surface at a distance y from the x-axis and of width dy so that area of strip, dA is at a depth h from the liquid free surface.

Pressure on the elementary strip, $p = \gamma h$

Force on the strip, $dP = p \times dA = \gamma h \times dA$

Total force on the plane surface, $P = \int \gamma h \cdot dA$

Now
$$\frac{h}{y} = \frac{\bar{h}}{\bar{y}} = \frac{h_c}{y_c} = \sin\theta$$

$$h = y \sin\theta$$

$\therefore$
$$P = \gamma \sin\theta \int y \cdot dA$$

But
$$\int y \cdot dA = A\bar{y}$$

$$P = \gamma \sin\theta \cdot \bar{y}A = \gamma A \bar{h} \qquad \text{... (2.17)}$$

(b) Centre of pressure, h_c

Force on the elementary strip, $dP = \gamma y \sin\theta .\ dA$

Moment of dP about x-axis, $dM = dp \times y = \gamma y^2 \sin\theta .\ dA$

Total moment, $M = \gamma \sin\theta \int y^2 dA$

$$\int y^2 dA = \text{moment of inertia of surface about } x\text{-axis}$$

$$= I_x$$

$\therefore$
$$M = \gamma \sin\theta .\ I_x$$
$$= P \times y_c$$

$\therefore$
$$y_c = \frac{\gamma \sin\theta \cdot I_x}{\gamma \sin\theta\, \bar{y}A} = \frac{I_x}{A\bar{y}} = \frac{I_x \sin\theta}{A\bar{h}}$$

By the theorem of parallel axis, we have

$$I_x = I_G + A\bar{y}^2 = I_G + A\left(\frac{\bar{h}}{\sin\theta}\right)^2$$

$$\therefore \qquad \frac{h_c}{\sin\theta} = \frac{\sin\theta}{A\bar{h}}\left[I_G + A\left(\frac{\bar{h}}{\sin\theta}\right)^2\right]$$

$$h_c = \frac{I_G \sin^2\theta}{A\bar{h}} + \bar{h} \qquad \text{... (2.18)}$$

2.7.4 Curved Submerged Surface

Consider the trace AB of the curved surface submerged in a static liquid as shown in Fig. 2.14. Consider an elementary area dA of the curved surface lying at a vertical depth *y* below the liquid free surface.

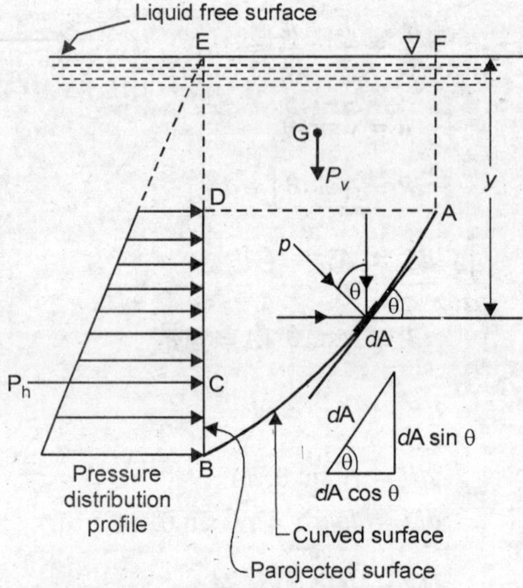

Fig. 2.14 Curved submerged surface.

Let $\qquad p =$ normal pressure on the elementary area.

Normal force on the elementary area,

$$dP = p.dA = \gamma y.\, dA$$

Total force on the curved surface,

$$P = \int \gamma y \cdot dA$$

Horizontal component of force on elementary area,

$$dP_x = dP \sin\theta = p \sin\theta . dA$$

Vertical component of force on elementary area,

$$dP_y = dP \cos\theta = p \cos\theta \cdot dA$$

Total forces on curved surface are:

$$P_h = \int dP_h = \gamma \int y\, dA \sin\theta$$

$$P_V = \int dP_V = \gamma \int y\, dA \cos\theta$$

Now $\gamma \int y\, dA \sin\theta$ = total force on the projected area of the curved surface on the vertical plane ... (2.19 a)

∴ P_h = total force on the projected area of curved surface on the vertical plane.

Now $\int y dA \cos\theta$ = total volume contained between the curved surface extended upto the free surface, i.e. volume *ABDEFA*.

∴ $P_V = \gamma \int y\, dA \cos\theta$ = total weight supported by the curved surface upto the liquid free surface. ... (2.19 b)

Resultant force, $$P = \sqrt{P_h^2 + P_V^2} \quad \text{... (2.20 a)}$$

$$\tan\phi = \frac{P_V}{P_h} \text{ with the horizontal} \quad \text{... (2.20 b)}$$

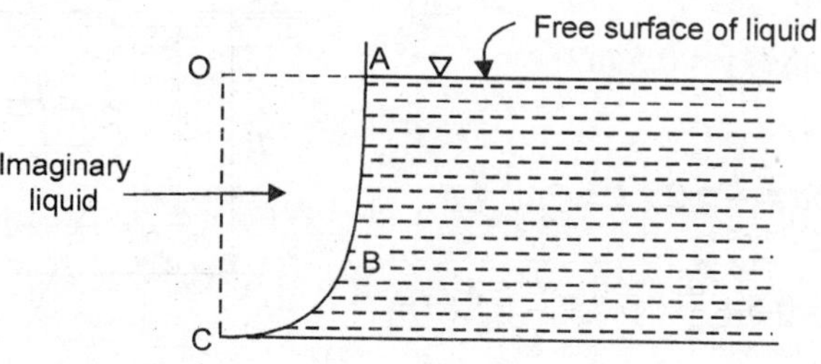

Fig. 2.15 Imaginary liquid

ote: If the liquid acts from below the curved surface then the vertical component of force acts upwards, and equals the weight of the imaginary column of liquid above the curved surface upto the free surface of liquid, as shown in Fig. 2.15.

Example 2.6 *A circular plate of diameter 1.5 is placed vertically in water such that the centre of the plate is 3 m below the free surface of water. Find the total pressure on the plate and the centre of pressure.*

Solution.

Given: $d = 1.5$ m

Refer to Fig. 2.16.

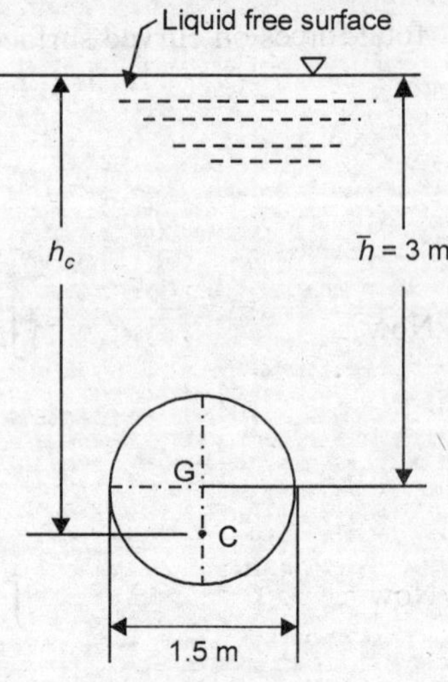

Fig. 2.16

Area of plate, $A = \frac{\pi}{4}d^2 = \frac{\pi}{4}(1.5)^2 = 1.767\text{m}^2$

$\bar{h} = 3$ m

Total pressure force, $P = \gamma A\bar{h}$

$= 9810 \times 1.767 \times 3 = = 52007$ N

Position of centre of pressure,

$$h_C = \bar{h} + \frac{I_G}{A\bar{h}}$$

$$I_G = \frac{\pi d^4}{64} = \frac{\pi(1.5)^4}{64} = 0.2485\,\text{m}^4$$

$$h_c = 3 + \frac{0.2485}{1.767 \times 3} = 3.047\,\text{m}$$

Example 2.7 *An isosceles triangular plate of base 2m is submerged vertically in an oil of specific gravity 0.82. The base of the plate is 0.5 m below the free surface of oil. Calculate the total pressure on the plate and centre of pressure.*

Solution.

Given: $b = d = 2$m, $s = 0.82$, $y = 0.5$ m

Refer to Fig. 2.17.

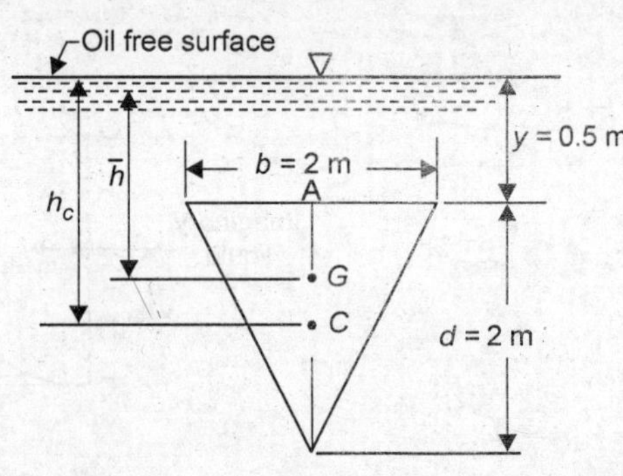

Fig. 2.17

Area of plate, $A = \frac{1}{2}bd = \frac{1}{2} \times 2 \times 2 = 2\text{m}^2$

$$\bar{h} = y + AG = 0.5 + \frac{d}{3} = 0.5 + \frac{2}{3} = 1.67\,\text{m}$$

Total pressure, $P = \gamma A\bar{h} = 0.82 \times 9810 \times 2 \times 1.67$

$= 18769.8$ N

$$I_G = \frac{bd^3}{36} = \frac{2 \times 2^3}{36} = 0.444\,\text{m}^4$$

Centre of pressure, $h_C = \bar{h} + \frac{I_G}{A\bar{h}} = 1.67 + \frac{0.444}{2 \times 1.67} = 1.357\,\text{m}$

Example 2.8 *A circular opening of 2 m diameter in the vertical side of a tank is closed by a disc of 2 m diameter which can rotate about a horizontal diameter. Calculate:*

(a) the force on the disc, and (b) the torque required to maintain the disc in equilibrium in vertical position when the head of water above the horizontal diameter is 3m.

Solution.

Given: $d = 2m,\ \bar{h} = 3\text{m}$

Refer to Fig. 2.18.

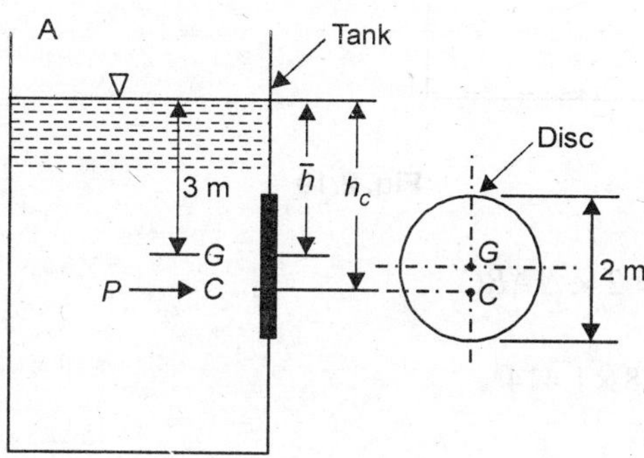

Fig. 2.18

Area of disc, $A = \frac{\pi}{4}d^2 = \frac{\pi}{4} \times 2^2 = \pi\,\text{m}^2$

(a) Force on the disc, $P = \gamma A\bar{h}$

$= 9810 \times \pi \times 3 = 92,457\ \text{N}$

(b) Centre of pressure, $h_c = \bar{h} + \frac{I_G}{A\bar{h}}$

$$= 3 + \frac{\frac{\pi d^4}{64}}{\pi \times 3} = 3 + \frac{16}{64 \times 3} = 3.083\,\text{m}$$

Moment of for ce P about horizontal diameter through $G = P(h_c - \bar{h})$

$= 92,457\ (3.083 - 3) = 7704.75$ N.m (ccw)

Torque to be applied on the disc to maintain equilibrium = 7704.75 N.m (cw)

Example 2.9 *A square opening in the vertical side of a tank has one diagonal vertical and is completely covered by a plane plate hinged along one of the upper sides of the opening. The diagonals of the opening are 2.828 m long. The tank contains a liquid of specific gravity 1.15. The centre of the opening is 2.5 m below the free surface of the liquid. Calculate: (a) the thrust exerted on the plate by the liquid, and (b) the position of its centre of pressure.*

Solution. Refer to Fig. 2.19:

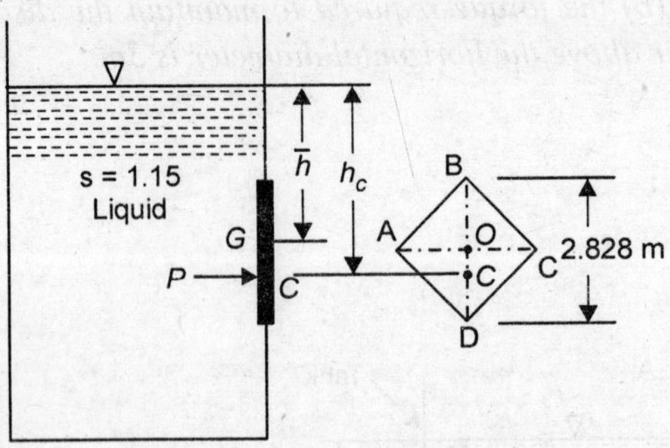

Fig. 2.19

Area of square opening, $A = 2 \times \Delta ABC$

$$= 2 \times \frac{1}{2} \times 2.828 \times 1.414$$

$$= 4 \text{ m}^2$$

$$\bar{h} = 2.5 \text{ m}$$

(a) Thrust exerted on plate = $\gamma A \bar{h}$

$$= 1.15 \times 9810 \times 4 \times 2.5$$

$$= 112,815 \text{ N}$$

I_G = moment of inertial of ABCD about diagonal AC

= 2 × moment of inertia of ΔABC about AC

$$= 2 \times \frac{2.828 \times (1.414)^3}{12} = 0.13325 \text{ m}$$

$$\frac{I_G}{A\bar{h}} = \frac{1.3325}{4 \times 2.5} = 0.13325 \text{ m}$$

$$h_c = \bar{h} + \frac{I_G}{A\bar{h}} = 2.5 + 0.13325 = 2.63325 \text{ m}$$

Example 2.10 *A trapezoidal gate plate 6 m wide at the top, 3m wide at the bottom and 1.5 m deep is used to close an opening of the same size. The water on the outside is just level with the top and dock is empty. Calculate total pressure and centre of pressure on the plate.*

Solution:

Given: AB = 6m, DC = 3m, ED = 1.5 m

Refer to Fig. 2.20.

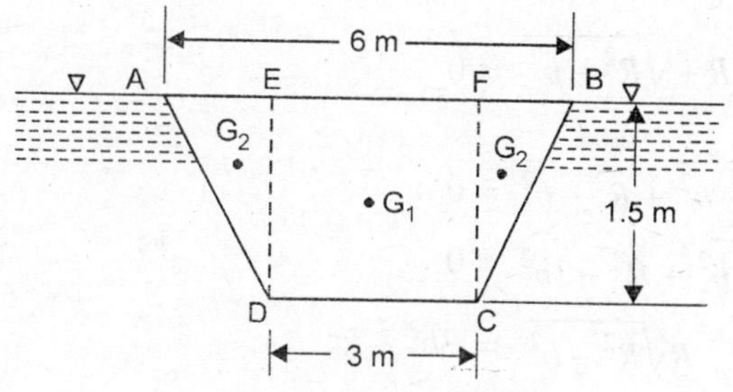

Fig. 2.20

Area, $A = A_1 + 2A_2 = 3 \times 1.5 + 2 \times 1.5 \times 2 = 6.75 \text{ m}^2$

$$\bar{h} = \left(4.5 \times 0.75 + 2 \times \frac{1}{2} \times 1.5 \times \frac{1.5}{3}\right) \times \frac{1}{6.75}$$

= 0.6667 m from free surface of water

Total pressure, $P = \gamma A\bar{h} = 9810 \times 6.75 \times 0.6667 = 44{,}145 \text{N}$

$$I_G = \frac{3 \times (1.5)^3}{12} + 4.5 \times (0.75 - 0.6667)^2 + 2\left[\frac{1.5 \times (1.5)^3}{12} + \frac{1}{2} \times 1.5 \times 1.5 \times (0.6667)^2\right]$$

$= 0.87497 + 1.84385 = 2.7188 \text{ m}^4$

$$h_c = \bar{h} + \frac{I_G}{A\bar{h}} = 0.6667 + \frac{2.7188}{6.75 \times 0.6667} = 1.2708 \text{m}$$

xample 2.11 *A circular lamina of radius R is kept immersed in a liquid such that its top most point A on the free surface. Determine the depth and width of the horizontal chord BC to that the total thrust e to hydrostatic pressure on the triangle ABC is maxium.*

olution. Refer to Fig. 2.21

The total pressure force on immersed triangle is ABC is,

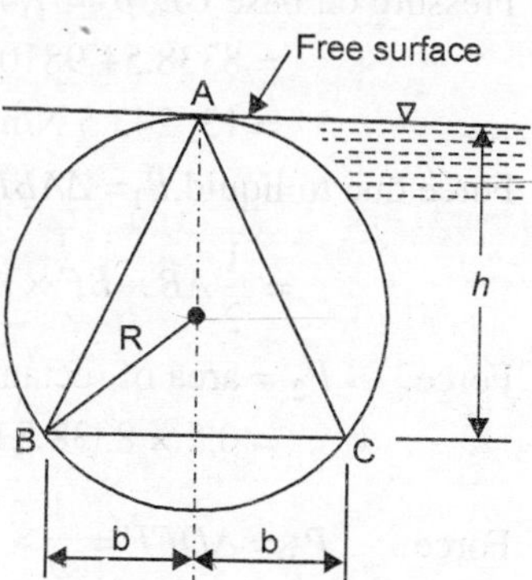

Fig. 2.21

$$P = \gamma A\bar{h} = \gamma \times \frac{1}{2} \times 2b \times h \times \frac{2h}{3} = \frac{2}{3}\gamma bh^2$$

$$h = R + \sqrt{R^2 - b^2}$$

$$P = \frac{2}{3}\gamma b[R + \sqrt{R^2 - b^2}]^2$$

For R to be maximum, $\frac{dP}{db} = 0$

$$\frac{d}{db}[b\{R + \sqrt{R^2 + b^2}\}^2] = 0$$

$$b \times 2\{R + \sqrt{R^2 - b^2}\} \times \frac{1}{2}(R^2 - b^2)^{-1/2} \times (-2b) + \{R + \sqrt{R^2 - b^2}\}^2 = 0$$

$$\frac{-2b^2}{\sqrt{R^2-b^2}} + R + \sqrt{R^2-b^2} = 0$$

$$-2b^2 + R\sqrt{R^2-b^2} + R^2 - b^2 = 0$$

$$R\sqrt{R^2-b^2} + R^2 - 3b^2 = 0$$

$$R\sqrt{R^2-b^2} = 3b^2 - R^2$$

Squaring both sides, we get

$$R^2(R^2-b^2) = 9b^4 + R^4 - 6b^2R^2$$

$$R^4 - R^2b^2 = 9b^4 + R^4 - 6b^2R^2$$

$$9b^4 - 5b^2R^2 = 0$$

$$9b^4 - 5R^2 = 0$$

$$b = \sqrt{\frac{5}{9}}R = \frac{\sqrt{5}R}{3}$$

$$h = R + \sqrt{R^2 - \frac{5}{9}R^2} = \frac{5}{3}R \quad \text{and width, } 2b = \frac{2}{3}\sqrt{5}R$$

Example 2.12 *A tank contains water and a liquid of specific gravity 0.85 upto a height of 0.5 m and 1 m respectively, as shown in Fig. 2.22 Calculate: (a) Total pressure on the side of the tank, and (b) the position of centre of pressure from one side of the tank, which is 1.5 m wide.*

Solution. (a) The pressure forces on the tank are shown in Fig. 2.23.

Pressure on BF, $p_B = \gamma_1 h_1 = 0.85 \times 9810 \times 1\ 8338.5$ N/m²

Pressure on base CE, $p_c = \gamma_1 h_1 + \gamma_2 h_2$

$= 8338.5 + 9810 \times 0.5$

$= 13,243.5$ N/m²

Force due to liquid, $P_1 = \Delta ABF \times$ width of tank

$$= \frac{1}{2} AB \times BF \times 1.5 = \frac{1}{2} \times 1 \times 8338.5 \times 1.5 = 6253.81\text{N}$$

Force, P_2 = area of rectangle $BCDF$ × width of tank

$= 0.5 \times 8338.5 \times 1.5 = 6253.87$ N

Force $P_3 = \Delta DFF = \frac{1}{2} \times FD \times DE$ × width of tank

$$= \frac{1}{2} \times 0.5 \times (13{,}243.5 - 8338.5) \times 1.5$$

$= 1839.4$ N

Total pressure, $P = P_1 + P_2 + P_3$

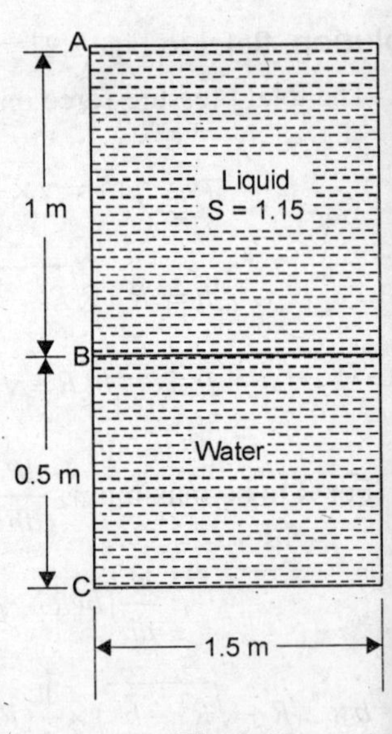

Fig. 2.22

$$= 6253.8 + 6253.8 + 1839.4$$
$$= 14347 \text{ N}$$

(b) Centre of pressure:

$$\bar{h}_1 = \frac{2}{3} \times AB = \frac{2}{3} \times 1 = \frac{2}{3} \text{ m}$$

$$\bar{h}_2 = AB + \frac{1}{2} BC = 1 + \frac{0.5}{2} = 1.25 \text{ m}$$

$$\bar{h}_3 = AB + \frac{2}{3} \times BC = 1 + \frac{2}{3} \times 0.5 = 1.333 \text{ m}$$

$$P \times h_c = P_1 \bar{h}_1 + P_2 \bar{h}_2 + P_3 \bar{h}_3$$

$$14347 \times h_c = 6253.87 \times \frac{2}{3} + 6253.87 \times 1.25 + 1839.4 \times 1.333$$
$$= 14438.3$$
$$h_c = 1.0064 \text{ m}$$

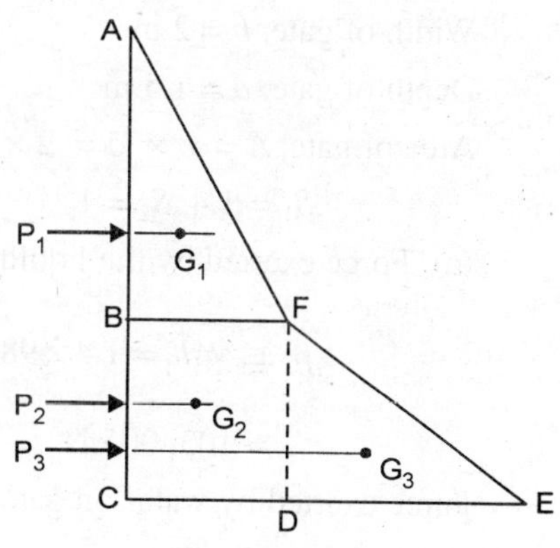

Fig. 2.23

Example 2.13 *An opening in a dam is covered by a vertical sluice gate. The opening is 2m wide and 1.5 m high. On the upstream of the gate the liquid of specific gravity 1.4, lies upto a height of 1.75 m above the top of the gate, whereas on the downstream side the water is available upto a height touching the top of the gate. Determine:*

(a) the resultant force acting on the gate and position of centre of pressure, and

(b) the force acting horizontally at the top of the gate which is capable of opening it.

Assume that the gate is hinged at the bottom.

Solution. Refer to Fig. 2.24

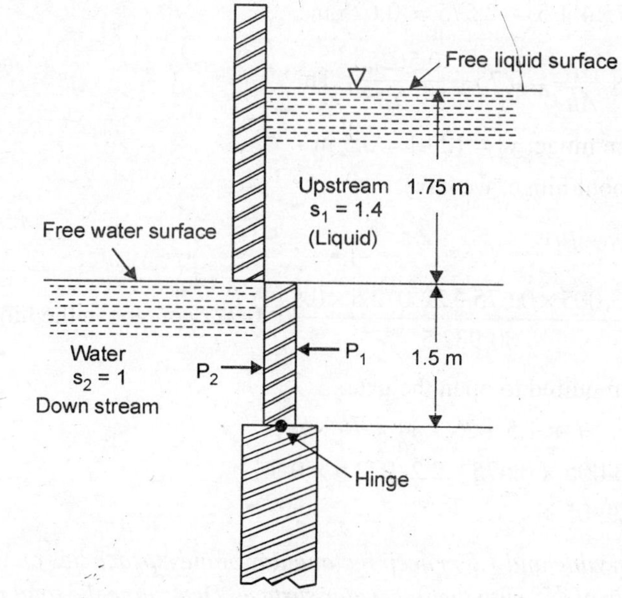

Fig. 2.24

Width of gate, $b = 2$ m

Depth of gate, $d = 1.5$ m

Area of gate, $A = b \times d = 2 \times 1.5 = 3\text{m}^3$

$$S_1 = 1.4, S_2 = 1$$

(a) Force exerted by the liquid on the gate,

$$P_1 = \gamma A\bar{h}_1 = 1.4 \times 9810 \times 3 \times \left(1.75 + \frac{1}{2} \times 1.5\right)$$

$$= 103, 005 \text{ N}$$

Force exerted by water on gate,

$$P_2 = \gamma A\bar{h}_2 = 9810 \times 3 \times \frac{1}{2} \times 1.5 = 22{,}072.5\text{N}$$

Resultant force, $P = P_1 - P_2 = 103{,}005 - 22{,}07.5 = 80{,}932.\ 5$ N

$$h_{c1} = \bar{h}_1 + \frac{I_G}{A\bar{h}_1}$$

$$I_G = \frac{bd^3}{12} = \frac{2 \times (1.5)^3}{12} = 0.5625\,\text{m}^4$$

$$\bar{h}_1 = 1.75 + 0.75 = 2.5\,\text{m}$$

$$h_{c1} = 2.5 + \frac{0.5625}{3 \times 2.5} = 2.575\,\text{m}$$

Distance of P_1 from hinge,

$$y_1 = (1.75 + 1.5) - 2.575 = 0.675 \text{ m}$$

$$h_{c2} = \bar{h}_2 + \frac{I_G}{A\bar{h}_2} = 0.75 + \frac{0.5625}{3 \times 0.75} = 1\,\text{m}$$

Distance of P_2 from hinge, $y_2 = 1.5 – 1 = 0.5$ m

Taking moments about hinge, we have

$$P \times y = P_1 y_1 - P_2 y_2$$

$$y = \frac{103{,}005 \times 0.675 - 22{,}072.5 \times 0.5}{80{,}932.5} = 0.7227\,\text{m} \text{ above the hinge.}$$

(b) Let F = force required to open the gate.

$$F \times 1.5 + P_2 \times y_2 = P_1 \times y$$

$$F \times 1.5 = 103{,}005 \times 0.675 - 2.2{,}072.5 \times 0.5$$

$$F = 38{,}9905 \text{ N}$$

Example 2.14 *A 1.2 m wide and 1.5 m deep rectangular plane surface lies on water in such a way th its plane makes an angle of 45° with the free water surface. Determine the total pressure and position centre of pressure when the upper edge is 1 m below the free surface of water.*

Solution. Refer to Fig. 2.25.

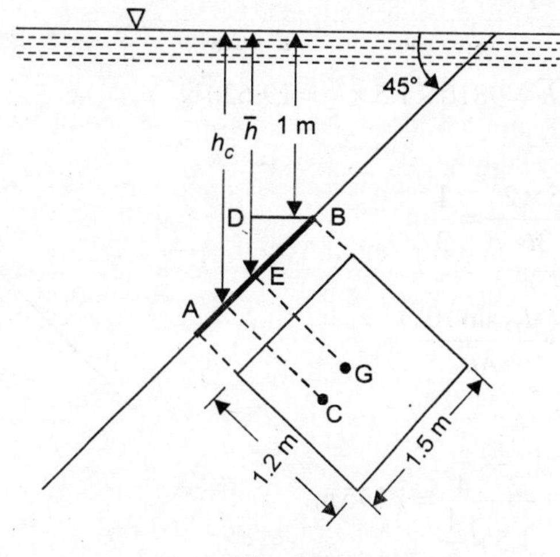

Fig. 2.25

Width of plane surface = 1.2 m

Depth of plane surface = 1.5 m

$\theta = 45°$

Distance of upper edge from free water surface = 1 m

(a) Total pressure, $P = \gamma A\bar{h}$

$$\bar{h} = 1 + BE \sin\theta = 1 + 0.75 \sin 45° = 1.53 \text{ m}$$

$$P = 9810 \times (1.2 \times 1.5) \times 1.53 = 27{,}017 \text{ N}$$

(b) Centre of pressure, $h_c = \dfrac{I_G \sin^2\theta}{A\bar{h}} + \bar{h}$

$$I_G = \frac{1.2 \times (1.5)^3}{12} = 0.3375 \text{ m}^4$$

$$h_c = \frac{0.3375 \times (\sin 45°)^2}{1.2 \times 1.5 \times 1.53} + 1.53 = 1.591 \text{m}$$

Example 2.15 *A triangular plate of 1.5 m base and 2 m height is immersed in water. The plane of the plate is inclined at 30° with free water surface. The base of plate is parallel to and at a depth of 1 m from water surface. Determine the total pressure on the plate and the position of centre of pressure.*

Solution. Refer to Fig. 2.26

Area of plate, $A = \dfrac{1}{2} \times 1.5 \times 2 = 1.5 \text{m}^2$

$\theta = 30°$

$$\bar{h} = 1 + \frac{2}{3}\sin 30° = \frac{4}{3}\text{ m}$$

Total pressure, $P = \gamma A\bar{h} = 9810 \times 1.5 \times \frac{4}{5} = 19{,}620\text{ N}$

$$I_G = \frac{1.5 \times 2^3}{36} = \frac{1}{3}\text{ m}^4$$

$$h_c = \bar{h} + \frac{I_G \sin^2\theta}{A\bar{h}}$$

$$= \frac{4}{3} + \frac{\frac{1}{3} \times \frac{1}{4}}{1.5 \times \frac{4}{3}} = 1.375\text{ m}$$

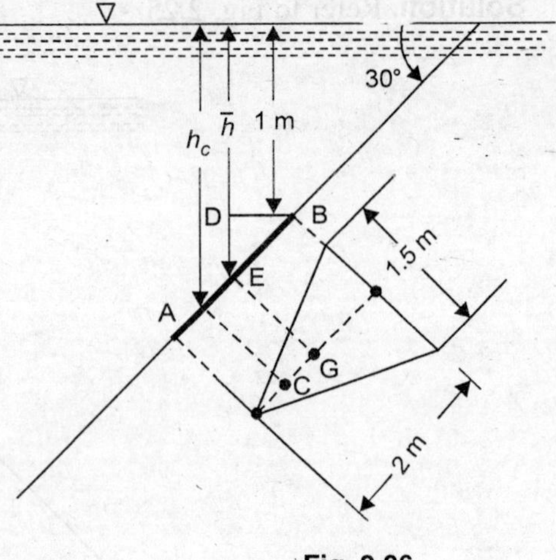

Fig. 2.26

Example 2.16 *A rectangular gate AB 3 m wide and 4 m long is hinged at A as shown in Fig. 2.27. It is kept closed by a weight fixed to the gate. The total weight of the gate and weight fixed to the gate is 450 kN. Find the height of water 'h' which will just cause the gate to open. The C.G. of weight and gate is at G.*

Solution.

Given: $b = 3$ m. $l = 4$ m, $W = 450$ kN, $\theta = 45°$

Let h = depth of water, m

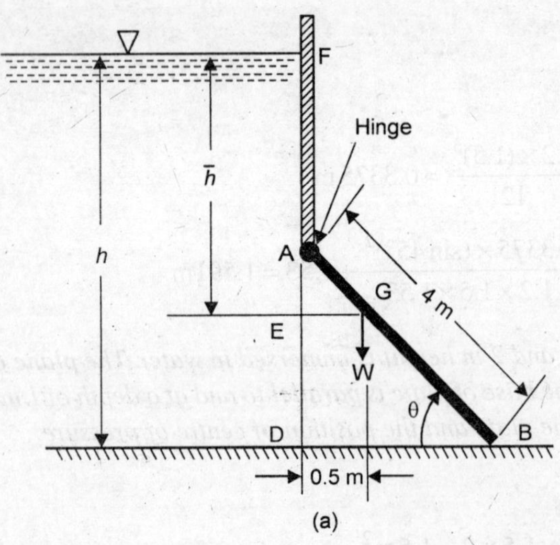

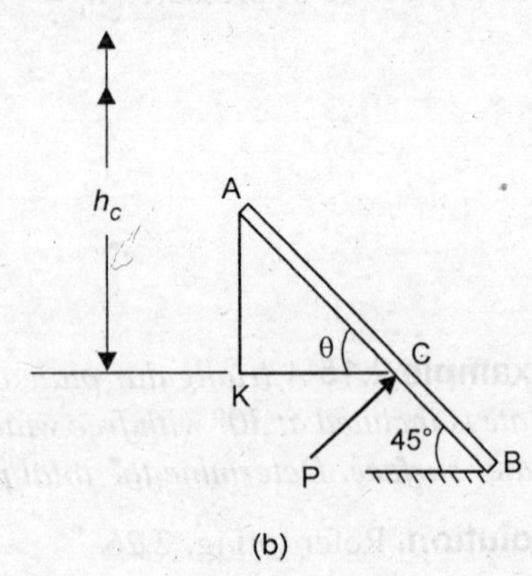

Fig. 2.27

$$\bar{h} = h - ED = h - (AD - AE) = h - (AB\sin\theta - EG\tan\theta)$$

$$= h - (4\sin 45° - 0.5\tan 45°)$$

$$= h - (2.8284 - 0.5) = (h - 2.3284)\ m$$

Total pressure force, $P = \gamma A\bar{h}$

$$= 9810 \times (3\times 4) \times (h - 2.3284)$$

$$= 117{,}720\ (h - 2.32\ 84)\ N$$

$$h_c = \frac{I_G \sin^2\theta}{A\bar{h}} + \bar{h}$$

$$I_G = \frac{I_G \sin^2\theta}{A\bar{h}} + \bar{h}$$

$$I_G = \frac{bl^3}{12} = \frac{3\times 4^3}{12} = 16\,\text{m}^4$$

$$h_c = \frac{16\times\sin^2 45°}{12\bar{h}} + \bar{h} = \frac{2}{3\bar{h}} + \bar{h}$$

Taking moments about A, we have

$$450 \times 10^3 \times EG = P \times AC$$

$$450 \times 10^3 \times 0.5 = 117,\,720\ (h - 2.3284) \times \frac{AK}{\sin 45°}$$

$$AK = \frac{1.3515}{h - 2.3284} \qquad \text{... (1)}$$

$$AK = h_C - AF = (h - 2.3284) + \frac{2}{3(h - 2.3284)} - AF$$

$$AF = FD - AD - h - AB\sin 45° = h - 4\sin 45° = h - 2.8284$$

$$\therefore \quad AK = (h - 2.3284) + \frac{2}{3(h - 2.3584)} - (h - 2.8284)$$

$$= 0.5 + \frac{2}{(h - 2.3284)} \qquad \text{... (2)}$$

Equating Eqs. (1) and (2), we have

$$\frac{1.3515}{h - 2.3284} = 0.5 + \frac{0.66667}{h - 2.3284}$$

$$\frac{0.68483}{h - 2.3284} = 0$$

$$0.68483 = 0.5\,h - 1.1642$$

$$h = 3.7 \text{ m}$$

Example 2.17: *A hemispherical projection of diameter 0.5 m exists on one the vertical sides of a tank. The tank contains water upto a height of 2m above the centre of the hemisphere. Calculate the vertical and horizontal forces acting on the projection.*

Solution. Refer to Fig. 2.28

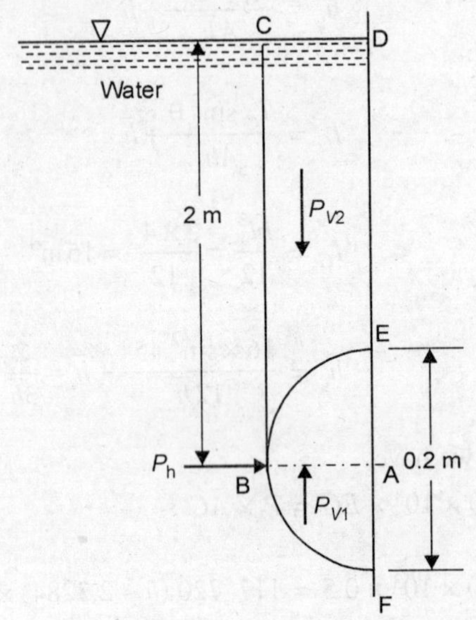

Fig. 2.28

Horizontal force, $P_h = \gamma A\bar{h} = 9810 \times \pi(0.25)^2 \times 2 = 3852.4\text{N}$

Vertical force, $P_V = P_{V1} - P_{V2}$

= Weight of volume of water *CDFBC*

– Weight of volume of water *BEDCB*

= Weight of water contained in hemisphere *EBFE*

$$= \gamma \times \frac{1}{2} \times \frac{4}{3} \times \pi r^3$$

$$= 9810 \times \frac{2}{3} \times \pi(0.25)^3 = 321\text{N}$$

Example 2.18 *Fig. 2.29 shows a curved surface AB in the form of a quadrant of a circle of radius 2 m immersed in water. The width of the gate is 1 m. Calculate the horizontal and vertical components of the forces acting on the curved surface.*

Solution: Refer to Fig. 2.29

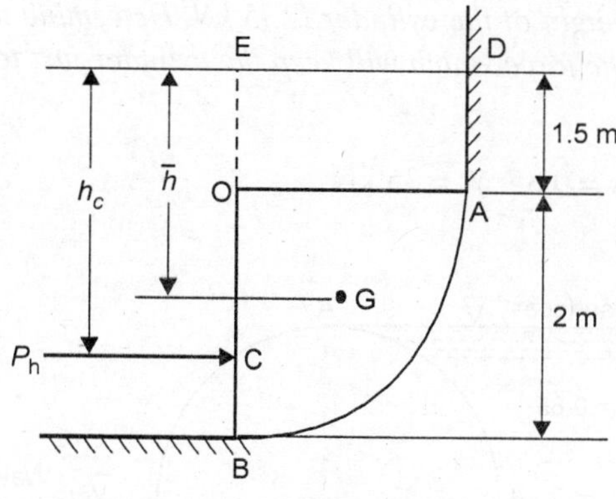

Fig. 2.29

Radius of gate, $r = 2\text{m}$

Width of gate = 1 m

Horizontal force exerted by water on gate,

P_h = Total pressure force on projected area of curved surface AB on vertical plane

$$= \gamma \times A \times \bar{h}$$

where $A = \text{OB} \times 1 = 2 \times 1 = 2\text{m}^2$

$$\bar{h} = 1.5 + \frac{2}{2} = 2.5\,\text{m}$$

$$P_h = 9810 \times 2 \times 2.5 = 49{,}050 \text{ N}$$

$$h_c = \bar{h} + \frac{I_G}{A\bar{h}}$$

where I_G = moment of inertia of OB about its C.G. = $\dfrac{1 \times 2^3}{12} = 0.667\,\text{m}^4$

$$h_c = 2.5 + \frac{0.667}{2.25} = 2.63\,\text{m} \text{ form water free surface}$$

Vertical component of total force,

P_v = Weight of water supported by AB upto the free surface.

= Weight of water *ABOEDA*

= γ [Volume *OABO* + Volume *OADEO*]

$$= \gamma\left[\frac{\pi \times 2^2}{2} \times 1 + 2 \times 1.5 \times 1\right]$$

= 9810 × 6.1416 = 60, 249 N

Example 2.19 *A cylinder having 3 m diameter and 2 m length is resting on the floor. On one side, water is filled upto half the depth while on the other side oil of specific gravity 0.85 is filled upto the top, as shown in Fig. 2.30. The weight of the cylinder is 35 kN. Determine the magnitude of the horizontal and vertical components of the force which will keep the cylinder just touching the floor.*

Solution.

Given: $d = 3\text{m}, l = 2\text{m}, s = 0.85, W = 35 \text{ kN}$

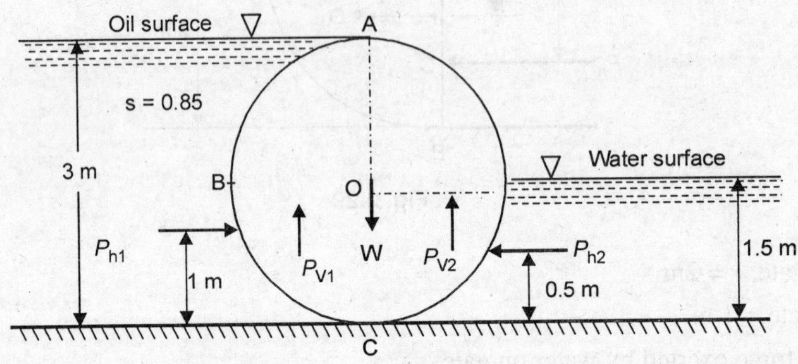

Fig. 2.30

Horizontal forces:

$$P_{h1} = s\gamma \times d \times l \times \frac{d}{2}$$

$$= 0.85 \times 9810 \times 3 \times 2 \times \frac{3}{2}$$

$$= 75{,}046.5 \text{ N}$$

This will act at $\frac{3}{3} = 1$ m from the bed.

$$P_{h2} = \gamma \times \frac{d}{2} \times l \times \frac{d}{4} = 9810 \times 1.5 \times 2 \times \frac{3}{4} = 22{,}072.5\text{N}$$

This will act at $\frac{1.5}{3} = 0.5\text{m}$ from the bed.

Resultant horizontal force = $P_{h1} - P_{h2} = 75{,}046.5 - 22{,}072.5 = 52{,}974$ N →

Free required to hold the cylinder stationary = 52,974 N to the left.

Vertical forces

$$P_{V1} = s\gamma \times \frac{\pi r^2}{2} \times l = 0.85 \times 9810 \times \frac{\pi \times (1.5)^2}{2} \times 2 = 58{,}945\text{N}$$

It will act at $\frac{4r}{3\pi} = \frac{4 \times 1.5}{3\pi} = 0.6366\,\text{m}$ to the left of AC.

$$P_{V2} = \gamma \times \frac{\pi r^2}{4} \times l = 9810 \times \frac{\pi (1.5)^2}{4} \times 12 = 34{,}611\,\text{N}$$

It will act at 0.6366 m to the right of AC.

For equilibrium of vertical forces, external force required

$$= 58941 + 34{,}611 - 35{,}000 = 58{,}552\ \text{N}\downarrow$$

Taking moments about P_{V2}, we have

$$W \times 0.6366 + 58{,}552 \times x = 58{,}941 \times 2 \times 0.6366$$

$$58{,}552\,x = 52{,}763$$

$$x = 0.901\ \text{m}$$

Example 2.20 *A cylinder 3 m in diameter and 4 m long retains water on one side as shown in Fig. 2.31. Determine the horizontal reaction at A and the vertical reaction at B. The weight of the cylinder is 200 kN.*

Solution.

Given: $d = 3\text{m}$, $l = 4$ m, $W = 200$ kN

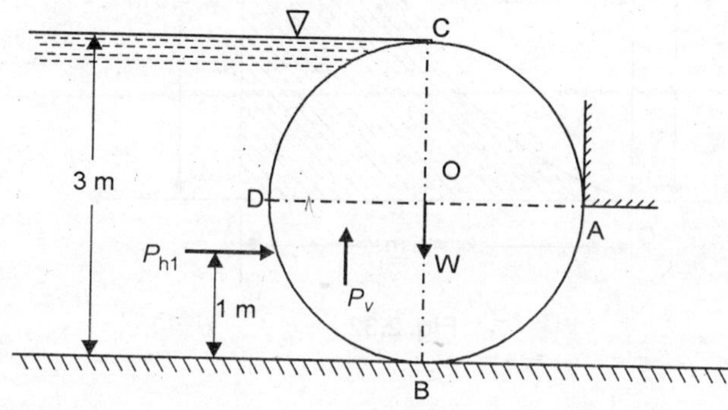

Fig. 2.31

Horizontal force exerted by water,

$$P_h = \text{Force on vertical area } BOC$$

$$= \gamma A\bar{h}$$

where

$$A = d \times l = 3 \times 4 = 12\text{m}^2$$

$$\bar{h} = \frac{d}{2} = \frac{3}{2} = 1.5\,\text{m}$$

$$P_h = 9810 \times 12 \times 1.5 = 176{,}580\ \text{N} \rightarrow$$

Vertical force exerted by water = Weight of imaginary water contained in volume *BDCOB*

$$P_v = r \times \frac{\pi r^2}{2} \times l = 9810 \times \frac{\pi \times (1.5)^2}{2} \times 4 = 138.686\,\text{N} \uparrow$$

For the equilibrium of cylinder,

$$(R_h)_A = P_h = 176580\text{ N}$$

$$(R_v)_B = \text{W} - P_v = 200{,}000 - 138,686 = 61{,}314\text{ N}$$

Example 2.21 *A water dam has a parabolic shape:* $y = y_0\left(\frac{x}{x_0}\right)^2$ *as shown in Fig. 2.32.* $x_0 = 5m$ *and* $y_0 = 8$ *m. Calculate the horizontal, vertical and the resultant thrust exerted by water per meter length of dam.*

Solution. Equation of curve *OA*,

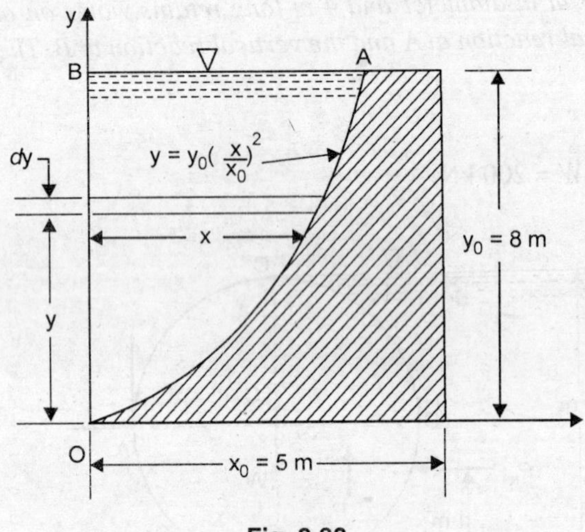

Fig. 2.32

$$y = y_0\left(\frac{x}{x_0}\right)^2 = 8\left(\frac{x}{5}\right)^2 = 0.32x^2$$

or

$$x^2 = 3.125\text{ y}$$

$$x = 1.768\sqrt{y}$$

Length of dam = 1 m

Horizontal thrust exerted by water,

P_h = Force exerted by water on projected surface *OB* of OA.

$$= \gamma A\bar{h} = 9810 \times (8 \times 1) \times 4 = 313{,}920\,\text{N}$$

Vertical thrust exerted by water,

P_V = Weight of water supported by curved surface *OA* upto the free surface of water

$= \gamma \times \text{Area } AOB \times \text{Length of dam}$

$$= 9810 \times [\int_0^8 x \cdot dy] \times 1.0$$

$$= 9810 \times 1.768 \int_0^8 y^{1/2} dy$$

$$= 17344 \times \frac{2}{3} | y^{3/2} |_0^8 = 115.72 \times 8^{3/2} = 261{,}634 \text{ N}$$

Resultant thrust, $P = \sqrt{P_h + P_V^2} = \sqrt{(313{,}920)^2 + (261{,}634)^2}$

$= 408{,}654$ N

$$\tan\theta = \frac{P_V}{P_h} = \frac{261{,}634}{313{,}920} = 0.83344$$

$\theta = 39.8°$

Example 2.22 *Calculate the horizontal and vertical components of the water pressure forces exerted on a tainter gate of radius 6m and unit width, as shown in Fig. 2.33.*

Solution. $AD = 6\ BC = \sin 30° = 3$ m

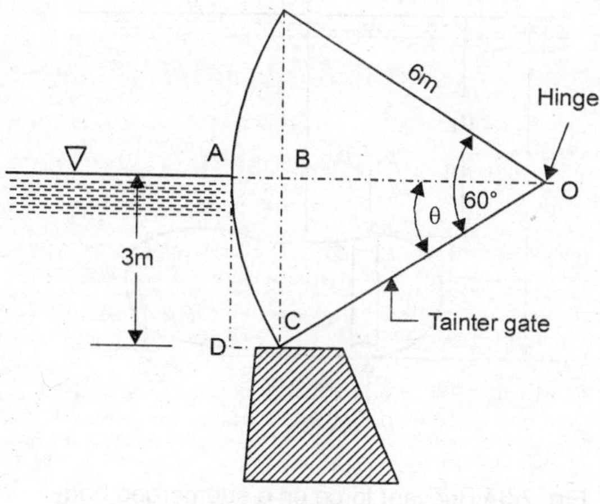

Fig. 2.33

Horizontal component of water pressure force = Force on projected area *AD*,

$$P_h = \gamma A\bar{h} = 9810 \times (3 \times 1) \times 1.5 = 44{,}145 \text{ N}$$

Vertical component of water pressure force,

$$P_v = \text{Weight of water enclosed by imaginary curved surface AC}$$
$$= \text{Weight of water in the portion } ACBA$$
$$= \gamma \times (\text{area of ACBA}) \times \text{Width of gate}$$
$$= \gamma \times (\text{area of sector } ACO - \text{area } \Delta\, \text{BCO})$$
$$= 9810 \times \left(\frac{30°}{360°} \times \pi r^2 - \frac{1}{2} \times BC \times BO \right)$$
$$= 9810 \times \left(\frac{\pi \times 36}{12} - \frac{1}{2} \times 3 \times 6 \cos 30° \right)$$
$$= 15{,}996 \text{ N}$$

2.8 BUOYANCY AND FLOATATION

Buoyancy: When a body is immersed in a fluid wholly or partially, an upward force is exerted by the fluid on the body which tends to lift it up. This tendency of an immersed body to be lifted up against gravity is called buoyancy.

Archimedes' Principle: This principle states that when a body is immersed in a fluid either wholly or partially, it is lifted up by a force, whose magnitude is equal to the weight of the fluid displaced by the body.

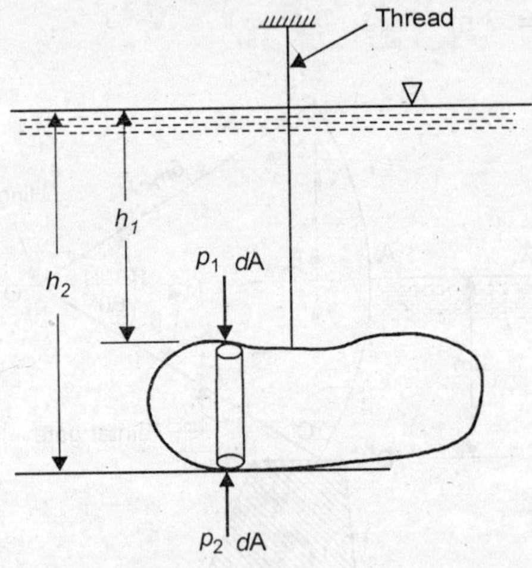

Fig. 2.34 Buoyant force on a submerged body

Consider a solid body suspended by a thread below the surface of a liquid. Consider a vertical cylindrical element of the body of cross-sectional area dA, as shown in Fig. 2.34. As dA is small, the pressure on the exposed ends of the cylinder may be taken as p_1 and p_2. Since p_2 is more than p_1, there will be an upward force $(p_2 - p_1)\, dA$ acting on the cylindrical element.

∴ The buoyant force on the element,

$$dF_B = (p_2 - p_1)dA = \gamma(h_2 - h_1)\,dA = \gamma.dV$$

where dV = volume of the cylindrical element.

Buoyant force on the whole body,

$$F_B = \int_A (p_2 - p_1)\,dA = \int_A \gamma(h_2 - h_1)\,dA = \gamma\int_V dV = \gamma V \quad ...(2.21)$$

where V = volume of the body

γ = specific weight of the liquid

If γ_b = specific weight of the body

Then, weight of the body, $W = \gamma_b V$

Tension is the thread = W – F_B = $V(\gamma_b - \gamma)$ ... (2.22)

Centre of Buoyancy: It is defined as the point through which the buoyant force acts. Since the buoyant force is the sum of a number of elementary forces, each equal in magnitude to the weight of cylindrical element of fluid, it follows that the buoyant force must pass through the centre of gravity of the displaced fluid mass.

Example 2.23 *A block of wood 6 m long × 2m wide × 1m thick is floating horizontally in water. The density of wood is 650 kg/m³. Find the volume of water displaced and position of centre of buoyancy.*

Solution. Volume of block, $V = 2 \times 9 \times 6 = 12$ m³

Weight of block, $W = \rho g V$

$= 650 \times 9.81 \times 12 = 76,518$ N

Refer to Fig. 2.35

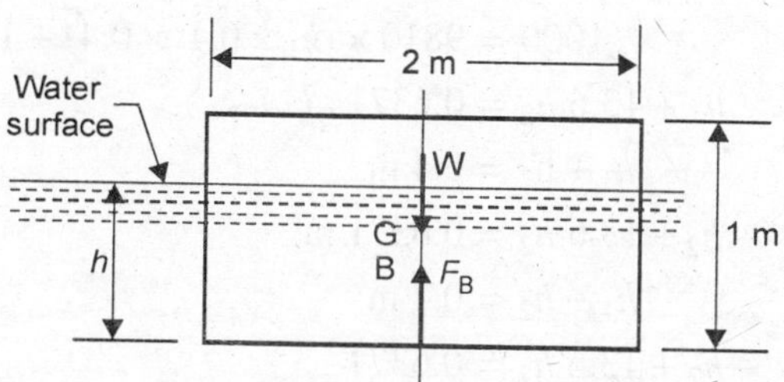

Fig. 2.35

Let h = depth of immersion of block in water

For equilibrium, $W = F_B$

$76,518 = (6 \times 2 \times h) \times 9810$

$h = 0.65$

Centre of buoyancy, $h_B = \dfrac{h}{2} = \dfrac{0.65}{2} = 0.325\,\text{m}$

$$\text{Volume of water displaced} = \frac{F_B}{\gamma} = \frac{76,518}{9810} = 7.8\,\text{m}^3$$

Example 2.24 *A metallic cube 400 mm side and weighing 1 kN is lowered into a tank containing a two-fluid layer of water and mercury. Determine the position of cube at mercury-water interface when it has reached equilibrium*

Solution. The position of the metallic cube, after immersion, is shown in Fig. 2.36. For equilibrium,

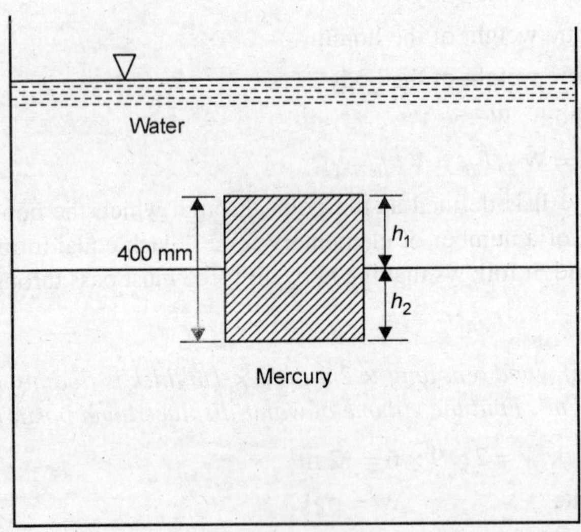

Fig. 2.36

$$W = F_B$$

$$1000 = 9810 \times (h_1 \times 0.4 \times 0.4) + 13.6 \times 9810 \times (h_2 \times 0.4 \times 0.4)$$

$$h_1 + 13.6\, h_2 = 0.6371 \text{ m}$$

$$h_1 + h_2 = 0.4 \text{ m}$$

$$0.4 - h_2 + 13.6\, h_2 = 0.6371 \text{ m}$$

$$h_1 + h_2 = 0.4 \text{ m}$$

$$0.4 - h_2 + 13.6\, h_2 = 0.6371$$

$$h_2 = 0.0188 \text{ m or } 18.8 \text{ mm}$$

Example 2.25 *A wooden log of 0.5 m diameter and 6 m long is floating in water. Find the depth of wooden log in water when the specific gravity of log is 0.65.*

Solution. Weight of wooden log, $W = \frac{\pi}{4} \times (0.5)^2 \times 6 \times 0.65 \times 10^3 \times 9.81 = 7512\,\text{N}$

For equilibrium, $W = F_B$

$$7512 = 9810 \times V_w$$

Volume of water displaced, $V_w = 0.7658 \text{ m}^3$

Let h = depth of immersion of log in water (see Fig. 2.37)

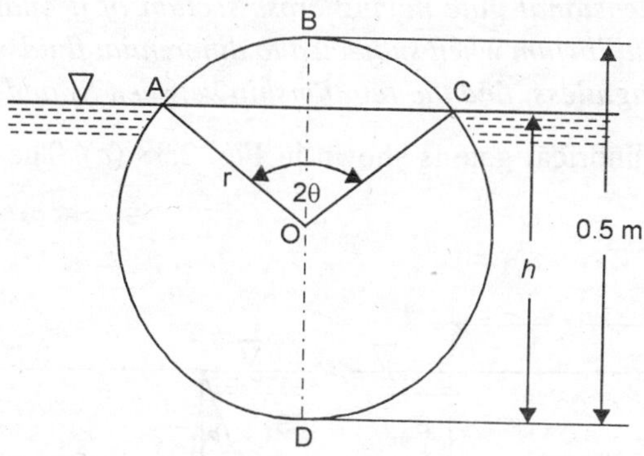

Fig. 2.37

$$\text{Volume of log immersed} = \text{Area ACDA} \times \text{length}$$
$$= \text{Volume of water displaced}$$
$$= 0.7658 \text{ m}^3$$

$$\text{Area ACDA} = \frac{0.7658}{6} = 0.1276 \text{ m}^2$$

$$\text{Area ACDA} = \text{Area of } \Delta AOC + \text{ Area } AOCDA$$

$$= \frac{1}{2} \times 2r \sin\theta \times r\cos\theta + \pi r^2 \left[1 - \frac{\theta}{180°}\right]$$

$$0.1276 = \frac{1}{2} \times 2 \times 0.25 \sin\theta \times 0.25 \cos\theta + \pi(0.25)^2 \left[1 - \frac{\theta}{180°}\right]$$

$$= 0.0625 \sin\theta \cos\theta + 0.1963 - 0.00109\,\theta$$

$$0.00109\,\theta - 0.0625 \sin\theta\cos\theta - \cos\theta - 0.0687 = 0$$

$$\theta - 57.3394 \sin\theta\cos\theta - 63.0275 = 0$$

$$\theta - 63.0275 = 57.3394 \sin\theta\cos\theta$$
$$= 28.6697 \sin 2\theta$$

θ, deg	**θ – 63.0275**	**28.6697 sin 2θ**
75	11.9725	14.3348
76	12.9725	13.4596
76.5	13.4725	13.0158
76.4	13.3725	13.1048
76.3	13.2725	13.1938
76.2	13.1725	13.2825

We take $\theta = 76.3°$

$$h = r + r\cos\theta \;= 0.25 + 0.25\cos 76°3 = 0.309 \text{ m}$$

Example 2.26 *A two dimensional gate having cross-section of a quarter cylinder is hinged at the bottom O and is in static equilibrium when subjected to differential fluids of specific gravities S_1 and S_2. Assuming the gate to be weightless, find the relationship between S_1 and S_2.*

Solution. The quarter-cylindrical gate is shown in Fig. 2.38 (*a*). The forces acting on the gate are shown in Fig. 2.38 (*b*).

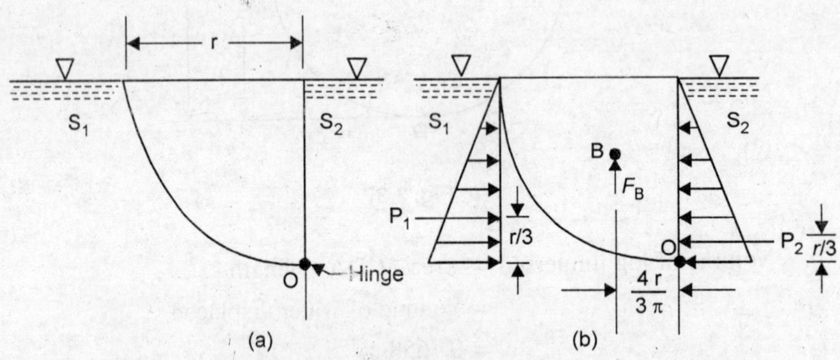

Fig. 2.38

Hydrostatic forces:

$$P_1 = \gamma_1 A\bar{h} = S_1\gamma \times r \times \frac{r}{2} = \frac{1}{2} S_1 \gamma r^2$$

$$P_2 = \gamma_2 A\bar{h} = S_2\gamma \times r \times \frac{r}{2} = \frac{1}{2} S_2 \gamma r^2$$

Buoyant force,

$$F_B = \frac{\pi r^2}{4} \times 1 \times \gamma_1 = \frac{\pi}{4} r^2 \times S_1 \gamma$$

For static equilibrium, taking moments about O, we have

$$(P_2 - P_1) \times \frac{r}{3} = F_B \times \frac{4r}{3\pi}$$

$$\frac{1}{2}\gamma r^2 (S_2 - S_1) \times \frac{r}{3} = \frac{\pi}{4} r^2 \times S_1 \gamma \times \frac{4r}{3\pi}$$

$$S_2 - S_1 = 2S_1$$

$$\frac{S_1}{S_2} = \frac{1}{3}$$

Example 2.27 *A cylindrical gate of diameter 2 m is 2.5 m long and has water on its both sides, as shown in Fig. 2.39. Determine:*

(a) the magnitude, location and direction of the resultant hydrostatic force exerted on the gate, and

(b) the minimum weight of the gate so that it will not float away from the floor.

Solution. Horizontal water force on projected vertical plane AB,

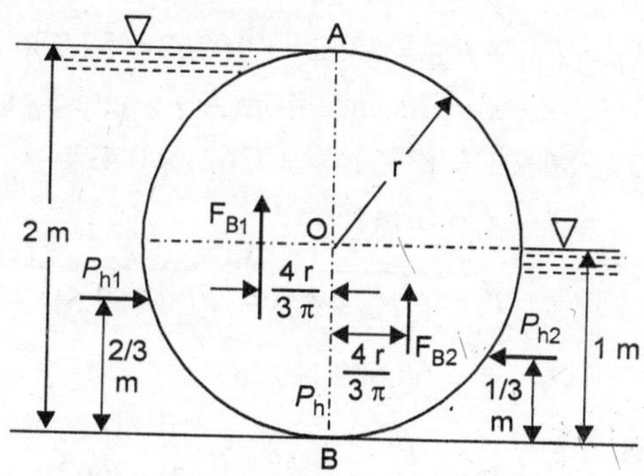

Fig. 2.39

$$Ph_1 = \gamma A h_1 = 9810 \times (2 \times 2.5) \times \frac{2}{2} = 49050\,\text{N}$$

$$h_{c1} = \frac{2}{3}\,\text{m from the bottom}$$

Buoyant force,
$$F_{B1} = \gamma \times \frac{\pi r^2}{2} \times 2.5 = 9810 \times \frac{\pi \times 1}{2} \times 2.5$$
$$= 38{,}524\ \text{N}$$

and acts at a distance,
$$x_1 = \frac{4r}{3\pi} = \frac{4 \times 1}{3\pi} = 0.4244\,\text{m from } AB.$$

Similarly,
$$P_{h2} = 9810 \times (1 \times 2.5) \times \frac{1}{2} = 12{,}262.5\,\text{N}$$

$$h_{c2} = \frac{1}{3}\,\text{m from the bottom.}$$

$$F_{B2} = 9810 \times \frac{\pi \times 1}{4} \times 2.5 = 19262\,\text{N}$$

$$x_2 = 0.4244\ \text{m from AB}$$

(*a*) Resultant horizontal force, $P_h = P_{h1} - P_{h2} = 49050 - 12262.5 = 36787.5$ N

Let y= distance from the bottom where P_h acts.

Then
$$36787.5 \times y = 49050 \times \frac{2}{3} - 12262.5 \times \frac{1}{3}$$
$$y = 0.778\ \text{m}$$

(*b*) Resultant vertical force, $F_B = F_{B1} + F_{B2} = 3852 + 19262 = 57{,}786$ N

Let x = distance from AB where F_B acts.

Then $$36787.5 \times y = 49050 \times \frac{2}{3} - 12262.5 \times \frac{1}{3}$$

$$y = 0.778 \text{ m}$$

(*b*) Resultant vertical force, $F_B = F_{B1} + F_{B2} = 38524 + 19262 = 57{,}786$ N

Let x = distance from AB where F_B acts.

Then $$57{,}786 \times x = (38524 - 19262) \times 0.4244$$

$$x = 0.1415 \text{ m}$$

Resultant water force, $$P = \sqrt{P_h^2 + F_B^2} = \sqrt{(36787.5)^2 + (57786)^2}$$

$$= 68{,}502 \text{ N}$$

$$\tan\theta = \frac{P_h}{F_B} = \frac{36787.5}{57786} = 0.63662$$

$$\theta = 32.48°$$

Example 2.28 *A cylinder of mass 10 kg and area of cross-section 0.1 m² is tied down with string in a vessel containing two liquids as shown in Fig. 2.40. Calculate the gauge pressure on the cylinder bottom and the tension in the string. Density of water = 1000 kg/m³, specific gravity of liquid A = 0.8 and that of liquid B = 1.0.*

Solution.

Given: $m = 10$ kg, $A = 0.1$ m², $S_A = 0.8$, $S_B = 1.0$

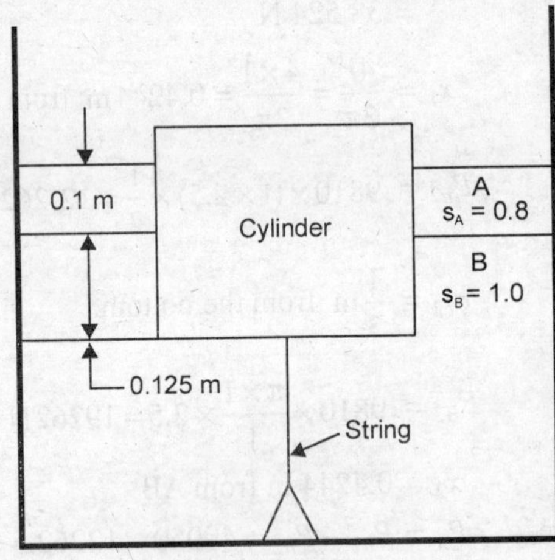

Fig. 2.40

Volume of liquid A displaced, $V_A = Ah_1 = 0.1 \times 0.1 = 0.01$ m³

$$m_A = s_A \rho V_A = 0.8\ 1000 \times 0.01 = 8 \text{ kg}$$

$$V_B = Ah_2 = 0.1 \times 0.125 = 0.0125 \text{ m}^3$$

$$m_B = s_B \rho V_B = 1.0 \times 1000 \times 0.0125 = 12.5 \text{ kg}$$

Total mass, $m = m_A + m_B = 8 + 12.5 = 20.5$ kg

Upward thrust, $F_B = mg = 20.5 \times 9.81 = 201.1$ N

Weight of cylinder, $W = 10 \times 9.81 = 98.1$ N

Net upward force $= F_B - W = 201.1 - 98.1 = 103$ N

Tension in the string, $T = 103$ N

Gauge pressure on the cylinder bottom $= \dfrac{T}{A} = \dfrac{103}{0.1} = 1030 \text{ N/m}^2$

Example 2.29 *A float valve shown is Fig. 2.41 in required to close an opening of water supply pipe feeding a cistern. The valve should close down when 60% of the volume of the spherical float is immersed in water. The valve has a diameter of 15 mm and the fulcrum of the operating lever is to be 150 mm from the valve and 500 mm from the centre of the float. Find the minimum diameter of the float if it is required to close the valve against a pressure of 150 kPa (gauge). Neglect the weight of the lever and the float.*

Solution. Force required to close the valve,

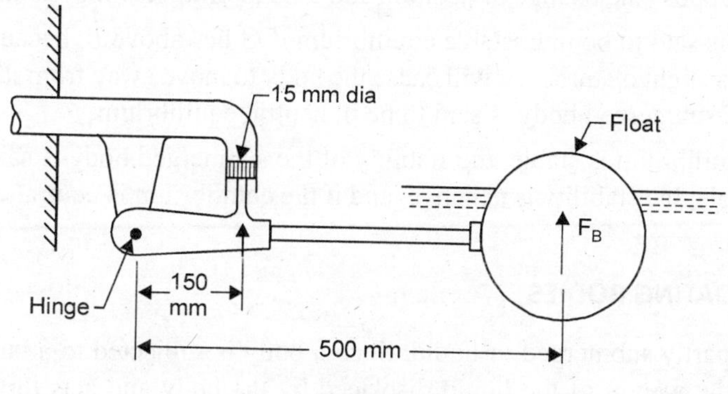

Fig. 2.41

$$F = \frac{\pi}{4} d^2 p = \frac{\pi}{4}(15)^2 \times 0.15$$

$$= 26.51 \text{ N}$$

Let F_B = buoyant force on the float

Taking moments about the hinge, we have

$$F_B \times 500 = 26.51 \times 150$$

$$F_B = 7.95 \text{ N}$$

Let V = Volume of the spherical float of diameter d_f.

Then $F_B = 0.6\, \gamma V$

$$7.95 = 0.6 \times 9810 \times \frac{\pi}{6} d_f^3$$

$$d_f = 0.137\, m \text{ or } 137 \text{ mm}$$

Example 2.30 *A stone weighs 400 N in air and 220 N in water. Find the volume of stone and its specific gravity.*

Solution.

$$\text{Buoyant force} = 400 - 220 = 180 \text{ N}$$

$$= \text{Weight of the liquid displaced}$$

$$= 9810 \times V$$

$$V = 0.01835 \text{ m}^3$$

$$= \text{Volume of stone immersed in water}$$

$$\text{Specific gravity of stone} = \frac{400}{9810 \times 0.01835} = 2.222$$

2.9 STABILITY OF SUBMERGED BODIES

A completely submerged body, such as a submarine, will remain in equilibrium at rest only if (*i*) the buoyant force F_B is equal to the weight W, and (*ii*) the centre of buoyancy (*B*) lies directly above the centre of gravity (G). A body is said to be in stable equilibrium, if a slight rotational displacement generates forces which oppose the change of position and tend to bring the body to its original position.

A submerged body is said to be in unstable equilibrium if G lies above *B*, because the overturning couple produced due to a slight disturbance will cause the body to move away from its original position. If B and G coincide, the submerged body is said to be in neutral equilibrium.

In general, if the equilibrium is stable, the stability of the submerged body is said to be positive, if the equilibrium is unstable, the stability is negative, and if the equilibrium is neutral, the body will have zero stability.

2.10 STABILITY OF FLOATING BODIES

A floating body is partly submerged in liquid. Such a body is subjected to a buoyant force that is equal in magnitude to the weight of the liquid displaced by the body and acts through the centre of gravity of the displaced liquid. The criterion of stability of a floating body is based on the position of a fictitious point, known as metacenter, with respect to the centre of gravity. The metacentre is denoted by the letter M.

2.10.1 Metacentre and Metacentric Height

Fig. 2.42 shows the plan and vertical cross-sections of a floating vessel. Fig. 2.42(a) shows the water line plan of a non-prismatic vessel in a state of equilibrium Let us consider an outside force applied causing the vessel to till or heel through a small angle θ, as shown in Fig. 2.42 (c). Due to change in the shape of displaced volume, the centre of buoyancy shifts to B′. The line of action of the buoyant force, acting through *B*′, cuts the axis of symmetry *c*–*d* at a point *M*, forming a ccw couple tending to restore the vessel to its original upright position. If M lies between *B* and *G* the cw couple is produced which results in progressive tilting of the vessel, thus rendering it unstable.

The position of *M* varies w.r.t *G* as the angle θ is increased or decreased. As the angle θ approaches zero, the point M approaches a limiting position above *G*. This position of *M* is known as the metacentre. The distance between *G* and *M* is called the metacentric height $\overline{GM}$.

The metacentre may also be defined as a point of intersection of the axis of a body passing through G and B and a vertical line passing through B′, the centre of buoyancy of the tilted position of the body.

The metacentric height will be positive when M lies above G, and represents the condition for stable equilibrium of the vessel. If M lies below G in between G and B, then the metacentric height is negative, which represents a condition of unstable equilibrium. If M and G coincide then it represents neutral equilibrium.

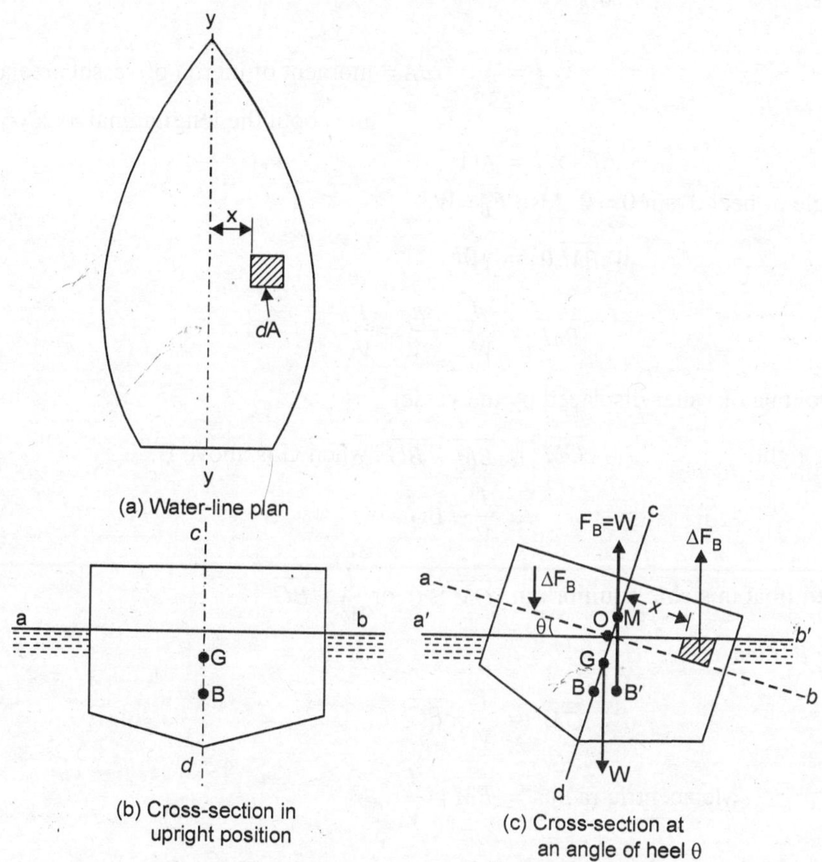

Fig. 2.42 Stability of non-prismatic floating vessel.

2.10.2 Determination of Metacentric Height

Consider the clockwise tilting of the vessel shown in Fig. 2.42 (c). The gain in vertically upwards buoyant force on right hand side is ΔF_B, while the loss is indicated by an equal vertically downwards force ΔF_B. These forces act through the centroids of wedge-shaped prisms *bob′* and *aoa′*. Taking moments of F_B and ΔF_B forces about B, we have

$$F_B \times \overline{BM} \sin\theta = \Delta F_B \times l$$

Consider an element of area dA in plan at a distance x from O, as shown in Fig. 2.42 (a) and (c). The buoyant force acting on this element is $\gamma x\theta.\ dA$.

Gain in buoyant force, $$\Delta F_B = \int \gamma x \theta . dA$$

Moment of ΔF_B about $y - y$ axis is,

$$\Delta F_B \times \frac{l}{2} = \int \gamma \theta x^2 \, dA$$

or $$\Delta F_B \times l = \int \gamma \theta x^2 dA = \gamma \theta \int_A x^2 \cdot dA = r\theta \cdot I$$

where $I = \int_A x^2 \cdot dA$ = moment of inertia of vessel area at the water line about the longitudinal axis *y-y*.

$$\therefore \quad \Delta F_B \times l = \gamma \theta \, \mathrm{I}$$

For small angle of heel θ, sin θ = θ. Also $F_B = W$.

Thus $$W.\overline{BM}.\theta = \gamma \theta I$$

$$\overline{BM} = \frac{\gamma I}{W} = \frac{\gamma I}{\gamma V} = \frac{I}{V} \qquad \text{... (2.23)}$$

where V = volume of water displaced by the vessel.

Metacentric height, $$\overline{GM} = \overline{BM} - \overline{BG}, \text{ when G is above B.}$$

$$= \frac{I}{V} - \overline{BG} \qquad \text{.... (2.24)}$$

For a vessel to float in stable equilibrium, $\overline{GM} > 0$ or $\frac{I}{V} > \overline{BG}$.

If G is below *B*, then

$$\overline{GM} = \frac{I}{V} + \overline{BG} \qquad \text{... (2.25)}$$

$$\text{Metacentric radius} = \overline{BM} = \frac{I}{V} \qquad \text{... (2.26)}$$

The metacentric height $\overline{GM}$ may be considered as a measure of stability of a vessel. In order to achieve high stability, I shall be large, a lower C.G. and smaller $\overline{BG}$.

2.10.3 Period of Rolling

The period of rolling oscillation T may be expressed as,

$$T = \frac{K}{\sqrt{g \times GM}} \cdot f(\theta) \qquad \text{... (2.27)}$$

where K = radius of gyration w.r.t axis of roll.

$f(\theta)$ = an unknown function, which is constant for small angles of roll.

A more stable vessel will have shorter rolling period.

2.10.4 Shape of Floating Body and its Stability

1. Cylinder floating with its axis parallel to the liquid surface:

   $\overline{GM} = 0$. Zero stability.
2. Vessel with vertical sides: good stability.
3. Vessel having diverging sides upwards:
   Better stability.
4. Vessel having converging sides upwards:
   Doubtful stability.

2.10.5 Experimental Determination of Metacentric Height

The experimental method for the determination of metacentric height $\overline{GM}$ consists of moving a weight w across the deck through a certain distance l and observing the corresponding angle of heel of roll θ. The moment wxl is balanced by the moment W.$\overline{GM}$ tan θ (see Fig. 2.43).

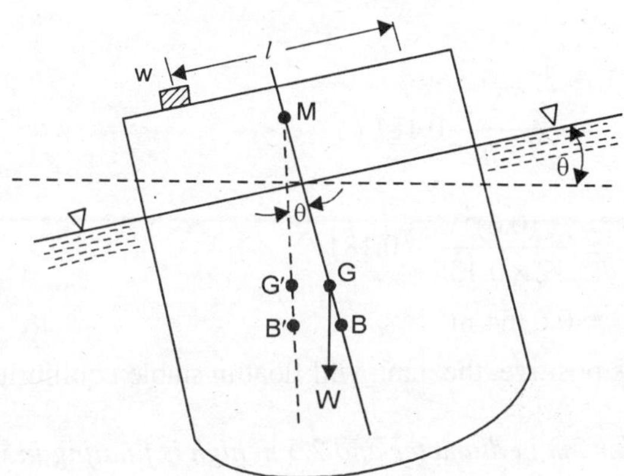

Fig. 2.43 Determination of metacentric height

$$\therefore \quad wl = W.\overline{GM} \tan\theta$$

$$\overline{GM} = \frac{wl}{W \tan\theta} \qquad \text{... (2.28)}$$

where W = total weight of the vessel including w. The angle θ may be measured by using a plumb bob.

Example 2.31 *An empty closed tank with sides 10 m long, 0.6 m wide and 0.5 high is allowed to float in fresh water with 0.5 m edges vertical. The sheet metal weights 360 N/m² of the surface area. Find its type of equilibrium.*

Solution. Weight of tank, $W = (0.6 \times 0.5 + 10 \times 0.6 + 10 \times 0.5) \times 2 \times 360$

$= 8136$ N

Weight of water displaced by tank, W = 8136 N

Volume of water displaced, $V = \frac{W}{\gamma} = \frac{8136}{9810} = 0.829\,\text{m}^3$

Depth of immersion of tank, $h = \frac{V}{\text{Bottom area of tank}}$

$$= \frac{0.829}{0.6 \times 10} = 0.138\,\text{m}$$

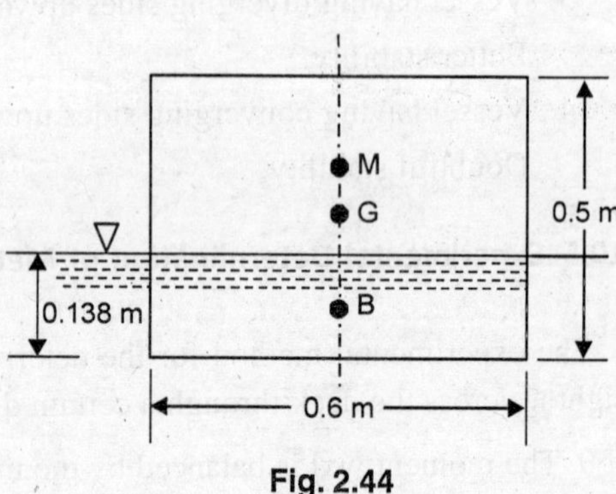

Fig. 2.44

$$\overline{BG} = \frac{0.5}{2} - \frac{0.138}{2} = 0.181\,\text{m}$$

$$\overline{MG} = \frac{I}{V} - \overline{BG}$$

$$= \frac{\frac{1}{12}bd^3}{bd \times h} - 0.181$$

$$= \frac{(0.6)^2}{12 \times 0.138} - 0.181$$

$$= 0.0364\text{ m}$$

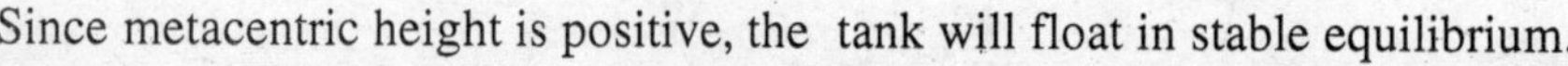

Since metacentric height is positive, the tank will float in stable equilibrium.

Example 2.32 *A solid cylinder 2m in diameter and 2.5 m high is floating in with its axis vertical. The specific gravity of cylinder is 0.65. Find whether a equilibrium is stable or not.*

Solution.

Given: $d = 2$m, h = 2.5 m, $s = 0.65$

Depth of cylinder in water, $h_1 = S \times h = 0.65 \times 2.5 = 1.625$ m

$$\overline{BG} = \frac{1}{2}(2.5 - 1.625) = 0.4375\,\text{m}$$

$$\overline{BG} = \frac{I}{V} = \frac{\frac{\pi}{64}d^4}{\frac{\pi}{4}d^2 \times h_1} = \frac{d^2}{16h_1} = \frac{4}{16 \times 1.625} = 0.1538\,\text{m}$$

$$\overline{GM} = \overline{BM} - \overline{BG} = 0.1538 - 0.4375 = -0.2837\,\text{m}$$

Since $\overline{GM}$ is negative the cylinder is in unstable equilibrium

Example 2.33 *A solid cylinder of 0.25 m diameter and 1 m length has its base 25 mm thick and of specific gravity 6. The remaining part of the cylinder is of specific gravity 0.65. State, if it can float vertically in water.*

Solution.

Given: $d = 0.25$n m, $l = 1$ m, $t = 25$ mm, $s_1 = 0.65$, $s_2 = 6$

Refer to Fig. 2.45.

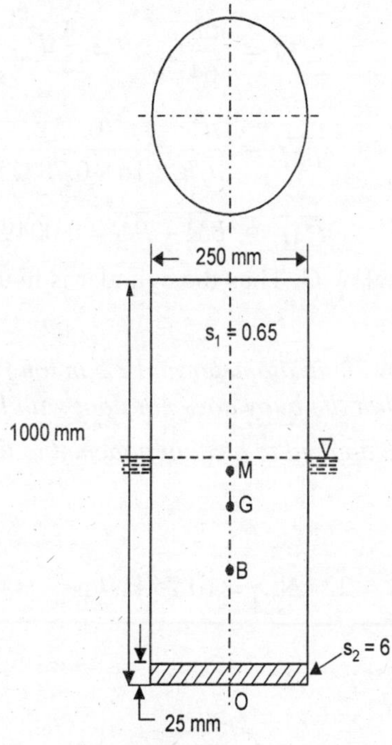

Fig.2.45

Cylinder area of cross-section, $A = \dfrac{\pi d^2}{4} = \dfrac{\pi(0.25)^2}{4} = 0.0491\text{m}^2$

Length of cylinder of specific gravity 0.65 $= 1000 - 25 = 975\,\text{mm}$

$$\overline{OG} = \frac{[A \times 0.975 \times 0.65\left(\dfrac{0.975}{2} + 0.025\right) + \left[A \times 0.025 \times 6 \times \dfrac{0.025}{2}\right]}{(A \times 0.975 \times 0.65) + (A \times 0.025 \times 6)}$$

$= 0.4168$ m

Combined specific gravity of cylinder,

$$s = \frac{0.975 \times 0.65 + 0.025 \times 6}{1.0} = 0.78375$$

Depth of immersion of cylinder, $h = 1 \times 0.78375 = 0.78375$ m

$$\overline{OB} = \frac{0.78375}{2} = 0.3919\,\text{m}$$

$$\overline{BG} = \overline{OG} - \overline{OB} = 0.4168 - 0.3919 = 0.0249\,\text{m}$$

$$\overline{BM} = \frac{I}{V}$$

$$I = \frac{\pi d^4}{64}, \quad V = \frac{\pi}{4} d^2 \times h$$

$$\overline{BM} = \frac{d^2}{16h} = \frac{(0.25)^2}{16 \times 0.78375} = 0.00498\,\text{m}$$

$$\overline{GM} = \overline{BM} - \overline{BG} = 0.00498 - 0.0249 = -0.0199\,\text{m}$$

–ve sign indicates that M is below G. Thus the cylinder is in unstable equilibrium and cannot float vertically in water.

Example 2.34 *A cylindrical buoy 2m in diameter and 2.5 m long weighs 22 kN. The specific weight of sea water is 10.25 kN/m³. Show that the buoy does not float with its axis vertical.*

What minimum pull should be applied to a chain attached to the centre of the base to keep the buoy vertical?

Solution.

Given: $d = 2\text{m}$, $l = 2.5$ m, $W_b = 22$ kN, $\gamma = 10.25$ kN/m³

Refer to Fig. 2.46

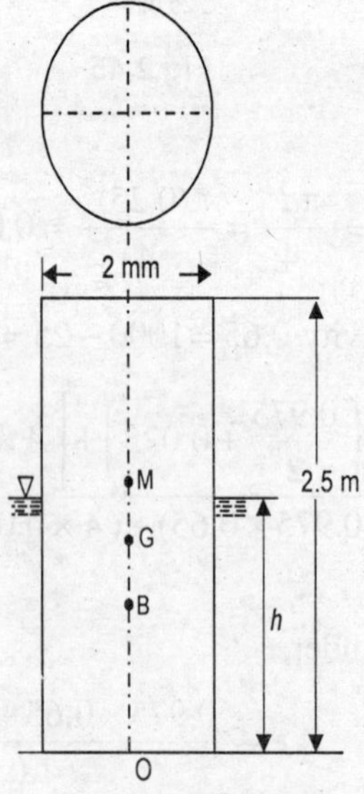

Fig. 2.46

Volume of sea water displaced, $V = \frac{W_b}{\gamma} = \frac{22}{10.25} = 2.146 \text{ m}^3$

Let h = depth of immersion

$$= \frac{V}{A} = \frac{4V}{\pi d^2} = \frac{4 \times 2.146}{\pi \times 4} = 0.6831 \text{m}$$

$$\overline{OB} = \frac{h}{2} = \frac{0.6831}{2} = 0.3415 \text{m}$$

$$\overline{BM} = \frac{I}{V} = \frac{\pi d^4}{64V} = \frac{\pi \times 2^4}{64 \times 2.146} = 0.366 \text{m}$$

$$\overline{OM} = \overline{OB} + \overline{BM} = 0.3415 + 0.366 = 0.7075 \text{m}$$

$$\overline{OG} = \frac{l}{2} = \frac{2.5}{2} = 1.25 \text{m}$$

Since $\overline{OM} < \overline{OG}$, the buoy is unstable with axis vertical .

Let T = minimum pull required to be applied to the chain to keep the buoy vertical.

Total downward force = $W_b + T$ = (22+ T) kN

$$\text{V} = \frac{22 + T}{10.25}$$

$$h' = \frac{22 + T}{10.25 \times \pi} = \frac{22 + T}{32.2} \text{m}$$

$$\text{OB}' = \frac{h'}{2} = \frac{22 + T}{64.4} \text{m}$$

$$\text{B}'\text{M}' = \frac{I}{V} = \frac{\pi \times 2^4}{64} \times \frac{10.25}{22 + T} = \frac{8.05}{22 + T}$$

Taking moments about O, we get

$$22 \times 1.25 = (22 + T) \times OG'$$

$$OG' = \frac{27.5}{22 + T}$$

$$B'G' = OG' - OB' = \frac{27.5}{22 + T} - \frac{22 + T}{64.4} = \frac{27.5 \times 64.44 - (22 + T)^2}{64.4(22 + T)}$$

$$= \frac{1287 - T^2 - 44T}{64.4(22 + T)}$$

For stable equilibrium, M′ must l is above G'.

$$\therefore \quad B'M' > B'G'$$

$$\frac{8.05}{22+T} = \frac{1287 - T^2 - 44T}{64.4(22+T)}$$

or
$$518.42 > (1287 - T^2 = 44T)$$

$$T^2 + 44T - 768.58 = 0$$

$$T = \frac{1}{2}[-44 \pm \sqrt{(44)^2 + 4 \times 768.58}]$$

$$= 13.39 \text{ kN}$$

Example 2.35 *A solid cone of base diameter 0.4 m and height 0.33 m floats with its vertex downwards in water as shown in Fig. 2.47. The specific gravity of cone is 0.8. Find the equilibrium status of cone.*

Solution. D = 0.4m, H = 0.33 m, s = 0.8

Refer to Fig. 2.47.

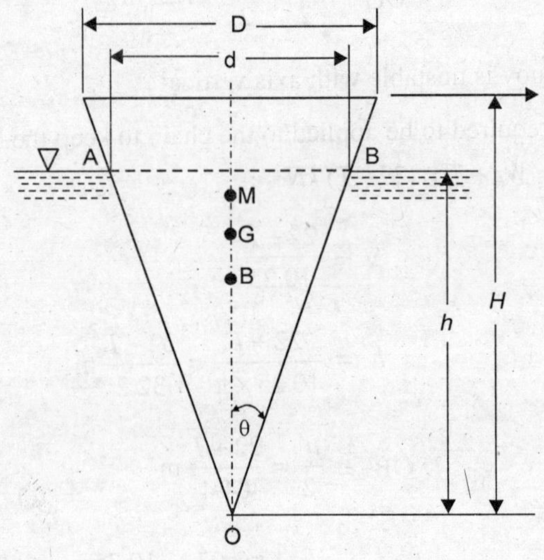

Fig. 2.47

Let θ = semi-vertex angle of cone

$$\tan\theta = \frac{D}{2H} = \frac{0.4}{2\times 0.33} = 0.606$$

$$\theta = 31.2°$$

Diameter of cone at water surface, $d = 2h \tan\theta$

Weight of cone = Weight of water displaced

$$\frac{1}{3}\times\pi\left(\frac{D}{2}\right)^2 \times H \times s \times \gamma = \frac{1}{3}\times\pi\left(\frac{d}{2}\right)^2 \times h \times \gamma$$

$$D^2HS = d^2h = 4h^3\tan^2\theta = 4h^3 \times \frac{D^2}{4H^2}$$

$$= \frac{h^3D^2}{H^2}$$

$$h^3 = H^3S$$

$$h = HS^{1/3} = 0.33 \times (0.8)^{1/3} = 0.306 \text{ m}$$

$$OB = \frac{3}{4}h = \frac{3}{4} \times 0.306 = 0.2295 \text{ m}$$

$$OG = \frac{3}{4}H = \frac{3}{4} \times 0.33 = 0.2475 \text{ m}$$

$$d = 2h\tan\theta = 2 \times 0.306 \times \tan 31.2° = 0.37 \text{ m}$$

$$\overline{BM} = \frac{I}{V} = \frac{\left(\frac{\pi d^4}{64}\right)}{\frac{1}{3} \times \pi\left(\frac{d}{2}\right)^2 \times h} = \frac{3}{16}\left(\frac{d^2}{h}\right) = \frac{3}{16} \times \frac{(0.37)^2}{0.306} = 0.0839 \text{m}$$

$$\overline{OM} = \overline{OB} + \overline{BM} = 0.2295 + 0.839 = 0.3134 \text{ m}$$

$$\overline{OG} = 0.2475 \text{ m}$$

$$\overline{GM} = \overline{OM} - \overline{OG} = 0.3134 - 0.2475 = 0.0659 \text{ m}$$

Since M is above G, hence the cone is under stable equilibrium.

Example 2.36 *Derive on expression for the time of oscillation of a floating body.*

Solution. Refer to Fig. 2.48.

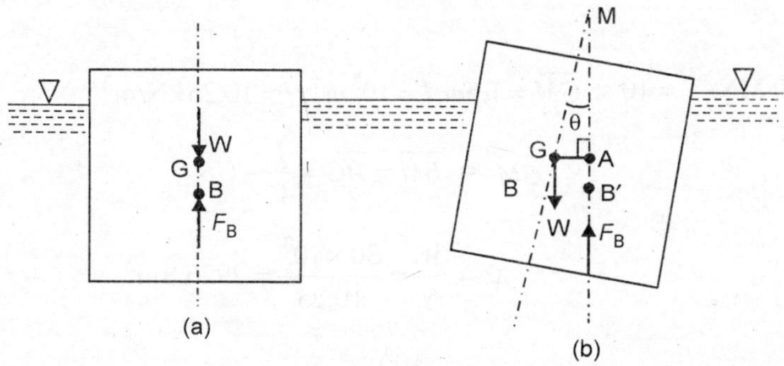

Fig. 2.48

Let W = weight of floating body

θ = angle of oscillation of body

T = time of rolling in seconds

$I_G = \frac{W}{g}K^2$ = moment of inertia of the body about its G.G.

K = radius of gyration.

$\overline{GM}$ = metacentric height of body

Disturbing couple = $I_G\ddot{\theta}$

Restoring couple = W × GA = $W \times \overline{GM}\tan\theta = W \times \overline{GM}\cdot\theta$

The equation of motion is,

$$I_G\ddot{\theta} + W.\overline{GM}.\theta = 0$$

$$\frac{W}{g}K^2\ddot{\theta} + W.\overline{GM}.\theta = 0$$

$$\ddot{\theta} + \frac{g}{K^2}\cdot\overline{GM}.\theta = 0$$

$$\omega_n = \sqrt{\frac{g}{K^2}.\overline{GM}}$$

$$T = 2\pi\sqrt{\frac{K^2}{g.\overline{GM}}} \quad \text{... (2.29)}$$

Example 2.37 *The time period of oscillation of a ship of weight 30 MN in the water is 10 s. The centre of buoyancy of the ship is 1.5 m below the centre of gravity Calculate the radius of gyration of the ship if the moment of inertia of the ship at the water line about fore and aft axis is 10^4 m^4. Take specific weight of sea water as 10.25 kN/m^3.*

Solution.

Given: W = 30 MN, T = 10 s, $\overline{GM} = 1.5m, I = 10^3 m^4, \gamma = 10.25\,\text{kN/m}^3$

$$\overline{GM} = \overline{BM} = \overline{BG} = \frac{I}{V} - \overline{GB}$$

$$V = \frac{W}{\gamma} = \frac{30\times10^3}{10.25} = 2926.8\,\text{m}^3$$

$$\overline{GM} = \frac{10^4}{2926.8} - 1.5 = 1.9167\,\text{m}$$

$$T = 2\pi\sqrt{\frac{K^2}{g \times GM}}$$

$$10 = 2\pi\sqrt{\frac{K^2}{9.81 \times 1.9167}}$$

$$K = 6.9 \text{ m}$$

Example 2.38 *Compare the stability of submerged and floating bodies.*

Solution. B = Centre of buoyancy

G = Centre of gravity

M = Metacentre

Type of stability	Submerged body	Floating body
1. Stable	B is above G	M is above G
2. Unstable	B is below G	M is below G
3. Neutral	B and G coincide	M and G coincide

2.11 RELATIVE EQUILIBRIUM

Consider a container filled with a liquid moving with uniform acceleration. Initially the liquid particles move relative to each other and to the boundaries of the container. After some time the liquid masses will not have any relative motion between its particles or between liquid particles and boundaries of container. In such a situation, the entire liquid mass moves as a single unit. Liquids in such a motion are said to be in *relative equilibrium*. Because of no relative motion, the liquid is free from shear stresses and is governed by the law of fluid statics.

2.11.1 Static Fluid Subjected to Uniform Acceleration

Consider a fluid element of size $dx \times dz$ and unit length, as shown in Fig. 2.49. It is subjected to acceleration components a_x and a_z along x and z directions. Let p be the pressure at the centre of the element.

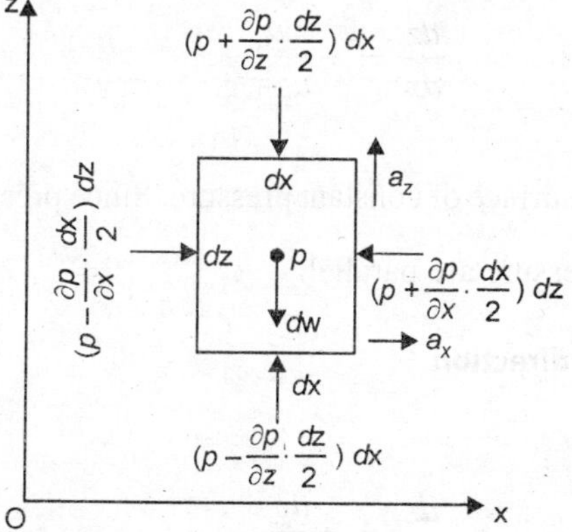

Fig. 2.49 Fluid mass subjected to acceleration

Applying Newton's second law of motion along the x and z directions, we have

$$\Sigma F_x = m \cdot a_x$$

$$= \left(p - \frac{\partial p}{\partial x} \cdot \frac{dx}{2}\right) dz - \left(p + \frac{\partial p}{\partial x} \cdot \frac{dx}{2}\right) dz$$

$$= -\frac{\partial p}{\partial x} \cdot dx . dz$$

$$= \left(\frac{\gamma}{g} \cdot dx \cdot dz\right) \cdot a_x$$

$$\therefore \quad \frac{-\partial p}{\partial x} = \frac{\gamma}{g} \cdot a_x$$

$$\Sigma F_z = \text{m} \cdot a_z$$

$$= \left(p - \frac{\partial p}{\partial z} \cdot \frac{dz}{2}\right) dx - \left(p + \frac{\partial p}{\partial z} \cdot \frac{dz}{2}\right) dx - \gamma . dx \cdot dz$$

$$= -\frac{\partial p}{\partial z} \cdot dx \cdot dz - \gamma \cdot dx . dz$$

$$= \frac{\gamma}{g} \cdot dx \cdot dz \cdot a_z$$

$$-\frac{\partial p}{\partial z} = \gamma \left(1 + \frac{a_z}{g}\right)$$

Now $$dp = \frac{\partial p}{\partial x} \cdot dx + \frac{\partial p}{\partial z} \cdot dz = -\frac{\gamma}{g} a_x \cdot dx - \gamma \left(1 + \frac{a_z}{g}\right) dz$$

At the surface of constant pressure, $dp = 0$.

$$\therefore \quad -\frac{\gamma}{g} a_x \cdot dx = \frac{\gamma}{g}\left(1 + \frac{a_z}{g}\right) dz$$

$$\frac{dz}{dx} = -\frac{a_x}{a_z + g} \qquad \text{... (2.30)}$$

Now $\frac{dz}{dx}$ is the slope of the surface of constant pressure. Since pressure variation follows hydrostatic law, the surfaces of constant pressure are parallel.

1. Acceleration in horizontal direction

For $a_z = 0$, we have

$$\frac{dz}{dx} = -\frac{a_x}{g} \qquad \text{... (2.31)}$$

The -ve sign shows that the surface is sloping downwards, if $a_x > 0$, i.e. the container is accelerating. The surfaces will have + *ve* slope, if $a_x < 0$, i.e. the container is decelerating. These two situations are shown in Fig. 2.50.

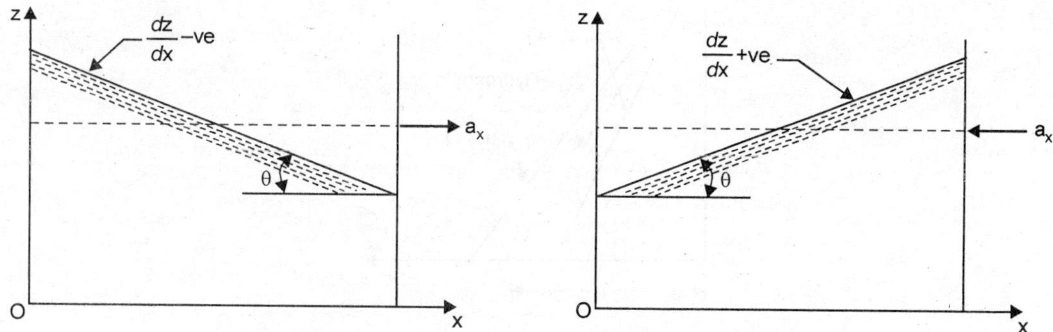

Fig. 2.50. Free surface during horizontal acceleration

If θ = angle made by the free surface with horizontal

Then $$\tan\theta = \frac{a_x}{g} \quad \text{... (2.32)}$$

2. Acceleration in vertical direction

For $a_x = 0$, $\frac{dz}{dx} = 0$, and the surfaces of constant pressure are horizontal Pressure gradient,

$$\frac{dp}{dz} = -\gamma\left(1+\frac{a_z}{g}\right)$$

The -ve sign indicates that the pressure decreases as *z* increases. If distance *h* is measured vertically downwards from the free surface, then

$$\frac{dp}{dh} = \gamma\left(1+\frac{a_z}{g}\right)$$

Upon integration ,we have

$$p = \gamma\left(1+\frac{a_z}{g}\right)h + C$$

At the free surface, $h = 0$, $p = 0$. Therefore, $C = 0$. Thus

$$p = \gamma\left(1+\frac{a_z}{g}\right)h \quad \text{... (2.33)}$$

If $a_z > 0$, then $(p - p_{\text{hyd}}) = \frac{\gamma a_z h}{g}$

If $a_z < 0$, then $(p - p_{\text{hyd}}) = -\frac{\gamma a_z h}{g}$

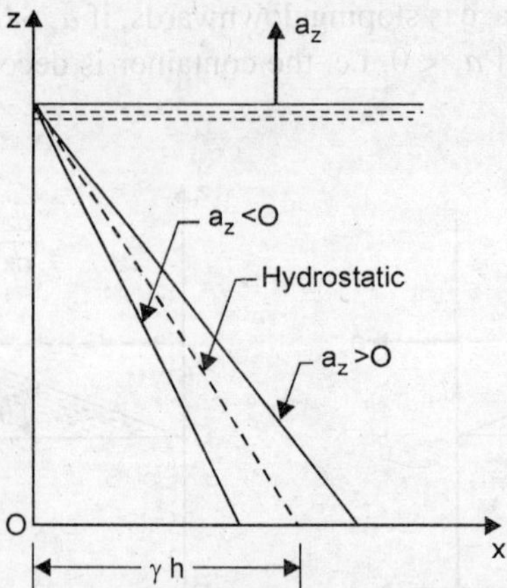

Fig. 2.51 Pressure variation in the fluid subjected to vertical acceleration.

If $a_z = -g$ (freely falling fluid), then $p = 0$ (atmospheric)

The above observations are shown graphically in Fig. 2.51.

Example 2.39 *A closed container containing water to a depth of 2 m is moved vertically downwards with an acceleration of 3m/s². Calculate the pressure on the bottom of the container.*

Solution.

Given: $h = 2\text{m}$, $a_z = -3\text{m/s}^2$

$$p = \gamma\left(1+\frac{a_z}{g}\right)h$$

$$= 9810\left(1-\frac{3}{9.81}\right)\times 2$$

$$= 13620 \text{ N/m}^2$$

Example 2.40 *An oil tanker of 2.4 m × 2.4 m cross-section is 4 m long. Oil is filled upto a depth of 2m. It is moved at uniform acceleration so that the front bottom corner is exposed. Calculate the net horizontal force acting on the tanker sides. Take specific gravity of oil as 0.85.*

Solution. Refer to Fig. 2.52

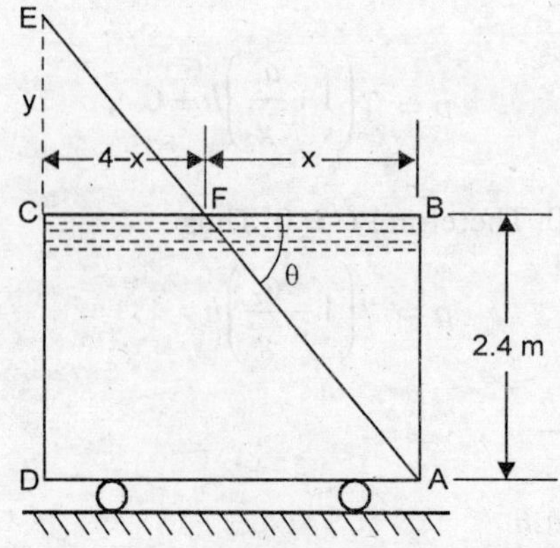

Fig. 2.52

Volume of oil inside the tanker, $V = 2.4 \times 2 \times 4 = 19.2\ \text{m}^3$

Volume of empty space inside the tanker

$$= 0.4 \times 2.4 \times 4 = 3.84\ \text{m}^3$$

$$\tan\theta = \frac{AB}{BF} = \frac{2.4}{x}$$

$$\text{Volume of empty space} = \frac{1}{2} \times AB \times BF \times 2.4$$

$$= \frac{1}{2} \times 2.4 \times x \times 2.4 = 2.88\,x$$

$$\therefore \quad 2.88\,x = 3.84$$

$$x = 1.333\ \text{m}$$

$$\tan\theta = \frac{a_x}{g} = \frac{2.4}{x}$$

$$a_x = \frac{2.4}{1.333} \times 9.81 = 17.66\ \text{m/s}^2$$

ΔS AED and FEC are similar.

$$\therefore \quad \frac{AD}{DE} = \frac{CF}{CE}$$

$$\frac{4}{2.4 + y} = \frac{4 - 1.333}{y}$$

$$4y = 2.667\ (2.4 + y) = 6.4 + 2.667\ y$$

$$1.333\ y = 6.4$$

$$y = 4.8\ \text{m}$$

Hydrostatic force on side $CD = \gamma h_c A = 0.85 \times 9810 \times \left(4.8 + \frac{2.4}{2}\right) \times 2.4 \times 2.4$

$$= 288{,}178\ \text{N}$$

Example 2.41 *An open rectangular tank 5 m long × 2.5 m wide is filled with water to a depth of 1.5 m. Find the slope of water surface when the tank moves with an acceleration of 3m/s² up an inclined plane of 30°.*

Solution.

Given: $\alpha = 30°$, $a = 3\ \text{m/s}^2$

Refer to Fig. 2.53.

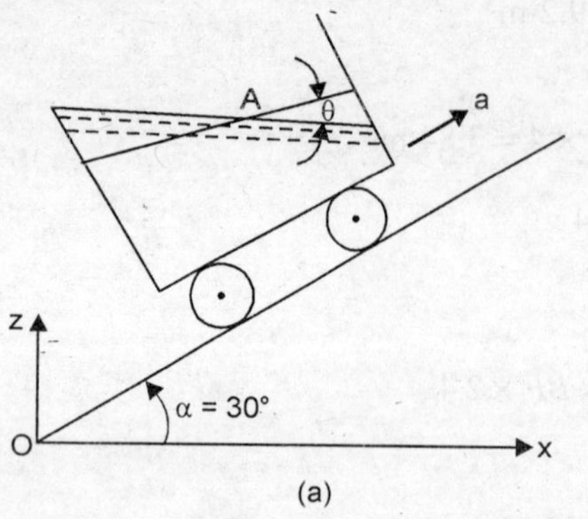

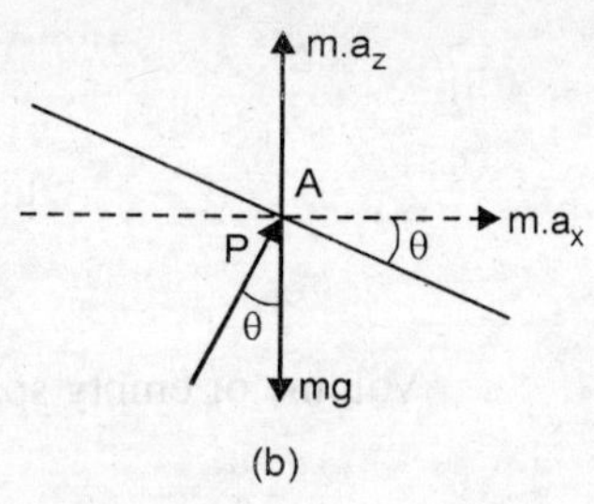

Fig. 2.53

Let P = Pressure force acting normal to the water surface.

Then
$$P \sin\theta = m.a_x$$
$$P \cos\theta = m \cdot a_z + mg$$

$$\tan\theta = \frac{a_x}{a_z + g} = \frac{a\cos\alpha}{a\sin\alpha + g}$$

$$= \frac{3\cos 30^\circ}{3\sin 30^\circ + 9.81} = 0.22971$$

$$\theta = 12.94^\circ$$

SUMMARY

1. The study of fluids at rest or undergoing uniform acceleration in a closed space is called fluid statics.
2. Hydrostatics means the study of pressure exerted by a liquid at rest.
3. The variation of static pressure is given by:

$$-\frac{dp}{dz} = \rho g = \gamma$$

4. Pressure head, $h = \dfrac{p}{\gamma}$
5. A barometer measures absolute pressure and Bourdon gauge measures gauge pressure.
6. The apparent contradiction in the hydrostatic force on the container base and the weight of liquid inside the container is known as the hydrostatic paradox.
7. The Pascal's law states that the pressure at any point in a liquid at rest is same in all directions.

8. Total pressure is defined as the force exerted by static fluid on a surface when it comes in contact with the surface.

9. Centre of pressure is defined as the point of application of the total pressure on the surface.

10. Hydrostatics deals with the study of pressure exerted by a fluid at rest

(a) Horizontal submerged plane surface.

$$p = \gamma h$$

$$P = \gamma \bar{h} A = \gamma h A$$

$$h_c = \bar{h} = h$$

(b) Vertical submerged plane surface.

$$P = \gamma \bar{h} A$$

$$h_c = \bar{h} + \frac{I_G}{A\bar{h}}$$

(c) Inclined submerged plane surface.

$$\frac{h}{y} = \frac{\bar{h}}{\bar{y}} = \frac{h_c}{y_c} = \sin\theta$$

$$P = \gamma \bar{h} A$$

$$h_c = \bar{h} + \frac{I_G \sin^2\theta}{A\bar{h}}$$

(d) Curved submerged surface

Horizontal force, P_h = Total force on the projected area of the curved surface on the vertical plane.

Vertical force, P_v = Weight of the liquid supported by the curved surface upto the free surface of liquid.

Resultant force, $P = \sqrt{P_h^2 + P_v^2}$

11. The tendency for an immersed body to be lifted up in the liquid due to an upward force opposite to the action of gravity is known as buoyancy.

12. Archimedes' principle states that when a body is immersed in a liquid either wholly or partially, it experiences an upward force, whose magnitude is equal to the weight of the liquid displaced by the body.

13. Centre of buoyance is the point through which the buoyant force acts. It is the C.G of the displaced fluid.

14. Stability of submerged bodies:

Stable equilibrium: *B* is directly above G.

Unstable equilibrium: *B* is directly below G.

Neutral equilibrium: B and G coincide.

15. Stability of floating bodies:

Metacentre: It is defined as a point of intersection of the axis of the body through G and B and a vertical line passing through the centre of buoyancy *B′* of the tilted position of the body. It is denoted by M. Metacentric height is the distance between G and M.

$$\overline{GM} = \frac{I}{V} \pm \overline{BG}$$

Take +Ve sign when G is below B.

$$\overline{GM} = \frac{wl}{W \tan\theta}$$

16. A vessel having diverging sides upwards has better stability.

17. Conditions of stability for floating body:

Stable: M is above G.

Unstable: M is below G.

Neutral : M and G coincide.

18. Time of rolling of a vessel, $T = 2\pi \frac{K}{\sqrt{g \cdot \overline{GM}}}$

19. Static fluid subjected to uniform acceleration.

$$\frac{dz}{dx} = \frac{a_x}{a_z + g}$$

Short Answer Type Questions

1. Define fluid statics.
2. What are the forces to which a fluid element at rest is subjected to?
3. How does the static pressure due to fluids vary?
4. Explain absolute, gauge and vacuum pressures.
5. Barometer and Bourdon gauge measure which type of pressure.
6. Define total pressure.
7. What is hydrostatic paradox?
8. State Pascal's law.
9. Define total pressure and centre of pressure.
10. Write the formula to determine centre of pressure on a vertical plane submerged surface.
11. Write the formula to determine centre of pressure on an inclined submerged plane surface.
12. How the horizontal and vertical force on a curved submerged surface are determined?
13. Define buoyancy and centre of buoyancy.

14. State the Archimedes' principle.
15. Explain the stability of a submerged body.
16. Define metacentre and metacentric height.
17. Discuss the stability of floating bodies.
18. How the shape of a floating body affect its stability?
19. How metacentric height can be determined experimentally?
20. How the rolling period of a ship affect its stability?

EXERCISES

1. Calculate the total pressure and location of centre of pressure on a plane rectangular surface 2m wide are 3 m deep when its upper edge is horizontal and (*a*) coincides with water surface, (*b*) 2m below the water surface. **[Ans.** 88290 N, 2m; 206010 N, 3.714 m]
2. A circular plate of diameter 2m is placed vertically in water in such a way that centre of plate is 3m below the surface of water. Determine:

   (a) total pressure on the plate, (b) The position of centre of pressure. **[Ans.** 92457 N, 3.08 m]
3. A circular opening of 3m diameter in the vertical side of a tank is closed by a disc of 3m diameter which can rotate about a horizontal diameter. Calculate: (a) the force on the disc, and (b) the torque required to maintain the disc in vertical position when the head of water above the horizontal diameter is 6m. **[Ans.** 416057 N, 39005 N.m]
4. A pipe of diameter 3 m contains a fluid at 0.3 N/mm^2. The pipe contains fluid of specific gravity 0.85 and is fitted with a gate valve. Calculate the force exerted by the fluid on the gate and the position of centre of pressure. **[Ans.** 2,120,575 N, 0.0156 M below centre of pipe]
5. Calculate the total pressure and centre of pressure on an isosceles triangular plate of base 4 m and height 4 m when it is immersed vertically in an oil of specific gravity 0.9. The base of the plate is 1 m below the free surface of water. **[Ans.** 164,808 N, 2.714 m]
6. A tank contains water upto a height of 1 m above the base. An immiscible liquid of specific gravity 0.8 is filled on the top of water upto 1.5 m height. Calculate:

   (a) the total pressure on one side of the tank, and (b) the position of centre of pressure for one side of the tank of 3 m width. **[Ans.** 76518 N, 1.686 m]
7. A rectangular plate 2m wide and 3m deep lies in water in such a way that its plane makes an angle of 30° with the free surface of water. Calculate the total pressure and position of centre of pressure when the upper edge of plate is 2m below the free surface of water.

   [Ans. 161,865 N, 2.818 m]
8. A circular plate of 3 m diameter is immersed in water in such a way that the plane of plate makes an angle of 60° with the free surface of water. Calculate: (a) the total pressure, and (b) the position of centre of pressure when the upper edge of the plate is 2 m below the free surface of water. **[Ans.** 228, 764 N, 3.47 m]

9. A circular opening of 2.5 m diameter in the vertical side of a tank is closed by a disc of 2.5 m diameter which can rotate about a horizontal diameter.

Calculate: (*a*) The forces on the disc, and

(b) the torque required to maintain the disc in equilibrium in vertical position when the head of water above horizontal diameter is 3.5 m.) **[Ans.** 168,541 N, 18,810 N.m]

10. A circular lamina of radius R is kept immersed in a liquid such that its top most point A is on the free surface. Calculate the depth and width of the horizontal chord BC so that the total thrust due to hydrostatic pressure on the triangle ABC is maximum **[Ans.** $\frac{5}{3}R, \frac{2}{3}\sqrt{5}R$]

11. A sliding gate 3m wide and 1.5 m high lies in a vertical plane and has a coefficient of friction of 0.2 between itself and guides. The gate weighs 30 kN. Find the vertical force required to raise the gate if its upper edge is at a depth of 9 m from the free surface of water. **[Ans.** 116,083 N]

12. A rectangular vertical door 2.4 m (height) × 1.2 m (wide) is fastened by two hinges situated 0.18 m below the top and 0.18 m above the bottom on one vertical edge, and by one clamp at the centre of other vertical edge. The door is subjected to water pressure on one side and the depth of water above the top of door is 1.2 m. Calculate the reactions at the hinges and at the clamp. **[Ans.** 10.3 kN, 23.6 kN, 33.9 kN]

13. An annular plate 2 m outside and 1.2 m inside diameter, with its maximum and minimum depths below the water surface being 1.5 m and 0.75 m respectively. Calculate the magnitude, direction and location of the force acting upon one side of the plate due to water pressure. **[Ans.** Ans. 26 kN at 68° with the vertical, 1.164 m]

14. A 6m × 2 m rectangular gate inclined at 60° with the horizontal is hinged at the base. The upper end of the gate is kept in position by a weight of 60 kN acting at an angle of 90°. Neglecting the weight of the gate, find the level of water when the gate begins to fall. **[Ans.** 4.36 m]

15. A cylindrical tank 1.5 m diameter and height 0.75 m has a hemispherical dome. The tank contains oil of specific gravity 0.84. The dome is joined to the cylinder position by four equally spaced bolts. If the pressure gauge at a point 0.3 m above the base of the tanks reads 50 kPa, determine the force on each bolt. **[Ans.** 18.64 kN]

16. The shape of a gate is in the form of a quadrant of radius 3m. Find the resultant force due to water per metre length of the gate and the angle at which this resultant force will act. **[Ans.** 82.2 kN, 57° – 31′]

17. A 8 cm side cube weighing 4 N is immersed in a liquid of specific gravity 0.8 contained in a rectangular tank of cross-sectional area 12 cm × 12 cm. The tank contains liquid to a height of 6.4 cm before the immersion. Determine the levels of the bottom of the cube and the liquid surface. **[Ans.** 1.975 cm, 9.94 cm]

18. A body has the cylindrical upper portion of 4 m diameter and 2.4 m deep. The lower portion, which is curved, displaces a volume of 800 litres of water and its centre of buoyancy is situated 2.6 m below the top of the cylinder. The C.G. of the whole body is 1.6 m below the top of the cylinder and the total displacement of water is 52 kN. Find the metacentric height of the body. **[Ans.** 1.693 m]

19. A hollow cylinder closed at both ends has an outside diameter of 1.25 m, length 3.5 m and specific weight 75 kN/m^3. If the cylinder is to float just in stable equilibrium in sea water (specific weight 10 kN/m^3), find its minimum permissible thickness. [**Ans.** 0.606 mm]

20. A buoy having a diameter of 2.4 m and length 1.95 m is floating with its axis vertical in sea water (specific weight = 10 kN.m^3). Its weight is 16.5 kN and a load of 1.65 kN is placed centrally at its top. If the buoy is to remain in stable equilibrium, find the maximum permissible height of the centre of gravity of the load above the top of the buoy. [**Ans.** 0.368 m]

21. A cone of specific gravity s, is floating in water with its apex downward. It has a radius R and vertical height H. Show that for stable equilibrium of cone,

$$\text{(a) } \sec^2\theta > \frac{H}{h}, \text{ and (b) } H < \left[\frac{R^2 s^{1/3}}{1 - s^{1/3}}\right]^{1/2}$$

where h = depth of immersion. and θ = semi– vertex angle of cone.

22. A vessel has a length of 60 m, width 12 m and a displacement of 19620 kN. When a weight of 294.3 kN is rolled off transversely across the deck through a distance of 6.5 m, the vessel tilts through 5°. The second moment of area of water line section about its fore–and aft axis is 0.75 percent of that of the circumscribing rectangle. The centre of buoyancy is 2.75 below the water line. Find:

(a) The metacentric height.

(b) The position of C.G. of the vessel.

Take specific weight of sea water = 10.104 kN/m^3.

[**Ans.** 1.1145 m , 0.53 m below water surface]

23. An open tank 10 long and 2 m deep is filled with 1.5 m of oil of specific gravity 0.82. The tank is accelerated uniformly from rest to speed of 90 m/s. What is the shortest time in which this speed may be attained without spilling any oil? [**Ans.** 20.4 s]

24. An open rectangular tank 1.5 m wide, 2 m deep and 3 m long is half filled with water. It is accelerated horizontally at 3.27 m/s^2 in the direction of its length. Determine the depth of water at each end of the tank and total force. [**Ans.** 0.5 m, 1.5 m, 1836 N, 16550 N]

3

FLUID KINEMATICS

3.1 INTRODUCTION

Fluid kinematics heals with the study of velocity and acceleration of fluid particles without taking into account the forces associated with them. The fluid motion can be described completely by an expression describing the location of a fluid particle in space at different times, thus enabling determination of the magnitude and direction of velocity and acceleration of the flow fluid at any instant of time.

3.2 METHOD OF DESCRIBING FLUID MOTION

There are two methods for describing the fluid motion.

1. Eulerian method, and 2. Lagrangian method.

1. Eulerian Method

In this method, we choose a coordinate system fixed in space and study the motion of fluid particles. For example, and observer sitting on the bank of a river and studying the movement of fluid particles or a floating body illustrates this method. This method is mostly used in fluid mechanics because of its simplicity. Actually, we are not concerned with the motion of each particle, but study the general state of motion at various points in the fluid system.

The velocities at any point (x, y, z) can be written as:

$$\left.\begin{aligned} u &= f_1(x, y, z, t) \\ v &= f_2(x, y, z, t) \\ w &= f_3(x, y, z, t) \end{aligned}\right\} \quad \text{... (3.1)}$$

Now

$$du = \frac{\partial u}{\partial x}\cdot dx + \frac{\partial u}{\partial y}\cdot dy + \frac{du}{dz}.dz + \frac{\partial u}{\partial t}\cdot dt$$

$$a_x = \frac{du}{dt} = \left(\frac{\partial u}{dx}\cdot\frac{dx}{dt} + \frac{\partial u}{\partial y}\cdot\frac{dy}{dt} + \frac{\partial u}{\partial z}\cdot\frac{dz}{dt}\right) + \frac{\partial u}{\partial t}\cdot\frac{dt}{dt}$$

$$a_x = u\frac{\partial u}{\partial x} + v\frac{\partial u}{\partial y} + w\frac{\partial u}{\partial z} + \frac{\partial u}{\partial t} \quad \text{... (3.2a)}$$

Similarly,
$$a_y = \frac{dv}{dt} = u\frac{\partial v}{\partial x} + \frac{v\partial v}{\partial y} + w\frac{\partial v}{\partial z} + \frac{\partial v}{\partial t} \quad \text{... (3.2 b)}$$

$$a_z = \frac{dw}{dt} = u\frac{\partial w}{\partial x} + v\frac{\partial w}{\partial y} + w\frac{\partial w}{\partial z} + \frac{\partial w}{dt} \quad \text{... (3.2 c)}$$

Resultant velocity,
$$V = \sqrt{u^2 + v^2 + w^2} \quad \text{... (3.3)}$$

Resultant acceleration,
$$a = \sqrt{a_x^2 + a_y^2 + a_z^2} \quad \text{... (3.4)}$$

2. Lagrangian Method

In this method, we choose a coordinate system attached to a particular particle and study the changes in its velocity and acceleration.

In the Cartesian system, the position of the fluid particle in space (x, y, z) at any time t from its position (a, b, c) at time $t = 0$, shall be given as:

$$\left.\begin{aligned} x &= f_1\,(a, b, c, t) \\ y &= f_2\,(a, b, c, t) \\ z &= f_3\,(a, b, c, t) \end{aligned}\right\} \quad \text{... (3.1)}$$

The velocity components are given by:

$$u = \frac{\partial x}{\partial t},\; v = \frac{\partial y}{\partial t},\; w = \frac{\partial z}{\partial t} \quad \text{... (3.6)}$$

The acceleration components are:

$$a_x = \frac{\partial^2 x}{\partial t^2},\; a_y = \frac{\partial^2 y}{\partial t^2},\; a_z = \frac{\partial^2 z}{\partial t^2} \quad \text{... (3.7)}$$

This method has the following shortcomings:

(i) The equations of motion are non-linear and very difficult to solve.

(*ii*) The method is cumbersome and complex.

(*iii*) The motion is hard to understand.

3.3 CLASSIFICATION OF FLUID FLOW

The fluids may be classified as follows:

1. Steady and unsteady flow
2. Uniform and non-uniform flow
3. One–, two–, and three-dimensional flows
4. Rotational and irrotational flow

5. Laminar and turbulent flow
6. Compressible and incompressible flow.

1. Steady and Unsteady Flow

A flow is said to be steady if the fluid and flow characteristic such as density, velocity, pressure, acceleration, etc. at a point do not change with time. Thus for a steady flow:

$$\left(\frac{\partial \rho}{\partial t}\right)_{x_0,y_0,z_0} = 0, \left(\frac{\partial v}{\partial t}\right)_{x_0,y_0,z_0} = 0, \left(\frac{\partial p}{\partial t}\right)_{x_0,y_0,z_0} = 0, \text{ and so on.}$$

In an unsteady flow, one or more of the above variables at a point may change with time. Thus, for unsteady flow:

$$\left(\frac{\partial \rho}{\partial t}\right)_{x_0,y_0,z_0} \neq 0, \left(\frac{\partial v}{\partial t}\right)_{x_0,y_0,z_0} \neq 0, \left(\frac{\partial p}{\partial t}\right)_{x_0,y_0,z_0} \neq 0, \text{ and so on.}$$

where (x_0, y_0, z_0) is a fixed point in the fluid.

2. Uniform and Non-Uniform Flow

A flow is said to be uniform if the velocity vector at all points in the flow is same at any instant of time. Thus for uniform flow

$$\left(\frac{\partial v}{\partial s}\right)_{t_0} = 0$$

in which v is the change in velocity corresponding to a displacement ∂s in any direction.

In a non-uniform flow, the velocity vector varies from point to point at any instant of time. Thus.

$$\left(\frac{\partial v}{\partial s}\right)_{t_0} \neq 0$$

3. One-,Two-, and Three-Dimensional Flow

In a one-dimensional flow, the flow parameters are functions of time and one space coordinate only. It may be expressed by:

$$u = f_1(x, t), \text{v} = 0, w = 0$$

In a two-dimensional flow, the flow parameters are functions of time and two rectangular space coordinates, say x and y. Thus

$$u = f_1(x, y, t), \text{v} = f_2,(x, y, t), w = 0$$

In a three-dimensional flow, the flow parameters are functions of time and three rectangular space coordinates. Thus for a 3-D unsteady flow:

$$u = f_1(x, y, z, 6), \text{v} = f_2(x, y, z, t), w = f_3(x, y, z, t)$$

4. Rotational and Irrotational Flow

A flow is said to be rotational if the fluid particles while moving in the direction of flow rotate about their mass centres.

A flow is said to be irrotational if the fluid particles while moving in the direction of flow do not rotate about their mass centres.

5. Laminar and Turbulent Flow

In a laminar flow, the fluid particles move along smooth and regular paths which are well defined in advance. The laminar flow is also called as streamline flow.

The turbulent flow is characterised by random and erratic movement of fluid particles resulting in the formation of eddies.

6. Compressible and Incompressible Flow

A compressible flow is one in which the density of the fluid changes from point to point, i.e. $\rho \neq$ const.

In an incompressible flow, the density of the fluid remains constant through the flow, i.e ρ = const.

3.4 EXAMPLES OF VARIOUS FLOWS

Steady flow – Flow through a prismatic or non-prismatic conduit at constant flow rate.

Unsteady flow – Flow in a pipe whose valve is being opened or closed gradually.

Uniform flow– Flow through a straight prismatic conduit, i.e a straight pipe of constant diameter.

Non-uniform flow – Flow through a non-prismatic conduit, around a uniform diameter pipe bend or a canal bend.

Laminar flow– Flow in a conduit of small size at low velocity, flow of blood in veins and arteries, flow through a capillary tube and ground water flow.

Turbulent flow– High velocity flow in conduit of large size.

Rotational flow– Flow near the solid boundaries, motion of liquid in a rotating tank

Irrotational flow– Flow of ideal fluid within parallel boundaries, flow above a drain hole of a stationary tank or a wash basin

One– dimensional flow – Flow in a pipe.

Two – dimensional flow – Flow between parallel plates of infinite extent, flow in a wide river.

Three– dimensional flow – Flow in a converging or diverging pipe or channel.

Compressible flow –Flow of gases through nozzles, compressors and gas turbines.

Incompressible flow – Subsonic aerodynamics.

3.5 TYPES OF FLOW LINES

The various types of flow lines are:

1. Pathline 2. Stream line 3. Stream tube, and 4. Streakline.

1. Pathline

It is a curve traced by a single fluid particle during its motion. Fig. 3.1 (*a*) shows a path line of a fluid practice. It is a zig-zag paths motion.

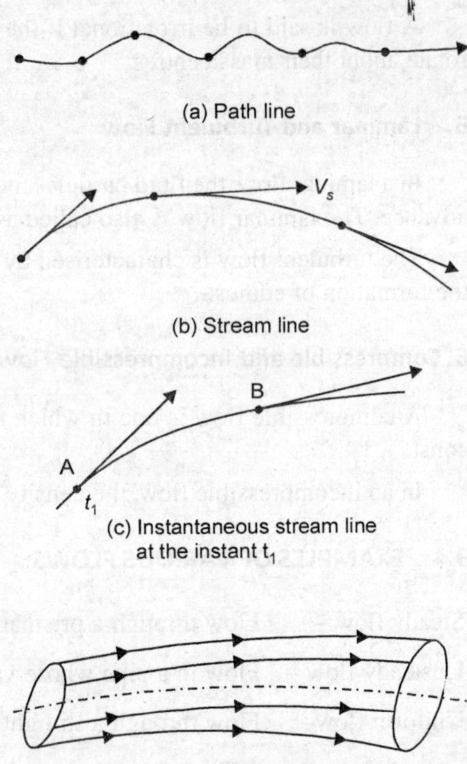

2. Stream line

It is an imaginary line drawn in a flow field such that a tangent drawn at any point on this line represents the direction of the velocity vector. Consider a fluid particle moving along a stream line, shown in Fig. 3.1 (b), for a very short distance *ds*. Let *dx*, *dy* and *dz* be the components of *ds* along the three rectangular coordinate axes and *u*, *v*, *w* those of V_s. The time taken by a fluid particle to move a distance *ds* along the streamline with a velocity V_s is,

$$t = \frac{ds}{V_s} = \frac{dx}{u} = \frac{dy}{v} = \frac{dz}{w}$$

Hence, the differential equation of the streamline may be written as:

$$\frac{dx}{u} = \frac{dy}{v} = \frac{dz}{w} \quad \text{... (3.8)}$$

The main characteristic of a stream line are:

(i) A streamline can neither intersect itself nor two streamlines can cross.

(ii) There can be no flow across a streamline.

(iii) The spacing between the streamlines varies inversely as the velocity of flow.

(iv) A stream line indicates the direction of velocity of a number of particles at the same instant.

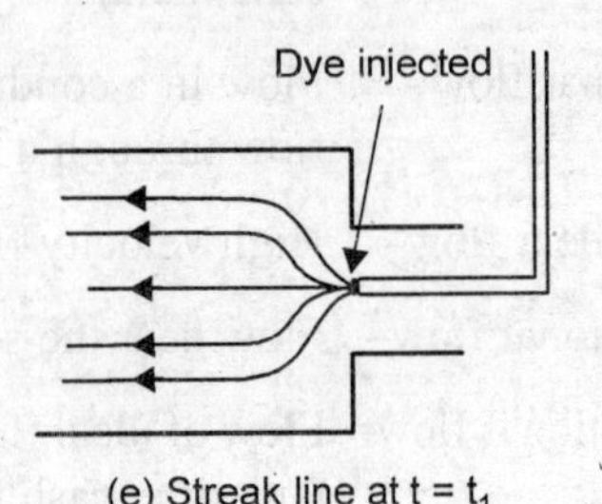

Fig. 3.1 Types of flow lines

3. Instantaneous stream line

In case of time-dependent flow, a fluid particle follows a streamline for only a short interval of time, before changing over to another streamline. The streamlines in such cases are not fixed in space, but change with time. The position of a stream line at a give instant of time is known as instantaneous stream line, as shown in Fig. 3.1 (c).

4. Stream tube

If stream lines are drawn through a closed curve, they form a boundary surface which fluid cannot penetrate. Such a surface bounded by streamlines is a sort of tube, and is known as a stream tube shown in Fig. 3.1. (d).

5. Streakline

A streak line is a curve which gives an instantaneous picture of the location of the fluid particles, which have passed through a given point. Thus a streakline connects all particles passing through a given point. Fig. 3.1 (e) shows streak lines formed due to dye injected at the outlet of a nozzle flow.

Example 3.1 *The velocity distribution for a three-dimensional flow is given by*

$$\vec{V} = 2x\hat{i} + 3y\hat{j} - 5z\hat{k}$$

Find the equation of streamline passing through the point (1, 2, 3).

Solution. The equation of a streamline is,

$$\frac{dx}{u} = \frac{dy}{v} = \frac{dz}{w}$$

$$\int \frac{dx}{2x} = \int \frac{dy}{3y} \quad \text{or} \quad \frac{3}{2}\int \frac{dx}{x} = \int \frac{dy}{y}$$

$$\frac{3}{2}\ln x = \ln y + \ln C_1 = \ln (yC_1)$$

$$\ln x^{3/2} = \ln(yC_1)$$

$$x = (yC_1)^{2/3}$$

$$C_1 = \frac{1}{2}$$

$$x = \left(\frac{y}{2}\right)^{2/3}$$

$$\int \frac{dx}{2x} = -\int \frac{dz}{5z}$$

$$\frac{1}{2}\ln x = -\frac{1}{5}\ln z + \ln C_2$$

$$\ln x^{1/2} = \ln z^{-1/5} + \ln C_2 = \ln(C_2 z^{-1/5})$$

$$x^{1/2} = C_2 z^{-1/5}$$

$$x^{1/2} = 3^{1/5} \times z^{-1/5} = \left(\frac{3}{z}\right)^{1/5}, \quad x = \left(\frac{3}{z}\right)^{2/5} \quad \therefore \quad x = \left(\frac{y}{2}\right)^{2/3} = \left(\frac{3}{z}\right)^{2/5}$$

3.6 EXPRESSIONS FOR ACCELERATION

3.6.1 One-Dimensional Flow

Consider a fluid particle at P undergoing a small displacement $ds = PQ$ in a short interval of time dt, as shown in Fig. 3.2 (a). The velocity V of this particle over the distance ds is,

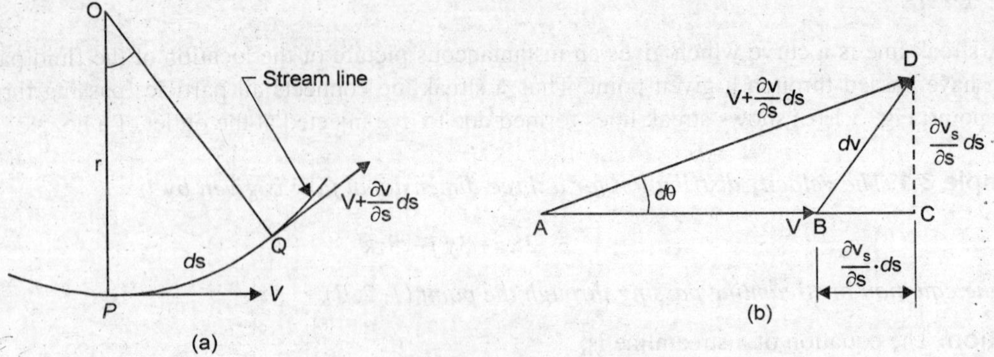

Fig. 3.2 Variation of velocity along a streamline

$$V = V_s = \frac{ds}{dt} \qquad \text{... (3.9)}$$

where V_s = velocity tangential to the streamline at P.

The acceleration of the particle will be the change in its velocity at Q w.r.t. its velocity at P, divided by dt.

$\therefore$ Acceleration,
$$a = \frac{dV}{dt} \qquad \text{... (3.10)}$$

Let the coordinate axes be along the tangential (s) and normal direction (n) to the streamline.

Now V_p = V

and
$$V_Q = V + \frac{\partial V}{\partial s} ds$$

Component of velocity at Q in the direction of tangent at P is (Fig. 3.2b),

$$V_s = \left(V + \frac{\partial V}{\partial s} ds\right)\cos d\theta = V + \frac{\partial Vs}{\partial s}\cdot ds \qquad (\because \cos(d\theta) \simeq 1)$$

Tangential acceleration,
$$a_s = \frac{dV_s}{dt} = \frac{V + \frac{\partial V_s}{\partial s} ds - V}{dt} = \frac{\partial Vs}{\partial s}\cdot\frac{ds}{dt}$$

$$= V\frac{\partial V}{\partial s} \qquad \text{... (3.11)}$$

Now
$$(V_P)_n = 0 \text{ and } (V_Q)n = \left(V + \frac{\partial V}{\partial s}\cdot ds\right)\sin(d\theta) = (V + \frac{\partial V}{ds}\cdot ds)\cdot\frac{ds}{r}$$

$$= V\frac{ds}{r} + \frac{1}{r}\frac{\partial V}{\partial s}(ds)^2 \qquad \left[\because \sin(d\theta) \simeq \frac{ds}{r}\right]$$

$$= V\frac{ds}{r} \qquad \text{(neglecting the second term, being small)}$$

Normal acceleration,
$$a_n = \frac{dV_n}{dt} = \frac{V\frac{ds}{r} - 0}{dt} = \frac{V}{r}\cdot\frac{ds}{dt} = \frac{V^2}{r} \qquad \text{... (3.12)}$$

For unsteady flow,

Total acceleration = Change in velocity with space + Change in velocity with time.

= Convective acceleration + Local acceleration

Then
$$a_s = \frac{\partial V}{\partial t} + V\frac{\partial V}{ds} \qquad \text{... (3.13)}$$

and
$$a_n = \frac{\partial V_n}{\partial t} + \frac{V^2}{r} \qquad \text{... (3.14)}$$

where $\frac{\partial V}{\partial t}, \frac{\partial V_n}{\partial t}$ = local accelerations due to unsteadiness of flow.

$V\frac{\partial V}{\partial s}, \frac{V^2}{r}$ = convective accelerations due to non-uniformity of flow.

(i) If the velocity is uniform (for constant streamlines spacing) then $\frac{\partial V}{ds} = 0$ and convective acceleration = 0.

(ii) If the flow is steady then local acceleration = 0.

(iii) If flow is uniform and steady then $a_s = 0$.

(iv) If the streamlines are straight the n $a_n = 0$.

(v) If the flow is steady and streamlines are straight and parallel, then $a_s = 0$ and $a_n = 0$.

Example 3.2 *The velocity along a circular streamline having its centre at the origin and radius 2 m is constant at 3m/s. Calculate the horizontal and vertical, tangential and normal accelerations at (2, 30°).*

Solution.

Given: $r = 2$m, $v_\theta = 3$m/s. Refer to Fig. 3.3.

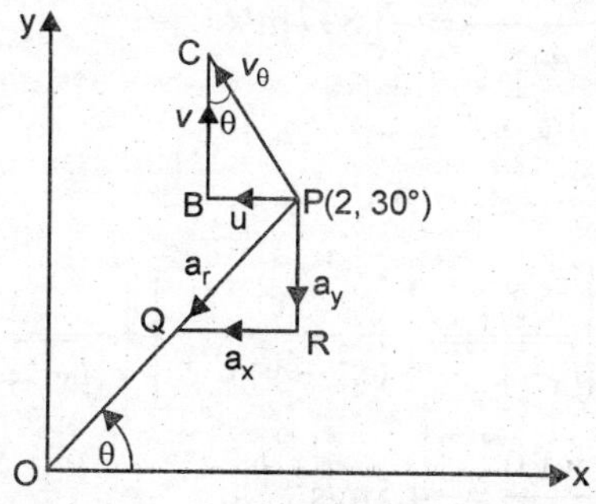

Fig. 3.3

Tangential acceleration, $a_\theta = 0$

Radial acceleration, $a_r = -\frac{v_\theta^2}{r} = -\frac{9}{2} = -4.5\,\text{m/s}^2$

$$\sin\theta = \frac{y}{\sqrt{x^2+y^2}} = -\frac{u}{v_\theta}$$

$$u = -\frac{yv_\theta}{\sqrt{x^2+y^2}} = -\frac{3y}{\sqrt{x^2+y^2}}$$

$$\cos\theta = \frac{x}{\sqrt{x^2+y^2}} = \frac{v}{v_\theta}$$

$$v = \frac{v_\theta\, x}{\sqrt{x^2+y^2}} = \frac{3x}{\sqrt{x^2+y^2}}$$

Now $\quad x = r\cos\theta = 2\cos 30^\circ = 1.732\,\text{m}$

$y = r\sin\theta = 2\sin 30^\circ = 1.0\ \text{m}$

and $\quad r^2 = x^2 + y^2$

$$u = -\frac{3y}{r} = -\frac{3\times 1}{2} = -1.5\,\text{m/s}$$

$$v = \frac{3x}{r} = \frac{3\times 1.732}{2} = 2.598\,\text{m/s}$$

$$a_x = u\frac{\partial u}{\partial x} + v\frac{\partial u}{\partial y}$$

$$= \left[-\frac{3y}{\sqrt{x^2+y^2}}\right]\times\left[\frac{3xy}{(x^2+y^2)^{3/2}}\right] + \left[\frac{3x}{\sqrt{x^2+y^2}}\right]\left[\frac{3y^2}{(x^2+y^2)^{3/2}} - \frac{3}{\sqrt{x^2+y^2}}\right]$$

$$= -\frac{9x}{x^2+y^2} = \frac{-9\times 1.732}{4} = -3.897\,\text{m/s}^2$$

$$a_y = u\frac{\partial v}{\partial x} + v\frac{\partial v}{\partial y}$$

$$= -\frac{3y}{\sqrt{x^2+y^2}}\times\left[\frac{3}{\sqrt{x^2+y^2}} - \frac{3x^2}{(x^2+y^2)^{3/2}}\right] + \frac{3x}{\sqrt{x^2+y^2}}\left[-\frac{3xy}{(x^2+y^2)^{3/2}}\right]$$

$$= -\frac{9y}{x^2+y^2} = -\frac{9\times 1.0}{2} = -4.5\,\text{m/s}^2$$

3.6.2 Two-Dimensional Flow

Let V = Velocity of a fluid particle at $P(x, y)$

u, v = components of V along x- and y-directions respectively.

a = acceleration of fluid particle at $P(x, y)$

a_x, a_y = components of 'a' along x-and y-directions respectively

$u = f_1(x, y, t)$

$v = f_2(x, y, t)$

(a) Cartesian coordinates (Fig. 3.4a)

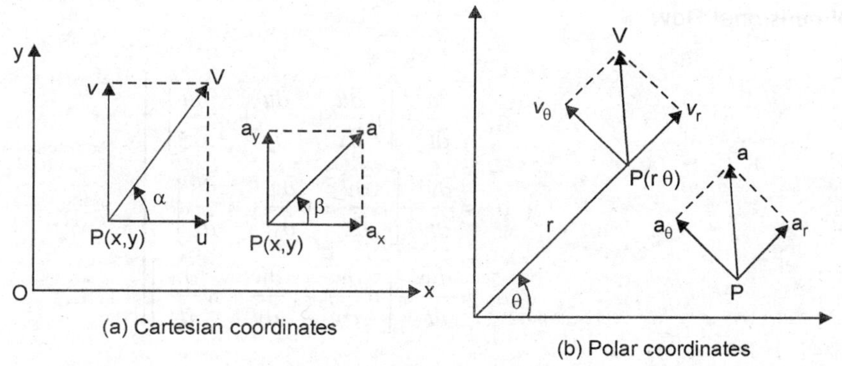

Fig 3.4 Two-dimensional flow

Now $$u = \frac{dx}{dt} \text{ and } v = \frac{dy}{dt} \quad \text{... (3.15)}$$

$$a_x = \frac{du}{dt} \text{ and } a_y = \frac{dv}{dt} \quad \text{... (3.16)}$$

Total derivative, $$du = \frac{\partial u}{\partial t}\cdot dt + \frac{\partial u}{\partial x}\cdot dx + \frac{\partial u}{\partial y}\cdot dy$$

$$a_x = \frac{du}{dt} = \frac{\partial u}{\partial t} + \frac{\partial u}{\partial x}\cdot\frac{dx}{dt} + \frac{\partial u}{\partial y}\cdot\frac{dy}{dt}$$

$$= \frac{\partial u}{\partial t} + u\frac{\partial u}{\partial x} + v\frac{\partial u}{\partial y} \quad \text{... (3.17a)}$$

Similarly $$a_y = \frac{\partial v}{\partial t} + u\frac{\partial v}{\partial x} + v\frac{\partial v}{dy} \quad \text{... (3.17 b)}$$

(b) Polar coordinates (Fig. 3.4b)

Radial velocity, $v_r = \frac{dr}{dt}$ ... (3.18 a)

Tangential velocity, $v_\theta = r\frac{d\theta}{dt}$ ... (3.18 b)

Radial acceleration, $a_r = \frac{\partial v_r}{\partial t} + v_r\frac{\partial v_r}{\partial r} + \frac{v_\theta}{r}\frac{\partial v_r}{\partial \theta} - \frac{v_\theta^2}{r}$ ... (3.19 a)

Tangential acceleration, $a_\theta = \frac{\partial v_\theta}{\partial t} + v_r\frac{\partial v_\theta}{\partial r} + \frac{v_\theta}{r}\frac{\partial v_\theta}{\partial \theta} + \frac{v_r v_\theta}{r}$ ... (3.19 b)

3.6.3 Three-Dimensional Flow

$$\left.\begin{aligned} a_x &= \frac{\partial u}{\partial t} + \left[u\frac{\partial u}{\partial x} + v\frac{\partial u}{\partial y} + w\frac{\partial u}{\partial z}\right] \\ a_y &= \frac{\partial v}{\partial t} + \left[u\frac{\partial v}{\partial x} + v\frac{\partial v}{\partial y} + w\frac{\partial v}{\partial z}\right] \\ a_z &= \frac{dw}{dt} + \left[u\frac{\partial w}{\partial x} + v\frac{\partial w}{\partial y} + w\frac{\partial w}{dz}\right] \end{aligned}\right\} \quad ...(3.20)$$

The expressions $\frac{\partial u}{\partial t}, \frac{\partial v}{\partial t}, \frac{\partial w}{\partial t}$ are known are *local acceleration*. The remaining terms within the square brackets are called *convective acceleration*.

3.7 CONTINUITY EQUATION

The continuity equation is based on the principle of conservation of mass, according to which fluid mass can neither be created nor destroyed.

3.7.1 One-Dimensional Flow

Consider a stream tube as shown in Fig. 3.5 having areas dA_1 and dA_2 of the closed curves. If these areas are sufficiently small, the velocities at their mid-point will represent the average velocities at those sections. As the fluid cannot cross the boundary surface of the stream tube, therefore the mass rate of steady flow passing any normal section remains constant. Thus

$$\rho_1\, dA_1 v_1 = \rho_2\, dA_2 v_2 \quad ... (3.21)$$

For incompressible fluids, $\rho_1 = \rho_2$, therefore

Volume flow rate, $dQ = v_1 \cdot dA_1 = v_2 \cdot dA_2$

or $Q = v_1 A_1 = v_2 A_2$ ... (3.22)

Eq. (3.22) is known as the continuity equation.

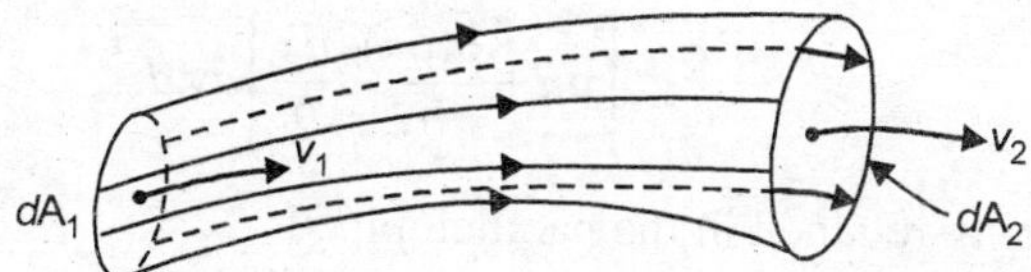

Fig. 3.5 Continuity of Flow through a stream tube

3.7.2 Two-Dimensional Flow

For a two-dimensional flow,

$$v_1 \cdot d_1 = v_2 \cdot d_2 \quad \text{... (3.23)}$$

where d_1, d_2 = normal distances between the streamlines at inlet and outlet respectively.

3.7.3 Cartesian coordinates

Consider an infinitesimal rectangular parallelopiped of size $dx \times dy \times dz$ with its centre $P(x, y, z)$ and velocity components u, v and w along x, y and z directions respectively, as shown in Fig. 3.6.

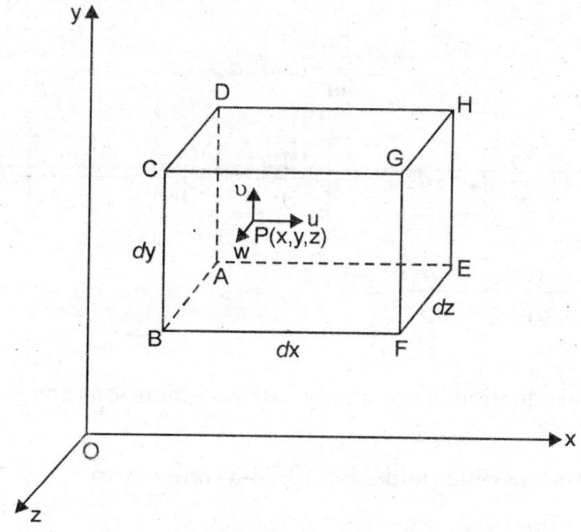

Fig. 3.6 Three-dimensional mass rate of flow

Rate of mass flow passing the central surface in x-direction = $\rho u \, dy. \, dz$

Rate of mass flow entering face ABCD

$$= \rho u \, dy.dz - \frac{\partial(\rho u)}{\partial x} \cdot dy.dz \times \frac{dx}{2}$$

$$= \left[\rho u - \frac{\partial(\rho u)}{\partial x} \cdot \frac{dx}{2}\right] dy \, dz$$

Rate of mass flow leaving the face EFGH

$$= \left[\rho u + \frac{\partial(\rho u)}{\partial x}\cdot\frac{dx}{2}\right]dy.dz$$

Net mass rate of flow being retained in the parallelopiped

$$= \left[\rho u + \frac{\partial(\rho u)}{\partial x}\cdot\frac{dx}{2}\right]dy.dz - \left[\rho u + \frac{\partial(\rho u)}{\partial x}\cdot\frac{dx}{2}\right]dy.dz$$

$$= -\frac{\partial(\rho u)}{\partial x}\cdot dx\,dy\,dz$$

Similar expressions may be obtained along the y- and z- directions.

Therefore, the total net gain of mass per unit time within the parallelopiped is,

$$-\left[\frac{\partial(\rho u)}{\partial x}+\frac{\partial(\rho v)}{\partial y}+\frac{\partial(\rho w)}{\partial z}\right]dx\,dy\,dz$$

Considering the time rate of variation of density, the change in the mass of the element per unit time will be

$$= \frac{\partial\rho}{\partial t}\cdot dx\,dy\,dz$$

$$\therefore \qquad \frac{\partial\rho}{\partial t}\cdot dx\,dy\,dz = -\left[\frac{\partial(\rho u)}{\partial x}+\frac{\partial(\rho v)}{\partial y}+\frac{\partial(\rho w)}{\partial z}\right]dx\,dy\,dz$$

$$\text{or} \qquad \frac{\partial\rho}{\partial t}+\frac{\partial(\rho u)}{\partial x}+\frac{\partial(\rho v)}{\partial y}+\frac{\partial(\rho w)}{\partial z} = 0 \qquad \text{... (3.24 a)}$$

Eq. (3.24 a) is the general equation of continuity in thee-dimensions and is applicable to compressible as well as incompressible fluids.

For steady flow of incompressible fluids, Eq. (3.24 a) reduces to

$$\frac{\partial u}{\partial x}+\frac{\partial v}{\partial y}+\frac{\partial w}{\partial z} = 0 \qquad \text{... (3.24 b)}$$

3.7.4 Polar coordinates

Consider a fluid element *ABCD*, as shown in Fig. 3.7, between θ and $\theta + d\theta$ and of radius r and $r + dr$. Let v_r and v_θ be the radial and tangential components of velocities. The thickness of the element perpendicular to the plane of the paper is taken as unity.

Flow in radial direction:

Mass flow rate of fluid entering the face $AD = \rho \times (v_r \times r d\theta)$

Mass flow rate of fluid leaving the face $BC = \left[\rho v_r + \frac{\partial}{\partial r}(\rho v_r)\cdot dr\right](r + dr)\cdot d\theta$

Mass flow rate retained in the element $= \rho \times (v_r \times r d\theta) - \left[\rho v_r + \frac{\partial}{\partial r}(\rho v_r) \cdot dr\right](r + dr) d\theta$

$$= -\left[\rho v_r \, dr \, d\theta + \frac{\partial}{\partial r}(\rho v_r) dr \cdot d\theta\right]$$

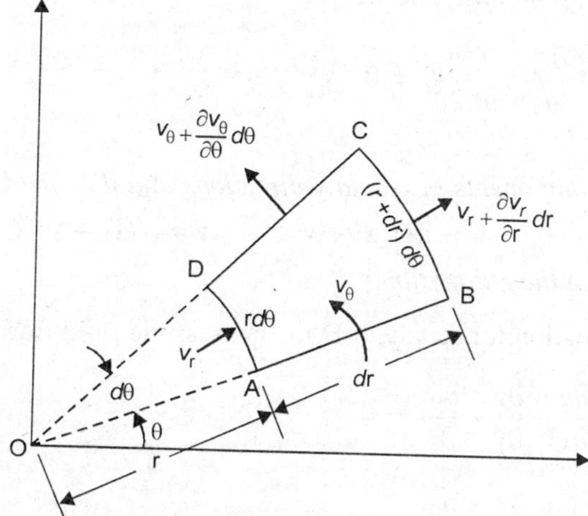

Fig. 3.7 Control volume of fluid in polar coordinates

Flow in tangential direction:

Mass flow rate retained in the element $= \rho \times (v_\theta \times dr) - \left[\rho v_\theta + \frac{\partial}{\partial \theta}(\rho v_\theta) d\theta\right] \cdot dr$

$$= -\left[\frac{\partial}{\partial \theta}(\rho v_\theta) dr \cdot d\theta\right]$$

Total gain in mass flow rate of fluid $= -\left[\rho v_r \, dr d\theta + \frac{\partial}{\partial r}(\rho v_r) r dr d\theta + \frac{\partial}{\partial \theta}(\rho v_\theta) dr \, d\theta\right]$

$$= -\left[\rho v_r + \frac{\partial}{\partial r}(\rho v_r) r + \frac{\partial}{\partial \theta}(\rho v_\theta)\right] dr \cdot d\theta$$

Also, the rate of change of fluid mass in the element *ABCD*

$$= \frac{\partial}{\partial t}\left[\rho \times \frac{r d\theta + (r + dr) d\theta}{2} + dr\right]$$

$$= \frac{\partial}{\partial t}(\rho r \, d\theta \cdot dr)$$

As per the principle of conservation of mass, we have

$$\frac{\partial}{\partial t}(\rho r\,d\theta \cdot dr) = -\left[\rho v_r + \frac{r\partial}{\partial r}(\rho v_r) + \frac{\partial}{\partial \theta}(\rho v_\theta)\right] dr \cdot d\theta$$

or $$\frac{\partial \rho}{\partial t} + \frac{\rho v_r}{r} + \frac{\partial}{\partial r}(\rho v_r) + \frac{1}{r}\frac{\partial}{\partial \theta}(\rho v_\theta) = 0 \qquad \text{... (3.25 a)}$$

For incompressible fluids, ρ = const, and we get

$$\frac{v_r}{r} + \frac{\partial v_r}{\partial_r} + \frac{1}{r}\frac{\partial v_\theta}{\partial \theta} = 0 \qquad \text{... (3.25 b)}$$

Example 3.3 *The velocity components in x- and y-directions in a flow field are given by:*

$$u = x^2 + y^2 z^3, \quad v = -(xy + yz + zx)$$

Find the velocity component in the z-direction.

Solution. The continuity equation for steady, 3-D incompressible fluid flow is:

$$\frac{\partial u}{\partial x} + \frac{\partial v}{\partial y} + \frac{\partial w}{\partial z} = 0$$

$$(2x + 0) + (-x - z) + \frac{\partial w}{\partial z} = 0$$

$$\frac{\partial w}{\partial z} = z - x$$

Integrating, we have

$$\int dw = \int (z - x)\,dz$$

$$w = \frac{z^2}{2} - xz + C(x, y)$$

where C = constant of integration

Example 3.4. *In an incompressible flow field, the velocity vector is given by*

$$\vec{V} = (4xt + 3yz^3)\hat{i} + (2t + xy^2)\hat{j} + (xy - 3tz)\hat{k}$$

(a) Verify whether the continuity equation is satisfied

(b) Determine the acceleration vector at (1, 2, 3) and t = 2.

Solution. (a) $u = 4xt + 3yz^3$, $v = 2t + xy^2$, $w = xy - 3tz$

$$\frac{\partial u}{\partial x} + \frac{\partial v}{\partial y} + \frac{\partial w}{\partial z} = 4t + 0 + 0 + 2xy - 3t = t + 2xy \neq 0$$

The continuity equation is not satisfied.

(b) $$a_x = \frac{\partial u}{\partial t} + u\frac{\partial u}{\partial x} + v\frac{\partial u}{\partial y} + w\frac{\partial u}{\partial z}$$

$$= 4x + (4xt + 3yz^3)(4t) + (2t + xy^2)(3z^3) + (xy - 3tz)(9yz^2)$$

$$= 4 \times 1 + (4 \times 1 \times 2 + 3 \times 2 \times 27)(4 \times 2) + (2 \times 2 + 1 \times 4)(3 \times 27)$$

$$+ (1 \times 2 - 3 \times 2 \times 3)(9 \times 2 \times 9)$$

$$= 4 + 1360 + 648 - 2592 = -580$$

$$a_y = \frac{\partial v}{\partial t} + u\frac{\partial v}{\partial x} + v\frac{\partial v}{\partial y} + w\frac{\partial v}{\partial z}$$

$$= 2 + 170 \times (y^2) + 8\,(2xy) + (-16)\,(0)$$

$$= 2 + 170 \times 4 + 8 \times 4 + 0 = 714$$

$$a_z = \frac{\partial w}{\partial t} + u\frac{\partial w}{\partial x} + v\frac{\partial w}{\partial y} + w\frac{\partial w}{\partial z}$$

$$= -3z + 170 \times y + 8 \times x + (-16) \times (-3t)$$

$$= 3 \times 3 + 170 \times 2 + 8 \times 1 + (-16) \times (3 \times 2)$$

$$= 435$$

$$\vec{a} = -580\hat{i} + 714\hat{j} + 435\hat{k}$$

Example 3.5 *The velocity components in x and y directions are given as*

$$u = \frac{2}{3}xy^3 - x^2y \text{ and } y = xy^2 - \frac{2}{3}yx^3$$

Find whether the given velocity distribution is:

(a) a possible flow field (b) not a possible flow field.

Solution. A possible flow field most satisfy the continuity equation,

$$\frac{\partial u}{\partial x} + \frac{\partial v}{\partial y} = 0$$

$$\frac{2}{3}y^3 - 2xy + 2xy - \frac{2}{3}x^3 \neq 0$$

Since the continuity equation is not satisfied, therefore it is not a possible flow field.

Example 3.6 *A pipe converges uniformly from 0.4 m to 0.2 m diameter over 2 m length. Find the convective acceleration at the middle of the pipe if the rate of flow is 0.02 m³/s.*

Solution. Here $v = 0$ and $w = 0$

$$\text{Convective acceleration} = u\frac{\partial u}{\partial x}$$

Diameter of pipe at a distance x from inlet,

$$d_x = 0.4\frac{(0.4-0.2)x}{2} = (0.4-0.1x)\,\text{m}$$

$$A_x = \frac{\pi}{4}d_x^2 = \frac{\pi}{4}(0.4-0.1x)^2$$

$$u = \frac{Q}{A_x} = \frac{0.02\times 4}{\pi(0.4-0.1x)^2} = \frac{0.02546}{(0.4-0.1x)^2} = 0.02546\,(0.4\text{–}0.1x)^{-2}$$

$$\frac{\partial u}{\partial x} = \frac{-0.051\times(-0.1)}{(0.4-0.1x)^3} = \frac{0.0051}{(0.4-0.1x)^3}$$

$$\text{Convective acceleration} = \frac{0.02546\times 0.0051}{(0.4-0.1x)^5} = \frac{0.0013}{(0.4-0.1x)^5}$$

$$= \frac{0.0013}{(0.4-0.1\times 1)^5} = 0.0534\,\text{m/s}^2$$

Example 3.7 *A 0.2 m diameter pipe carries oil of specific gravity 0.9 at a velocity of 2.5 m/s. At another section the pipe diameter is 0.15 m. Find the velocity at this section and the mass flow rate of oil.*

Solution. Applying continuity equation, we have

$$A_1 v_1 = A_2 v_2$$

$$v_2 = \left(\frac{d_1}{d_2}\right)^2 v_1 = \left(\frac{0.2}{0.15}\right)^2 \times 2.5 = 4.44\,\text{m/s}$$

$$\text{Mass flow rate, } \dot{m} = \rho A_1 V_1 = 0.9\times 10^3 \times \frac{\pi}{4}(0.2)^2 \times 2.5$$

$$= 70.686\ \text{kg/s}$$

3.8 SHEAR DEFORMATION, ROTATION AND VORTICITY

Shear deformation is defined as the average change in the angle contained by two adjacent sides of a fluid element. For example, for velocities along x- and y-directions,

$$\text{Shear strain rate} = \frac{1}{2}\left(\frac{\partial v}{\partial x} - \frac{\partial u}{\partial y}\right)$$

Rotation is defined as the movement of a fluid element in such a way that both of its axis (horizontal as well as vertical) rotate in the same direction. For a two-dimensional fluid element in x-y plane, rotation is equal to, $\frac{1}{2}\left(\frac{\partial v}{\partial x} - \frac{\partial u}{\partial y}\right)$.

Thus the total deformation matrix may be written as:

$$\begin{bmatrix} \frac{\partial u}{\partial x} & \frac{\partial u}{\partial y} & \frac{\partial u}{\partial z} \\ \frac{\partial v}{\partial x} & \frac{\partial v}{\partial y} & \frac{\partial v}{\partial z} \\ \frac{\partial w}{\partial x} & \frac{\partial w}{\partial y} & \frac{\partial w}{\partial z} \end{bmatrix} = \begin{bmatrix} \frac{\partial u}{\partial x} & \frac{1}{2}\left(\frac{\partial u}{\partial y}+\frac{\partial v}{\partial x}\right) & \frac{1}{2}\left(\frac{\partial u}{\partial z}+\frac{\partial w}{\partial x}\right) \\ \frac{1}{2}\left(\frac{\partial v}{\partial x}+\frac{\partial u}{\partial y}\right) & \frac{\partial v}{\partial y} & \frac{1}{2}\left(\frac{\partial v}{\partial z}+\frac{\partial w}{\partial y}\right) \\ \frac{1}{2}\left(\frac{\partial w}{\partial x}+\frac{\partial u}{\partial z}\right) & \frac{1}{2}\left(\frac{\partial w}{\partial y}+\frac{\partial v}{\partial z}\right) & \frac{\partial w}{\partial z} \end{bmatrix}$$

$$+\begin{bmatrix} 0 & \frac{1}{2}\left(\frac{\partial u}{\partial y}-\frac{\partial v}{\partial x}\right) & \frac{1}{2}\left(\frac{du}{dz}-\frac{\partial w}{\partial x}\right) \\ \frac{1}{2}\left(\frac{\partial v}{\partial x}-\frac{\partial u}{\partial y}\right) & 0 & \frac{1}{2}\left(\frac{du}{dz}-\frac{\partial w}{\partial x}\right) \\ \frac{1}{2}\left(\frac{\partial w}{\partial x}-\frac{\partial u}{\partial z}\right) & \frac{1}{2}\left(\frac{\partial w}{\partial y}-\frac{\partial v}{\partial z}\right) & 0 \end{bmatrix} \quad \text{... (3.26)}$$

= Shear strains rate matrix + Rotation matrix

The rotation components are:

$$\left.\begin{aligned} \omega_x &= \frac{1}{2}\left(\frac{\partial w}{\partial y}-\frac{\partial v}{\partial z}\right) \\ \omega_y &= \frac{1}{2}\left(\frac{\partial u}{\partial z}-\frac{\partial w}{\partial x}\right) \\ \omega_z &= \frac{1}{2}\left(\frac{dv}{dx}-\frac{\partial u}{\partial y}\right) \end{aligned}\right\} \quad \text{... (3.27)}$$

$$\vec{\omega} = \omega_x\hat{i}+\omega_y\hat{j}+\omega_z\hat{k} \quad \text{... (3.28)}$$

Vorticity: It is defined as twice the value of rotation.

Thus, vorticity, $\xi = 2\omega$. ... (3.29)

where ξ is Greek letter *xi*.

3.9 CIRCULATION AND VORTICITY

Consider a closed curve ABC within the fluid medium (Fig. 3.8) The circulation around such a curve is defined as the summation of product of velocity component along the element (such as PQ) of the curve and the elemental length *ds*.

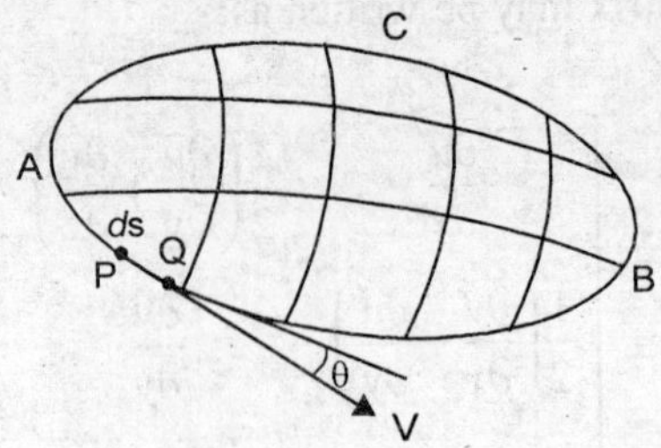

Fig. 3.8 Circulation around a closed curve.

Let V = velocity at point P

ds = length of element PQ

θ = angle between ds and V

Then the circulation denoted by Γ (Greek capital "gamma") is given by,

$$\Gamma = \oint V \cos\theta . ds \qquad \text{... (3.30a)}$$

Vectorially it is written as the dot product of the velocity vector and the length vector,

$$\Gamma = \oint V . ds \qquad \text{... (3.30 b)}$$

$$= \oint (u\hat{i} + v\hat{j} + w\hat{k}) \cdot (dx\hat{i} + dy\hat{j} + dz\hat{k})$$

$$= \oint (u\,dx + v dy + w\,dz) \qquad \text{... (3.30 c)}$$

The integral is taken around the closed curve in an anti-clockwise direction.

The circulation around a large closed curve may be shown to be equal to the sum of circulations around small meshes into which the closed curve can be sub-divided.

Consider an elementary rectangle ABCD of fluid element of size $dx \times dy$ as shown in fig. 3.9

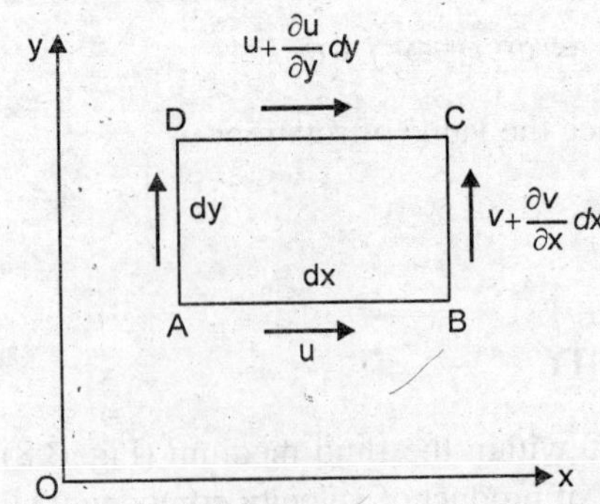

Fig. 3.9 Circulation around an elementary rectangle

$$d\Gamma = udx + \left(v + \frac{\partial v}{dx}dx\right)dy - \left(u + \frac{\partial u}{\partial y}dy\right)dx - v\,dy$$

$$= \frac{\partial v}{dx}dx\,dy - \frac{\partial u}{\partial y}dx\,dy$$

$$= \left(\frac{\partial v}{dx} - \frac{\partial u}{\partial y}\right)dx\,dy \qquad \text{... (3.31)}$$

The vorticity ξ at a point is defined as the ratio of the differential circulation around an infinitesimal closed curve enclosing the point to the area of the closed curve.

Thus from Eq. (3.31), the z-component of vorticity,

$$\xi_z = \frac{d\Gamma}{dx\,dy} = \frac{dv}{dx} - \frac{du}{\partial y} \qquad \text{... (3.32 a)}$$

In polar coordinates,

$$\xi_z = \frac{v_\theta}{r} + \frac{\partial v_\theta}{\partial r} - \frac{1}{r}\frac{\partial v_r}{\partial \theta} \qquad \text{... (3.32 b)}$$

Vorticity is a vector quantity having its direction perpendicular to the plane of the closed curve around which the circulation is measured.

If a fluid element is rotating at constant angular velocity ω around a circular path of radius r, then

$$d\Gamma = \int_0^{2\pi} vr\,d\theta = \int_0^{2\pi} r^2\omega\,d\theta = 2\pi r^2\omega$$

$$\xi_z = \frac{d\Gamma}{\pi r^2} = 2\omega$$

$\therefore$ Rotation, $\omega = \frac{1}{2}\xi_z$ ... (3.33)

For a flow to be irrotational, the vorticity is zero.

Example 3.8 The velocity components in a fluid flow are given as:

$$u = -4ax(x^2 - 3y^2), v = 4ay(3x^2 - y^2)$$

Find whether (i) these velocity components represent a physically possible two-dimension flow, and (ii) the flow is rotational or irrotational.

Solution. (i) A possible flow should satisfy the continuity equation,

$$\frac{\partial u}{\partial x} + \frac{\partial v}{\partial y} = 0$$

$$-4a(3x^2 - 3y^2) + 4a(3x^2 - 3y^2) = 0$$

Hence, it is a physically possible two-dimensional flow.

(*ii*) The flow will be irrotational if

$$\omega_z = \frac{1}{2}\xi_z = \frac{1}{2}\left(\frac{\partial v}{\partial x} - \frac{\partial u}{\partial y}\right) = 0$$

or
$$\frac{\partial v}{\partial x} = \frac{\partial u}{\partial y}$$

$$\frac{\partial v}{\partial x} = 4ay(6x) = 24axy$$

$$\frac{\partial u}{\partial y} = -4ax(0 - 6y) = 24axy$$

Hence the flow is irrotational.

Example 3.9 *The velocity field is given by:*

$$u = (8y - 5x),\ v = (6y - 3x)$$

Calculate the circulation around the closed curve defined by: x = 3, y = 2; x = 6, y= 4.

Solution. Refer Fig. 3.10.

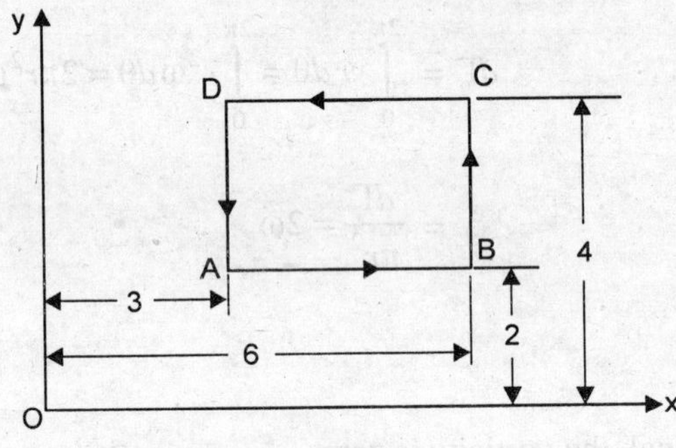

Fig. 3.10

$$\Gamma_{ABCD} = \oint_{ABCD} (udx + vdy)$$

$$= \int_{AB} (u\,dx + vdy) + \int_{BC} (udx + vdy) + \int_{CD} (udx + vdy) + \int_{DA} (udx + vdy)$$

$$= \int_3^6 (8y - 5x)\,dx + \int_2^4 (6y - 3x)\,dy + \int_6^3 (8y - 5x)\,dx + \int_4^2 (6y - 3x)\,dy$$

$$= \left|8yx - 2.5x^2\right|_3^6 + \left|3y^2 - 3xy\right|_2^4 + \left|8xy - 2.5x^2\right|_6^3 + \left|3y^2 - 3xy\right|_4^2$$

$$= [8 \times 2\times 3 - 2.5 \times 27]_{y=2} + [3 \times 12 - 3\times 6 \times 2]_{x=6} + [8 \times (-3) \times 4 - 2.5\,(-27)]_{y=4}$$
$$+ [3 \times (-12) - 3\times (-2) \times 3]_{x=3}$$

$$= -19.5 + 0 - 28.5 - 18 = -66$$

Example 3.10 *The velocity components in a three-dimensional flow are:*

$$u = xy^3z,\ v = -y^2z^2,\ w = yz^2 - \frac{y^3z^2}{2}$$

Determine the rotation at (1, 1, 2).

Solution. $w_x = \frac{1}{2}\left(\frac{\partial w}{\partial y} - \frac{\partial v}{dz}\right) = \frac{1}{2}\left[\left(z^2 - \frac{3}{2}y^2z^2\right) - (-2y^2z)\right] = \frac{1}{2}\left[\left(4 - \frac{3}{2}\times 1\times 4\right) + 2\times 1\times 2\right] = 3.25$

$$\omega_y = \frac{1}{2}\left(\frac{\partial u}{\partial z} - \frac{\partial w}{\partial x}\right) = \frac{1}{2}[xy^3 - 0] = \frac{1}{2}(1\times 1) = 0.5$$

$$\omega_z = \frac{1}{2}\left(\frac{\partial v}{\partial x} - \frac{\partial u}{\partial y}\right) = \frac{1}{2}[0 - 3xy^2z] = \frac{1}{2}(-3\times 1\times 1\times 2) = -3$$

$$\vec{\omega} = 3.25\hat{i} + 0.5\hat{j} - 3\hat{k}$$

$$= \sqrt{(3.25)^2 + (0.5)^2 + (-3)^2} = \sqrt{19.8125} = 4.451\,\text{rad.}$$

3.10 VELOCITY POTENTIAL

The velocity potential is a scalar quantity dependent upon space and time. Its negative derivative with respect to any direction gives the velocity in that direction. It is denoted by ϕ (Greek letter phi).

If u, v and w are the components of velocity in the x, y and z directions respectively, then

$$u = -\frac{\partial \phi}{\partial x},\ v = -\frac{\partial \phi}{\partial y},\ w = -\frac{\partial \phi}{dz} \qquad \text{... (3.34)}$$

The negative sign signifies that ϕ decreases with an increase in the values of x, y, and z

In polar coordinates (r, θ, z), the velocity components are:

$$v_r = -\frac{\partial \phi}{\partial r},\ v_\theta = -\frac{1}{r}\frac{\partial \phi}{\partial \theta},\ v_z = -\frac{\partial \phi}{\partial z} \qquad \text{... (3.35)}$$

For an incompressible steady flow the continuity equation is,

$$\frac{\partial u}{\partial x} + \frac{\partial v}{\partial y} + \frac{\partial w}{\partial z} = 0$$

$$\frac{\partial}{\partial x}\left(-\frac{\partial \phi}{\partial x}\right)+\frac{\partial}{\partial y}\left(-\frac{\partial \phi}{\partial y}\right)+\frac{\partial}{\partial z}\left(-\frac{\partial \phi}{\partial z}\right) = 0$$

$$\frac{\partial^2 \phi}{\partial x^2}+\frac{\partial^2 \phi}{\partial y^2}+\frac{\partial^2 \phi}{dz^2} = 0$$

or $$\nabla^2 \phi = 0 \qquad \text{... (3.36)}$$

Eq. (3.36) is the Laplace equation.

Thus any function ϕ that satisfies the Laplace equation will represent some fluid flow.

In case of two-dimensional flow in xy-plane, $\omega_x = \omega_y = 0$,

$$\omega_z = \frac{1}{2}\left(\frac{\partial v}{\partial x}-\frac{\partial u}{\partial y}\right)$$

and $$\xi_z = \frac{\partial v}{\partial x}-\frac{\partial u}{\partial y}$$

$$\omega_z = \frac{1}{2}\left[-\frac{\partial^2 \phi}{\partial x \partial y}+\frac{\partial^2 \phi}{\partial y \partial x}\right] = 0 \qquad \text{... (3.37)}$$

$\therefore$ $$\omega = \sqrt{\omega_x^2+\omega_y^2+\omega_z^2} = 0$$

The flow is irrotational if there exists velocity potential. The incompressible, irrotational flow is called *potential flow.*

3.10.1 Equipotential Line

A line along which the velocity potential ϕ is constant is called an equipotential line.

For an equipotential line, ϕ = const.

$\therefore$ $$d\phi = 0$$

For steady flow, $$\phi = f(x, y)$$

$\therefore$ $$d\phi = \frac{\partial \phi}{dx}dx+\frac{\partial \phi}{\partial y}dy$$

$$= -udx - vdy = -(udx + vdy) = 0$$

$\therefore$ $$\frac{dy}{dx} = -\frac{u}{v} \qquad \text{... (3.38)}$$

where $\frac{dy}{dx}$ = slope of equipotential line

3.11 STREAM FUNCTION

The stream function is defined as a function of space and time such that its partial derivative with respect to any direction gives the velocity component at right angles to this direction. It is denoted by ψ (Greek letter psi).

In case of two-dimensional steady flow, the stream function may be defined as:

$$\psi = f(x, y)$$

$$u = \frac{\partial \psi}{\partial y}, v = -\frac{\partial \psi}{\partial x} \quad \text{... (3.39)}$$

Eq. (3.39) states that the partial derivative of the stream function with respect to any direction gives the velocity component 90° clockwise with respect to that direction.

The continuity.

The continuity equation in two-dimensions is,

$$\frac{\partial u}{\partial x} + \frac{\partial v}{\partial y} = 0$$

$$\frac{\partial^2 \psi}{\partial x \partial y} - \frac{\partial^2 \psi}{\partial y \partial x} = 0$$

Since ψ is a continuous function of x and y, it satisfies the continuity equation

$$d\psi = \frac{\partial \psi}{\partial x} \cdot dx + \frac{\partial \psi}{dy} \cdot dy$$

$$= -v.dx + u.dy$$

If $d\psi = 0$, then

$$\frac{dy}{dx} = \frac{v}{u}$$

which is the equation of a streamline. Thus, lines of constant ψ represent stream lines since there is no flow across these lines. Conversely, a streamline is that line along which the stream function ψ is constant.

For irrotational flow, $\omega_z = 0$

or

$$\frac{\partial v}{\partial x} = \frac{\partial u}{\partial y}$$

$$-\frac{\partial^2 \psi}{\partial x^2} = \frac{\partial^2 \psi}{\partial y^2}$$

or

$$\frac{\partial^2 \psi}{\partial x^2} + \frac{\partial^2 \psi}{\partial y^2} = 0$$

or

$$\nabla^2 \psi = 0 \quad \text{... (3.40)}$$

Eq. (3.40) shows that stream function must satisfy the Laplace's equation provided that the flow is irrotational.

3.11.1 Relationship between Flux and Stream Function

Consider a two-dimensional incompressible flow (Fig. 3.11) in an open channel. From the principle of continuity, the rate of flow per unit width of the channel is the same at all cross sections. Thus

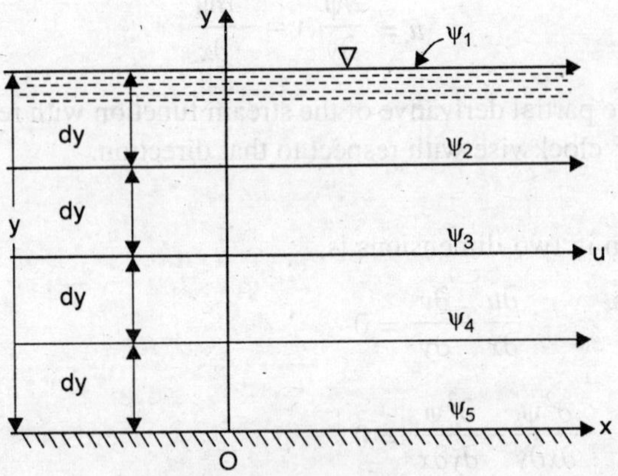

Fig. 3.11 Two-dimensional flow in an open channel

$$q = \int_0^y u.dy$$

where dy = spacing between the adjacent streamlines measured perpendicular to u.

As the flow is along the x- direction, $v = 0$.

Thus $$d\psi = \text{u}dy$$

$$\therefore \quad \int_5^1 d\psi = \int_0^y u.dy$$

$$\psi_1 - \psi_5 = q \qquad \text{... (3.41)}$$

Eq. (3.41) shows that the numerical difference in ψ between two streamlines is equal to the flow rate or flux. The unit of ψ are m^3/s, *i.e.* discharge per unit thickness of flow.

3.11.2 Properties of Stream Function

The properties of stream function are:

1. ψ = const. on any streamline.
2. If the flow is continuous, the flow around any path in the fluid is zero.
3. The rate of change of ψ with distance in any arbitrary direction is proportional to the component of velocity normal to that direction.
4. The algebraic sum of stream function for two incompressible flow patterns superimpose.

3.11.3 Cauchy-Riemann Equations

1. Velocity potential function ϕ exists only for irrotational flow.
2. Stream function ψ applies to both rotational and Irrotational flows.
3. Both the velocity potential function and stream function for irrotational flow, satisfy Laplace equation.

For irrotational incompressible flow,

$$\left.\begin{aligned} u &= -\frac{\partial \phi}{\partial x} = \frac{\partial \psi}{\partial y} \\ v &= -\frac{\partial \phi}{\partial y} = -\frac{\partial \psi}{\partial x} \end{aligned}\right\} \qquad \text{... (3.42)}$$

Eqs. (3.42) are sometimes called Cauchy-Riemann equations.

3.11.4 Relation between Velocity Potential and stream Function

We have observed that:

ψ = const. represents a streamline

and ϕ = const. represents an equipotential line.

Let the two curves, ϕ = const. and ψ = const. intersect each other at any point.

At the point of intersection, the slopes are:

$$m_1 = \frac{\partial y}{\partial x} = \frac{\partial \phi / \partial x}{\partial \phi / \partial y} = \frac{-u}{-v} = \frac{u}{v} \quad \text{for } \phi = \text{const.}$$

$$m_2 = \frac{\partial y}{\partial x} = \frac{\partial \psi / \partial x}{\partial \psi / \partial y} = \frac{-v}{u} = -\frac{v}{u} \quad \text{for } \psi = \text{const.}$$

$$m_1 \times m_2 = \frac{u}{v} \times \left(-\frac{v}{u}\right) = -1 \qquad \text{... (3.43)}$$

Eq. (3.43) shows that ϕ and ψ curves intersect each other orthogonally at all points of intersection.

Example 3.11 *Verify whether the following functions represent irrotational flow:*

(a) $\psi = A(x^2 - y^2)$ *(b)* $\phi = Ur\cos\theta + \frac{U}{r}\cos\theta$

Solution. For irrotational flow, both ϕ and ψ satisfy Laplace equation.

(a)

$$\psi = A(x^2 - y^2)$$

$$\frac{\partial \psi}{\partial x} = 2Ax, \quad \frac{\partial \psi}{\partial y} = -2Ay$$

$$\frac{\partial^2 \psi}{\partial x^2} = 2A, \quad \frac{\partial^2 \psi}{\partial y^2} = -2A$$

$$\frac{\partial^2 \psi}{\partial x^2} + \frac{\partial^2 \psi}{\partial y^2} = 2A - 2A = 0$$

Hence, it is a possible irrotational flow.

(*b*) $\phi = Ur\cos\theta + \frac{U}{r}\cos\theta$

$$\frac{\partial \phi}{\partial r} = U\cos\theta - \frac{U}{r^2}\cos\theta, \quad \frac{\partial^2 \phi}{\partial r^2} = \frac{2U}{r^3}\cos\theta$$

$$\frac{\partial \phi}{\partial \theta} = -Ur\sin\theta - \frac{U}{r}\sin\theta, \quad \frac{\partial^2 \phi}{\partial \theta^2} = -Ur\cos\theta - \frac{U}{r}\cos\theta$$

The Laplace equation in polar coordinates is,

$$\frac{1}{r}\frac{\partial \phi}{\partial r} + \frac{\partial^2 \phi}{\partial r^2} + \frac{1}{r^2}\frac{\partial^2 \phi}{\partial \theta^2} = 0$$

$$\frac{1}{r}(U\cos\theta - \frac{U}{r^2}\cos\theta) + \frac{2U}{r^3}\cos\theta + \frac{1}{r^2}(-Ur\cos\theta - \frac{U}{r}\cos\theta)$$

$$\frac{U}{r}\cos\theta - \frac{U}{r^3}\cos\theta + \frac{2U}{r^3}\cos\theta - \frac{U}{r}\cos\theta - \frac{U}{r^3}\cos\theta = 0$$

The Laplace equation is satisfied and hence it is a possible irrotational flow.

Example 3.12 *The velocity components for a steady incompressible flow are:*

$$u = -\frac{ax}{y}, v = a\ln xy$$

Find a relevant stream function.

Solution. The continuity equation is:

$$\frac{\partial u}{\partial x} + \frac{\partial v}{\partial y} = 0$$

$$-\frac{a}{y} + \frac{a}{xy}\cdot x = -\frac{a}{y} + \frac{a}{y} = 0$$

Hence the flow is possible and stream function ψ exists.

$$u = \frac{\partial \psi}{\partial y} = -\frac{ax}{y}$$

$$\psi = -ax \ln y + f_1(x)$$

$$v = -\frac{\partial \psi}{\partial x} = a \ln xy = a \ln x + a \ln y$$

$$-\psi = a\left[\ln x \times x - \int \frac{1}{x} \times x\,dx\right] + (a \ln y) \times x + f_2(y)$$

$$= a(x \ln x - x) + ax \ln y + f_2(y)$$

$$\psi = -ax \ln xy + ax - f_2(y) \text{ is the relevant stream function.}$$

Example 3.13 *In a two dimensional flow field, $\phi = x(2y - 1)$. Determine the stream function and its value at the point (2, 3).*

Solution.

$$u = -\frac{\partial \phi}{\partial x} = -(2y - 1) = 1 - 2y$$

$$v = -\frac{\partial \phi}{\partial y} = -2x$$

$$d\psi = \frac{\partial \psi}{\partial x}dx + \frac{\partial \psi}{\partial y}dy$$

$$= -v\,dx + u\,dy$$

$$= 2x\,dx + (1 - 2y)\,dy$$

Integrating both side, we get

$$\psi = x^2 + y - y^2 + C$$

At $x = y = 0$, $\psi = 0$, then $C = 0$

$$\therefore \quad \psi = x^2 - y^2 + y$$

At (2, 3), $\psi = 4 - 9 + 3 = -2$ units.

Example 3.14 *For a two-dimensional flow the velocity potential, $\phi = x^2 - y^2$. Determine the stream function and the discharge between the streamlines (2, 0) and (2, 2).*

Solution.

$$\phi = x^2 - y^2$$

$$u = \frac{-\partial \phi}{\partial x} = -2x, \; v = -\frac{\partial \phi}{\partial y} = 2y$$

$$d\psi = \frac{\partial \psi}{\partial x}dx + \frac{\partial \psi}{\partial y}dy$$

$$= -v\,dx + u\,dy$$

$$= -2y\,dx - 2x\,dy$$

Integrating, we get

$$\psi = -2yx - 2xy + C = -4xy + C$$

where $C =$ constant of integration.

At $x = y = 0$, $\psi = 0$ and $C = 0$

$$\psi(2, 0) = -4 \times 2 \times 0 = 0$$

$$\psi(2, 2) = -4 \times 2 \times 2 = -16$$

$$\text{Discharge} = 0 - (-16) = 16 \text{ m}^3/\text{s}$$

Example 3.15 *In a two -dimensional flow, the velocity components are: $u = cy$ and $v = 0$ where c is a constant. Find the circulation about the circle: $x^2 + y^2 - 2ay = 0$, situated in the flow, where a = radius of the circle.*

Solution. Circulation, $\Gamma = \xi_z A$

$$\xi_z = \frac{\partial v}{\partial x} - \frac{\partial u}{\partial y} = 0 - c = -c$$

$$A = \pi a^2$$

$$\Gamma = -c\pi a^2$$

Example 3.16 *The flow field of a fluid is given by:*

$$\vec{V} = xy\hat{i} + 2yz\hat{j} - (2yz + z^2)\hat{k}$$

(a) *Verify whether it is a possible three-dimensional steady incompressible continuous flow*

(b) *Is this flow rotational or irrotational at (2, 3, 4)?*

(c) *Calculate angular velocity, vorticity, shear strain, and dilatency at (2, 3, 4).*

Solution. (a) The continuity equation must be satisfied for a steady incompressible flow.

$$\frac{\partial u}{\partial x} + \frac{\partial v}{\partial y} + \frac{\partial w}{\partial z} = 0$$

$$2y + 2z - (2y + 2z) = 0$$

Hence the given flow is steady incompressible.

(b)
$$\omega_z = \frac{1}{2}\left(\frac{\partial v}{\partial x} - \frac{\partial u}{\partial y}\right)$$

$$= \frac{1}{2}(0 - 2x) \neq 0$$

Hence flow is rotational

(c) Angular velocity ,

$$\omega = \frac{1}{2}\begin{bmatrix} \hat{i} & \hat{j} & \hat{k} \\ \frac{\partial}{\partial x} & \frac{\partial}{\partial y} & \frac{\partial}{\partial z} \\ 2xy & 2yz & -(2yz + z^2) \end{bmatrix}$$

$$= \frac{1}{2}\left[\hat{i}\left\{\frac{\partial}{\partial y}(-2yz - z^2) - \frac{\partial}{\partial z}(2yz)\right\} - \hat{j}\left\{\frac{\partial}{\partial x}(-2yz - z^2) - \frac{\partial}{\partial z}(2xy)\right\}\right.$$

$$\left. + \hat{k}\left\{\frac{\partial}{\partial x}(2yz) - \frac{\partial}{\partial y}(2xy)\right\}\right]$$

$$= \frac{1}{2}[\hat{i}\{-2z - 2y\} - \hat{j}\{0 - 0\} + \hat{k}\{0 - 2x\}]$$

$$= \frac{1}{2}[-2(y+z)\hat{i} - 2x\hat{k}]$$

$$= \frac{1}{2}[-2(3+4)\hat{i} - 2\times 2\hat{k}]$$

$$= -7\hat{i} - 2\hat{k} = -(7\hat{i} + 2\hat{k})$$

Vorticity, $\xi = 2\omega = -(14\hat{i} + 4\hat{k})$

Shear strains:

$$\gamma_{xy} = \frac{1}{2}\left(\frac{\partial v}{\partial x} + \frac{\partial u}{\partial y}\right) = \frac{1}{2}(0 + 2x) = \frac{1}{2}(0 + 2\times 2) = 2$$

$$\gamma_{yz} = \frac{1}{2}\left(\frac{\partial v}{\partial z} + \frac{\partial w}{\partial y}\right) = \frac{1}{2}(2y - 2z) = y - z = 3 - 4 = -1$$

$$\gamma_{xz} = \frac{1}{2}\left(\frac{\partial u}{\partial z} + \frac{\partial w}{\partial x}\right) = \frac{1}{2}(0 + 0) = 0$$

Dilatency or linear strains:

$$\xi_x = \frac{\partial u}{\partial x} = 2y = 2\times 3 = 6$$

$$\xi_y = \frac{\partial v}{\partial y} = 2z = 2\times 4 = 8$$

$$\xi_z = \frac{\partial w}{\partial z} = -(2y + 2z) = -2(3+4) = -14$$

Example 3.17 *The stream function $\psi = 4xy$, in which ψ is in cm²/s and x and y are in metres, describe the incompressible flow between the boundaries shown below in Fig. 3.12. Calculate:*

(a) Velocity at B. (b) convective acceleration at B. (c) Flow rate per unit width across AB.

Solution. $\psi = 4xy$

Coordinates of A(3, 0)

Coordinates of B:

$$x = 3\text{m},\ y = \frac{3}{x} = \frac{3}{3} = 1\,\text{m}$$

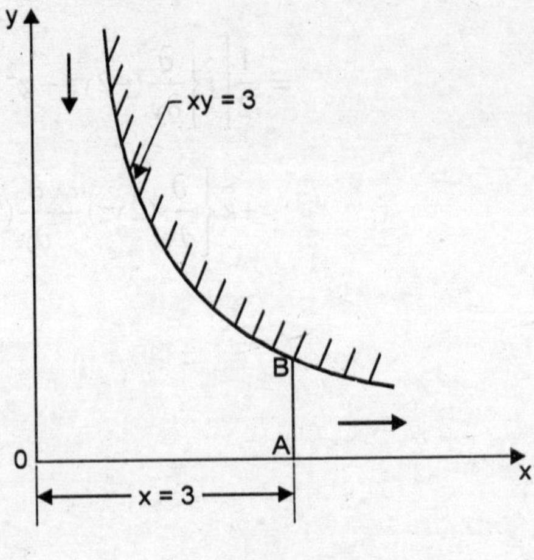

Fig. 3.12

$B(3, 1)$

Velocity components:

$$u = \frac{\partial \psi}{\partial y} = 4x$$

$$v = -\frac{\partial \psi}{\partial x} = -4y$$

$$u_B = 4 \times 3 = 12 \text{ cm/s}$$

$$v_B = -4 \times 1 = -4 \text{ cm/s}$$

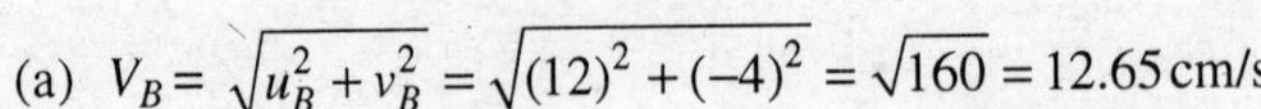

(a) $V_B = \sqrt{u_B^2 + v_B^2} = \sqrt{(12)^2 + (-4)^2} = \sqrt{160} = 12.65\,\text{cm/s}$

(b) Convective acceleration.

$$a_x = u\frac{\partial u}{\partial x} + v\frac{\partial u}{\partial y} = 4x \times 4 - 4y \times 0 = 16x$$

$$a_y = u\frac{\partial v}{\partial x} + v\frac{\partial v}{\partial y} = 4x \times 0 - 4y \times (-4) = 16y$$

$$a = \sqrt{a_x^2 + a_y^2} = 16\sqrt{x^2 + y^2}$$

$$a_B = 16\sqrt{9+1} = 50.6\,\text{cm/s}^2$$

(c) $q_{AB} = \psi_B - \psi_A = 4 \times 3 \times 1 - 4 \times 3 \times 0 = 12\,\text{cm}^3/\text{s/cm}$

Example 3.18 *The stream function in a two-dimensional flow is:*

$$\psi = 6x - 4y + 8xy.$$

(a) Verify whether the flow is irrotational or rotational.

(b) Determine the direction of streamline at a point (1, –1)

(c) Find the expression for velocity potential.

Solution. (a) For irrotational flow, the vorticity component must the zero.

$$\xi_z = \frac{\partial v}{\partial x} - \frac{\partial u}{\partial y}$$

Now
$$u = \frac{\partial \psi}{\partial y} = -4 + 8x, \frac{\partial u}{\partial y} = 0$$

$$v = -\frac{\partial \psi}{\partial x} = -6 - 8y, \frac{\partial v}{\partial x} = 0$$

$\therefore$
$$\xi_z = 0$$

Hence, the flow is irrotation.

(b) $\tan\theta = \frac{v}{u} = \frac{-6-8y}{-4+8x} = \frac{-6+8}{-4+8} = \frac{2}{4} = \frac{1}{2}$

$\theta = 26.56°$ (see Fig. 3.13)

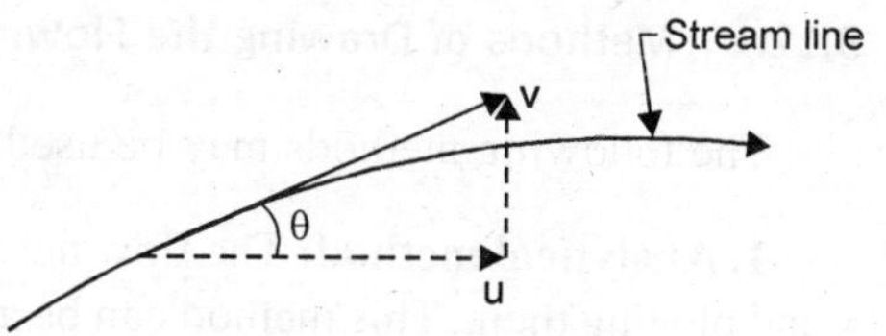

Fig. 3.13

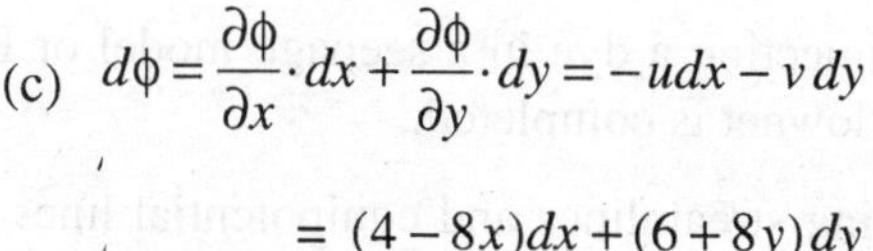

(c) $d\phi = \frac{\partial \phi}{\partial x} \cdot dx + \frac{\partial \phi}{\partial y} \cdot dy = -u dx - v dy$

$$= (4 - 8x)dx + (6 + 8y)dy$$

Integrating, we have

$$\phi = (4x - 4x^2) + (6y + 4y^2)$$
$$= 4(y^2 - x^2) + 4x + 6y$$

3.12 FLOWNET

The flownet is a graphical representation of two-dimensional irrotation flow. It consists of a family of streamlines intersecting orthogonally forming a family of equipotential lines. In this process they form small curvilinear squares. Fig. 3.14 shows a family of streamlines represented by ψ, $\psi + \Delta\psi$ and $\psi + 2\Delta\psi$, which are intersecting orthogonally a series of equipotential lines represented by ϕ, $\phi - \Delta\phi$, $\phi - 2\Delta\phi$ and $\phi - 3\Delta\phi$. It may be noted that ϕ decreases in the direction of flow.

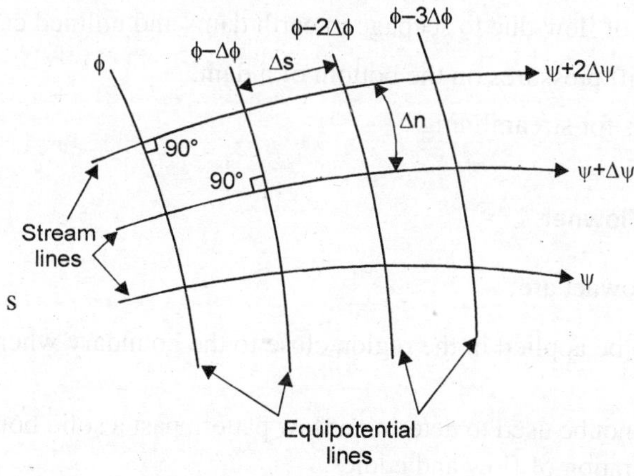

Fig. 3.14 Flownet

Let Δn = distance between the stream lines

Δs = distance between the equipotential lines.

Now
$$\frac{\partial \phi}{\partial n} = \frac{\partial \psi}{\partial s}$$

If $\Delta\phi = \Delta\psi$, then $\Delta n = \Delta s$

3.12.1 Methods of Drawing the Flownet

The following methods may be used to draw the flow net:

1. Analytical method: The flow net can be drawn from the expression for ϕ and ψ in terms of x and y and plotting them. This method can be applied to problems with simple and ideal boundary conditions.

2. Hydraulic models: Streamlines can be traced by injecting a dye in a seepage model or Hele-Shaw apparatus. Then by drawing equipotential lines the flownet is completed.

3. Graphical method: This method consists of drawing streamlines and equipotential lines such that they cut orthogonally and form curvilinear squares. This method consumes lot of time and requires lot of erasing to obtain orthogonal and square shaped net.

4. Electrical analogy method: This method is based on the fact that the flow of fluids and flow of electricity through a conductor are analogous. These two systems are similar in the respect that (i) electric potential is analogous to the velocity potential, (ii) the electric current is analogous to the velocity of flow, and (iii) the homogeneous conductor is analogous to the homogeneous fluid.

3.12.2 Uses of Flownet

The following are some of the important uses of flownet analysis:

1. To determine the streamlines and equipotential lines.
2. To determine the velocity and pressure distribution for given boundaries of flow.
3. To evaluate loss of flow due to seepage in earth dams and unlined canals.
4. To work out uplift pressures on the bottom of a dam.
5. Design of outlets for streamlining.

3.12.3 Limitations of Flownet

The limitations of flownet are:

1. The method can be applied in the region close to the boundary where the viscosity effects are predominant.
2. The method cannot be used to determine flow pattern past a solid boundary on the downstream side due to separation of flow and eddies.

SUMMARY

1. Fluid kinematics deals with the study of velocity and acceleration of fluid particles without taking into account the forces associated with them.
2. In the Eulerian method, we chose a coordinate system fixed in space and study the motion of fluid particles. This method is preferred over Lagrangian method.
3. In the Lagrangian method, a coordinate system attached to a particular particle is chosen and the change in its velocity and acceleration are studied.
4. A path line is a curve traced by a single fluid particle during its motion.
5. A stream line is an imaginary line drawn in a fluid flow field such that a tangent drawn at any point on this line represents the direction of the velocity vector.
6. A stream tube is a surface bounded by streamlines through which fluid cannot penetrate.
7. A streak line is a curve which gives an instantaneous picture of the location of fluid particles.
8. The equation of stream line is:

$$\frac{dx}{u} = \frac{dy}{v} = \frac{dz}{w}$$

9. Expressions for acceleration:

(a) One-dimensional flow.

Tangential acceleration, $a_s = V\frac{\partial V}{\partial s}$

Normal acceleration, $a_n = \frac{V^2}{r}$

(b) Unsteady flow.

$$\left.\begin{aligned} a_s &= \frac{\partial V}{\partial t} + V\frac{\partial V}{\partial s} \\ a_n &= \frac{\partial V_n}{\partial t} + \frac{V^2}{r} \end{aligned}\right\} = \text{local acceleration + convective acceleration}$$

The local acceleration is due to unsteadiness of flow and convective acceleration is due to non-uniformity of flow.

(c) Two dimensional flow.

(i) Cartesian coordinates:

$$u = \frac{dx}{dt}, v = \frac{dy}{dt}$$

$$a_x = \frac{du}{dt} = \frac{d^2x}{dt^2}, a_y = \frac{dv}{dt} = \frac{d^2y}{dt^2}$$

$$\left.\begin{aligned} a_x &= \frac{\partial u}{\partial t} + \left(u\frac{\partial u}{\partial x} + v\frac{\partial u}{\partial y} \right) \\ a_y &= \frac{\partial v}{\partial t} + \left(u\frac{\partial v}{\partial x} + v\frac{\partial v}{\partial y} \right) \end{aligned}\right\} = \text{local acceleration + convective acceleration}$$

(ii) Polar coordinates:

$$v_r = \frac{dr}{dt},\ v_\theta = r\frac{d\theta}{dt}$$

$$a_r = \frac{\partial v_r}{\partial t} + \left(v_r\frac{\partial v_r}{\partial r} + \frac{v_\theta}{r}\cdot\frac{\partial v_r}{\partial \theta} - \frac{v_\theta^2}{r} \right)$$

$$a_\theta = \frac{\partial v_\theta}{\partial t} + \left(v_r\frac{\partial v_\theta}{\partial r} + \frac{v_\theta}{r}\frac{\partial v_\theta}{\partial \theta} + \frac{v_r \cdot v_\theta}{r} \right)$$

(d) Three-dimensional flow.

$$\left.\begin{aligned} a_x &= \frac{\partial u}{\partial t} + \left(u\frac{\partial u}{\partial x} + v\frac{\partial u}{\partial y} + w\frac{\partial u}{\partial z} \right) \\ a_y &= \frac{\partial v}{\partial t} + \left(u\frac{\partial v}{\partial x} + v\frac{\partial v}{\partial y} + w\frac{\partial v}{\partial z} \right) \\ a_z &= \frac{\partial w}{\partial t} + \left(u\frac{\partial w}{\partial x} + v\frac{\partial w}{\partial y} + w\frac{\partial w}{\partial z} \right) \end{aligned}\right\} = \text{local acceleration + convective acceleration}$$

10. Continuity equation.

(i) one-dimensional flow.

Discharge, $Q = \rho_1 v_1 A_1 = \rho_2 v_2 A_2$

(ii) Two-dimensional flow.

$$v_1 \cdot d_1 = v_2 \cdot d_2$$

where d_1, d_2 = normal distance between the streamlines at inlet and outlet respectively.

(iii) Cartesian coordinates.

For steady flow of incompressible fluids,

$$\frac{\partial u}{\partial x} + \frac{\partial v}{\partial y} + \frac{\partial w}{\partial z} = 0$$

(iv) Polar coordinates.

For incompressible fluids,

$$\frac{v_r}{r} + \frac{\partial v_r}{\partial r} + \frac{1}{r}\frac{\partial v_\theta}{\partial \theta} = 0$$

11. Total deformation matrix = Shear strain matrix + Rotation matrix.

Rotation components are:

$$\omega_x = \frac{1}{2}\left(\frac{\partial w}{\partial y} - \frac{\partial v}{\partial z}\right)$$

$$\omega_y = \frac{1}{2}\left(\frac{\partial u}{\partial z} - \frac{\partial w}{\partial x}\right)$$

$$\omega_z = \frac{1}{2}\left(\frac{\partial v}{\partial x} - \frac{\partial u}{\partial y}\right)$$

$$\vec{\omega} = \omega_x \hat{i} + \omega_y \hat{j} + \omega_z \hat{k}$$

12. Vorticity, $\xi = 2\omega$

13. Circulation, $\Gamma = \oint V.ds = \oint V.\cos\theta\, ds = \oint (udx + v\,dy + w\,dz)$

Vorticity,

$$\xi = \frac{d\Gamma}{dA}$$

$$\xi_z = \frac{\partial v}{\partial x} - \frac{\partial u}{\partial y}$$

$$= \frac{v_\theta}{r} + \frac{\partial v_\theta}{\partial r} - \frac{1}{r}\frac{\partial v_r}{\partial \theta}$$

For a flow to be irrotational, the vorticity is zero.

14. Velocity potential, ϕ:

$$u = -\frac{\partial \phi}{\partial x}, v = -\frac{\partial \phi}{\partial y}, w = -\frac{\partial \phi}{\partial z}$$

The velocity potential decreases in the direction of flow.

Velocity potential is defined as a scalar function of space and time such that its negative derivative w.r.t any direction gives the fluid velocity components in that direction.

In polar cylindrical coordinates,

$$u_r = -\frac{\partial \phi}{\partial r}, v_\theta = -\frac{1}{r}\frac{\partial \phi}{\partial \theta}, v_z = -\frac{\partial \phi}{\partial z}$$

The velocity potential satisfies the Laplace's equation: $\nabla^2\phi = 0$, for a steady and incompressible flow.

(i) The flow is irrotational if there exists velocity potential.

(ii) The incompressible, irrotational flow is called potential flow.

(iii) A line along which ϕ = const, is called an equipotential line.

Slope of equipotential line, $\frac{dy}{dx} = -\frac{u}{v}$

15. Stream function, ψ : It is defined as a scalar function of space and time, such that its partial derivative w.r.t. any direction gives the velocity components at right angles to this direction in (cw direction).

$$u = \frac{\partial \psi}{\partial y}, v = -\frac{\partial \psi}{\partial x}$$

(i) The stream function satisfies the Laplace's equation: $\nabla^2\psi = 0$, provided that the flow is irrotational.

(ii) The lines of constant ψ represent streamlines.

(iii) The slope of a streamline is, $\frac{dy}{dx} = \frac{v}{u}$

(iv) Flow rate or flux = numerical difference between two streamlines.

(v) ϕ and ψ curves intersect each other orthogonally at all points.

(vi) The existence of ψ means a possible case of flow.

16. Flownet is a graphical representation of two-dimensional irrotational flow.

17. Cauchy–Riemann equations.

$$u = -\frac{\partial \phi}{\partial x} = \frac{\partial \psi}{\partial y}$$

$$v = -\frac{\partial \phi}{\partial y} = -\frac{\partial \psi}{\partial x}$$

Short Answer Type Questions

1. What is fluid kinematics?
2. Differentiate between Eulerian and Lagrangian methods of describing fluid motion.
3. Differentiate between steady and uniform flow.
4. What is the difference between rotational and irrotational flow?
5. Differentiate between laminar and turbulent flow
6. Explain path line, stream line and streak line.
7. What is a stream tube? Explain with a sketch.
8. What is the difference between local and convective accelerations?
9. What is the basis for continuity equation?
10. What is a streamline? Write equation for a streamline.
11. Write expressions for acceleration in 3-D Cartesian coordinates.
12. Define shear deformation and rotation.
13. Define vorticity, rotation and circulation.

14. What is velocity potential? How components of velocity can be evaluated from velocity potential?
15. Define stream function. How stream function can be determined from velocity potential?
16. List the properties of stream function
17. Show that for ideal fluids the flow is irrational.
18. Show that the streamlines are equipotential lines form a net of orthogonal lines.
19. What is a flownet? What are its uses?
20. Why flownet analysis is not applicable to rotational flow?

EXERCISES

1. The velocity components of an incompressible fluid flow are;

   $u = x^2 y, v = xyz - xy^2, w = x^2 - z^2$

   Show that this flow is kinematically possible.

2. Does the velocity potential exists for two-dimensional incompressible flow prescribed by?

   $u = x + 4y, v = -(4x + y)$

   If so, determine its form and that of stream function.

   [**Ans.** $\phi = \frac{1}{2}(x^2 - y^2) - 4xy + C_1, \psi = 2(x^2 - y^2) + xy + C_2$]

3. A two-dimensional flow is described by the velocity components:

   $u = 5x^3, v = -15x^2 y$

   Determine the stream function, velocity and acceleration at (1, 2)m.

   [**Ans.** 10m^3/s, 30.41 m/s, 167.7m/s^2]

4. The velocity field in a flow is given by: $u = x^2 + 2xy$ and $v = y^2 + 2xy$. Determine the circulation around a closed curve defined by $x = 1$, $x = 3$, $y = 1$, $y = 4$. [**Ans.** –86]

5. A stream function is given by: $\psi = 3x^2y + (3 + t)\, y^2$

   Determine the flow rates across the faces of a triangular prism having thickness of 2.5 m in the *z*-direction at $t = 3$s. The length of the triangular prism along *x*-axis is 1.5 m and 1.0 m along y-axis.

   [**Ans.** 15m^3/s]

6. The velocity components in a fluid flow are given by:

   $u = \frac{2x}{t}, v = -\frac{y}{t}, w = -\frac{x}{t}$

   (a) Does this represent a possible fluid flow?

   (b) Is the flow steady and irrotational? [**Ans.** Possible, unsteady, irrotational]

7. A stream function is defined by : $\psi = x^3 - y^3$. Show that this cannot be a potential flow.

8. The flow of a fluid has velocity components given by:

$u = 3x + y,\ v = 2x - 3y$

Determine whether or not the flow is irrotational **[Ans.** Rotational]

9. A certain flow pattern has a velocity potential, $\phi = \frac{x^3}{3} - x^2 + xy^2 + y^2$.

Find the stream function

10. A 3-D flow field is described by: $\vec{V} = (x + y)\hat{i} + (y + z)\hat{j} + (x^2 + y^2 + z^2)\hat{k}$.

Find the rotation at (1, 1, 1) **[Ans.** $\sqrt{1.5}\, s^{-1}$]

11. Water flows through a pipe AB of 1.2 m diameter at 3m/s and then passes through a pipe BC of 1.5 m diameter. At C the pipe branches into two pipes CD and CE. Branch CD is of 0.8 m diameter and carries one-third of the flow in AB. The flow velocity in branch CE is 2.5 m/s. Find the volume rate of flow in AB, velocities in BC and CD and diameter of CE.

[Ans. 3.39 m^3/s, 1.92 m/s, 2.25 m/s, 1.073 m]

12. The velocity field in a fluid flow is given by:

$u = x + y, v = x^3 - y$

Calculate the circulation around a closed contour defined by:

$x = 1, x = 0, y = 0,\ y = 1$ **[Ans.** zero]

13. Sketch the streamlines represented by: $\psi = x^2 + y^2$. Also calculate the velocity and its direction at point (1, 2). **[Ans.** 4.47 units/s, 26°–34′]

14. The velocity components in a two-dimensional flow are:

$u = \frac{y^3}{3} + 2x - x^2 y$ and $v = xy^2 - 2y - \frac{x^3}{3}$

Show that these velocity components represent a possible case of an irrotational flow.

15. Find the convective acceleration at the middle of a pipe which converges uniformly from 0.6 m diameter to 0.3 m diameter over 3m length. The rate of flow is 40 lit/s. If the rate of flow changes uniformly from 40 lit/s to 80 lit/s in 40 s, find the total acceleration at the middle of the pipe at 20th second. **[Ans.** 0.05 m/s^2, 0.1187 m/s^2]

4

FLUID DYNAMICS

4.1 INTRODUCTION

Fluid dynamics deals with the study of forces which cause motion of fluids in producing velocity and acceleration Dynamics of fluid motion is governed by the momentum principle (Euler's equation) and energy principle (Bernoulli's equation). The fluid is assumed to be in compressible and non-viscous.

4.2 SYSTEM, CONTROL VOLUME AND CONTROL SURFACE

A *system* may be defined as that region of space occupied by the quantity of fluid under consideration. The fluid contained within the system is separated from the surroundings by a boundary. The system may contain either a constant or a variable mass. Its boundaries may be fixed or deformable. It may be at rest or in motion with respect to a chosen coordinate system. The region outside the boundary of the system is known as surroundings.

The *control volume* is an arbitrary volume fixed in space and bounded by a closed surface which is known as the *control surface*. The fluid may enter and leave control volume by crossing the bounding surface enabling transfer of mass,, momentum and energy.

In fluid flow problems, part of the control surface coincides with some physical boundary such as the wall of the container. The remaining part of the control surface is called *cross-section,* which is a hypothetical and fictitious surface through which the fluid can flow.

The concept of control volume and control surface is shown in Fig. 4.1

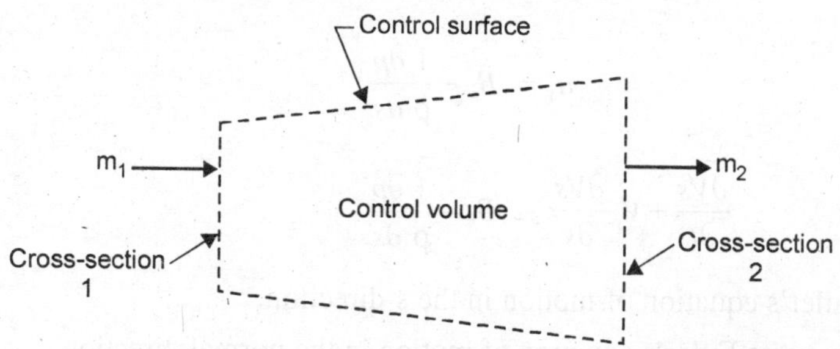

Fig. 4.1 Concept of control volume and control surface

4.3 EULER'S EQUATIONS OF MOTION

Consider a stream line of steady flow of an ideal fluid along the stream tube, shown in Fig. 4.2, with the flow taking place in s-direction. Consider a cylindrical element of cross-sectional area dA and length *ds* from the stream tube as a free body from the moving fluid.

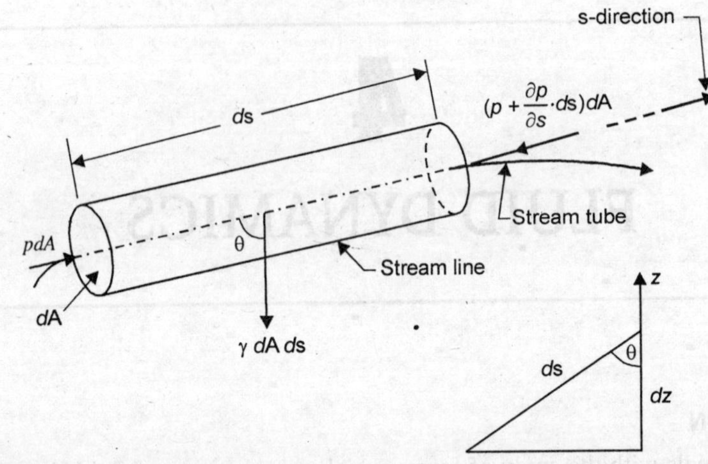

Fig. 4.2 Pressure and gravity forces on a cylindrical element along a streamline

Let $\qquad p$ = pressure on the upstream face

$p + \frac{\partial p}{\partial s} ds$ = pressure on the downstream face

B_s, B_n = body forces per unit mass in the direction of the stream line and normal to it respectively

According to Newton's second law of motion, the equation of motion in the *s*-direction may be written as:

$$(\rho\ dA\ ds)\ a_s = pdA - \left(p + \frac{\partial p}{\partial s} ds\right) dA + \rho dA.ds.B_s$$

$$= -\frac{\partial p}{\partial s} dA.ds + \rho dA.ds.B_s$$

$$a_s = B_s - \frac{1}{\rho}\frac{\partial p}{\partial s}$$

$$\frac{\partial Vs}{\partial t} + V_s \frac{\partial Vs}{\partial s} = B_s - \frac{1}{\rho}\frac{\partial p}{\partial s} \qquad \text{... (4.1)}$$

Eq. (4.1) is the Euler's equation of motion in the s-direction.

Similarly, we may write Euler's equation of motion in the normal direction.

$$a_n = \frac{\partial V_n}{\partial t} + V_s \frac{\partial V_n}{\partial s} = B_n - \frac{1}{\rho}\frac{\partial p}{\partial n} \quad \text{... (4.2)}$$

Euler's equation of motion in Cartesian coordinates may be written as:

$$\left.\begin{aligned}
\frac{Du}{Dt} &= \frac{\partial u}{\partial t} + u\frac{\partial u}{\partial x} + v\frac{\partial u}{\partial y} + w\frac{\partial u}{\partial z} = B_x - \frac{1}{\rho}\frac{\partial p}{\partial x} \\
\frac{Dv}{Dt} &= \frac{\partial v}{\partial t} + u\frac{\partial v}{\partial x} + v\frac{\partial v}{\partial y} + w\frac{\partial v}{\partial z} = B_y - \frac{1}{\rho}\frac{\partial p}{\partial y} \\
\frac{Dw}{Dt} &= \frac{\partial w}{\partial t} + u\frac{\partial w}{\partial x} + v\frac{\partial w}{\partial y} + w\frac{\partial w}{\partial z} = B_z - \frac{1}{\rho}\frac{\partial p}{\partial z}
\end{aligned}\right\} \quad \text{... (4.3)}$$

For steady flow, the local acceleration terms $\frac{\partial u}{\partial t}, \frac{\partial v}{\partial t}$ and $\frac{\partial w}{\partial t}$ are zero.

4.4 INTEGRATION OF EULER'S EQUATION

Let Ω be the body force potential, such that

$$B_s = -\frac{\partial \Omega}{\partial s} \text{ and } B_n = -\frac{\partial \Omega}{\partial n}$$

Then for steady flow, we have

$$V_s \frac{\partial V_s}{\partial s} = -\frac{\partial \Omega}{\partial s} - \frac{1}{\rho}\frac{\partial p}{\partial s}$$

$$\frac{1}{2}\frac{\partial}{\partial s}(V_s^2) = -\frac{\partial}{\partial s}\left(\frac{p}{\rho} + \Omega\right)$$

or

$$\frac{\partial}{\partial s}\left(\frac{V_s^2}{2} + \frac{p}{\rho} + \Omega\right) = 0$$

For incompressible fluid, integrating above expression, we get

$$\frac{V_s^2}{2} + \frac{p}{\rho} + \Omega = \text{const.} \quad \text{... (4.4)}$$

Body force component in s-direction due to gravity = $\rho\, dA\, ds.B_s$

Component of weight in s-direction = $\gamma\, dA\, ds \cos\theta$

$$= -\gamma dA\, ds \cdot \frac{dz}{ds}$$

$\therefore$

$$\rho dA.ds.B_s = -\gamma\, dA.ds.\frac{dz}{ds}$$

$$\therefore \qquad B_s = -g\frac{dz}{ds}$$

$$= -\frac{\partial \Omega}{\partial s}$$

$$B_s = -\frac{\partial \Omega}{\partial s} = -g\frac{dz}{\partial s}$$

Integrating, we get

$$\Omega = gz$$

Eq. (4.4) becomes,

$$\frac{V_s^2}{2} + \frac{p}{\rho} + gz = \text{const} \qquad \text{... (4.5)}$$

The conditions to be satisfied by Eq. (4.5) for its applicability are:

1. The flow is steady
2. The fluid is incompressible.
3. The fluid is non-viscous.
4. It is applicable to points along a streamline.
5. The flow is irrotational

Eq. (4.5) may be written as:

$$\frac{v^2}{2g} + \frac{p}{\gamma} + z = \text{const.} \qquad \text{... (4.6 a)}$$

or

$$\frac{\rho v^2}{2} + p + \gamma z = \text{const.} \qquad \text{... (4.6 b)}$$

Eqs. (4.5) and (4.6) are known as Bernoulli's equation.

4.5 VARIOUS HEADS OF A LIQUID IN MOTION

In Eq. (4.6 a), the various heads are:

$$\text{Velocity head} = \frac{v^2}{2g}$$

$$\text{Pressure head} = \frac{p}{\gamma}$$

$$\text{Potential head} = z$$

$$\text{Total head} = \text{Velocity head} + \text{Pressure head} + \text{Potential head} \qquad \text{... (4.7)}$$

The various heads the shown in Fig. 4.3.

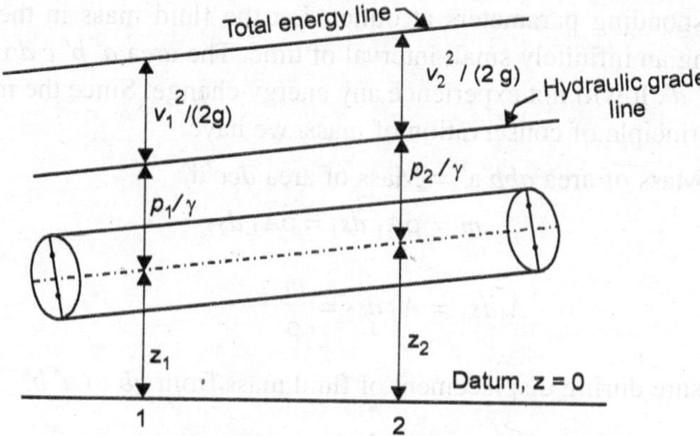

Fig. 4.3 Various heads in fluid flow

The hydraulic grade line represents the sum of $\left(z+\frac{p}{\gamma}\right)$. The energy line represents the total head.

4.6 BERNOULLI'S THEOREM

The Bernoulli's theorem states that in an ideal incompressible fluid when the flow is steady and continuous, the sum of pressure energy, kinetic energy and potential energy is constant along a stream line. Thus,

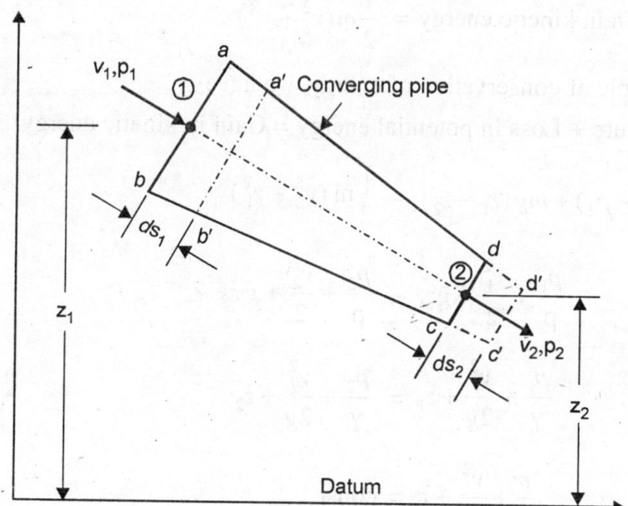

Fig. 4.4 Energy conservation during fluid flow

$$\frac{p}{\gamma}+\frac{v^2}{2g}+z = \text{const.}$$

Proof: Consider an ideal incompressible liquid flowing through a converging pipe as shown in Fig. 4.4. Let p_1, v_1, A_1 be the pressure, average flow velocity and area of cross-section respectively at inlet and p_2 v_2, A_2 the corresponding parameters at outlet. Let the fluid mass in the region *abcd* shift to position *a′ b′ c′ d′* during an infinitely small interval of time. The area *a′ b′ c d* is common to both the regions *abcd* and *a′ b′ c′ d′*. It will not experience any energy change. Since the mass flow rate remains constant, applying the principle of conservation of mass, we have

$$\text{Mass of area } abb'a' = \text{Mass of area } dcc'd'$$

$$\therefore \quad m = \rho A_1\, ds_1 = \rho A_2\, ds_2$$

$$A_1 ds_1 = A_2\, ds_2 = \frac{m}{\rho}$$

Work done by pressure during displacement of fluid mass from *ab* to *a′ b′*

$$= p_1 A_1\, ds_1$$

$$\text{Similarly work done by pressure at } cd = -\,p_2\, A_2\, ds_2$$

$$\text{Net work done by pressure} = p_1\, A_1\, ds_1 - p_2\, A_2\, ds_2$$

$$= A_1\, ds_1\, (p_1 - p_2) \qquad [\because A_1\, ds_1 = A_2\, ds_2\,]$$

$$= \frac{m}{\rho}(p_1 - p_2)$$

$$\text{Loss of potential energy} = mg\,(z_1 - z_2)$$

$$\text{Gain in kinetic energy} = \frac{1}{2}m(v_2^2 - v_1^2)$$

Applying the principle of conservation of energy, we have

Work done by pressure + Loss in potential energy = Gain in kinetic energy

$$\frac{m}{\rho}(p_1 - p_2) + mg(z_1 - z_2) = \frac{1}{2}\mathrm{m}\,(v_2^2 - v_1^2)$$

$$\frac{p_1}{\rho} + \frac{v_1^2}{2} + gz_1 = \frac{p_2}{\rho} + \frac{v_2^2}{2} + g\,z_2$$

or

$$\frac{p_1}{\gamma} + \frac{v_1^2}{2g} + z_1 = \frac{p_2}{\gamma} + \frac{v_2^2}{2g} + z_2$$

or

$$\frac{p}{\gamma} + \frac{v^2}{2g} + z = \text{const}$$

Which is Bernoulli's equation.

Example 4.1 *Water is flowing through a horizontal converging pipe whose diameter varies from 100 mm to 50 mm over a length of 0.5 m. Calculate the velocity heads at the two ends and at the middle of the pipe. Also calculate the mass flow rate if the velocity of flow at the bigger diameter is 2m/s.*

Solution.

Given: $d_1 = 100$ mm, $d_2 = 50$ mm, $l = 0.5$ m, $v_1 = 2$m/s.

Area across flow, $$A_1 = \frac{\pi}{4}(0.1)^2 = 7.854 \times 10^{-3}\, m^2$$

$$A_2 = \frac{\pi}{4}(0.05)^2 = 1.963 \times 10^{-3}\, \text{m}^2$$

$$Q = A_1 v_1 = 7.854 \times 10^{-3} \times 2 = 0.0157 \text{ m}^3/\text{s}$$

Mass flow rate, $\dot{m} = pQ = 10^3 \times 0.0157 = 15.7$ kg/s

Velocity head at bigger diameter $= \dfrac{v_1^2}{2g} = \dfrac{2^2}{2 \times 9.81} = 0.2039\,\text{m}$

$$A_1 v_1 = A_2 v_2$$

$$v_2 = 2 \times \left(\frac{0.1}{0.05}\right)^2 = 8\,\text{m/s}$$

Velocity head at smaller diameter $= \dfrac{v_2^2}{2g} = \dfrac{64}{5 \times 9.81} = 3.262\,\text{m}$

$$d_m = \frac{1}{2}(100 + 50) = 75\text{mm}$$

Velocity at the middle of pipe, $v_m = \left(\dfrac{100}{75}\right)^2 \times 2 = 3.566$ m/s

Velocity head at the middle of pipe $= \dfrac{(3.556)^2}{2 \times 9.81} = 0.644\,\text{m}$

Example 4.2 *Brine of specific gravity 1.5 is draining from the bottom of a large open tank through a 80 mm pipe. The drain pipe ends at a point 10 m below the surface of brine in the tank. Calculate the velocity of flow along the streamline flow at the point of discharge from the pipe.*

Solution.

Given : $s = 1.15$, $d_2 = 80$ mm, $p_1 = p_{\text{atm}} = p_2$, $z_1 - z_2 = 10$ m, $v_1 = 0$

Refer to Fig. 4.5

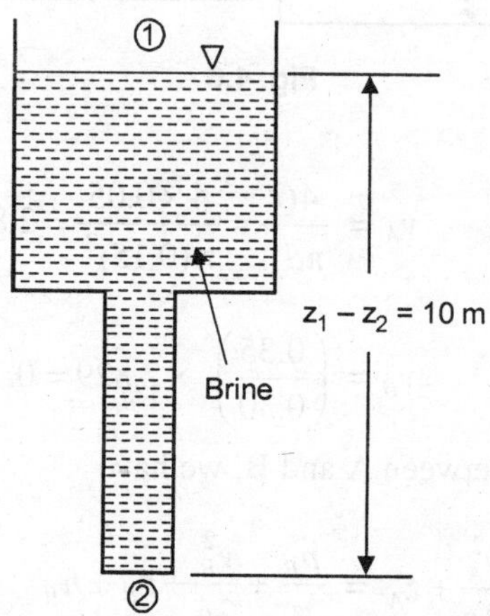

Fig. 4.5

Applying Bernoulli's equation between points 1 and 2, we have

$$\frac{p_1}{\gamma}+\frac{v_1^2}{2g}+z_1=\frac{p_2}{\gamma}+\frac{v_2^2}{2g}+z_2$$

$$v_2^2 = 2g(z_1-z_2)$$

$$= 2\times 9.81\times 10$$

$$= 196.2$$

$$v_2 = 14 \text{ m/s}$$

Example 4.3 *A pipe line is 0.15 m in diameter and it is at an elevation of 100 m at section A (Fig. 4.6). At a section B it is at an elevation of 107 m and has diameter of 0.3 m. When a discharge of 50 litres/s of water is passed through this pipeline, pressure at A is 35 kPa. The energy loss in pipe is 2 m of water. Calculate pressure at B if flow is from A to B.*

Solution.

Given: $d_A = 0.15$ m, $d_B = 0.30$ m, $z_A = 100$ m, $z_B = 107$ m, $Q = 50$ lps $= 0.05$ m^3/s, $p_A = 35$ kPa, $h_{fl} = 2$m.

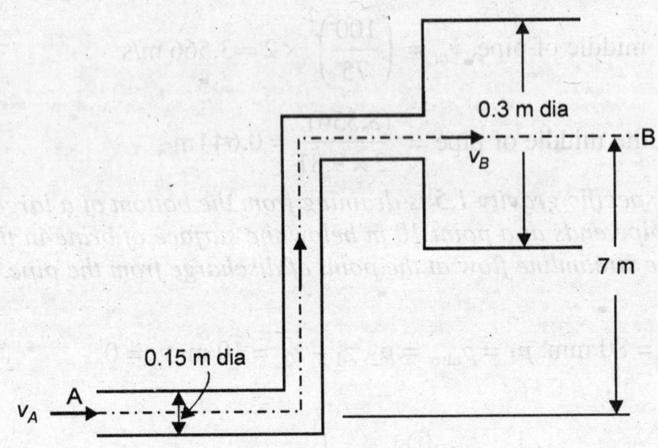

Fig. 4.6

$$v_A = \frac{4Q}{\pi d_A^2}=\frac{4\times 0.05}{\pi(0.15)^2}=2.829\,\text{m/s}$$

$$v_B = \left(\frac{0.35}{0.30}\right)^2\times 2.829 = 0.707\,\text{m/s}$$

Applying Bernoulli's theorem between A and B, we have

$$\frac{p_A}{\gamma}+\frac{v_A^2}{2g}+z_A=\frac{p_B}{\gamma}+\frac{v_B^2}{2g}+z_B+h_{fl}$$

$$\frac{35\times10^3}{9.81\times10^3}+\frac{(2.829)^2}{2\times9.81}+100 = \frac{p_B}{9810}+\frac{(0.707)^2}{2\times9.81}+107+2$$

$$p_B = -49.25 \text{ kPa}$$

Example 4.4 *At one section of a circular water pipe the diameter is 0.3 m, static pressure 260 kPa gauge, velocity of flow 3m/s and elevation 10 m above ground level. The elevation at a section downstream is zero metre and pipe diameter 0.15 m. Find the gauge pressure at the downstream section Take* γ = *9810 N/m³.*

Solution.

Given: $d_1 = 0.1\text{m}$, $d_2 = 0.15$ m, $p_1 = 206$ kPa (gauge), $z_1 = 10$ m, $z_2 = 0$, $v_1 = 3$ m/s

Refer to Fig. 4.7.

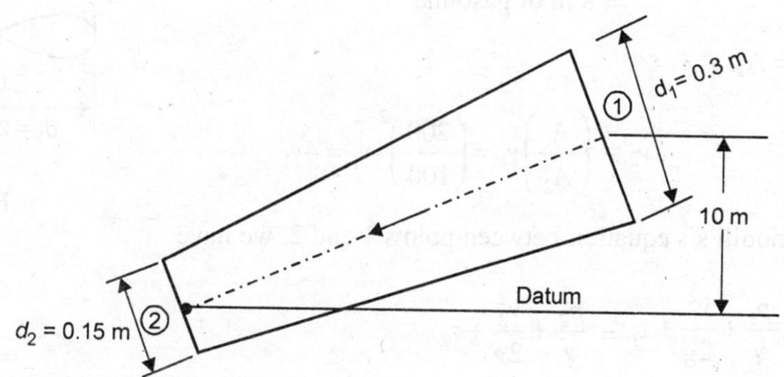

Fig. 4.7

$$v_2 = \left(\frac{d_1}{d_2}\right)^2 v_1 = \left(\frac{0.3}{0.15}\right)^2 v_1 = 4v_1 = 12\text{m/s}$$

Applying Bernoulli's equation between points 1 and 2, we have

$$\frac{p_1}{\gamma}+\frac{v_1^2}{2g}+z_1 = \frac{p_2}{\gamma}+\frac{v_2^2}{2g}+z_2$$

$$\frac{260\times10^3}{9810}+\frac{9}{2\times9.81}+10 = \frac{p_2}{9810}+\frac{144}{2\times9.81}+0$$

$$p_2 = 290.6 \text{ kPa (gauge)}$$

Example 4.5 *Gasoline (sp. gr. = 0.8) is flowing upwards in a vertical pipe line which tapes from 200 mm to 100 mm diameter. A gasoline mercury differential manometer is connected to the pipe to measure the rate of flow. The distance between the manometer tappings is 1 m and gauge reading is 500 mm of mercury. Calculate : (a) Differential gauge reading in terms of gasoline head, and*

(b) Rate of flow

Solution. Refer to Fig. 4.8

$$A_1 = \frac{\pi}{4}d_1^2 = \frac{\pi}{4}(0.2)^2 = 0.0314\text{m}^2$$

$$A_2 = \frac{\pi}{4}d_2^2 = \frac{\pi}{4}(0.1)^2 = 0.00785\text{m}^2$$

$$z_1 = 0\text{m}, \quad z_2 = 1\text{ m}$$

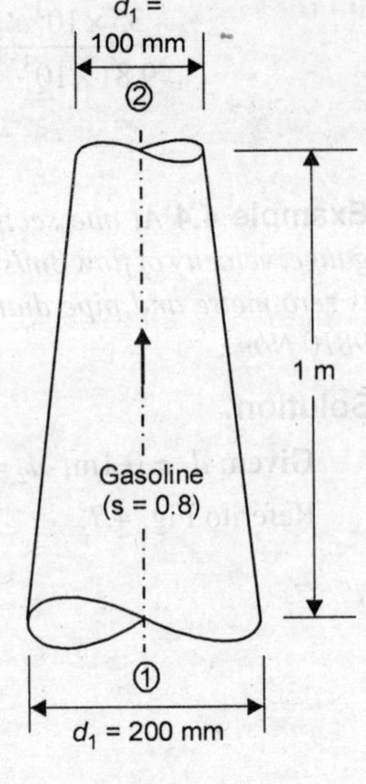

Fig. 4.8

(a) Gauge reading = 0.5 m of mercury

$$= \left[\frac{13.6 - 0.8}{0.8}\right] \times 0.5$$

$$= 8 \text{ m of gasoline}$$

(b) $Q = A_1 v_1 = A_2 v_2$

$$v_2 = \left(\frac{A_1}{A_2}\right) v_1 = \left(\frac{200}{100}\right)^2 v_1 = 4v_1$$

Applying Bernoulli's's equation between points 1 and 2, we have

$$\frac{p_1}{\gamma} + \frac{v_1^2}{2g} + z_1 = \frac{p_2}{\gamma} + \frac{v_2^2}{2g} + z_2$$

$$\frac{p_1 - p_2}{\gamma} = \frac{v_2^2 - v_1^2}{2g} + (z_2 - z_1)$$

$$8 = \frac{v_1^2}{2 \times 9.81}(16 - 1) + 1$$

$$7 = 0.7645 \; v_1^2$$

$$v_1 = 3.026 \text{ m/s}$$

Rate of flow, $Q = 0.314 \times 3.026 = 0.095$ m³/s

Example 4.6 *Two stationary, horizontal flat plates with an external diameter of 0.4 m are placed 10 mm apart. A vertical pipe 50 mm in diameter delivers 0.005 m³/s of water to the centre of the plates. The water is discharged to the periphery of the plates at atmospheric pressure of 98 kN/m². Assuming radial flow and neglecting losses, determine the absolute pressure at the entrance of the flow.*

Solution.

Given: $d_0 = 0.4$ m, $t = 10$ mm, $d_i = 80$ mm, Q = 0.005 m³/s, $p_0 = 98$ kN/m²

Refer to Fig. 4.9

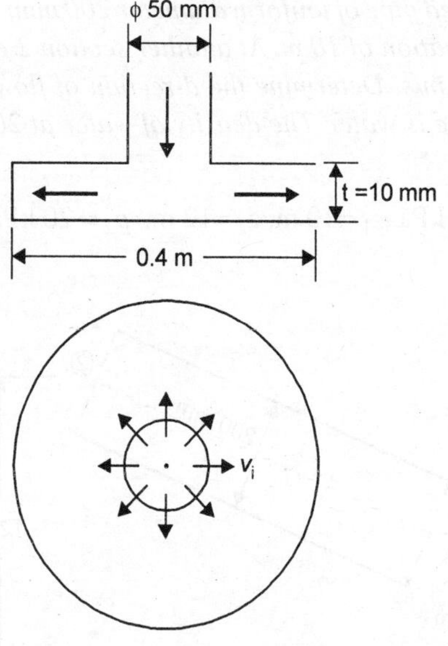

Fig. 4.9

$$v_i = \frac{Q}{\pi d_i t} = \frac{0.005}{\pi \times 0.05 \times 0.01} = 3.183\,\text{m/s}$$

$$v_0 = \frac{Q}{\pi d_0 t} = \frac{0.005}{\pi \times 0.4 \times 0.01} = 0.398\,\text{m/s}$$

Applying Bernoulli's equation between the inlet and exit from the plates, we have

$$\frac{p_i}{\gamma} + \frac{v_i^2}{2g} = \frac{p_0}{\gamma} + \frac{v_0^2}{2g}$$

$$\frac{p_i}{\gamma} + \frac{(3.183)^2}{2 \times 9.81} = \frac{98 \times 10^3}{9810} + \frac{(0.398)^2}{2 \times 9.81}$$

$$\frac{p_i}{\gamma} = 9.481 \text{ m of water absolute}$$

$$= 93 \text{ kN/m}^2$$

4.7 BERNOULLI'S EQUATION FOR REAL FLUIDS

All real fluids are viscous and as such there are always some losses in fluid flows. There are also friction losses due to friction of the pipe carrying fluid. This is accounted for by considering frictional head loss in Bernoulli's equation.

$$\frac{p_1}{\gamma} + \frac{v_1^2}{2g} + z_1 = \frac{p_2}{\gamma} + \frac{v_2^2}{2g} + z_2 + h_{fl} \qquad \text{... (4.8)}$$

where h_{fl} = frictional head loss

Example 4.7 *In a smooth inclined pipe of uniform diameter 200 mm, a pressure of 50 kPa was observed at section 1 which was at an elevation of 10 m. At another section 2 at elevation 12m, the pressure was 20 kPa and the velocity was 1.5 m/s. Determine the direction of flow and the head loss between these two sections. The fluid in the pipe is water. The density of water at 20° C is 998 kg/m³.*

Solution.

Given: $d = 200$ mm, $p_1 = 50$ kPa, $z_1 = 10$ m, $z_2 = 12$ m, $p_2 = 20$ kPa, $v_1 = v_2 = 1.5$ m/s, $\rho = 998$ kg/m^3

Refer to Fig. 4.10

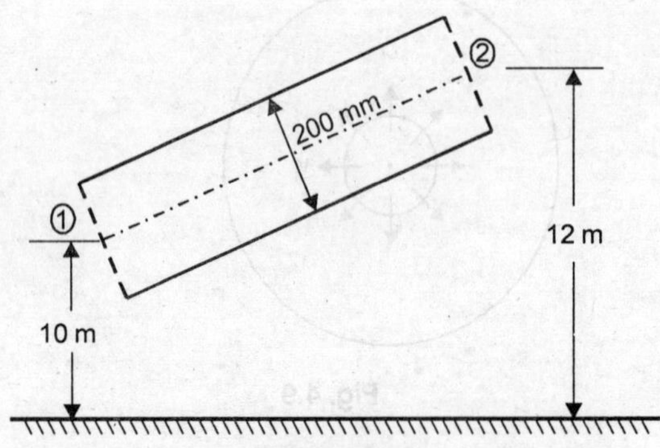

Fig. 4.10

Total head at section 1,

$$h_1 = \frac{p_1}{\gamma} + \frac{v_1^2}{2g} + z_1$$

$$= \frac{50 \times 10^3}{998 \times 9.81} + \frac{(1.5)^2}{2 \times 9.81} + 10 = 15.22 \text{ m}$$

Total head at section 2,

$$h_2 = \frac{p_2}{\gamma} + \frac{v_2^2}{2g} + z_2$$

$$= \frac{20 \times 10^3}{998 \times 9.81} + \frac{(1.5)^2}{2 \times 9.81} + 12 = 14.16 \text{ m}$$

Head loss, $h_{fl} = h_1 - h_2 = 15.22 - 14.16 = 1.06$ m

$h_1 > h_2$, the flow is from section 1 to 2.

Example 4.8 *The centre line of a tapered pipe AB slopes upwards from A to B at an angle of 30° to the horizontal. The distance AB is 6 m and the diameter increases uniformly from 150 mm at A to 200 mm at B. The pipe carries petrol of specific gravity 0.74 and pressure gauges are installed at A and B. Find (a) the pressure difference across AB for the same rate of flow when the direction of taper is reversed.*

Solution.

Given: $d_A = 0.15$ m, $d_B = 0.20$ m, $\theta = 30°$, $l = 6$m, $s = 0.74$

Refer to Fig. 4.11

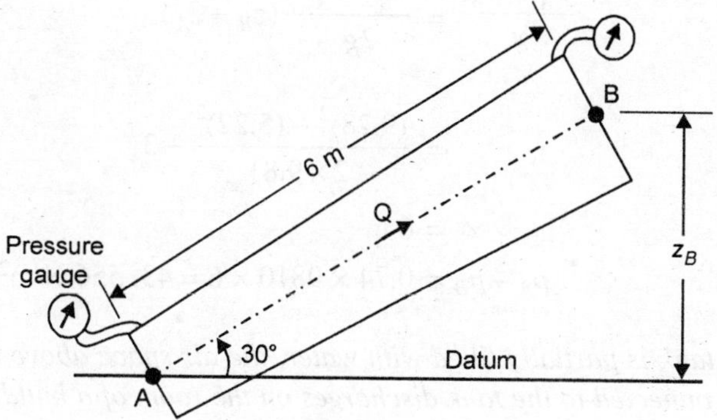

Fig. 4.11

$$v_A = \frac{4Q}{\pi d_1^2} = \frac{4Q}{\pi(0.15)^2} = 56.59Q \text{ m/s}$$

$$v_B = \frac{4Q}{\pi d_2^2} = \frac{4Q}{\pi(0.2)^2} = 31.83Q \text{ m/s}$$

Applying Bernoulli's equation between A and B, we have

$$\frac{p_A}{\gamma} + \frac{v_A^2}{2g} + z_A = \frac{p_B}{\gamma} + \frac{v_B^2}{2g} + z_B$$

(a) $p_A = p_B$

$$\frac{(56.59Q)^2}{2g} + 0 = \frac{(31.832Q)^2}{2g} + 6\sin 30°$$

$$\left[\frac{(56.59)^2 - (31.83)^2}{2 \times 9.81}\right] Q^2 = 3$$

$$111.584\ Q^2 = 3$$

$$Q = 0.164 \text{ m}^3\text{/s or 164 litres/s}$$

(b) When the direction of taper is reversed.

$$v_A = 31.83 \times 0.164 = 5.22 \text{ m/s}$$

$$v_B = 56.59 \times 0.164 = 9.28 \text{ m/s}$$

$$\frac{p_A}{\gamma} + \frac{v_A^2}{2g} + z_A = \frac{p_B}{\gamma} + \frac{v_B^2}{2g} + z_B$$

$$\frac{p_A - p_B}{\gamma} = \frac{v_B^2 - v_A^2}{2g} + (z_B - z_A)$$

$$= \frac{(9.28)^2 - (5.22)^2}{2 \times 9.81} + 3$$

$$= 6 \text{ m}$$

$$p_A - p_B = 0.74 \times 9810 \times 6 = 43,556 \text{ N/m}^2 \text{ or } 43.556 \text{ kPa}$$

Example 4.9 *A large tank is partially filled with water, the air space above being under pressure. A 0.05 m diameter hose connected to the tank discharges on the room of a building 20 m high above the level of tank. The frictional head loss is 0.06 m Calculate the air pressure which must be maintained in the tank to deliver 0.001 m³/s on the roof.*

Solution. Velocity of water in hose, $v = \frac{4Q}{\pi d^2}$

$$= \frac{4 \times 0.001}{\pi (0.05)^2} = 0.509 \text{ m/s}$$

Refer to Fig. 4.12.

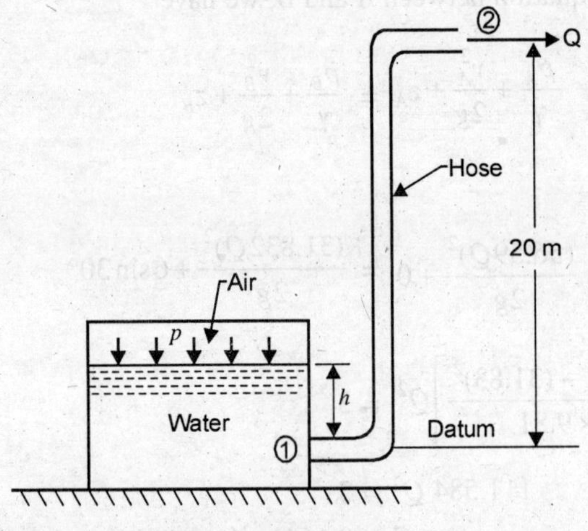

Fig. 4.12

Applying Bernoulli's equation between points 1 and 2,
we have,

$$\frac{p}{\gamma} + h = \frac{v^2}{2g} + z + h_{fl}$$

Assuming $h << z$, we get

$$\frac{p}{\gamma} = \frac{(0.509)^2}{2 \times 9.81} + 20 + 0.06$$

$$= 20.0732$$

$$p = 9810 \times 20.0732 = 196{,}918.14 \text{ N/m}^2$$

4.7 GRAPHICAL REPRESENTATION OF BERNOULLI'S EQUATION

The Bernoulli's equation considering friction head loss may be written as:

$$\frac{p_1}{\gamma} + \frac{v_1^2}{2g} + z_1 = \frac{p_2}{\gamma} + \frac{v_2^2}{2g} + z_2 + h_{fl} \qquad \text{... (4.7)}$$

Fig. 4.13 shows the graphical representation of Bernoulli's equation considering friction head loss.

The hydraulic gradient (ϕ) and energy gradient (θ) are the slopes of the hydraulic grade line (H.G.L.) and the total energy line (T.E.L.)

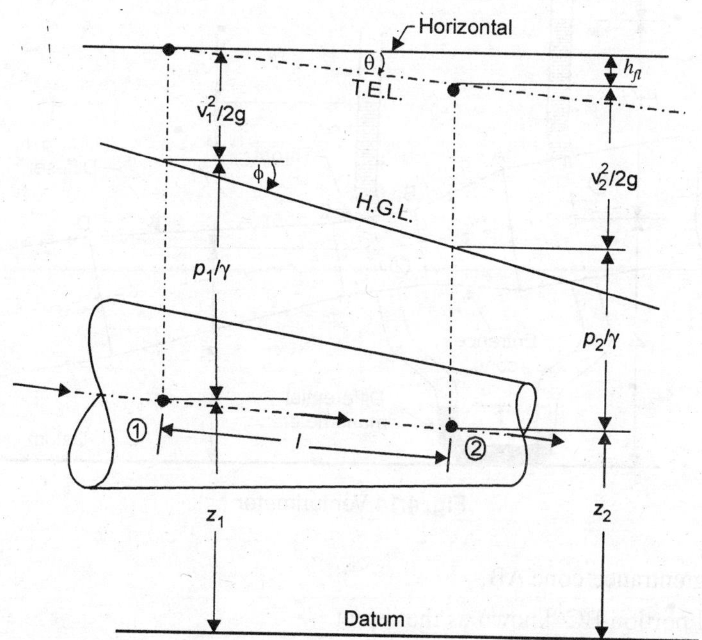

Fig. 4.13 Graphical representation of Bernoulli's equation

Energy gradient, $$\theta = \sin^{-1}\left[\frac{h_{fl}}{l}\right] = \sin^{-1}\left[\frac{1}{l}\left\{\frac{p_1 - p_2}{\gamma} + \frac{v_1^2 - v_2^2}{2g} + (z_1 - z_2)\right\}\right] \qquad \text{... (4.8)}$$

Hydraulic gradient, $$\phi = \sin^{-1}\left[\frac{1}{l}\left\{\frac{p_1 - p_2}{\gamma} + (z_1 - z_2)\right\}\right] \qquad \text{... (4.9 a)}$$

$$= \sin^{-1}\left[\frac{1}{l}\left\{\frac{v_2^2 - v_1^2}{2g} + h_{fl}\right\}\right] \quad \text{... (4.9b)}$$

4.8 VENTURIMETER

A Venturimeter is a device for measuring rate of flow in a pipeline. It is based on the Bernoulli's principle, that is, when the velocity head increases in an accelerated flow, there is a corresponding reduction in the piezometric head. Fig. 4.14 shows a venturimeter. It may be divided into three component parts:

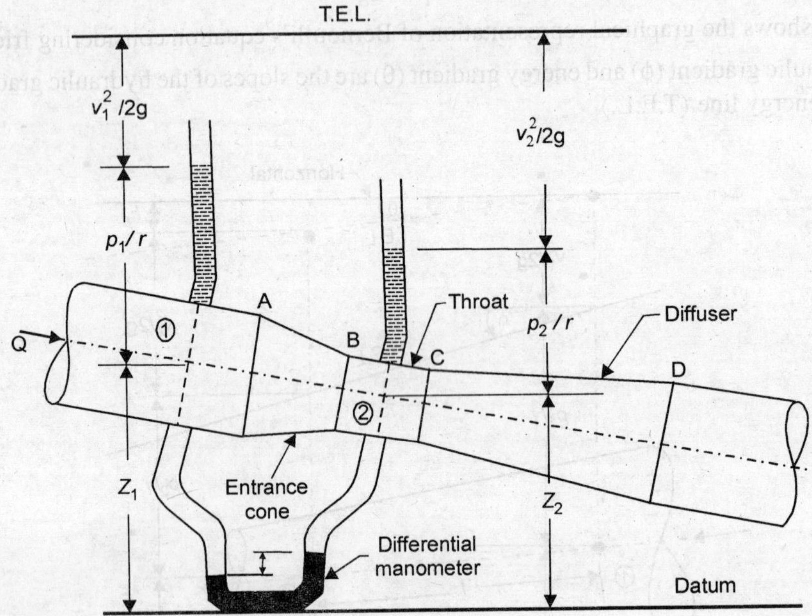

Fig. 4.14 Venturimeter

1. Converging entrance cone AB.
2. Cylindrical portion BC, known as the throat.
3. Diverging cone CD, known as the diffuser.

The converging cone converts the pressure energy into kinetic energy. Even for steady flow the converging cone produces an accelerating flow. The velocity in the throat is maximum. The diffuser converts the kinetic energy into pressure energy.

Applying Bernoulli's equation between points 1 and 2, we have

$$\frac{v_1^2}{2g} + \left(\frac{p_1}{\gamma} + z_1\right) = \frac{v_2^2}{2g} + \left(\frac{p_2}{\gamma} + z_2\right) \quad \text{... (1)}$$

For an incompressible fluid, the continuity equation is,

$$Q = A_1 v_1 = A_2 v_2$$

or $$v_2 = \frac{A_1}{A_2} \times v_1$$

Substituting Eq. (2) in (1), we get

$$v_1 = \frac{A_2\sqrt{2g\left[\left(\frac{p_1}{\gamma}+z_1\right)-\left(\frac{p_2}{\gamma}+z_2\right)\right]}}{\sqrt{A_1^2 - A_2^2}} \quad \text{... (4.10 a)}$$

$$= \frac{A_2\sqrt{2g \times \Delta h}}{\sqrt{A_1^2 - A_2^2}} \quad \text{... (4.10. b)}$$

where $$\Delta h = \left(\frac{p_1}{\gamma}+z_1\right)-\left(\frac{p_2}{\gamma}+z_2\right)$$

= Change in piezometric head

The piezometric head may be measured by a differential manometer.

Now $$\Delta h = \left[\frac{s_2 - s_1}{s_1}\right] \times h = \left(\frac{s_2}{s_1} - 1\right)h \quad \text{... (4.11)}$$

where s_2 = specific gravity of manometer liquid

s_1 = specific gravity of flowing liquid.

h = difference of level in the differential manometer.

$$Q_{\text{ideal}} = A_1 v_1 = \frac{A_1 A_2}{\sqrt{A_1^2 - A_2^2}}\sqrt{2gh\left(\frac{s_2}{s_1} - 1\right)} \quad \text{... (4.12)}$$

Let C_d = coefficient of discharge for real fluids.

Then $$Q = C_d \times Q_{\text{ideal}} = \frac{C_d A_1 A_2}{\sqrt{A_1^2 - A_2^2}}\sqrt{2gh\left(\frac{s_2}{s_1} - 1\right)} \quad \text{... (4.13)}$$

Example 4.10 *A venturimeter having inlet diameter of 0.2 m and throat diameter of 0.1 m is used for measuring flow of oil of specific gravity 0.85. The oil-mercury differential manometer shows a reading of 0.5 m. Assuming the coefficient of discharge as 0.98, calculate the discharge.*

Solution.

Given: $d_1 = 0.2$ m, $d_2 = 0.1$ m, $s_1 = 0.85$, $s_2 = 13.6$, $h = 0.3$ m $C_d = 0.98$

$$Q = \frac{C_d A_1 A_2}{\sqrt{A_1^2 - A_2^2}}\sqrt{2gh\left(\frac{s_2}{s_1} - 1\right)}$$

$$= \frac{C_d A_2}{\sqrt{1-\left(\frac{A_2}{A_1}\right)^2}} \sqrt{2gh\left(\frac{s_2}{s_1}-1\right)}$$

$$= \frac{0.98 \times \frac{\pi}{4}(0.1)^2}{1-\left(\frac{0.1}{0.2}\right)^4} \sqrt{2 \times 9.81 \times 0.3\left(\frac{13.6}{0.85}-1\right)}$$

$$= 0.0741 \text{ m}^3\text{/s or } 74.7 \text{ lit/s}$$

Example 4.11 *A vertical venturimeter shown in Fig. 4.15 has an area ratio of 4. It has a throat diameter of 10 mm. When an oil of specific gravity 0.9 flows through it the mercury differential manometer indicates a difference in height of 100 mm. Calculate the discharge through the manometer.*

Solution.

Given: $\frac{A_1}{A_2} = 4,\ d_2 = 0.01 \text{ m},\ s_1 = 0.9,\ s_2 = 13.6,\ h = 0.1 \text{ m}$

$$\Delta h = \left(\frac{p_1}{\gamma} + z_1\right) - \left(\frac{p_2}{\gamma} + z_2\right)$$

$$= \left(\frac{s_2}{s_1} - 1\right) h$$

$$= \left(\frac{13.6}{0.9} - 1\right) \times 0.1 = 1.41 \text{m of oil}$$

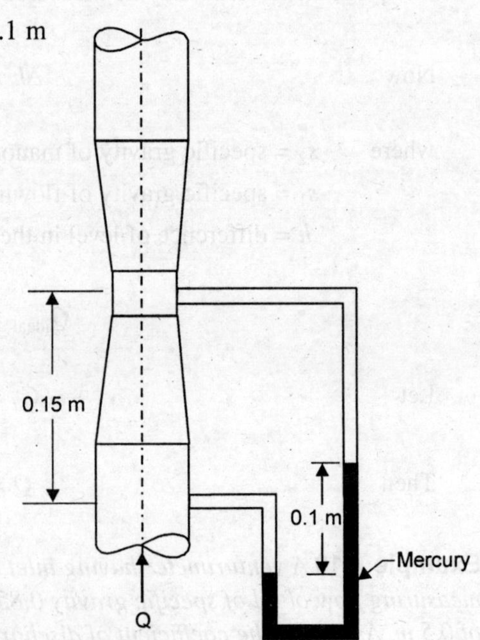

Fig 4.15

From continuity equation, we have

$$Q = A_1 v_1 = A_2 v_2$$

$$v_1 = \frac{v_2}{4}$$

Applying Bernoulli's equation to the inlet and throat of venturimeter, we have

$$\frac{v_2^2 - v_1^2}{2g} = \Delta h$$

$$\frac{v_2^2 - \frac{v_2^2}{16}}{2 \times 9.81} = 1.41$$

$$v_2 = \sqrt{29.51} = 5.43 \text{m/s}$$

$$Q = \frac{\pi}{4} d_2^2 v_2 = \frac{\pi}{4}(0.01)^2 \times 5.43$$

$$= 4.265 \times 10^{-4} \text{m}^3\text{/s or } 25.59 \text{ lit/min}$$

Example 4.12 *A venturimeter of throat diameter 0.1 m is fitted to a water pipe of 0.2 m diameter as shown in Fig. 4.16. The mercury manometer shows a difference of 0.8 m. Calculate the velocity in the throat and the discharge.*

Solution.

Given: $d_2 = 0.1$ m, $d_1 = 0.2$ m, $h = 0.8$ m

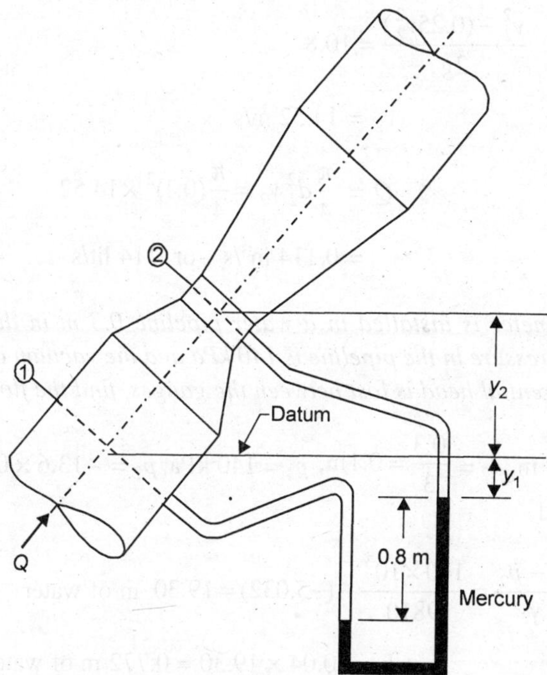

Fig. 4.16

Applying Bernoulli's theorem between points 1 and 2, we have

$$\frac{p_1}{\gamma} + \frac{v_1^2}{2g} = \frac{p_2}{\gamma} + \frac{v_2^2}{2g} + y_2$$

From the principles of manometry, we have

$$\frac{p_1}{\gamma} + y_1 + h = \frac{p_2}{\gamma} + y_1 + y_2 + s_2 \times h$$

$$\frac{p_1 - p_2}{\gamma} - y_2 = s_2 h - h$$

$$= (13.6 - 1) \times 0.8 = 10.08 \text{ m of water}$$

$$= \frac{v_2^2 - v_1^2}{2g}$$

From continuity equation, we have

$$A_1 v_1 = A_2 v_2$$

$$v_1 = \left(\frac{0.1}{0.2}\right)^2 v_2 = \frac{v_2}{4} = 0.25 v_2$$

$$\frac{v_2^2 - (0.25 v_2^2)}{2g} = 10.8$$

$$v_2 = 14.52 \text{ m/s}$$

$$Q = \frac{\pi}{4} d_2^2 v_2 = \frac{\pi}{4} (0.1)^2 \times 14.52$$

$$= 0.114 \text{ m}^3\text{/s} \quad \text{or} \quad 114 \text{ lit/s}$$

Example 4.13 *A venturimeter is installed in a water pipeline 0.3 m in diameter. The throat pipe diameter ratio is 1:3. The pressure in the pipeline is 140 kPa and the vacuum in the throat is 370 mm of mercury. If 4 % of the differential head is lost between the gauges, find the flow in the pipeline.*

Solution. Given : $d_1 = 0.3$ m, $d_2 = \frac{0.3}{3} = 0.1$ m, $p_1 = 140$ kPa, $p_2 = -13.6 \times 0.37 = -5.032$ m of water, $h_{fl} = 4\%$ of differential head

$$\text{Differential head} = \frac{p_1 - p_2}{\gamma} = \frac{140 \times 10^3}{9810} - (-5.032) = 19.30 \text{ m of water}$$

$$h_{fl} = 0.04 \times 19.30 = 0.772 \text{ m of water}$$

Applying Bernoulli's theorem, we have

$$\frac{v_1^2}{2g} + \frac{p_1}{\gamma} + z_1 = \frac{v_2^2}{2g} + \frac{p_2}{\gamma} + z_2 + h_{fl}$$

For horizontal pipeline, $z_1 = z_2$

$$\frac{v_1^2 - v_2^2}{2g} = \frac{p_2 - p_1}{\gamma} + h_{fl}$$

$$= -19.30 + 0.772 = -18.528$$

Now
$$Q = A_1 v_1 = A_2 v_2$$

$$v_2 = \left(\frac{d_1}{d_2}\right)^2 v_1 = \left(\frac{0.3}{0.1}\right)^2 v_1 = 9v_1$$

$$\frac{v_1^2 - 81v_1^2}{2 \times 9.81} = -18.528$$

$$v_1 = 2.132 \text{ m/s}$$

$$Q = A_1 v_1 = \frac{\pi}{4}(0.3)^2 \times 2.132 = 0.1501 \text{m}^3/\text{s or } 150.7 \text{ lit/s}$$

Example 4.14 *A venturimeter of throat diameter 10.5 cm is fitted to a water pipe of internal diameter 21 cm in order to monitor flow rate. To improve accuracy of measurement, pressure difference across the venturimeter is measured with the help of an inclined tube manometer, the angle of inclination being 30°, as shown in Fig. 4.17. For a manometer reading of 9.5 cm of mercury, find the flow rate. Take discharge coefficient of ventuimeter as 0.984.*

Solution. Given: $d_1 = 21$ cm, $d_2 = 10.5$ cm, $\theta = 30°$, $C_d = 0.984$

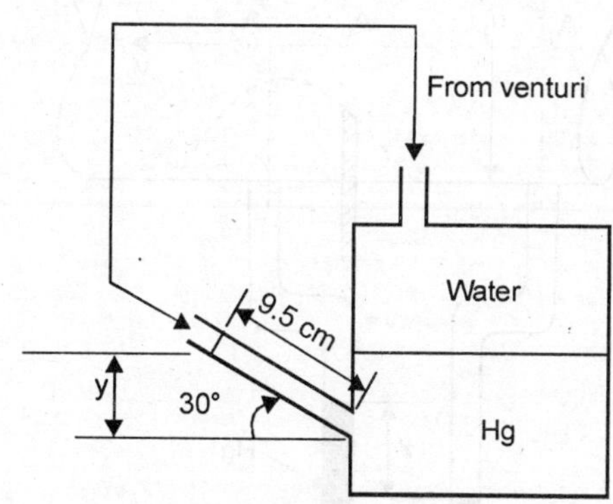

Fig. 4.17

$$h = y\left[\frac{s_2}{s_1} - 1\right] = 9.5 \sin 30° \left[\frac{13.6}{1} - 1\right] = 59.85 \text{ cm}$$

$$Q = \frac{C_d A_1}{\sqrt{\left(\frac{A_1}{A_2}\right)^2 - 1}} \sqrt{2gh}$$

$$= \frac{0.984 \times \frac{\pi}{4}(0.21)^2}{\sqrt{\left(\frac{21}{10.5}\right)^4 - 1}} = \sqrt{2 \times 9.81 \times 0.5985} = 0.3016 \text{m}^3/\text{s}$$

4.9 ORIFICEMETER

Orificemeter or orifice plate is a device used for measuring discharge of fluid through a pipe. The orifice is a sharp edged circular hole in a flat plate which is clamped between the flanges at a joint in the pipe line so that its plane is perpendicular to the axis of the pipe and the hole is concentric with the pipe. A differential manometer is used to measure the difference in the fluid pressure by having pressure tappings on both sides of the orifice. This difference in pressure or differential head is used to measure the fluid flow. The main feature of the orificemeter is that most of the potential energy of liquid is converted into kinetic energy of the free jet issuing through the orifice.

A diagramatic sketch of an orificemeter is shown in Fig. 4.18. The differential manometer is connected on the upstream side (section 1) at a distance of about two times the pipe diameter from the orifice plate, and at section 2 which is at a distance of about half the diameter of orifice from the orifice plate on the downstream side.

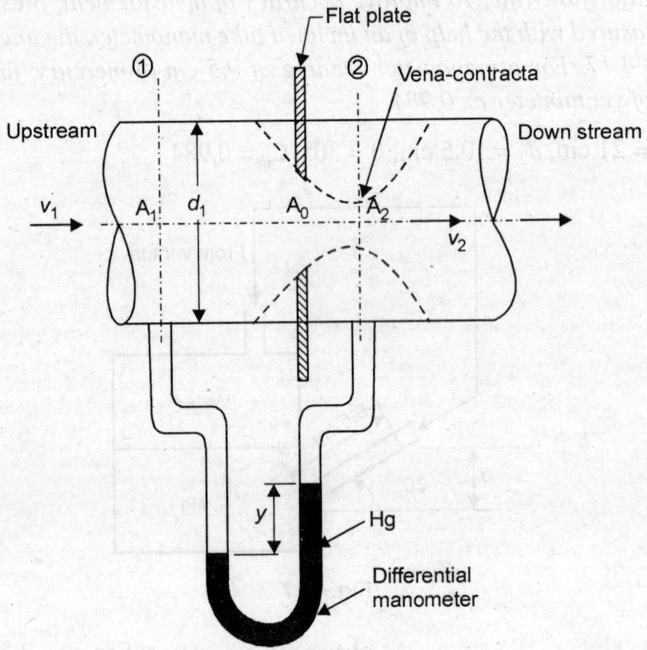

Fig. 4.18 Orificemeter

Let A_1 = cross-sectional area of pipe at section 1

v_1 = Velocity of fluid flow at section 1

p_1 = pressure at section 1

A_2, v_2, p_2 are the corresponding values at section 2, the vena-contracta.

Applying Bernoulli's equation between sections 1 and 2, we get

$$\frac{p_1}{\gamma}+\frac{v_1^2}{2g}+z_1 = \frac{p_2}{\gamma}+\frac{v_2^2}{2g}+z_2$$

$$\left[\left(\frac{p_1}{\gamma}+z_1\right)-\left(\frac{p_2}{\gamma}+z_2\right)\right] = \frac{v_2^2 - v_1^2}{2g}$$

or

$$h = \frac{v_2^2 - v_1^2}{2g}$$

$$v_2 = \sqrt{2g\left(h+\frac{v_1^2}{2g}\right)} = \sqrt{2gh+v_1^2}$$

If A_0 = area of orifice, then

$$A_2 = C_c A_0$$

where C_c = coefficient of contraction

Using continuity equation, we have

$$A_1 v_1 = A_2 v_2 = C_c A_0 v_2$$

$$v_1 = \frac{C_c A_0 v_2}{A_1}$$

$$\therefore \quad v_2 = \sqrt{2gh+\left(\frac{C_c A_0 v_2}{A_1}\right)^2}$$

$$v_2^2 = 2gh + \frac{C_c^2 A_0^2}{A_1^2} v_2^2$$

$$v_2^2\left[1-\left(\frac{A_0}{A_1}\right)^2 C_c^2\right] = 2\,gh$$

$$v_2 = \frac{\sqrt{2\,gh}}{\sqrt{1-\left(\frac{A_0}{A_1}\right)^2 C_c^2}} \quad \text{... (4.14)}$$

Discharge,

$$Q = A_2 v_2 = C_c A_0 v_2$$

$$= C_c A_0 \frac{\sqrt{2gh}}{\sqrt{1-\left(\frac{A_0}{A_1}\right)^2 C_c^2}} \quad \text{... (4.15)}$$

Coefficient of discharge, $$C_d = C_c \frac{\sqrt{1-\left(\frac{A_0}{A_1}\right)^2}}{\sqrt{1-\left(\frac{A_0}{A_1}\right)^2 C_c^2}}$$

or $$C_c = C_d \frac{\sqrt{1-\left(\frac{A_0}{A_1}\right)^2 C_c^2}}{\sqrt{1-\left(\frac{A_0}{A_1}\right)^2}}$$

Eq. (4.15) becomes,

$$Q = A_0\, C_d \frac{\sqrt{1-\left(\frac{A_0}{A_1}\right)^2 C_c^2}}{\sqrt{1-\left(\frac{A_0}{A_1}\right)^2}} \times \frac{\sqrt{2gh}}{\sqrt{1-\left(\frac{A_0}{A_1}\right)^2 C_c^2}}$$

$$= C_d \frac{A_0\, A_1}{\sqrt{A_1^2 - A_0^2}} \cdot \sqrt{2gh}$$

$$= C_d \frac{A_0}{\sqrt{1-\left(\frac{A_0}{A_1}\right)^2}} \cdot \sqrt{2gh} \qquad \text{... (4.16)}$$

Example 4.15 *Water flows at the rate of 0.15 m³/s through a 100 mm diameter orifice used in a 200 mm pipe. What is the difference of pressure head between the upstream section and the vena-contracta section? Take coefficient of contraction, C_c = 0.60 and coefficient of velocity, C_V = 1.0.*

Solution.

Given: $Q = 0.015\ \text{m}^3/\text{s}$, $d_0 = 100$ mm, $d_1 = 200$ mm

$$A_1 = \frac{\pi}{4} d_1^2 = \frac{\pi}{4}(0.2)^2 = 0.03141\,\text{m}^2$$

$$A_0 = \frac{\pi}{4} d_0^2 = \frac{\pi}{4}(0.1)^2 = 0.00785\ \text{m}^2$$

$$C_d = C_c \times C_v = 0.6 \times 1 = 0.6$$

$$Q = \frac{C_d A_0}{\sqrt{1-\left(\frac{A_0}{A_1}\right)^2}} \times \sqrt{2gh}$$

$$0.015 = \frac{0.6 \times 0.00785}{\sqrt{1-\left(\frac{0.00785}{0.03141}\right)^2}} \times \sqrt{2 \times 9.81 \times h}$$

$$h = 0.4846 \text{ m of water}$$

Example 4.16 *Water is flowing through a pipeline of 0.5 m inside diameter. An orifice of diameter 0.2 m is placed in the pipeline to measure the flow rate. The differential manometer reads 0.3 m of mercury. Calculate:*

(a) Flow rate of water, and (b) Velocity of fluid through the pipe.

Take $C_0 = 0.6$, $\rho_{water} = 990$ kg/m³, $\rho_{Hg} = 13600$ kg/m³.

Solution. Given $d_1 = 0.5$ m, $d_0 = 0.2$ m, $y = 0.3$ m of Hg, $C_d = 0.6$, $\rho_{H_2o} = 990\,\text{kg/m}^3$, $\rho_{Hg} = 13600$ kg/m³.

$$A_1 = \frac{\pi}{4} d_1^2 = \frac{\pi}{4}(0.5)^2 = 0.1963\,\text{m}^2$$

$$A_0 = \frac{\pi}{4} d_0^2 = \frac{\pi}{4}(0.2)^2 = 0.03141\,\text{m}^2$$

$$h = y\left[\frac{\rho_{Hg}}{\rho_{H_2O}} - 1\right] = 0.3\left[\frac{13600}{990} - 1\right] = 3.821 \text{ m of water}$$

(a)

$$Q = \frac{C_d A_0}{\sqrt{1-\left(\frac{A_0}{A_1}\right)^2}} \cdot \sqrt{2gh}$$

$$= \frac{0.6 \times 0.03141}{\sqrt{1-\left(\frac{0.03141}{0.1963}\right)^2}} \cdot \sqrt{2 \times 9.81 \times 3.821}$$

$$= 0.1653 \text{ m}^3/\text{s}$$

(b) $v_1 = \dfrac{Q}{A_1} = \dfrac{0.1653}{0.1963} = 0.842\,\text{m/s}$

Example 4.17 *An orificemeter with orifice diameter 0.1 m is inserted in a pipe of 0.2 m diameter. The water pressure on the upstream and downstream side of the orifice are 1.5 bar and 1.0 bar respectively. Calculate:*

(a) The discharge, if coefficient of discharge = 0.6, and (b) coefficient of contraction of the jet.

Solution. Given: $d_0 = 0.1$ m, $d_1 = 0.2$ m, $p_1 = 1.5$ bar, $p_2 = 1.0$ bar, $C_d = 0.6$

Differential head,
$$h = \frac{p_1 - p_2}{\gamma} = \frac{(1.5-1)\times 10^5}{9810} = 5.097\,\text{m}$$

$$A_0 = \frac{\pi}{4} d_0^2 = \frac{\pi}{4}(0.1)^2 = 0.00\,7854\ \text{m}^2$$

$$A_1 = \frac{\pi}{4} d_1^2 = \frac{\pi}{4}(0.2)^2 = 0.031416\,\text{m}^2$$

(a)
$$Q = \frac{C_d A_0}{\sqrt{1-\left(\dfrac{A_0}{A_1}\right)^2}}\sqrt{2gh}$$

$$= \frac{0.6\times 0.007854}{\sqrt{1-\left(\dfrac{0.007854}{0.03146}\right)^2}}\sqrt{2\times 9.81\times 5.097}$$

$$= 0.04867\ \text{m}^3/\text{s}$$

(b)
$$C_d = \frac{C_c\sqrt{1-\left(\dfrac{A_0}{A_1}\right)^2}}{\sqrt{1-\left(\dfrac{A_0}{A_1}\right)^2 C_c^2}}$$

$$0.6 = \frac{C_c\sqrt{1-\left(\dfrac{1}{2}\right)^4}}{\sqrt{1-\left(\dfrac{1}{2}\right)^4 C_c^2}}$$

$$= \frac{0.9682\,C_c}{\sqrt{1-0.0625\,C_c^2}}$$

$$0.36(1-0.0625C_c^2) = 0.9375C_c^2$$

$$0.36-0.0225C_c^2 = 0.9375C_c^2$$

$$0.36 = 0.96C_c^2$$

$$C_c = 0.61$$

4.10 ORIFICES

An orifice is an opening, usually round, through which the fluid flows. It may be used for measuring flow from a reservoir or through a pipe. The main feature of orifice flow is that most of the potential energy of fluid is converted into the kinetic energy of the free jet issued through the orifice. An orifice in a tank or reservoir may be located in the side wall or in the bottom.

4.10.1 Classification of Orifices

The orifices may be classified as follows:

1. According to size

(i) **Small orifice:** In a small orifice, the dimensions of the orifice are small as compared to the head causing flow. The flow velocity from top to bottom edge of orifice is assumed to be uniform.

(ii) **Large orifice :** In such an orifice, the dimensions of orifice are comparable with the head causing flow. The variation in velocity from top to bottom edge of orifice is quite appreciable.

2. According to Shape

(i) Circular (ii) Rectangular (iii) Square (iv) Triangular.

3. Shape of upstream edge

(i) Sharp-edged (ii) Bell-mouthed.

4. According to discharge condition

(i) Free discharge (ii) Drowned or submerged-fully or partially submerged

4.10.2 Flow Through an Orifice

Consider a large tank filled with a liquid having a small circular sharp-edged orifice located in the wall at a fairly large depth H below the free surface as shown in Fig. 4.19. The liquid flows out through the orifice into the atmosphere. As the liquid flows through the orifice, it contracts due to inability of streamlines to take a sharp turn and the jet size becomes minimum at the vena-contracta. Beyond this section, the jet diverges and is attracted in the downward direction by gravity.

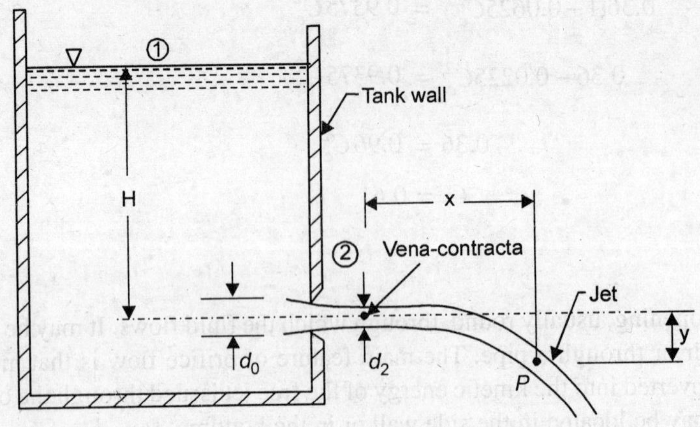

Fig. 4.19 Flow through an orifice

Applying Bernoulli's theorem between points 1 and 2, we have

$$\frac{p_1}{\gamma}+\frac{v_1^2}{2g}+z_1 = \frac{p_2}{\gamma}+\frac{v_2^2}{2g}+z_2$$

Now $p_1 = p_2 = p_a$, the atmospheric pressure.

$$z_1 = z_2 + H \text{ and } v_1 = 0$$

$$\therefore \quad H = \frac{v_2^2}{2g}$$

or $$v_2 = \sqrt{2gh} \qquad \text{... (4.17)}$$

Eq. (4.17) is known as Torricelli''s theorem,

4.10.3 Hydraulic Coefficients

1. Coefficient of contraction, C_c

The ratio of the area of the jet at vena-contracta to the area of the orifice is known as coefficient of contraction, C_c.

Let d_0 = diameter of orifice.

d_2 = diameter of jet at vena-contracta.

Then $A_0 = \frac{\pi}{4}d_0^2$ and $A_2 = \frac{\pi}{4}d_2^2$

$$C_c = \frac{A_2}{A_0} = \frac{\text{Area of jet at vena-contracta}}{\text{Area of orifice}}$$

The value of C_c varies from 0.61-0.69, but the average value is taken as 0.64.

2. Coefficient of Velocity, C_v

The ratio between the actual velocity at the vena-contracta and the theoretical velocity expressed by Eq. (4.17) is known as the coefficient of velocity, C_V. Thus

$$C_v = \frac{\text{Actual velocity of jet at vena-contracta}, v_a}{\text{Theoretical velocity}, v_2} = \frac{v_a}{\sqrt{2gh}} \quad \text{... (4.19)}$$

The value of C_v varies from 0.95 to 0.99.

3. Coefficient of Discharge, C_d

It is defined as the ratio of actual discharge (Q) through an orifice to the theoretical discharge (Q_{th})

$$C_d = \frac{\text{Acutal discharge}, Q}{\text{Theoretical discharge}, Q_{th}}$$

$$= \frac{\text{Actual area of jet}}{\text{Theoretical area of jet}} \times \frac{\text{Actual velocity}}{\text{Theoretical velocity}}$$

$$= C_c \times C_v \quad \text{... (4.20)}$$

The value of C_d varies from 0.62 to 0.65.

4. Coefficient of Resistance, C_r

It is defined as the ratio of loss of head (or kinetic energy) in orifice to the head of water (or actual kinetic energy) available at the exit of the orifice.

$$C_r = \frac{\text{Loss of head in the orifice}}{\text{Head of water}} \quad \text{... (4.21)}$$

4.10.4 Experimental Determination of Hydraulic Coefficients

1. Coefficient of velocity, C_v

Consider the coordinates (x, y) of a point P on the free trajectory of jet from the vena-contracta, as shown in Fig. 4.19.

$$x = v_a \cdot t \text{ or } t = \frac{x}{v_a}$$

$$y = \frac{1}{2} g t^2$$

$$\therefore \quad y = \frac{1}{2} g \left(\frac{x}{v_a}\right)^2 = \frac{g x^2}{2 v_a^2}$$

$$v_a = \sqrt{\frac{gx^2}{2y}}$$

$$v_{\text{th}} = \sqrt{2gh}$$

$$C_v = \frac{v_a}{v_{th}} = \sqrt{\frac{gx^2}{2y}} \times \frac{1}{\sqrt{2gh}} = \sqrt{\frac{x^2}{4yH}} \quad \text{... (4.22)}$$

2. Coefficient of discharge, C_d

$$C_d = \frac{Q}{Q_{th}}$$

where $Q = \dfrac{\text{Area of measuring tank} \times \text{Rise of water level in measuring tank}}{\text{time}}$

$$Q_{\text{th}} = A_0 v_2 = A_0 \times \sqrt{2gH}$$

$$A_0 = \text{area of orifice}$$

$$C_d = \frac{Q}{A_0\sqrt{2gh}} \quad \text{... (4.23)}$$

3. Coefficient of contraction, C_c

$$C_c = \frac{C_d}{C_v} \quad \text{... (4.24)}$$

4. Loss of head, h_{fl}

$$\frac{p_1}{\gamma} + \frac{v_1^2}{2g} + z_1 = \frac{p_2}{\gamma} + \frac{v_2^2}{2g} + z_2 + h_{fl}$$

$$0 + 0 + H = 0 + \frac{v_2^2}{2g} + 0 + h_{fl}$$

$$h_{fl} = H - \frac{v_2^2}{2g} = H\left(1 - \frac{v_2^2}{2gH}\right)$$

$$= H(1 - C_v^2)$$

$$= \frac{v_2^2}{2g}\left(\frac{2gH}{v_2^2} - 1\right)$$

$$= \frac{v_2^2}{2g}\left(\frac{1}{C_v^2} - 1\right) \quad \text{... (4.25)}$$

Example 4.18 *An orifice 20 mm in diameter is discharging water under a head of 5m. If $C_d = 0.62$ and $C_v = 0.97$, Calculate:*

(a) Actual discharge, and (b) Actual velocity of jet at vena-contracta.

Solution.

Given: $d_0 = 20$ mm, $H = 5$ m

$$A_0 = \frac{\pi}{4}(0.2)^2 = 3.1416 \times 10^{-4}\ \text{m}^2$$

(*a*) Actual discharge,

$$Q = C_d A_0 \sqrt{2gh}$$

$$= 0.62 \times 3.1416 \times 10^{-4} \sqrt{2 \times 9.81 \times 5}$$

$$= 1.93 \times 10^{-3}\ \text{m}^3/\text{s}$$

(b) Actual velocity,

$$v_a = C_v \sqrt{2gH} = 0.97\sqrt{2 \times 9.81 \times 5} = 9.6\ \text{m/s}$$

Example 4.19 *Oil of specific gravity 0.9 issues from a 30 mm diameter orifice under a pressure of 0.5 bar (gauge). The diameter of jet at the vena-contracta is 20 mm and the discharge is 0.5 m³/min. Determine the coefficient of velocity.*

Solution.

Given: $s = 0.9$, $d_0 = 30$ mm, $p_1 = 0.5$ bar (gauge), $d_2 = 20$ mm, $Q = 0.17$, m³/min.

$$H = \frac{p_1}{\gamma} = \frac{0.5 \times 10^5}{0.9 \times 9810} = 4.53\,\text{m}\ \text{of oil}$$

$$v_2 = \sqrt{2gH} = \sqrt{2 \times 9.81 \times 4.53} = 9.43\ \text{m/s}$$

Actual velocity at vene-contracta,

$$v_a = \frac{Q}{A_2} = \frac{(0.17/60)}{\frac{\pi}{4}(0.02)^2} = 9\ \text{m/s}$$

$$C_v = \frac{9}{9.43} = 0.954$$

Example 4.20 *The head loss in flow through a 50 mm diameter orifice under certain head is 0.15 m of water and the velocity of jet is 6 m/s. If the coefficient of discharge is 0.6, calculate:*

(a) head on the orifice causing the flow,

(b) diameter of the jet, and

(c) coefficient of velocity.

Solution.

Given: $d_0 = 50$ mm, $h_{fl} = 0.15$ m of water, $v_2 = 6$m/s, $C_d = 0.6$

(a) Applying Bernoulli's equation between the reservoir surface and the vena-contracta, we have

$$H = \frac{v_2^2}{2g} + h_{fl} = \frac{6^2}{2 \times 9.81} + 0.15 = 1.985\,\text{m}$$

(c) $$v_2 = C_V\sqrt{2gH}$$

or $$C_v = \frac{6}{\sqrt{2\times 9.81\times 1.985}} = 0.961$$

$$C_c = \frac{C_d}{C_v} = \frac{0.6}{0.961} = 0.624$$

(b) $$C_c = \left(\frac{d_2}{d_0}\right)^2$$

$$0624 = \left(\frac{d_2}{0.05}\right)^2$$

Diameter of jet

$$d_2 = 0.0395 \text{ m or } 39.5 \text{ mm}$$

Example 4.21 *A tank has two identical orifices, each 50 mm in diameter, in one of its vertical sides and are located one above the other. The upper orifice is located 5 m below the water surface and the lower one 6 m. The coefficients of contraction and velocity are 0.62 and 0.97 for both the orifices. Calculate:*

(a) The combined rate of discharge from the two orifices, and

(b) the distance of point of intersection of the two jets from the vertical side.

Solution. Given: $d_{OA} = d_{OB} = 50$ mm, $H_A = 5$m, $H_B = 6$ m, $C_c = 0.62$, $C_v = 0.97$

Refer to Fig. 4.20.

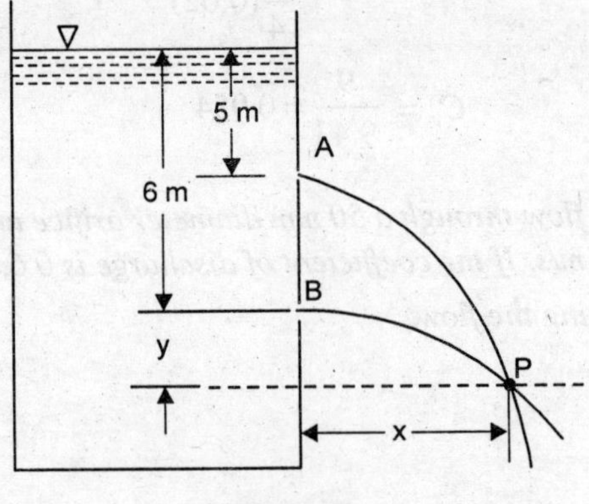

Fig. 4.20

(a) $$v_{2A} = C_v\sqrt{2gH_A} = 0.97\sqrt{2\times 9.81\times 5} = 9.6 \text{ m/s}$$

$$v_{2A} = C_v\sqrt{2gH_B} = 0.97\sqrt{2\times 9.81\times 6} = 10.52 \text{ m/s}$$

$$Q_A = C_c \times \frac{\pi}{4} d_{OA}^2 \times v_{2A} = 0.97 \times \frac{\pi}{4}(0.05)^2 \times 9.6 = 0.01828 \text{m}^3/\text{s}$$

$$Q_B = C_c \times \frac{\pi}{4} d_{OB}^2 \times v_{2B} = 0.02004 \, \text{m}^3/\text{s}$$

$$Q = Q_A + Q_B = 0.01828 + 0.02004 = 0.03832 \text{ m}^3/\text{s}$$

Let P be the point of intersection of the two jets. Then for the lower jet,

$$y = \frac{1}{2} g t^2$$

or

$$t = \sqrt{\frac{2y}{g}}$$

$$x = v_{2B} \times t = 10.52\sqrt{\frac{2y}{g}} m \quad \text{... (1)}$$

For the upper jet,

$$(H_B - H_A) + y = \frac{1}{2} g t^2$$

$$(6 - 5) + y = \frac{1}{2} g t^2$$

$$t = \sqrt{\frac{2(1+y)}{g}}$$

$$x = v_{2A} t = 9.6\sqrt{\frac{2(1+y)}{g}} \quad \text{... (2)}$$

Equating Eqs. (1) and (2), we get

$$10.52\sqrt{\frac{2y}{g}} = 9.6\sqrt{\frac{2(1+y)}{g}}$$

$$\left(\frac{10.52}{96}\right)^2 \times \frac{2y}{g} = \frac{2(1+y)}{g}$$

$$1.2\, y = 1 + y$$

$$y = 5 \text{ m}$$

and

$$x = 10.52\sqrt{\frac{2 \times 5}{9.81}} = 10.62 \text{ m}$$

Example 4.22 *A tank shown in Fig. 4.21 has a nozzle of exit diameter D_1 at a depth H_1 below the* ***free*** *surface of water. At the side opposite to this nozzle, another nozzle at a depth $H_1/2$ of diameter D_2* ***is to*** *be provided so that the net horizontal force on the tank is zero. Find D_2 in terms of D_1.*

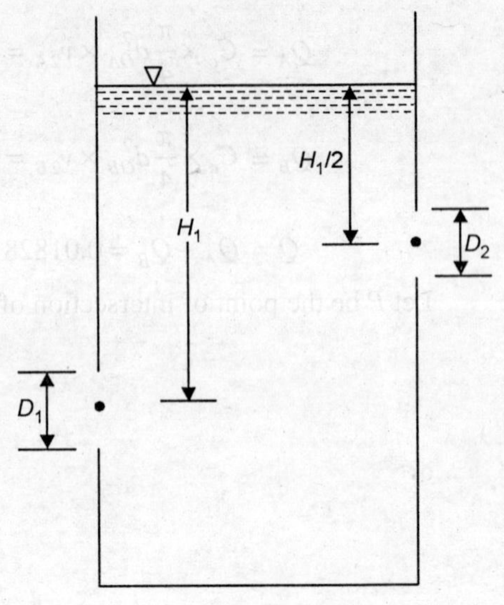

Fig 4.21

Solution. Jet 1: $v_1 = \sqrt{2g H_1}$

$$\text{Mass flow rate}, m_1 = \rho A_1 v_1 = \rho \times \frac{\pi}{4} D_1^2 \times \sqrt{2g H_1}$$

$$\text{Horizontal force}, \; F_1 = m_1 v_1 = \rho \times \frac{\pi}{4} D_1^2 \times 2g H_1$$

Jet2: $v_2 = \sqrt{2g H_2}, m_2 = \rho A_2 v_2 = \rho \times \frac{\pi}{4} D_2^2 \times \sqrt{2g H_2}$

$$F_2 = m_2 v_2 = \rho \times \frac{\pi}{4} D_2^2 \times 2g H_2$$

For $F_1 = F_2$

$$D_1^2 H_1 = D_2^2 H_2$$

$$D_2 = D_1 \sqrt{\frac{H_1}{H_2}} = \sqrt{2}\, D_1 = 1.414\, D_1$$

4.10.5 Large Rectangular Orifice

Consider a large rectangular orifice in one side of a tank discharging water freely into the atmosphere, as shown in Fig. 4.22.

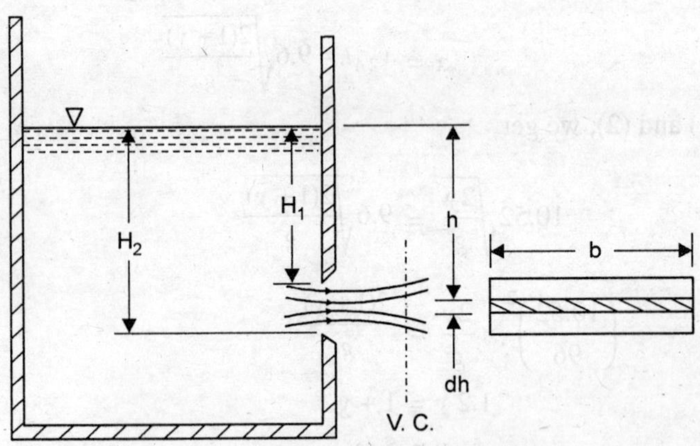

Fig. 4.22 Large rectangular orifice

Let H_1, H_2 = height of liquid above the top and bottom of orifice respectively.

b = breadth of orifice

C_d = coefficient of discharge

Consider an elementary horizontal strip of depth dh at depth h below the water surface.

Area of strip, $dA = b \cdot dh$

Theoretical velocity of water through the strip, $v = \sqrt{2gh}$

Discharge through the strip, $d_Q = C_d \times dA \times v$

$$= C_d \times b \times dh \times \sqrt{2gh}$$

Total discharge, $$Q = \int_{H_1}^{H_2} C_d \times b \times \sqrt{2gh} \cdot dh$$

$$= C_d \times b \times \sqrt{2g} \int_{H_1}^{H_2} h^{1/2}\, dh$$

$$= C_d \times b \times \sqrt{2g} \left| \frac{h^{3/2}}{(3/2)} \right|_{H_1}^{H_2}$$

$$= \frac{2}{3} C_d \cdot b \cdot \sqrt{2g}[H_2^{3/2} - H_1^{3/2}] \quad \text{... (4.26)}$$

Example 4.23 *A rectangular orifice 0.5 m wide and 0.6 m deep is discharging water from a vessel. The top edge of the orifice is 0.4 m below the water surface. Calculate:*

(a) The discharge through the vessel if $C_d = 0.6$.

(b) The percentage error if the orifice is treated as a small orifice.

Solution.

Given: $b = 0.5$ m, d = 0.6m, $H_1 = 0.4$ m, $C_d = 0.6$

(a) $H_2 = H_1 + d = 0.4 + 0.6 = 1.0$ m

$$Q = \frac{2}{3} C_d b \sqrt{2g}\ [H_2^{3/2} - H_1^{3/2}]$$

$$= \frac{2}{3} \times 0.6 \times 0.5 \times \sqrt{2 \times 9.81}\,[(1.0)^{3/2} - (0.4)^{3/2}]$$

$= 0.6618$ m³/s

(b) For a small orifice, $H = H_1 + \frac{d}{2} = 0.4 + \frac{0.6}{2} = 0.7\,\text{m}$

$$Q' = C_d, A_0 \sqrt{2gh} = 0.6 \times 0.5 \times 0.6 \times \sqrt{2 \times 9.81 \times 0.7} = 0.667\ \text{m}^3/\text{s}$$

$$\%\ \text{error} = \frac{Q'}{Q} - 1 = \frac{0.667}{0.6618} - 1 = 0.00786 \text{ or } 0.786\ \%$$

4.10.6 Fully Submerged Orifice

When the orifice discharges liquid such that the orifice is fully submerged (or drowned) into the liquid of same kind, then it is known as a fully submerged orifice

Consider a fully submerged orifice as shown in Fig. 4.23

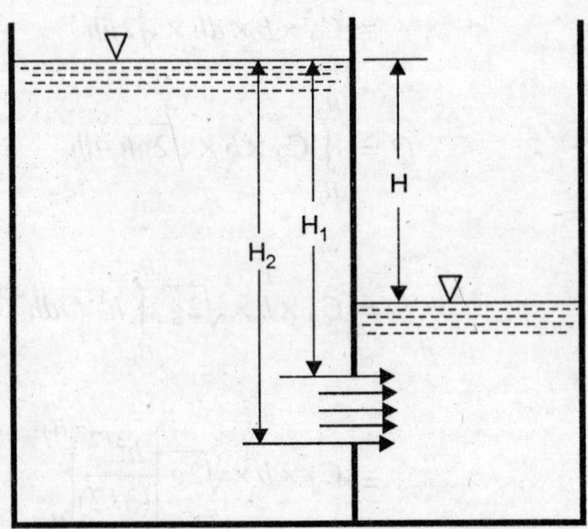

Fig 4.23. Fully submerged orifice

Let H_1, H_2 = height of water above the top and bottom of the orifice respectively

H = difference between the two water surfaces.

b = width of orifice

C_d = coefficient of discharge.

Area of orifice, $A = b\,(H_2 - H_1)$

Actual velocity of water through the orifice, $v = C_v\sqrt{2gH}$

In this case, $C_c = 1.0$

$\therefore$ $C_d = C_v$

and $v = C_d\sqrt{2gh}$

Discharge, $$Q = Av = C_d b(H_1 - H_1)\sqrt{2gH} \qquad \text{... (4.27 a)}$$

$$= C_d\, bd\sqrt{2gh} \qquad \text{... (4.27 b)}$$

where $d = H_2 - H_1 =$ depth of orifice.

4.10.7 Partially Submerged Orifice

If the outlet side of an orifice is partly drowned under liquid then it is known as partially submerged orifice, as shown in Fig. 4.24.

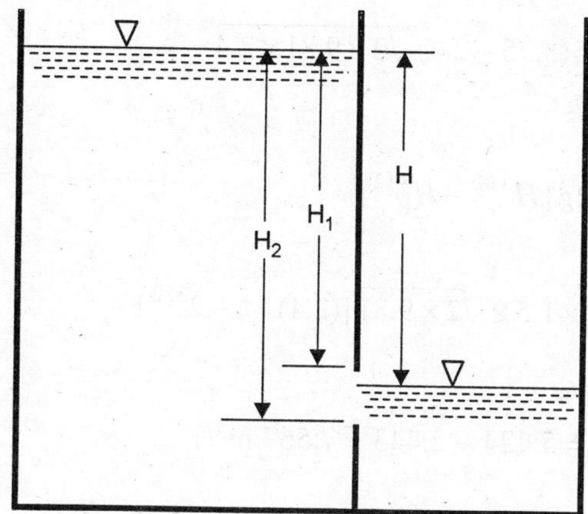

Fig. 4.24 Partially submerged orifice

Discharge through the submerged portion,

$$Q_1 = C_d b(H_2 - H)\sqrt{2gH}$$

Discharge through the unsubmerged portion,

$$Q_2 = \frac{2}{3} C_d b\sqrt{2g}\,[H^{3/2} - H_1^{3/2}]$$

Total discharge,

$$Q = Q_1 + Q_2$$

$$= C_d b\sqrt{2g}\left[(H_2 - H)\sqrt{H} + \frac{2}{3}(H^{3/2} - H_1^{3/2})\right] \quad \text{... (4.28)}$$

Example 4.24 *Calculate the discharge through a totally submerged orifice 1 m wide and 0.5 m deep, if the difference of water levels on both sides of the orifice is 2m. Take $C_d = 0.62$.*

Solution. Given: $b = 1$m, $d = 0.5$ m, $H = 2$ m, $C_d = 0.62$

$$Q = C_d \cdot b \cdot d\sqrt{2gH}$$

$$= 0.62 \times 1 \times 0.5 \times \sqrt{2 \times 9.81 \times 2}$$

$$= 1.942 \text{ m}^3\text{/s}$$

Example 4.25 *A rectagular orifice 1.5 m wide and 1.25 m deep has water level 2 m above the top edge of the orifice on one side. On the other side of orifice, the water level is 0.4 m below its top edge. Calculate the discharge through the orifice, if $C_d = 0.62$.*

Solution.

Given: $b = 1.5$ m, $d = 1.25$ m, $H_1 = 2$ m, $H = 2 + 0.4 = 2.4$ m, $C_d = 0.62$,

$H_2 = 2 + 1.25 = 3.25$ m

$$Q_1 = C_d \cdot b(H_2 - H)\sqrt{2gH}$$

$$= 0.62 \times 1.5(3.25 - 2.4)\sqrt{2 \times 9.81 \times 2.4}$$

$$= 5.424 \text{ m}^3/\text{s}$$

$$Q_2 = \frac{2}{3} C_d b \sqrt{2g} [H^{3/2} - H_1^{3/2}]$$

$$= \frac{2}{3} \times 0.62 \times 1.5 \times \sqrt{2 \times 9.81} [(2.4)^{3/2} - 2^{3/2}]$$

$$= 2.443 \text{ m}^3/\text{s}$$

$$Q = Q_1 + Q_2 = 5.424 + 2.443 = 7.567 \text{ m}^3/\text{s}$$

4.10.8 Time Required for Emptying a Tank

(a) Side Orifice

Consider a tank of uniform area of cross-section, containing some liquid and having a side orifice, as shown in Fig. 4.25.

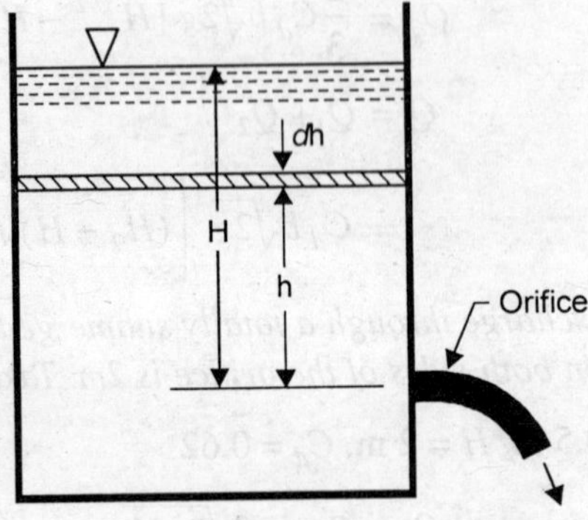

Fig. 4.25 Emptying a tank through side orifice

Let A_1 = area of cross-section of tank

A_2 = area of cross-section of orifice.

H = initial height of liquid surface above the orifice at any time t

dh = fall in liquid surface in time dt.

Discharge through the orifice, $Q = C_d \cdot A_2 \sqrt{2gh}$

Provided that the drop in liquid surface is slow enough for Bernoulli's equation to be applicable without appreciable error

During time dt, the volume of liquid discharged by the orifice = $Q \cdot dt$

Reduction in volume of liquid in tank in time $dt = -A_1 \times dh$

$$\therefore \quad -A_1 \cdot dh = Q \cdot dt$$

$$= C_d A_2 \sqrt{2gh}.dt$$

$$dt = \frac{-A_1}{C_d A_2 \sqrt{2gh}} \cdot \frac{dh}{\sqrt{h}}$$

Integrating both sides, we get

$$\int_0^t dt = \frac{A_1}{C_d A_2 \sqrt{2g}} \int_H^h \frac{dh}{\sqrt{h}}$$

$$t = \frac{-A_1}{C_d A_2 \sqrt{2g}} 2 \left| \sqrt{h} \right|_H^h = \frac{2A_1}{C_d A_2 \sqrt{2g}} [\sqrt{H} - \sqrt{h}] \quad \text{... (4.29)}$$

(b) Bottom orifice

Let A_1 = uniform cross-sectional area of tank

A_2 = area of orifice

H_1 = initial height of liquid in tank

t = time required to empty the tank from H_1 height of liquid to H_2.

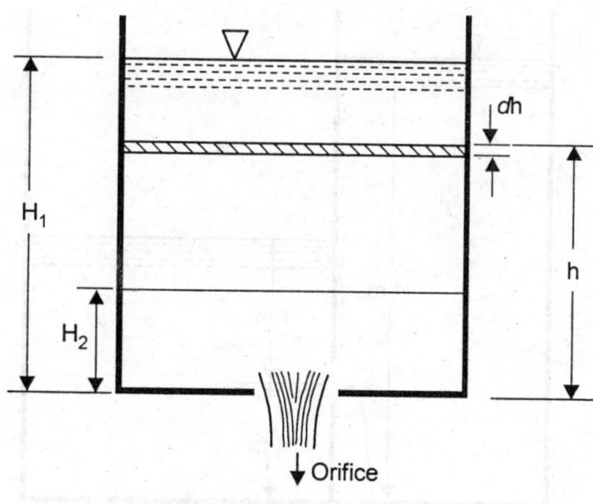

Fig. 4.26 Emptying a tank through orifice

Let at some instant the liquid height be h above the orifice and and in time dt it falls by dh.

Then $$-A_1 dh = Q \cdot dt$$

$$= C_d A_2 \sqrt{2gh} \cdot dt$$

$$dt = \frac{-A_1 dh}{C_d A_2 \sqrt{2gh}} = \frac{-A_1}{C_d A_2 \sqrt{2g}} \cdot \frac{dh}{\sqrt{h}}$$

$$\int_0^t dt = \frac{-A_1}{C_d A_2 \sqrt{2g}} [\sqrt{H_1} - \sqrt{H_2}] \quad \text{... (4.30)}$$

For emptying the tank completely, $H_2 = 0$.

$$t = \frac{2A_1 \sqrt{H_1}}{C_d A_2 \sqrt{2g}} \quad \text{... (4.31)}$$

Example 4.26 *A vessel has compartments A and B communicating by an orifice of area 0.015 m². Its centre is 1 m above the bottom of the vessel. The cross-section of compartment A is 10 m². Its centre is 1 m above the bottom of the vessel. The cross-section of comparment A is 10m² and that of B is 20 m². At a certain time the water stands at 4m in A and 2m in B. How soon thereafter water will attain common level? Assume $C_d = 0.62$.*

Solution.

Given: $A_0 = 0.015 m^2$, $A_A = 10\ m^2$, $A_B = 20\ m^2$, $H_A = 4 - 1 = 3m$, $H_B = 1\ m$, $C_d = 0.62$

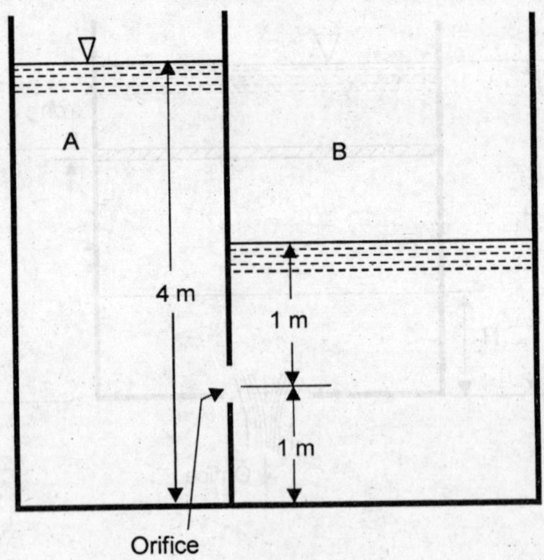

Fig. 4.27

Let y = depth of water in compartment A at any instant.

dy = change in depth of water during time dt.

Rise in the level of water in compartment B during time dt.

$$= \frac{A_A}{A_B} \times dy = \frac{10}{20} \times dy = 0.5\,dy$$

Net change in head, $\qquad dh = dy - (-0.5\,dy) = 1.5\,dy$

or $\qquad dy = \frac{2}{3} dh$

Discharge through orifice in time dt $= Q \cdot dt = C_d A_0 \sqrt{2gh} \cdot dt$ ... (1)

Decrease in volume of water in compartment A in time dt = $10\,dy = 10\frac{2}{3}dh = \frac{20}{3}dh$ ... (2)

Equating Eqs. (1) and (2), we get

$$\frac{-20}{3} dh = C_d\, A_0 \sqrt{2gh} \cdot dt$$

$$dt = -\frac{20}{3} \times \frac{1}{C_d A_0 \sqrt{2g}} \cdot \frac{dh}{\sqrt{h}}$$

$$= -\frac{20}{3} \times \frac{1}{0.62 \times 0.015 \times \sqrt{2 \times 9.81}} \times \frac{dh}{\sqrt{h}}$$

$$= -161.84 \times \frac{dh}{\sqrt{h}}$$

Integrating both sides, we have

$$\int_0^t dt = -161.84 \int_2^0 \frac{dh}{\sqrt{h}}$$

$$t = 161.84\, | 2\sqrt{h}\, |_0^2 = 457.7\,s \text{ or } 7.63 \text{ min}$$

Example 4.27 *A cylindrical tank 3 m in diameter and 6m high has an orifice 150 mm in diameter at the bottom centre of the tank. A constant discharge of 85 lit/s of water is fed into the tank. At the same time water is being discharged through the orifice. Determine the time taken to lower the water surface level in the tank from 5 m to 2.5 m above the centre of orifice. The top of tank is open to the atmosphere. Take* $C_d = 0.72$.

Solution.

Given: $D = 3$m, $d_0 = 150$ mm, $Q_i = 0.085$ m^3/s, $H_1 = 5m$, $H_2 = 2.5\ m$

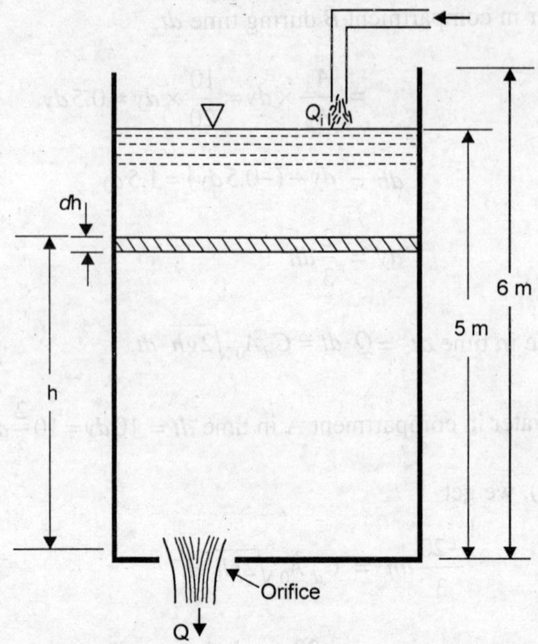

Fig. 4.28

Let dh = fall in water surface in the tank in time dt

Area of tank, $$A = \frac{\pi}{4}D^2 = \frac{\pi}{4}\times 9 = 7.068 \text{ m}^2$$

Area of orifice, $$A_0 = \frac{\pi}{4}(0.15)^2 = 0.01767 \text{ m}^2$$

Applying continuity equation, we have

$$-A \cdot dh = (Q - Q_i)\,dt$$

$$= (C_d\, A_0 \sqrt{2gh} - Q_i)\cdot dt$$

$$dt = -\frac{A \cdot dh}{(C_d A_0 \sqrt{2gh - Q_i})}$$

$$= -\frac{7.068\, dh}{(0.72 \times 0.01767) \times \sqrt{2 \times 9.81} \times \sqrt{h} - 0.085)}$$

$$= \frac{-7.068 \times dh}{(0.0563)\sqrt{h} - 0.085)} = \frac{-125.42\, dh}{(\sqrt{h} - 1.51)}$$

Let $x = \sqrt{h} - 1.51$

$$dx = \frac{1}{2} h^{-1/2}\, dh = \frac{dh}{2\sqrt{h}}$$

or

$$dh = 2\sqrt{h}\, dx$$

$$dt = \frac{-125.42}{x} \times 2\sqrt{h}\, dx = \frac{-125.42 \times 2}{x} \times (x + 1.51)\, dx$$

$$= -250.84 \left(1 + \frac{1.51}{x}\right) dx$$

Integrating, we get

$$t = 250.84 \,|\, x + 1.51 \ln x \,|_{H_2}^{H_1}$$

when $h = \mathrm{H}_1 = 5\text{m},\ x = 0.726$ m

and when $h = \mathrm{H}_2 = 2.5$ m, $x = 0.071$

$$t = 250.84 \left[(0.726 - 0.071) + 1.51 \ln \left(\frac{0.726}{0.071} \right) \right]$$

$$= 1044.9 \text{ s or } 17.4 \text{ min}$$

4.10.9 Emptying a Swimming Pool with Two Outlets

Refer to Fig. 4.29.

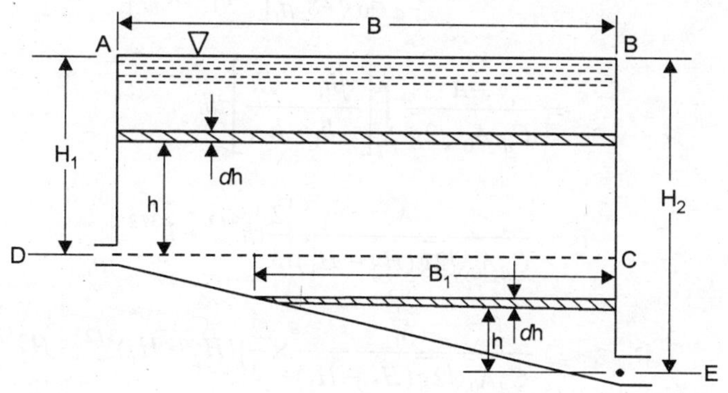

Fig. 4.29 Emptying a swimming pool.

Let L = Length of swimming pool

B = width of swimming pool

H_1, H_2 = depth of water on shallow and deep sides respectively

A = area of pool,

A_0 = area of orifice

d = diameter of each orifice.

t_1 = time required to bring water surface from AB to DC.

t_2 = time required to empty the triangle DCE.

Then, time required to empty the tank, $t = t_1 + t_2$

(a) Time t_1:

Now $CE = BE - BC = DE - AD = H_2 - H_1$

Consider an instant, when the height of water above the centre of top orifice is h. At that instant, the height of water above the bottom orifice is $h_1 = h + (H_2 - H_1)$. If during a small time interval dt the water level falls by dh, then

Volume of water leaving the pool in time dt.

= Discharge through the two orifices in time dt.

$$-A \cdot dh = C_d\, A_0\sqrt{2gh_1}\cdot dt + C_d A_0\sqrt{2gh}\cdot dt$$

$$dt = \frac{-A\cdot dh}{C_d\cdot A_0\sqrt{2g}\,[\sqrt{h_1}+\sqrt{h}]}$$

$$\int_0^{t_1} dt = \frac{-A}{C_d\, A_0\sqrt{2g}}\int_{h=H_1}^{h=0}\frac{dh}{[\sqrt{h_1}+\sqrt{h}]}$$

$$= \frac{-A}{C_d\, A_0\sqrt{2g}}\int_{H_1}^{0}\left[\frac{\sqrt{h_1}-\sqrt{h}}{h_1-h}\right]dh$$

$$= \frac{-A}{C_d\, A_0\sqrt{2g}}\int_{H_1}^{0}\left[\frac{\sqrt{h_1}-\sqrt{h}}{h_1-h}\right]dh$$

$$t_1 = \frac{-A}{C_d A_0\sqrt{2g}(H_2-H_1)}\left|\frac{2}{3}(h_1^{3/2}-h^{3/2})\right|_{H_1}^{0}$$

$$= \frac{-A_1}{C_d A_0\sqrt{2g}(H_2-H_1)}\times\frac{2}{3}[(H_2-H_1)^{3/2}-H_2^{3/2}+H_1^{3/2}]$$

$$= \frac{2}{3}\times\frac{A_1}{C_d\cdot A_0\sqrt{2g}(H_2-H_1)}[H_2^{3/2}-H_1^{3/2}-(H_2-H_1^{3/2})] \quad \text{... (4.32)}$$

(b) Time t_2:

Within the triangle DCE, let h be the height of water above the bottm orifice at any instant. Let the water level fall by dh during a small time interval dt. At that instant,

Width of water surface, $B_1 = \dfrac{hB}{(H_2 - H_1)}$

Area of water surface, $A_1 = B_1 \times L = \dfrac{hBL}{(H_2 - H_1)} = \left(\dfrac{BL}{H_2 - H_1}\right) h = kh$

where $k = \dfrac{BL}{(H_2 - H_1)}$

$$\therefore \quad -A_1 \cdot dh = C_d A_0 \sqrt{2gh} \cdot dt$$

$$dt = \frac{-A_1 \cdot dh}{C_d \cdot A_0 \sqrt{2gh}} = \frac{-k\,h\,dh}{C_d\,A_0 \sqrt{2gh}} = \frac{-k\sqrt{h}\,dh}{C_d A_0 \sqrt{2g}}$$

$$\int_0^{t_2} dt = \frac{-k}{C_d A_0 \sqrt{2g}} \int_{(H_2 - H_1)}^{0} \sqrt{h}\,dh$$

$$t_2 = \frac{-k}{C_d\,A_0 \sqrt{2g}} \left| \frac{2}{3} h^{3/2} \right|_{(H_2 - H_1)}^{0} = \frac{2}{3} \frac{k}{C_d A_0 \sqrt{2g}} \times (H_2 - H_1)^{3/2}$$

$$= \frac{2}{3} \times \frac{BL}{C_d\,A_0 \sqrt{2g}} \times (H_2 - H_1)^{1/2} \qquad \text{... (4.33)}$$

4.10.10 Emptying a Truncated Conical Tank

Refer to Fig. 4.30

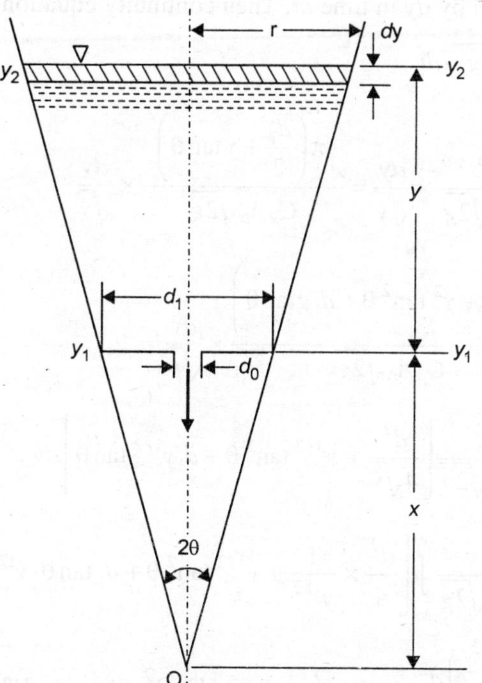

Fig. 4.30 Emptying a truncated conical tank

Let d_0 = diameter of orifice

2θ = vertex angle of cone

d_1 = bottom diameter of cone

Then $\tan\theta = \dfrac{d_1}{2x}$

$$x = \frac{d_1}{2\tan\theta}$$

Let r = radius of water surface at a distance $(x + y)$ from the vertex, O.

Then $\dfrac{r}{x+y} = \tan\theta$

or $r = (x+y)\tan\theta$

$$= \left[\frac{d_1}{2\tan\theta} + y\right]\tan\theta$$

$$= \frac{d_1}{2} + y\tan\theta$$

Let the liquid surface fall by dy in time dt. Then continuity equation gives,

$$-\pi r^2 \times dy = C_d A_0 \sqrt{2gy}\cdot dt$$

$$dt = -\frac{\pi r^2}{C_d A_0\sqrt{2g}} \times \frac{dy}{\sqrt{y}} = \frac{-\pi\left(\dfrac{d_1}{2} + y\tan\theta\right)^2}{C_d A_0\sqrt{2g}} \times \frac{dy}{\sqrt{y}}$$

$$= -\frac{\pi\left(\dfrac{d^2}{4} + y^2\tan^2\theta + d_1 y\tan\theta\right)}{C_d A_0\sqrt{2g}} \times \frac{dy}{\sqrt{y}}$$

$$= -\frac{\pi}{C_d A_0\sqrt{2g}}\left[\frac{d^2}{4\sqrt{y}} + y^{3/2}\tan^2\theta + d_1 y^{1/2}\tan\theta\right]dy$$

$$t = -\frac{\pi}{C_d A_0\sqrt{2g}}\int_{y_2}^{y_1}\left[\frac{d^2}{4} \times \frac{1}{y^{1/2}} + y^{3/2}\tan^2\theta + d_1\tan\theta\cdot y^{1/2}\right]dy$$

$$= \frac{\pi}{C_d A_0\sqrt{2g}}\left|\frac{d^2}{2}\cdot y^{1/2} + \frac{2}{5}\tan^2\theta\cdot y^{5/2} + \frac{2}{3}d_1\tan\theta\cdot y^{3/2}\right|_{y_1}^{y_2}$$

$$= \frac{\pi}{C_d A_0 \sqrt{2g}}\left[\frac{d^2}{2}(y_2^{1/2} - y_1^{1/2}) + \frac{2}{5}\tan\theta \cdot (y_2^{5/2} - y_1^{5/2}) + \frac{2}{3} d_1 \tan\theta\,(y_2^{3/2} - y_1^{3/2})\right] \quad \text{... (4.34)}$$

4.10.11 Emptying a Hemispherical Tank

Refer to Fig. 4.31.

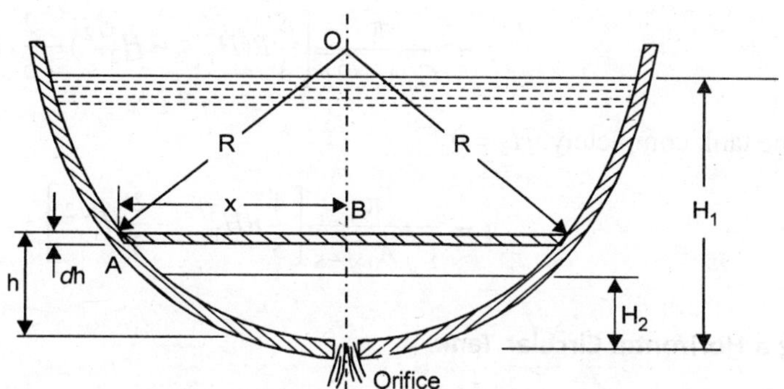

Fig 4.31 Emptying a hemispherical tank

Let R = radius of tank

A_0 = area of orifice

H_1, H_2 = initial and final height of liquid respectively

t = time for the liquid to fall from H_1 to H_2

Let at any instant of time, h be the height of the liquid over the orifice and x the radius of liquid surface.

Let dh be the decrease in height of the liquid surface in time dt.

Then
$$-\pi x^2 \times dh = C_d A_0 \sqrt{2gh} \times dt$$

$$dt = \frac{-\pi x^2 \cdot dh}{C_d A_0 \sqrt{2gh}}$$

Now $OB = R - h$

and
$$AB = x = \sqrt{R^2 - (R-h)^2} = \sqrt{2Rh - h^2}$$

$$x^2 = 2RH - h^2$$

$$dt = \frac{-\pi(2Rh - h^2)\cdot dh}{C_d A_0 \sqrt{2g\,h}}$$

$$= \frac{-\pi}{C_d A_0 \sqrt{2g}}[2Rh^{1/2} - h^{3/2}]dh$$

$$\int_0^t dt = \int_{H_1}^{H_2} \frac{-\pi}{C_d A_0 \sqrt{2g}} [2Rh^{1/2} - h^{3/2}] dh$$

$$= \frac{\pi}{C_d A_0 \sqrt{2g}} \left| \frac{4}{3} Rh^{3/2} - \frac{2}{5} h^{5/2} \right|_{H_2}^{H_1}$$

$$t = \frac{\pi}{C_d A_0 \sqrt{2g}} \left[\frac{4}{3} R(H_1^{3/2} - H_2^{3/2}) - \frac{2}{5}(H_1^{5/2} - H_2^{5/2}) \right]$$

For emptying the tank completely, $H_2 = 0$

$$\therefore \quad t = \frac{\pi}{C_d A_0 \sqrt{2g}} \left[\frac{4}{3} RH_1^{3/2} - \frac{2}{5} H_1^{5/2} \right] \quad \text{... (4.36)}$$

4.10.12 Emptying a Horizontal Circular Tank

Refer to Fig. 4.32.

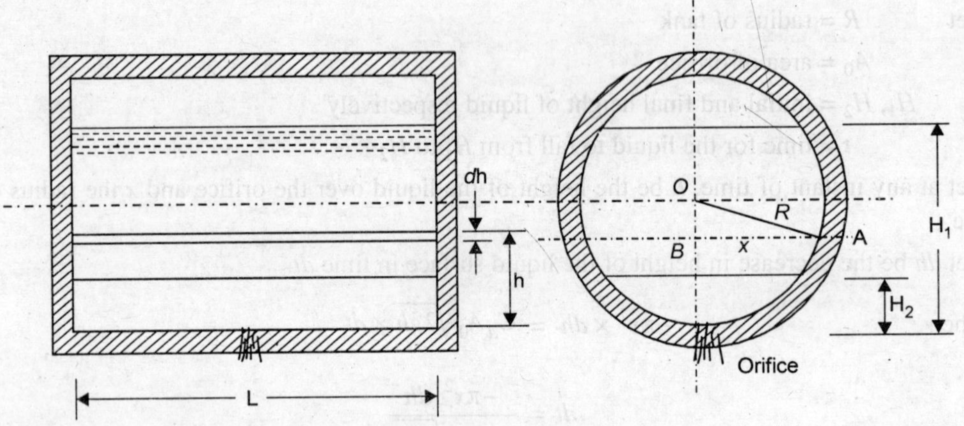

Fig. 4.32 Emptying a Horizontal Circular Tank

Let R = radius of the tank

L = length of the tank

H_1, H_2 = initial and final height of liquid

t = time of emptying the tank

h = height of liquid at any time

dh = decrease in h in time dt

x = radius of liquid surface

Then $$-A \cdot dh = C_d A_0 \sqrt{2gh} \cdot dt$$

Now $$A = 2x \times L$$

$$x = \sqrt{R^2 - (R-h)^2} = \sqrt{2Rh - h^2}$$

$$dt = -\frac{2\sqrt{2Rh - h^2} \times L \times dh}{C_d A_0 \sqrt{2gh}}$$

$$= -\frac{2L}{C_d A_0 \sqrt{2g}} \times \sqrt{2R - h} \cdot dh$$

$$\int_0^t dt = -\frac{2L}{C_d A_0 \sqrt{2g}} \int_{H_1}^{H_2} \sqrt{2R - h}\ dh$$

$$t = \frac{-2L}{C_d A_0 \sqrt{2g}} \left| \frac{2}{3}(2R-h)^{3/2}(-1) \right|_{H_1}^{H_2}$$

$$= \frac{4L}{3C_d A_0 \sqrt{2g}} [(2R - H_2)^{3/2} - (2R - H_1)^{3/2}] \qquad \text{... (4.37)}$$

For emptying the tank completely, $H_2 = 0$.

$$t = \frac{4L}{C_d A_0 \sqrt{2g}} [(2R)^{3/2} - (2R - H_1)^{3/2}] \qquad \text{... (4.38)}$$

4.11 MOUTHPIECES

A mouthpiece is a short pipe equal to the size of the orifice connected to the orifice and extended on the downstream side upto a length usually not more than 2.5 times the diameter. It is used to measure the discharge. The mountpieces may by cylindrical, converging or diverging. Cylindrical or parallel moutpieces with sharp edged entrance have considerable friction loss because of its subsequent re-expansion to fill the mouthpiece Rounding the entrance greatly reduces the losses.

A mouthpiece is said to be running free if the jet of liquid after contraction does not touch the sides of the mouthpiece, But if the jet after contraction expands and fills the whole mouthpiece, it is known as running full.

4.11.1 EXTERNAL MOUTHPIECE

An external mouthpiece is attached to the vessel such that it projects outside, as shown in Fig. 4.33.

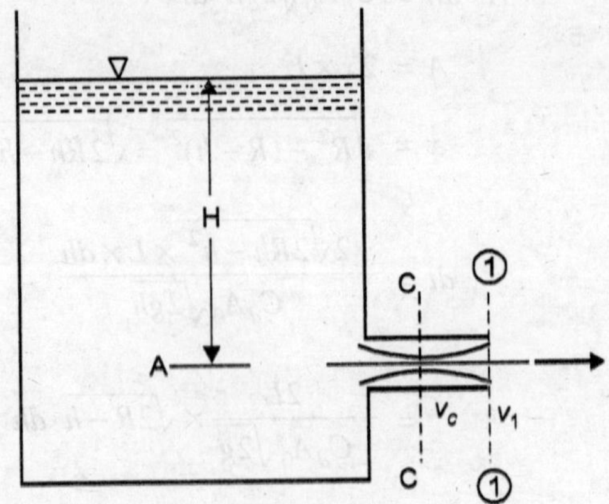

Fig. 4.33 External moutpiece

Let A_1 = area of mouthpiece at outlet

v_1 = velocity of liquid at outlet section 1–1

A_C = area of flow at vena-contracta C-C

v_C = velocity of flow at $C - C$.

H = height of liquid above the centre of moutpiece

C_c = coefficient of contraction

Applying continuity equation at $C - C$ and $1 - 1$, we have

We get

$$A_C v_C = A_1 v_1$$

$$v_C = \frac{V_1}{(A_C / A_1)} = \frac{V_1}{C_c}$$

From section C-C the liquid jet suddenly enlarges at section 1–1. The loss of head due to sudden enlargement is,

$$h_l = \frac{(v_c - v_1)^2}{2g} = \frac{1}{2g}\left[\frac{v_1}{C_c} - v_1\right]^2 = \frac{v_1^2}{2g}\left(\frac{1-C_c}{C_c}\right)^2$$

Applying Bernoulli's equation between point A and section 1 – 1, we get

$$\frac{p_A}{\gamma} + \frac{v_A^2}{2g} + z_A = \frac{p_1}{\gamma} + \frac{v_1^2}{2g} + z_1 + h_l$$

But $z_A = z_1$, $\frac{p_1}{\gamma} = 0$, and $v_A \simeq 0$.

$$\therefore \qquad H+0 = 0+\frac{v_1^2}{2g}+\frac{v_1^2}{2g}\left(\frac{1-C_c}{C_c}\right)^2$$

$$= \frac{v_1^2}{2g}\left[\left\{\frac{1-C_c}{C_c}\right\}^2+1\right] = \frac{v_1^2}{2g}\left[\frac{1+C_c^2-2C_c+C_c^2}{C_c^2}\right]$$

$$= \frac{v_1^2}{2g}\left[\frac{1+2C_c^2-2C_c}{C_c^2}\right]$$

For $C_c = 0.62$

$$H = \frac{1.3756\, v_1^2}{2g}$$

$$v_1 = \left[\frac{2gH}{1.3756}\right]^{1/2} = 0.852\sqrt{2gH} \qquad \text{... (4.39)}$$

Theoretical velocity at outlet, $\quad v_{\text{th}} = \sqrt{2gH}$

Coefficient of velocity, $\quad C_v = \frac{v_1}{v_{th}} = \frac{0.852\sqrt{2gH}}{\sqrt{2gH}} = 0.852$

Since the area of jet of liquid at outlet for a mouthpiece is equal to the area of mouthpiece at outlet, therefore, $C_C = 1$.

Hence $\quad C_d = C_c \times C_v = 1 \times 0.852 = 0.852$

Since $\quad (C_d)_{\text{mouthpiece}} > (C_d)_{\text{orifice}}$

$\therefore \quad Q_{\text{mouth piece}} > Q_{\text{orifice}}$

4.11.2 Convergent divergent Mouthpiece

In a convergent divergent mouthpiece, there is no sudden enlargement of the jet, as shown in Fig. 4.34. Therefore, the loss of energy due to sudden enlargement is eliminated

Thus $\quad C_d = 1.0$

Let $\quad H$ = head of liquid pressure head

H_a = atomospheric pressure head

H_c = absolute pressure head at vena-contracta.

Applying Bernoulli's equation at free liquid surface and section C–C, we get

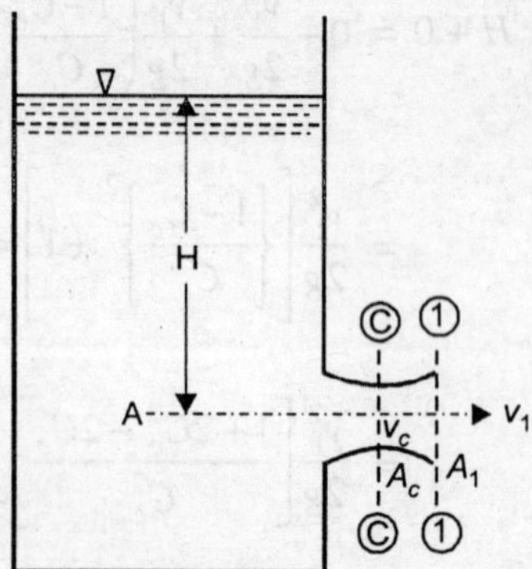

Fig. 4.34. Convergent-divergent mouthpiece

$$\frac{p}{\gamma}+\frac{v^2}{2g}+z = \frac{p_c}{\gamma}+\frac{v_c^2}{2g}+z_c$$

Let centre of the mouthpiece be the datum. Then

$$\frac{p}{\gamma} = H_a, v=0, z=H, \frac{p_c}{\gamma}=H_c, z_c=0$$

$$H_a+0+H = H_c+\frac{v_c^2}{2g}+0 \quad \text{... (1)}$$

$$\frac{v_c^2}{2g} = H_a+H-H_c$$

$$v_c = \sqrt{2g(H_a+H-H_c)} \quad \text{... (2)}$$

Now applying Bernoulli's equation at sections C-C and 1–1, we get

$$\frac{p_c}{\gamma}+\frac{v_c^2}{2g}+z_c = \frac{p_c}{\gamma}+\frac{v_1^2}{2g}+z_1$$

But $z_c = z_1$ and $\frac{p_1}{\gamma}=H_a$

From Eq. (1), we have

$$H_c+\frac{v_c^2}{2g} = H_a+H$$

$$H_a+\frac{v_1^2}{2g} = H_a+H$$

$$v_1 = \sqrt{2gH}$$

But continuity equation, we have

$$A_c\, v_c = A_1\, v_1$$

$$\frac{A_1}{A_c} = \frac{v_c}{v_1} = \frac{\sqrt{2g(H_a + H - H_c)}}{\sqrt{2gH}} = \sqrt{\frac{H_a}{H} + 1 - \frac{H_c}{H}} = \sqrt{1 + \frac{H_a - H_c}{H}} \quad \text{... (4.40)}$$

$$Q = A_1\sqrt{2gH} \quad \text{... (4.41)}$$

4.11.3 Internal (or re-entrant or Borda's) Mouthpiece

Internal or re-entrant mouthpieces are those which project from the wall into the tank as shown in Fig. 4.35. For a cylindrical re-entrant mouthpiece, if the length is between 2 to 3 times its diameter then the flow shall be full at the outlet.

Borda's mouthpiece is a short cylindrical re-entrant tube which projects into the tank. Its length is about one diameter and has sharp edges to ensure perfect contraction so that the jet does not touch the sides of the tube.

(a) Mouthpiece running free

For a mouthpiece running free, the length of the tube is equal to diameter and the jet of liquid comes out without touching the sides of the tube (see Fig. 4.35).

Let H = height of liquid above the mouth piece,

A_0 = area of orifice or moutpiece.

A_c = area of contracted jet

v_c = velocity through moutpieces.

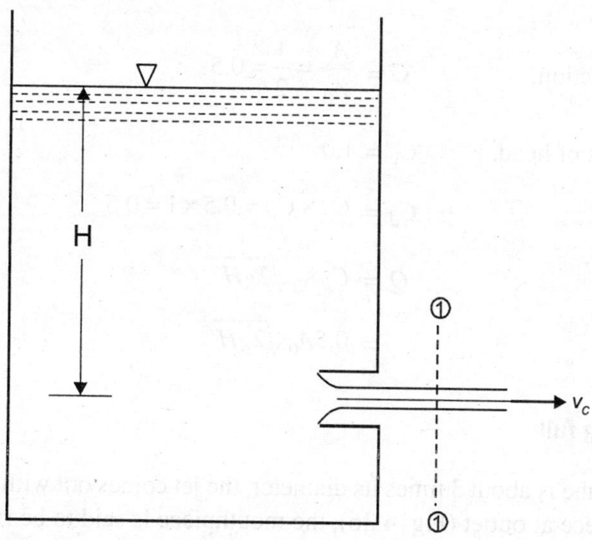

Fig.4.35 Mouthpiece running free

Pressure of liquid on mouthpiece, $p = \gamma H$

Force acting on mouthpiece, $F = p \times A_0 = \gamma H A_0$ ... (1)

Mass rate of flow of liquid, $\dot{m} = \dfrac{\gamma A_c V_c}{g}$

Momentum of flowing liquid per second $= \dot{m} \times v_c = \dfrac{\gamma A_c v_c^2}{g}$

Initial momentum of water $= 0$

Change in momentum $= \dfrac{\gamma}{g} A_c v_c^2$ ... (2)

According to Newton's second law of motion,

Force = Rate of change of momentum.

$$\gamma H A_0 = \frac{\gamma}{g} A_c v_c^2$$

$$H A_0 = \frac{A_c v_c^2}{g}$$

$$\frac{v_c^2}{2g} \times A_0 = A_c \times \frac{v_c^2}{g}$$

Coefficient of contraction, $C = \dfrac{A_c}{A_0} = \dfrac{1}{2} = 0.5$

Since there is no loss of head, $C_v = 1.0$

$\therefore$ $C_d = C_c \times C_v = 0.5 \times 1 = 0.5$

Discharge, $Q = C_d A_0 \sqrt{2gH}$ ... (4.42)

$$= 0.5 A_0 \sqrt{2gH}$$

(b) Mouthpiece running full

If the length of the tube is about 3 times its diameter, the jet comes out with its diameter equal to the diameter of the mouthpiece at outlet (Fig. 4.36), the mouthpiece is said to be running full.

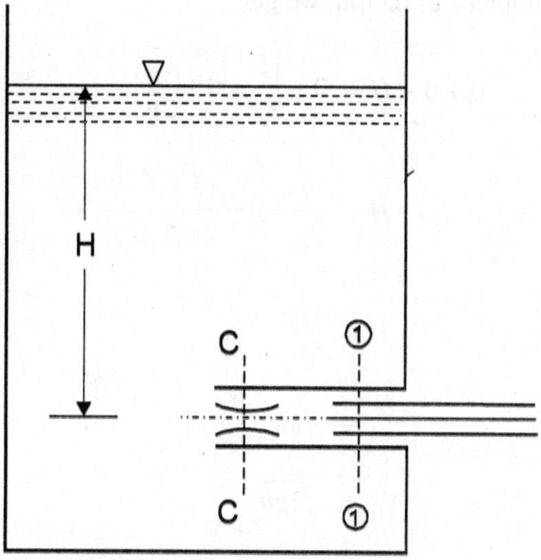

Fig. 4.36 Mouthpiece running full

Let A_C = area of vena-contracta

$A = A_1$ = area of orifice or mouthpiece

v_c = velocity of liquid at vena-contracta, C-C

v_1 = velocity of liquid at outlet, 1–1

H = height of liquid above the mouthpiece.

From continuity equation, we have.

$$A_c\, v_c = A_1\, v_1$$

$$v_c = \frac{A_1 v_1}{A_c} \qquad \text{... (1)}$$

For internal mouthpiece, $C_c = \dfrac{A_c}{A_1} = 0.5$

Eq. (1) becomes,

$$v_c = 2v_1$$

The jet of liquid after passing though C–C, suddenly enlarges at section 1–1. Therefore, there will be loss of head due to sudden enlargement.

$$h_l = \frac{(v_c - v_1)^2}{2g} = \frac{(2v_1 - v_1)^2}{2g} = \frac{v_1^2}{2g}$$

Applying Bernoulli's equation to free liquid surface in tank and section 1 – 1 (or outlet), we have

$$\frac{p}{\gamma} + \frac{v^2}{2g} + z = \frac{p_1}{\gamma} + \frac{v_1^2}{2g} + z_1 + h_l$$

Taking centreline of mouthpiece as datum, we get

$$0+0+H = 0+\frac{v_1^2}{2g}+0+\frac{v_1^2}{2g}$$

$$H = \frac{v_1^2}{2g}+0+\frac{v_1^2}{2g}$$

$$H = \frac{v_1^2}{g}$$

$$V_1 = \sqrt{gh}$$

Now $$V_{th} = \sqrt{2gh}$$

∴ $$C_v = \frac{v_1}{v_{th}} = \frac{1}{\sqrt{2}} = 0.707$$

Here $$C_c = 1.0, \text{ because, } A = A_1$$

∴ $$C_d = C_c \times C_v = 1.0 \times \frac{1}{\sqrt{2}} = 0.707$$

$$Q = C_d A \times \sqrt{2gh} = 0.707 \times \sqrt{2gh} \quad \text{... (4.43)}$$

4.12 IMPULSE-MOMENTUM RELATIONSHIP

The force facting on a fluid mass m moving with acceleration 'a' is given by the Newton's second law of motion, $F = ma$. The direction of F is the same as that of "a". As the mass of the fluid is constant, we may write,

$$F = ma = m\frac{dv}{dt} = \frac{d}{dt}(mv) \quad \text{... (4.44)}$$

Eq. (4.44) embodies what is known as the *momentum principle*. The momentum principle expresses that the time rate of change of momentum in any direction is equal to the net force acting on the fluid mass in that direction. It is usually known as the momentum equation.

Eq. (4.44) may be expressed as:

$$F \cdot dt = d\,(mV) \quad \text{... (4.45)}$$

Eq. (4.45) is known as the *inpulse-momentum equation*. It states that the impulse of a force F acting over a short interval of time dt is equal to the change in momentum in the direction of the force. Impulse and momentum are both vector quantities.

The momentum principle provides a bulk solution to determine the hydro-dynamic force arising from change in momentum instead of analysing the details of pressure and velocity distribution. Some examples involving application of the momentum principle are:

- Force on a pipe-bend caused by the fluid flowing through it.
- Force exerted by jet of fluid striking against a solid surface.
- Thrust on propeller.
- Jet propulsion

4.12.1 Steady Flow Momentum Equation

Consider steady flow of fluid through a diverging streamtube as shown in Fig. 4.37. The cross-section of the streamtube is sufficiently small so that the velocity may be considered uniformly distributed over the flow section. The flow is steady so that the streamtube remains stationary w.r.t the fixed coordinate axes.

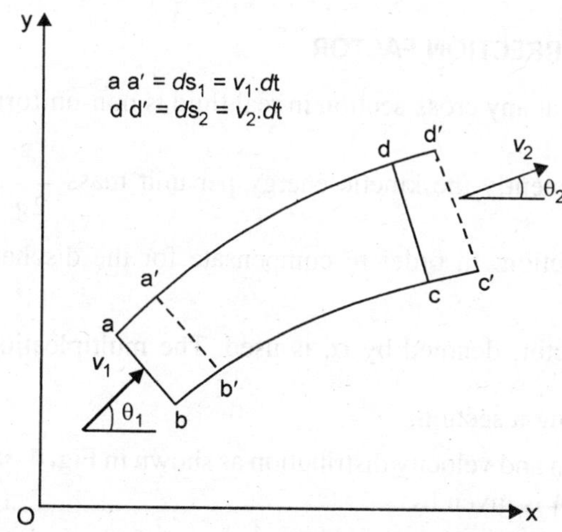

Fig. 4.37 Momentum change in a streamtube

Let v_1 = Velocity of flow at *ab*

v_2 = Velocity of flow at *cd*.

Let the mass of fluid in the region *abcd* shift to new position *a′b′c′d′* under the action of external forces in a short interval of time *dt*.

Since the area *a′b′c d* is common to both the regions *abcd* and *a′b′c′d′*, it will not experience any momentum change. There shall be momentum change due to fluid mass in *a bb′a′* and *cc′d′d.*

According to the continuity equation, for steady incompressible flow, we have

$$A_1 \cdot ds_1 = A_2 \cdot ds_2 \qquad (\because \rho_1 = \rho_2)$$

Momentum of fluid mass in region $a\,bb'a' = \rho A_1\, ds_1 v_1 = \rho(A_1 v_1 \cdot dt)\, v_1$

Momentum of fluid mass in region $cc'd'd = \rho A_2\, ds_2 v_2 = \rho(A_2 v_2 \cdot dt)\, v_2$

$$\text{Change in momentum} = \rho[A_2 v_2 \cdot dt \cdot v_2 - A_1 v_1 \cdot dt \cdot v_1]$$

Now $$A_1 v_1 = A_2 v_2 = Q$$

$\therefore$ $$\text{Change in momentum} = \rho Q(v_2 - v_1)\, dt$$

By impulse – momentum principle, we have

$$F \cdot dt = \rho Q(v_2 - v_1)\,dt$$

or

$$F = \rho Q(v_2 - v_1) \qquad \text{... (4.47)}$$

where $\rho Q = \dot{m}$ = mass flux.

Component of force along x-and y-axes are:

$$\left.\begin{aligned} F_x &= \rho Q(v_2 \cos\theta_2 - v_1 \cos\theta_1) \\ F_y &= \rho Q(v_2 \sin\theta_2 - v_1 \sin\theta_1) \end{aligned}\right\} \qquad \text{... (4.48)}$$

4.13 KINETIC ENERGY CORRECTION FACTOR

The velocity distribution at any cross-section in real fluid is non-uniform on account of viscous and boundary resistance. Consequently the kinetic energy per unit mass $\frac{v^2}{2g}$ does not represent the true kinetic energy across the section. In order to compensate for the discharge, a coefficient known as kinetic energy correction factor, denoted by α, is used. The multiplcation of α with $\frac{v^2}{2g}$ yields the kinetic energy actually passing a section.

Consider the flow-section and velocity distribution as shown in Fig. 4.38. The kinetic energy passing through an elemetary area dA is given by:

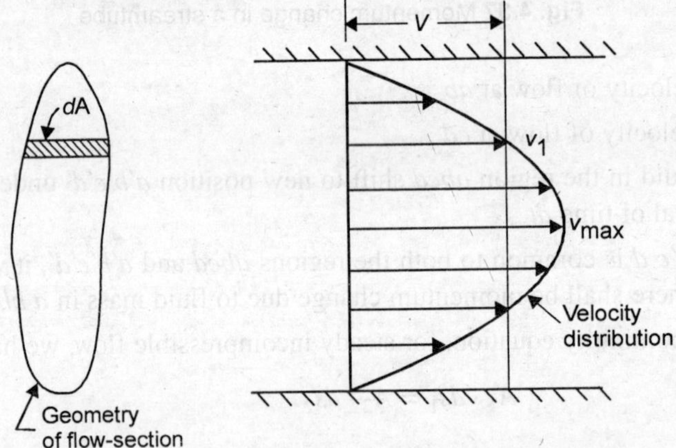

Fig. 4.38 Flow-section and velocity distribution

$$= \frac{1}{2}(\rho \cdot dA \cdot v)v^2 = \frac{1}{2}\rho dA v^3$$

Total kinetic energy passing the section

$$= \int_A \frac{1}{2}\rho v^3 \cdot dA \qquad \text{... (1)}$$

Actual kinetic energy based on average velocity V passing the section

$$= \alpha \cdot \frac{1}{2}\rho V^3 A \qquad \text{... (2)}$$

Equating Eqs. (1) and (2), we get

$$\alpha \cdot \frac{1}{2}\rho V^3 A = \int_A \frac{1}{2}\rho v^3 \cdot dA$$

$$\alpha = \frac{1}{A}\int \left(\frac{v}{V}\right)^3 dA \qquad \text{... (4.49)}$$

α is a measure of viscous resistance, generated in a given flow, the effect of which is reflected in the non-uniform nature of velocity distribution. Practical values of α are:

α = 2.0 for liminar flow with parabolic velocity distribution.

= 1.01 to 1.15 for fully developed turbulent flow in pipes.

Incorporating the kinetic energy correction factor, the Bernoulli's equation may be modified as:

$$\frac{p_1}{\gamma} + \alpha_1 \frac{v_1^2}{2g} + z_1 = \frac{p_2}{\gamma} + \alpha_2 \frac{v_2^2}{2g} + z_2 + h_l \qquad \text{... (4.50)}$$

4.14 MOMENTUM CORRECTION FACTOR

The non-uniformity of velocity distribution at the inlet and outlet sections affect the accuracy of momentum flux (*i.e.* momentum per second) based on the average velocity of flow.

To correct it, a coefficient β, known as momentum correction factor is used to represent the true momentum flux passing a section.

Consider an elementary area dA, as shown in Fig. 4.39, where the velocity is v. Then

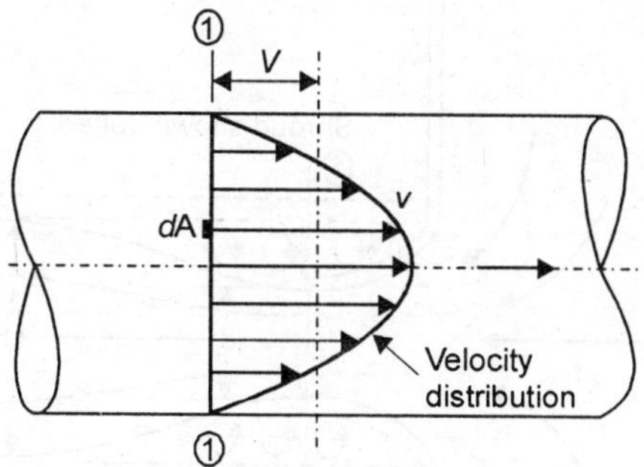

Fig. 4.39 Velocity distribution a pipe.

Momentum of fluid passing this area = $\rho v^2 \cdot dA$

Momentum flux passing section 1–1 = $\int_A \rho v^2 \cdot dA$

Actual momentum flux = $\beta \rho A V^2$

$$\therefore \qquad \beta \rho A V^2 = \int_A \rho v^2 \cdot dA$$

$$\beta = \frac{1}{A}\int_A \left(\frac{v}{V}\right)^2 dA \qquad \text{... (4.51)}$$

Practical values of β are:

$\beta = \frac{4}{3}$ for laminar flow in a circular pipe.

= 1.01–1.05 for turbulent flow in a circular pipe.

The impulse–Momentum equations may be modified as:

$$\left.\begin{aligned} F_x &= \rho Q(\beta_1 v_1 \cos\theta_1 - \beta_2 v_2 \cos\theta_2) \\ F_y &= \rho Q(\beta_1 v_1 \sin\theta_1 - \beta_2 v_2 \sin\theta_2) \end{aligned}\right\} \qquad \text{... (4.52)}$$

Example 4.28 *A Streamlined nozzle of diameter d is supplied at a constant head, the magnitude of the head being large compared to d. The nozzle discharges directly into the atmosphere and is so shaped that the issuing jet is parallel at the nozzle exit. To increase the flow rate a shroud of diameter D is firmly secured to the nozzle as shown in Fig. 4.40. The jet expands to fill the shroud and the shroud is long enough to ensure that the flow leaving is steady and parallel.*

Determine what the diameter of shroud should be to maximise the flow rate. Neglect shear stresses at the wall of the shroud.

Solution. Given that $H > d$.

Writing the Bernoulli's equation between sections (1) and (2), including the loss due to sudden expansion (see Fig. 4.40),

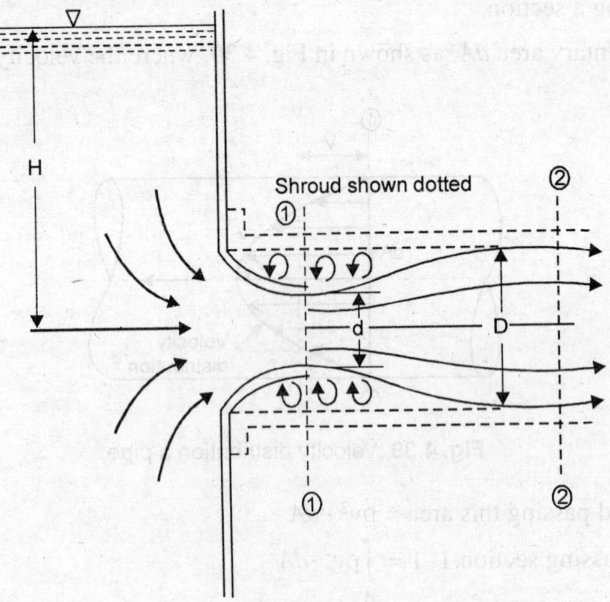

Fig. 4.40

$$\frac{p_1}{\gamma}+\frac{v_1^2}{2g} = \frac{v_2^2}{2g}+\frac{(v_1-v_2)^2}{2} \qquad \text{... (1)}$$

and between the water-surface in the tank and section 1–1,

$$H = \frac{p_1}{\gamma}+\frac{v_1^2}{2g} \qquad \text{... (2)}$$

From the continuity equation,

$$Q = \frac{\pi}{4}d^2 v_1 = \frac{\pi}{4}D^2 v_2 \qquad \text{... (3)}$$

Eliminating p_1 between Eqs. (1) and (2) and using (3), one can write

$$H = \frac{v_2^2}{2g}+\frac{(v_1-v_2)^2}{2g}$$

$$= \frac{1}{2g}\left(\frac{4Q}{\pi D^2}\right)^2+\frac{1}{2g}\left(\frac{4Q}{\pi d^2}-\frac{4Q}{\pi D^2}\right)$$

$$= \frac{8Q^2}{\pi^2 g}\left[\frac{2}{D^4}+\frac{1}{d^4}-\frac{2}{D^2 d^2}\right]$$

from which $$Q = \frac{\sqrt{\pi^2 gH}}{8}\frac{D^2 d^2}{\sqrt{d^4+(D^2-d^2)^2}}.$$

For maximum discharge, the shroud diameter can be determined from the condition $dQ/dD = 0$. Differentiating the discharge equation,

$$\frac{dQ}{dD} = \frac{1}{\sqrt{d^4+(D^2-d^2)^2}}\{2Dd^2\}+D^2 d^2\left[-\frac{1}{2}\{d^4+(D^2-d^2)^2\}^{-3/2}\right]2(D^2-d^2).2D$$

equating it to zero and simplifying, one gets

$$D = \sqrt{2}\,d$$

The maximum discharge,

$$Q_{\max} = \sqrt{\frac{\pi^2 g}{8}}\frac{2d^2\cdot d^2}{\sqrt{d^4+(2d^2-d^2)^2}} = \frac{\pi d^2}{2\sqrt{2}}\sqrt{2gH}$$

Percentage increase in discharge

$$= \frac{Q_{\max}-Q}{Q}\times 100 = \frac{\frac{\pi d^2}{2\sqrt{2}}\sqrt{2gH}-\frac{\pi d^2}{4}\sqrt{2gH}}{\frac{\pi d^2}{4}\sqrt{2gH}}$$

$$= (\sqrt{2}-1)\times 100 = 41.4\%$$

Example 4.29 *In a 45° bend, a rectanular duct of 1m² cross-sectional area used for conveying air is gradually reduced to 0.5 m² area. Find the magnitude and direction of force required to hold the duct in postion if the velocity of air flow at 1m² section is 10 m/s, and pressure is 30 kPa. Take specific weight of air 11.6 N/m³.*

Solution.

Given: $\theta = 45°$, $A_1 = 1\text{m}^2$, $A_2 = 0.5\text{ m}^2$, $v_1 = 10\text{ m/s}$, $p_1 = 30\text{ kPa}$, $\gamma = 11.6\text{ N/m}^3$.

Refer to Fig. 4.41

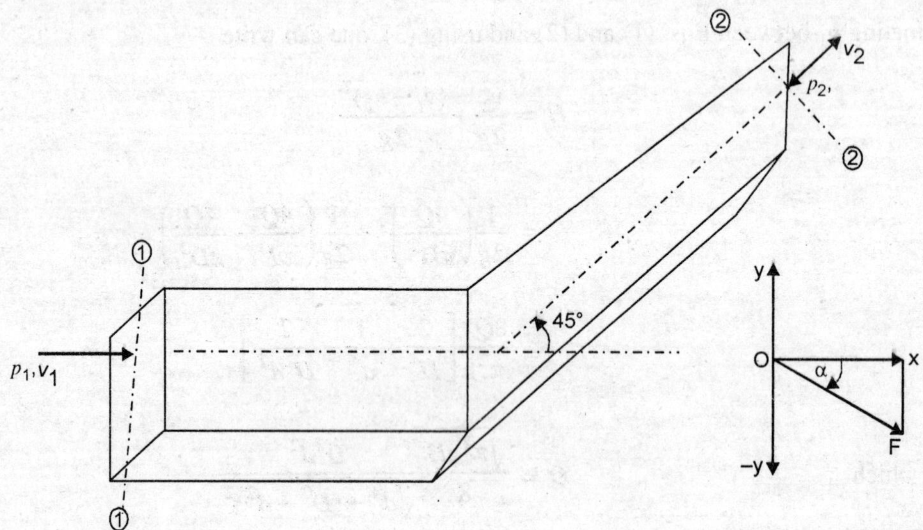

Fig. 4.41

From continuity equation,

$$A_1 v_1 = A_2 v_2$$

$$v_2 = \frac{1\times 10}{0.5} = 20\,\text{m/s}$$

Applying Bernoulli's equation at sections 1 and 2, we get

$$\frac{p_1}{\gamma}+\frac{v_1^2}{2g}+z_1 = \frac{p_2}{\gamma}+\frac{v_2^2}{2g}+z_2$$

Here $z_1 = z_2$.

$$\frac{30\times 10^3}{11.6}+\frac{10^2}{2\times 9.81} = \frac{p_2}{11.6}+\frac{20^2}{2\times 9.81}$$

$$p_2 = 29.823\text{ kPa}$$

$$Q = A_1 v_1 = 1\times 10 = 10\text{m}^3/\text{s}$$

$$F_x = \frac{\gamma Q}{g}[v_{1x}-v_{2x}]+(p_1A_1)_x+(p_2A_2)_x$$

where $v_{1x} = 10$ m/s, $v_{2x} = 20 \cos 45° = 14.14$ m/s

$$F_x = \frac{11.6 \times 10}{9.81}(10 - 14.14) + 30 \times 10^3 \times 1 - 29.823 \times 10^3 \times 0.5 \times \cos 45°$$

$$= -48.954 + 30 \times 10^3 - 10.544 \times 10^3$$

$$= 19.41 \text{ kN } (\rightarrow)$$

$$F_y = \frac{\gamma Q}{g}(v_{1y} - v_{2y}) + (p_1 A_1)y + (p_2 A_2)_y$$

$$= \frac{11.6 \times 10}{9.81}(0 - 20 \sin 45°) + 0 - 29.823 \times 10^3 \times 0.5 \times \sin 45°$$

$$= -167.226 - 10.544 \times 10^3 = -10.711 \text{ kN} \downarrow$$

Resultant force, $F = \sqrt{F_x^2 + F_y^2} = \sqrt{(19.41)^2 + (-10.711)^2} = 22.169$ kN

$$\tan \alpha = \frac{F_y}{F_x} = \frac{-10.711}{19.41} = -0.55183$$

$$\alpha = -28.89° \text{ or } (28° - 53')$$

Since numerator is –ve and denominator is +ve, α lies in the 4th quadrant

Example 4.30 *360 lit/s of water is flowing in a pipe which is bent by 120°. The pipe bend measures 400 mm x 300 mm and volume of bend is 0.15m³. The pressure at the enttance is 75 kPa and the exit is 2.5 m above the entrance section. Find the force exerted on the bend.*

Solution.

Given: $Q = 0.36$ m³/s, $\theta = 120°$, $d_1 = 0.4$ m, $d_2 = 0.3$ m, $p_1 = 75$ kPa, $z_2 = 2.5$ m, $V = 0.15$ m³.

Refer to Fig. 4.42

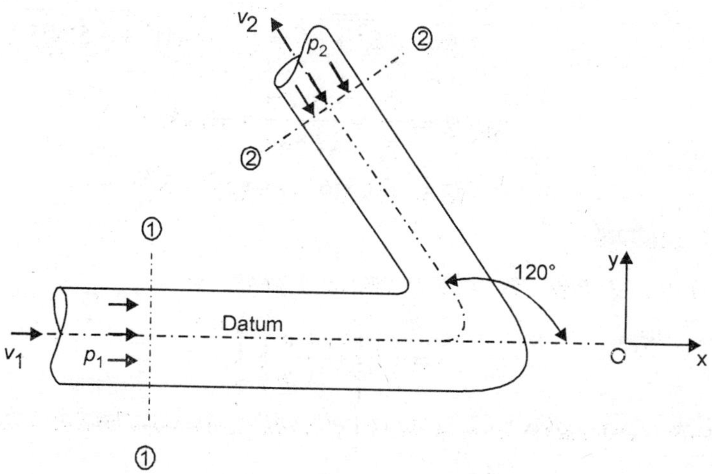

Fig. 4.42

$$v_1 = \frac{Q}{A_1} = \frac{0.36 \times 4}{\pi(0.4)^2} = 2.865 \text{ m/s}$$

$$v_2 = \frac{Q}{A_2} = \frac{0.36 \times 4}{\pi(0.3)^2} = 5.093 \text{ m/s}$$

Applying Bernoulli's theorem at sections 1–1 and 2–2, we get

$$\frac{p_1}{\gamma} + \frac{v_1^2}{2g} + z_1 = \frac{p_2}{\gamma} + \frac{v_2^2}{2g} + z_2$$

$$\frac{75}{9.810} + \frac{(2.865)^2}{2 \times 9.81} + 0 = \frac{p_2}{9.810} + \frac{(5.093)^2}{2 \times 9.81} + 2.5$$

$$p_2 = 41.61 \text{ kPa}$$

$$F_x = \frac{\gamma Q}{g}[v_1 - (-v_2 \cos 60°)] + p_1 A_1 + p_2 A_2 \cos 60°$$

$$= 10^3 \times 0.36[2.865 + 5.093 \cos 60°] + 75 \times 10^3 \times \frac{\pi}{4}(0.4)^2 + 41.61 \times 10^3 \times \frac{\pi}{4}(0.3)^2 \cos 60°$$

$$= (1.948 + 9.425 + 1.471)10^3 = 12.844 \text{ kN}$$

$$F_y = \frac{\gamma Q}{g}[0 - v_2 \sin 60°] + 0 - p_2 A_2 \sin 60° - V \times 9.81 \times 10^3$$

$$= -10^3 \times 0.36 \times 5.093 \sin 60° - 41.61 \times \frac{\pi}{4}(0.3)^2 \times \sin 60° - 0.15 \times 9.18 \times 10^3$$

$$= 10^3 [-1.587 - 2.547 - 1.471]$$

$$= -5.605 \text{ kN}$$

$$F = \sqrt{F_x^2 + F_y^2} = \sqrt{(12.844)^2 + (-5.605)^2} = 14.014 \text{ kN}$$

$$\tan \alpha = \frac{F_y}{F_x} = \frac{-5.605}{12.844} = -0.4364$$

$$\alpha = -23.576° \text{ or } -(23° - 34')$$

α lies in the 4th quadrant.

Example 4.31 *In a circular pipe, the velocity profile is given as:*

$$u = U_m \left[1 - \left(\frac{r}{R}\right)^2\right]$$

where u is the velocity at any radius r, U_m is the velocity at the pipe axis, and R is the radius of the pipe Find:

(a) Average velocity, (b) Kinetic energy correctin factor, and (c) Momentum corrrection factor..

Solution. Refer to Fig. 4.43.

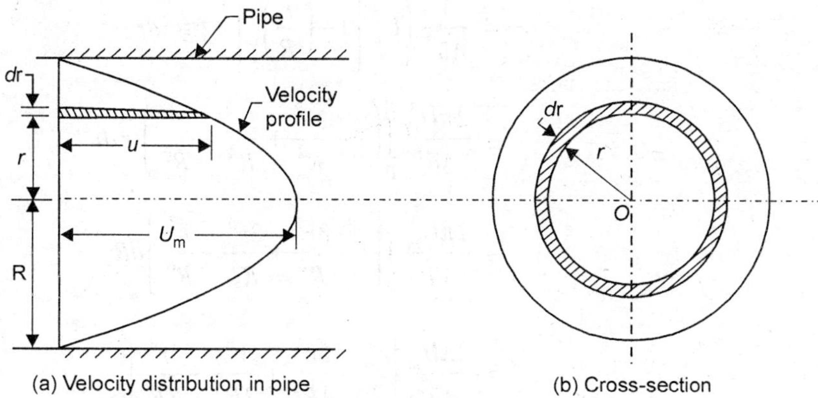

(a) Velocity distribution in pipe

(b) Cross-section

Fig. 4.43

Consider an elementary area dA of the ring between radii r and $r + dr$. Then $dA = 2\pi r.\ dr$

Flow through the elementary ring,

$$dQ = 2\pi r \cdot dr \cdot u$$

Total flow,

$$Q = \int_0^R 2\pi r.u.dr$$

$$= \int_0^R 2\pi r \cdot U_m \left[1 - \left(\frac{r}{R}\right)^2\right] dr$$

$$= 2\pi U_m \int_0^R \left[r - \frac{r^3}{R^2}\right] dr$$

$$= 2\pi U_m \left|\frac{r^2}{2} - \frac{r^4}{4R^2}\right|_0^R = 2\pi U_m \left[\frac{R^2}{2} - \frac{R^2}{4}\right] = \frac{\pi}{2} U_m R^2 \quad \text{... (1)}$$

(a) Let $\overline{u}$ = average velocity of flow

Then $Q = A\overline{u} = \pi R^2 \overline{u} = \frac{\pi}{2} U_m R^2$

$$\overline{u} = \frac{U_m}{2}$$

(b) Kinetic energy correction factor, $\alpha = \frac{1}{A\overline{u}^3} \int_0^R u^3 dA$

$$= \frac{1}{A\bar{u}^3}\int_0^R U_m^3\left[1-\left(\frac{r}{R}\right)^2\right]^3 \cdot 2\pi r\cdot dr$$

$$= \frac{2\pi U_m^3}{A\bar{u}^3}\int_0^R\left(1-\frac{3r^2}{R^2}+\frac{3r^4}{R^4}-\frac{r^6}{R^6}\right)\cdot r\,dr$$

$$= \frac{2\pi U_m^3}{A\bar{u}^3}\int_0^R\left[r-\frac{3r^3}{R^2}+\frac{3r^5}{R^4}-\frac{r^7}{R^6}\right]dR$$

$$= \frac{2\pi U_m^3}{A\bar{u}^3}\left|\frac{r^2}{2}-\frac{3r^4}{4R^2}+\frac{r^6}{2R^4}-\frac{r^8}{8R^6}\right|_0^R$$

$$= \frac{2\pi U_m^3}{A\bar{u}^3}\left[\frac{R^2}{2}-\frac{3}{4}R^2+\frac{R^2}{2}-\frac{R^2}{8}\right]$$

$$= \frac{2\pi U_m^3 R^2}{A\bar{u}^3}\left[\frac{4-6+4-1}{8}\right]$$

$$= \frac{2\pi U_m^3 R^2}{8A\bar{u}^3}$$

Now $$A = \pi R^2, \bar{u} = \frac{U_m}{2}$$

$$\alpha = \frac{2\pi U_m^3 R^2}{8\times\pi R^2\times\frac{U_m^3}{8}} = 2$$

(c) Momentum correction factor, $$\beta = \frac{1}{A\bar{u}^2}\int_0^R u^2\,dA$$

$$= \frac{1}{A\bar{u}^2}\int_0^R U_m^2\left[1-\left(\frac{r}{R}\right)^2\right]^2 2\pi r\cdot dr$$

$$= \frac{2\pi U_m^2}{A\bar{u}^2}\int_0^R\left[1-\left(\frac{r}{R}\right)^2\right]^2 r.dr$$

$$= \frac{2\pi U_m^2}{A\bar{u}^2}\int_0^R\left[1+\frac{r^4}{R^4}-\frac{2r^2}{R^2}\right]^2 r.dr$$

$$= \frac{2\pi U_m^2}{A\bar{u}^2}\int_0^R \left[r + \frac{r^5}{R^4} - \frac{2r^3}{R^2}\right]^2 dr$$

$$= \frac{2\pi U_m^2}{A\bar{u}^2}\left|\frac{r^2}{2} + \frac{r^6}{6R^4} - \frac{r^4}{2R^2}\right|_0^R$$

$$= \frac{2\pi U_m^2}{A\bar{u}^2}\left[\frac{R^2}{2} + \frac{R^2}{6} - \frac{R^2}{2}\right]$$

$$= \frac{\pi U_m^2 R^2}{3A\bar{u}^2}$$

$$= \frac{\pi U_m^2 R^2}{3 \times \pi R^2 \times \frac{U_m^2}{4}} = \frac{4}{3} = 1.33$$

SUMMARY

1. Fluid dyanmics deals with the study of forces which cause motion of fluids in producing velocity and acceleration.
2. A system may be defined as that region of space occupied by the quantity of fluid
3. A control volume is an arbitrary volume fixed in space and bounded by a closed surface called control surface.
4. Euler's equation of motion:

$$\frac{\partial V_s}{\partial t} + V_s\frac{\partial V_s}{\partial s} = B_s - \frac{1}{\rho}\frac{\partial p}{\partial s}$$

$$\frac{\partial V_s}{\partial t} + V_s\frac{\partial V_n}{\partial s} = B_n - \frac{1}{\rho}\frac{\partial p}{\partial n}$$

5. Integration of Euler's equation leads to Bernoulli's equation.

$$\frac{V_s^2}{2g} + \frac{p}{\gamma} + z = \text{const.}$$

6. Different types of head are: potential head, velocity head, and pressure head.
7. The hydraulic grade line (HGL) represents the sum of potential head (z) and pressure head $\left(\frac{p}{\gamma}\right)$
8. Total energy line (TEL) represents the sum of potential head, velocity head and pressure head.

9. Bernoulli's equation:

$$\frac{p}{\gamma}+\frac{v^2}{2g}+z = \text{const. for ideal, incompressible fluids neglecting losses.}$$

$$\frac{p_1}{\gamma}+\frac{v_1^2}{2g}+z_1 = \frac{p_2}{\gamma}+\frac{v_2^2}{2g}+z_2+h_l, \text{ for real fluids}$$

10. Venturimeter is a device to measure rate of flow in a pipeline.

$$Q = \frac{C_d A_1 A_2}{\sqrt{A_1^2 - A_2^2}}\sqrt{2gh\left(\frac{s_2}{s_1}-1\right)}$$

11. Orificemeter is a device to measure the rate of flow in a pipe line.

$$Q = \frac{C_d A_0 A_1}{\sqrt{A_1^2 - A_0^2}}\sqrt{2gh}$$

12. Flow through an orifice.

Torricelli's theorem: $v_2 = \sqrt{2gH}$

13. Hydraulic coefficients:

$$\text{Coefficient of contraction, } C_c = \frac{\text{area of jet at vena-contracta}}{\text{area of orifice}} = 0.61 \text{ to } 0.69$$

$$\text{Coefficient of velocity, } C_v = \frac{\text{actual velocity of jet at vena-contracta}}{\text{theoretical velocity}} = 0.95 \text{ to } 0.99$$

$$\text{Coefficient of discharge, } C_d = C_c \times C_v = \frac{Q_a}{Q_{th}} = 0.62 \text{ to } 0.65$$

$$C_v = \sqrt{\frac{x^2}{4yH}}, C_d = \frac{Q}{A_0\sqrt{2gH}}, h_{fl} = \frac{v_2^2}{2g}\left(\frac{1}{C_v^2}-1\right)$$

14. A mouthpiece is a short pipe, equal to the size of the orifice, connected to the orifice and extending on the downstream side upto a length usually not more than 2.5 times the diameter. It is used to measure the discharge.

15. Inpulse -momentum relationship:

$$F \cdot dt = d\,(mv)$$

$$\text{Impulse} = \text{change in momentum.}$$

16. Impulse-momentum principle:

$$F = \rho Q\,(v_2 - v_1)$$

17. Kinetic energy correction factor, $\alpha = \frac{1}{A}\int_A \left(\frac{v}{V}\right)^3 dA$

where V = average velocity.

18. Momentum correction factor, $\beta = \frac{1}{A}\int_A \left(\frac{v}{V}\right)^2 dA$

19. Discharge from a large rectangular orifice.

$$Q = \frac{2}{3}C_d\, b\sqrt{2g}\,[H_2^{3/2} - H_1^{3/2}]$$

20. Discharge from a fully submerged orifice.

$$Q = C_d b(H_2 - H_1)\sqrt{2gH} = C_d\, bd\sqrt{2gH}$$

where $d = H_2 - H_1$

21. Discharge from a partially submerged orifice.

$$Q = C_d b\sqrt{2g}\,[H_2 - H]\sqrt{H} + \frac{2}{3}(H^{3/2} - H_1^{3/2})$$

22. Time required for emptying a tank of uniform area of cross-section.

$$t = \frac{2A_1}{C_d\, A_2\sqrt{2g}}[\sqrt{H} - \sqrt{h}] \text{ with side orifice}$$

$$= \frac{2A_1\sqrt{H_1}}{C_d\, A_2\sqrt{2g}} \text{ with bottom orifice}$$

23. Time to empty a hemispherical tank

$$t = \frac{\pi}{C_d\, A_0\sqrt{2g}}\left[\frac{4}{3}R(H_1^{3/2} - H_2^{3/2}) - \frac{2}{5}(H_1^{5/2} - H_2^{5/2})\right]$$

24. Time to empty a horizontal circular tank.

$$t = \frac{4L}{3C_d A_0\sqrt{2g}}[(2R)^{3/2} - (2R - H_1^{3/2})]$$

25. Coefficient of discharge for mouthpiece.

C_d = 0.852 for external mouthpiece
= 0.707 for internal mouthpiece, running full
= 0.50 for internal mouthpiece running free
= 1.0 for convergent-divergent mouthpiece

26. Ratio of areas for convergent-divergent mouthpiece,

$$\frac{A_1}{A_c} = \sqrt{1 + \frac{H_a - H_c}{H}}$$

Short Answer Type Questions

1. Define a system, control volume and control surface.
2. What is Euler's equation of motion?
3. What are the assumptions made while deriving Bernoulli's equation?
4. What are the limitations of Bernoulli's equation?
5. What is a Venturimeter?
6. Differentiate between an orificemeter and Venturimeter.
7. Why the coefficient of discharge of an orificemeter is much smaller than that of a venturimeter?
8. How orifices can be classfied?
9. Differentiate between an orifice and mouthpiece.
10. What is Torricelli's theorem?
11. Explain various hydraulic coefficients.
12. Define vena-contracta.
13. What is the difference between a small and a large orifice?
14. Explain the difference between a wholly submerged and partially submerged orifice.
15. How mouthpieces are classified? What is their main advantage?
16. What is a Borda's mouthpiece?
17. State the momentum principle.
18. State the impulse-momentum principle.
19. Write the steady flow momentum equation.
20. Define kinetic energy correction factor and momentum correction factors.
21. List applications of momentum principle.

EXERCISES

1. A pipe P of 400 mm diameter, conveying water, branches into two pipes *Q* and *R* of diameter 300 mm and 200 mm respectively. Calculate:
   (a) discharge in pipe *P* if average velocity of water in this pipe is 3 m/s, and
   (b) Velocity of water in pipe *R* if average velocity in pipe *Q* is 2m/s

   [**Ans.** 0.377 m^3/s, 7.5 m/s]]
2. A pipe 300 m long has a slope of 1 in 100 and tapers from 1.0 m diameter at higher end to 0.5 m diameter at the lower end. Discharge through the pipe is 90 lit/s. Calculate the pressure at the lower end if the pressure at the higher end is 70 kPa. [**Ans.** 100 kPa]
3. A 2 m long concical tube is fixed vertically with its smaller end upwards. It carries liquid in downward direction. The flow velocities at the smaller and large ends are 5m/s and 2 m/s respectively. The pressure head at the smaller end is 2.5 m of liquid. If the loss of head in the tube is $\dfrac{0.35(v_1^2 - v_2^2)}{2g}$, determine the pressure head at the large end [**Ans.** 5.4 m of liquid]

4. The following data relate to an orificemeter:

   Diameter of pipe = 0.30 m

   Diameter of orifice = 0.15 m

   Differential manometric head = 0.5 m of mercury

   Specific gravity of liquid = 0.9

   Coefficient of discharge of orificemeter = 0.64

   Calculate the rate of flow. **[Ans.** 0.137 m^3/s]

5. A venturimeter with 0.15 m diameter at inlet and 0.10 m diameter at throat is laid with its axis horizontal. It is used to measure the flow of oil of specific gravity 0.9. The oil-mercury differential manometer shows a gauge difference of 0.2 m. Calculate the discharge of coefficient if discharge is 0.98 **[Ans.** 63.93 lit/s]

6. A venturimeter has its axis vertical, the inlet and throat diameters being 150 mm and 75 mm respectively. The throat is 0.225 m above inlet and venturimeter constant is 0.96. A liquid of specific gravity 0.78 flows up through the meter at a rate of 0.029 m^3/s. Calculate the pressure difference between inlet and throat. **[Ans.** 18.9 kPa]

7. A venturimeter is used to measure the flow of a liquid in a pipe line inclined at 35° to the horizontal. The specific gravity of liquid is 0.81 and throat area ratio is 4. The differential mercury manometer reading is 50 mm. Calculate the rate of flow if pipe diameter is 300 mm. Take venturimeter constant = 0.975. **[Ans.** 0.07 m^3/s]

8. A 0.3 m diameter, 150° bend discharges 0.35 m^3/s of water in the atmosphere. If the pressure of water entering the bend is 150 kPa (gauge), Calculate the force required to hold the bend in place. Assume the bend to be in horizontal plane. **[Ans.** 10.86 kN, 4.57°]

9. A pipeline 0.6 m diameter, carrying oil of specific gravity 0.85 at the flow rate of 1.8 m^3/s has a bend of 90° in the horizontal plane. The pressure at entrance to bend is 1.471 bar and the loss of head in the bend is 2 m of oil. Find the magnitude and direction of force exerted by the oil on the bend. **[Ans.** 69.357 kN, 42.3°]

10. An orifice 60 mm in diameter is discharging water under a head of 9 m. If $C_d = 0.6$ and $C_v = 0.9$, Calculate:

    (a) Actual discharge, and (b) Actual velocity of jet at vena-contracta

    [Ans. 22.54 lit/s, 11.26 m/s]

11. A jet of water issues from a circular orifice of 25 mm diameter under a constant head of 1.0 m. It falls 35 mm vertically down and strikes the ground at a distance of 350 mm from the centre of vena-contracta. If the discharge throgh the jet is 1.35 lit/s, find: (a) Coefficient of discharge, and (b) coefficient of velocity **[Ans.** 0.625, 0.935]

12. A tank has two identical orifices in one of its vertical sides. The upper orifice is 2m below the water surface and the lower one is 4 m below the water surface. If $C_v = 0.9$ for each orifice, find the point of intersection of the two jets. **[Ans.** 5.1 m]

13. A rectangular orifice 1.5 m wide and 1.0 m deep is discharging water from a tank. If the water level in the tank is 3 m above the top edge of the orifice, find the discharge through the orifice. Take $C_d = 0.6$. **[Ans.** 7.45 m^3/s]

14. A circular tank of diameter 4 m contains water upto a height of 5m. The tank is provided with an orifice of diameter 0.5 m at the bottom. Calculate the time taken by water to fall from 5 m to 2m, and for completely emptying the tank. Take $C_d = 0.6$. **[Ans.** 39.6 s, 107.7 s]

15. A hemispherical cistern of 6 m radius is full of water. It is fitted with a 75 mm diameter sharp-edged orifice at the bottom. Calculate the time required to lower the level of water in the cistern by 2 m Take $C_d = 0.6$ for the orifice. **[Ans.** 2h–18 min–42s]

16. A horizontal boiler drum 5 m long and 3 m in diameter is half full of water. It is provided with a 0.1 m diameter orifice at its bottom. Calculate the time taken to empty the boiler drum. Take $C_d = 0.6$. **[Ans.** 1070 s]

17. Find the discharge from a 0.1 m diameter external mouthpiece fitted to the side of a vessel, if the head over the mouthpiece is 4 m **[Ans.** 0.595 m^3/s]

18. An internal mouthpiece of 80 mm diameter is discharging under a constant head of 8 m. Find the discharge from the mouthpiece, when (a) the mouthpiece is running free, and (b) the mouthpiece is running full. **[Ans.** 22.26 *lps*. 31.47 *lps*]

19. A convergent-divergent mouthpiece is fitted into the vertical side of a tank containing water. Neglect losses in the convergent part and losses in the divergent part are equivalent to 0.2 times the velocity head at exit. The maximum absolute pressure head at the throat is 2.44 m of water for a barometric pressure of 760 mm of mercury. Determine the throat and exit diameters of the mouthpiece when the discharge is 4.25 lit/s for a head fo 1.52 m. **[Ans.** 20 mm, 33 mm]

20. In open channel flow, the velocity distribution is given by:

$$\frac{u}{U} = \left(\frac{y}{d}\right)^{1/4}$$

where U = velocity at the water surface

D = depth of the channel

y = depth of water measured from the channel bottom.

Determine the kinetic energy correction factor and momentum correction factor.

[Ans. $\frac{125}{112}, \frac{25}{24}$]

21. For turbulent flow of an incompressible fluid through a circular pipe of radius R, the velocity distribution is given by:

$$u = U\left[1 - \frac{r}{R}\right]^{1/7}$$

where u = velocity at radius r

U = velocity on the pipe axis.

Determine the kinetic energy correction factor. **[Ans.** 1.06]

5

COMPREESSIBLE FLUID FLOW

5.1 INTRODUCTION

A fluid is said to be compressible if its density changes with a change in pressure or temperature or velocity. All real fluids are compressible to some extent. If the relative change in density $\frac{d\rho}{\rho}$ is small, the fluid can be treated as incompressible. When the pressure changes produced by flow are large, the assumption of fluid being incompressible is no longer valid and the corresponding changes in density must be considered.

5.2 THERMODYNAMIC CONCEPTS

1. Equation of state.

$$pV = mRT$$

or

$$p = \rho RT \qquad \text{... (5.1)}$$

where density, $\rho = \frac{m}{V}$, V = volume

Taking log of both sides, we get

$$\ln p = \ln \rho + \ln R + \ln T$$

Differentiating, we get

$$\frac{dp}{p} = \frac{d\rho}{\rho} + \frac{dR}{R} + \frac{dT}{T}$$

Since R = const., therefore, $dR = 0$.

$$\therefore \quad \frac{dp}{p} - \frac{d\rho}{\rho} - \frac{dT}{T} = 0 \qquad \text{... (5.2)}$$

2. Internal energy and enthalpy.

Enthalpy, $$h = u + pv = u + \frac{p}{\rho} \quad \text{(per unit mass)} \qquad \text{... (5.3)}$$

where u = internal energy per unit mass.

3. Specific heats

$$c_v = \left(\frac{dq}{dT}\right)_{V=\text{const}} \qquad \text{... (5.4 a)}$$

$$c_p = \left(\frac{dq}{dT}\right)_{p=\text{const}} \qquad \text{... (5.4 b)}$$

Ratio of specific heats, $$k = \frac{c_p}{c_v} \qquad \text{... (5.5)}$$

4. Thermodynamic processes.

(i) Isothermal process.

$$\frac{p}{\rho} = \text{cont.} \qquad \text{... (5.6)}$$

(ii) Reversible Adiabatic (or isentropic) process.

$$\frac{p}{\rho^k} = \text{const.} \qquad \text{... (5.7)}$$

(iii) Polytropic process.

$$\frac{p}{\rho^n} = \text{const.} \qquad \text{... (5.8)}$$

(a) If $n = 0$, $p =$ const., the process is isobaric

(b) if $n = 1$, $T =$ const., the process is isothermal

(c) if $n = k$, Entropy = const., the process is isentropic.

5.3 BASIC EQUATIONS GOVERNING COMPRESSIBLE FLUID FLOW

The basic equations governing compressible fluid flow are:

1. Continuity equation.
2. Momentum equation.
3. Energy equation.
4. Equation of state.

1. Continuity equation

The continuity equation in Cartesian coordinates for compressible fluid flow may be expressed as,

$$\frac{\partial \rho}{\partial t}+\frac{\partial}{\partial x}(\rho u)+\frac{\partial}{\partial y}(\partial v)+\frac{\partial}{\partial z}(\rho w) = 0 \quad \text{... (5.9)}$$

The mass rate of flow along a narrow stream tube may be expressed as,

$$\rho A v = \text{const} \quad \text{... (5.10)}$$

where v = velocity.

Taking log of both sides, we get

ln $\ln \rho + \ln A + \ln v = \ln \text{const.}$

Differentiating, we get

$$\frac{d\rho}{\rho}+\frac{dA}{A}+\frac{dv}{v} = 0 \quad \text{...(5.11)}$$

Eq. (5.11) is the continuity equation in differential form.

2. Momentum equation

Momentum flux = mass flux × velocity = $\rho A v \times v$ ... (5.12)

But mass flux, $\rho A v$ = const., by continuty equation.

Thus the momentum equation is independent of the compressibility effects.

3. Energy equation

The Bernoulli's equation for incompressible fluids is given by,

$$\frac{p}{\rho}+\frac{v^2}{2}+gz = \text{const.} \quad \text{... (5.13)}$$

As the flow of compressible fluid is steady, Eq. (5.13) can be written as:

$$\frac{dp}{\rho}+v.dv+g.dz = 0 \quad \text{... (5.14)}$$

Integrating both sides, we get

$$\int\frac{dp}{\rho}+\int v.dv+\int g.dz = \text{const.}$$

or

$$\int\frac{dp}{\rho}+\frac{v^2}{2}+gz = \text{const.} \quad \text{... (5.15)}$$

(a) Energy equation for isohermal process

For isothermal process, $\qquad pV = \text{const.} \quad \text{or} \quad \dfrac{p}{\rho} = \text{const.}\ C_1$ (say)

where V = specific volume $= \dfrac{1}{\rho}$

$$\therefore \qquad \rho = \frac{p}{C_1}$$

Hence
$$\int \frac{dp}{\rho} = \int \frac{C_1}{p} dp = C_1 \ln p = \frac{p}{\rho} \ln p$$

Eq. (5.15) becomes,

$$\frac{p}{\rho} \ln p + \frac{v^2}{2} + gz = \text{const.}$$

Dividing both sides by g, we get

$$\frac{p}{\rho g} \ln p + \frac{v^2}{2g} + z = \text{const.} \qquad \text{... (5.16)}$$

(b) Energy equation for adiabatic process.

Let
$$\frac{p}{\rho^k} = C_2 \quad \text{or} \quad \rho^k = \frac{p}{C_2} \text{ or } \rho = \left(\frac{p}{C_2}\right)^{1/k}$$

Hence
$$\int \frac{dp}{\rho} = \int \left(\frac{C_2}{p}\right)^{1/k} dp = C_2^{1/k} \int \frac{dp}{p^{1/k}} = C_2^{1/k} \int p^{-1/k} \cdot dp$$

$$= C_2^{1/k} \frac{p^{\frac{k-1}{k}}}{\left(\frac{k-1}{k}\right)} = \left(\frac{k}{k-1}\right) C_2^{1/k} p^{\left(\frac{k-1}{k}\right)}$$

$$= \left(\frac{k}{k-1}\right) \cdot \frac{p^{1/k}}{\rho} \cdot \rho^{\left(\frac{k-1}{k}\right)} = \left(\frac{k}{k-1}\right) \frac{p}{\rho}$$

Eq. (5.15) become,

$$\left(\frac{k}{k-1}\right) \frac{p}{\rho} + \frac{v^2}{2} + gz = \text{const.}$$

or
$$\left(\frac{k}{k-1}\right) \frac{p}{\rho g} + \frac{v^2}{2g} + z = \text{const.} \qquad \text{... (5.17)}$$

Example 5.1 *A gas is flowing through a horizontal pipe. On a section where cross-sectional area is 40 cm², the pressure and temperature are 2.5 bar (gauge) and 17°C respectively. At another section the area of cross-section is 20 cm² and pressure is 2 bar (gauge). The mass flow rate of gas is 0.5 kg/s through the pipe. Calculate the velocities of gas at these sections, assuming isothermal process. Take R = 0.287 kJ/kg.K and atmospheric pressure = 1 bar.*

Solution. $P_{abs} = p_g + p_a$

Given: $A_1 = 40$ cm², $p_1 = 3.5$ bar, $T_1 = 290$ K, $A_2 = 20$ cm², $p_2 = 3$ bar, $\dot{m} = 0.5$ kg/s

$R = 0.287$ kJ/kg. K, $p_a = 1$ bar, $T_2 = T_1$

Characterstic equation:

$$p = \rho RT$$

$$p_1 = \rho_1 RT_1, \rho_1 = \frac{p_1}{RT_1} = \frac{3.5 \times 10^5}{0.287 \times 10^3 \times 290} = 4.205 \text{ kg/m}^3$$

$$\dot{m} = \rho_1 A_1 v_1$$

$$v_1 = \frac{0.5}{4.205 \times 40 \times 10^{-4}} = 29.72 \text{ m/s}$$

$$\rho_2 = \frac{p_2}{RT_2} = \frac{3 \times 10^5}{0.287 \times 10^3 \times 290} = 3.604 \text{ kg/m}^3$$

$$v_2 = \frac{0.5}{3.604 \times 20 \times 10^{-4}} = 69.37 \text{ m/s}$$

Example 5.2 *A gas is flowing in a horizontal pipe at a temperature of 5°C as shown in Fig. 5.1. The pressure at sections 1 and 2 are 5 bar (gauge) and 4 bar (gauge) respectively. Calculate the velocities of gas at these sections. Take R = 0.287 kJ/kg.K and p_a = 1 bar.*

Solution.

Given: $T_1 = 273 + 5 = 278$ k, $p_1 = 5 + 1 = 6$ bar,

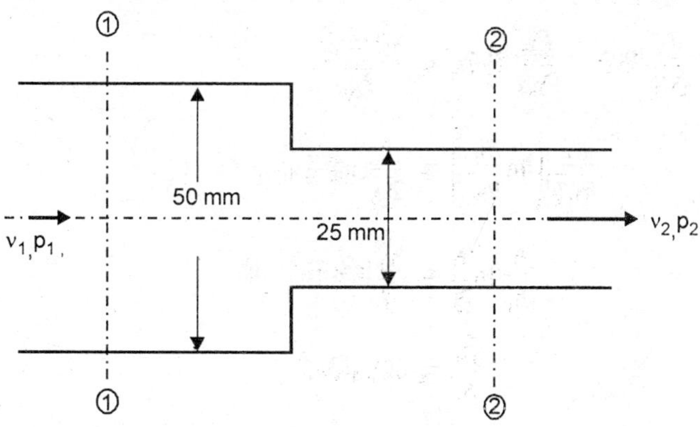

Fig. 5.1

$p_2 = 4 + 1 = 5$ bar, $R = 0.287$ kJ/kg. K

$p_a = 1$ bar, $d_1 = 50$ mm, $d_2 = 25$ mm

$$A_1 = \frac{\pi}{4} d_1^2 = \frac{\pi}{4}(0.050)^2 = 1.9635 \times 10^{-3} \text{m}^2,$$

$$A_2 = \frac{\pi}{4} d_2^2 = \frac{\pi}{4}(0.025)^2 = 4.9087 \times 10^{-4} \text{m}^2$$

Continuity equation:

$$\rho_1 A_1 v_1 = \rho_2 A_2 v_2$$

$$\frac{v_2}{v_1} = \frac{\rho_1 A_1}{\rho_2 A_2} = \frac{\rho_1}{\rho_2}\left(\frac{d_1}{d_2}\right)^2 = \frac{\rho_1}{\rho_2}\left(\frac{50}{25}\right)^2 = \frac{4\rho_1}{\rho_2} \quad \text{... (1)}$$

For isothermal process, we have

$$\frac{p_1}{\rho_1} = \frac{p_2}{\rho_2}, \frac{\rho_1}{\rho_2} = \frac{p_1}{p_2} = \frac{6}{5} = 1.2$$

Substituting the value of $\frac{\rho_1}{\rho_2}$ in Eq. (1), we get

$$\frac{v_2}{v_1} = 4 \times 1.2 = 4.8$$

Applying energy equation at sections 1 and 2 for isothermal process, we have

$$\frac{p_1}{\rho_1 g} \ln p_1 + \frac{v_1^2}{2g} + z_1 = \frac{p_2}{\rho_2 g} \ln p_2 + \frac{v_2^2}{2g} + z_2$$

Here $z_1 = z_2$, since the pipe is horizontal.

$$\frac{p_1}{\rho_1 g} \ln p_1 - \frac{p_2}{\rho_2 g} \ln p_2 = \frac{v_2^2 - v_1^2}{2g}$$

$$\frac{p_1}{\rho_1 g}\left(\ln \frac{p_1}{p_2}\right) = \frac{1}{2g}(v_2^2 - v_1^2)$$

$$\frac{p_1}{\rho_1} \ln \frac{6}{5} = \frac{v_1^2}{2}[(4.8)^2 - 1]$$

$$\frac{p_1}{\rho_1} = 60.443 v_1^2 \quad \text{... (3)}$$

Equation of state:

$$\frac{p_1}{\rho_1} = RT_1 = 0.287 \times 10^3 \times 278 = 79{,}786$$

$$\therefore \quad 79{,}786 = 60.443\ v_1^2$$

$$v_1 = 36.33 \text{ m/s}$$

$$v_2 = 4.8 \times 36.33 = 174.4 \text{ m/s}$$

Example 5.3 *A gas is flowing through a horizontal pipe with a velocity of 300 m/s. At a section the pressure and temperature are 60 kPa absolute and 30°C. The pipe changes in diameter and at this section the pressure is 90 kPa absolute. Calculate the velocity of gas at this section if the flow of gas is adiabatic. Take R = 0.287 kJ/kg.K, and k =1.4.*

Solution.

Given: $v_1 = 300$ m/s, $p_1 = 60$ kPa, $T_1 = 303$ K, $p_2 = 90$ kPa, R = 0.287 kJ/kg.K, $k = 1.4$.

Applying energy equation at sections, 1 and 2 for adiabatic process, we have

$$\left(\frac{k}{k-1}\right)\frac{p_1}{\rho_1 g} + \frac{v_1^2}{2g} + z_1 = \left(\frac{k}{k-1}\right)\frac{p_2}{\rho_2 g} + \frac{v_2^2}{2g} + z_2$$

For a horizontal pipe, $z_1 = z_2$.

$$\left(\frac{k}{k-1}\right)\left(\frac{p_1}{\rho_1} - \frac{p_2}{\rho_2}\right) = \frac{1}{2}(v_2^2 - v_1^2)$$

$$\left(\frac{k}{k-1}\right)\frac{p_1}{\rho_1}\left[1 - \frac{p_2}{p_1} \times \frac{\rho_1}{\rho_2}\right] = \frac{1}{2}(v_2^2 - v_1^2)$$

For an adiabatic process, $$\frac{p_1}{\rho_1^k} = \frac{p_2}{\rho_2}, \frac{\rho_1}{\rho_2} = \left(\frac{p_1}{p_2}\right)^{1/k}$$

$$\therefore \quad \left(\frac{k}{k-1}\right)\frac{p_1}{\rho_1}\left[1 - \frac{p_2}{p_1} \times \left(\frac{p_1}{p_2}\right)^{1/k}\right] = \frac{1}{2}(v_2^2 - v_1^2)$$

$$\left(\frac{k}{k-1}\right)\frac{p_1}{\rho_1}\left[1 - \left(\frac{p_2}{p_1}\right)^{\frac{k-1}{k}}\right] = \frac{1}{2}(v_2^2 - v_2^2)$$

Now $$\frac{p_1}{\rho_1} = RT_1 = 287 \times 303 = 86{,}961$$

$$\frac{p_2}{p_1} = \frac{90}{60} = 1.5 \quad \text{and} \quad k = 1.4, \quad \frac{k}{k-1} = \frac{1.4}{0.4} = 3.5$$

$$3.5 \times 86{,}961 \times [1 - (1.5)^{1/3.5}] = \frac{1}{2}[v_2^2 - (300)^2]$$

$$-37{,}383.2 = \frac{1}{2}(v_2^2 - 90000)$$

$$v_2 = 123.4 \text{ m/s}$$

5.4 PROPAGATION OF ELASTIC WAVES IN FLUIDS

In fluids the molecules are relatively apart as compared to the molecules in solids. Due to distrubance the fluid molecules change their position. The propagation of disturbance depends upon the elastic properties of the fluid. The velocity of disturbance depens upon the changes of pressure and density of the fluid.

The propagation of elastic waves in fluids is similar to the propagation of sound waves through a media. The speed of propagation of sound waves in a media is known as acoustic (or sonic) velocity.

5.4.1 Sonic Velocity

Consider a one-dimensional flow through a long straight rigid pipe of uniform cross-sectional area fitted with a frictionless piston at one end, as shown in Fig. 5.2. The pipe is filled with a compressible fluid initially at rest. If the piston is moved suddenly to the right with velocity v, a pressure wave would be propagated through the fluid with a velocity of sound wave.

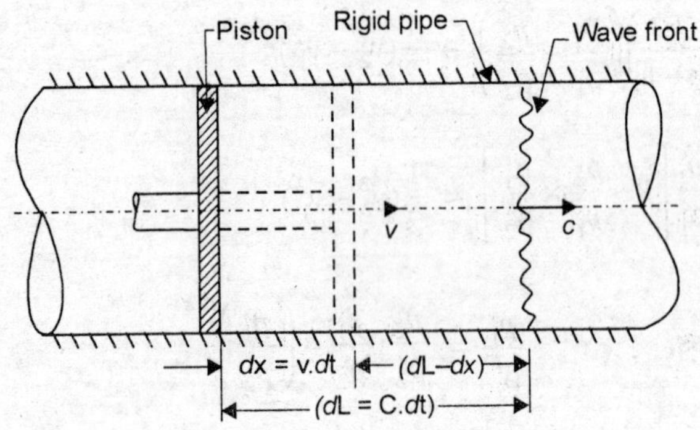

Fig. 5.2 One-dimensional pressure wave propagation

Let A = area of cross-sectin of pipe

v = velocity of piston

c = sonic velocity

p = fluid pressure in the pipe initially

ρ = fluid density initially

dt = elementary time interval for piston movement

Mass of the fluid in length dL and and density ρ before the movement of the piston

$$= \rho \times A \times dL$$

When the piston moves through a distance dx, the fluid density in the compressed region of length $(dL - dx)$ shall be increased to $(\rho + d\rho)$. The mass of the fluid becomes

$$= (\rho + d\rho) \times A \times (dL - dx)$$

Using continuity equation, we have

$$\rho \times A \times dL = (\rho + d\rho) \times A \times (dL - dx)$$

Now $$dL = c \times dt \quad \text{and } dx = v \times dt$$

$$\rho \times A \times c \times dt = (\rho + d\rho) \times A \times (c \times dt - v \times dt)$$

$$\rho c = (\rho + d\rho)(c - v)$$

$$= \rho c + c \times d\rho - v\rho - v \times d\rho$$

∴ $$0 = c \times d\rho - vp - v \times dp$$

Neglecting the term $v \times dp$ as compared to $c \times d\rho$, we get

$$c \times d\rho = v\rho$$

$$c = v\frac{\rho}{d\rho} \qquad \text{... (5.18)}$$

Applying impulse-momentum equation to the fluid in the compressed region for time dt, we get

$$dp \times A \times dt = \rho \times dL \times A(v - 0)$$

or $$dp = \rho \times \frac{dL}{dt} \times v = \rho \times \frac{c \times dt}{dt} \times v = \rho c v$$

or $$c = \frac{dp}{\rho v} \qquad \text{... (5.19)}$$

Multiplying Eqs. (5.18) and (5.19), we get

$$c^2 = v \times \frac{\rho}{d\rho} \times \frac{dp}{\rho v} = \frac{dp}{d\rho}$$

$$c = \sqrt{\frac{dp}{d\rho}} \qquad \text{... (5.20)}$$

5.4.2 Sonic Velocity in term of Bulk Modulus

Bulk modulus of elasticity of a fluid (K) is defined as:

$$K = \frac{dp}{-\left(\frac{dV}{V}\right)}$$

where V = original volume, and dV = decrease in volume.

-ve sign indicates that volume decreases as pressure increases.

Alos $$V\rho = \text{const.}$$

Differentiating both sides, we get

$$V \cdot d\rho + \rho \cdot dV = 0$$

or $$-\frac{dV}{V} = \frac{d\rho}{\rho}$$

Eq. (1) becomes, $$K = \frac{dp}{\left(\frac{d\rho}{\rho}\right)}$$

or $$\frac{dp}{d\rho} = \frac{K}{\rho}$$

Eq. (5.20) then becomes, $$c = \sqrt{\frac{K}{\rho}} \quad \text{... (5.21)}$$

5.4.3 Sonic Velocity for Isothermal process

For an isothermal process, we have

$$\frac{p}{\rho} = \text{const.}$$

Differenting both sides, we get

$$\frac{\rho \cdot dp - p \cdot d\rho}{\rho^2} = 0$$

or $$\frac{dp}{d\rho} = \frac{p}{\rho} = RT$$

$\therefore$ $$c = \sqrt{\frac{p}{\rho}} = \sqrt{RT} \quad \text{... (5.22)}$$

5.4.4 Sonic velocity for Adiabatic process

For reversible adiabatic (isentropic) process, we have

$$\frac{p}{\rho^k} = \text{const.}$$

Differentiating both sides, we have

$$p(-k)\rho^{-k-1} . d\rho + \rho^{-k} \cdot dp = 0$$

Dividing by ρ^{-k}, we get

$$-pk\rho^{-1} \cdot d\rho + dp = 0$$

$$\frac{dp}{d\rho} = \frac{pk}{\rho} = k\,RT \qquad \left[\because \frac{p}{\rho} = RT\right]$$

$$c = \sqrt{\frac{dp}{d\rho}} = \sqrt{kRT} \qquad \text{... (5.23)}$$

where $k = \frac{c_p}{c_v}$

5.5 MACH NUMBER

The Mach number is defined as the ratio of free stream velocity and the correponding sonic velocity in the fluid medium. It is expressed as:

$$M = \frac{v}{c}$$

The mach number is a measure of the fluid compressibility. It may also be defined as the square root of the ratio of inertia force of a fluid to the elastic force.

$$M = \sqrt{\frac{\text{Inertia force}}{\text{Elastic force}}} = \sqrt{\frac{\rho A v^2}{KA}} = \sqrt{\frac{v^2}{(K/\rho)}} = \frac{v}{c}$$

5.5.1 Compressible Flow Regimes

Based on the value of mach number, flow regimes are generally defined as:

1. Incompressible flow regime: $0 < M < 0.3$. In this range the effects of compressiblity are considered negligible.
2. Subsonic flow regime: $0.3 < M < 1.0$.
3. Transonic flow regime: $0.9 < M < 1.1$.
4. Sonic flow: $M = 1.0$.
5. Supersonic flow regime: $M > 1$, upto $\simeq 5$.
6. Hypersonic flow regime: $M >> 1$.

Example 5.4 *Calculate the sonic velocity for the following fluid:*

(a) Crude oil of specific gravity 0.85 and bulk modulus of 1.6 GPa.

(b) Mercury having a bulk modulus of 26.5 GPa.

Solution.

(a) $c = \sqrt{\frac{K}{\rho}} = \sqrt{\frac{1.6 \times 10^9}{0.85 \times 10^3}} = 1372 \text{ m/s}$

(b) $c = \sqrt{\frac{26.5 \times 10^9}{13.6 \times 10^3}} = 1395.9 \text{ m/s}$

Example 5.5. *An aircraft is flying at a height of 15 km where temperature is –50°C. The speed of the aircraft corresponds to M = 2. Find the speed of the aircraft taking R = 287 J/kg k and K = 1.4.*

Solution. $T = 273 - 50 = 223$ k

$$c = \sqrt{kRT} = \sqrt{1.4 \times 287 \times 223} = 299.335 \text{ m/s}$$

$$M = \frac{v}{c} = 2$$

Speed of aircraft ,

$$v = 2\times 299.335 = 598.67 \text{ m/s or } 2155.2 \text{ km/h}$$

Example 5.6 *A supersonic aircraft flies at an altitude of 2 km where the air temperature is 3°C. Calculate the speed of the aircraft if its sound is heard 5 seconds after passage over the head of the observer.*

Solution. Let O be the position of observer and A the position of the aircraft just vertically over the observer. After 5 seconds, the aircraft reaches a position B, as shown in Fig. 5.3 OB represents the wavefront and $\angle ABO = \alpha$ = Mach angle.

$T = 273 + 3 = 276$ K, $R = 287$ J/kg.K, $k = 1.4$.

Sonic velocity, $c = \sqrt{kRT} = \sqrt{1.4 \times 287 \times 276} = 333$ m/s

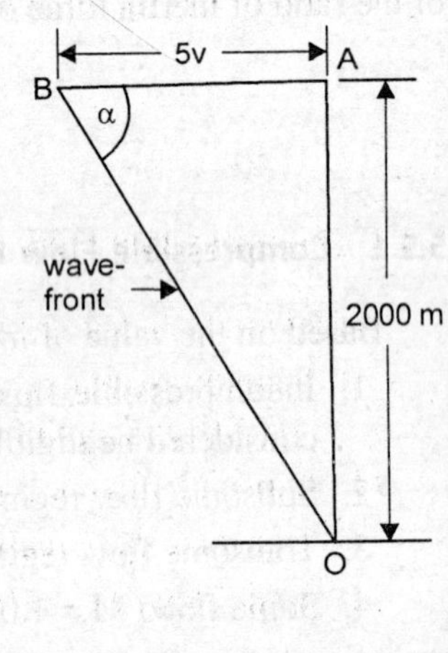

$$\tan \alpha = \frac{2000}{5v}$$

$$M = \frac{v}{c} = \frac{1}{\sin \alpha}$$

$$v = \frac{c}{\sin \alpha}$$

$$\frac{\sin \alpha}{\cos \alpha} = \tan \alpha = \frac{2000 \sin \alpha}{5 \times c} = \frac{2000 \sin \alpha}{5 \times 333} = 1.2012 \sin \alpha$$

$\therefore$ $\cos \alpha = 0.8325$

$\alpha = 33.64°$

$$\text{Speed of aircraft}, v = \frac{333}{\sin 33.64°} = 601 \text{ m/s}$$

5.6 PROPAGATION OF ELASTIC WAVES DUE TO DISTURBANCE IN FLUID

Consider a small projectile moving from right to left in a stationary fluid at a velocity smaller than the velocity of sound ($0 < v < c$). The motion of the projectile creates pressure disturbances which are propaged spherically outward with the sonic speed. If the projectile was stationary, the wave fronts would speed out spherically, as shown in Fig 5.4(a), for successive time intervals, $dt = (t_2 - t_1) = (t_3 - t_2)$.

Case I: When $M < 1$ (i.e $0 < v < c$)

In this case, since $v < c$, the projectile lags behind the disturbance (or pressure wave) so that at $t = 0$ lies inside the sphere of radius ct and also inside other spheres formed by the waves started at intermediate points, shown in Fig. 5.4 (b).

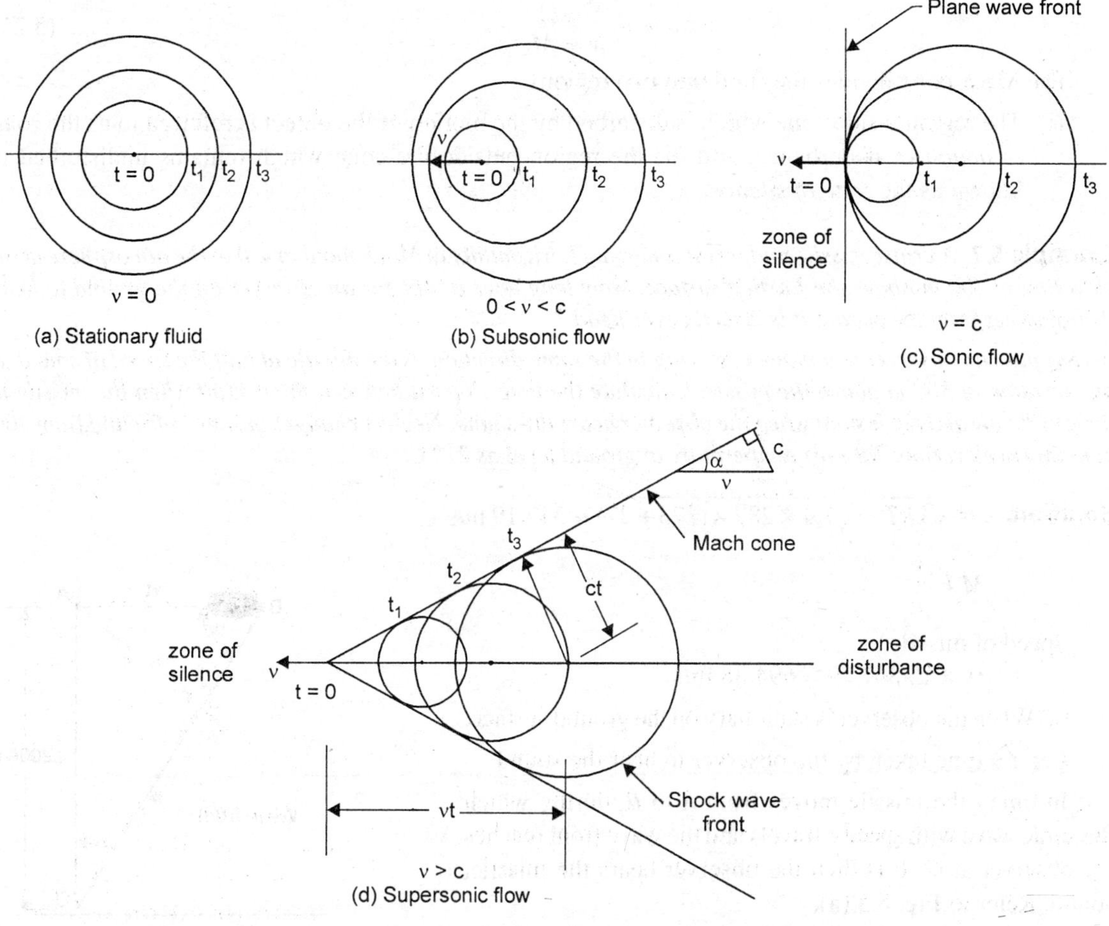

Fig. 5.4 Wave patterns for a small object moving with different velocities

CaseII: When M = 1 (i.e v = c)

In this case, the waves do not move ahead of the object but appear to remain stationary (see Fi.g 5.4 c). The wave fronts merge to form a plane wave front which is tangential to the upstream side shown, and the downstream wave moves at a speed of $(v + c)$. In This case, the pressure wave is unable to move upstream against the oncoming flow and the fluid ahead of this plane wave front is not influenced by the motion of the object.

CaseIII: When M> 1 (i.e. v > c)

In this case, the individual pressure waves combine to form a conical wave front, as shown in Fig. 5.4 (d). This conical pattern of wave fronts is known as the *mach cone*. The fluid ahead of the cone is undisturbed but suddenly undergoes changes in pressure, temperature and density, as it passes through the Mach cone. An actual wave through which the flow experiences these sudden changes is called a *shock wave.* The semi-vertex angle of Mach cone is known as the *Mach angle,* which is given by:

$$\sin \alpha = \frac{c}{v} = \frac{1}{M} \qquad \text{... (5.25)}$$

The Mach cone divides the fluid into two regions:

(i) The region inside cone which is disturbed by the motion of the object is referred to as the *zone of action* or disturbance, and (ii) the region outside the cone which remains undisturbed is known as the *zone of silence*.

Example 5.7 *A cruise missile under test is moving horizontally at Mach number 2.0 in the atmosphere at an elevation of 300 m above the Earth's surface. How long does it take for an observer on the ground to hear distrubances from the instant it is directly over head ?*

(b) Suppose an observer in a plane is moving in the same direction as the missile at half the speed of sound at an elevation of 300 m above the missle. Calculate the time elapsed between the instant when the missile is directly below and the instant when the observer hears the sound. Neglect changes in speed of sound from 300 m to 600 m elevation. Take air temperature at ground level as 27° C.

Solution. $c = \sqrt{kRT} = \sqrt{1.4 \times 287 \times (273 + 27)} = 347.19$ m/s

$$M = \frac{v}{c}$$

Speed of missile,

$$v = 2 \times 347.19 = 694.38 \text{ m/s}$$

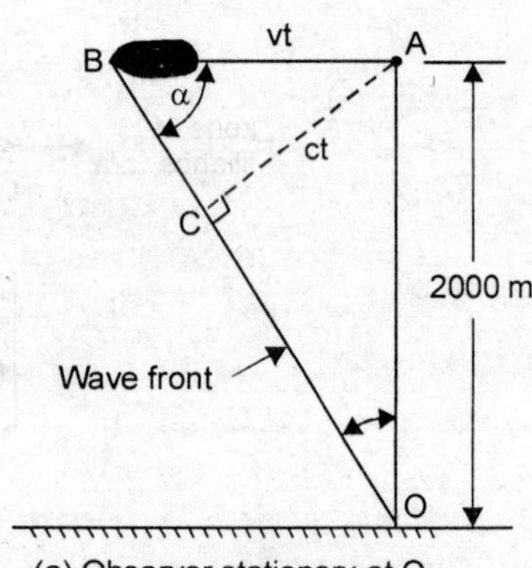

(a) Observer stationary at O

(a) When the observer is stationary on the ground surface.

Let t = time taken by the observer to hear the sound.

In time t the missile moves from A to B, during which the sonic wave with speed c travels and the wave front reaches the observer at O. It is then the observer hears the missile, sound. Refer to Fig. 5.5 (a).

In ΔABC, we have

$$\sin \alpha = \frac{AC}{AB} = \frac{ct}{vt} = \frac{c}{v} = \frac{347.19}{644.38} = 0.5$$

$$\alpha = 30°$$

$$\angle AOB = 60°$$

$$\sin 60° = \frac{AC}{OA} = \frac{ct}{300} = 0.866$$

$$t = \frac{300 \times 0.866}{347.19} = 0.748 \text{ s}$$

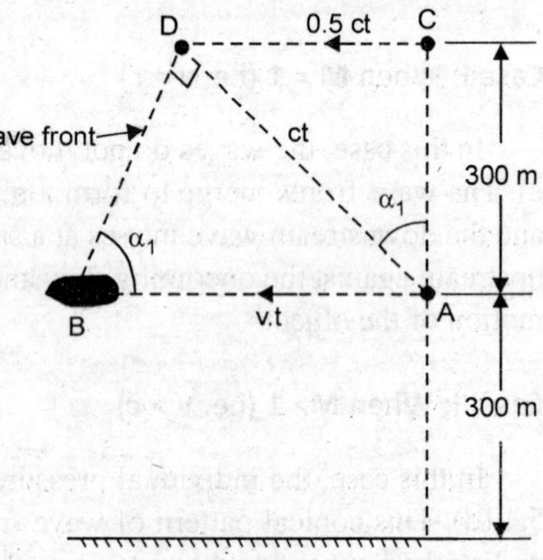

(b) Observer moving in a plane at 600 m elevation

Fig. 5.5

(b) When the observer is moving in a plane in the same direction as the missile.

Fig. 5.5 (b) shows the position of the observer C travelling in a plane and the missile is at A, directly below C. In time t, the wave front reaches the observer as both move to their new positions D and B respectively. From ΔABD, we have

$$\sin \alpha_1 = \frac{AD}{AB} = \frac{ct}{vt} = \frac{1}{2} = 0.5$$

$$\alpha_1 = 30°, \angle CAD = \alpha_1 = 30°$$

From ΔCAD, we have

$$\tan \alpha_1 = \frac{CD}{AC} = \frac{0.5\,ct}{300}$$

$$t = \frac{300 \tan 30°}{0.5c} = \frac{300}{\sqrt{3} \times 0.5 \times 347.19} = 0.998\,s$$

5.7 STAGNATION PROPERTIES

A point on an immersed body in a fluid flow, where the velocity is zero, is called the *stagnation point.* At the stagnation point, the velocity head is converted into pressure head. In a compressible fluid flow, since the pressure, temperature and density are inter-related, a change in pressure also, affects the temperature. At a stagnation point where the velocity is zero, the pressure is called the *stagnation pressure* and the temperature there is called the *stagnation temperature.*

5.7.1 Expression for stagnation pressure

Consider the compressible flow past a body in which the velocity becomes zero at some point. Let us consider two points 0 and 1 in the flow such that the point 0 lies on the body where the velocity is zero.

Let p_0, v_0, ρ_0, T_0 = pressure velocity, density and temperature at the stagnation point 0.

p_1, v_1, ρ_1, T_1 = corresponding quantities at point 1.

Applying Bernoulli's equation at points 0 and 1 for isentropic flow, we have

$$\left(\frac{k}{k-1}\right)\frac{p_0}{\rho_0 g} + \frac{v_0^2}{2g} + z_0 = \left(\frac{k}{k-1}\right)\frac{p_1}{\rho_1 g} + \frac{v_1^2}{2g} + z_1$$

For $z_0 = z_1$, we have

$$\left(\frac{k}{k-1}\right)\frac{p_0}{\rho_0 g} + \frac{v_0^2}{2g} = \left(\frac{k}{k-1}\right)\frac{p_1}{\rho_1 g} + \frac{v_1^2}{2g}$$

or

$$\left(\frac{k}{k-1}\right)\frac{p_0}{\rho_0} + \frac{v_0^2}{2} = \left(\frac{k}{k-1}\right)\frac{p_1}{\rho_1} + \frac{v_1^2}{2}$$

At point 0, $v_0 = 0$, the above equation becomes,

$$\left(\frac{k}{k-1}\right)\left(\frac{p_0}{\rho_0} - \frac{p_1}{\rho_1}\right) = \frac{v_1^2}{2}$$

or $$\left(\frac{k}{k-1}\right)\frac{p_1}{\rho_1}\left(1-\frac{p_0}{\rho_1}\times\frac{\rho_1}{\rho_0}\right) = -\frac{v_1^2}{2} \quad \text{... (1)}$$

For adiabatic process, we have

$$\frac{p_1}{\rho_1^k} = \frac{p_0}{\rho_0^k} \text{ or } \frac{p_1}{p_0} = \frac{\rho_1^k}{\rho_0^k} = \left(\frac{\rho_1}{\rho_0}\right)^k$$

or $$\frac{\rho_1}{\rho_0} = \left(\frac{p_1}{p_0}\right)^{1/k} \quad \text{... (2)}$$

Substituting the value of $\frac{\rho_1}{\rho_0}$ from Eq. (2) in (1), we get

$$\left(\frac{k}{k-1}\right)\frac{p_1}{\rho_1}\left[1-\frac{p_0}{p_1}\times\left(\frac{p_1}{p_0}\right)^{1/k}\right] = -\frac{v_1^2}{2}$$

$$\left(\frac{k}{k-1}\right)\frac{p_1}{\rho_1}\left[1-\left(\frac{p_0}{p_1}\right)^{1-\frac{1}{k}}\right] = -\frac{v_1^2}{2}$$

$$1-\left(\frac{p_0}{p_1}\right)^{\frac{k-1}{k}} = -\frac{v_1^2}{2}\left(\frac{k-1}{k}\right)\frac{\rho_1}{p_1}$$

or $$1+\frac{v_1^2}{2}\left(\frac{k-1}{k}\right)\frac{\rho_1}{p_1} = \left(\frac{p_0}{p_1}\right)^{\frac{k-1}{k}} \quad \text{... (3)}$$

For addiabatic process, sonic velocity

$$c = \sqrt{kRT} = \sqrt{k\frac{p}{\rho}} \qquad \left[\because \frac{p}{\rho} = RT\right]$$

At point 1, $$c_1 = \sqrt{k\frac{p_1}{\rho_1}}$$

or $$c_1^2 = k\frac{p_1}{\rho_1}$$

Substituting the value of $\frac{kp_1}{\rho_1}$ in Eq. (3), we get

$$1+\frac{v_1^2}{2}(k-1)\times\frac{1}{c_1^2} = \left(\frac{p_0}{p_1}\right)^{\frac{k-1}{k}}$$

$$1+\frac{v_1^2}{2c_1^2}(k-1) = \left(\frac{p_0}{p_1}\right)^{\frac{k-1}{k}}$$

$$1+\frac{M_1^2}{2}(k-1) = \left(\frac{p_0}{p_1}\right)^{\frac{k-1}{k}}$$

or

$$\left(\frac{p_0}{p_1}\right)^{\frac{k-1}{k}} = 1+\left(\frac{k-1}{2}\right)M_1^2$$

$$\frac{p_0}{p_1} = \left[1+\left(\frac{k-1}{2}\right)M_1^2\right]^{\frac{k}{k-1}}$$

$$p_0 = p_1\left[1+\left(\frac{k-1}{2}\right)M_1^2\right]^{\frac{k}{k-1}} \qquad \text{... (5.26)}$$

Eq. (5.26) gives the value of stagnation pressure.

5.7.2 Compressibility Correction Factor

Expanding R.H.S of Eq. (5.26) by Binomial theorem, we have

$$p_0 = p_1\left[1+\frac{k}{2}M_1^2+\frac{k}{8}M_1^4+\frac{k(2-k)}{48}M_1^6+...\right]$$

$$= p_1\left[1+\frac{k}{2}M_1^2\left\{1+\frac{M_1^2}{4}+\left(\frac{2-k}{24}\right)M_1^4+...\right\}\right]$$

$$= p_1+\frac{p_1k}{2}M_1^2\left\{1+\frac{M_1^2}{4}+\left(\frac{2-k}{24}\right)M_1^4+...\right\} \qquad \text{... (5.27)}$$

Now

$$M_1^2 = \left(\frac{v_1}{c_1}\right)^2 = \frac{v_1^2\rho_1}{k\,p_1} \qquad \left[\because c_1=\frac{kp_1}{\rho_1}\right]$$

Substituting the value of M_1^2 in Eq. (5.27), we get

$$p_0 = p_1 + \frac{p_1 k}{2} \times \frac{v_1^2 \rho_1}{k p_1}\left[1 + \frac{M_1^2}{4} + \left(\frac{2-k}{24}\right) M_1^4 + \ldots\right]$$

$$= p_1 + \frac{\rho_1 v_1^2}{2}\left[1 + \frac{M_1^2}{4} + \left(\frac{2-k}{24}\right) M_1^4 + \ldots\right] \qquad \ldots (5.28)$$

Also $p_0 = p_1 + \frac{1}{2}\rho_1 v_1^2$, when compressibility effects are neglected. ... (5.29)

where $1 + \frac{M_1^2}{4} + \left(\frac{2-k}{24}\right) M_1^4$ = compressibility correction factor (CCF) ... (5.30)

CCF is taken into account when M_1 exceeds a value of about 0.3

$$\text{CCF} = \frac{(p_0 - p_1)}{\left[\frac{1}{2}\rho_1 v_1^2\right]} \qquad \ldots (5.30\text{ a})$$

5.7.3 Expression for Stagnation Density

Now

$$\frac{\rho_1}{\rho_0} = \left(\frac{p_1}{p_0}\right)^{\frac{1}{k}}$$

or

$$\frac{\rho_0}{\rho_1} = \left(\frac{p_0}{p_1}\right)^{\frac{1}{k}}$$

Substituting the value of $\frac{p_0}{p_1}$ from Eq. (5.26), have

$$\frac{\rho_0}{\rho_1} = \left[1 + \left(\frac{k-1}{2}\right) M_1^2\right]^{\frac{1}{k-1}}$$

Stagnation density,

$$\rho_0 = \rho_1\left[1 + \left(\frac{k-1}{2}\right) M_1^2\right]^{\frac{1}{k-1}} \qquad \ldots (5.31)$$

5.7.4 Expression for Stagnation Temperature

The equation of state is given by:

$$\frac{p}{\rho} = RT$$

$$\frac{p_0}{\rho_0} = RT_0 \quad \text{or} \quad T_0 = \frac{p_0}{R\,\rho_0}$$

$$\therefore \qquad T_0 = \frac{1}{R}\left[\frac{p_1\left\{1+\left(\frac{k-1}{2}\right)M_1^2\right\}^{k/(k-1)}}{\rho_1\left\{1+\left(\frac{k-1}{2}\right)M_1^2\right\}^{1/(k-1)}}\right]$$

$$= \frac{p_1}{R\rho_1}\left[1+\left(\frac{k-1}{2}\right)M_1^2\right]^{\left(\frac{k}{k-1}-\frac{1}{k-1}\right)}$$

$$= \frac{p_1}{R\rho_1}\left[1+\left(\frac{k-1}{2}\right)M_1^2\right]$$

Stagnation temperature, $$T_0 = T_1\left[1+\left(\frac{k-1}{2}\right)M_1^2\right] \qquad \text{... (5.32)}$$

Example 5.8 *Air has at velocity of 1000 km/h at a pressure of 100 kPa vacuum and a temperature of 47°C. Calculate its stagnation properties and the local Mach number. Take atomospheric pressure = 10 kPa and R =287 J/kg.K Also calculate the compressibility correct factor when M = 0.8.*

Solution.

Given: $v = \frac{1000\times10^3}{3600} = 277.78$ m/s, $T = 273 + 47 = 320\ K$, $p = 100$ kPa, (vacuum),

$p_a = 10$ kPa

Sonic velocity, $$c = \sqrt{kRT} = \sqrt{1.4\times287\times320} = 358.57 \text{ m/s}$$

Mach number, $$\text{M} = \frac{v}{c} = \frac{277.78}{358.57} = 0.7747$$

$$\frac{p_0}{p} = \left[1+\left(\frac{k-1}{2}\right)M^2\right]^{k/(k-1)} = \left[1+\left(\frac{1.4-1}{2}\right)(0.7747)^2\right]^{1.4/0.4} = 1.487$$

$$p_0 = p = 1.487\,(100-10) = 133.\,87 \text{ kPa}$$

Stagnation temperature, $$T_0 = \left[1+\left(\frac{k-1}{2}\right)M^2\right] = 320\left[1+\left(\frac{1.4-1}{2}\right)(0.7747)^2\right] = 358.4\,K$$

$$\rho_0 = \frac{p_0}{RT_0} = \frac{133.87}{0.287 \times 358.4} = 1.301\,\text{kg/m}^3$$

$$\text{CCF} = \frac{p_0 - p}{\frac{1}{2}\rho v^2} = 1 + \frac{M^2}{4} + \left(\frac{2-k}{24}\right) M^4$$

$$= 1 + \frac{(0.8)^2}{4} + \left(\frac{2-1.4}{24}\right)(0.8)^4$$

$$= 1.17$$

Example 5.9 *An aeroplane is flying at 900 km/h through still air having pressure of 80 kPa (abs) and temperature –10°C. Calculate the following on the nose of the plane:*

(a) Stagnation pressure, (b) stagnation temperature, and (c) stagnation density.

Take for air : R = 0.287 kJ/kg.K, k = 1.4

Solution. Given: $v = \frac{900 \times 10^3}{3600} = 250$ m/s , $T = 273\text{-}10 = 263$ K, $p = 80$ kPa, $R = 0.287$ kJ/kg. K, $k = 1.4$.

(a) Stagnation pressure, $p_0 = p\left[1 + \left(\frac{k-1}{2}\right) M^2\right]^{\frac{k}{k-1}}$

Sonic velocity, $c = \sqrt{kRT} = \sqrt{1.4 \times 287 \times 263} = 325.07$ m/s

Mach number, $M = \frac{v}{c} = \frac{250}{325.07} = 0.769$

$$p_0 = 80\left[1 + \left(\frac{1.4-1}{2}\right)(0.769)^2\right]^{\frac{1.4}{0.4}} = 80 \times 1.4788 = 118.3\,\text{kPa}$$

(b) Stagnation temperature, $T_0 = R\left[1 + \left(\frac{k-1}{2}\right) M^2\right]$

$$= 263\left[1 + \left(\frac{1.4-1}{2}\right)(0.769)^2\right] = 294.1\,\text{K or}\,21.1°C$$

(c) Stagnation density, $\rho_0 = \frac{p_0}{RT_0} = \frac{118.3}{0.287 \times 294.1} = 1.401\,\text{kg/m}^3$

Example 5.10 *An aircraft is flying at an altitude where the pressure is 40 kPa and temperature –37°C. The stagnation pressure measured was 65 kPa. Calculate the speed of the aircraft. Take molecular weight of air as 28.*

Solution.

Given: $p = 40$ kPa, $T = 273 - 37 = 236$ K, $p_0 = 65$ kPa, $M_a = 28$

$$R = \frac{R_u}{M_a} = \frac{8314}{28} = 296.928 \text{ J/kg.K}$$

$$\rho = \frac{p}{RT} = \frac{40 \times 10^3}{296.928 \times 236} = 0.5708 \text{ kg/m}^3$$

$$p_0 = p + \frac{1}{2}\rho v^2$$

$$v = \sqrt{\frac{2(p_0 - p)}{\rho}} = \sqrt{\frac{2(65 - 40) \times 10^3}{0.5708}} = 295.96 \text{ m/s}$$

5.8 EFFECTS OF AREA CHANGES ON FLOW PROPERTIES

For a one-dimensional flow across an area of cross-section A with velocity v and constant density ρ, the continuity equation for steady flow is,

$$\rho A v = \text{const.}$$

Taking log of both sides, we have

$$\ln \rho + \ln A + \ln v = \ln \text{const.}$$

Differentiating, we have

$$\frac{d\rho}{\rho} + \frac{dA}{A} + \frac{dv}{v} = 0 \quad \ldots (1)$$

If signficant changes in cross-sectional area and, therefore, in v and ρ occur over only a short length of conduit, such as in a nozzle, frictional effects may be neglected. Euler's equation of motion for steady, frictionaless flow neglecting the gravity and body forces may be written as:

$$v dv + \frac{dp}{p} = 0 \quad \ldots (2)$$

Now sonic velocity, $c = \sqrt{\frac{dp}{d\rho}}$ or $c^2 = \frac{dp}{d\rho}$

Eq. (2) may be writen as:

$$\therefore \quad v dv + \frac{dp}{\rho} \times c^2 \times \frac{d\rho}{dp} = 0$$

$$\frac{v dv}{c^2} + \frac{d\rho}{\rho} = 0 \quad \ldots (3)$$

From Eq. (1), we have

$$\frac{d\rho}{\rho} = -\frac{dA}{A} - \frac{dv}{v}$$

Eq. (3) becomes

$$\frac{dA}{A} + \frac{dv}{v} = \frac{v\,dv}{c^2}$$

or

$$\frac{dA}{A} = \frac{dv}{v}\left(\frac{v^2}{c^2} - 1\right) = \frac{dv}{v}(M^2 - 1) \quad \text{... (5.33)}$$

Eq. (5.33) is known as the Hugoniot equation.

Also $$\frac{dv}{v} = -\frac{dp}{\rho v^2} \text{ from Eq. (2)}$$

Substituting in Eq. (1), we get

$$\frac{d\rho}{\rho} - \frac{dp}{\rho v^2} + \frac{dA}{A} = 0$$

or

$$\frac{dA}{A} = \frac{dp}{\rho v^2} - \frac{d\rho}{\rho} = \frac{dp}{\rho v^2}\left(1 - v^2\frac{d\rho}{dp}\right) = \frac{dp}{\rho v^2}\left(1 - \frac{v^2}{c^2}\right) = \frac{dp}{\rho v^2}(1 - M^2)$$

$$\therefore \quad dp = \frac{pv^2}{A}\left(\frac{1}{1-M^2}\right)dA = pv^2\left(\frac{1}{1-M^2}\right)\frac{dA}{A} \quad \text{... (5.34)}$$

5.8.1 Flow Through Ducts of Varying Area

The following conclusions of practical significance may be drawn from Eqs. (5.33) and 5.34):

1. Subsonic flow: (M < 1)

(i) For a convergent nozzle Fig. 5.6 (a)

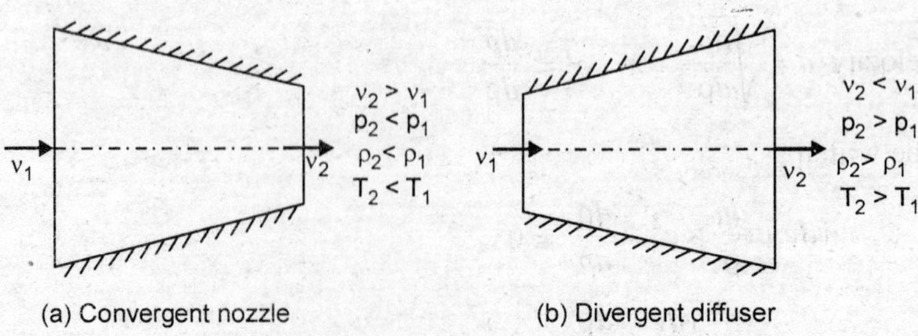

(a) Convergent nozzle (b) Divergent diffuser

Fig. 5.6 Subsonic flow (M < 1)

$$\frac{dv}{v} > 0, \frac{dA}{A} < 0, dp < 0$$

(ii) For a divergent diffuser, Fig. 5.6 (b),

$$\frac{dv}{v} < 0, \frac{dA}{A} > 0, dp > 0$$

Eq. (5.33) shows that for $M < 1$ $(M^2 - 1)$ is negative. Therefore dA and dv must be of opposite sign. In Eq. (5.34) likewise dp and dA must be of same sign.

2. Supersonic flow: (M >1)

(i) For a divergent nozzle, Fig. 5.7 (a),

$$\frac{dv}{v} > 0, \frac{dA}{A} > 0, dp < 0$$

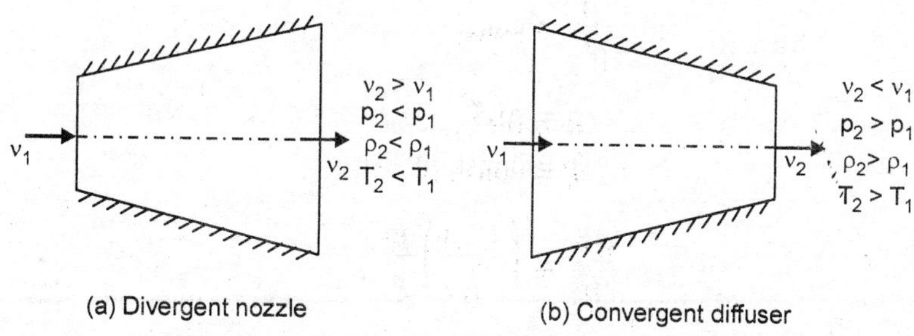

Fig. 5.7 Supersonic flow (M > 1)

(ii) For a convergent diffuser, Fig. 5.7 (b),

$$\frac{dv}{v} < 0, \frac{dA}{A} < 0, dp > 0$$

For M > 1, dA and dv are of same sign but dp and dA are of opposite sign.

3. Sonic flow (M = 1): Fig. 5.8

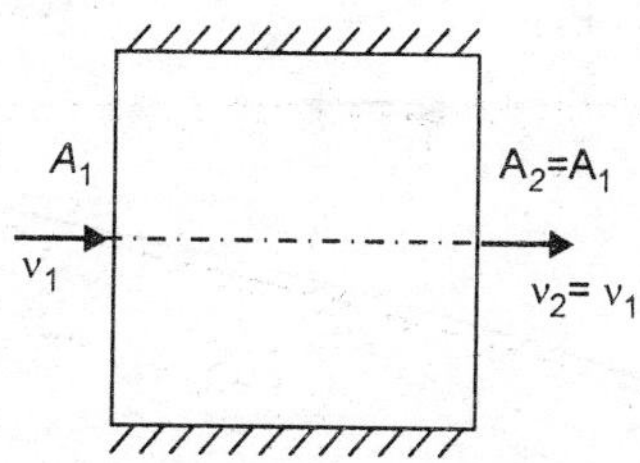

Fig. 5.6 Sonic flow

$$\frac{dA}{A} = 0$$

$dp = 0$ since $dA = 0$, the flow is frictionless.

5.8.2 Variatin of Density and Temperature with Area

Sonic Velocity, $c = \sqrt{\frac{dp}{d\rho}}$

$$d\rho = \frac{dp}{c^2} \quad \text{... (5.35)}$$

Eq. (5.35) shows that density varies in the same manner as the pressure,

Now for isentropic flow,

$$\frac{p}{\rho^k} = \text{const.}$$

Also $p = \rho RT$

$\therefore$ $T = \text{const. } p^{k-1/k}$

$$\frac{dT}{T} = \left(\frac{k-1}{k}\right)\frac{dp}{p} \quad \text{... (5.36)}$$

Eq. (5.36) shows that temperature varies in the same manner as pressure.

5.9 ISENTROPIC FLOW THROUGH A CONVERGENT NOZZLE

Consider a convergent nozzle fitted to a large vessel containing a compressible fluid as shown in Fig. 5.9. Consider two points 1 and 2 inside the vessel and exit of the nozzle. For insentropic flow from the vessel, let

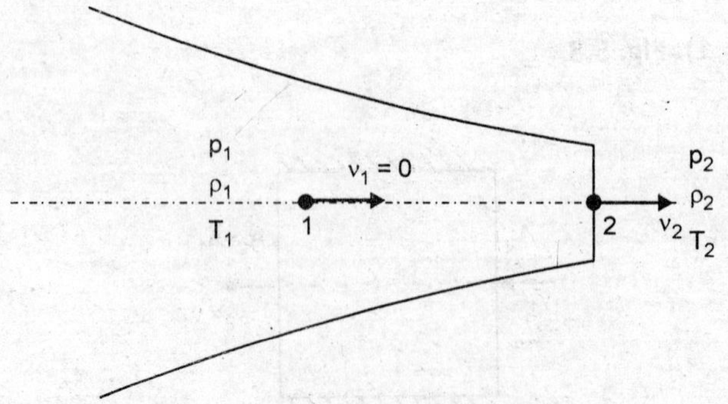

Fig. 5.9 Flow through a convergent nozzle

$p_1, v_1, \rho_1\ T_1$ = pressure, velocity, density and temperature at point 1 respectively.

p_2, v_2, ρ_2, T_2 = corresponding parameters at point 2.

Bernoulli's equation for insentropic flow gies,

$$\left(\frac{k}{k-1}\right)\frac{p_1}{\rho_1 g} + \frac{v_1^2}{2g} + z_1 = \left(\frac{k}{k-1}\right)\frac{p_2}{\rho_2 g} + \frac{v_2^2}{2g} + z_2$$

With $z_1 = z_2$ and $v_1 = 0$, we get

$$\left(\frac{k}{k-1}\right)\left[\frac{p_1}{\rho_1 g} - \frac{p_2}{\rho_2 g}\right] = \frac{v_2^2}{2g}$$

or

$$\left(\frac{k}{k-1}\right)\left[\frac{p_1}{\rho_1} - \frac{p_2}{\rho_2}\right] = \frac{v_2^2}{2}$$

$$v_2 = \sqrt{\left(\frac{2k}{k-1}\right)\left(\frac{p_1}{\rho_1} - \frac{p_2}{\rho_2}\right)}$$

$$= \sqrt{\left(\frac{2k}{k-1}\right)\frac{p_1}{\rho_1}\left(1 - \frac{p_2}{p_1} \times \frac{\rho_1}{\rho_2}\right)} \qquad \text{... (1)}$$

For isentropic flow,

$$\frac{p_1}{\rho_1^k} = \frac{p_1}{\rho_2^k} \text{ or } \frac{p_2}{p_1} = \left(\frac{\rho_2}{\rho_1}\right)^k \text{ or } \frac{\rho_2}{\rho_1} = \left(\frac{p_2}{p_1}\right)^{1/k} \qquad \text{... (2)}$$

Substituting Eq. (2) in (1), we get

$$v_2 = \sqrt{\left(\frac{2k}{k-1}\right)\frac{p_1}{\rho_1}\left[1 - \frac{p_2}{p_1} \times \left(\frac{p_1}{p_2}\right)^{1/k}\right]}$$

$$= \sqrt{\left(\frac{2k}{k-1}\right)\frac{p_1}{\rho_1}\left[1 - \left(\frac{p_2}{p_1}\right)^{\frac{k-1}{k}}\right]} \qquad \text{... (3)}$$

Mass flow rate of compressible fluid,

$$\dot{m} = \rho_2 A_2 v_2, \text{ where } A_2 = \text{exit area of nozzle.}$$

$$\dot{m} = \rho_2 A_2 \sqrt{\left(\frac{2k}{k-1}\right)\frac{p_1}{\rho_1}\left[1 - \left(\frac{p_2}{p_1}\right)^{\frac{k-1}{k}}\right]} \qquad \text{... (5.37)}$$

Now $\quad \dfrac{p_2}{\rho_2^k} = \dfrac{p_1}{\rho_1^k}$ or $\dfrac{p_2}{p_1} = \left(\dfrac{\rho_2}{\rho_1}\right)^k$ or $\dfrac{\rho_2}{\rho_1} = \left(\dfrac{p_2}{p_1}\right)^{1/k}$ or $\rho_2^2 = \rho_1^2\left(\dfrac{p_2}{p_1}\right)^{2/k}$

$$\dot{m} = A_2\sqrt{\left(\frac{2k}{k-1}\right)p_1\rho_1\left(\frac{p_2}{p_1}\right)^{2/k}\left[1-\left(\frac{p_2}{p_1}\right)^{\frac{k-1}{k}}\right]}$$

$$= A_2\sqrt{\left(\frac{2k}{k-1}\right)p_1\,\rho_1\left[\left(\frac{p_2}{p_1}\right)^{2/k}-\left(\frac{p_2}{p_1}\right)^{\frac{k+1}{k}}\right]} \qquad \text{... (5.38)}$$

For maximum mass flow rate, $\dfrac{d\dot{m}}{d\left(\dfrac{p_2}{p_1}\right)} = 0$

As all other quantities except the ratio p_2/p_1 are constant, therefore

$$\frac{d}{d\left(\dfrac{p_2}{p_1}\right)}\left[\left(\frac{p_2}{p_1}\right)^{2/k}-\left(\frac{p_2}{p_1}\right)^{\frac{k+1}{k}}\right] = 0$$

$$\frac{2}{k}\left(\frac{p_2}{p_1}\right)^{\frac{2}{k}-1}-\left(\frac{k+1}{k}\right)\left(\frac{p_2}{p_1}\right)^{1/k} = 0$$

$$\left(\frac{p_2}{p_1}\right)^{\frac{2}{k}-1}-\left(\frac{k+1}{2}\right)\left(\frac{p_2}{p_1}\right)^{\frac{1}{k}} = 0$$

$$\left(\frac{p_2}{p_1}\right)^{\frac{2}{k}-1-\frac{1}{k}} = \frac{k+1}{2}$$

$$\left(\frac{p_2}{p_1}\right)^{\frac{1-k}{k}} = \frac{k+1}{2}$$

Critical pressure ratio, $\quad \dfrac{p_2}{p_1} = \left(\dfrac{k+1}{2}\right)^{\frac{k}{1-k}} = \left(\dfrac{2}{k+1}\right)^{\frac{k}{k-1}} \qquad \text{... (5.39)}$

For air with $k = 1.4$, the critical pressure ratio,

$$\left(\frac{p_2}{p_1}\right)_{critical} = \left(\frac{2}{2.4}\right)^{3.5} = 0.528 \qquad \text{... (5.40)}$$

A convergent nozzle is used when the exit pressure is equal to or greater than the critical pressure, and a convergent – divergent nozzle is used when the discharge pressure is less than the critical pressure.

Critical density ratio, $$\frac{\rho_2}{\rho_1} = \left(\frac{2}{k+1}\right)^{\frac{1}{k+1}} = 0.634 \text{ for } k = 1.4 \qquad \text{... (5.41)}$$

Critical temperature ratio, $$\frac{T_2}{T_1} = \frac{2}{k+1} = 0.833 \text{ for } k = 1.4 \qquad \text{... (5.42)}$$

5.9.1 Exit velocity for maximum mass rate of flow

Substituting the value of $\frac{p_2}{p_1}$ from Eq. (5.39) in Eq. (3), we get

$$(v_2)_{max} = \sqrt{\left(\frac{2k}{k-1}\right)\left(\frac{p_1}{\rho_1}\right)\left[1-\left(\frac{2}{k+1}\right)^{\frac{k}{k-1}\times\frac{k-1}{k}}\right]}$$

$$= \sqrt{\left(\frac{2k}{k-1}\right)\frac{p_1}{\rho_1}\left[1-\frac{2}{k+1}\right]}$$

$$= \sqrt{\left(\frac{2k}{k+1}\right)\frac{p_1}{\rho_1}} \qquad \text{... (5.43)}$$

5.9.2 Maximum Mass flow rate

$$(\dot{m})_{max} = A_2\sqrt{\left(\frac{2k}{k-1}\right)p_1\rho_1\left[\left(\frac{2}{k+1}\right)^{\frac{k}{k-1}\times\frac{2}{r}} - \left(\frac{2}{k+1}\right)^{\frac{k}{k-1}\times\frac{k+1}{k}}\right]}$$

$$= A_2\sqrt{\left(\frac{2k}{k-1}\right)p_1\,\rho_1\left[\left(\frac{2}{k+1}\right)^{\frac{k}{k-1}} - \left(\frac{2}{k+1}\right)^{\frac{k+1}{k-1}}\right]}$$

For air, $k = 1.4$

$$(\dot{m})_{max} = A_2\sqrt{\left(\frac{2\times1.4}{0.4}\right)p_1\rho_1\left[\left(\frac{2}{2.4}\right)^{\frac{2}{0.4}}-\left(\frac{2}{2.4}\right)^{\frac{2.4}{0.4}}\right]}$$

$$= 0.685\, A_2\sqrt{p_1\,\rho_1} \quad \text{... (5.44)}$$

5.9.3 Variation of Mass Flow Rate with Pressure Ratio

It may be observed that the downstream velocity v_2 will increase due to decrease in downstream pressure p_2. So long as the stream remains subsonic, the jet velocity v_2 will increase due to reduction of downstream pressure p_2. Once however, the fluid reaches sonic velocity at the throat ($M = 1$), any further reduction in downstream pressure cannot be propagated upstream. The flow through the nozzle then becomes independent of p_2, and is termed as *choked flow.* Thus a passage in which the sonic velocity has been reached and the mass flow rate is maximum is called the choked flow.

Fig. 5.10 Shows the variation of mass flow rate v's p_2/p_1. The following observation may be made from Fig. 5.10.

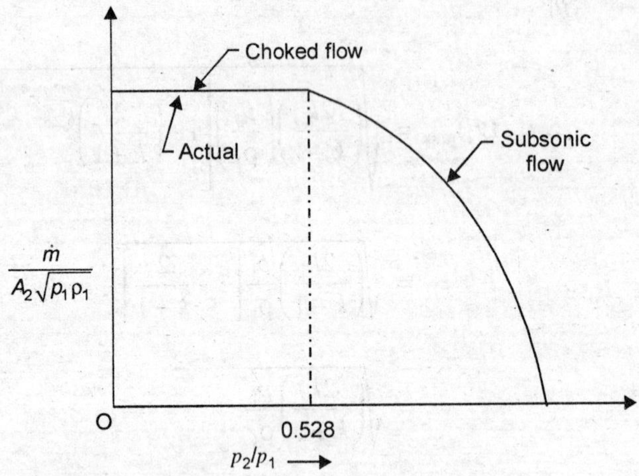

Fig. 5.10 Mass flow rate through a convergent nozzle

1. The mass flow rate increases with decreases in pressure ratio p_2/p_1 and becomes maximum at critical pressure ratio of 0.528.
2. The actual mass flow rate remains constant after attaining the maximum value.

5.9.4 Velocity at Outlet of Nozzle for Maximum Flow Rate

From Eq. (5.43) the velocity at outlet of nozzle for maximum flow rate is given by,

$$v_2 = \sqrt{\left(\frac{2k}{k+1}\right)\frac{p_1}{\rho_1}}$$

Now
$$\frac{p_2}{p_1} = \left(\frac{2}{k+1}\right)^{\frac{k}{k-1}}$$

$$p_1 = \frac{p_2}{\left(\frac{2}{k+1}\right)^{k/(k-1)}} = p_2\left(\frac{2}{k+1}\right)^{\left(\frac{k}{1-k}\right)}$$

For adiabatic flow, we have

$$\frac{p_1}{p_2} = \left(\frac{\rho_1}{\rho_2}\right)^k \text{ or } \frac{\rho_1}{\rho_2} = \left(\frac{p_1}{p_2}\right)^{1/k} = \left(\frac{p_2}{p_1}\right)^{-1/k}$$

$$\rho_1 = \rho_2\left(\frac{p_2}{p_1}\right)^{-1/k} = \rho_2\left(\frac{2}{k+1}\right)^{\frac{k}{k-1}\times\left(\frac{-1}{k}\right)} = \rho_2\left(\frac{2}{k+1}\right)^{\frac{1}{1-k}}$$

∴
$$v_2 = \left[\left(\frac{2k}{k+1}\right)\left\{\frac{p_2\left(\frac{2}{k+1}\right)^{\frac{k}{1-k}}}{\rho_2\left(\frac{2}{k+1}\right)^{\frac{1}{1-k}}}\right\}\right]^{1/2}$$

$$= \left[\left(\frac{2k}{k+1}\right)\frac{p_2}{\rho_2}\left(\frac{2}{k+1}\right)^{\frac{k}{1-k}-\frac{1}{1-k}}\right]^{1/2}$$

$$= \left[\left(\frac{2k}{k+1}\right)\times\frac{p_2}{\rho_2}\left(\frac{2}{k+1}\right)^{-1}\right]^{1/2}$$

$$= \sqrt{\frac{k\,p_2}{\rho_2}} = c_2 \qquad \text{... (5.45)}$$

∴ As maximum flow rate, velocity at the outlet of nozzle = sonic velocity.

Example 5.11 *A large container holds carbon dioxide at a temperature of 10°C and pressure 5 atmosphere. The gas is discharged through an oifice 10 mm in diameter. Determine the velocity of jet and the mass flow rate. Take k = 1.30, and atmospheric pressure 101.25 kPa, R = 189, R = J/kg.K.*

Solution.

Given: $T_1 = 273 + 10 = 283\ K$, $p_1 = 5 \times 101.25 = 506.25$ kPa, $d = 10$ mm, $k = 1.30$, $R = 189$ J/kg.K.

Density of carbon dioxide in cotainer, $\rho_1 = \dfrac{p_1}{RT_1} = \dfrac{506.25 \times 10^3}{189 \times 283} = 9.465 \text{ kg/m}^3$

Critical pressure ratio for choked flow,

$$\frac{p_c}{p_1} = \left(\frac{2}{k+1}\right)^{\frac{k}{k-1}} = \left(\frac{2}{1.3+1}\right)^{\frac{1.3}{0.3}} = 0.5457$$

$$\frac{p_2}{p_1} = \frac{1}{5} = 0.2$$

Since $\dfrac{p_2}{p_1} < \dfrac{p_c}{p_1}$, the flow is choked and $v_c = c_c$ at $T = T_c$.

$$\frac{T_c}{T_1} = \left(\frac{p_c}{p_1}\right)^{\frac{k-1}{k}} = (0.5457)^{\frac{0.3}{1.3}} = 0.8696$$

$$T_c = 0.8696 \times 283 = 246.1\ K$$

$$v_c = c_c = \sqrt{kRT_c} = \sqrt{1.3 \times 189 \times 246.1} = 245.9 \text{ m/s}$$

∴ Velocity of jet = 245.9 m/s

Density of CO_2 at the orifice, $\rho_c = \dfrac{p_c}{RT_c} = \dfrac{0.5457 \times 5 \times 101.25 \times 10^3}{189 \times 246.1} = 5.94 \text{ kg/m}^3$

Mass flow rate, $\dot{m} = \rho_c A v_c = 5.94 \times \dfrac{\pi}{4}(0.01)^2 \times 245.9 = 0.1147 \text{ kg/s}$

5.10 FLOW VARIBALES IN TERMS OF MACH NUMBER

In order to express changes in velocity, pressure, temperature and density in terms of Mach number, use is made of the continuity equation, perfect gas equation, energy equation and isentropic flow.

Continuity Equation:

$$\rho A v = \text{const.}$$

Taking log of both sides, we have

$$\ln \rho + \ln A + \ln v = \ln \text{const.}$$

Differentiating , we get

$$\frac{d\rho}{\rho} + \frac{dA}{A} + \frac{dv}{v} = 0$$

Isentropic flow equation:

$$\frac{p}{\rho^k} = \text{const.} \quad \text{or} \quad \frac{dp}{p} = k\frac{d\rho}{\rho}$$

Perfect gas equation:

$$p = \rho RT \text{ or } \frac{dp}{p} = \frac{d\rho}{\rho} + \frac{dT}{T}$$

Energy equation:

$$c_p T + \frac{v^2}{2} = \text{const.}\ ,\ c_p = \left(\frac{k}{k-1}\right)R$$

$$c_p \cdot dT + v \cdot dv = 0$$

$$\left(\frac{k}{k-1}\right)R \cdot dT + v.dv = 0$$

or $$\left(\frac{k}{k-1}\right)R \cdot \frac{dT}{v^2} + \frac{dv}{v} = 0$$

$$c = \sqrt{kRT}$$

∴ $$kR = \frac{c^2}{T}$$

or $$\left(\frac{1}{k-1}\right)\frac{c^2}{T} \cdot \frac{dT}{v^2} + \frac{dv}{v} = 0$$

$$\left(\frac{k}{k-1}\right) \cdot \frac{1}{M^2} \cdot \frac{dT}{T} + \frac{dv}{v} = 0 \qquad \text{... (5.46)}$$

$$M = \frac{v}{\sqrt{kRR}}$$

$$\frac{dM}{M} = \frac{dv}{v} - \frac{1}{2}\frac{dT}{T} \qquad \text{... (5.47)}$$

Substituting for $\frac{dT}{T}$ from Eq. (5.46), we have

$$\frac{dM}{M} = \frac{dv}{v}\left[1 + \left(\frac{k-1}{2}\right)M^2\right]$$

or
$$\frac{dv}{v} = \frac{1}{\left[1+\left(\frac{k-1}{2}\right)M^2\right]} \cdot \frac{dM}{M} \quad \text{... (5.48)}$$

Sustiuting Eq. (5.48) in Eq. (5.46) we get

$$\frac{dT}{T} = \left[\frac{-(k-1)M^2}{1+\left(\frac{k-1}{2}\right)M^2}\right]\frac{dM}{M} \quad \text{... (5.49)}$$

Since right hand side is negative, the temperatre changes follow an opposite trend to that of Mach number. Similar trends are obtained for pressure and density.

$$\frac{dp}{p} = \left[\frac{-kM^2}{1+\left(\frac{k-1}{2}\right)M^2}\right]\frac{dM}{M} \quad \text{... (5.50)}$$

and
$$\frac{d\rho}{\rho} = \left[\frac{-M^2}{1+\left(\frac{k-1}{2}\right)M^2}\right]\frac{dM}{M} \quad \text{... (5.51)}$$

For area changes, we have

$$\frac{dA}{A} = \left[\frac{-(1-M\ )}{1+\left(\frac{k-1}{2}\right)M^2}\right]\frac{dM}{M} \quad \text{... (5.52)}$$

Eq. (5.52) can be integrated to obtain a relationship between the critical throat area A_c, where $M = 1$ and the area A at any section.

$$\frac{A}{A_c} = \frac{1}{M}\left[\frac{2+(k-1)M^2}{k+1}\right]^{\frac{k+1}{2(k-1)}} \quad \text{... (5.53)}$$

Example 5.12 *A supersonic nozzle is to be designed for air flow with Mach number 3 at the exit section which is 0.2 m in diameter. The pressure and temperature of air at the nozzle exit are to be 8 kPa and 200 K respectively. Caluculate the reservoir pressure and temperature and throat area. Take k = 1.4.*

Solution. The stagnation conditions in the reservoir are:

$$p_0 = p\left[1+\left(\frac{k-1}{2}\right)M^2\right]^{\frac{k}{k-1}} = 8\left[1+\left(\frac{1.4-1}{2}\right)\times 3^2\right]^{\frac{1.4}{0.4}} = 293.86 \text{ kPa}$$

$$T_0 = T\left[1+\left(\frac{k-1}{2}\right)M^2\right] = 200\left[1+\left(\frac{1.4-1}{2}\right)\times 3^2\right] = 560 \text{ K}$$

$$\frac{A}{A_c} = \frac{1}{M}\left[\frac{2+(k-1)M^2}{k+1}\right]^{\frac{k+1}{2(k-1)}}$$

$$= \frac{1}{3}\left[\frac{2+(1.4-1)\times 3^2}{1.4+1}\right]^{\frac{2.4}{2\times 0.4}} = 4.2346$$

$$Ac = \frac{\frac{\pi}{4}(0.2)^2}{4.2346} = 0.00742 \text{ m}^2$$

Example 5.13 *The pressure leads from a Pitot-static tube mounted on an aircraft were connected to a pressure gauge in the cockpit. The dial of the pressure gauge is calibrated to read the aircraft speed in m/s. The calibration is done on the ground by applying a known pressure across the gauge and calculating the equivalent velocity using incompressible Bernoulli's equation and assuming that density is 1.224 kg/m³.*

The gauge having been calibrated in this way, the aircraft is flown at 9200 m altitude, where the density is 0.454 kg/m³ and ambient pressure is 30 kPa. The gauge indicates a velocity of 152 m/s. What is the true speed of aircraft?

Solution.

Given: $\rho_0 = 1.224$ kg/m^3, $\rho = 0.451$ kg/m³, $p = 30$ kPa, $v = 152$ m/s

Bernoulli's equation for an incompressible flow in given by,

$$p + \frac{\rho v^2}{2} = \text{const.}$$

Stagnation pressure created at Pitot-static tube,

$$p_0 = p + \frac{\rho_0 v^2}{2} = 30 + \frac{1.224\times 152^2}{2\times 10^3} = 44.139 \text{ kPa}$$

Neglecting compressibility effects, the speed of the aircraft when $\rho = 0.454$ kg/m^3,

$$44.139 = 30 + \frac{0.454\times v^2}{2\times 10^3}$$

$$v = 249.57 \text{ m/s}$$

Sonic velocity,

$$c = \sqrt{kRT} = \sqrt{k\frac{p}{\rho}} = \sqrt{\frac{1.4\times 30\times 10^3}{0.454}} = 304.16 \text{ m/s}$$

$$\text{Mach number, } M = \frac{v}{c} = \frac{249.57}{304.16} = 0.82$$

$$\text{Compressibility correction factor, CCF} = 1 + \frac{M^2}{4} = 1 + \frac{(0.82)^2}{4} = 1.168$$

$$\text{True speed of aircraft} = \frac{v}{\sqrt{CCF}} = \frac{249.57}{\sqrt{1.168}} = 230.9\,\text{m/s}$$

Example 5.14 *A large vessel contains air at 300 kPa gauge pressure and 27°C temperature. The air flows from the vessel to the atmosphere through a convergent nozzle. If the diameter at the nozzle outlet is 20 mm, calculate the maximum flow rate of air. Take R = 287 J/kg.K, k = 1.4, and atmospheric pressure 100 kPa.*

Solution.

Given: $p_1 = 300 + 100 = 400$ kPa (*abs*), $T_1 = 273 + 27 = 300$ *K*, $d = 20$ mm, $R = 287$ J/kg.K, $k = 1.4$, $p_a = 100$ kPa

Equation of state,

$$p_1 = \rho_1 R T_1$$

$$\rho_1 = \frac{400 \times 10^3}{287 \times 300} = 4.646\ \text{kg/m}^3$$

Maximum mass flow rate,

$$(\dot{m})_{\max} = 0685 A_2 \sqrt{p_1 \rho_1}$$

$$= 0.685 \times \frac{\pi}{4} (0.02)^2 \sqrt{400 \times 10^3 \times 4.646}$$

$$= 0.293\ \text{kg/s}$$

Example 5.15 *A large vessel contains air at a pressure of 2500 kPa (abs) and 25°C. A nozzle is fitted to the vessel at the outlet of which the pressure is 1800 kPa (abs). Find the velocity of air flowing out from the nozzle.* Take $R = 287$ J/kg.K and $k = 1.4$

Solution.

Given: $p_1 = 2500$ kPa (abs). $T_1 = 273 + 25 = 298$ *K*, $p_2 = 1800$ kPa (*abs*), $R = 287$ J/kg.K, $k = 1.4$

Velocity of air at nozzle exit,

$$v_2 = \sqrt{\left(\frac{2k}{k-1}\right)\frac{p_1}{\rho_1}\left[1 - \left(\frac{p_2}{p_1}\right)^{\frac{k-1}{k}}\right]}$$

$$\rho_1 = \frac{p_1}{RT_1} = \frac{2500 \times 10^3}{287 \times 298} = 29.23\ \text{kg/m}^3$$

$$v_2 = \sqrt{\left(\frac{2 \times 1.4}{1.4 - 1}\right) \times \frac{2500 \times 10^3}{29.23}\left[1 - \left(\frac{1800}{2500}\right)^{\frac{0.4}{1.4}}\right]}$$

$$= 231.6\ \text{m/s}$$

Example 5.16 *A tank fitted with a convergent nozzle contains air at a temperature of 25°C. The diameter at the outlet of the nozzle is 30 mm. Assuming isontropic flow, calculate the mass rate of flow of air through the nozzle to the atmosphere when the pressure in the tank is 125 kPa (abs.).*

Take for air: R = 0.287 kJ/kg.K, k = 1.4. Barometric pressure = 100 kPa.

Solution.

Given: $T_1 = 273 + 25 = 298\ K$, $p_2 = 100$ kPa. $d = 30$ mm, $p_1 = 125$ kPa, $R = 0.287$ kJ/kg.K, $k = 1.4$

$$\rho_1 = \frac{p_1}{RT_1} = \frac{125}{0.287 \times 298} = 1.4615 \text{ kg/m}^3$$

$$r_p = \frac{p_2}{p_1} = \frac{100}{125} = 0.8$$

$\frac{p_2}{p_1} > 0.528,$ the flow in the nozzle will be subsonic

$$\text{Mass flow rate of air, } \dot{m} = A_2\sqrt{\left(\frac{2k}{k-1}\right)p_1\rho_1\left[r_p^{2/k} - r_p^{(k+1)/k}\right]}$$

$$= \frac{\pi}{4}(0.03)^2\sqrt{\left(\frac{2\times1.4}{1.4-1}\right)\times125\times10^3\times1.4615[(0.8)^{2/1.4} - (0.8)^{2.4/1.4}]}$$

$$= 0.1694 \text{ kg/s}$$

SUMMARY

1. In a compressible flow the density of fluid changes during flow.
2. Thermodynamic concepts:

   (i) Equation of state: $\frac{p}{\rho} = RT$

   $$\frac{dp}{p} - \frac{d\rho}{\rho} - \frac{dT}{T} = 0$$

   (*ii*) Continuity equation:

   $\rho A v = \text{const.}$

   $$\frac{d\rho}{\rho} + \frac{dA}{A} + \frac{dv}{v} = 0$$

   (iii) Bernoulli's equation:

   $$\frac{p}{\rho g}\ln p + \frac{v^2}{2g} + z = \text{const. for isothermal process.}$$

   $$\left(\frac{k}{k-1}\right)\frac{p}{\rho g} + \frac{v^2}{2g} + z = \text{const. for adiabatic process.}$$

(iv) $\dfrac{p}{\rho}$ = const. for isothermal process

$\dfrac{p}{\rho^k}$ = const. for adiabatic process

3. Sonic velocity: $c = \sqrt{\dfrac{dp}{d\rho}} = \sqrt{\dfrac{K}{\rho}}$ in terms of bulk moduls k.

$$= \sqrt{\frac{p}{\rho}} = \sqrt{RT} \text{ for isothermal process}$$

$$= \sqrt{\frac{kp}{\rho}} = \sqrt{kRT} \text{ for adiabatic process.}$$

4. Mach number, $M = \dfrac{v}{c}$

Subsonic flow: $0.3 < M < 1.0$

Transonic flow: $0.9 < M < 1.1$

Sonic flow: $M = 1.0$

Supersonic flow: $M > 1.0$ and upto $\simeq 5.0$

Hypersonic flow: $M >> 1.0$

5. Mach angle, $\alpha = \sin^{-1}\left(\dfrac{1}{M}\right)$

6. Stagnation properties:

$$p_0 = p_1\left[1+\left(\frac{k-1}{2}\right)M_1^2\right]^{\frac{k}{k-1}}$$

$$\rho_0 = \rho_1\left[1+\left(\frac{k-1}{2}\right)M_1^2\right]^{\frac{1}{k-1}}$$

$$T_0 = T_1\left[1+\left(\frac{k-1}{2}\right)M_1^2\right]$$

7. Compressiblity correctrin factor, $\text{CCF} = 1+\dfrac{M^2}{4}+\left(\dfrac{2-k}{24}\right)M^4$

8. Area-velocity relationship:

$$\frac{dA}{A} = \frac{dv}{v}(M^2 - 1)$$

$$dp = \rho v^2 \left(\frac{1}{1-M^2}\right)\frac{dA}{A}$$

(i) Subsonic flow ($M < 1$):

(a) $\frac{dv}{v} > 0, \frac{dA}{A} < 0, dp < 0,$ for convergent nozzle

(b) $\frac{dv}{v} < 0, \frac{dA}{A} > 0, dp > 0,$ for divergent diffuser

(ii) Supersonic flow (M > 1):

(a) $\frac{dv}{v} > 0, \frac{dA}{A} > 0, dp < 0$ fro divergent nozzle

(b) $\frac{dv}{v} < 0, \frac{dA}{A} < 0, dp > 0,$ for convergent diffuser

(iii) Sonic flow ($M = 1$):

$$\frac{dA}{A} = 0$$

$dp = \frac{\text{zero}}{\text{zero}}$ i.e. indetrminate. When evaluated, $dp = 0$, $dA = 0$ and the flow is frictionless.

9. Flow through a convergent nozzle:

(i) Nozzle exit velocity, $v_2 = \sqrt{\left(\frac{2k}{k-1}\right)\frac{p_1}{\rho_1}\left[1-\left(\frac{p_2}{p_1}\right)^{\frac{k-1}{k}}\right]}$

(ii) Mass flow rate, $\dot{m} = A_2\sqrt{\left(\frac{2k}{k-1}\right)p_1\,\rho_1\left[\left(\frac{p_2}{p_1}\right)^{2/k} - \left(\frac{p_2}{p_1}\right)^{\frac{k+1}{k}}\right]}$ for $\left(\frac{p_2}{p_1}\right) > 0.528$

(iii) For maximum mass flow rate:

Critical pressure ratio $= \frac{p_2}{p_1} = \frac{p_c}{p_1} = \left(\frac{2}{k+1}\right)^{\frac{k}{k-1}} = 0.528$ for $k = 1.4$

(iv) Value of v_2 for maximum mass flow rate:

$$v_2 = \sqrt{\left(\frac{2k}{k+1}\right)\frac{p_1}{\rho_1}} = \sqrt{\frac{kp_2}{\rho_2}} = c_2$$

(v) Maximum rate of mass flow:

$$(\dot{m})_{max} = A_2\sqrt{\left(\frac{2k}{k-1}\right)p_1\,\rho_1\left[\left(\frac{2}{k+1}\right)^{\frac{2}{k-1}} - \left(\frac{2}{k+1}\right)^{\left(\frac{k+1}{k-1}\right)}\right]}$$

$$= 0685\, A_2\sqrt{p_1\,\rho_1} \quad \text{for } k = 1.4$$

(vi) Critical density ratio, $\dfrac{\rho_2}{\rho_1} = \dfrac{\rho_c}{\rho_1} = \left(\dfrac{2}{k+1}\right)^{\frac{1}{k-1}} = 0.634$ for $k = 1.4$

(vii) Critical temperature ratio, $\dfrac{T_2}{T_1} = \dfrac{T_c}{T_1} = \dfrac{2}{k+1} = 0.833$ for $k = 1.4$

10. Flow variables in term of Mach number:

(i) $$\frac{dv}{v} = \frac{1}{\left[1+\left(\frac{k-1}{2}\right)M^2\right]}\cdot\frac{dM}{M}$$

(ii) $$\frac{dT}{T} = \left[\frac{-(k-1)M^2}{1+}\right]\cdot\frac{dM}{M}$$

(iii) $$\frac{dp}{p} = \left[\frac{-kM^2}{1+\left(\frac{k-1}{2}\right)M^2}\right]\frac{dM}{M}$$

(iv) $$\frac{d\rho}{\rho} = \left[\frac{-M^2}{1+\left(\frac{k-1}{2}\right)M^2}\right]\frac{dM}{M}$$

(v) $$\frac{dA}{A} = \left[\frac{-(1-M^2)}{1+\left(\frac{k-1}{2}\right)M^2}\right]\frac{dM}{M}$$

(vi) $$\frac{A}{A_c} = \frac{1}{M}\left[\frac{2+(k-1)M^2}{k+1}\right]^{\frac{k+1}{2(k-1)}}$$

Short Answer Type Questions

1. What is the difference between the incompressible and compressible flows?
2. What is the relation between pressure and density of a compressible fluid for (i) insothermal process, and (ii) adiabatic process?
3. Explain the mechanism of propagation of disturbances in compressible fluids.
4. Define sonic velcity. On what factors does it depend?
5. Define Mach number and Mach cone.
6. Differntiate between subsonic flow, supersonic flow, and sonic flow.
7. Define stagnation point.
8. What is static pressure and stagnation pressure?
9. Differentiate between static and stagnation temperatures.
10. What is the critical pressure ratio for a compressible flow throug a nozzle?
11. Define compressibility correction factor.
12. What is choked flow?
13. Explain zone of disturbance and zone of silence.
14. Define Mach cone angle.
15. How does the pressure and velocity vary with area for (a) subsonic flow, and (b) supersonic flow?

EXERCISES

1. A gas is flowing through a horizontal pipe of 80 mm diameter at a temperature of 4°C and 3 bar (gauge) pressure. The diameter of the pipe suddenly changes to 40 mm where pressure is 2 bar (gauge). Calculate the velocities of the gas in these diameters of pipe assuming isothermal process. Take $R = 0.287$ kJ/kg.K, and atmospheric pressure = 1 bar

   [**Ans.** 40.82 m/s, 217.73 m/s]

2. A gas is flowing through a horizontal pipe having area of cross-section as 20 cm^2 and pressure 3 bar (gauge), The mass flow rate through the pipe is 0.5 kg/s. Calculate the velocities of gas through these sections, assuming isothermal process. Take $R = 292$ J/kg.K and atmospheric pressure = 1 bar. [**Ans.** 21.02 m/s, 52.56 m/s]

3. A gas is flowing through a horizontal pipe with a velocity of 300 m/s, where the pressure is 0.6 bar (abs.) and temperature 40°C. At another section, the pressure is 0.9 bar (abs.). Find the velocity of gas in this section if the flow of gas is adiabatic. Take $R = 0.287$ kJ/kg. K and $k = 1.4$. [**Ans.** 113 m/s]

4. Calculate the speed of sound wave in air at sea-level where the temperature is 20°C. Take $R = 0.287$ kJ/kg.K and $k = 1.4$. [**Ans.** 343.1 m/s]

5. A jet propelled aircraft is flying at a speed of 1000 km/h at an altitude of 2 km where the air temperature is –5°C. Calculate the Mach number. Take $k = 1.4$ and $R = 0.287$ kJ/kg. K.

   [**Ans.** 0.846]

6. A projectile travels in air of pressure 1.01325 bar at –10°C at a speed of 1200 km/h. Calculate the Mach number and Mach angle. Take $k = 1.4$ and $R = 287$ J/kg.K. [**Ans.** 1.025, 77.22°]
7. An aeroplane is flying at 1050 km/h through still air having a pressure of 0.7 bar and temperature –5°C. Calculate the pressure, temperature and density at stagnation point on the nose. Take $R = 0.287$ kJ/kg.K and $k = 1.4$ [**Ans.** 1.169 bar, 37.3°C, 1.313 kg/m^3]
8. A nozzle is fitted to a large vessel containing air at 30 bar (*abs.*) and 20°C. The air is flowing out from the nozzle outlet at a pressure of 20 bar (abs.). Calculate the air velocity at the nozzle outlet. Take $R = 287$ J/kg. K and $k = 1.4$. [**Ans.** 253.75 m/s]
9. A tank contains air at a temperature of 30°C. Air flows from the tank into atmosphere through a convergent nozzle. The outlet diameter of nozzle is 25 mm. Assuming adiabatic flow, calculate the mass rate of flow of air through the nozzle when the pressure of air in tank is (a) 0.4 bar (gauge), and (b) 3.5 bar (gauge). Take R= 287 J/kg.K, $k = 1.4$ and atmospheric pressure = 1 bar (abs.) [**Ans.** 0.091 kg/s, 0.507 kg/s]
10. Calculate the mass rate of flow of air through a nozzle of diameter 20 mm fitted to a large tank containing air at 20°C. The air flows into atmosphere from tank at (i) 0.6 bar (gauge), and (ii) 3 bar (gauge), Assume adiabatic flow, R = 287 kg.K, and $k = 1.4$, and atmospheric pressure = 1 bar. [**Ans.** 0.116 kg/s, 0.2436 kg/s]

6

VISCOUS FLOW

6.1 INTRODUCTION

The viscous flow deals with the flow of fluid which are viscous and flowing at very low velocity . At low velocity the fluid mouses in layers such that each layers such that each layer of fluid slides over the adjacent layer. Due to relative velocity between two layers the velocity gradient $\frac{du}{dy}$ exises and according to Newton's law a shear $I = \mu \frac{du}{dy}$ acts on the layers.

6.2 REYNOLDS NUMBER AND FLOW REGIMES

It was observed by Qsborne Reynolds from his experiments that the nature of flow in a closed conduit depends on the following parameters:

1. Diameter of conduict, d.
2. Density of liquid, ρ.
3. Viscosity of the liquid μ, and
4. Velocity of flow, v.

Reynolds number is defined as the ratio of inertia force to the viscous force.

$$\text{Reynolds number, } Re = \frac{\text{Inetia force}}{\text{Viscous force}}$$

$$= \frac{\rho v d}{\mu} = \frac{vd}{\nu} \qquad \text{... (6.1)}$$

where $\nu = \frac{\mu}{\rho}$ = kinetic viscoity.

In general, *d* is replaced by *l*, known as the characteristic length. Thus,

$$Re = \frac{vl}{\nu}$$

When Re < 2000, the flow in a pipe is laminar (or viscous)

2000 < *Re* < 4000, indicates transistion from laminar to turbulent flow

Re > 4000, the flow is turbulent.

Some examples of laminar (or viscous) flow are:

1. Flow past tiny bodies.
2. Underground flow.
3. Blood flow in the arteries of a human body.
4. Flow of oil in measuring instruments.
5. Rise of water in plants through their roots.

6.3 NAVIER-STOKES EQUATIONS OF MOTION

The Navier-Stokes equations of motion take into account the pressure force, body force and force due to viscosity.

Consider an infinitesimal small mass of fluid element enclosed in an elementary parallelopiped of sides *dx*, *dy* and *dz* shown in Fig. 6.1. The various forces acting on the fluid element are:

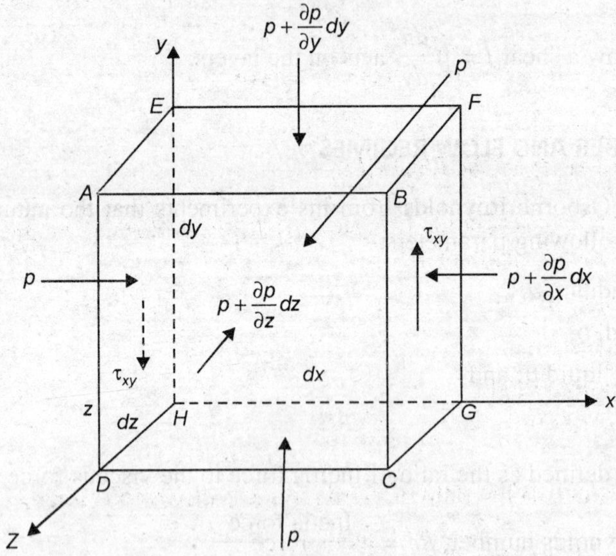

Fig. 6.1 Forces acting on elementary parallelopied

1. Normal forces due to pressure

The net force in the *x*-direction due to pressure,

$$F_{px} = p\cdot dy\cdot dz - \left(p + \frac{\partial p}{\partial x}dx\right)dy\cdot dz$$

$$= -\frac{\partial p}{\partial x}dx\cdot dy\cdot dz$$

2. Body forces due to gravity

Let B_x, B_y, and B_z be the component of the body force per unit mass of fluid along the coordinate axes. Then

$$F_{gx} = B_x\cdot\rho\cdot dx\cdot dy\cdot dz$$

3. Inertia forces due to acceleratin

$$F_{ix} = (\rho.dx\,dy\,dz)\cdot\frac{du}{dt}$$

4. Shear forces due to viscosity

Let S_x, S_y, and S_z be the components of shear force per unit mass due to viscosity of fluid element along the coordinate axes. Then

$$F_{vx} = S_x.\rho.dx.dy.dz$$

According to Newton's second law of motion, we have

$$F_{ix} = F_{px} + F_{gx} + F_{vx}$$

$$\rho.dx\,dy\,dz\cdot\frac{du}{dt} = -\frac{\partial p}{\partial x}\cdot dx\,dy\,dz + B_x\cdot\rho.dx\,dy\,dz + S_x\cdot\rho\cdot dx\,dy\,dz$$

or

$$B_x - \frac{1}{\rho}\frac{\partial p}{\partial x} = \frac{du}{dt} - S_x$$

Similarly

$$B_y - \frac{1}{\rho}\frac{\partial p}{\partial y} = \frac{dv}{dt} - S_y \qquad \text{... (6.2)}$$

$$B_z - \frac{1}{\rho}\frac{\partial p}{\partial z} = \frac{dw}{dt} - S_z$$

The shear forces can be evaluated as follows:

Let τ_{xy} be the shear stress on the x faces of the parallelopiped.

Now

$$\tau_{xy} = \mu\frac{\partial u}{\partial x}$$

Shear force on the -ve x–face $AEHD$ $= -\mu \frac{\partial u}{\partial x} \cdot dy \cdot dz$

Shear force on the +ve x-face $BFGC$ $= \mu \frac{\partial}{\partial x}\left(u + \frac{\partial u}{\partial x} dx \right) dy \cdot dz$

$$= \mu \left(\frac{\partial u}{\partial x} + \frac{\partial^2 u}{\partial x^2} dx \right) dy\ dz$$

Net shear force on x-faces $= -\mu \frac{\partial u}{\partial x} \cdot dydz + \mu \left(\frac{\partial u}{\partial x} + \frac{\partial^2 u}{\partial x^2} \cdot dx \right) dy \cdot dz$

$$= \mu \frac{\partial^2 u}{\partial x^2} dx\, dy\, dz$$

Similarly net shear forces on y- and z-faces in the x-direction are:

$$\mu \frac{\partial^2 u}{\partial y^2} \cdot dx\, dy\, dz \ \ and \ \ \mu \frac{\partial^2 u}{\partial z^2} \cdot dx\, dy\, dz$$

The total viscous force parallel to x-axis

$$= \mu \left(\frac{\partial^2 u}{\partial x^2} + \frac{\partial^2 u}{\partial y^2} + \frac{\partial^2 u}{\partial z^2} \right) dx\, dy\, dz$$

$$\therefore \quad S_x = \frac{\mu}{\rho} \left(\frac{\partial^2 u}{\partial x^2} + \frac{\partial^2 u}{\partial y^2} + \frac{\partial^2 u}{\partial z^2} \right) = \frac{\mu}{\rho} \cdot \nabla^2 u = \nu \nabla^2 u$$

Similarly $S_y = \frac{\mu}{\rho} \nabla^2 v = \nu \nabla^2 v$ and $S_z = \frac{\mu}{\rho} \nabla^2 w = \nu \nabla^2 w$

Eqs. (6.2) become

$$\left. \begin{aligned} B_x - \frac{1}{\rho} \frac{\partial p}{\partial x} &= \frac{du}{dt} - \nu \nabla^2 u \\ B_y - \frac{1}{\rho} \frac{\partial p}{\partial y} &= \frac{dv}{dt} - \nu \nabla^2 v \\ B_z - \frac{1}{\rho} \frac{\partial p}{\partial z} &= \frac{dw}{dt} - \nu \nabla^2 w \end{aligned} \right\} \quad \text{... (6.3)}$$

Eqs. (6.3) are known as Navier-Stokes equations.

Some of the important applications of N-S equations are:

1. Laminar flow in circular pipes.
2. Laminar flow between concentric rotating cylinders.
3. Laminar one-dimensional flow between stationary parallel plates.
4. Laminar one-dimensional flow between parallel plates having relative motion.

6.4 RELATIONSHIP BETWEEN SHEAR STRESS AND PRESSURE GRADIENT

Consider a fluid element in the form of an infinitesimal parallelopied shown in Fig. 6.2 (a). The velocity profile is shown in Fig. 6.2 (b). The velocity profile is non-uniform due to relative motion between different layers of fluid. The motion of the fluid element is resisted by shearing forces which must be overcome by maintaining a pressure gradient in the direction of flow.

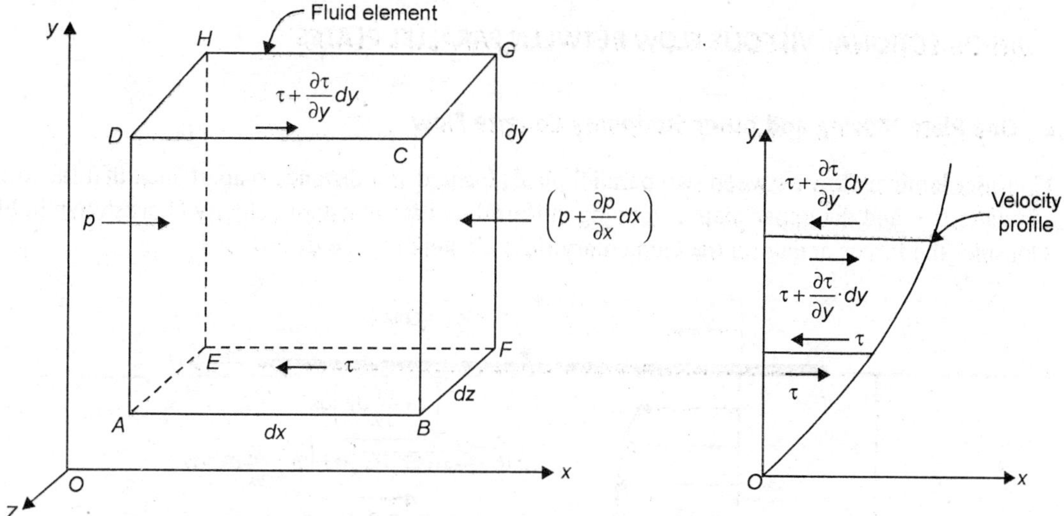

Fig. 6.2 Forces on a fluid element in laminar flow

Let τ = shear stress on the lower face *ABFE*.

$\tau + \frac{\partial \tau}{\partial y} dy$ = shear stress on the upper face *DCGH*

For two-dimensional steady flow there shall be no shear stresses on the vertical faces *ABCD* and *EFGH*. Thus the only forces acting on the element in the *x*-direction of flow shall be the pressure and shear forces.

$$\text{Net shear force on the element} = \left(\tau + \frac{\partial \tau}{\partial y} dy\right) dx \cdot dz - \tau \cdot dx \cdot dz$$

$$= \frac{\partial \tau}{\partial y} \cdot dx\, dy\, dz \qquad \text{... (1)}$$

$$\text{Net pressure force on the element} = p.dy.dz - \left(p + \frac{\partial p}{\partial x} \cdot dx\right) dy.dz$$

$$= -\frac{\partial p}{\partial x} \cdot dx\,dy\,dz \qquad \text{... (2)}$$

From Eqs. (1) and (2), we get

$$\frac{\partial \tau}{\partial y} \cdot dx\,dy\,dz - \frac{\partial p}{\partial x} \cdot dx\,dy\,dz = 0$$

$$\therefore \qquad \frac{\partial \tau}{\partial y} = \frac{\partial p}{\partial x} \qquad \text{... (6.4)}$$

Eq. (6.4) indicates that the pressure gradient in the direction of flow is equal to the shear gradient in the direction normal to the direction of flow.

6.5 UNI-DIRECTIONAL VISCOUS FLOW BETWEEN PARALLEL PLATES

6.5.1 One Plate Moving and other Stationary-Couette Flow

Consider laminar flow between two parallel plates located at a distance b apart such that the lower plate is stationary and the upper plate is moving uniformly with a constant velocity U, as shown in Fig. 6.3. Consider the forces acting on the elementary fluid element of size $dx \times dy \times 1$.

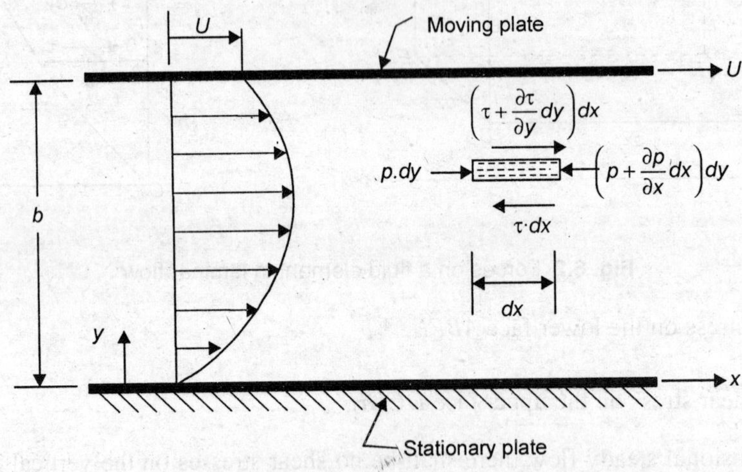

Fig. 6.3 Couette flow

Resultant force on the fluid element in the direction of flow

$$= pdy - \left(p + \frac{\partial p}{\partial x} dx\right) dy - \tau . dx + \left(\tau + \frac{\partial \tau}{\partial y} \cdot dy\right) dx = 0$$

or $$-\frac{\partial p}{\partial x} dx\,dy + \frac{\partial \tau}{\partial y} dx\,dy = 0$$

or
$$\frac{\partial p}{\partial x} = \frac{\partial \tau}{\partial y} \qquad \text{... (6.5)}$$

According to Newton's law of viscosity, $\tau = \mu \frac{du}{dy}$

Substituting in Eq. (6.5), we get

$$\frac{\partial p}{\partial x} = \mu \frac{d^2 u}{dy^2}$$

Integrating twice, we get

$$u = \frac{1}{2\mu}\frac{\partial p}{\partial x} y^2 + C_1 y + C_2$$

where C_1 and C_2 are constants of integration.

The boundary conditions are;

At $\quad y = 0,\ u = 0$ and at $y = b,\ u = U.$

$$\therefore \quad C_2 = 0 \quad \text{and} \quad C_1 = \frac{U}{b} - \frac{1}{2\mu}\left(\frac{\partial p}{\partial x}\right) b$$

$$\therefore \quad u = \frac{U}{b} y - \frac{1}{2\mu}\frac{\partial p}{\partial x}(by - y^2) \qquad \text{... (6.6)}$$

Eq. (6.6) is the velocity distribution in generalised *Couette flow.*

For
$$\frac{\partial p}{\partial x} = 0, \quad u = \frac{U}{b} y \qquad \text{... (6.7)}$$

Which gives linear velocity distribution. This is called simple (or plain) Couette flow or simple shear flow.

Discharge per unit width,
$$q = \int_0^b u.dy = \int_0^b \left[\frac{U}{b} y - \frac{1}{2\mu}(by - y^2)\right] dy$$

$$= \left| \frac{U}{b}\cdot\frac{y^2}{2} - \frac{1}{2\mu}\frac{\partial p}{\partial x}\left(\frac{by^2}{2} - \frac{y^3}{3}\right)\right|_0^b$$

$$= \frac{Ub}{2} - \frac{1}{2\mu}\frac{\partial p}{\partial x}\left(\frac{b^3}{2} - \frac{b^3}{3}\right)$$

$$= \frac{Ub}{2} - \frac{b^3}{12\mu}\frac{\partial p}{\partial x} \qquad \text{... (6.8)}$$

Shear stress, $$\tau = \mu\frac{du}{dy} = \mu\left[\frac{U}{b} - \frac{1}{2\mu}\frac{\partial p}{\partial x}(b-2y)\right]$$

$$= \mu\cdot\frac{U}{b} - \frac{1}{2}\cdot\frac{\partial p}{\partial x}(b-2y) \quad \text{... (6.9)}$$

6.5.2 Both Plates Stationary

The velocity distribution and shear stress distribution between parallel stationary plates are shown in Fig. 6.4.

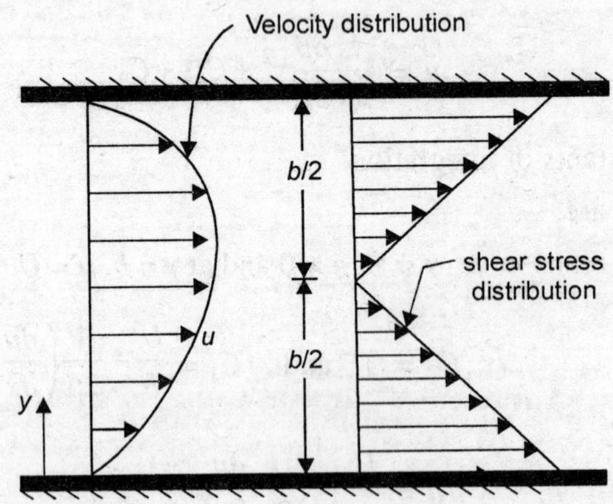

Fig. 6.4 Fllow between stationary parallel plates.

Putting $U = 0$ in Eq. (6.6), we get

$$u = -\frac{1}{2\mu}\frac{\partial p}{\partial x}(by - y^2) \quad \text{... (6.10)}$$

Eq. (6.10) represents plane Poiseville flow. Discharge per unit width from Eq. (6.8) for $U = 0$ becomes,

$$q = -\frac{b^3}{12\mu}\cdot\frac{\partial p}{\partial x} \quad \text{... (6.11)}$$

Shear stress from Eq. (6.9) for $U = 0$ becomes,

$$\tau = -\frac{1}{2}\frac{\partial p}{\partial x}(b-2y) \quad \text{... (6.12)}$$

At $y = 0$, $$\tau_0 = -\frac{1}{2}\frac{\partial p}{\partial x}\times b \quad \text{... (6.12 a)}$$

6.5.3 Both Plates Moving in Opposite Directions

Fig 6.5 shows the velocity distribution in flow between parallel plates moving in opposite directions.

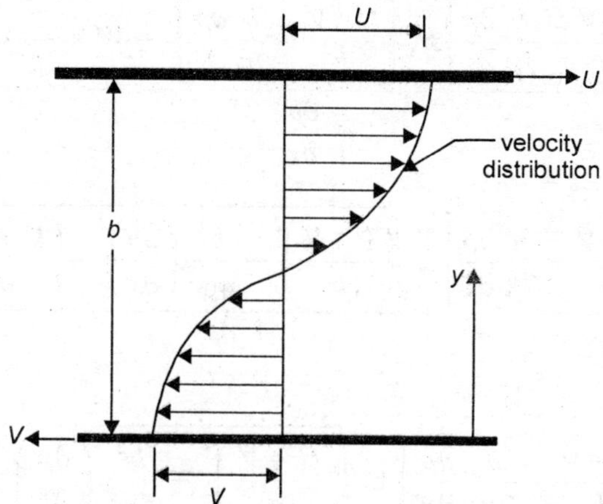

Fig. 6.5 Flow between parallel plates moving in opposite directions

Let U, V = velocities of upper and lower plates respectively.

The velocity distribution for flow between parallel plates is given by,

$$u = \frac{1}{2\mu}\frac{\partial p}{\partial x}y^2 + C_1 y + C_2$$

The boundary conditions are:

At $y = 0$, $u = -V$ and at $y = b$, $u = U$.

$$\therefore \qquad -V = C_2 \quad \text{or } C_2 = -V$$

$$U = \frac{1}{2\mu}\frac{\partial p}{\partial x}b^2 + C_1\, b - V$$

$$C_1 = \frac{1}{b}\left(U + V - \frac{1}{2\mu}\frac{\partial p}{\partial x}b^2\right) = \left(\frac{U+V}{b}\right) - \frac{b}{2\mu}\frac{\partial p}{\partial x}$$

$$\therefore \qquad u = \frac{1}{2\mu}\frac{\partial p}{\partial x}y^2 + \left[\left(\frac{U+V}{b}\right) - \frac{b}{2\mu}\frac{\partial p}{\partial x}\right]y - V$$

$$= \left(\frac{U+V}{b}\right)y - \frac{1}{2u}\frac{\partial p}{\partial x}(by - y^2) - V \qquad \text{... (6.13)}$$

For $u = 0$, we have

$$\left(\frac{U+V}{b}\right)y - \frac{2}{2u}\frac{\partial p}{\partial x}(by - y^2) - V = 0$$

or
$$\left(\frac{1}{2u}\frac{\partial p}{\partial x}\right)y^2+\left[\frac{U+V}{b}-\frac{b}{2\mu}\frac{\partial p}{\partial x}\right]y-V=0$$

$$y=\frac{-\left[\frac{U+V}{b}-\frac{b}{2\mu}\frac{\partial p}{\partial x}\right]\pm\sqrt{\left[\frac{U+V}{b}-\frac{b}{2\mu}\frac{\partial p}{\partial x}\right]^2+4V\times\frac{1}{2\mu}\times\frac{\partial p}{\partial x}}}{\frac{1}{\mu}.\frac{\partial p}{\partial x}}$$

$$=\frac{-\left[\frac{U+V}{b}-\frac{b}{2\mu}\frac{\partial p}{\partial x}\right]\pm\sqrt{\left(\frac{U+V}{b}\right)^2+\frac{b^2}{4\mu^2}\left(\frac{\partial p}{\partial x}\right)^2-\left(\frac{U+V}{\mu}\right)\frac{\partial p}{\partial x}+\frac{2V}{\mu}\cdot\frac{\partial p}{\partial x}}}{\frac{1}{\mu}\frac{\partial p}{\partial x}}$$

$$=\frac{-\left[\frac{U+V}{b}-\frac{b}{2\mu}\cdot\frac{\partial p}{\partial x}\right]\pm\sqrt{\left(\frac{U+V}{b}\right)^2+\frac{b^2}{4\mu^2}\left(\frac{\partial p}{\partial x}\right)^2-\left(\frac{U-V}{\mu}\right)\frac{\partial p}{\partial x}}}{\frac{1}{\mu}\frac{\partial p}{\partial x}} \quad \text{... (6.14)}$$

Discharge per unit width of plates,

$$q=\int_0^b u\,dy=\int_0^b\left[\left(\frac{U+V}{b}\right)y-\frac{1}{2\mu}\frac{\partial p}{\partial x}(by-y^2)-V\right]dy$$

$$=\left|\left(\frac{U+V}{2b}\right)y^2-\frac{1}{2\mu}\frac{\partial p}{\partial x}\left(\frac{by^2}{2}-\frac{y^3}{3}\right)-Vy\right|_0^b$$

$$=(U+V)\frac{b}{2}-\frac{1}{2\mu}\frac{\partial p}{\partial x}\left(\frac{b^3}{2}-\frac{b^3}{3}\right)-Vb$$

$$=(U-V)\frac{b}{2}-\frac{b^3}{12\mu}\cdot\frac{\partial p}{\partial x} \quad \text{... (6.15)}$$

Shear stress,
$$\tau=\mu\frac{du}{dy}=\mu\frac{d}{dy}\left[\left(\frac{U+V}{b}\right)y-\frac{1}{2\mu}\frac{\partial p}{\partial x}(by-y^2)-V\right]$$

$$=\mu\left[\frac{U+V}{b}-\frac{1}{2u}\frac{\partial p}{\partial x}(b-2y)\right]$$

$$=(U+V)\frac{\mu}{b}-\frac{1}{2}\frac{\partial p}{\partial x}\left(\frac{b}{2}-y\right) \quad \text{... (6.16)}$$

For $\tau = 0$, $\quad (U+V)\dfrac{\mu}{b} - \dfrac{\partial p}{\partial x}\left(\dfrac{b}{2} - y\right) = 0$

$$\frac{b}{2} - y = \frac{(U+V)\dfrac{\mu}{b}}{\left(\dfrac{\partial p}{\partial x}\right)}$$

$$y = \frac{b}{2} - \frac{\mu}{b}\left(\frac{U+V}{\dfrac{\partial p}{\partial x}}\right) \quad \text{... (6.17)}$$

6.5.4 Ratio of Maximum Velocity to Average Velocity

1. One plate Stationary and the other moving

The velocity distribution from Eq. (6.6) is given by:

$$u = \frac{U}{b}y - \frac{1}{2\mu}\frac{\partial p}{\partial x}(by - y^2)$$

For u to be maximum, $\dfrac{du}{dy} = 0$.

$$\frac{U}{b} - \frac{1}{2u}\frac{\partial p}{\partial x}(b-2y) = 0$$

$$b - 2y = \frac{2\mu U}{b\dfrac{\partial p}{\partial x}}$$

$$y = \frac{1}{2}\left[b - \frac{2\mu U}{b\dfrac{\partial p}{\partial x}}\right] = \frac{b}{2} - \frac{\mu U}{b} \times \frac{1}{\dfrac{\partial p}{\partial x}}$$

$$u_{max} = \frac{U}{b}\left[\frac{b}{2} - \frac{\mu U}{b} \times \frac{1}{\dfrac{\partial p}{\partial x}}\right] - \frac{1}{2\mu}\frac{\partial p}{\partial x}\left[\frac{b^2}{2} - \mu U \times \frac{1}{\dfrac{\partial p}{\partial x}} - \frac{b^2}{4} - \frac{\mu^2 U^2}{b^2} \times \frac{1}{\left(\dfrac{\partial p}{\partial x}\right)^2} + \mu U + \frac{1}{\dfrac{\partial p}{\partial x}}\right]$$

$$= \frac{U}{2} - \frac{\mu U^2}{b^2} \times \frac{1}{\dfrac{\partial p}{\partial x}} - \frac{b^2}{8\mu}\frac{\partial p}{\partial x} + \frac{\mu U^2}{2b^2} \times \frac{1}{\dfrac{\partial p}{\partial x}}$$

$$= \frac{U}{2} - \frac{\mu U^2}{2b^2} \times \frac{1}{\frac{\partial p}{\partial x}} - \frac{b^2}{8\mu}\frac{\partial p}{\partial x} \quad \text{... (6.18)}$$

$$\text{Average velocity, } \bar{u} = \frac{q}{b} = \frac{U}{2} - \frac{b^2}{12\mu}\frac{\partial p}{\partial x} \quad \text{... (6.19)}$$

$$\frac{u_{max}}{\bar{u}} = \frac{\frac{U}{2} - \frac{\mu U^2}{2b^2} \times \frac{1}{(\partial p/\partial x)} - \frac{b^2}{8\mu}\frac{\partial p}{\partial x}}{\frac{U}{2} - \frac{b^2}{12\mu}\frac{\partial p}{\partial x}} \quad \text{... (6.20)}$$

2. Both plates stationary

Putting $U = 0$ in Eq. (6.18), we get

$$u_{max} = -\frac{b^2}{8\mu}\frac{\partial p}{\partial x} \quad \text{... (6.21)}$$

$$\bar{u} = -\frac{b^2}{12\mu}\frac{\partial p}{\partial x} \quad \text{... (6.22)}$$

$$\frac{u_{max}}{\bar{u}} = \frac{-\frac{b^2}{8\mu}\frac{\partial p}{\partial x}}{\left(-\frac{b^2}{12\mu}\cdot\frac{\partial p}{\partial x}\right)} = \frac{3}{2} = 1.5 \quad \text{... (6.23)}$$

3. Plates moving in opposite directions

From Eq. (6.3), we have

$$u = (U+V)\frac{y}{b} - \frac{1}{2\mu}\frac{\partial p}{\partial x}(by - y^2) - V$$

For u to be maximum, $\frac{du}{dy} = 0$

$$\frac{U+V}{b} - \frac{1}{2\mu}\frac{\partial p}{\partial x}(b - 2y) = 0$$

$$y = \frac{b}{2} - \frac{\mu(U+V)}{b} \times \frac{1}{\frac{\partial p}{\partial x}}$$

$$u_{\max} = \left(\frac{U+V}{b}\right)\left[\frac{b}{2} - \frac{\mu(U+V)}{b} \times \frac{1}{\frac{\partial p}{\partial x}}\right] - \frac{1}{2u} \cdot \frac{\partial p}{\partial x}\left[\frac{b^2}{2} - u(U+V) \times \frac{1}{\frac{\partial p}{\partial x}} - \left\{\frac{b}{2} - \frac{\mu(U+V)}{b} \times \frac{1}{\frac{\partial p}{\partial x}}\right\}^2\right] - V$$

$$= \left(\frac{U+V}{2}\right) - \frac{\mu(U+V)^2}{b^2} \times \frac{1}{\partial p/\partial x} - \frac{1}{2\mu} \cdot \frac{\partial p}{\partial x}\left[\frac{b^2}{2} - \mu(U+V) \times \frac{1}{\frac{\partial p}{\partial x}} - \frac{b^2}{4} - \frac{\mu^2(U+V)^2}{b^2} \times \frac{1}{\left(\frac{\partial p}{\partial x}\right)^2} + \mu(U+V) \times \frac{1}{\frac{\partial p}{\partial x}}\right] - V$$

$$= \left(\frac{U+V}{2}\right) - \frac{\mu(U+V)^2}{b^2} \times \frac{1}{\partial p/\partial x} - \frac{1}{2\mu} \times \frac{\partial p}{\partial x}\left[\frac{b^2}{4} - \frac{\mu^2(U+V)^2}{b^2} \times \frac{1}{\left(\frac{\partial p}{\partial x}\right)^2}\right] - V$$

$$= \left(\frac{U+V}{2}\right) - \frac{\mu(U+V)^2}{b^2} \times \frac{1}{\partial p/\partial x} - \frac{b^2}{8\mu}\frac{\partial p}{\partial x} + \frac{\mu}{2b^2}(U+V)^2\frac{1}{\frac{\partial p}{\partial x}} - V$$

$$= \left(\frac{U+V}{2}\right) - \frac{\mu(U+V)^2}{2b^2} \times \frac{1}{\frac{\partial p}{\partial x}} - \frac{b^2}{8\mu}\frac{\partial p}{\partial x} - V$$

$$= \frac{U-V}{2} - \frac{\mu(U+V)^2}{2b^2} \times \frac{1}{\frac{\partial p}{\partial x}} - \frac{b^2}{8\mu} \cdot \frac{\partial p}{\partial x} \qquad \text{... (6.24)}$$

$$\bar{u} = \frac{q}{b} = \frac{U-V}{2} - \frac{b^2}{12\mu}\frac{\partial p}{\partial x} \qquad \text{... (6.25)}$$

$$\frac{u_{max}}{\bar{u}} = \frac{\frac{U-V}{2} - \frac{\mu(U+V)^2}{2b^2} \times \frac{1}{\partial p/\partial x} - \frac{b^2}{8\mu}\frac{\partial p}{\partial x}}{\frac{U-V}{2} - \frac{b^2}{12\mu}\frac{\partial p}{\partial x}} \quad \text{... (6.26)}$$

6.5.5 Drop of Pressure Head for a given Length

For both plates stationary,

$$\bar{u} = -\frac{b^2}{12\mu}\frac{\partial p}{\partial x}$$

or

$$\frac{\partial p}{\partial x} = -\frac{12\mu\bar{u}}{b^2}$$

$$\int_2^1 dp = \int_2^{j01} -\frac{12\mu\bar{u}}{b^2} dx$$

$$p_1 - p_2 = \frac{12\mu\bar{u}}{b^2}(x_1 - x_2) = \frac{12\mu\bar{u}}{b^2}(x_2 - x_1)$$

$$= \frac{12\mu\bar{u}L}{b^2} \quad \text{for } L = x_2 - x_1$$

Drop in pressure head,

$$h_l = \frac{p_1 - p_2}{\rho g} = \frac{12\mu\bar{u}L}{\rho g b^2} \quad \text{... (6.27)}$$

Example 6.1 *Determine the directin and amount of flow per metre width between two parallel plates when one is moving relative to the other with a velocity of 3m/s in the negative direction, if* $\frac{\partial p}{\partial x} = -1 \times 10^8\ \text{N/m}^3$ *and* $\mu = 0.04$ *Pa. s and distance between the plates is 1 mm.*

Solution.

Given : $U–V = -3\text{m/s}$, $\frac{\partial p}{\partial x} = -1 \times 10^8\ \text{N/m}^3$, $\mu = 0.04$ Pa.s, $b = 1$ mm

Discharge per meter width of plate,

$$q = (U-V)\frac{b}{2} - \frac{b^3}{12\mu}\cdot\frac{\partial p}{\partial x}$$

$$= -3 \times \frac{1 \times 10^{-3}}{2} - \frac{(1 \times 10^{-3})^3}{12 \times 0.04} \times (-1 \times 10^8)$$

$$= 0.2068\ \text{m}^3/\text{s}$$

The direction of flow is opposite to that of moving plate.

Example 6.2 *Two parallel plates kept 100 mm apart have laminar flow of oil between them with a maximum velocity of 1.8 m/s. Calculate the following:*
(a) Discharge per meter width, (b) Shear stress at the plates, (c) Difference in pressure between two points 10 m apart, (d) Velocity gradient at the plates, and (e) Velocity at 30 mm from the plate. Assume viscosity of oil = 2.5 Pa.s.

Solution.

Given : $b = 0.1$ m, $u_{max} = 1.8$ m/s, $y = 0.03$ m, $\mu = 2.5$ Pa.s

(a) $\bar{u} = \frac{2}{3}u_{max} = \frac{2}{3}\times 1.8 = 1.2$ m/s

$$q = \bar{u}\times b = 1.2\times 0.1 = 0.12 \text{ m}^3\text{/s}$$

(b) $\tau_0 = -\frac{1}{2}\frac{\partial p}{\partial x}\cdot b$

$$q = \frac{b^3}{12\mu}\left(-\frac{\partial p}{\partial x}\right)$$

$$0.12 = \frac{(0.1)^3}{12\times 2.5}\left(-\frac{\partial p}{\partial x}\right)$$

$$-\frac{\partial p}{\partial x} = 3600 \text{ N/m}^3$$

$$\tau_0 = \frac{1}{2}\times 3600\times 0.1 = 180 \text{ N/m}^2$$

(c) $p_1 - p_2 = 3600(x_2 - x_1)$

$$= 3600\times 10 = 36000 \text{ N/m}^2 \text{ or } 36 \text{ kPa}$$

(d) $\tau_0 = \mu\left(\frac{\partial u}{\partial y}\right)_{y=0}$

$$\left(\frac{\partial u}{\partial y}\right)_{y=0} = \frac{180}{2.5} = 72 \text{ m}^2\text{/s}$$

(e) $u = -\frac{1}{2\mu}\frac{\partial p}{\partial x}(by - y^2)$

$$= -\frac{1}{2\times 2.5}\times(-3600)[0.1\times 0.03 - (0.03)^2]$$

$$= 1.512 \text{ m/s}$$

Example 6.3 *A liquid of viscosity 0.85 Poise is fitted between two horizontal plates 10 mm apart. The upper plate is moving at 1.5 m/s w.r.t the lower plate which is stationary. The pressure difference between two sections 50 m apart is 50 kPa. Determine:*

(a) The velocity distribution, (b) Discharge per unit width, and (c) Shear stress on the upper plate.

Solution.

Given : $\mu = 0.85$ Poise $= 0.085$ Pa.s, $b = 0.01$ m, $U = 1.5$ m/s, $p_1 - p_2 = 50$ kPa, $x_2 - x_1 = 50$ m,

$$-\frac{\partial p}{\partial x} = \frac{50\times10^3}{50} = 10^3\,\text{N/m}^3$$

(a)
$$u = \frac{U}{b}y + \frac{1}{2\mu}\left(-\frac{\partial p}{\partial x}\right)(by - y^2)$$

$$= \left(\frac{1.5}{0.01}\right)y + \frac{1}{2\times0.085}\times10^3(0.01y - y^2)$$

$$= 150\,y + 5882.35\,(0.01\,y - y^2)$$

$$= 150y + 58.8235y - 5882.35y^2$$

$$= 208y - 5882.35\,y^2$$

$$= y\,(208 - 5882.35\,y)$$

(b)
$$q = \int_0^b u.dy = \int_0^{0.01}(208y - 5882.35y^2)\,dy$$

$$= |104y^2 - 1960.784y^3\,|_0^{0.01} = 0.00844\,\text{m}^3/\text{s}$$

(c)
$$\tau = \mu\frac{du}{dy} = \mu(208 - 11764.70\,y)$$

At $y = 0.01$ m

$$\tau_u = 0.085\,(208 - 11764.70\times0.01) = 7.68\,\text{N/m}^2$$

Example 6.4 *There is a horizontal crack 50 mm wide and 3 mm deep in a wall of thickness 100 mm. Water leaks through the crack. Calculate the rate of leakage of water through the crack if the difference of pressure between the two ends of the crack is 295 Pa. Take viscosity of water equal to 0.001 Pa.s.*

Solution.

Given : $w = 50$ mm, $b = 3$ mm, $L = 100$ mm, $\Delta p = 295$ Pa, $\mu = 0.001$ Pa.s.

$$p_1 - p_2 = \frac{12\mu\bar{u}L}{b^2}$$

$$295 = \frac{12\times0.001\times\bar{u}\times0.1}{(0.003)^2}$$

$$\bar{u} = 2.215 \text{ m/s}$$

Rate of leakege, $q = \bar{u} \times w \times b = 2.2125 \times 0.05 \times 0.003 = 3 \cdot 318 \times 10^{-4} \text{ m}^3/\text{s}$ or 0.3318 *lps*

6.6 FLOW OF VISCOUS FLUID IN CIRCULAR PIPES: HAGEN-POISEUILLE LAW

Consider a horizontal circular pipe of radius *R* having laminar flow of fluid through it. Consider a small cylindrical fluid element of radius *r*, thickness *dr* and length *dx*, as shown in Fig. 6.6.

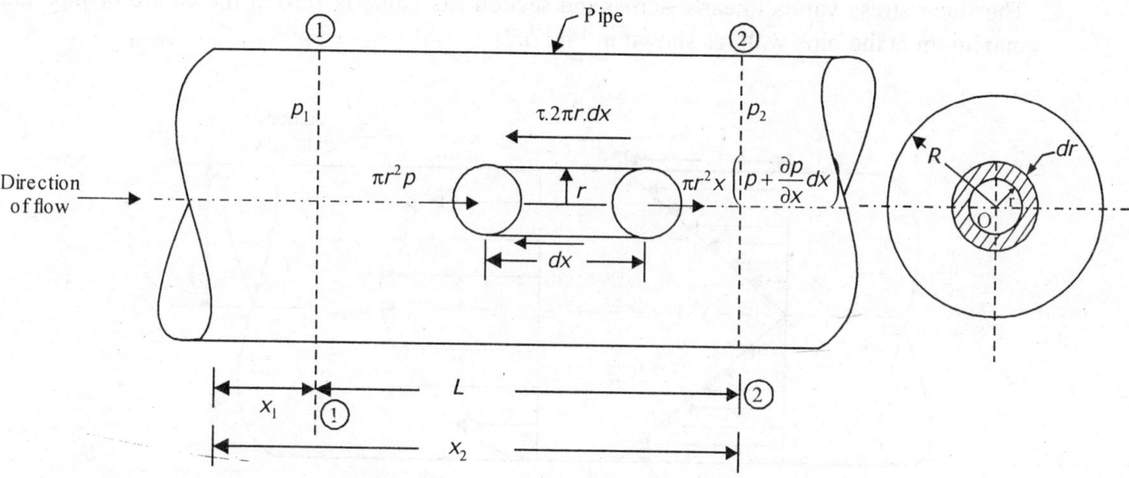

Fig.6.6 Viscous laminar flow through a circular pipe

The following assumpation are made;

1. Fluid obeys Newton's law of viscosity.
2. There is no slip of fluid paritcles at the boundary.

Let $\quad p$ = pressure at left end of fluid element.

$p + \frac{\partial p}{\partial x} dx$ = pressure at right end of fluid element.

The forces acting on fluid element are:

1. Shear force on the surface of fluid element = $\tau \times 2\pi r \times dx$
2. Force due to pressure on left end = $p \times \pi r^2$
3. Force due to pressure on right end = $\left(p + \frac{\partial p}{\partial x} \cdot dx\right) \pi r^2$

For steady flow, the net force on the fluid element must be zero.

$$\therefore \; p \times \pi r^2 - \left(p + \frac{\partial p}{\partial x} \cdot dx\right) \pi r^2 - \tau \times 2\pi r \times dx = 0$$

or $$-\frac{\partial p}{\partial x}\cdot dx \times \pi r^2 - \tau \times 2\pi r \times dx = 0$$

or $$\tau = -\frac{\partial p}{\partial x}\frac{r}{2} \quad \text{... (6.28)}$$

The following observation can be made from Eq. (6.28).

1. The flow will occur only if pressure gradient exists in the direction of flow.
2. The shear stress varies linearly across the section. Its value is zero at the centre of pipe and maximum at the pipe wall, as shown in Fig. 6.7.

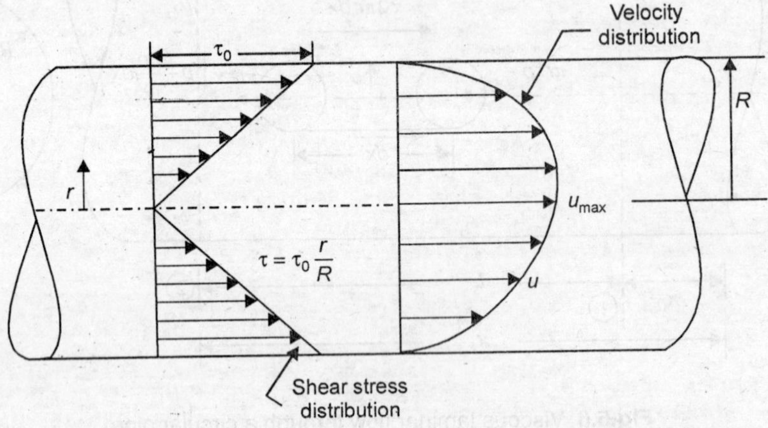

Fig. 6.7. Velocity and shear stress distribution across a section in pipe flow.

At $r = R$,

$$\tau_0 = -\frac{\partial p}{\partial x}\cdot\frac{R}{2} \quad \text{... (6.29)}$$

Thus $$\tau = \tau_0 \frac{r}{R} \quad \text{... (6.29 a)}$$

From Newton's law of viscosity

$$\tau = \mu \frac{du}{dy}$$

In this equation, the distance y is measured from the pipe boundary.

$$y = R - r$$

or $$dy = -dr$$

$\therefore$ $$\tau = -\mu \frac{du}{dr} \quad \text{... (6.30)}$$

Comparing Eqs. (6.28) and (6.30), we have

$$-\mu\frac{du}{dr} = -\frac{\partial p}{\partial x}\cdot\frac{r}{2}$$

or

$$du = \frac{1}{2\mu}\left(\frac{\partial p}{\partial x}\right)r\cdot dr$$

Integrating w.r.t r, we get

$$u = \frac{1}{4\mu}\frac{\partial p}{\partial x}\cdot r^2 + C \quad \text{... (6.31)}$$

where C = a constant of integration.

At $r = R,\ u = 0$

$$\therefore \quad 0 = \frac{1}{4u}\cdot\frac{\partial p}{\partial x}\cdot R^2 + C$$

$$C = -\frac{1}{4\mu}\cdot\frac{\partial p}{\partial x}R^2$$

$$\therefore \quad u = \frac{1}{4\mu}\cdot\frac{\partial p}{\partial x}r^2 - \frac{1}{4\mu}\frac{\partial p}{\partial x}R^2$$

or

$$u = -\frac{1}{4\mu}\frac{\partial p}{\partial x}(R^2 - r^2) \quad \text{... (6.32)}$$

Eq. (6.32) shows that the velocity distribution curve is a parabola (see Fig. 6.7). The maximum flow velocity occurs at the centre and is given by,

$$u_{max} = -\frac{1}{4\mu}\frac{\partial p}{\partial x}\cdot R^2 \quad \text{... (6.33)}$$

From Eqs. (6.32) and (6.33), we get

$$u = u_{max}\left[1 - \left(\frac{r}{R}\right)^2\right] \quad \text{... (6.34)}$$

The discharge through an elementary ring of thickness dr at radial distance r is given by,

$$dQ = u \times 2\pi r \times dr$$

$$= u_{max}\left[1 - \left(\frac{r}{R}\right)^2\right]\cdot 2\pi r\cdot dr$$

Total discharge, $$Q = \int_0^R u_{max}\left[1-\left(\frac{r}{R}\right)^2\right]\cdot 2\pi r\cdot dr$$

$$= 2\pi\, u_{max}\int_0^R\left[1-\left(\frac{r}{R}\right)^2\right]r\cdot dr$$

$$= 2\pi\, u_{max}\int_0^R\left[r-\frac{r^3}{R^2}\right]dr$$

$$= 2\pi\, u_{max}\left|\frac{r^2}{2}-\frac{r^4}{4R^2}\right|_0^R = 2\pi u_{max}\left(\frac{R^2}{2}-\frac{R^2}{4}\right)$$

$$= \frac{\pi}{2}u_{max}\cdot R^2 \qquad \text{... (6.35)}$$

Average velocity of flow , $$\bar{u} = \frac{Q}{A} = \frac{\frac{\pi}{2}u_{max}R^2}{\pi R^2} = \frac{u_{max}}{2} \qquad \text{... (6.36)}$$

Substituting the value of u_{max} from Eq. (6.33), we have

$$\bar{u} = -\frac{1}{8\mu}\cdot\frac{\partial p}{\partial x}\cdot R^2$$

or $$-\partial p = \frac{8\mu\bar{u}}{R^2}\cdot\partial x$$

The pressure difference between two sections 1 and 2 at distance x_1 and x_2 (see Fig. 6.6), is given by,

$$-\int_{p_1}^{p_2}\partial p = \frac{8\mu\bar{u}}{R^2}\int_{x_1}^{x_2}\partial x$$

$$p_1 - p_2 = \frac{8\mu\bar{u}}{R^2}(x_2 - x_1) = \frac{8\mu\bar{u}L}{R^2} = \frac{32\mu\bar{u}L}{D^2} \qquad \text{... (6.37)}$$

where D = diameter of pipe

L = length of pipe

Eq. (6.37) is known as Hagen-Poiseuille equation.

Example 6.5 *An oil of viscosity 0.86 Pa. s and specific gravity 0.92 is flowing through a horizontal pipe of 50 mm diameter. The pressure drop in 80 m length of pipe is 1500 kPa. Calculate:*

(a) Flow rate of oil (b) Maximum flow velocity (c) Total friction drag (d) Power required to maintain the flow (e) Velocity gradient at the pipe wall, and (f) Velocity and shear stress at 10 mm from the wall.

Solution.

Given : $\mu = 0.86$ Pa.s, s = 0.92, $D = 50$ mm, $L = 80$ m, $\Delta p = 1500$ kPa, $y = 10$ mm.

(a) $$\Delta p = \frac{32\mu \bar{u} L}{D^2}$$

$$1500 \times 10^3 = \frac{32 \times 0.86 \times \bar{u} \times 80}{(0.05)^2}$$

$$\bar{u} = 1.7 \text{ m/s}$$

$$Re = \frac{\rho \bar{u} D}{\mu} = \frac{0.92 \times 10^3 \times 1.7 \times 0.05}{0.86} = 90.93$$

As $Re < 2000$, the flow is laminar.

Rate of flow, $Q = A\bar{u} = \frac{\pi}{4}(0.05)^2 \times 1.7 = 3.338 \times 10^{-3}\,\text{m}^3/\text{s}$ or 3.338 *lps*.

(b) $u_{max} = 2\bar{u} = 2 \times 1.7 = 3.4$ m/s

(c) Wall shear stress, $\tau_0 = -\frac{\partial p}{\partial x} \cdot \frac{R}{2}$

$$-\frac{\partial p}{\partial p} = \frac{\Delta p}{L} = \frac{1500 \times 10^3}{80} = 18{,}750$$

$$\tau_0 = 18750 \times \frac{0.05}{4} = 234.375 \text{ N/m}^2$$

Frictional drag, $F_D = \tau_0 \times \pi D L = 234.375 \times \pi \times 0.05 \times 80 = 2945.2$ N

(d) Power required to maintain flow,

$$P = F_D \times \bar{u} = 2945.2 \times 1.7 \times 10^{-3} = 5 \text{ kW}$$

$$P = Q \times \Delta p = 3.338 \times 10^{-3} \times 1500 \times 10^3 \times 10^{-3} = 5 \text{ kW}$$

(e) $\tau_0 = \mu \left(\frac{du}{dy}\right)_{y=0}$

$$\left(\frac{du}{dy}\right)_{y=0} = \frac{\tau_0}{\mu} = \frac{234.375}{0.86} = 272.53\ s^{-1}$$

(f) $u = -\frac{1}{4\mu}\frac{\partial p}{\partial x}(R^2 - r^2)$

$$y = R - r, r = 25 - 10 = 15 \text{ mm}$$

$$u = -\frac{1}{4\times 0.86}\times\frac{1500\times 10^3}{80}\times[(0.025)^2-(0.015)^2]$$

$$= 2.18 \text{ m/s}$$

$$\tau = \tau_0\frac{r}{R} = 234.375\times\frac{15}{25} = 140.625 \text{ N/m}^2$$

Example 6.6 *A pipe is of 200 mm diameter and the maximum velocity of flow is 2m/s. If the flow in the pipe is laminar, calculate:*

(a) The average velocity and the radius it which it occurs, and

(b) The velocity at 40 mm from the wall of the pipe.

Solution.

Given : $D = 200$ *mm*, $u_{max} = 2$ *m/s*, $y = 40$ *mm.*

(a) $$\bar{u} = \frac{u_{max}}{2} = \frac{2}{2} = 1 \text{ m/s}$$

$$u = u_{max}\left[1-\left(\frac{r}{R}\right)^2\right]$$

$$u = \bar{u} = 1 \text{ m/s}$$

$$1 = 2\left[1-\left(\frac{r}{100}\right)^2\right]$$

$$r = 100\sqrt{\frac{1}{2}} = 70.71 \text{ mm}$$

(b) $$y = R - r$$

$$40 = 100 - r \text{ or } \quad r = 60 \text{ mm}$$

$$u = 2\left[1-\left(\frac{60}{100}\right)^2\right] = 1.28 \text{ m/s}$$

Example 6.7 *A crude oil of viscosity 0.15 Pa.s and specific gravity 0.85 flows through a 20 mm diameter vertical pipe. Two pressure gauges fixed 15 m apart read 500 kPa and 200 kPa. Find the direction and rate of flow through the pipe.*

Solution.

Given : $\mu = 0.15$ Pa.s, s = 0.85, $D = 20$ mm, $L = 15$m, $p_1 = 500$ kPa, $p_2 = 200$ kPa.

Specfic weight of oil, $\gamma = g\, s\rho_w = 9.81 \times 0.85 \times 10^3 = 8338.5 \text{ N/m}^3$

$$h_1 = \frac{p_1}{\gamma} + z_1 = \frac{500\times 10^3}{8338.5} + 0 = 59.963 \text{ m}$$

$$h_2 = \frac{p_2}{\gamma} + z_2 = \frac{200 \times 10^3}{8338.5} + 15 = 38.985 \text{ m}$$

$h_1 > h_2$, therefore flow takes place from sectin 1 to 2.

Loss of head $h_l = h_1 - h_2 = 59.963 - 38.985 = 20.978$ m

$$h_l = \frac{32\mu\bar{u}L}{\gamma D^2}$$

$$20.978 = \frac{32 \times 0.15 \times \bar{u} \times 15}{8338.5 \times (0.02)^2}$$

$$\bar{u} = 0.972 \text{ m/s}$$

Flow rate of oil, $Q = \bar{u} \times \frac{\pi}{4} D^2 = 0.972 \times \frac{\pi}{4}(0.02)^2 = 3.054 \times 10^{-4} \text{ m}^3\text{/s}$ or 0.3054 lit/s

Example 6.8 *A total of 15 litres per second of an oil is pumped through two pipes in parallel, one of 100 mm is diameter and the other of 120 mm diameter, Both pipes are 800 m long. The specific gravity of the oil is 0.96 and kinematic viscosity 9 cm²/s. Calcualte the flow rate through each pipe and the power of the pump if its efficiency is 60 percent.*

Solution.

Given : $Q = 0.015$ m³/s, $D_1 = 0.1$ m, $D_2 = 0.12$m, L =800 m, $s = 0.96$, $\nu = 9$ cm²/s or 9×10^{-4} m²/s, $\eta_p = 0.6$.

For two pipes in parallel,

$$Q = Q_1 + Q_2$$

$$0.015 = Q_1 + Q_2 \quad \text{... (1)}$$

For laminar flow,

$$\Delta p = \frac{32\mu u_1 L}{D_1^2} = \frac{32\mu u_2 L}{D_2^2} \quad \text{... (2)}$$

$$u_1 = \frac{4Q_1}{\pi D_1^2}, u_2 = \frac{4Q_2}{\pi D_2^2}$$

From Eq. (2), we have

$$\frac{u_1}{D_1^2} = \frac{u_2}{D_2^2}$$

$$\frac{4Q_1}{\pi D_1^4} = \frac{4Q_2}{\pi D_2^4}$$

$$\frac{Q_1}{Q_2} = \left(\frac{D_1}{D_2}\right)^4 \quad \text{... (3)}$$

Substituting in Eq. (1), we get

$$0.015 = Q_1\left[1+\left(\frac{D_2}{D_1}\right)^4\right] = Q_1\left[1+\left(\frac{0.12}{0.1}\right)^4\right] = 3.0736\,Q_1$$

$$Q_1 = 4.8803 \times 10^{-3}\ \text{m}^3/\text{s or } 4.8803\ \text{lit/s}$$

$$u_1 = \frac{4\times 4.8803\times 10^{-3}}{\pi(0.9)^2} = 0.621\ \text{m/s}$$

$$u_2 = \frac{4\times 10.1197\times 10^{-3}}{\pi(0.12)^2} = 0.895\ \text{m/s}$$

Loss of head, $$h_l = \frac{\Delta p}{\gamma} = \frac{32\mu u_1 L}{\gamma D_1^2}$$

$$\nu = \frac{\mu}{\rho},\ u = \rho\nu$$

$$h_l = \frac{\Delta p}{\gamma} = \frac{32\mu u_1 L}{g D_1^2}$$

$$= \frac{32\times 9\times 10^{-4}\times 0.621\times 800}{9.81\times(0.1)^2} = 145.85\,\text{m of oil}$$

Power loss, $P = \gamma Q \,.\, h_l = 0.96\times 9.81\times 0.015\times 145.85 = 20.6\ \text{kW}$

Required pump power $= \dfrac{P}{\eta_p} = \dfrac{20.6}{0.6} = 34.34\ \text{kW}$

Example 6.9 *The velocity distrbution in a pipe is given by*

$$\frac{u}{u_{\max}} = 1-\left(\frac{r}{R}\right)^n$$

Obtain an expression for mean velocity in terms of u_{max} and n.

Solution.

Given : Consider a small fluid ring of a radius r and thickness dr. The discharge through this ring is,

$$dQ = u\times 2\pi r\,dr = u_{\max}\left[1-\left(\frac{r}{R}\right)^n\right]\times 2\pi r\,.dr$$

Integrating, we get

$$Q = \int_0^R u_{max}\left[1-\left(\frac{r}{R}\right)^n\right]\times 2\pi r.dr$$

$$= 2\pi\, u_{max}\int_0^R\left[r-\frac{r^{n+1}}{R^n}\right]dr$$

$$= 2\pi u_{max}\left|\frac{r^2}{2}-\frac{r^{n+2}}{(n+2)R^n}\right|_0^R = 2\pi u_{max}\left[\frac{R^2}{2}-\frac{R^2}{n+2}\right]$$

$$= 2\pi u_{max} R^2\left[\frac{n+2-2}{2(n+2)}\right] = \frac{\pi n}{(n+2)}R^2\cdot u_{max}$$

Mean velocity, $$\bar{u} = \frac{Q}{\pi R^2} = \left(\frac{n}{n+1}\right)u_{max}$$

6.6.1 Kinetic Energy Correction Factor, α

Kinetic energy of fluid flowing through an elementary ring of radius r and thickness *dr* per second is given by

$$= \frac{1}{2}\times \text{mass}\times(\text{velocity})^2 = \frac{1}{2}(\rho.dQ)\times u^2$$

$$= \frac{1}{2}(\rho\times u\times 2\pi r\times dr)\times u^2 = \pi\rho u^3 rdr$$

Total kinetic energy of flow per second $= \int_0^R \pi\rho u^3\, rdr$

$$= \pi\rho\int_0^R r\left[-\frac{1}{4\mu}\left(\frac{\partial p}{\partial x}\right)(R^2-r^2)\right]^3 dr$$

$$= \pi\rho\left[-\frac{1}{4\mu}\left(\frac{\partial p}{\partial x}\right)\right]^3\int_0^R (R^2-r^2)^3\, r\,dr$$

$$= \pi\rho\left[-\frac{1}{4\mu}\left(\frac{\partial p}{\partial x}\right)\right]^3\int_0^R (R^6-r^6-3R^4r^2+3R^2r^4)r\,dr$$

$$= \pi\rho\left[-\frac{1}{4\mu}\left(\frac{\partial p}{\partial x}\right)\right]^3\int_0^R (R^6-r^7-3R^4r^3+3R^2r^5)\,dr$$

$$= \pi\rho\left[-\frac{1}{4\mu}\left(\frac{\partial p}{\partial x}\right)\right]^3\left|\frac{R^6r^2}{2}-\frac{r^8}{8}-\frac{3}{4}R^4r^4+\frac{1}{2}R^2r^6\right|_0^R$$

$$= \pi\rho\left[-\frac{1}{4\mu}\left(\frac{\partial p}{\partial x}\right)\right]^3\left[\frac{R^8}{2}-\frac{R^8}{8}-\frac{3}{4}R^8+\frac{1}{2}R^8\right]$$

$$= \pi\rho\left[-\frac{1}{4\mu}\left(\frac{\partial p}{\partial x}\right)\right]^3\frac{R^8}{8}(4-1-6+4]$$

$$= \frac{\pi\rho}{8}\left[-\frac{1}{4\mu}\left(\frac{\partial p}{\partial x}\right)\right]^3 R^8$$

$$= \frac{-\pi\rho}{512\mu^3}\left(\frac{\partial p}{\partial x}\right)^3 R^8$$

Kinetic energy of flow based on average velocity

$$= \frac{1}{2}\times \text{mass}(\text{averag velocity})^2$$

$$= \frac{1}{2}\times\rho A\bar{u}\times\bar{u}^2 = \frac{1}{2}\times\rho A\bar{u}^3$$

$$= \frac{1}{2}\times\rho\times\pi R^2\times\left[-\frac{1}{8\mu}\left(\frac{\partial p}{\partial x}\right)R^2\right]^3$$

$$= -\frac{\pi\rho}{1024\mu^3}\left(\frac{\partial p}{\partial x}\right)^3 R^8$$

Kinetic energy correction factor, $\alpha = \dfrac{\text{K.E. per second based on actual velocity}}{\text{K.E. per second based on average velocity}}$

$$= \frac{\dfrac{\pi\rho}{512\mu^3}\left(\dfrac{\partial p}{\partial x}\right)^3 R^8}{-\dfrac{\pi\rho}{1024\mu^3}\left(\dfrac{\partial p}{\partial x}\right)^3 R^8} = 2 \qquad \text{... (6.38)}$$

6.6.1 Momentum Correction Factor, β

$$u = -\frac{1}{4\mu}\left(\frac{\partial p}{\partial x}\right)(R^2 - r^2)$$

Consider an elementary ring of radius r and thickness dr of pipe fluid.

$$dA = 2\pi r \cdot dr$$

$$dQ = u \times dA = u \times 2\pi r . dr$$

Momentum of fluid passing through the ring per second

$$= m \times u = (\rho . dQ) \times u = \rho \times u \times 2\pi r \times dr \times u$$

$$= 2\pi \rho u^2 r \cdot dr$$

Total actual momentum of fluid per second across the section

$$= \int_0^R 2\pi \rho u^2 r \cdot dr$$

$$= 2\pi\rho \int_0^R \left[-\frac{1}{4\mu}\left(\frac{\partial p}{\partial x}\right)(R^2 - r^2)\right]^2 r \cdot dr$$

$$= \frac{\pi\rho}{8\mu^2}\left(\frac{\partial p}{\partial x}\right)^2 \int_0^R (R^4 r + r^5 - 2R^2 r^3)\, dr$$

$$= \frac{\pi\rho}{8\mu^2}\left(\frac{\partial p}{\partial x}\right)^2 \left| \frac{R^4 r^2}{2} + \frac{r^6}{6} - \frac{R^2 r^4}{2} \right|_0^R$$

$$= \frac{\pi\rho}{8\mu^2}\left(\frac{\partial p}{\partial x}\right)^2 R^6 \left(\frac{1}{2} + \frac{1}{6} - \frac{1}{2}\right)$$

$$= \frac{\pi\rho}{48\mu^2}\left(\frac{\partial p}{\partial x}\right)^2 R^6$$

Momentum of fluid per second based on average velocity

$$= \bar{m} \times \bar{u} = \rho A \bar{u} \times \bar{u} = \rho A \bar{u}^2$$

$$\bar{u} = \frac{u_{\max}}{2} = -\frac{1}{8\mu}\left(\frac{\partial p}{\partial x}\right) R^2$$

Momenutum of fluid per second based on average velocity

$$= \rho \times \pi R^2 \times \left[-\frac{1}{8\mu}\left(\frac{\partial p}{\partial x}\right) R^2 \right]^2$$

$$= \frac{\pi\rho}{64\mu^2}\left(\frac{\partial p}{\partial x}\right)^2 \cdot R^6$$

$$\beta = \frac{\dfrac{\pi\rho}{48\mu^2}\left(\dfrac{\partial p}{\partial x}\right)^2 R^6}{\dfrac{\pi\rho}{64\mu^3}\left(\dfrac{\partial p}{\partial x}\right) R^6} = \frac{4}{3} \qquad \text{... (6.39)}$$

6.7 FLOW OF VISCOUS FLUID THROUGH AND ANNULUS

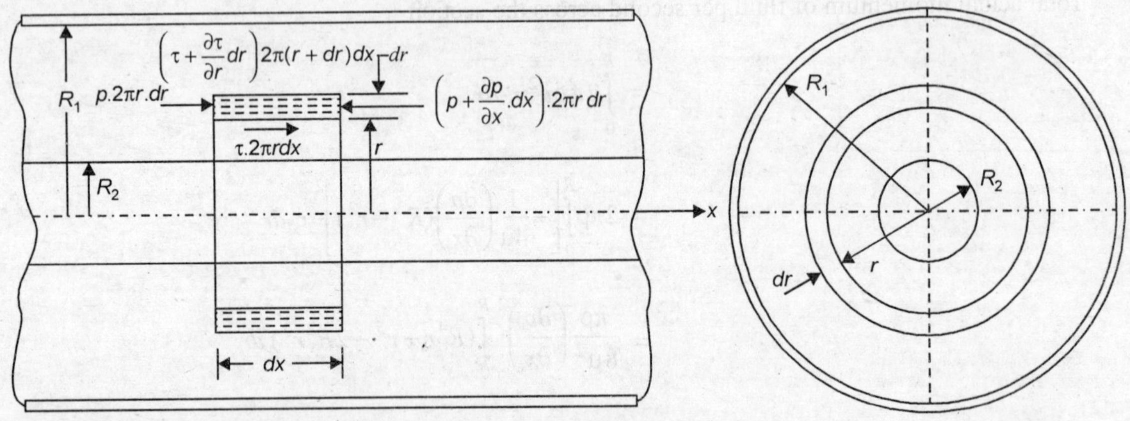

Fig. 6.8 Laminar flow through an annulus.

Consider an annulus of outer radius R_1 and inner radius R_2 through which steady laminar flow of an incompressible fluid is taking place. Consider a fluid element having shape of a small concentric cylindrical sleeve of length dx between radial distance r and $r + dr$, as shown in Fig. 6.8. The forces acting on the fluid element in the direction of flow are:

1. Normal pressure forces over the end areas,

$$F_p = p.2\pi r\cdot dr - \left(p + \frac{\partial p}{\partial x}\cdot dx\right)\cdot 2\pi r\cdot dr$$

$$= -\frac{\partial p}{\partial x}\cdot dx.2\pi r.dr$$

2. Shear forces over the inner and outer curved surfaces.

$$F_s = \tau.2\pi r.dx - \left(\tau + \frac{\partial \tau}{\partial r}dr\right)\cdot 2\pi(r+dr)\cdot dx$$

$$= \tau.2\pi r.dx - \tau\times 2\pi\,(r+dr).dx - 2\pi(r+dr)\cdot dx.\frac{\partial\tau}{\partial r}dr$$

$$= -2\pi\tau.dr.dx - 2\pi r\frac{\partial\tau}{\partial r}\cdot dr.dx \quad \text{(neglecting small quantities)}$$

Resultant force in x-direction = $F_p + F_s$

$$= -2\pi r\frac{\partial p}{\partial x}\cdot dr.dx - 2\pi\tau.dr.dx - 2\pi r\frac{\partial \tau}{\partial r}dr\,dx = 0$$

$$-\frac{\partial p}{\partial x} - \frac{\tau}{r} - \frac{\partial \tau}{\partial r} = 0$$

or
$$\frac{\partial p}{\partial x} + \frac{\tau}{r} + \frac{\partial \tau}{\partial r} = 0 \quad \text{... (6.40a)}$$

or
$$\frac{\partial p}{\partial x} + \frac{1}{r}\frac{\partial}{\partial r}(\tau r) = 0 \quad \text{... (6.40 b)}$$

Integrating w.r.t. r, we get

$$\frac{\partial p}{\partial x}\int r.dr + \int \frac{d}{dr}(\tau r).dr = C_1$$

where C_1 = a constant of integration.

$$\left(\frac{\partial p}{\partial x}\right)\frac{r^2}{2} + \tau r = C_1$$

Now
$$\tau = -\mu\frac{\partial u}{\partial r}$$

$\therefore$
$$\left(\frac{\partial p}{\partial x}\right)\frac{r^2}{2} - \mu\frac{\partial u}{\partial r}\cdot r = C_1$$

Dividing throughout by r, we get

$$\left(\frac{\partial p}{\partial x}\right)\frac{r}{2} - \mu\frac{\partial u}{\partial r} = \frac{C_1}{r}$$

Integrating w.r.t. r, we get

$$\left(\frac{\partial p}{\partial x}\right)\frac{r^2}{4} - \mu u = C_1 \ln r + C_2$$

where C_2 = another constant of integration.

The boundary conditions are:

At $r = R_1$, $u = 0$ and at $r = R_2$, $u = 0$

$$\left(\frac{\partial p}{\partial x}\right)\frac{R_1^2}{4} = C_1 \ln R_1 + C_2 \quad \text{... (i)}$$

$$\left(\frac{\partial p}{\partial x}\right)\frac{R_2^2}{4} = C_1 \ln R_2 + C_2 \quad \text{... (ii)}$$

Substracting Eq. (ii) from (i), we get

$$C_1 \ln\frac{R_1}{R_2} = \frac{1}{4}\left(\frac{\partial p}{\partial x}\right)(R_1^2 - R_2^2)$$

$$C_1 = \frac{1}{\ln\frac{R_1}{R_2}}\left[\frac{1}{4}\left(\frac{\partial p}{\partial x}\right)(R_1^2 - R_2^2)\right]$$

$$C_2 = \left(\frac{\partial p}{\partial x}\right)\frac{R_1^2}{4} - \frac{\ln R_1}{\ln\frac{R_1}{R_2}}\left[\frac{1}{4}\left(\frac{\partial p}{\partial x}\right)(R_1^2 - R_2^2)\right]$$

$$u = \frac{1}{\mu}\left[\left(\frac{\partial p}{\partial x}\right)\frac{r^2}{4} - \frac{\ln r}{\ln\frac{R_1}{R_2}} \times \frac{1}{4}\left(\frac{\partial p}{\partial x}\right)(R_1^2 - R_2^2) - \left(\frac{\partial p}{\partial x}\right)\frac{R_1^2}{4} + \frac{\ln R_1}{\ln\frac{R_1}{R_2}} \times \left(\frac{\partial p}{\partial x}\right)(R_1^2 - R_2^2)\right]$$

or

$$u = \frac{-1}{4\mu}\left(\frac{\partial p}{\partial x}\right)\left[R_1^2 - r^2 - \frac{R_1^2 - R_2^2}{\ln\frac{R_1}{R_2}}\ln\frac{R_1}{r}\right] \quad \text{... (6.41)}$$

For u to be maximum, $\frac{\partial u}{\partial r} = 0$

$$-2r + \frac{R_1^2 - R_2^2}{\ln\frac{R_1}{R_2}} \times \frac{1}{r} = 0$$

$$r = \left[\frac{R_1^2 - R_2^2}{2\ln\frac{R_1}{R_2}}\right]^{1/2} \quad \text{... (6.42)}$$

Discharge through the annulus,

$$Q = \int_{R2}^{R_1} 2\pi r.dr.u$$

$$= 2\pi \int_{R_2}^{R_1} -\frac{1}{4\mu}\left(\frac{\partial p}{\partial x}\right)\left[R_1^2 - r^2 - \frac{R_1^2 - R_2^2}{\ln\frac{R_1}{R_2}}\cdot \ln\frac{R_1}{r}\right] r.dr$$

$$= -\frac{2\pi}{4\mu}\left(\frac{\partial p}{\partial x}\right)\int_{R_2}^{R_1}\left[R_1^2 r - r^3 - \frac{R_1^2 - R_2^2}{\ln\frac{R_1}{R_2}}(r\ln R_1 - r\ln r)\right] dr$$

$$= -\frac{\pi}{2\mu}\left(\frac{\partial p}{\partial x}\right)\left|\frac{R_1^2 r^2}{2} - \frac{r^4}{4} - \frac{R_1^2 - R_2^2}{\ln\frac{R_1}{R_2}}\left(\frac{r^2}{2}\ln R_1 - \frac{r^2}{2}\ln r + \frac{r^2}{4}\right)\right|_{R_2}^{R_1}$$

$$= -\frac{\pi}{2\mu}\left(\frac{\partial p}{\partial x}\right)\left[\frac{R_1^2}{2}(R_1^2 - R_1^2) - \frac{1}{4}(R_1^4 - R_2^4) - \frac{R_1^2 - R_2^2}{\ln\frac{R_1}{R_2}}\left\{\frac{\ln R_1}{2}(R_1^2 - R_2^2)\right.\right.$$

$$\left.\left. - \frac{1}{2}(R_1^2 \ln R_1 - R_2^2 \ln R_2) + \frac{1}{4}(R_1^2 - R_2^2)\right\}\right]$$

$$= -\frac{\pi}{2\mu}\left(\frac{\partial p}{\partial x}\right)\left[\frac{1}{2}R_1^4 - \frac{1}{2}R_1^2 R_2^2 - \frac{1}{4}R_1^4 + \frac{1}{4}R_2^4 - \frac{R_1^2 - R_2^2}{\ln\frac{R_1}{R_2}}\left\{\ln R_1\left(\frac{R_1^2}{2} - \frac{R_2^2}{2} - \frac{R_1^2}{2}\right)\right.\right.$$

$$\left.\left. + \frac{R_2^2}{2}\ln R^2 + \frac{1}{4}R_1^2 - \frac{1}{4}R_2^2\right\}\right]$$

$$= -\frac{\pi}{2\mu}\left(\frac{\partial p}{\partial x}\right)\left[\frac{R_1^4}{4} - \frac{1}{2}R_1^2 R_2^2 + \frac{1}{4}R_2^4 - \frac{R_1^2 - R_2^2}{\ln\frac{R_1}{R_2}}\left\{-\frac{R_2^2}{2}\ln R_1 + \frac{R_2^2}{2}\ln R_2 + \frac{1}{4}(R_1^2 - R_2^2)\right\}\right]$$

$$= -\frac{\pi}{2\mu}\left(\frac{\partial p}{\partial x}\right)\left[R_1^4 + R_2^4 - 2R_1^2 R_2^2 - \frac{R_1^2 - R_2^2}{\ln\frac{R_1}{R_2}}\left\{-2R_2^2 \ln\frac{R_1}{R_2}(R_1^2 - R_2^2)\right\}\right]$$

$$= -\frac{\pi}{8\mu}\left(\frac{\partial p}{\partial x}\right)\left[R_1^4 + R_2^4 - 2R_1^2R_2^2 - (R_1^2 - R_2^2)\left\{-2R_2^2 + \frac{(R_1^2 - R_2^2)}{\ln\frac{R_1}{R_2}}\right\}\right]$$

$$= -\frac{\pi}{8\mu}\left(\frac{\partial p}{\partial x}\right)\left[R_1^4 - R_2^4 - \frac{(R_1^2 - R_2^2)^2}{\ln\frac{R_1}{R_2}}\right]$$

Average velocity, $\bar{u} = \dfrac{Q}{\pi(R_1^2 - R_2^2)} = -\dfrac{1}{8\mu}\left(\dfrac{\partial p}{\partial x}\right)\left[R_1^2 + R_2^2 - \dfrac{R_1^2 - R_2^2}{\ln\frac{R_1}{R_2}}\right]$

Shear stress, $\tau = -\mu\dfrac{du}{dr}$

$$\frac{du}{dr} = -\frac{1}{4\mu}\left(\frac{\partial p}{\partial x}\right)\left[-2r - \frac{R_1^2 - R_2^2}{\ln\frac{R_1}{R_2}}\left(-\frac{1}{r}\right)\right]$$

$$= -\frac{1}{4\mu}\left(-\frac{\partial p}{\partial x}\right)\left[2r - \frac{1}{r}\frac{R_1^2 - R_2^2}{\ln\frac{R_1}{R_2}}\right]$$

$$\tau = \frac{1}{4}\left(-\frac{\partial p}{\partial x}\right)\left[2r - \frac{1}{r}\frac{R_1^2 - R_2^2}{\ln\frac{R_1}{R_2}}\right] \quad \ldots (6.45$$

6.8 POWER ABSORBED IN BEARINGS

A very thin film of lubricating oil is maintained between the journal and the bearing. The lubricatin oils are viscous and hence the theory of laminar viscous flow can be applied to bearings.

6.8.1 Journal Bearing

Fig 6.9 shows a journal bearing and shaft. A lubricating oil is filled in the clearance space between the bearing and the journal.

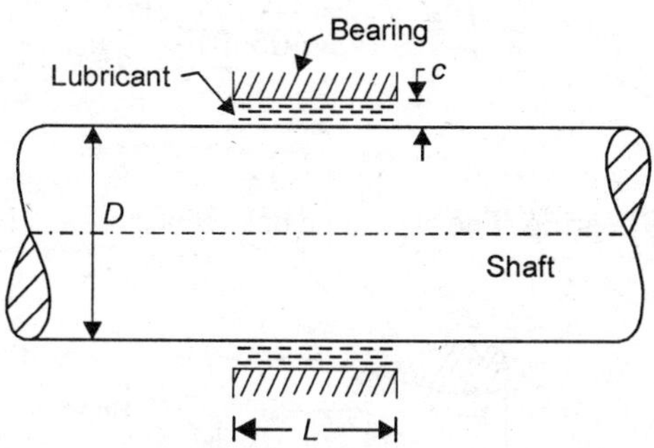

Fig. 6.9 Journal bearing

Let D = diameter of journal, m

L = Length of bearing, m

c = clearance between bearing and journal, m

N = speed of journal, rpm.

h_0 = oil film thickness, $m = c$

Angular speed of journal, $\omega = \dfrac{2\pi N}{60}$ rad/s

Tangential speed of journal, $u = \dfrac{\pi DN}{60}$ m/s

As the oil film thickness is very small, a linear velocity distribution can be presumed.

Hence
$$\frac{du}{dy} = \frac{u-0}{h} = \frac{u}{h_0} = \frac{\pi DN}{60 \times h_0}$$

Shear stress,
$$\tau = \mu\frac{du}{dy} = \mu \times \frac{\pi DN}{60 \times h_0}$$

Shear force or viscous resistance,
$$F = \tau \times \text{surface area of journal}$$
$$= \tau \times \pi DL$$
$$= \frac{\mu \pi^2 D^2 NL}{60 \times h_0}$$

Torque required to overcome viscous resistance,

$$T = F \times \frac{D}{2} = \frac{\mu \pi^2 D^3 NL}{120 \times h_0} \quad \text{... (6.46)}$$

Power absorbed ,
$$P = T\omega = \frac{2\pi NT}{60 \times 10^3} kW \quad \text{... (6.47)}$$

6.8.2 Foot-step Bearing

Fig. 6.10. shows a foot-step bearing. The clearance between shaft and bearing is filled with a lubricant of viscosity μ.

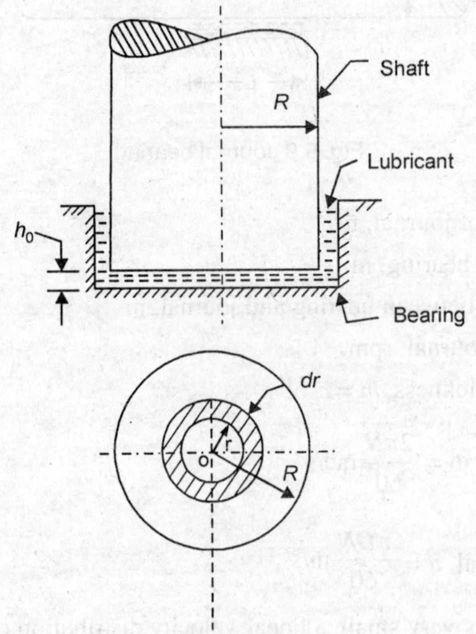

Fig. 6.10 Foot-step bearing

Let R = radius of shaft, m

N = speed of shaft, rpm.

h_0 = thickness of oil film.

Consider an elementary circular ring of radius r and thickness dr of oil film.

Area of elementary ring, $dA = 2\pi r \cdot dr$

Shear stress,
$$\tau = \mu \frac{du}{dy} = \frac{\mu u}{h_0}$$

where u = tangential velocity of shaft at radius r.

$$u = \omega r = \frac{2\pi\, Nr}{60} \text{ m/s}$$

Shear force on elementary ring, $dF = \tau \times dA$

$$= \mu \times \frac{2\pi NR}{60 \times h_0} \times 2\pi r . dr$$

$$= \frac{\mu}{15} \times \frac{\pi^2 N}{h_0} \times r^2 \cdot dr$$

Total torque required to overcome viscous resistance,

$$T = \frac{\mu \pi^2 N}{15 \times h_0} \int_0^R r^3 \cdot dh$$

Total torque required to overcome viscous resistance,

$$T = \frac{\mu\pi^2 N}{15 \times h_0} \int_0^R r^3 \cdot dh$$

$$= \frac{\mu\pi^2 N}{15 \times h_0} \left| \frac{r^4}{4} \right|_0^R = \frac{\mu\pi^2 NR^4}{60 \times h_0} \qquad \text{... (6.48)}$$

Power absorbed by the bearing, $P = T\omega = \frac{2\pi NT}{60}$ Watts ... (6.49)

6.8.3 Collar Bearing

The collar bearing is shown in Fig. 6.11. It is used to take up axial thrust in the shaft.

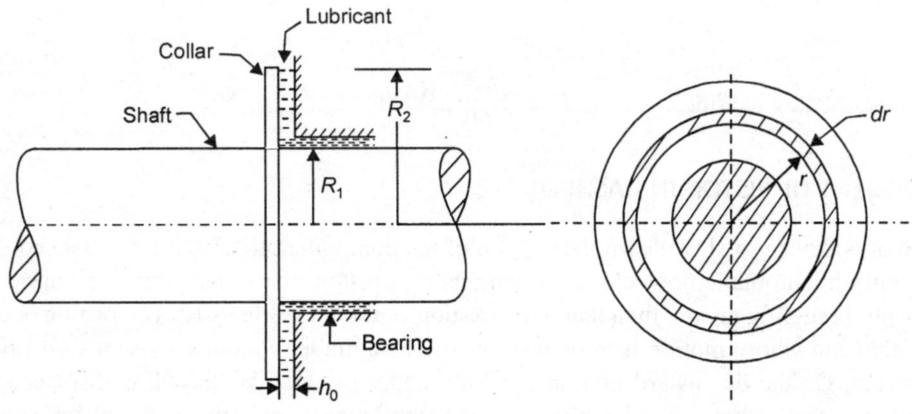

Fig. 6.11 Collar bearing

Let R_1, R_2 = inner and outer radii of collar respectively

h_0 = thickness of oil film.

N = speed of shaft, rpm.

Consider an elementary ring of radius r and thickness dr.

Area of elementary ring, $dA = 2\pi r.dr$

Shear stress, $$\tau = \mu\frac{du}{dy} = \mu\frac{u}{h_0}$$

where $$u = \frac{2\pi r N}{60}$$

Shear force on elementary ring, $$dF = \tau \times dA = \mu \times \frac{2\pi r N}{60 \times h_0} \times 2\pi r\, dr$$

$$= \frac{\mu}{15} \times \frac{\pi^2 N}{h_0} \times r^2 \cdot dr$$

Torque on elementary ring, $$dT = dF \times r = \frac{\mu}{15} \times \frac{\pi^2 N}{h_0} \times r^3 \cdot dr$$

Total torque required to overcome viscous resitance,

$$T = \int_{R_1}^{R_2} \left(\frac{\mu\pi^2 N}{15 \times h_0}\right) \times r^3 \cdot dr$$

$$= \frac{\mu\pi^2 N}{60 \times h_0} \left| r^4 \right|_{R_1}^{R_2} = \frac{\mu\pi^2 N}{60 \times h_0}(R_2^4 - R_1^4) \quad \text{... (6.50)}$$

Power obsorbed by bearing, $$P = \frac{2\pi NT}{60} \text{ Watts.} \quad \text{... (6.51)}$$

6.9 MOVEMENT OF PISTON IN DASHPOT

A dashpot is device to slow down the motion of the component to which it is attached. It is used most commonly in automatic door closer. It consists of a piston that moves in a concentric cylinder, whose diameter is slightly greater than that of the pistion as shown in Fig. 6.12. The pistion is connected with the component whose motion is to be restrained. The cylinder contains a viscous oil sufficient to cover the piston. On the downward motion of piston under the load W, the oil is displaced from the underneath to the space above the piston through the small clearance between the piston and cylinder wall. During the upward motion of the piston the oil is displaced downwards. This is how the motion of the component attached to it is restricted.

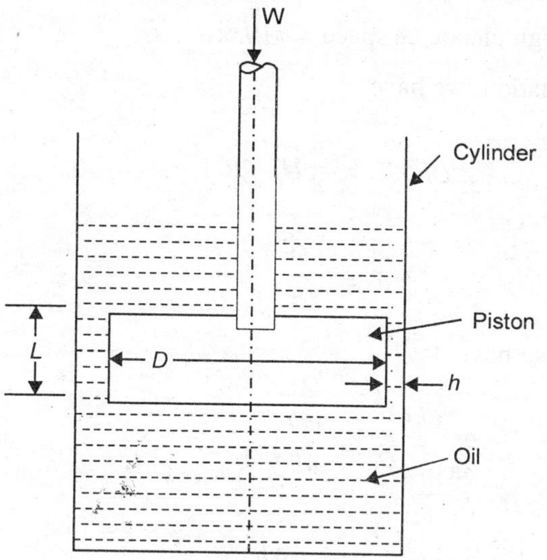

Fig 6.12 Dashpot mechanism

Let D = piston diameter

L = Piston length

μ = oil viscosity

u = velocity of pisiton

$\bar{u}$ = average velocity of oil in the clearance space.

h = clearance between piston and cylinder.

Δp = difference of pressure between two ends of piston.

$$= \frac{4W}{\pi D^2} \quad \text{... (1)}$$

The flow of oil through the clearance space is similar to the laminar flow between parallel plates. Thus

$$\Delta p = \frac{12\mu\bar{u}L}{h^2} \quad \text{... (2)}$$

$$\therefore \quad \frac{4W}{\pi D^2} = \frac{12\mu\bar{u}L^2}{h^2}$$

or

$$\bar{u} = \frac{Wh^2}{3\pi uLD^2} \quad \text{...(3)}$$

The rate of flow of oil in dashpot $= \frac{\pi}{4}D^2 \times u$

Rate of flow of oil through clearance space = $\pi Dh \times \bar{u}$

Applying continuity equation, we have

$$\pi Dh \times \bar{u} = \frac{\pi}{4} D^2 \times u$$

$$\bar{u} = \frac{uD}{4h} \qquad \text{... (4)}$$

From Eqs. (3) and (4), we have

$$\frac{Wh^2}{3\pi\mu LD^2} = \frac{uD}{4h}$$

$$\mu = \frac{4Wh^3}{3\pi LD^3 u} \qquad \text{... (6.52)}$$

Example 6.9 *A shaft of diameter 40 mm rotates in a journal bearing of diameter 40.2 mm and length 100 mm. The clearance space between the shaft and bearing is filled with an oil of viscosity 0.09 Pa.s. Calculate the power absorbed in bearing when the speed of rotation is 750 rpm.*

Solution.

Given : $D = 40$ mm, $L = 100$ mm, $h_0 = \frac{1}{2}(40.2 - 40) = 0.1$mm, $\mu = 0.09$ Pa.s, $N = 750$ rpm.

$$u = \frac{\pi DN}{60 \times 10^3} = \frac{\pi \times 40 \times 750}{60 \times 10^3} = 1.57 \text{m/s}$$

$$\tau = \mu \frac{du}{dy} = \frac{\mu u}{h_0} = \frac{0.09 \times 1.57}{0.1 \times 10^{-3}} = 1413 \text{ N/m}^2$$

Shear force, $F = \tau \times A = 1413 \times \pi \times 400 \times 100 \times 10^{-6} = 17.76$ N

Torque, $T = F \times \frac{D}{2} = 17.76 \times \frac{40}{2} \times 10^{-3} = 0.355$ N.m

Power absorbed, $P = \frac{2\pi NT}{60} = \frac{2\pi \times 750 \times 0.355}{60} = 27.88$ W

Example 6.10 *The shaft of a foot-step bearing is 100 mm in diameter and rotates at 600 rpm. The oil film thickness is 0.5 mm and its viscosity is 0.15 Pa.s. Calculate the power absorbed.*

Solution.

Given : $D = 100$ mm, $N = 600$ rpm, $h_0 = 0.5$ mm, $\mu = 0.15$ Pa.s

Torque, $T = \frac{\mu\pi^2 NR^4}{60 \times h_0} = \frac{0.15 \times \pi^2 \times 600 \times (0.05)^4}{60 \times 0.5 \times 10^{-3}} = 0.185 \text{ N.m}$

Power absorbed, $P = \frac{2\pi NT}{60} = \frac{2\pi \times 600 \times 0.185}{60} = 11.62 \text{ W}$

Example 6.11 *The inner and outer diameters of a collar bearing are 80 mm and 120 mm respectively. An oil film of thickness 0.2 mm is maintained between the collar surface and bearing. The oil viscosity is 0.09 Pa.s. Calculate the power absorbed in overcoming viscous resistance when the shaft rotates at 300 rpm.*

Solution.

Given : $R_1 = 40$ mm, $R_2 = 60$ mm, $h_0 = 0.2$ mm, $\mu = 0.09$ Pa.s, N = 300 rpm.

Torque, $T = \frac{\pi^2 \mu N}{60 \times h_0}(R_1^4 - R_1^4)$

$$= \frac{\pi^2 \times 0.09 \times 300}{60 + 0.2 \times 10^{-3}}(60^4 - 40^4) \times 10^{-12} = 0.2309 \text{ N.m}$$

Power obsorbed, $P = \frac{2\pi NT}{60} = \frac{2\pi \times 300 \times 0.2309}{60} = 7.254 \text{ W}$

Example 6.12 *The piston of a dashpot moves with a speed of 0.5 mm/s. If an additional weight of 1.2 N is placed on the top of piston, it moves with 0.6 mm/s. The diameter of the piston is 80 mm and length 10 mm. The clearance between the piston and cylinder is 1.2 mm. Calculate the viscosity of oil.*

Solution. Given : $u_1 = 0.5$ mm/s, $W_a = 1.2$ N, $u_2 = 0.6$ mm/s, D = 80 mm, L = 100 mm, $h = 1.2$ mm

$$\mu = \frac{4Wh^3}{3\pi LD^3 \mu_1} = \frac{4(W + W_a)h^3}{3\pi LD^3 u_2}$$

$$\frac{W}{u_1} = \frac{W + W_a}{u_2}$$

$$\frac{u_2}{u_1} = \frac{W + W_a}{W}$$

$$1.2\ W = W + 1.2$$

$$W = 6 \text{ N}$$

$$\mu = \frac{4 \times 6 \times (1.2 \times 10^{-3})^3}{3\pi \times 0.1 \times (0.08)^3 \times 0.5 \times 10^{-3}} = 0.172 \text{ Pa.s}$$

SUMMARY

1. In a viscous flow the fluid flows in layers at a very low velocity.
2. The Reynolds number is the ratio of inertia force to viscous force. For flow in pipe, Reynolds number is given by,

$$Re = \frac{\rho v d}{\mu} = \frac{vd}{\nu}$$

$Re < 2000$ for laminar flow

$2000 < Re < 4000$ for transition from laminar to turbulent flow

$Re > 4000$ for turbulent flow

3. The Navier-Stokes equations of motion takes into account the pressure forces, body forces, viscous forces, and inertia forces. These equations are of the form:

$$B_x - \frac{1}{\rho}\frac{\partial p}{\partial x} = \frac{du}{dt} - \nu\nabla^2 u, \text{ etc.}$$

4. The pressure gradient in the direction of flow is equal to the shear gradient in the direction normal to the direction of flow.

$$\frac{\partial p}{\partial x} = \frac{\partial \tau}{\partial y}$$

5. Viscous flow between parallel plates:

(i) Couette flow-one plate moving and other stationary.

$$u = \frac{U}{b}y - \frac{1}{2\mu}\frac{\partial p}{\partial x}(by - y^2)$$

$$q = \frac{Ub}{2} - \frac{b^3}{12\mu}\cdot\frac{\partial p}{\partial x}$$

$$\tau = \mu\frac{U}{b} - \frac{1}{2}\frac{\partial p}{\partial x}(b - 2y)$$

$$\bar{u} = \frac{U}{2} - \frac{b^2}{12\mu}\cdot\frac{\partial p}{\partial x}$$

(iii) Both plates stationary.

$$u = -\frac{1}{2\mu}\cdot\frac{\partial p}{\partial x}(by - y^2)$$

$$q = -\frac{b^3}{12\mu}\cdot\frac{\partial p}{\partial x}$$

$$\tau = -\frac{1}{2}\cdot\frac{\partial p}{\partial x}(b - 2y)$$

$$\tau_0 = -\frac{b}{2}\cdot\frac{\partial p}{\partial x}$$

$$\bar{u} = -\frac{b^2}{12\mu}\cdot\frac{\partial p}{\partial x}$$

$$\frac{u_{max}}{\bar{u}} = 1.5$$

(iii) Both plates moving in opposite directions.

$$u = \left(\frac{U+V}{b}\right)y - \frac{1}{2\mu}\cdot\frac{\partial p}{\partial x}(by - y^2) - V$$

$$q = (U - V)\frac{b}{2} - \frac{b^3}{12\mu}\cdot\frac{\partial p}{\partial x}$$

$$\tau = (U + V)\frac{\mu}{b} - \frac{1}{2}\frac{\partial p}{\partial x}\left(\frac{b}{2} - y\right)$$

6. Drop of pressure head in flow between parallel plates.

$$h_l = \frac{\Delta p}{\rho g} = \frac{12\mu\bar{u}L}{\rho g b^2}$$

7. Viscous flow through circular pipes.

(i) $\tau = -\frac{\partial p}{\partial x}\cdot\frac{r}{2}$, $\tau = \tau_0\frac{r}{R}$

$$u = -\frac{1}{4\mu}\left(\frac{\partial p}{\partial x}\right)(R^2 - r^2)$$

$$u_{max} = -\frac{1}{4\mu}\left(\frac{\partial p}{\partial x}\right)R^2$$

$$u = u_{max}\left[1 - \left(\frac{r}{R}\right)^2\right]$$

$$Q = \frac{\pi}{2}U_{max}R^2$$

$$\bar{u} = \frac{u_{max}}{2}$$

$$\Delta p = \frac{32\mu\bar{u}L}{D^2}, h_l = \frac{32\mu\bar{u}L}{\rho g D^2}$$

(iii) $\alpha = 2.0, \beta = \frac{4}{3}$

8. Viscous flow through an annulus.

$$u = -\frac{1}{4\mu}\left(\frac{\partial p}{\partial x}\right)\left[R_1^2 - r^2 - \frac{R_1^2 - R_2^2}{\ln\frac{R_1}{R_2}} \times \ln\frac{R_1}{r}\right]$$

$$Q = -\frac{\pi}{8\mu}\left(\frac{\partial p}{\partial x}\right)\left[R_1^4 - R_2^4 - \frac{(R_1^2 - R_2^2)^2}{\ln\frac{R_1}{R_2}}\right]$$

$$\bar{u} = -\frac{1}{8\mu}\left(\frac{\partial p}{\partial x}\right)\left[R_1^2 + R_2^2 - \frac{R_1^2 - R_2^2}{\ln\frac{R_1}{R_2}}\right]$$

$$\tau = -\frac{1}{4}\left(\frac{\partial p}{\partial x}\right)\left[2r - \frac{1}{r} \times \frac{R_1^2 - R_2^2}{\ln\frac{R_1}{R_2}}\right]$$

9. Viscous resistance of journal bearing.

$$u = \frac{\pi DN}{60}, \tau = \frac{\mu\pi DN}{60 \times h_0}, F = \frac{\mu\pi^2 D^2 NL}{60 \times h_0}$$

$$T = \frac{\mu\pi^2 D^3 NL}{120 \times h_0}, P = \frac{2\pi NT}{60}$$

10. Foot-step bearing.

$$u = \frac{\pi DN}{60}, \tau = \frac{\mu u}{h_0}, T = \frac{\mu\pi^2 NR^4}{60 \times h_0}$$

$$p = \frac{2\pi NT}{60}$$

11. Collar bearing

$$T = \frac{\mu\pi^2 N}{60 \times h_0}(R_2^4 - R_1^4),\; P = \frac{2\pi NT}{60}$$

12. Movement of piston in dashpot.

$$\mu = \frac{4Wh^3}{32LD^3u},\; \bar{u} = \frac{Wh^2}{3\pi\mu LD^2} = \frac{uD}{4h}$$

Short Answer Type Questions

1. Define laminar flow.
2. Give few examples of laminar flow.
3. Define Reynolds number.
4. What are Navier-Stokes equations of motion?
5. What is the relationship between shear gradient and pressure gradient?
6. What is Couette flow?
7. What is plain Couette flow?
8. What is Poiseuille flow?
9. For laminar flow between two stationary plates, what is the ratio of maximum velocity to average velocity?
10. What is Hagen-Poiseville equation?
11. How much are the values for α and β for laminar flow through a circular pipe?
12. What do you understand by 'no slip' condition?

EXERCISES

1. An oil of dynamic viscosity 0.09 Pa.s is flowing between two stationary horizontal plates with a maximum velocity of 1.5 m/s. If the distance between the plates is 80 mm, Calculate: (a) the pressure gradient, (b) the maximum shear stress, and (c) the discharge per metre width.

   [**Ans.** 168.75 N/m^3, 6.75 N/m^3, 0.08 m^3/s]
2. Water is flowing between two large parallel plates which are 2 mm apart. Calculate: (a) maximum velocity, (b) the pressure drop per unit length, and (c) shear stress at wall of the plate. Average velocity is 0.4 m/s and viscosity of water is 0.001 Pa.s. [**Ans.** 0.6 m/s, 1200 N/m^3, 1.2 N/m^2]
3. There is a horizontal crack 50 mm wide and 3 mm deep in a wall of thickness 150 mm. Water leaks through the crack. Calculate the rate of leakage of water through the crack if the difference of pressure between the two ends of the crack is 250 N/m^2. Take viscosity of water =0.001 Pa.s

   [**Ans.** 1.875 $\times 10^{-4}$ m^3/s]

4. A fluid of viscosity 0.5 Pa.s and specific gravity 1.2 is flowing through a circular pipe of diameter 100 mm. The maximum shear stress at the pipe wall is 150 N/m^2. Calculate: (a) the pressure gradient, (b) the average velocity, and (c) the Reynolds number of pipe.

[**Ans.** 6000 N/m^3, 3.75 m/s, 900]

5. A crude oil of viscosity 0.09 Pa.s and specfic gravity 0.8 is flowing through a horizontal circular pipe of diameter 80 mm and length 15m. Calculate the difference of pressure at the two ends of the pipe, if 50 kg of the oil is collected in a tank in 15 s.

[**Ans.** 5595.75 N/m^2]

6. A shaft of diameter 100 mm rotates in a journal bearing of diameter 100.2 mm and length 200 mm. The annular space between the shaft and bearing is filled with an oil of viscosity 0.08 Pa.s. Calculate the power absorbed in the bearing when the speed of rotation is 600 rpm.

[**Ans.** 496.1 W]

7. A vertical shaft of diameter 100 mm rotates at 750 rpm in a foot-step bearing. The oil film thickness is 0.75 mm and viscosity of oil is 0.12 Pa.s Calculate the power absorbed in the bearing. [**Ans.** 9.69 W]

8. A collar bearing having external and internal radii of 10 mm and 50 mm respectively is used to take the thrust of a shaft. The oil film thickness is 0.3 mm between the collar surface and bearing. Calculate the power lost in overcoming the viscous resistance when the shaft rotates at 300 rpm. Take oil viscosity = 0.09 Pa.s [**Ans.** 43.6 W]

9. An oil dashpot consists of piston moving in a cylinder having oil. The piston falls with uniform speed and covers 50 mm in 100 s. It falls through 50 mm in 80 s when an additonal weight of 1.5 N is placed on the top of the piston. The diameter of piston is 100 mm and its length is 150 mm. The clearance between the piston and the cylinder is 1.5 mm. Calculate the viscosity of oil.

[**Ans.** 0.1146 Pa.s]

10. An oil of specific gravity 0.9 and viscosity 1.0 Pa.s is flowing through a pipe of diameter 100 mm. The velocity at the centre is 2m/s. Calculate:

    (a) the pressure gradient in the flow direction, (b) shear stress at the pipewall, (c) Reynolds number, and (d) velocity at a distance of 30 mm from the wall.

[**Ans.** 3200 N/m^3, 80 N/m^2, 90, 1.28 m/s]

7
FLOW THROUGH PIPES

7.1 INTRODUCTION

A pipe is a conduit, generally of circular cross-secion, which is used for transporting fluids under pressure. The flow in a pipe is termed pipe flow only when the fluid completely fills the cross-section and there is no free surface of fluid. We shall study the laminar flow through pipes in this chapter.

7.2 LOSS OF ENERGY IN PIPES

When a fluid is flowing through a pipe, the fluid experiences some resistance due to which some of the energy of fluid is lost. This loss of energy may be classified as follows:

1. Major Energy Losses.
   These energy losses are due to friction of the pipes.
2. Minor energy losses are due to the following:
   - (i) Sudden expansion of pipe.
   - (ii) Sudden contraction of pipe.
   - (iii) Bend of pipe.
   - (iv) Pipe fittings, etc.
   - (v) An obstruction in pipe.
   - (v) At the entrance and exit of pipe.

7.3 DETERMINATION OF MAJOR ENERGY LOSSES

These losses which are due to friction are calculated by the following formulae:

1. Darcy-Weisbach Formula
2. Chezy's Formula.

1. Darcy–Weisbach Formula

The loss of head (or energy) in pipes due to friction is calculatesd from Darcy-Weisbach formula, which is given by,

$$h_f = \frac{4 f L v^2}{D \times 2g} \quad \text{... (7.1)}$$

where h_f = loss of head due to friction.

f = coefficient of friction, which is a function of Reynolds number.

$$= \frac{16}{\text{Re}} \text{ for Re} < 2000$$

$$= \frac{0.079}{(\text{Re})^{0.25}} \text{ for Re} = 4000 \text{ to } 10^6$$

L = length of pipe

v = mean velocity of flow

D = diameter of pipe.

This formula is generally used for flow through pipes.

2. Chezy's Formula.

An equilibrium between the propelling force due to pressure differnce and the frictional resistance gives,

$$(p_1 - p_2)A = f' \, PLV^2$$

$$\frac{(p_1 - p_2)}{\gamma} A = \frac{f'}{\gamma} PLV^2$$

$$h_f = \frac{f'}{\gamma} \frac{P}{A} Lv^2$$

Mean velocity,

$$v = \sqrt{\frac{\gamma}{f'}} \times \sqrt{\frac{A}{P} \times \frac{h_f}{L}}$$

where $\sqrt{\dfrac{\gamma}{f'}}$ = Chezy's constant, C

$$\frac{A}{P} = \frac{\text{area of flow}}{\text{wetted perimeter}} = \text{Hydraulic mean depth} = \text{m}$$

$$= \frac{\frac{\pi}{4} D^2}{\pi D} = \frac{D}{4}$$

$$\frac{h_f}{L} = \text{loss of head per unit length of pipe} = i$$

$$\therefore \quad v = C\sqrt{mi} \qquad \text{... (7.2)}$$

Eq. (7.2) is known as Chezy's formula.

This formula is generally used for the flow through open channels.

Example 7.1 *Water is flowing at a velocity of 3m/s in a pipe of diameter 300 mm and length 50 m. Calculate the head lost due to friction using: (a) Darcy-Weisbach formula, and (b) Chezy's formula taking C = 55. Assume kinematic viscosity of water = 0.012 Stokes.*

Solution.

Given: $v = 3\text{m/s}$, $D = 0.3$ m, $L = 50$ m, $C = 55$, $\nu = 0.012 \times 10^{-4}$ m²/s

(a) Darcy-Weisbach formula:

$$h_f = \frac{4FLv^2}{D \times 2g}$$

$$Re = \frac{vD}{\nu} = \frac{3 \times 0.3}{0.012 \times 10^{-4}} = 7.5 \times 10^5$$

$$f = \frac{0.079}{(\text{Re})^{0.25}} = \frac{0.079}{(7.5 \times 10^5)^{0.25}} = 0.00268$$

$$h_f = \frac{4 \times 0.00268 \times 50 \times 3^2}{0.3 \times 2 \times 9.81} = 0.819\text{m}$$

(b) Chezy'formula:

$$v = C\sqrt{mi}$$

$$m = \frac{A}{P} = \frac{D}{4} = \frac{0.3}{4} = 0.075 \text{ m}$$

$$3 = 55\sqrt{0.075 \times i}$$

$$i = 0.039667$$

$$i = \frac{h_f}{l}, h_f = 0.03967 \times 50 = 1.98 \text{ m}$$

Example 7.2 *An oil of specific gravity 0.85 and kinematic viscosity 0.3 Stokes is flowing at the rate of 0.5 m³/s in a pipe of diameter 300 mm and length 900 m. Calculate: (a) Head lost due to friction, and (b) power required to maintain the flow.*

Solution.

Given: $s = 0.85$ $\nu = 0.3 \times 10^{-4}$ m²/s, $Q = 0.5$ m³/s, $D = 0.3$ m, $L = 900$ m

$$(a)\ v = \frac{4Q}{\pi D^2} = \frac{4 \times 0.5}{\pi (0.3)^2} = 7.07 \text{ m/s}$$

$$\text{Re} = \frac{vD}{\nu} = \frac{7.07 \times 0.3}{0.3 \times 10^{-4}} = 7.07 \times 10^4$$

$$f = \frac{0.079}{(\text{Re})^{0.25}} = \frac{0.079}{(7.07 \times 10^4)^{0.25}} = 0.00484$$

$$h_f = \frac{4 f L v^2}{D \times 2g} = \frac{4 \times 0.00484 \times 900 \times (7.07)^2}{0.3 \times 2 \times 9.81} = 147.97 \text{ m}$$

(b) Power required, $P = \gamma Q h_f = 0.85 \times 9.81 \times 0.5 \times 147.97 = 616.9$ kW

Example 7.3 *Water flows through a tapering pipeline of diameter varying from 300 mm to 200 mm in a length of 10 m. Calculate the loss of head due to frictin when the discharge through the pipe is 0.05 m³/s. The Darcy-Weisbach friction factor for the whole pipe may be taken as 0.02.*

Solution.

Given: $D_1 = 0.3$ m, $D_2 = 0.2$ m, L = 10 m, $f = 0.02$, $Q = 0.05$ m³/s

Consider an elementary length dx of pipe at a distance x from $D_1 = 0.3$m, as shown in Fig 7.1.

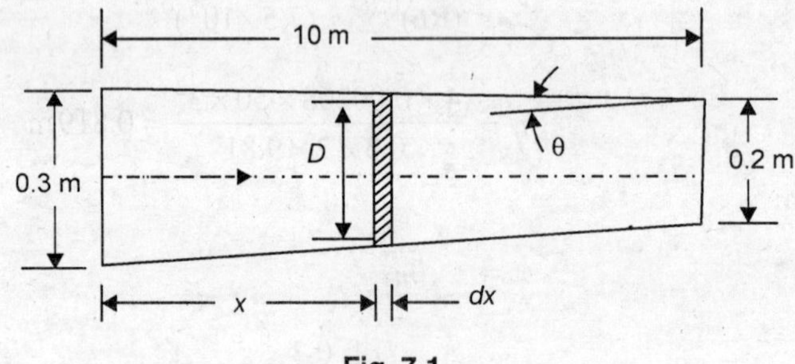

Fig. 7.1

$$\tan \theta = \frac{0.3 - 0.2}{2 \times 10} = 0.005$$

$$D = 0.3 - 2 \times 0.005x = 0.3 - 0.01\,x$$

$$dh_{fx} = \frac{4f \times dx \times \left(\frac{4Q}{\pi D^2}\right)^2}{D \times 2g} = \frac{4 \times 0.02 \times dx \times \left(\frac{4 \times 0.05}{\pi D^2}\right)^2}{D \times 2 \times 9.81}$$

$$= \frac{1.6525 \times 10^{-5}}{D^5} \times dx$$

$$= \frac{1.6525\times10^{-5}}{\left[\frac{1}{100}(30-x)\right]^5}dx = \frac{1.6525\times10^{5}}{(30-x)^5}dx$$

$$h_f = 1.6525\times10^5\int_0^{10}(30-x)^{-5}dx$$

$$= 1.6525\times10^5\left|\frac{(30-x)^{-4}}{4}\right|_0^{10}$$

$$= \frac{1.6525\times10^5}{4}\left[\frac{1}{20^4}-\frac{1}{30^4}\right]$$

$$= 0.207 \text{ m}$$

7.4 DETERMINATION OF MINOR ENERGY LOSSES

7.4.1 Losss of Head Due to Sudden Enlargement

Consider a liquid flowing through a pipe having sudden enlargement as shown in Fig. 7.2. Consider two sections 1–1 and 2–2 before and after the enlargement.

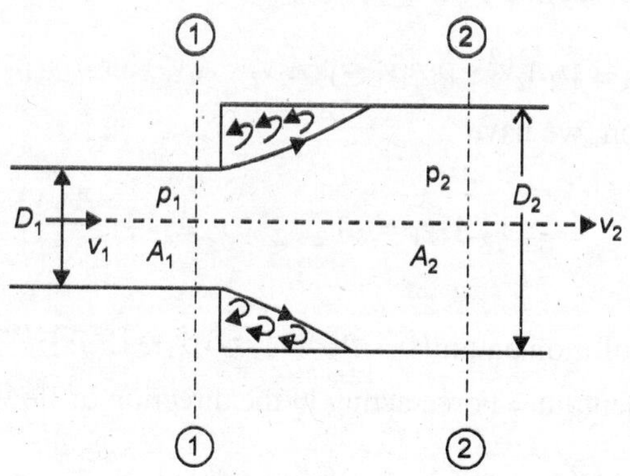

Fig. 7.2 Sudden enlargement

Let p_1, v_1, A_1 = pressure, velocity and area of pipe at section 1–1 respectively.

p_2, v_2, A_2 = corresponding parameters at section 2–2.

Due to sudden change of diameter of pipe from D_1 to D_2, the flowing liquid is not able to follow the abrupt change of pipe bondary. Thus the flow separates from the boundary and turbulent eddies are formed as shown in Fig. 7.2. The loss of head occurs due to the formation of these eddies.

Let p_0 = pressure of the liquid edidies on the area $(A_2 - A_1)$

h_e = loss of head due to sudden enlargement.

Applying Bernoulli's equation at sections 1–1 and 2–2, we have

$$\frac{p_1}{\gamma} + \frac{v_1^2}{2g} + z_1 = \frac{p_2}{\gamma} + \frac{v_2^2}{2g} + z_2 + h_e$$

For horizontal pipe, $z_1 = z_2$

$$\therefore \quad h_e = \frac{p_1 - p_2}{\gamma} + \frac{v_1^2 - v_2^2}{2g} \qquad \text{... (1)}$$

Consider the control volume of liquid between sections 1–1 and 2–2. The net force acting on the liquid in the control volume in the direction of flow is given by,

$$F_x = p_1 A_1 + p_0 (A_2 - A_1) - p_2 A_2$$

Assumming $p_0 = p_1$ we have

$$F_x = p_1 A_1 + p_1 (A_2 - A_1) - p_2 A_2$$

$$= (p_1 - p_2) A_2$$

Momentum of liquid/s at section 1–1 = $\rho_1 A_1 v_1 \times v_1 = \rho_1 A_1 v_1^2$

Momentum of liquid/s at section 2–2 = $\rho_2 A_2 v_2^2$

Change of momentum/s = $\rho_2 A_2 v_2^2 - \rho_1 A_1 v_1^2 = \rho(A_2 v_2^2 - A_1 v_1^2)$ for $\rho_1 = \rho_2 = \rho$

From continuity equation, we have

$$A_1 v_1 = A_2 v_2 \text{ or } A_1 = \frac{A_2 v_2}{v_1}$$

$\therefore$ Change of momentum/s = $\rho[A_2 v_2^2 - A_2 v_1 v_2] = \rho A_2 (v_2^2 - v_1 v_2)$

Rate of change of momentum = Force acting in the direction of flow

$$(p_1 - p_2) A_2 = \rho A_2 (v_2^2 - v_1 v_2)$$

$$\frac{p_1 - p_2}{\rho g} = \frac{v_2^2 - v_1 v_2}{g} \qquad \text{... (2)}$$

Substituting Eq. (2) in (1), we get

$$h_e = \frac{v_2^2 - v_1 v_2}{g} + \frac{v_1^2 - v_2^2}{2g} = \frac{2v_2^2 - 2v_1 v_2 + v_1^2 - v_2^2}{2g} = \frac{(v_1 - v_2)^2}{2g}$$

$$\therefore \qquad h_e = \frac{(v_1 - v_2)^2}{2g} \qquad \text{... (7.3)}$$

Example 7.4 *At a sudden enlargement of a water main from 250 mm to 500 mm diameter, the hydraulic gradient rises by 15 mm. Calculate the flow rate.*

Solution.

Given: $D_1 = 250$ mm, $D_2 = 500$ mm, $\left[\left(\frac{p_2}{\gamma} + z_2\right) - \left(\frac{p_1}{\gamma} + z_1\right)\right] = 0.015$ m

Applying Bernoulli's theorem at sections 1 and 2, we have

$$\frac{p_1}{\gamma} + \frac{v_1^2}{2g} + z_1 = \frac{p_2}{\gamma} + \frac{v_2^2}{2g} + z_2 + h_e$$

where

$$h_e = \frac{(v_1 - v_2)^2}{2g}$$

From continuity equation, we have

$$v_1 A_1 = v_2 A_2$$

$$v_1 = \left(\frac{D_2}{D_1}\right)^2 v_2 = \left(\frac{500}{250}\right)^2 v_2 = 4v_2$$

$$\therefore \qquad h_e = \frac{(4v_2 - v_2)^2}{2g} = \frac{9}{2} \times \frac{v_2^2}{g}$$

$$\frac{p_1}{\gamma} + \frac{16v_2^2}{2g} + z_1 = \frac{p_2}{\gamma} + \frac{v_2^2}{2g} + z_2 + \frac{9v_2^2}{2g}$$

$$\left(\frac{p_2}{\gamma} + z_2\right) - \left(\frac{p_1}{\gamma} + z_1\right) = \frac{6}{2}\frac{v_2^2}{g} = \frac{3v_2^2}{g} = 0.015$$

$$v_2 = \sqrt{\frac{0.015 \times 9.81}{3}} = 0.2215 \text{ m/s}$$

Flow rate, $Q = A_2 V_2 = \frac{\pi}{4}(0.5)^2 \times 0.2215 = 0.0434 \text{ m}^3/\text{s}$

Example 7.5 *In an 80 mm diameter pipeline an oil of specific gravity 0.8 is flowing at the rate of 0.0125 m³/s. A sudden expansion takes place into a second pipeline of such diameter that maximum pressure is obtained. Find: (a) Loss of energy in sudden expansion, and (b) differential gauge length indicated by an oil-mercury manometer connected between the two pipes.*

Solution.

Given: $D_1 = 80$ mm, $s = 0.8$, $Q = 0.0125$ m³/s,

(a) $v_1 = \dfrac{4Q}{\pi D_1^2} = \dfrac{4 \times 0.0125}{\pi (0.08)^2} = 2.487$ m/s

The pressure rise will be maximum, when $\dfrac{D_1}{D_2} = \dfrac{1}{\sqrt{2}}$ (see art 7.4.2)

$$D_2 = \sqrt{2} \times 80 = 113.4 \text{mm}$$

$$v_2 = \frac{4Q}{\pi D_2^2} = \frac{4 \times 0.0125}{\pi (0.11314)^2} = 1.243 \text{ m/s}$$

$$h_e = \frac{(v_1 - v_2)^2}{2g} = \frac{(2.487 - 1.243)^2}{2 \times 9.81} = 0.0787 \text{ m}$$

(b) Applying Bernoulli's equation, we have

$$\frac{p_1}{\gamma} + \frac{v_1^2}{2g} + z_1 = \frac{p_2}{\gamma} + \frac{v_2^2}{2g} + z_2 + h_e$$

Here $z_1 = z_2$

$$\frac{p_2 - p_1}{\gamma} = \frac{v_1^2 - v_2^2}{2g} - h_e = \frac{(2.487)^2 - (1.243)^2}{2 \times 9.81} - 0.0787 = 01578 \text{ m of oil}$$

Let h = reading of U-tube oil-mercury manometer.

$$\frac{p_2 - p_1}{\gamma} = h\left(\frac{13.6}{5} - 1\right) = h\left(\frac{13.6}{0.8} - 1\right) = 16h = 0.1578$$

$$h = 0.00986 \text{ m or } 9.86 \text{ mm}$$

7.4.2 Optimum Diameter Ratio of Pipes for Maximum Pressure Rise

Bernoulli's equation at sections 1 and 2 gives,

$$\frac{p_1}{\gamma} + \frac{v_1^2}{2g} + z_1 = \frac{p_2}{\gamma} + \frac{v_2^2}{2g} + z_2 + h_e$$

where
$$h_e = \frac{(v_1 - v_2)^2}{2g}$$

$$\therefore \quad \frac{p_1}{\gamma} + \frac{v_1^2}{2g} = \frac{p_2}{\gamma} + \frac{v_2^2}{2g} + \frac{(v_1 - v_2)^2}{2g}$$

Pressure rise,
$$\Delta p = p_2 - p_1 = \gamma\left[\frac{v_1^2 - v_2^2 - (v_1 - v_2)^2}{2g}\right]$$

$$= \frac{\gamma}{g}(v_1 v_2 - v_2^2)$$

From continuity equation, we have

$$A_1 v_1 = A_2 v_2$$

$$v_2 = \left(\frac{D_1}{D_2}\right)^2 v_1$$

$$\Delta p = \frac{\gamma}{g}\left[v_1^2\left(\frac{D_1}{D_2}\right)^2 - v_1^2\left(\frac{D_1}{D_2}\right)^4\right]$$

$$= \frac{\gamma v_1^2}{g}\left[\left(\frac{D_1}{D_2}\right)^2 - \left(\frac{D_1}{D_2}\right)^4\right]$$

For Δp to be maximum, $\dfrac{d(\Delta p)}{d\left(\dfrac{D_1}{D_2}\right)} = 0$

$$2\left(\frac{D_1}{D_2}\right) - 4\left(\frac{D_1}{D_2}\right)^3 = 0$$

$$\frac{D_1}{D_2} = \frac{1}{\sqrt{2}} \qquad \text{... (7.4)}$$

and $(\Delta p)_{max} = \dfrac{\gamma v_1^2}{g}\left[\dfrac{1}{2} - \dfrac{1}{4}\right] = \dfrac{\gamma v_1^2}{4g}$ ... (7.5)

7.4.3 Loss of Head due to Sudden Contraction

Consider a liquid flowing in a pipe having sudden contraction in area as shown in Fig. 7.3. As the liquid flows from larger pipe to smaller pipe, the flow area becomes minimum at the vena-contracta section C-C. After section C-C, a sudden enlargement of area takes place. The loss of head due to sudden cotraction is actually due to sudden enlargement from vena contracta to smaller pipe.

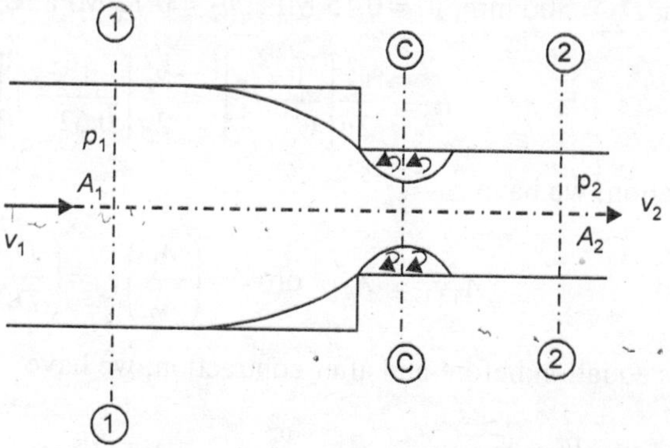

Fig. 7.3 Sudden contraction

Let A_c, v_c = area and velocity of flow respecitively at C-C.

A_2, v_2 = area and velocity of flow at section 2-2.

h_c = loss of head due to sudden contraction.

= loss of head due to sudden enlargement from C-C to 2–2.

$$= \frac{(v_c - v_2)^2}{2g} = \frac{v_2^2}{2g}\left[\frac{v_c}{v_2} - 1\right]^2 \quad \text{... (1)}$$

From continuity equation, we have

$$A_c v_c = A_2 v_2 \text{ or } \frac{v_c}{v_2} = \frac{A_2}{A_c} = \frac{1}{C_c}$$

where C_c = coefficient of contraction.

$$\therefore \quad h_c = \frac{v_2^2}{2g}\left[\frac{1}{C_c} - 1\right]^2 = \frac{k v_2^2}{2g} \quad \text{... (7.6)}$$

where $k = \left[\frac{1}{C_c} - 1\right]$

For $C_c = 0.62$, $k = 0.375$

In general,

$$h_c = \frac{0.5 v_2^2}{2g} \quad \text{... (7.7)}$$

Example 7.6 *A horizontal pipe of diameter 600 mm is suddenly contracted to a diameter 300 mm. The pressure in the larger and smaller pipes are 0.15 MPa and 0.12 MPa respectively. Calculate the loss of head due to contraction and flow rate of water. Take $C_c = 0.62$.*

Solution.

Given: $D_1 = 600$ mm, $D_2 = 300$ mm, $p_1 = 0.15$ MPa, $p_2 = 0.12$ MPa, $C_c = 0.62$

$$h_c = \frac{v_2^2}{2g}\left[\frac{1}{C_c} - 1\right]^2 = \frac{v_2^2}{2g}\left[\frac{1}{0.62} - 1\right]^2 = 0.375\frac{v_2^2}{2g}$$

From continuity equation, we have

$$A_1 v_1 = A_2 v_2 \text{ or } v_2 = \left(\frac{A_1}{A_2}\right) v_1 = \left(\frac{D_1}{D_2}\right)^2 v_1 = \left(\frac{600}{300}\right)^2 v_1 = 4v_1$$

Applying Bernoulli's equation before and after contraction, we have

$$\frac{p_1}{\gamma} + \frac{v_1^2}{2g} + z_1 = \frac{p_2}{\gamma} + \frac{v_2^2}{2g} + z_2 + h_c$$

$$\frac{0.15\times10^3}{g}+\frac{v_2^2}{16\times2g} = \frac{0.12\times10^3}{g}+\frac{v_2^2}{2g}+0.375\frac{v_2^2}{2g} \qquad (\because z_1 = z_2)$$

$$\frac{v_2^2}{2g}[1+0.375-0.0625] = \frac{0.03\times10^3}{g}$$

$$0.65625v_2^2 = 30$$

$$v_2 = 6.76 \text{ m/s}$$

$$h_c = 0.375\times\frac{(6.76)^2}{2\times9.81}=0.873\text{m}$$

$$Q = A_2v_2 = \frac{\pi}{4}(0.3)^2\times6.76 = 0.478 \text{ m}^3\text{/s}$$

Example 7.7 *In a horizontal pipe line there is a sudden contraction from 400 mm diameter to 200 mm diameter and then sudden enlagement to 400 mm diameter. The pressure changes from 102 kPa in larger diameter to 70 kPa in smaller diameter. What is the pressure in the pipe after sudden enlargement and rate of flow? Take $C_C = 0.65$.*

Solution.

Given: $D_1 = 0.4$ m, $D_2 = 0.2$ m, $p_1 = 102$ kPa, $p_2 = 70$ kPa, $C_c = 0.65$

Refer to Fig. 7.4.

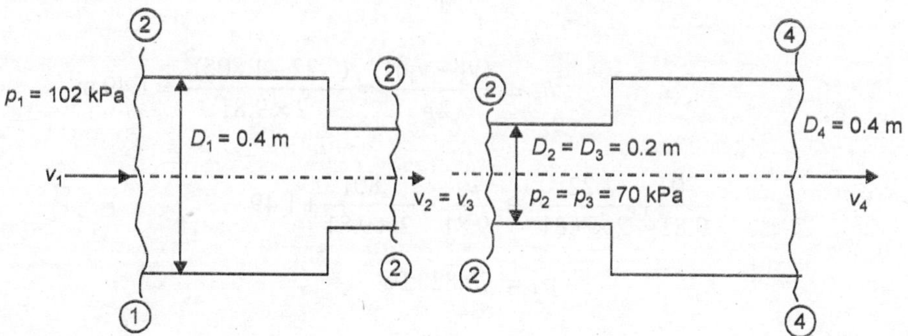

Fig. 7.4

$$h_c = \frac{v_2^2}{2g}\left[\frac{1}{C_c}-1\right]^2 = \frac{v_2^2}{2g}\left[\frac{1}{0.65}-1\right]^2 = 0.28994\frac{v_2^2}{2g}$$

From continuity equation, we have

$$A_1v_1 = A_2v_2, v_1 = \frac{A_2}{A_1}v_2 = \left(\frac{D_2}{D_1}\right)^2 v_2$$

$$v_1 = \left(\frac{0.2}{0.4}\right)^2 v_2 = \frac{v_2}{4}$$

Applying Bernoulli's equation at sections 1–1 and 2–2, we have

$$\frac{p_1}{\gamma} + \frac{v_1^2}{2g} = \frac{p_2}{\gamma} + \frac{v_2^2}{2g} + h_c$$

$$\frac{102}{g} + \frac{v_2^2}{16 \times 2g} = \frac{70}{g} + \frac{v_2^2}{2g} + 0.28994\frac{v_2^2}{2g}$$

$$32 = v_2^2\left[\frac{1}{2} + 0.14497 - \frac{1}{32}\right] = 0.61372\, v_2^2$$

$$v_2 = 7.22 \text{ m/s}$$

$$Q = A_2 v_2 = \frac{\pi}{4}(0.2)^2 \times 7.22 = 0.2268 \text{m}^3/\text{s}$$

Applying Bernoulli's equation to sections 3–3 and 4–4, we have

$$\frac{p_3}{\gamma} + \frac{v_3^2}{2g} = \frac{p_4}{\gamma} + \frac{v_4^2}{2g} + h_e$$

$$p_3 = p_2 = 70 \text{ kPa},\ v_3 = v_2 = 7.22 \text{ m/s},\ v_4 = v_1 = \frac{v_2}{4} = \frac{7.22}{4} = 1.805 \text{ m/s}$$

$$h_e = \frac{(v_3 - v_4)^2}{2g} = \frac{(7.22 - 1.805)^2}{2 \times 9.81} = 1.49 \text{ m}$$

$$\frac{70}{9.81} + \frac{(7.22)^2}{2 \times 9.81} = \frac{p_4}{9.81} + \frac{(1.80)^2}{2 \times 9.81} + 1.49$$

$$p_4 = 79.82 \text{ kPa}$$

7.4.4 Loss of head at the Entrance of a Pipe

This loss of head occurs when a liquid enters a pipe which is connected to a large tank or reservoir. This loss is similar to the loss of head due to sudden contraction In practice, the value of loss of head at the entrance of a pipe is taken as,

$$h_i = 0.5\frac{v^2}{2g} \qquad \text{... (7.8)}$$

where v = velocity of liquid in pipe.

7.4.5 Loss of Head at the Exit of Pipe

This is the loss of head due to the velocity of liquid at the outlet of pipe, which is dissipated either in the form of a free jet or in the tank or reservoir, This loss is given by,

$$h_0 = \frac{v^2}{2g} \quad \text{... (7.9)}$$

7.4.6 Loss of Head due to obstruction in a Pipe

The loss of head due to obstruction takes place due to reduction of area of cross-section of pipe at the location of obstruction. Beyond the obstruction there is sudden enlargement of area due to which loss of head takes place.

Consider a pipe having an obstruction as shown in Fig. 7.5.

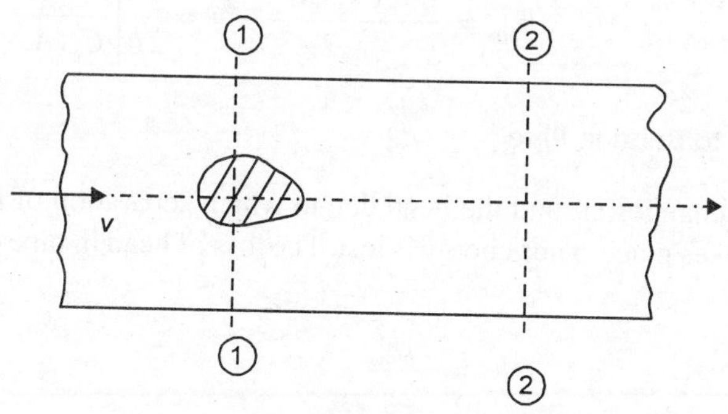

Fig. 7.5 An obstruction in a pipe

Let

A = area of pipe

a = maximum area of obstruction

v = velocity of liquid in pipe

Then $A - a$ = area of flow and passes through section 1–1. As the liquid flows and passes through section 1–1, a vena -contracta is formed beyond section 1–1, after which the stream of liquid widens again and velocity of flow at section 2–2 becomes uniform and becomes equal to velocity v in the pipe. This situation is similar to the flow of liquid through sudden enlargement.

Let v_c = velocity of liquid at vena-contracta.

Then loss of head due to obstruction is equal to the loss of head due to enlargement from vena-contracta to section 2–2.

$$h_{obs} = \frac{(v_c - v)^2}{2g} \quad \text{... (1)}$$

From continuity equation, we have

$$a_c \times v_c = A \times v \quad \text{... (2)}$$

where a_c = area at vena contracta

Coefficient of contraction, $$C_c = \frac{a_c}{A-a}$$

or $$a_c = C_c(A-a)$$

Substituting this value in Eq. (2), we get

$$C_e(A-a)\times v_c = A\times v$$

$$\therefore \quad v_c = \frac{A\times v}{C_c(A-a)}$$

Eq. (1). then becomes,

$$h_{obs} = \frac{\left[\frac{Av}{C_c(A-a)} - v\right]^2}{2g} = \frac{v^2}{2g}\left[\frac{A}{C_c(A-a)} - 1\right]^2 \quad \text{... (7.10)}$$

7.4.7 Loss of Head due to Bend in Pipe

The velocity of flow changes around the bend due to which separation of flow from the boundary and formation of eddies takes place. Thus energy is lost. The loss of head in pipe due to bend is expressed as:

$$h_b = \frac{k_b v^2}{2g}$$

where k_b= coefficient of bend, which depends on angle of bend, radius of curvature of bend, and diameter of pipe.

7.4.8 Loss of Head in Pipe Fittings

The loss of head in pipe fittings such as valves, couplings etc. is expressed as:

$$h_{pf} = \frac{k_{pf} v^2}{2g}$$

where k_{pf} = coefficient of pipe fitting

Example 7.8 *Water is flowing through a horizontal pipe of diameter 250 mm at a velocity of 2.5 m/s. A cirular solid plate of diameter 100 mm is placed in the pipe to obstruct the flow. Calculate the loss of head due to obstruction in the pipe, if $C_c = 0.62$.*

Solution.

Given: $D = 0.25$ m, $v = 2.5$ m/s, $d = 0.1$ m, $C_c = 0.62$

$$h_{obs} = \frac{v^2}{2g}\left[\frac{A}{C_c(A-a)} - 1\right]^2$$

$$= \frac{(2.5)^2}{2\times 9.81}\left[\frac{(0.25)^2}{0.62(0.25)^2-(0.1)^2]}-1\right]^2$$

$$= 0.27 \text{ m}$$

Example 7.9 *Water is flowing through a horizontal pipe of diameter 250 mm and length 60 m when one end of the pipe is connected to a large tank and the other end of the pipe is open to the atmosphere. The height of water in the tank is 6 m above the centre of pipe. Considering all losses, determine the rate of flow of water. Take f = 0.008 in Darcy–Weisbach formula.*

Solution.

Given: $D = 0.25$ m, $L = 60$ m, $H = 6$ m, $f = 0.008$

Applying Bernoulli's equation at the water surface in the tank and the outlet of pipe, we have

$$\frac{p_1}{\gamma}+\frac{v_1^2}{2g}+z_1 = \frac{p_2}{\gamma}+\frac{v_2^2}{2g}+z_2+(h_l)_{all}$$

Let us take datum at the centre of pipe.

$$0+0+6 = 0+\frac{v_2^2}{2g}+0+(h_i+h_f)$$

$$6 = \frac{v_2^2}{2g}+0.5\frac{v_2^2}{2g}+\frac{4fLv_2^2}{2gD}$$

$$= \frac{1.5v_2^2}{2g}+\frac{4\times 0.008\times 60\times v_2^2}{2g\times 0.25}$$

$$= \frac{1.5v^2}{2g}+\frac{7.68v_2^2}{2g} = \frac{9.18v_2^2}{2g}$$

$$v_2 = \sqrt{\frac{6\times 2\times 9.81}{9.18}} = 3.58 \text{ m/s}$$

$$Q = Av_2 = \frac{\pi}{4}(0.25)^2\times 3.58 = 0.1757 \text{ m}^3/\text{s}$$

Example 7.10 *Two tanks are connected by a horizontal pipe of diameter 300 mm and length 300 m. The rate of water flow through the pipe is 0.3 m³/s. Considering all losses, determine the difference in elevation of water surfaces in the two tanks. Take f = 0.008.*

Solution.

Given: $D = 0.3$m, $L = 300$ m, $Q = 0.3$ m³/s, $f = 0.008$

$$v = \frac{4Q}{\pi D^2} = \frac{4\times 0.3}{\pi(0.3)^2} = 4.244 \text{ m/s}$$

Let H_1, H_2 = elevation of water surfaces from the centre of pipe in the two tanks.

$$H_1 = H_2 + (h_l)_{all}$$

$$= H_2 + h_i + h_f + h_0$$

where $$h_i = 0.5\frac{v^2}{2g} = \frac{0.5(4.544)^2}{2 \times 9.81} = 0.459\,\text{m}$$

$$h_f = \frac{4fLv^2}{2gD} = \frac{4 \times 0.008 \times 300(4.244)^2}{2 \times 9.81 \times 0.3} = 29.377\text{ m}$$

$$h_0 = \frac{v^2}{2g} = \frac{(4.244)^2}{2 \times 9.81} = 0.918\,\text{m}$$

$$H_1 - H_2 = 0.459 + 29.377 + 0.918 = 30.754\text{ m}$$

7.5 HYDRAULIC GRADIENT LINE AND TOTAL ENERGY LINE

Hydraulic Gradient Line

It is defined as the line which gives the sum of pressure head $\left(\frac{p}{\gamma}\right)$ and datum head (z) of a flowing fluid in a pipe with resepct to some reference line. It is the line which is obtained by joining the top of all vertical ordinates, showing the pressure head $\left(\frac{p}{\gamma}\right)$ of a flowing fluid in a pipe from the centre of the pipe. It is briefly written as H.G.L (Hydraulic Gradient Line).

Total Energy Line

It is defined as the line which gives the sum of pressure head, datum head and kinetic head of a flowing fluid on a pipe with respect to some reference line. It is also defined as the line which is obtained by joining the tops of all vertical ordiantes showing the sum of pressure head and kinetic head from the centre of the pipe. It is briefly written as T.E.L. (Total Energy Line) or E.G.L (Energy Gradient Line).

Example 7.11 *A horizontal pipeline 50 m long is connected to a water tank at one end and discharges freely into the atmosphere at the other end. For the first 30 m of its length from the tank, the pipe is 100 mm diameter and its diameter is suddenly enlarged to 200 mm. The height of water level on the tank is 6 m above the centre of the pipe. Considering all losses of head, (a) Calculate the rate of flow, and (b) Draw the hydraulic gradient line and total energy line. Take f = 0.01 for both sections of the pipe.*

Solution.

Given: $L_1 = 30$ m, $D_1 = 0.1$ m, $L_2 = 20$ m, $D_2 = 0.2$ m, $H = 6$ m, $f = 0.01$

Applying Bernoulli's theorem to the free surface of water in the tank and outlet of pipe as shown in Fig. 7.6 and taking datum line passing through the centre of the pipe, we have

$$H = 0 + \frac{v_2^2}{2g} + h_i + h_{f1} + h_e + h_{f2}$$

$$= \frac{v_2^2}{2g} + \frac{0.5v_1^2}{2g} + \frac{4fL_1v_1^2}{2gD_1} + \frac{(v_1 - v_2)^2}{2g} + \frac{4fL_2v_2^2}{2gD_2}$$

From continuity equation, we have

$$A_1v_1 = A_2v_2$$

$$v_1 = \frac{A_2}{A_1}v_2 = \left(\frac{D_2}{D_1}\right)^2 v_2 = \left(\frac{0.2}{0.1}\right)^2 v_2 = 4v_2$$

$$H = \frac{v_2^2}{2g} + \frac{8v_2^2}{2g} + \frac{64fL_1v_2^2}{2gD_1} + \frac{9v_2^2}{2g} + \frac{4fL_2v_2^2}{2gD_2}$$

$$6 = \frac{v_2^2}{2g}\left[1 + 8 + \frac{64 \times 0.01 \times 30}{0.1} + 9 + \frac{4 \times 0.01 \times 20}{0.2}\right]$$

$$= 107\frac{v_2^2}{g}$$

$$v_2 = \sqrt{\frac{6 \times 9.81}{107}} = 0.7417 \text{ m/s}$$

$$v_1 = 4 \times 0.7417 = 2.9668 \text{ m/s}$$

(a) $Q = A_2v_2 = \frac{\pi}{4}(0.2)^2 \times 0.7417 = 0.0233 \text{m}^3\text{/s or } 23.3 \text{ lit/s}$

(b) $h_i = 0.5\frac{v_1^2}{2g} = \frac{0.5 \times (2.9668)^2}{2 \times 9.81} = 0.2243 \text{ m}$

$$h_{f1} = \frac{4fL_1v_1^2}{2gD_1} = \frac{4 \times 0.01 \times 30 \times (2.9668)^2}{2 \times 9.81 \times 0.1} = 5.3834 \text{ m}$$

$$h_e = \frac{(v_1 - v_2)^2}{2g} = \frac{(2.9668 - 0.7417)^2}{2 \times 9.81} = 0.2523 \text{m}$$

$$h_{f2} = \frac{4fL_2v_2^2}{2gD_2} = \frac{4 \times 0.01 \times 20 \times (0.7417)^2}{2 \times 9.81 \times 0.2} = 0.1120 \text{m}$$

$$h_0 = \frac{v_2^2}{2g} = \frac{(0.7417)^2}{2 \times 9.81} = 0.0280 \text{m}$$

Also $\frac{v_1^2}{2g} = \frac{(2.9668)^2}{2 \times 9.81} = 0.4486 \text{m}$

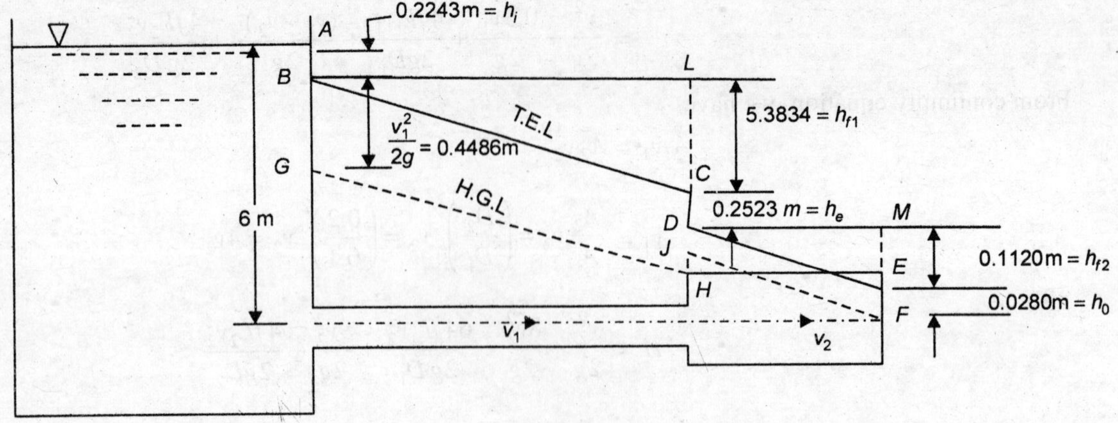

Fig. 7.6

Total energy Line

1. Point *A* lies on free surface of water.
2. Take $AB = h_i = 0.2243$ m.
3. From *B* draw a horizontal line $BL = L_1$. From *L* draw a vertical line downwards
4. Cut the line $LC = h_{f1} = 5.3834$ m.
5. Join points *B* and *C* by a straight line. From *C*, take a line $CD = h_e = 0.2523$ m vertically downwards.
6. From *D*, draw *DM* horizontal and from point *F*, which is lyaing on the centre of the pipe, draw is vertical line in the upward direction, meeting at *M*. Take $ME = h_{f2} = 0.1120$ m. Join DE. $EF = h_0 = 0.0280$ m

   Then line *ABCDE* represents the T.E.L.

Hydraulic Gradient Line

1. From *B*, take $BG = \frac{v_1^2}{2g} = 0.4486$ m.
2. Draw line *GH* parallel to *BC*.
3. From *F* draw a line FJ parallel to *ED*.
4. Join the points *H* and *J*.

Then GHJF is the H.G.L.

Example 7.12 *Two reservoirs are connected by a pipeline consisting of two pipes, one of 150 mm diameter and length 6 m and the other of diameter 300 mm and 12 m length. If the difference of water levels in the two reservoirs is 6m, calculate the discharge and draw the energy gradient line. Take f = 0.04.*

Solution.

Given: $D_1 = 0.15$ m, $L_1 = 6$ m, $D_2 = 0.3$ m, $L_2 = 12$ m, $H = 6$ m, $f = 0.04$.

From the continuity equation, we have

$$A_1 v_1 = A_2 V_2, v_1 = \frac{A_2}{A_1} v_2 = \left(\frac{D_2}{D_1}\right)^2 v_2 = \left(\frac{0.3}{0.15}\right)^2 v^2 = 4v_2$$

The various head losses are:

$$h_i = \frac{0.5v_1^2}{2g} = \frac{8v_2^2}{2g}$$

$$h_{f1} = \frac{4fL_1v_1^2}{2gD_1} = \frac{4\times 0.04\times 6\times 4v_2^2}{2g\times 0.15} = \frac{102.4\,v_2^2}{2g}$$

$$h_e = \frac{(v_1 - v_2)^2}{2g} = \frac{(4v_2 - v_2)^2}{2g} = \frac{9v_2^2}{2g}$$

$$h_{f2} = \frac{4fL_2v_2^2}{2gD_2} = \frac{4\times 0.04\times 12\times v_2^2}{2g\times 0.3} = \frac{6.4v_2^2}{2g}$$

$$h_0 = \frac{v_2^2}{2g}$$

Applying Bernoulli's theorem to free water surfaces in the two reservoirs, we have

$$\frac{p_1}{\gamma} + \frac{v_1^2}{2g} + z_1 = \frac{p_2}{\gamma} + \frac{v_2^2}{2g} + z_2 + (h_i + h_{f1} + hl + h_{f2})$$

$$0 + 0 + H = 0 + \frac{v_2^2}{2g} + 0 + \frac{8v_2^2}{2g} + \frac{102.4\,v_2^2}{2g} + \frac{9v_2^2}{2g} + \frac{6.4v_2^2}{2g}$$

$$6 = \frac{v_2^2}{2g}[1 + 8 + 102.4 + 9 + 9 + 6.4] = 63.4\frac{v_2^2}{g}$$

$$v_2 = \sqrt{\frac{6\times 9.81}{63.4}} = 0.9635 \text{ m/s}$$

$$v_1 = 4 \times 0.9635 = 3.8541 \text{ m/s}$$

$$Q = A_1 v_1 = \frac{\pi}{4}(0.15)^2 \times 3.8541 = 0.681 \text{m}^3\text{/s or } 68.1 \text{lit/s}$$

The E.G.L is shown in Fig. 7.7

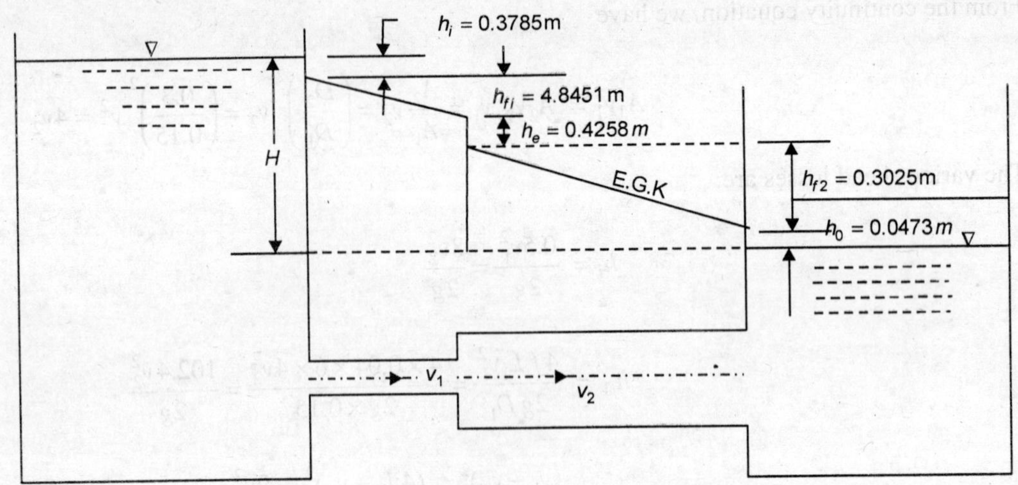

Fig 7.7

$$h_i = \frac{8v_2^2}{2g} = \frac{8(0.9635)^2}{2\times 9.81} = 0.3785\,\text{m}$$

$$h_{f1} = \frac{102.4\,v_2^2}{2g} = \frac{102.4(0.9635)^2}{2\times 4.81} = 4.8451\,\text{m}$$

$$h_e = \frac{9v_2^2}{2g} = \frac{9(0.9635)^2}{2\times 9.81} = 0.4258\,\text{m}$$

$$h_{f2} = \frac{6.4v_2^2}{2g} = \frac{6.4(0.9635)^2}{2\times 9.81} = 0.3028\,\text{m}$$

$$h_0 = \frac{v_2^2}{2g} = \frac{(0.9635)^2}{2\times 9.81} = 0.0473\,\text{m}$$

Example 7.13 *Two reservoirs having a difference of level of 16 m are connected by a pipe line ABCD. AB = 300 m, BC = 450 m, CD = 200 m. The head of water in first reservoir is 4 m above AB. CD is 2 m below AB. The diameter of pipe is 200 mm throughout. There is a partially closed valve in CD at which the head loss is 3m. The flow through the pipeline is 4m³/min. Calculate:*

(a) the friction coefficient of pipeline, and

(b) draw the energy gradient line of pipe line.

Solution.

Given: $H = 16$ m, $L_1 = 300$ m, $L_2 = 450$ m, $L_3 = 200$ m, $H_1 = 4$ m, $H_2 = 2$m, $D = 0.2$ m, $h_v = 3$m, $Q = 4\text{m}^3/\text{min}$

$$v = \frac{4Q}{\pi D^2} = \frac{4\times 4}{60\times\pi(0.2)^2} = 2.122\,\text{m/s}$$

Applying Bernoulli's equation at the free water surfaces in the reservoirs, we have

$$\frac{p_1}{\gamma}+\frac{v_1^2}{2g}+z_1 = \frac{p_2}{\gamma}+\frac{v_2^2}{2g}+z_2+\frac{0.5v^2}{2g}+\frac{4f(L_1+L_2+L_3)v^2}{2gD}+h_v+\frac{v^2}{2g}$$

$$0+0+16 = 0+0+0+\frac{v^2}{2g}\left(0.5+\frac{4f\times 950}{0.2}+1\right)+3$$

$$13 = \frac{v^2}{2g}\times(1.5+19{,}000f)$$

$$= \frac{(2.122)^2}{2\times 9.81}(1.5+19{,}000f) = 0.3442+9{,}360.6\times f$$

$$f = 0.0029$$

The various losses are:

$$h_i = \frac{0.5v^2}{2g} = 0.1148\,\text{m}$$

$$h_f = \frac{4fLv^2}{2gD} = \frac{4\times 0.0029\times 950\times(2.122)^2}{2\times 9.81\times 0.2} = 12.646\,\text{m}$$

$$h_0 = \frac{v^2}{2g} = \frac{(2.122)^2}{2\times 9.81} = 0.2295\,\text{m}$$

The E.G.L. is shown in Fig. 7.8.

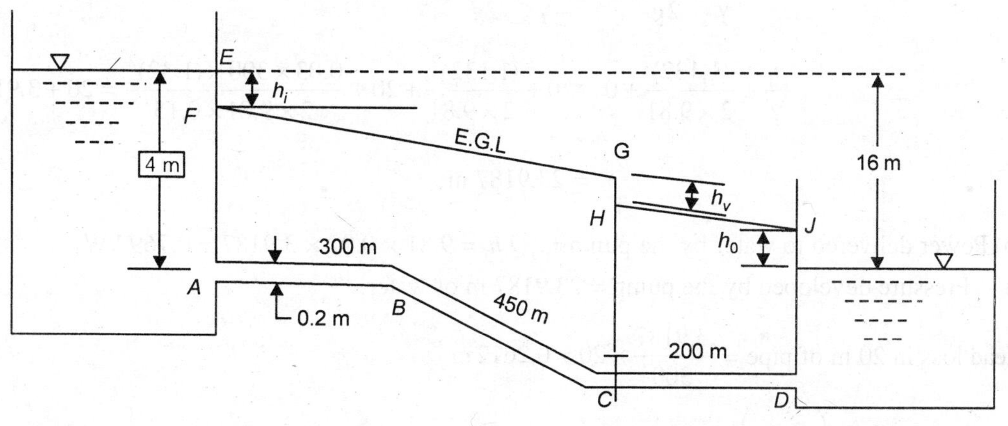

Fig. 7.8

Example 7.14 *Water is being pumped at the rate of 0.02 m³/s to an overhead tank through a 150 mm diameter 300 m long delivery pipe. The pipe discharges freely in the tank at a height of 15 m above the pump. The Dacy-Weisbach friction factor = 0.03 for the pipe. Calculate: (a) the pressure developed by the pump on its delivery side, and (b) the power delivered to water by the pump.*

Draw the hydraulic gradient line from the pump to the tank. Assume that first 280 m of the pipe is horizontal and the rest is vertical.

Solution.

Given: $Q = 0.02$ m³/s, $D = 0.15$ m, $L = 300$ mm, $4f = 0.03$

(a) $v = \dfrac{4Q}{\pi D^2} = \dfrac{4 \times 0.02}{\pi (0.15)^2} = 1.132\,\text{m/s}$

Applying Bernoulli's equation at sections 1 and 2, we have

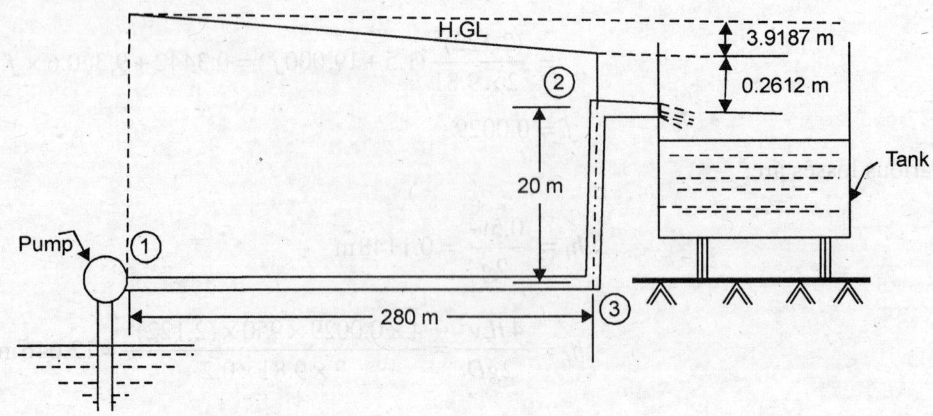

Fig. 7.9

$$\frac{p_1}{\gamma} + \frac{v_1^2}{2g} + z_1 = \frac{p_2}{\gamma} + \frac{v_2^2}{2g} + z_2 + h_f$$

$$\frac{p_1}{\gamma} + \frac{(1.132)}{2 \times 9.81} + 0 = 0 + \frac{(1.132)^2}{2 \times 9.81} + 20 + \frac{0.03 \times 300 \times (1.32)^2}{2 \times 9.81 \times 0.15} = 20 + 3.9187$$

$$\frac{p_1}{\gamma} = 23.9187 \text{ m}$$

(b) Power delivered to water by the pump = $\gamma Q\, h_f = 9.81 \times 0.02 \times 3.9187 = 0.769$ kW

$\therefore$ Pressure developed by the pump = 23.9187 m of water.

Head loss in 20 m of pipe = $\dfrac{3.9187}{300} \times 20 = 0.2612\,\text{m}$

Piezometric head $\left(\dfrac{p}{\gamma} + z\right)$ at:

Section 1 = 23.9187 + 0.2612 = 24.1799 m

Section 2 = 24.1799 – 3.9187 = 20.2612 m

Section 3 = 20 m

The H.G.L. is shown in Fig. 7.9.

7.6 PIPES IN SERIES

Pipes in series or compound pipes are defined as the pipes of different length and diameters connected end to end to form a pipe line as shown in Fig. 7.10

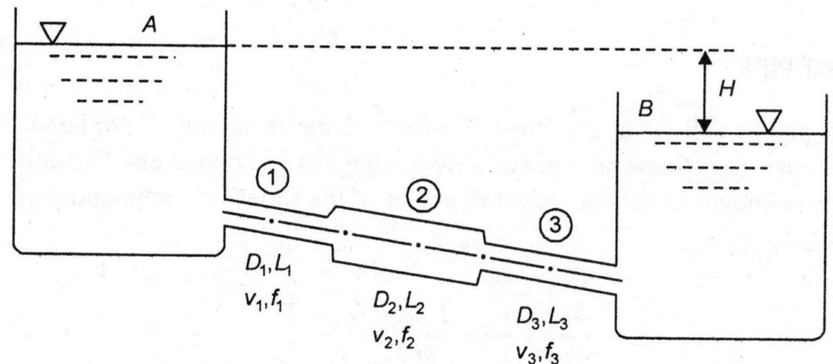

Fig. 7.10 Pipes in series

Let D_1, L_1, v_1, f_1 = diameter, length velocity of flow, coefficient of friction respectively for pipe 1. and similar parameters for other pipes.

H = difference of water level in the two tanks A and B.

The rate of flow Q of water through each pipe is same.

$$\therefore \quad Q = A_1 v_1 = A_2 v_2 = A_3 v_3$$

Now
$$H = h_i + h_{f1} + h_e + h_{f2} + h_c + h_{f3} + h_0 \quad \text{... (7.13)}$$

wher
$$h_i = \frac{0.5 v_1^2}{2g}, h_{f1} = \frac{4 f_1 L_1 v_1^2}{D_1 \times 2g}, h_c = \frac{0.5 v_2^2}{2g},$$

$$h_{f2} = \frac{4 f_2 L_2 v_2^2}{D_2 \times 2g}, h_e = \frac{(v_2 - v_3)^2}{2g}, h_{f3} = \frac{4 f_3 L_3 v_3^2}{D_3 \times 2g}, h_0 = \frac{v_3^2}{2g}$$

If minor losses are neglected and $f_1 = f_2 = f_3$, then

$$H = \frac{4f}{2g}\left[\frac{L_1 v_1^2}{D_1} + \frac{L_2 v_2^2}{D_2} + \frac{L_3 v_3^2}{D_3}\right] \quad \text{... (7.14 a)}$$

Also $$v_2 = \left(\frac{A_1}{A_2}\right) v_1 = \left(\frac{D_1}{D_2}\right)^2 v_1 \text{ and } v_3 = \left(\frac{D_1}{D_3}\right)^2 v_1$$

$$\therefore \quad H = \frac{4 f v_1^2}{2g}\left[\frac{L_1}{D_1} + \frac{L_2 \times D_1^4}{D_2^5} + \frac{L_3 D_1^4}{D_3^5}\right]$$

$$= \frac{4 f v_1^2}{D_1 \times 2g}\left[L_1 + \left(\frac{D_1}{D_2}\right)^5 L_2 + \left(\frac{D_1}{D_3}\right)^5 L_3\right] \quad \text{... (7.14.b)}$$

7.7 EQUIVALENT PIPE

An equivalent pipe is defined as the pipe of uniform diameter having loss of head and discharge of a compound pipe consisting of several pipes of different lengths and diameters. The uniform diameter of the equivalent pipe is known as the equivalent diameter of the series or compound pipe.

From Eq. (7.14 a), we have

$$H = \frac{4f}{2g}\left[\frac{L_1 v_1^2}{D_1} + \frac{L_2 v_2^2}{D_2} + \frac{L_3 v_3^2}{D_3}\right]$$

Now $$Q = \frac{\pi}{4} D_1^2 v_1 \text{ or } v_1 = \frac{4Q}{\pi D_1^2} \text{ and so on.}$$

$$H = \frac{4f}{2g}\left[\frac{L_1}{D_1} \times \left(\frac{4Q}{\pi D_1^2}\right)^2 + \frac{L_2}{D_2}\left(\frac{4Q}{\pi D_2^2}\right)^2 + \frac{L_3}{D_3}\left(\frac{4Q}{\pi D_3^2}\right)^2\right]$$

$$= \frac{4f}{2g} \times \left(\frac{4Q}{\pi}\right)^2 \left[\frac{L_1}{D_1^5} + \frac{L_2}{D_2^5} + \frac{L_3}{D_3^5}\right]$$

$$= \frac{64 fQ^2}{\pi^2 \times 2g}\left[\frac{L_1}{D_1^5} + \frac{L_2}{D_2^5} + \frac{L_3}{D_3^5}\right] \quad \text{... (1)}$$

Let D, L = diameter and length of equivalent pipe respectively.

Then $$H = \frac{4 fLv^2}{D \times 2g}$$

$$v = \frac{4Q}{\pi D^2}$$

$$\therefore \qquad H = \frac{4fL}{D \times 2g} \times \left(\frac{4Q}{\pi D^2}\right)^2 = \frac{64 f Q^2}{\pi^2 \times 2g} \times \left[\frac{L}{D^5}\right] \qquad \text{... (2)}$$

From Eqs (1) and (2), we have

$$\frac{L}{D^5} = \frac{L_1}{D_1^5} + \frac{L_2}{D_2^5} + \frac{L_3}{D_3^5} \qquad \text{... (7.15)}$$

In general, for n pipes in series,

$$\frac{L}{D^5} = \sum_{i=1}^{n} \frac{L_i}{D_i^5} \qquad \text{... (7.16)}$$

Eq. (7.16) is known as Dupit's equation. In this equation,
$L = L_1 + L_2 + L_3$ and D_1, D_2, D_3, are known. Hence D can be calculated.

7.8 PIPES IN PARALLEL

The pipes are said to be in parallel when a main line divides into two or more parallel pipes which again join together downstream and continues as a main line. Fig. 7.11 shows two pipes in parallel. For pipes in parallel, we have

1. Total discharge = Sum of the discharge in each parallel pipe.

$$Q = \Sigma Q_i = Q_1 + Q_2 + \ldots \qquad \text{... (7.17)}$$

2. The loss of head in each pipe is same.

$$h_{f1} = h_{f2} = \ldots$$

or

$$\frac{4 f_1 L_1 v_1^2}{2g D_1} = \frac{4 f_2 L_2 v_2^2}{2g D_2} = \ldots$$

when $f_1 = f_2 = \ldots$, then

$$\frac{L_1 v_1^2}{D_1} = \frac{L_2 v_2^2}{D_2} \qquad \text{... (7.19)}$$

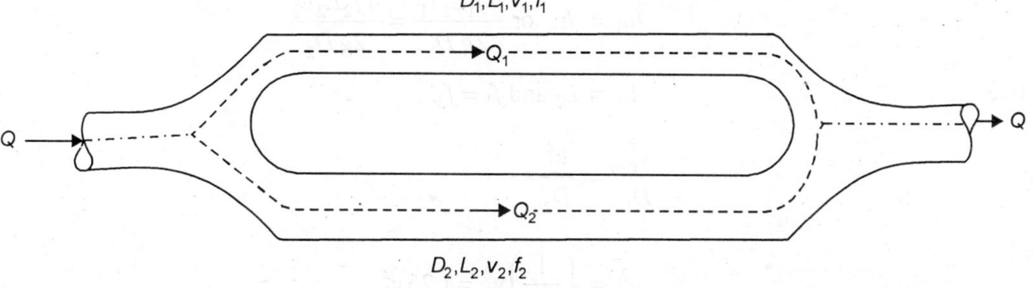

Fig. 7.11 Pipes in parallel

Example 7.15 *A piping system consists of three pipes arranged in series. The length of the pipes are: 1000 m, 750 m and 600 m and diameter 800 mm, 600 mm and 450 mm respectively.*

(a) Transform the system to an equivalent 400 mm diameter pipe, and

(b) Determine an equivalent diameter of pipe, 2500 m long.

Solution.

Given: $L_1 = 1000$ m, $L_2 = 750$ m, $L_3 = 600$ m, $D_1 = 0.8$ m, $D_2 = 0.6$ m, $D_3 = 0.45$ m

(a) $$\frac{L}{D^5} = \frac{L_1}{D_1^5} + \frac{L_2}{D_2^5} + \frac{L_3}{D_3^5}$$

or $$L = \left(\frac{D}{D_1}\right)^5 L_1 + \left(\frac{D}{D_2}\right)^5 L_2 + \left(\frac{D}{D_3}\right)^5 L_3$$

$$= \left[\frac{0.4}{0.8}\right]^5 \times 1000 + \left(\frac{0.4}{0.6}\right)^5 \times 750 + \left(\frac{0.4}{0.45}\right)^5 \times 600$$

$$= 31.25 + 98.76 + 332.96$$
$$= 462.97 \text{ m}$$

(b) $$\frac{2500}{D^5} = \frac{1000}{(0.8)^5} + \frac{750}{(0.6)^5} + \frac{600}{(0.45)^5} = 5212.2$$

$$D = 0.560 \text{ m} \quad \text{or} \quad 560 \text{ mm}$$

Example 7.16 *A main pipe divides into two parallel pipes which again forms one pipe. The given data is as follows:*

$L_1 = 1000$ m, $D_1 = 1$ m, $L_2 = 1000$ m, $D_2 = 0.8$ m, $f_1 = f_2 = 0.005$, $Q = 2.5$ m³/s.

Calculate the rate of flow in each pipe.

Solution.

Given: $$Q = Q_1 + Q_2$$

$$h_{f1} = h_{f_2} \text{ or } \frac{4 f_1 L_1 v_1^2}{2g\,D_1} = \frac{4 f_2 L_2 v_2^2}{2g\,D_2}$$

Here $$L_1 = L_2 \text{ and } f_1 = f_2.$$

$\therefore$ $$\frac{v_1^2}{D_1} = \frac{v_2^2}{D_2}$$

$$v_1^2 = \left(\frac{1}{0.8}\right) v_2^2 = 1.25\, v_2^2$$

or $$v_1 = 1.118\, v_2$$

$$Q_1 = \frac{\pi}{4} D_1^2 v_1 = \frac{\pi}{4} \times 1 \times v_1 = 0.7854\, v_1$$

$$Q_2 = \frac{\pi}{4} D_2^2 v_2 = \frac{\pi}{4} (0.8)^2 \times v_2 = 0.5026\, v_2$$

$$2.5 = 0.7854\, v_1 + 0.5026\, v_2$$

$$= 0.7854 \times 1.11\, v_2 + 0.5026\, v_2 = 1.3807\, v_2$$

$$v_2 = 1.81 \text{ m/s and } v_1 = 2.024 \text{ m/s}$$

$$Q_1 = A_1 v_1 = \frac{\pi}{4} \times 1 \times 2.024 = 1.59 \text{ m}^3\text{/s}$$

$$Q_2 = 2.5 - 1.59 = 0.91 \text{ m}^3\text{/s}$$

Example 7.17 *A 200 mm diameter, 3 km long straight pipe runs between two reservoirs of surface elevations 140 m and 70 m. A 1.5 km long, 250 mm diameter pipe is laid parallel to the 200 mm diameter pipe from its mid-point to the lower reservoir. Neglecting all minor losses and assuming a friction factor of 0.02 for both pipes, find the increase in discharge casued by addition of 250 mm diameter pipe.*

Solution.

Given: $D_1 = D_2 = 0.2$ m, $L_1 = L_2$ 1.5 km, $D_3 = 0.25$ m, $L_3 = 1.5$ km, $f_1 = f_3 = 0.02$, $H_1 = 140$ m, $H_2 = 70$ m

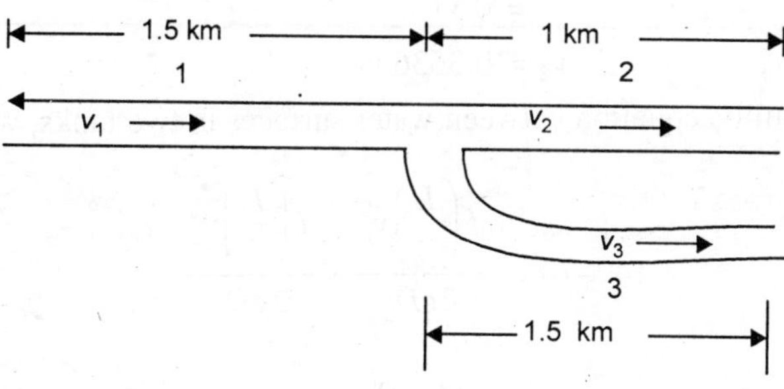

Fig. 7.12

Applying Bernoulli's equation between the free water surfaces in two tanks, we have

$$H_1 - H_2 = \frac{fLv^2}{2gD}$$

$$140 - 70 = \frac{0.02 \times 3000 \times v^2}{2 \times 9.81 \times 0.2}$$

$$v = 2.14 \text{ m/s}$$

$$Q = \frac{\pi}{4} (0.2)^2 \times 2.14 = 0.06723 \text{ m}^3\text{/s}$$

The altered pipeline is shown in Fig. 7.12.

$$Q_1 = Q_2 + Q_3$$

$$D_1^2 v_1 = D_2^2 v_2 + D_3^2 v_3$$

$$= D_1^2 v_2 + D_3^2 v_3 \quad \text{... (1)}$$

$$h_{f2} = h_{f3}$$

$$\frac{f_2 \times \frac{L}{2} \times v_2^2}{2gD_2} = \frac{f_3 \times \frac{L}{2} \times v_3^2}{2gD_3}$$

$$\frac{v_2^2}{D_2} = \frac{v_3^2}{D_3}$$

$$v_3 = \left(\frac{D_3}{D_2}\right)^{1/2} v_2 = \left(\frac{0.25}{0.2}\right)^{1/2} v_2 = 1.118\, v_2 \quad \text{... (2)}$$

Substituting in Eq. (1), we get

$$(0.2)^2\, v_1 = (0.2)^2\, v_2 + (0.25)^2\,(1.118\, v_2)$$

$$0.04\, v_1 = 0.04\, v_2 + 0.07\, v_2$$

$$= 0.11\, v_2$$

$$v_2 = 0.3636\, v_1$$

Again applying Bernoulli's equation between water surfaces in two tanks, we get

$$140\text{–}70 = \frac{f\left(\frac{L}{2}\right)v_1^2}{2gD_1} + \frac{f\left(\frac{L}{2}\right)v_2^2}{2gD_2}$$

$$70 = \frac{f(L/2)}{2gD_1}(v_1^2 + v_2^2) \qquad (\because D_1 = D_2)$$

$$= \frac{0.02 \times 1500}{2 \times 9.81 \times 0.2}[v_1^2 + (0.3636)^2\, v_1^2]$$

$$= 8.656\, v_1^2$$

$$v_1 = 2.844 \text{ m/s}$$

$$Q = \frac{\pi}{4}(0.2)^2 \times 2.844 = 0.08935 \text{ m}^3\text{/s}$$

Increase in discharge = 0.08935 – 0.6723 = 0.02212 m^3/s

$$\% \text{ increase in discharge} = \frac{0.02212}{0.06723} \times 100 = 32.89\%$$

Example 7.18 *When two pipes of diameter D and d are arranged in parallel, the loss of head for either pipe for a flow of Q is h. When the pipes are arranged in series, the loss of head is H for the same discharge Q. If d = 0.3 D, Calculate the total flow through each pipe when placed in parallel and the ratio of H to h neglecting minor losses and assuming friction coefficient to be constant.*

Solution.

Given: $L_1 = L_2 = L$, $D_1 = D$, $D_2 = d$, $h_{lp} = \text{h}$, $h_{ls} = H$, $d = 0.5D$

1. Pipes in parallel:

$$Q = Q_1 + Q_2 \qquad \text{... (1)}$$

$$(h_{lp})_1 = (h_{lp})_2 = h$$

$$\frac{4f\,L_1 v_1^2}{2gD_1} = \frac{4f\,L_2 v_2^2}{2g\,D_2} = h \qquad \text{... (2)}$$

$$\frac{v_1^2}{D} = \frac{v_2^2}{d}$$

$$\frac{\left(\frac{4Q_1}{\pi D^2}\right)^2}{D} = \frac{\left(\frac{4Q_2}{\pi d^2}\right)^2}{d}$$

$$\frac{Q_1^2}{D^5} = \frac{Q_2^2}{d^5}$$

$$\frac{Q_1}{Q_2} = \left(\frac{D}{0.5D}\right)^{5/2} = 2^{5/2} = 5.657$$

$$Q = (5.657 + 1)Q_2 = 6.657\,Q_2$$

$$Q_2 = 0.15\,Q$$

$$Q_1 = 0.85\,Q$$

2. Pipes in series:

$$H = \frac{4fL\,v_1^2}{2gD} + \frac{4fL\,v_2^2}{2gD} + \frac{4fL}{2gD}\left(v_1^2 + \frac{v_2^2}{0.5}\right)$$

$$= \frac{4fL}{2gD}\left[\left(\frac{4Q}{\pi D^2}\right)^2 + \frac{1}{2}\left\{\frac{4Q}{\pi \times (0.5D)^2}\right\}^2\right]$$

$$= \frac{4fL}{2gD} \times \frac{16Q^2}{\pi^2 D^4}[1+8] = \frac{32fLQ^2}{\pi^2 g D^5} \quad \text{... (3)}$$

From Eq. (2), we have

$$\frac{32fL}{\pi^2 gD^5} = \frac{h}{Q_1^2} \text{ and } \frac{32fL}{\pi^2 g d^5} = \frac{h}{Q_2^2}$$

Substituting these values in Eq. (3), we get

$$H = Q^2 \times \frac{h}{Q_1^2} + Q^2 \times \frac{h}{Q_2^2} = h\left[\left(\frac{Q}{Q_1}\right)^2 + \left(\frac{Q^2}{Q_2}\right)^2\right]$$

$$= h\left[\left(\frac{1}{0.85}\right)^2 + \left(\frac{1}{0.15}\right)^2\right] = 45.828\,h$$

$$\frac{H}{h} = 45.828$$

7.9 FLOW THROUGH BRANCHED PIPES

When three or more reservoirs are connected by means of pipes, having one or more junctions, the system is called a branching pipe system. Fig.7.13 shows three reservoirs connected to a single junction, by means of pipes which are called brached pipes. It is desired to find the discharge and direction of flow in each pipe.

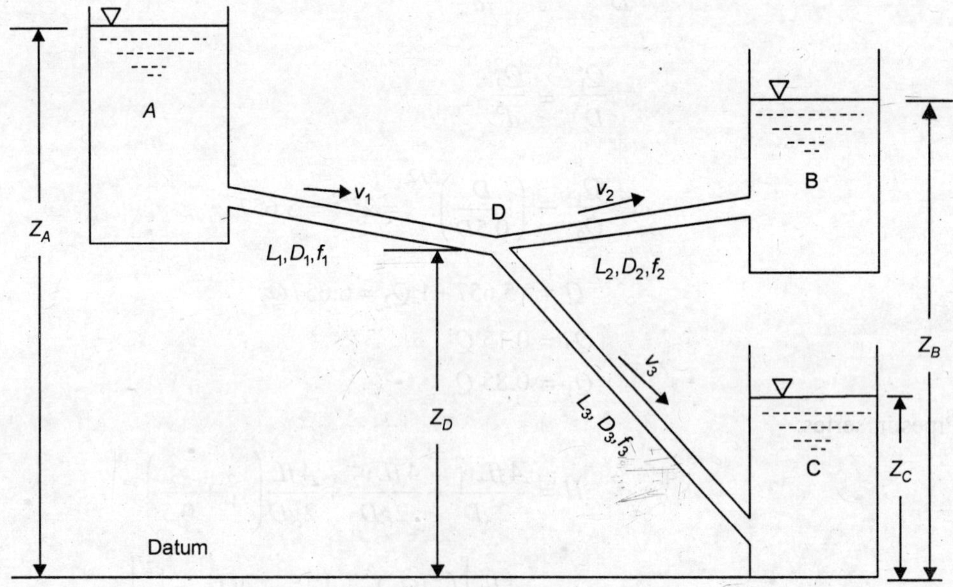

Fig. 7.13 Branched pipes

The basic equations for solving such problems are:

1. Continuity equation 2. Bernoulli's equation, and 3. Darcy-Weisbach equation.

The following assumpations are made:

(i) The reservoirs are very large so that the water levels in the reservoirs are constant to keep steady conditions in the pipes.

(ii) All minor losses are assumed very small and neglected.

(iii) The flow is from reservoir *A* to junction *D*, and from *D* to reservoir *C*.

(iv) The flow from junction *D* to reservoir *B* will take place only if piezometric head at *D* is more than the piezometric head at *B*.

Let us assume that flow is from *D* to *B*.

Applying Bernoulli's equation between *A* and *D*, D and B, and D and C, we have

$$z_A = z_D + \frac{p_D}{\gamma} + h_{f1} \quad \text{... (1)}$$

$$z_D + \frac{p_D}{\gamma} = z_B + h_{f2} \quad \text{... (2)}$$

$$z_D + \frac{p_d}{\gamma} = z_C + h_{f3} \quad \text{... (3)}$$

From continuity equation, we have

$$\frac{\pi}{4}D_1^2 v_1 = \frac{\pi}{4}D_2^2 v_2 + \frac{\pi}{4}D_3^2 v_3 \quad \text{... (4)}$$

There are four unknows i.e. v_1, v_2, v_3 and $\frac{p_D}{\gamma}$ and four equations to solve the problem.

Example 7.19 *Fig 7.14 shows three reservoirs connected by pipes. Each is 300 mm in diameter and1600 m long. Assuming coefficient of friction for each pipe equal to 0.01, find the discharge in each pipe.*

Solution.

Given: $D_1 = D_2 = D_3 = 0.3$ m, $L_1 = L_2 = L_3 = 1600$ m, $f_1 = f_2 = f_3 = 0.1$

Let us assume that no flow occurs in pipe 2, i.e. the piezonmetric level is 30 m.

Head loss in pipe 1, $h_{f1} = 70 - 30 = 40$ m

$$h_{f1} = \frac{4fL_1 v_1^2}{2gD_1}$$

$$40 = \frac{4 \times 0.01 \times 1600 \times v_1^2}{2 \times 9.81 \times 0.3}$$

$$v_1 = 1.91.8 \text{ m/s}$$

$$Q_1 = \frac{\pi}{4}(0.3)^2 \times 1.918$$

$$= 0.1356 \text{ m}^3\text{/s}$$

$$h_{f3} = 30 - 15 = 15 \text{ m}$$

$$= \frac{4 f L_3 v_3^2}{2g D_3}$$

$$15 = \frac{4 \times 0.01 \times 1600 \times v^2 3}{2 \times 9.81 \times 0.3}$$

$$v_3 = 1.174 \text{ m/s}$$

$$Q_3 = \frac{\pi}{4}(0.3)^2 \times 1.174$$

$$= 0.083 \text{ m}^3\text{/s}$$

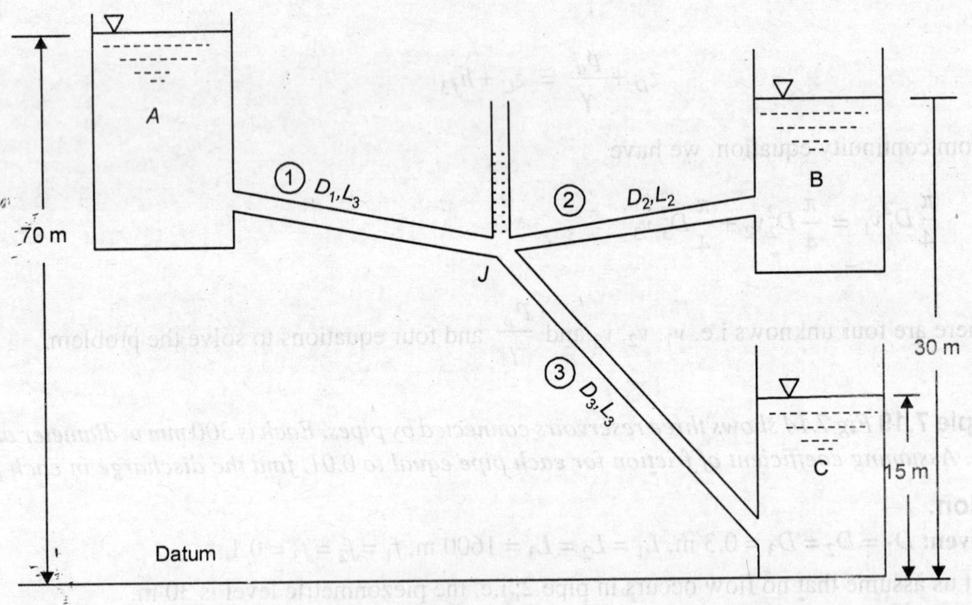

Fig. 7.14

Since $Q_1 > Q_3$, the direction of flow *is from J* to *B*.

Considering the flow from reservoirs *A* and *B*, we have

$$(70 - 30) = h_{f1} + h_{f2} = \frac{4 f L_1 v_1^2}{2g D_1} + \frac{4 f L_2 v_2^2}{2g D_2}$$

Mass rate of fluid entering through AD, $\dot{m}_1 = \int_0^\delta \rho u\, dy$

Mass rate of fluid leaving through BC, $\dot{m}_2 = \int_0^\delta \rho u\, dy + \frac{d}{dx}\left[\int_0^\delta \rho u\, dy\right] dx$

∴ Mass rate of fluid entering the control volume through the face DC = $\dot{m}_2 - \dot{m}_1$

$$= \frac{d}{dx}\left[\int_0^\delta \rho u\, dy\right] dx \quad \text{... (1)}$$

The fluid is entering through DC with a uniform velocity U_∞.

Momentum rate of fluid entering the control volume in x-direction through AD

$$= \int_0^\delta \rho u^2\, dy$$

Momentum rate of fluid leaving the central volume in x-direction through BC

$$= \int_0^\delta \rho u^2 dy + \frac{d}{dx}\left[\int_0^\delta \rho u^2 dy\right] dx \quad \text{... (2)}$$

Momentum rate of fluid entering the control volume through DC in x-direction

$$= \frac{d}{dx}\left[\int_0^\delta \rho u\, dy\right] dx \times U_\infty$$

$$= \frac{d}{dx}\left[\int_0^\delta \rho u U_\infty\, dy\right] dx \quad \text{... (3)}$$

Rate of change of momentum of control volume

$$= \text{Eq. (2)} - \text{Eq. (1)} - \text{Eq. (3)}$$

$$= \frac{d}{dx}\left[\int_0^\delta \rho u^2 dy - \int_0^\delta \rho u U_\infty dy\right] dx$$

$$= \rho \frac{d}{dx}\left[\int_0^\delta (u^2 - u U_\infty) dy\right] dx \quad \text{... (8.9)}$$

According to the momentum principle, the rate of change of momentum on the control volume $ABCD$ must be equal to the total force on the control volume in the same direction. The only external force acting on the control volume is the shear force acting on the side AB in the direction B to A, due to boundary shear stress, τ_0 (see Fig. 8.4b). The value of this shear drag force is given by,

$$\Delta F_D = \tau_0 \times dx$$

$$40 = \frac{4 \times 0.01 \times 1600 \times v_1^2}{2 \times 9.81 \times 0.3} + \frac{4 \times 0.01 \times 1600 \times v_2^2}{2 \times 9.81 \times 0.3}$$

$$= 10.873(v_1^2 + v_2^2)$$

$$v_1^2 + v_2^2 = 3.678$$

or $v_2 = \sqrt{3.678 - v_1^2}$ ... (1)

Now consider the flow from reservoir A to C, we have

$$70 - 15 = h_{f1} + h_{f3}$$

$$55 = 10.873(v_1^2 + v_3^2)$$

or $v_1^2 + v_3^2 = 5.058$ ... (2)

$$v_3 = \sqrt{5.058 - v_1^2}$$

From continuity of flow, we have

$$Q_1 = Q_2 + Q_3$$

Here $A_1 = A_2 = A_3$

∴ $v_1 = v_2 + v_3$ ... (3)

Using Eqs. (1), (2) and (3), we get

$$v_1 = \sqrt{3.678 - v_1^2} + \sqrt{5.058 - v_1^2}$$

$$[v_1 - \sqrt{3.678 - v_1^2}]^2 = 5.058 - v_1^2$$

$$v_1^2 + 3.678 - v_1^2 - 2v_1\sqrt{3.678 - v_1^2} = 5.058 - v_1^2$$

$$3.678 - 2v_1\sqrt{3.678 - v_1^2} = 5.058 - v_1^2$$

$$v_1^2 - 1.38 = 2v_1\sqrt{3.678 - v_1^2}$$

$$v_1^4 + 1.904 - 2.76v_1^2 = 4v_1^2(3.678 - v_1^2)$$

$$= 14.712v_1^2 - 4v_1^4$$

$$5v_1^4 - 17.472v_1^2 + 1.904 = 0$$

$$v_1^4 - 3.394 + 0.380 = 0$$

$$v_1^2 = \frac{1}{2}[3.394 \pm \sqrt{(3.394)^2 - 4 \times 0.38}]$$

Thus the total external force in the direction of rate of change of momentum

$$= -\tau_0 \times dx \qquad \text{... (8.10)}$$

Equating Eqs. (8.9) and (8.10), we have

$$-\tau_0 \times dx = \rho \frac{d}{dx}\left[\int_0^{\delta}(u^2 - uU_\infty)\,dy\right]dx$$

$$\tau_0 = -\rho \frac{d}{dx}\left[\int_0^{\delta}(u^2 - uU_\infty)\,dy\right]$$

$$= \rho \frac{d}{dx}\left[\int_0^{\delta}(uU_\infty - u^2)\,dy\right]$$

$$= \rho \frac{d}{dx}\left[\int_0^{\delta}U_\infty^2\left(\frac{u}{U_\infty} - \frac{u^2}{U_\infty^2}\right)dy\right]$$

$$= \rho U_\infty^2 \frac{d}{dx}\left[\int_0^{\delta}\frac{u}{U_\infty}\left(1 - \frac{u}{U_\infty}\right)dy\right]$$

or

$$\frac{\tau_0}{\rho U_\infty^2} = \frac{d}{dx}\left[\int_0^{\delta}\frac{u}{U_\infty}\left(1 - \frac{u}{U_\infty}\right)dy\right] \qquad \text{... (8.11a)}$$

Now momentum thickness, $\theta = \int_0^{\delta}\frac{u}{U_\infty}\left(1 - \frac{u}{U_\infty}\right)dy$

$$\therefore \quad \frac{\tau_0}{\rho U_\infty^2} = \frac{d\theta}{dx} \qquad \text{... (8.11b)}$$

Eq. (8.11) is known as von-Karman momentum integral equation for boundary layer flow. This is applicable to:

(i) Laminar boundary layer, (ii) Transition boundary layer, and (iii) Turbulent boundary layer.

Boundary conditions

The following boundary conditions must be satisfied by any assumed velocity distribution:

1. At the surface of plate:

   $y = 0, u = 0$, and $\frac{du}{dy}$ = a finite value

2. At the outer edge of boundary layer:

   $y = \delta, u = U_\infty$

$$\frac{u}{U_\infty} = 1-\frac{y^2}{R^2}$$

$$\therefore \quad \delta^* = \frac{1}{R}\int_0^R \frac{y^2}{R^2}\cdot y dy = \frac{1}{R^3}\int_0^R y^3 dy = \frac{1}{R^3}\left|\frac{y^4}{4}\right|_0^R = \frac{R}{4}$$

From the definition for momentum thickness, we have

$$[\rho\{\pi R^2 - \pi(R-\theta)^2\}]U_\infty^2 = \int_0^R \rho.2\pi y\,.dy.u(U_\infty - u)$$

$$2\pi R\theta U_\infty^2 = \int_0^R 2\pi y u(U_\infty - u)dy$$

$$\theta = \frac{1}{R}\int_0^R \frac{u}{U_\infty}\left(1-\frac{u}{U_\infty}\right)y\,dy$$

$$= \frac{1}{R}\int_0^R \left(1-\frac{y^2}{R^2}\right)\frac{y^2}{R^2}\cdot y\,dy = \frac{1}{R}\int_0^R\left(\frac{y^3}{R^2}-\frac{y^5}{R^4}\right)dy$$

$$= \frac{1}{R}\left|\frac{y^4}{4R^2}-\frac{y^6}{6R^4}\right|_0^R = \frac{R}{12}$$

$$\frac{\delta^*}{\theta} = \frac{R}{4}\times\frac{12}{R} = 3$$

Example 8.6 *The velocity distribution in the boundary layer over the face of a high spillway is given by:*

$$\frac{u}{U_\infty} = \left(\frac{y}{\delta}\right)^{0.22}$$

Calculate: (a) the displacement thickness, (b) the energy thickness, and (c) the loss of energy, for the following data:

$U_\infty = 25$ m/s, $\delta = 30$ mm, $q = 5$ m³/s per metre length of spillway.

Solution. (a) $\delta^* = \int_0^\delta\left(1-\frac{u}{U_\infty}\right)dy = \int_0^\delta\left[1-\left(\frac{y}{\delta}\right)^{0.22}\right]dy$

$$= \left|y - \frac{y^{1.22}}{1.22\delta^{0.22}}\right|_0^\delta = \left[\delta - \frac{\delta}{1.22}\right] = 0.18\delta$$

$$= 0.18\times 30 = 5.4 \text{ mm}$$

Solution. (b) $\delta_e = \int_0^{\delta} \frac{u}{U_\infty}\left[1-\left(\frac{u}{U_\infty}\right)^2\right] dy = \int_0^{\delta} \left(\frac{y}{\delta}\right)^{0.22}\left[1-\left(\frac{y}{\delta}\right)^{0.44}\right] dy$

$$= \int_0^{\delta}\left[\left(\frac{y}{\delta}\right)^{0.22} - \left(\frac{y}{\delta}\right)^{0.66}\right] dy = \left|\frac{y^{1.22}}{1.22\,y^{0.22}} - \frac{y^{1.66}}{1.66\,\delta^{0.66}}\right|_0^{\delta}$$

$$= \frac{\delta}{1.22} - \frac{\delta}{1.66} = 0.2173\,\delta$$

$$= 0.2173 \times 30 = 6.518 \text{ mm}$$

(c) Energy loss per metre length of spillway,

$$E_l = \frac{1}{2}\rho\delta_e U_\infty^3 = \frac{1}{2}\times 10^3 \times 6.518\times 10^{-3} \times 25^3 = 50922 \text{ N.m/s}$$

$$h_l = \frac{E_l}{\gamma q} = \frac{50922}{9810\times 5} = 1.038 \text{ m}$$

8.3 VON-KARMAN MOMENTUM INTEGRAL EQUATION

Von-Karman suggested a method based on the momentum equation by which the growth of boundary layer along a flat plate, wall shear stress, and drag force can be determined, when the velocity distribution in the boundary layer is known. This method can be used for both laminar and turbulent boundary layers.

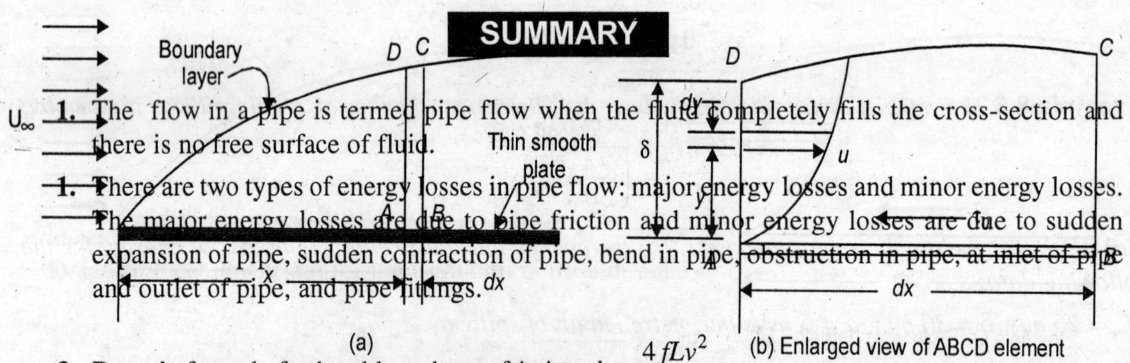

Fig. 8.4 Von-Karman boundary layer

The fluid flowing over a thin flat smooth plate, with free-stream velocity U_∞, is shown in Fig. 8.4 (a). Consider a small length dx of the plate at a distance x from the leading edge. The enlarged view of the elementary length dx of the plate is shown in Fig. 8.4 (b). Consider unit width of the plate perpendicular to the direction of flow. Let ABCD be a small element of the boundary layer

Given: $L = 2000$ m, $P = 120$ kW, $p_1 = 4$ MPa, $p_2 = 0.8$ MPa, $f = 0.005$

(a) $H = \frac{p}{\gamma} = \frac{4\times 10^6}{9810} = 407.747$ m

$$h_f = \frac{p_2}{\gamma} = \frac{0.8\times 10^6}{9810} = 81.5 \text{ m}$$

$$P = \frac{\gamma Q(H-h_f)}{10^3}$$

$$120 = \frac{9810\times Q(407.747 - 81.549)}{10^3}$$

$$Q = 0.0375 \text{ m}^3/\text{s}$$

$$v = \frac{4Q}{\pi D^2} = \frac{4\times 0.0375}{\pi D^2} = \frac{0.04775}{D^2}$$

$$h_f = \frac{4fLv^2}{2gD}$$

$$81.549 = \frac{4\times 0.005\times 2000\times (0.04775)^2}{2\times 9.81\times D^5}$$

$$D = 141.6 \text{ mm}$$

(b) $\frac{h_f}{H} = \frac{81.549}{407.747} = 0.8$ or 80%

SUMMARY

1. The flow in a pipe is termed pipe flow when the fluid completely fills the cross-section and there is no free surface of fluid.
2. There are two types of energy losses in pipe flow: major energy losses and minor energy losses. The major energy losses are due to pipe friction and minor energy losses are due to sudden expansion of pipe, sudden contraction of pipe, bend in pipe, obstruction in pipe, at inlet of pipe and outlet of pipe, and pipe fittings.
3. Darcy's formula for head loss due to friction, $h_f = \frac{4fLv^2}{2gD}$

   where f = coefficient of friction, and $4f$ = friction factor.
4. Chezy's formula for head loss due to friction, $v = C\sqrt{mi}$

   where C = Chezy's constant, $m = \frac{A}{P} = \frac{D}{4}$, $i = \frac{h_f}{L}$

5. Minor head losses:

(i) Sudden expansion of pipe, $h_e = \frac{(v_1 - v_2)^2}{2g}$

(ii) Sudden contraction of pipe, $h_c = \frac{0.5 v_2^2}{2g}$

(iii) Entrance of pipe, $h_i = \frac{0.5 v^2}{2g}$

(iv) Exit of pipe, $h_0 = \frac{v^2}{2g}$

(v) Obstruction in a pipe, $h_{obs} = \frac{v^2}{2g}\left[\frac{A}{C_c(A-a)} - 1\right]^2$

(vi) Pipe bend, $h_b = \frac{k_b v^2}{2g}$

(vii) pipe fittings, $h_{pf} = \frac{k_{pf} v^2}{2g}$

6. The line representing the sum of pressure head and datum head w.r.t some reference line is called hydraulic gradient line (H.G.L.), and the line representing the sum of pressure head, datum head, and velocity head w.r.t some reference line is called energy gradient line (E.G.L) or total energy line (T.E.L.)

7. The pipes of different lengths and diameters connected end to end are called pipes in series or compound pipes.

$$Q = A_1 v_1 = A_2 v_2 = \ldots$$

$$H = (h_i + h_{f1}) + h_e + h_{f2} + h_c + h_{f3} + h_0$$

$$= \frac{4 f v_1^2}{2 g D_1}\left[L_1 + \left(\frac{D_1}{D_2}\right)^5 L_2 + \left(\frac{D_1}{D_3}\right)^5 L_3 + \ldots\right]$$

8. An equivalent pipe is defined as the pipe of uniform diameter having loss of head and discharge of a compound pipe consisting of several pipes of different lengths and diameters. The uniform diameter of the equivalent pipe is known as the equivalent diameter of compound pipe.

$$\frac{L}{D^5} = \sum \frac{L_i}{D_i^5}$$

Equating Eqs. (8.15) and (8.16), we get

$$\frac{2}{15}\rho U_\infty^2 \frac{d\delta}{dx} = \frac{2\mu U_\infty}{\delta}$$

$$\delta \cdot \frac{d\delta}{dx} = \frac{15\mu}{\rho U_\infty}$$

or

$$\delta \cdot d\delta = \frac{15\mu}{\rho U_\infty} \cdot dx$$

Integrating both sides, we get

$$\frac{\delta^2}{2} = \frac{15\mu}{\rho U_\infty} x + C$$

where C is a constant of integration.

At $x = 0$, $\delta = 0$, thus $C = 0$.

$$\frac{\delta^2}{2} = \frac{15\mu x}{\rho U_\infty}$$

or

$$\delta = \sqrt{\frac{30\mu x}{\rho U_\infty}} = 5.48\sqrt{\frac{\mu x}{\rho U_\infty}} = 5.48\sqrt{\frac{\mu x \times x}{\rho U_\infty x}}$$

$$= 5.48\sqrt{\frac{x^2}{Re_x}} = \frac{5.48x}{\sqrt{Re_x}} \quad \left[\because Re_x = \frac{\rho U_\infty x}{\mu}\right] \quad \ldots (8.17)$$

Eq. (8.17) shows that for laminar boundary layer over a flat plate,

$$\delta \propto \sqrt{x} \propto \frac{1}{\sqrt{Re_x}} \quad \ldots (8.18)$$

i.e. thickness of laminar boundary layer over a flat plate is proporation to the square root of the distance from the leading edge and inversely proportional to the square root of Reynolds number.

2. Shear stress-τ_0

We have from Eq. (8.16),

$$\tau_0 = \frac{2\mu U_\infty}{\delta}$$

and

$$\delta = \frac{5.48x}{\sqrt{Re_x}}$$

∴

$$\tau_0 = \frac{2\mu U_\infty \sqrt{Re_x}}{5.48 x} = 0.635 \frac{\mu U_\infty}{x} \cdot \sqrt{Re_x} \quad \ldots (8.19)$$

3. Local coefficient of drag-C^*_D

$$\tau_0 = C_D^* \frac{\rho U_\infty^2}{2}$$

$$\therefore \quad 0.365 \frac{\mu U_\infty}{x} \sqrt{\text{Re}_x} = C_D^* \frac{\rho U_\infty^2}{2}$$

$$C_D^* = \frac{0.73\sqrt{\text{Re}_x}}{\left(\dfrac{\rho U_\infty x}{\mu}\right)} = \frac{0.73\sqrt{\text{Re}_x}}{\text{Re}_x} = \frac{0.73}{\sqrt{\text{Re}_x}} \qquad \text{... (8.20)}$$

4. Coefficient of drag-C_D

$$C_D = \frac{FD}{\frac{1}{2}\rho A U_\infty^2}$$

where $F_D = \int_0^L \tau_0 \times b \times dx$

$$= \int_0^L 0.365 \frac{\mu U_\infty}{x} \sqrt{\text{Re}_x} \times b \times dx$$

$$= \int_0^L 0.365 \frac{\mu U_\infty}{x} \sqrt{\frac{\rho U_\infty x}{\mu}} \times bx\, dx$$

$$= 0.365 \mu U_\infty \sqrt{\frac{\rho U_\infty}{\mu}} \times b \times \int_0^L x^{-1/2} dx$$

$$= 0.365 \mu U_\infty \sqrt{\frac{\rho U_\infty}{\mu}} \times b \left| \frac{x^{1/2}}{(1/2)} \right|_0^L = 0.73 \mu U_\infty \sqrt{\frac{\rho U_\infty}{\mu}} \times bL^{1/2}$$

$$F_D = 0.73 \mu U_\infty b \sqrt{\frac{\rho U_\infty L}{\mu}} = 0.73\, uU_\infty \sqrt{\text{Re}_L}$$

$$C_D = \frac{0.73 \mu U_\infty b \sqrt{\text{Re}_L}}{\frac{1}{2}\rho L b U_\infty^2} = \frac{1.46\mu}{\rho L U_\infty} \sqrt{\text{Re}_L} = \frac{1.46}{\sqrt{\text{Re}_L}} \qquad \text{... (8.21)}$$

Drag force on elemental length dx of plate is given by,

$$\Delta F_D = \tau_0 \times b \times dx$$

where b = width of plate.

Total drag force on one side of plate of length L,

$$F_D = \int \Delta F_D = \int_0^L \tau_0 \times b \times dx \qquad \text{... (8.12)}$$

8.3.1 Local Coefficient of Drag – C_D^*

It is defined as the ratio of the shear stress τ_0 to the quantity $\frac{1}{2}\rho U_\infty^2$. It is denoted by C_D^*. It is also called *coefficient of skin friction.*

Hence

$$C_D^* = \frac{\tau_0}{\frac{1}{2}\rho U_\infty^2} \qquad \text{...(8.13)}$$

8.3.2 Average Coefficient of Drag, C_D

It is defined as the ratio of the total drag force to the quantity $\frac{1}{2}\rho A U_\infty^2$. It is also called coefficient of drag and is denoted by C_D. Thus

$$C_D = \frac{F_D}{\frac{1}{2}\rho A U_\infty^2} \qquad \text{... (8.14)}$$

where ρ = mass density of fluid

A = area of surface (or plate)

U_∞ = free-stream velocity.

8.4 LAMINAR BOUNDARY LAYER

Case I: $= \dfrac{u}{U_\infty} = 2\left(\dfrac{y}{\delta}\right) - \left(\dfrac{y}{\delta}\right)^2$

1. Boundary layer thickness-δ

   Momentum integral equation is,

$$\frac{\tau_0}{\rho U_\infty^2} = \frac{d}{dx}\left[\int_0^\delta \frac{u}{U_\infty}\left(1 - \frac{u}{U_\infty}\right)dy\right]$$

$$= \frac{d}{dx}\left[\int_0^\delta \left\{2\left(\frac{y}{\delta}\right) - \left(\frac{y}{\delta}\right)^2\right\}\left\{1 - \left(\frac{2y}{\delta} - \frac{y^2}{\delta^2}\right)\right\} dy\right]$$

$$= \frac{d}{dx}\left[\int_0^\delta \left\{\frac{2y}{\delta} - \frac{y^2}{\delta^2}\right\}\left\{1 - \frac{2y}{\delta} + \frac{y^2}{\delta^2}\right\} dy\right]$$

8

$$= \frac{d}{dx}\left[\int_0^\delta \left\{\frac{2y}{\delta} - \frac{4y^2}{\delta^2} + \frac{2y^3}{\delta^3} - \frac{y^2}{\delta^2} + \frac{2y^3}{\delta^3} - \frac{y^4}{\delta^4}\right\} dy\right]$$

BOUNDARY LAYER FLOW

$$= \frac{d}{dx}\left[\int_0^\delta \left\{\frac{2y}{\delta} - \frac{5y^2}{\delta^2} + \frac{4y^3}{\delta^3} - \frac{y^4}{\delta^4}\right\} dy\right]$$

8.1 INTRODUCTION

$$= \frac{d}{dx}\left[\frac{y^2}{\delta} - \frac{5}{3}\frac{y^3}{\delta^2} + \frac{y^4}{\delta^3} - \frac{y^5}{5\delta^4}\right]_0^\delta$$

When a real (or viscous) fluid flows passed a solid boundary a condition of no slip occurs, which implies that the velocity of fluid must be same as that of boundary itself. The layer of fluid which cannot slip away from the boundary surface undergoes retardation. This retarded layer further causes retardation in the adjacent layes of fluid. This develops a small region in the immediate vicinity of the boundary surface in which the velocity of the flowing fluid increases rapidly from zero at the boundary surface and approaches the free-stream velocity U_∞. The layer adjacent to the boundary is nown as *boundary layer*. The theory dealing with boundary layer flow is called boundary layer theory. Boundary layer is formed whenever there is relative motion between the boundary and the fluid.

$$= \frac{d}{dx}\left[\delta - \frac{5}{3}\delta + \delta - \frac{\delta}{5}\right] = \frac{d}{dx}\left(\frac{2}{15}\delta\right)$$

$$\tau_0 = \rho U_\infty^2 \times \frac{d}{dx}\left(\frac{2}{15}\delta\right) = \frac{2}{15}\rho U_\infty^2 \frac{d\delta}{dx} \quad \text{... (8.15)}$$

According to Newton's law of viscosity,

According to boundary layer theory, the flow of fluid in the neighbourhood of the solid boundary may be divided into following two regions:

$$\tau_0 = \mu\left(\frac{du}{dy}\right)_{y=0}$$

1. A very thin layer of fluid, called the boundary layer, in the immediate vicinity of the solid boundary, where the variation of velocity from zero at the solid boundary to free-stream velocity in the direction normal to the boundary takes place. In this region, the velocity gradient $\frac{du}{dy}$ exists and the fluid exerts a shear stress on the boundary in the direction of flow. The value of shear stress is given by: $\tau_0 = \mu\left(\frac{du}{dy}\right)_{y=0}$

Now $u = U_\infty\left[\frac{2y}{\delta} - \frac{y^2}{\delta^2}\right]$

$$\frac{du}{dy} = U_\infty\left(\frac{2}{\delta} - \frac{2y}{\delta^2}\right)$$

$$\left(\frac{du}{dy}\right)_{y=0} = \frac{2U_\infty}{\delta}$$

2. A region outside the boundary layer, where the velocity of fluid is the free stream velocity, the fluid flow is like that of an ideal fluid and the potential flow theory is applicable. In this region, $\therefore\ \tau_0 = 0$ and $\frac{du}{dy} = 0$.

$$\tau_0 = \frac{2\mu U_\infty}{\delta} \quad \text{... (8.16)}$$

8.2 DEFINITIONS FOR BOUNDARY LAYER FLOW OVER A FLAT PLATE

Consider the flow of a fluid over a smooth thin flat plate placed parallel to the direction of flow as shown in Fig. 8.1. The edge of the plate facing the direction of flow is called leading edge and the rear edge is called the trailing edge. Let the flow be with zero pressure gradient on one side of the plate, when stationary.

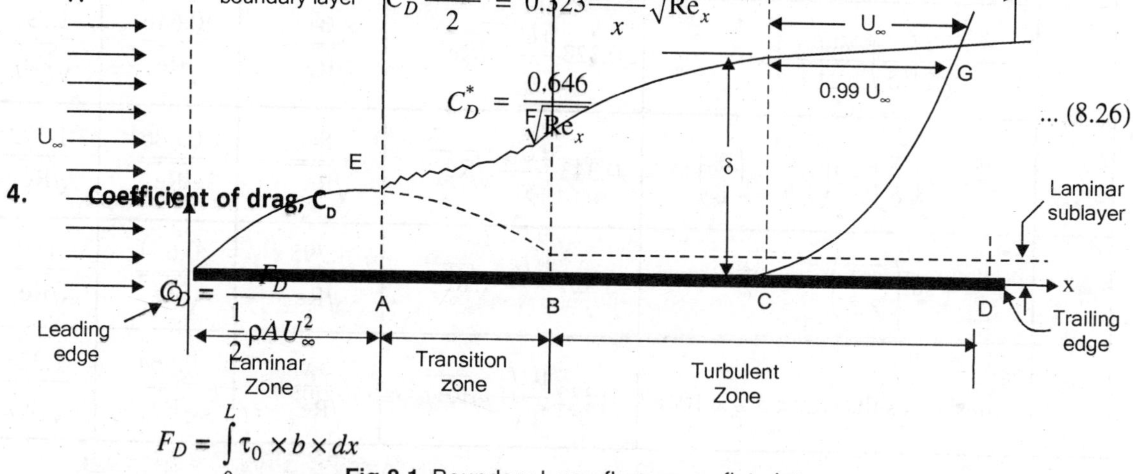

Fig 8.1 Boundary Layer flow over a flat plate

1. Laminar Boundary Layer

Consider the flow with zero pressure gradient on one side of the stationary plate. As the plate is stationary hence the velocity of fluid on the surface of plate is zero. At a distance from the plate surface a velocity gradient is set up in the fluid. This velocity gradient develops shear resistance, which retards the fluid. Thus the boundary layer region begins at the sharp leading edge of the plate. The boundary layer region increases because the retarded fluid is further retarded. This is referred to as the growth of boundary layer. Where the thickness is small, the flow in the boundary layer is laminar. This layer of the fluid is said to be laminar boundary layer, as shown by *OE* in Fig. 8.1. The length of the plate from the leading edge to the end of the laminar boundary layer is called the laminar zone, as shown by *OA*. The distance *OA* is obtained from Reynolds number, $Re = 5 \times 10^5$ for a plate, where

$$Re_x = \frac{U_\infty x}{\nu} \qquad \text{... (8.1)}$$

and x = distance from leading edge

U_∞ = free-stream velocity

ν = kinematic viscosity of fluid

3. Local coefficient of drag C_D^*

Also

$$\tau_0 = C_D^* \frac{\rho U_\infty^2}{2}$$

$\therefore$

$$C_D^* \frac{\rho U_\infty^2}{2} = 0.323 \frac{\mu U_\infty}{x} \sqrt{Re_x}$$

$$C_D^* = \frac{0.646}{\sqrt{Re_x}} \qquad \text{... (8.26)}$$

4. Coefficient of drag C_D

$$C_D = \frac{F_D}{\frac{1}{2}\rho A U_\infty^2}$$

$$F_D = \int_0^L \tau_0 \times b \times dx$$

$$= \int_0^L \frac{0.323 U_\infty}{x} \sqrt{Re_x} \times b \times dx = 0.323 \int_0^L \frac{\mu U_\infty}{x} \times \sqrt{\frac{\rho U_\infty x}{\mu}} \times b \times dx$$

$$= 0.323\, b \mu U_\infty \sqrt{\frac{\rho U_\infty}{\mu}} \int_0^L x^{-1/2}\, dx = 0.646\, b \mu U_\infty \sqrt{\frac{\rho U_\infty L}{\mu}}$$

$$C_D = \frac{0.646\, b \mu U_\infty \sqrt{\frac{\rho U_\infty L}{\mu}}}{\frac{1}{2}\rho \times b \times L \times U_\infty^2} = \frac{1.292 \sqrt{\frac{\rho U_\infty L}{\mu}}}{\left(\frac{\rho L U_\infty}{\mu}\right)}$$

$$= \frac{1.292}{\sqrt{Re_L}} \qquad \text{... (8.27)}$$

Table 8.1 Shows the values of δ, τ_0, C^*_D and C_D for various velocity profiles

Table 8.1 Values of δ, τ_0, C^*_D and C_D for various velocity profiles

S.No	Velocity profile	τ_0	δ	C^*_D	CD
1.	$\frac{u}{U_\infty} = 2\left(\frac{y}{\delta}\right) - \left(\frac{y}{\delta}\right)^2$	$0.365\frac{\mu U_\infty}{x}\sqrt{Re_x}$	$\frac{5.48x}{\sqrt{Re_x}}$	$\frac{0.73}{\sqrt{Re_x}}$	$\frac{1.46}{\sqrt{Re_L}}$
2.	$\frac{u}{U_\infty} = \frac{3}{2}\left(\frac{y}{\delta}\right) - \frac{1}{2}\left(\frac{y}{\delta}\right)^3$	$0.323\frac{\mu U_\infty}{x}\sqrt{Re_x}$	$\frac{4.64x}{\sqrt{Re_x}}$	$\frac{0.646}{\sqrt{Re_x}}$	$\frac{1.292}{\sqrt{Re_L}}$
3.	$\frac{u}{U_\infty} = 2\left(\frac{y}{\delta}\right) - 2\left(\frac{y}{\delta}\right)^3 + \left(\frac{y}{\delta}\right)^4$	$0.343\frac{\mu U_\infty}{x}\sqrt{Re_x}$	$\frac{5.84x}{\sqrt{Re_x}}$	$\frac{0.686}{\sqrt{Re_x}}$	$\frac{1.372}{\sqrt{Re_L}}$
4.	$\frac{u}{U_\infty} = \sin\left(\frac{\pi}{2}\frac{y}{\delta}\right)$	$0.327\frac{\mu U_\infty}{x}\sqrt{Re_x}$	$\frac{4.795x}{\sqrt{Re_x}}$	$\frac{0.654}{\sqrt{Re_x}}$	$\frac{1.31}{\sqrt{Re_L}}$
5.	Blasius results (Re < 3.2 × 10^5)	$0.332\frac{\mu U_\infty}{x}\sqrt{Re_x}$	$\frac{5x}{\sqrt{Re_x}}$	$\frac{0.664}{\sqrt{Re_x}}$	$\frac{1.328}{\sqrt{Re_L}}$

$$\sqrt{Re_x} = \sqrt{\frac{\rho U_\infty x}{\mu}}, \sqrt{Re_L} = \sqrt{\frac{\rho U_\infty L}{\mu}}$$

Example 8.7 *The boundary layer thickness at a distance of 1.0 m from the leading edge of a flat plate kept at zero angle of incidence to the flow direction is 1 mm. The free-stream velocity is 20 m/s. Find the boundary layer thickness at a distance of 3m.*

Solution. Given: $x_1 = 1$ m, δ_1 = m, $\delta_1 = 1.0$ mm, $U_\infty = 20$ m/s, $x_2 = 3$ m

$$\delta \propto \sqrt{x}$$

$$\frac{\delta_2}{\delta_1} = \sqrt{\frac{x_2}{x_1}}$$

$$\delta_2 = 1.0\sqrt{\frac{3}{1}} = 1.732 \text{ mm}$$

Example 8.8 *A smooth plate 3 m long and 2 m wide is towed in oil of specific gravity 0.8 at a velocity of 2m/s along its length. Calculate the boundary layer thickness and shear stress at the trailing edge of the plate. Kinematic viscosity of oil is 10^{-4} m^2/s*

where $\mathrm{Re}_L = \dfrac{\rho L U_\infty}{\mu}$ is the Reynolds number at the trailing edge of plate.

$A = L \times b$

Case 2: $\dfrac{u}{U_\infty} = \dfrac{3}{2}\left(\dfrac{y}{\delta}\right) - \dfrac{1}{2}\left(\dfrac{y}{\delta}\right)^3$

1. Boundary layer thickness-δ

$$\frac{\tau_0}{\tau U_\infty^2} = \frac{d}{dx}\left[\int_0^\delta \frac{u}{U_\infty}\left(1 - \frac{u}{U_\infty}\right)dy\right]$$

$$= \frac{d}{dx}\left[\int_0^\delta \left\{\frac{3}{2}\left(\frac{y}{\delta}\right) - \frac{1}{2}\left(\frac{y}{\delta}\right)^3\right\}\left\{1 - \frac{3}{2}\left(\frac{y}{\delta}\right) + \frac{1}{2}\left(\frac{y}{\delta}\right)^3\right\}dy\right]$$

$$= \frac{d}{dx}\left[\int_0^\delta \left\{\frac{3}{2}\frac{y}{\delta} - \frac{9}{4}\frac{y^2}{\delta^2} + \frac{3}{4}\frac{y^4}{\delta^4} - \frac{1}{2}\frac{y^3}{\delta^3} + \frac{3}{4}\frac{y^4}{\delta^4} - \frac{1}{4}\frac{y^6}{\delta^6}\right\}dy\right]$$

$$= \frac{d}{dx}\left[\int_0^\delta \left\{\frac{3}{2}\frac{y}{\delta} - \frac{9}{4}\frac{y^2}{\delta^2} + \frac{3}{2}\frac{y^4}{\delta^4} - \frac{1}{2}\frac{y^3}{\delta^3} - \frac{1}{4}\frac{y^6}{\delta^6}\right\}dy\right]$$

$$= \frac{d}{dx}\left[\left|\frac{3}{4}\frac{y^2}{\delta} - \frac{3}{4}\frac{y^3}{\delta^2} + \frac{3}{10}\frac{y^5}{\delta^4} - \frac{1}{8}\frac{y^4}{\delta^3} - \frac{1}{28}\frac{y^7}{\delta^6}\right|_0^\delta\right]$$

$$= \frac{d}{dx}\left[\frac{3}{4}\delta - \frac{3}{4}\delta + \frac{3}{10}\delta - \frac{1}{8}\delta - \frac{1}{28}\delta\right]$$

$$= \frac{d}{dx}\left[\frac{39}{280}\delta\right] = \frac{39}{280}\frac{d\delta}{dx}$$

$$\tau_0 = \rho U_\infty^2 \times \frac{39}{280}\frac{d\delta}{dx} = \frac{39}{280}\rho U_\infty^2 \frac{d\delta}{dx} \qquad \text{... (8.22)}$$

Also $\qquad \tau_0 = \mu\left(\dfrac{du}{dy}\right)_{y=0}$

Let θ = distance by which plate be displaced when the fluid is flowing with constant velocity U_∞.

Then rate of momentum loss of fluid flowing through distance θ with velocity U_∞

$$= (\rho\theta b U_\infty)\times U_\infty = \rho\theta b U_\infty^2 \quad \text{... (8.5)}$$

Equating Eqs. (8.4) and (8.5), we get

$$\rho\theta b U_\infty^2 = \rho b\int_0^\delta u(U_\infty - u)\,dy$$

or

$$\theta = \int_0^\delta \frac{u}{U_\infty}\left(1-\frac{u}{U_\infty}\right)dy \quad \text{... (8.6)}$$

8.2.4 Energy Thickness-δ_e

It is defined as the distance measured perpendicular to the boundary of the solid body by which the boundary should be displaced to compensate for reduction in kinetic energy of the flowing fluid on account of boundary laryer formation. It is denoted by δ_e.

Kinetic energy of this fluid $= \frac{1}{2}(\rho u b dy)u^2$

Kinetic energy of this fluid in the absence of boundary layer $= \frac{1}{2}(\rho u b dy)U_\infty^2$

∴ Loss of kinetic energy through element strip $= \frac{1}{2}\rho u b\,(U_\infty^2 - u^2)\,dy$

Total loss of kinetic energy of fluid passing through AB

$$= \frac{1}{2}\rho b\int_0^\delta u(U_\infty^2 - u^2)\,dy$$

Let δ_e = distance by which the plate be displaced to compensate for reduction in kinetic energy.

Loss of kinetic energy through δ_e of flid flowing with U_∞

$$= \frac{1}{2}(\rho b\delta_e U_\infty)U_\infty^2 = \frac{1}{2}\rho b\,\delta_e U_\infty^3$$

∴

$$\frac{1}{2}\rho b\,\delta_e U_\infty^3 = \frac{1}{2}\rho b\int_0^\delta u(U_\infty^2 - u^2)\,dy$$

But

$$u = U_\infty\left[\frac{3}{2}\left(\frac{y}{\delta}\right)-\frac{1}{2}\left(\frac{y}{\delta}\right)^3\right]$$

$$\frac{du}{dy} = U_\infty\left[\frac{3}{2}\times\frac{1}{\delta}-\frac{3}{2}\frac{y^2}{\delta^3}\right]$$

$$\left(\frac{du}{dy}\right)_{y=0} = \frac{3}{2}\frac{U_\infty}{\delta}$$

$$\tau_0 = \frac{3}{2}\frac{\mu U_\infty}{\delta} \quad \text{... (8.23)}$$

Equating Eqs. (8.22) and (8.23), we get

$$\frac{39}{280}\rho U_\infty^2\frac{d\delta}{dx} = \frac{3}{2}\frac{\mu U_\infty}{\delta}$$

From Eq. (1) of art. 8.2.2, we have

$$\delta.d\delta = \frac{3}{2}\mu U_\infty\,dx\times\frac{280}{39}\times\frac{1}{\rho U_\infty^2}\cdot dx = \frac{140}{13}\frac{\mu}{\rho U_\infty}\cdot dx$$

Integrating both sides, we get

$$\frac{\delta^2}{2} = \frac{140}{13}\times\frac{\mu}{\rho U_\infty}\times x + C$$

where C = a constant of integration

At $\alpha = 0$ $\delta = 0$, therefore, $C = 0$

$$\delta^2 = \frac{280}{13}\frac{\mu x}{\rho U_\infty}$$

$$\delta = \sqrt{\frac{280}{13}\times\frac{\mu x}{\rho U_\infty}} = \frac{4.64\,x}{\sqrt{Re_x}} \quad \text{... (8.24)}$$

2. Shear stress-τ_0

$$\tau_0 = \frac{3\mu U_\infty}{2\delta}$$

But

$$\tau = \frac{4.64x}{\sqrt{Re_x}}$$

∴

$$\tau_0 = \frac{3\mu U_\infty}{2\times 4.64x} = 0.323\frac{\mu U_\infty}{x}\sqrt{Re_x} \quad \text{... (8.25)}$$

Boundary Layer Flow || 353

Eq. (8.28) is not applicable in the thin laminar sub-layer. In this laminar sub-layer, the velocity distribution is influenced only by viscous effects.

Wall shear stress in the turbulent boundary layer is taken as;

$$\tau_0 = 0.0225\rho U_\infty^2\left(\frac{\mu}{\rho\delta U_\infty}\right)^{1/4} \quad \text{... (8.29)}$$

Substituting the value of $\frac{u}{U_\infty}$ in Von-Karman momentum-integral equation (8-11a), we have

$$\frac{\tau_0}{\rho U_\infty^2} = \frac{d}{dx}\left[\int_0^\delta \frac{u}{U_\infty}\left(1-\frac{u}{U_\infty}\right)dy\right]$$

$$= \frac{d}{dx}\left[\int_0^\delta\left\{\left(\frac{y}{\delta}\right)^{1/7}\left\{1-\left(\frac{y}{\delta}\right)^{1/7}\right\}dy\right\}\right]$$

$$= \frac{d}{dx}\left[\int_0^\delta\left\{\left(\frac{y}{\delta}\right)^{1/7}-\left(\frac{y}{\delta}\right)^{2/7}\right\}dy\right]$$

$$= \frac{d}{dx}\left|\frac{7y^{8/7}}{8\delta^{1/7}}-\frac{7}{9}\frac{y^{9/7}}{\delta^{2/7}}\right|_0^\delta = \frac{d}{dx}\left[\frac{7}{8}\delta-\frac{7}{9}\delta\right] = \frac{7}{72}\frac{d\delta}{dx}$$

$$\tau_0 = \frac{7}{72}\rho U_\infty^2\frac{d\delta}{dx} \quad \text{.. (8.30)}$$

Also

$$\tau_0 = 0.0225\rho U_\infty^2\left(\frac{\mu}{\rho U_\infty\delta}\right)^{1/4} \quad \text{Eq. (8.29)}$$

Equating Eqs. (8.29) and (8.30), we get

$$\frac{7}{72}\rho U_\infty^2\frac{d\delta}{dx} = 0.0225\rho U_\infty^2\left(\frac{\mu}{\rho U_\infty\delta}\right)^{1/4}$$

$$\delta^{1/4}.d\delta = 0.0225\times\frac{72}{7}\times\left(\frac{\mu}{\rho U_\infty}\right)^{1/4}dx$$

$$= 0.2314\left(\frac{\mu}{\rho U_\infty}\right)^{1/4}dx$$

$$\delta_e = \int_0^\delta \frac{u}{U_\infty}\left[1-\left(\frac{u}{U_\infty}\right)^2\right]dy \quad \text{... (8.7)}$$

8.2.5 Shape Factor

1. Boundary Layer Thickness-δ

Shape Factor,

$$H = \frac{\delta^*}{\theta} \quad \text{... (8.8)}$$

Example 8.1 *The velocity distribution in boundary layer is given by $\frac{u}{U_\infty}=\frac{y}{\delta}$. Calcualte : (a) Displacement thickness, (b) Momentum thickness, (c) Energy thickness, and (d) Shape Factor*

Solution. (a) $\delta^* = \int_0^\delta\left(1-\frac{u}{U_\infty}\right)dy = \int_0^\delta\left(1-\frac{y}{\delta}\right)dy$

$$= \left|y-\frac{y^2}{2\delta}\right|_0^\delta = \delta-\frac{\delta}{2}=\frac{\delta}{2}$$

(b) $\theta = \int_0^\delta \frac{u}{U_\infty}\left(1-\frac{u}{U_\infty}\right)dy$

$$= \int_0^\delta\frac{y}{\delta}\left(1-\frac{y}{\delta}\right)dy = \int_0^\delta\left[\frac{y}{\delta}-\frac{y^2}{\delta^2}\right]dy$$

$$= \left|\frac{y^2}{2\delta}-\frac{y^3}{3\delta^2}\right| = \frac{\delta}{2}-\frac{\delta}{3}=\frac{\delta}{6}$$

(c) $\delta_e = \int_0^\delta\frac{u}{U_\infty}\left[1-\left(\frac{u}{U_\infty}\right)^2\right]dy = \int_0^\delta\frac{y}{\delta}\left[1-\left(\frac{y}{\delta}\right)^2\right]dy$

$$= \int_0^\delta\left[\frac{y}{\delta}-\frac{y^3}{\delta^3}\right]dy = \left|\frac{y^2}{2\delta}-\frac{y^4}{4\delta^3}\right|_0^\delta$$

$$= \frac{\delta}{2}-\frac{\delta}{4}=\frac{\delta}{4}$$

(d) $H = \frac{\delta^*}{\theta}=\frac{\delta}{2}\times\frac{6}{\delta}=3$

Integrating both sides, we get

$$\frac{4}{5}\delta^{5/4} = 0.2314\left(\frac{\mu}{\rho U_\infty}\right)x + C$$

where C is a constant of integration.

At $x = 0$, $\delta = 0$, so that $C = 0$

$$\frac{4}{5}\delta^{5/4} = 0.2314\left(\frac{\mu}{\rho U_\infty}\right)^{1/4} x$$

$$\delta = 0.371\left(\frac{\mu}{\rho U_\infty}\right)^{1/5} \times x^{4/5} \qquad \text{... (8.31 a)}$$

$$= 0.371\left(\frac{\mu}{\rho U_\infty}\right)^{1/5} \times x^{4/5}$$

$$\therefore \qquad \delta = \frac{0.371\, x}{(\mathrm{Re}_x)^{1/5}} \qquad \text{... (8.31 b)}$$

From Eq. (8.31 a), we observe that for for turbulent boundary layer, $\delta_x\ x^{4/5}$.

2. Wall shear stress, τ_0

$$\tau_0 = 0.225\rho U_\infty^2\left(\frac{\mu}{\rho U_\infty \delta}\right)^{1/4}$$

$$= 0.0225\,\rho U_\infty^2\left[\frac{\mu}{\rho U_\infty \times \dfrac{0.371\,x}{(\mathrm{Re}_x)^{1/5}}}\right]^{1/4}$$

$$= \frac{0.0225}{(0.371)^{1/4}} \times \rho U_\infty^2\left[\frac{\mu}{\rho U_\infty x} \times (\mathrm{Re}_x)^{1/5}\right]^{1/4}$$

$$= 0.0288\,\rho U_\infty^2\left[\frac{(\mathrm{Re}_x)^{1/5}}{\mathrm{Re}_x}\right]^{1/4}$$

or

$$\tau_0 = \frac{\rho U_\infty^2}{2} \times \frac{0.0576}{(\mathrm{Re}_x)^{1/5}} \qquad \text{... (8.32)}$$

Solution.

Given: $L = 3\text{m}$, $b = 2\text{m}$, $s = 0.8$, $U_\infty = 2\text{m/s}$, $\nu = 10^{-4}\ \text{m}^2/\text{s}$

$$\text{Re}_L = \frac{U_\infty L}{\gamma} = \frac{2\times 3}{10^{-4}} = 60{,}000 < 5\times 10^5$$

Therefore, the boundary layer at the trailing edge is laminar.

$$\delta_L = \frac{5L}{\sqrt{\text{Re}_L}} = \frac{5\times 3}{60{,}000} = 0.6124\,\text{m or } 61.24\,\text{mm}$$

According to Blasius, average coefficient of drag,

$$C_D^* = \frac{0.664}{\sqrt{\text{Re}_L}} = \frac{0.664}{\sqrt{60{,}000}} = 2.71\times 10^{-3}$$

Shear stress,
$$\tau_L = C_D^* \times \frac{1}{2}\rho U_\infty^2$$

$$= 2.71\times 10^{-3} \times \frac{1}{2}\times 0.8\times 10^3 \times 2^2 = 4.336\,\text{N/m}^2$$

Example 8.9 *A plate 500 mm × 200 mm has been placed longitudinally in a stream of crude oil of specific gravity 0.925 and kinetmatic viscosity of 0.9 Stoke. The free-stream velocity of oil is 5 m/s. Calculate:*

(a) The friction drag on the plate, (b) The boundary layer thickness at the trailing edge, and

(c) shear stress at the trailing edge.

Solution.

Given: $L = 0.5$ m, $b = 0.2$ m, $s = 0.925$, $\nu = 0.9\times 10^{-4}\ \text{m}^2/\text{s}$, $U_\infty = 5$ m/s

(a) $$\text{Re}_L = \frac{U_\infty L}{\nu} = \frac{5\times 0.5}{0.9\times 10^{-4}} = 27{,}778 < 5\times 10^5$$

Hence the flow over the plate is laminar.

$$C_D = \frac{1.328}{\sqrt{\text{Re}_L}} = \frac{1.328}{\sqrt{27{,}778}} = 0.007968$$

$$F_D = C_D \times \frac{1}{2}\rho A U_\infty^2 = 0.07968\times \frac{1}{2}\times 0.925\times 10^3 \times 0.5\times 0.2\times 5^2 = 9.213\,\text{N}$$

(b) $$\delta_L = \frac{5L}{\sqrt{\text{Re}_L}} = \frac{5\times 0.5}{\sqrt{27{,}778}} = 0.015\,\text{m or } 15\text{ mm}$$

(c) $$C_D^* = \frac{0.664}{\sqrt{\text{Re}_x}} = \frac{0.664}{\sqrt{27{,}778}} = 0.03984$$

$$\tau_0 = C_D^* \times \frac{1}{2}\rho U_\infty^2 = 0.003984\times \frac{1}{2}\times 0.925\times 10^3 \times 5^2 = 46.06\ \text{N/m}^2$$

Example 8.10 *Air is flowing over a flat plate 5m ×3m with a velocity of 4m/s. If ρ = 1.2 kg/m³ and ν = 0.15 × 10⁻⁴ m²/s, Calculate:*

$$= \frac{1}{2}(\rho\,\delta_e U_\infty)U_\infty^2 = \frac{1}{2}\times 1.2\times 3.5\times 10^{-3}\times 10^3 = 2.1\ \text{N.m/s}$$

(a) Length of plate over which the boundary layer thickness is laminar, and thickness of boundary layer,

Let h_l = energy loss per unit width in terms of metres of head.

(b) Shear stress at the location where boundary layer ceases to be laminar, and

$$\rho q h_l = 2.1$$

(c) Total drag force on both sides of the plate for the laminar layer.

Solution.

$$h_l = \frac{2.1}{1.2\times 5} = 0.35\,\text{m}$$

Given: $L = 5$ m, $b = 3$ m, $U_\infty = 4$ m/s, $\rho = 1.2$ kg/m³, $\nu = 0.15\times 10^{-4}$ m²/s

Example 8.5 *For steady Poiseuille flow in a pipe of radius R, obtain an expression for ratio of displacement thickness to momentum thickness.*

(a) $$\text{Re}_L = \frac{U_\infty L}{\nu} = \frac{4\times 5}{0.15\times 10^{-4}} = 13.33\times 10^5 > 5\times 10^5$$

Solution. For steady Poiseuille flow in a circular pipe, the velocity distribution is given by (see Fig. 8.3),

Hence the flow is turbulent near the trailing edge of the plate.

For laminar flow, $\text{Re}_x = 5\times 10^5$

$$\therefore\quad \frac{U_\infty x}{\nu} = \frac{4\times x}{0.15\times 10^{-4}} = 5\times 10^5$$

$$x = 1.875\ \text{m}$$

$$\delta = \frac{5x}{\sqrt{\text{Re}_x}} = \frac{5\times 1.875}{\sqrt{5\times 10^5}} = 0.01326\,\text{m or } 13.26\ \text{mm}$$

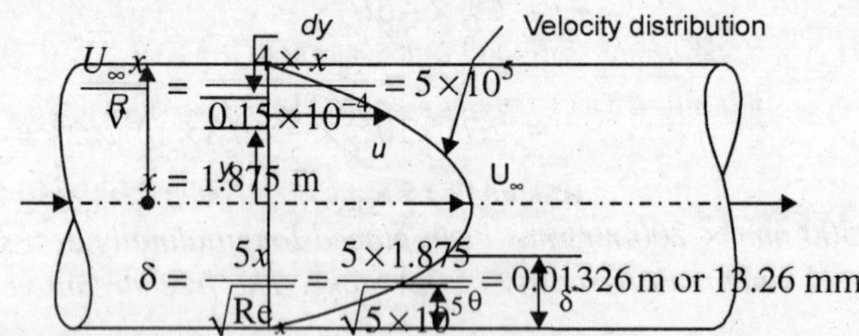

(b) $$C_D^* = \frac{0.644}{\sqrt{5\times 10^5}} = 9.39\times 10^{-4}$$

Fig. 8.3 Flow in a pipe

$$u = -\frac{1}{4\mu}\cdot\frac{\partial p}{\partial x}(R^2 - y^2) \quad \text{... (1)}$$

$$\tau_0 = C_D^*\times\frac{1}{2}\rho U_\infty^2 = 9.39\times 10^{-4}\times\frac{1}{2}\times 1.2\times 4^2 = 0.00901\ \text{N/m}^2$$

At $y = 0$, $u = U_\infty = u_{max}$

(c) $$F_D = 2\times C_D\times\frac{1}{2}\rho A U_\infty^2$$

$$\therefore\quad U_\infty = -\frac{1}{4\mu}\cdot\frac{\partial p}{\partial x}R^2 \quad \text{... (2)}$$

$$C_D = \frac{1.328}{\sqrt{5\times 10^5}} = 1.878\times 10^{-3}$$

From the definition of displacement thickness, we have

$$A = 1.875\times 3 = 5.625\ \text{m}^2$$

$$[\pi R^2 - \pi(R-\delta^*)^2]U_\infty = \int_0^R 2\pi y(U_\infty - u)\,dy$$

$$F_D = 2\times 1.878\times 10^{-3}\times\frac{1}{2}\times 1.2\times 5.625\times 4^2 = 0.2028\ \text{N}$$

$$2\pi R\delta^* U_\infty = \int_0^R 2\pi y(U_\infty - u)\,dy \quad [\text{neglecting } (\delta^*)^2 \text{ term, being small}]$$

8.5 TURBULENT BOUNDARY LAYER OVER A FLAT PLATE

According to Blasius, the velocity profile for turbulent boundary layer is given by:

$$\delta^* = \frac{1}{R}\int_0^R\left[1-\frac{u}{U_\infty}\right]y.dy$$

$$\frac{u}{U_\infty} = \left(\frac{y}{\delta}\right)^{1/7} \quad \text{for } 5\times 10^5 < \text{Re} < 10^7 \quad \text{... (8.28)}$$

Dividing Eq (1) by (2), we have

$$\frac{u}{U_\infty} = 1 - \frac{y^2}{R^2}$$

$$\therefore \quad \delta^* = \frac{1}{R}\int_0^R \frac{y^2}{R^2}\cdot y dy = \frac{1}{R^3}\int_0^R y^3 dy = \frac{1}{R^3}\left|\frac{y^4}{4}\right|_0^R = \frac{R}{4}$$

From the definition for momentum thickness, we have

$$[\rho\{\pi R^2 - \pi(R-\theta)^2\}]U_\infty^2 = \int_0^R \rho.2\pi y\,.dy.u(U_\infty - u)$$

$$2\pi R\theta U_\infty^2 = \int_0^R 2\pi y u(U_\infty - u)\,dy$$

$$\theta = \frac{1}{R}\int_0^R \frac{u}{U_\infty}\left(1-\frac{u}{U_\infty}\right)y\,dy$$

$$= \frac{1}{R}\int_0^R\left(1-\frac{y^2}{R^2}\right)\frac{y^2}{R^2}\cdot y\,dy = \frac{1}{R}\int_0^R\left(\frac{y^3}{R^2}-\frac{y^5}{R^4}\right)dy$$

$$= \frac{1}{R}\left|\frac{y^4}{4R^2} - \frac{y^6}{6R^4}\right|_0^R = \frac{R}{12}$$

$$\frac{\delta^*}{\theta} = \frac{R}{4}\times\frac{12}{R} = 3$$

Example 8.6 *The velocity distribution in the boundary layer over the face of a high spillway is given by:*

$$\frac{u}{U_\infty} = \left(\frac{y}{\delta}\right)^{0.22}$$

Calculate: (a) the displacement thickness, (b) the energy thickness, and (c) the loss of energy, for the following data:

U_∞ = 25 m/s, δ = 30 mm, q = 5 m³/s per metre length of spillway.

Solution. (a) $\delta^* = \int_0^\delta\left(1-\frac{u}{U_\infty}\right)dy = \int_0^\delta\left[1-\left(\frac{y}{\delta}\right)^{0.22}\right]dy$

$$= \left|y - \frac{y^{1.22}}{1.22\delta^{0.22}}\right|_0^\delta = \left[\delta - \frac{\delta}{1.22}\right] = 0.18\delta$$

$$= 0.18 \times 30 = 5.4 \text{ mm}$$

$$\text{(b) } \delta_e = \int_0^\delta \frac{u}{U_\infty}\left[1-\left(\frac{u}{U_\infty}\right)^2\right]dy = \int_0^\delta \left(\frac{y}{\delta}\right)^{0.22}\left[1-\left(\frac{y}{\delta}\right)^{0.44}\right]dy$$

$$= \int_0^\delta \left[\left(\frac{y}{\delta}\right)^{0.22} - \left(\frac{y}{\delta}\right)^{0.66}\right]dy = \left|\frac{y^{1.22}}{1.22y^{0.22}} - \frac{y^{1.66}}{1.66\,\delta^{0.66}}\right|_0^\delta$$

$$= \frac{\delta}{1.22} - \frac{\delta}{1.66} = 0.2173\,\delta$$

$$= 0.2173 \times 30 = 6.518 \text{ mm}$$

(c) Energy loss per metre length of spillway,

$$E_l = \frac{1}{2}\rho\delta_e U_\infty^3 = \frac{1}{2}\times 10^3 \times 6.518\times 10^{-3} \times 25^3 = 50922 \text{ N.m/s}$$

$$h_l = \frac{E_l}{\gamma q} = \frac{50922}{9810\times 5} = 1.038 \text{ m}$$

8.3 VON-KARMAN MOMENTUM INTEGRAL EQUATION

Von-Karman suggested a method based on the momentum equation by which the growth of boundary layer along a flat plate, wall shear stress, and drag force can be determined, when the velocity distribution in the boundary layer is known. This method can be used for both laminar and turbulent boundary layers.

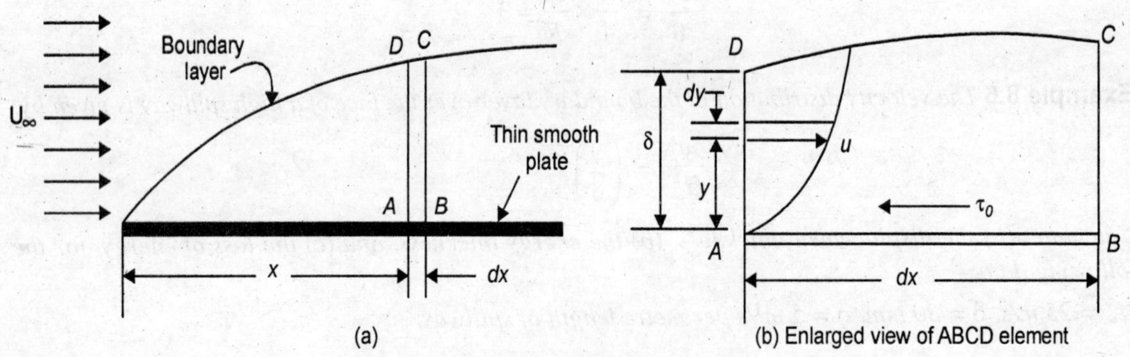

Fig. 8.4 Von-Karman boundary layer

The fluid flowing over a thin flat smooth plate, with free-stream velocity U_∞, is shown in Fig. 8.4 (a). Consider a small length dx of the plate at a distance x from the leading edge. The enlarged view of the elementary length dx of the plate is shown in Fig. 8.4 (*b*). Consider unit width of the plate perpendicular to the dirction of flow. Let ABCD be a small element of the boundary layer

Mass rate of fluid entering through AD, $\dot{m}_1 = \int_0^\delta \rho u \, dy$

Mass rate of fluid leaving through BC, $\dot{m}_2 = \int_0^\delta \rho u \, dy + \frac{d}{dx}\left[\int_0^\delta \rho u \, dy\right] dx$

∴ Mass rate of fluid entering the control volume through the face DC = $\dot{m}_2 - \dot{m}_1$

$$= \frac{d}{dx}\left[\int_0^\delta \rho u \, dy\right] dx$$

The fluid is entering through DC with a uniform velocity U_∞.
Momentum rate of fluid entering the control volume in x-direction through AD

$$= \int_0^\delta \rho u^2 \, dy \qquad \text{... (1)}$$

Momentum rate of fluid leaving the central volume in x-direction through BC

$$= \int_0^\delta \rho u^2 dy + \frac{d}{dx}\left[\int_0^\delta \rho u^2 dy\right] dx \qquad \text{... (2)}$$

Momentum rate of fluid entering the control volume through DC in x-direction

$$= \frac{d}{dx}\left[\int_6^\delta \rho u \, dy\right] dx \times U_\infty$$

$$= \frac{d}{dx}\left[\int_0^\delta \rho u U_\infty \, dy\right] dx \qquad \text{... (3)}$$

Rate of change of momentum of control volume

= Eq. (2) – Eq. (1) – Eq. (3)

$$= \frac{d}{dx}\left[\int_0^\delta \rho u^2 dy - \int_0^\delta \rho u U_\infty dy\right] dx$$

$$= \rho \frac{d}{dx}\left[\int_0^\delta (u^2 - uU_\infty) dy\right] dx \qquad \text{... (8.9)}$$

According to the momentum principle, the rate of change of momentum on the control volume $ABCD$ must be equal to the total force on the control volume in the same direction. The only external force acting on the control volume is the shear force acting on the side AB in the direction B to A, due to boundary shear stress, τ_0 (see Fig. 8.4b). The value of this shear drag force is given by,

$$\Delta F_D = \tau_0 \times dx$$

Thus the total external force in the direction of rate of change of momentum

$$= -\tau_0 \times dx \quad \text{... (8.10)}$$

Equating Eqs. (8.9) and (8.10), we have

$$-\tau_0 \times dx = \rho \frac{d}{dx}\left[\int_0^\delta (u^2 - uU_\infty)dy\right]dx$$

$$\tau_0 = -\rho \frac{d}{dx}\left[\int_0^\delta (u^2 - uU_\infty)dy\right]$$

$$= \rho \frac{d}{dx}\left[\int_0^\delta (uU_\infty - u^2)dy\right]$$

$$= \rho \frac{d}{dx}\left[\int_0^\delta U_\infty^2\left(\frac{u}{U_\infty} - \frac{u^2}{U_\infty^2}\right)dy\right]$$

$$= \rho U_\infty^2 \frac{d}{dx}\left[\int_0^\delta \frac{u}{U_\infty}\left(1 - \frac{u}{U_\infty}\right)dy\right]$$

or

$$\frac{\tau_0}{\rho U_\infty^2} = \frac{d}{dx}\left[\int_0^\delta \frac{u}{U_\infty}\left(1 - \frac{u}{U_\infty}\right)dy\right] \quad \text{... (8.11a)}$$

Now momentum thickness, $\theta = \int_0^\delta \frac{u}{U_\infty}\left(1 - \frac{u}{U_\infty}\right)dy$

$\therefore$

$$\frac{\tau_0}{\rho U_\infty^2} = \frac{d\theta}{dx} \quad \text{... (8.11b)}$$

Eq. (8.11) is known as von-Karman momentum integral equation for boundary layer flow. This is applicable to:

(i) Laminar boundary layer, (ii) Transition boundary layer, and (iii) Turbulent boundary layer.

Boundary conditions

The following boundary conditions must be satisfied by any assumed velocity distribution:

1. At the surface of plate:

   $y = 0, u = 0$, and $\frac{du}{dy}$ = a finite value

2. At the outer edge of boundary layer:

   $y = \delta, u = U_\infty$

Drag force on elemental length dx of plate is given by,

$$\Delta F_D = \tau_0 \times b \times dx$$

where b = width of plate.

Total drag force on one side of plate of length L,

$$F_D = \int \Delta F_D = \int_0^L \tau_0 \times b \times dx \qquad \text{... (8.12)}$$

8.3.1 Local Coefficient of Drag $-C_D^*$

It is defined as the ratio of the shear stress τ_0 to the quantity $\frac{1}{2}\rho U_\infty^2$. It is denoted by C_D^*. It is also called *coefficient of skin friction.*

Hence
$$C_D^* = \frac{\tau_0}{\frac{1}{2}\rho U_\infty^2} \qquad \text{...(8.13)}$$

8.3.2 Average Coefficient of Drag, C_D

It is defined as the ratio of the total drag force to the quantity $\frac{1}{2}\rho A U_\infty^2$. It is also called coefficient of drag and is denoted by C_D. Thus

$$C_D = \frac{F_D}{\frac{1}{2}\rho A U_\infty^2} \qquad \text{... (8.14)}$$

where ρ = mass density of fluid

A = area of surface (or plate)

U_∞ = free-stream velocity.

8.4 LAMINAR BOUNDARY LAYER

Case I: $= \dfrac{u}{U_\infty} = 2\left(\dfrac{y}{\delta}\right) - \left(\dfrac{y}{\delta}\right)^2$

1. Boundary layer thickness-δ

   Momentum integral equation is,

$$\frac{\tau_0}{\rho U_\infty^2} = \frac{d}{dx}\left[\int_0^\delta \frac{u}{U_\infty}\left(1 - \frac{u}{U_\infty}\right) dy\right]$$

$$= \frac{d}{dx}\left[\int_0^\delta \left\{2\left(\frac{y}{\delta}\right)-\left(\frac{y}{\delta}\right)^2\right\}\left\{1-\left(\frac{2y}{\delta}-\frac{y^2}{\delta^2}\right)\right\}dy\right]$$

$$= \frac{d}{dx}\left[\int_0^\delta \left\{\frac{2y}{\delta}-\frac{y^2}{\delta^2}\right\}\left\{1-\frac{2y}{\delta}+\frac{y^2}{\delta^2}\right\}dy\right]$$

$$= \frac{d}{dx}\left[\int_0^\delta \left\{\frac{2y}{\delta}-\frac{4y^2}{\delta^2}+\frac{2y^3}{\delta^3}-\frac{y^2}{\delta^2}+\frac{2y^3}{\delta^3}-\frac{y^4}{\delta^4}\right\}dy\right]$$

$$= \frac{d}{dx}\left[\int_0^\delta \left\{\frac{2y}{\delta}-\frac{5y^2}{\delta^2}+\frac{4y^3}{\delta^3}-\frac{y^4}{\delta^4}\right\}dy\right]$$

$$= \frac{d}{dx}\left[\left|\frac{y^2}{\delta}-\frac{5}{3}\frac{y^3}{\delta^2}+\frac{y^4}{\delta^3}-\frac{y^5}{5\delta^4}\right|_0^\delta\right]$$

$$= \frac{d}{dx}\left[\delta-\frac{5}{3}\delta+\delta-\frac{\delta}{5}\right] = \frac{d}{dx}\left(\frac{2}{15}\delta\right)$$

$$\tau_0 = \rho U_\infty^2 \times \frac{d}{dx}\left(\frac{2}{15}\delta\right) = \frac{2}{15}\rho U_\infty^2 \frac{d\delta}{dx} \qquad \text{... (8.15)}$$

According to Newton's law of viscosity,

$$\tau_0 = \mu\left(\frac{du}{dy}\right)_{y=0}$$

Now

$$u = U_\infty\left[\frac{2y}{\delta}-\frac{y^2}{\delta^2}\right]$$

$$\frac{du}{dy} = U_\infty\left(\frac{2}{\delta}-\frac{2y}{\delta^2}\right)$$

$$\left(\frac{du}{dy}\right)_{y=0} = \frac{2U_\infty}{\delta}$$

$\therefore$

$$\tau_0 = \frac{2\mu U_\infty}{\delta} \qquad \text{... (8.16)}$$

Equating Eqs. (8.15) and (8.16), we get

$$\frac{2}{15}\rho U_\infty^2 \frac{d\delta}{dx} = \frac{2\mu U_\infty}{\delta}$$

$$\delta \cdot \frac{d\delta}{dx} = \frac{15\mu}{\rho U_\infty}$$

or

$$\delta \cdot d\delta = \frac{15\mu}{\rho U_\infty} \cdot dx$$

Integrating both sides, we get

$$\frac{\delta^2}{2} = \frac{15\mu}{\rho U_\infty} x + C$$

where C is a constant of integration.

At $x = 0$, $\delta = 0$, thus $C = 0$.

$$\frac{\delta^2}{2} = \frac{15\mu x}{\rho U_\infty}$$

or

$$\delta = \sqrt{\frac{30\mu x}{\rho U_\infty}} = 5.48\sqrt{\frac{\mu x}{\rho U_\infty}} = 5.48\sqrt{\frac{\mu x \times x}{\rho U_\infty x}}$$

$$= 5.48\sqrt{\frac{x^2}{\text{Re}_x}} = \frac{5.48x}{\sqrt{\text{Re}_x}} \qquad \left[\because \text{Re}_x = \frac{\rho U_\infty x}{\mu}\right] \qquad \text{... (8.17)}$$

Eq. (8.17) shows that for laminar boundary layer over a flat plate,

$$\delta \propto \sqrt{x} \propto \frac{1}{\sqrt{\text{Re}_x}} \qquad \text{... (8.18)}$$

i.e. thickness of laminar boundary layer over a flat plate is proporation to the square root of the distance from the leading edge and inversely proportional to the square root of Reynolds number.

2. Shear stress-t_0

We have from Eq. (8.16),

$$\tau_0 = \frac{2\mu U_\infty}{\delta}$$

and

$$\delta = \frac{5.48x}{\sqrt{\text{Re}_x}}$$

$\therefore$

$$\tau_0 = \frac{2\mu U_\infty \sqrt{\text{Re}_x}}{5.48} = 0.635\,\frac{uU_\infty}{x} \cdot \sqrt{\text{Re}_x} \qquad \text{... (8.19)}$$

3. Local coefficient of drag-C^*_D

$$\tau_0 = C_D^* \frac{\rho U_\infty^2}{2}$$

$$\therefore \quad 0.365 \frac{\mu U_\infty}{x}\sqrt{\text{Re}_x} = C_D^* \frac{\rho U_\infty^2}{2}$$

$$C_D^* = \frac{0.73\sqrt{\text{Re}_x}}{\left(\frac{\rho U_\infty x}{\mu}\right)} = \frac{0.73\sqrt{\text{Re}_x}}{\text{Re}_x} = \frac{0.73}{\sqrt{\text{Re}_x}} \quad \text{... (8.20)}$$

4. Coefficient of drag-C_D

$$C_D = \frac{FD}{\frac{1}{2}\rho A U_\infty^2}$$

where $F_D = \int_0^L \tau_0 \times b \times dx$

$$= \int_0^L 0.365 \frac{\mu U_\infty}{x}\sqrt{\text{Re}_x} \times b \times dx$$

$$= \int_0^L 0.365 \frac{\mu U_\infty}{x}\sqrt{\frac{\rho U_\infty x}{\mu}} \times bx\, dx$$

$$= 0.365 \mu U_\infty \sqrt{\frac{\rho U_\infty}{\mu}} \times b \times \int_0^L x^{-1/2} dx$$

$$= 0.365 \mu U_\infty \sqrt{\frac{\rho U_\infty}{\mu}} \times b \left| \frac{x^{1/2}}{(1/2)} \right|_0^L = 0.73 \mu U_\infty \sqrt{\frac{\rho U_\infty}{\mu}} \times bL^{1/2}$$

$$F_D = 0.73 \mu U_\infty b \sqrt{\frac{\rho U_\infty L}{\mu}} = 0.73\, uU_\infty \sqrt{\text{Re}_L}$$

$$C_D = \frac{0.73 \mu U_\infty b\sqrt{\text{Re}_L}}{\frac{1}{2}\rho L b U_\infty^2} = \frac{1.46\mu}{\rho L U_\infty}\sqrt{\text{Re}_L} = \frac{1.46}{\sqrt{\text{Re}_L}} \quad \text{... (8.21)}$$

where $Re_L = \frac{\rho L U_\infty}{\mu}$ is the Reynolds number at the trailing edge of plate.

$$A = L \times b$$

Case 2: $\frac{u}{U_\infty} = \frac{3}{2}\left(\frac{y}{\delta}\right) - \frac{1}{2}\left(\frac{y}{\delta}\right)^3$

1. Boundary layer thickness-δ

$$\frac{\tau_0}{\tau U_\infty^2} = \frac{d}{dx}\left[\int_0^\delta \frac{u}{U_\infty}\left(1 - \frac{u}{U_\infty}\right)dy\right]$$

$$= \frac{d}{dx}\left[\int_0^\delta \left\{\frac{3}{2}\left(\frac{y}{\delta}\right) - \frac{1}{2}\left(\frac{y}{\delta}\right)^3\right\}\left\{1 - \frac{3}{2}\left(\frac{y}{\delta}\right) + \frac{1}{2}\left(\frac{y}{\delta}\right)^3\right\}dy\right]$$

$$= \frac{d}{dx}\left[\int_0^\delta \left\{\frac{3}{2}\frac{y}{\delta} - \frac{9}{4}\frac{y^2}{\delta^2} + \frac{3}{4}\frac{y^4}{\delta^4} - \frac{1}{2}\frac{y^3}{\delta^3} + \frac{3}{4}\frac{y^4}{\delta^4} - \frac{1}{4}\frac{y^6}{\delta^6}\right\}dy\right]$$

$$= \frac{d}{dx}\left[\int_0^\delta \left\{\frac{3}{2}\frac{y}{\delta} - \frac{9}{4}\frac{y^2}{\delta^2} + \frac{3}{2}\frac{y^4}{\delta^4} - \frac{1}{2}\frac{y^3}{\delta^3} - \frac{1}{4}\frac{y^6}{\delta^6}\right\}dy\right]$$

$$= \frac{d}{dx}\left[\left|\frac{3}{4}\frac{y^2}{\delta} - \frac{3}{4}\frac{y^3}{\delta^2} + \frac{3}{10}\frac{y^5}{\delta^4} - \frac{1}{8}\frac{y^4}{\delta^3} - \frac{1}{28}\frac{y^7}{\delta^6}\right|_0^\delta\right]$$

$$= \frac{d}{dx}\left[\frac{3}{4}\delta - \frac{3}{4}\delta + \frac{3}{10}\delta - \frac{1}{8}\delta - \frac{1}{28}\delta\right]$$

$$= \frac{d}{dx}\left[\frac{39}{280}\delta\right] = \frac{39}{280}\frac{d\delta}{dx}$$

$$\tau_0 = \rho U_\infty^2 \times \frac{39}{280}\frac{d\delta}{dx} = \frac{39}{280}\rho U_\infty^2 \frac{d\delta}{dx} \qquad \text{... (8.22)}$$

Also $\qquad \tau_0 = \mu\left(\frac{du}{dy}\right)_{y=0}$

But $$u = U_\infty\left[\frac{3}{2}\left(\frac{y}{\delta}\right) - \frac{1}{2}\left(\frac{y}{\delta}\right)^3\right]$$

$$\frac{du}{dy} = U_\infty\left[\frac{3}{2}\times\frac{1}{\delta} - \frac{3}{2}\frac{y^2}{\delta^3}\right]$$

$$\left(\frac{du}{dy}\right)_{y=0} = \frac{3}{2}\frac{U_\infty}{\delta}$$

$$\tau_0 = \frac{3}{2}\frac{\mu U_\infty}{\delta} \quad \text{... (8.23)}$$

Equating Eqs. (8.22) and (8.23), we get

$$\frac{39}{280}\rho U_\infty^2 \frac{d\delta}{dx} = \frac{3}{2}\frac{\mu U_\infty}{\delta}$$

$$\delta.d\delta = \frac{3}{2}\mu U_\infty \times \frac{280}{39} \times \frac{1}{\rho U_\infty^2}\cdot dx = \frac{140}{13}\frac{\mu}{\rho U_\infty}\cdot dx$$

Integrating both sides, we get

$$\frac{\delta^2}{2} = \frac{140}{13} \times \frac{\mu}{\rho U_\infty} \times x + C$$

where C = a constant of integration

At $\alpha = 0$ $\delta = 0$, therefore, $C = 0$

$$\therefore \qquad \delta^2 = \frac{280}{13}\frac{\mu}{\rho U_\infty}$$

$$\delta = \sqrt{\frac{280}{13} \times \frac{\mu x}{\rho U_\infty}} = \frac{4.64\,x}{\sqrt{\text{Re}_x}} \quad \text{... (8.24)}$$

2. Shear stress-τ_0

$$\tau_0 = \frac{3\mu U_\infty}{2\delta}$$

But $$\tau = \frac{4.64\,x}{\sqrt{\text{Re}_x}}$$

$$\tau_0 = \frac{3\mu U_\infty}{2 \times 4.64 x} = 0.323\frac{\mu U_\infty}{x}\sqrt{\text{Re}_x} \quad \text{... (8.25)}$$

3. Local coefficient of drag-C^*_D

$$\tau_0 = 0.323\frac{\mu U_\infty}{x}\sqrt{\text{Re}_x}$$

Also
$$\tau_0 = C_D^*\frac{\rho U_\infty^2}{2}$$

∴
$$C_D^*\frac{\rho U_\infty^2}{2} = 0.323\frac{\mu U_\infty}{x}\sqrt{\text{Re}_x}$$

$$C_D^* = \frac{0.646}{\sqrt{\text{Re}_x}} \qquad \text{... (8.26)}$$

4. Coefficient of drag, C_D

$$C_D = \frac{F_D}{\frac{1}{2}\rho A U_\infty^2}$$

$$F_D = \int_0^L \tau_0 \times b \times dx$$

$$= \int_0^0 \frac{0.323 U_\infty}{x}\sqrt{\text{Re}_x} \times b \times dx = 0.323\int_0^L \frac{\mu U_\infty}{x} \times \sqrt{\frac{\rho U_\infty x}{\mu}} \times b \times dx$$

$$= 0.323\,\mu U_\infty \times \sqrt{\frac{\rho_{U_\infty}}{\mu}} \times b\,|\,2x^{1/2}\,|_0^L = 0.646\,\mu U_\infty\sqrt{\frac{\rho U_\infty L}{\mu}} \times b$$

$$= 0.646\, b\,\mu U_\infty\sqrt{\frac{\rho U_\infty L}{\mu}}$$

$$C_D = \frac{0.646\, b\mu U_\infty\sqrt{\frac{\rho U_\infty L}{\mu}}}{\frac{1}{2}\rho \times b \times L \times U_\infty^2} = \frac{1.292\sqrt{\frac{\rho U_\infty L}{\mu}}}{\left(\frac{\rho L U_\infty}{\mu}\right)}$$

$$= \frac{1.292}{\sqrt{\text{Re}_L}} \qquad \text{... (8.27)}$$

Table 8.1 Shows the values of δ, τ_0, C^*_D and C_D for various velocity profiles

Table 8.1 Values of δ, τ_0, C^*_D and C_D for various velocity profiles

S.No	Velocity profile	τ_0	δ	C_D^*	CD
1.	$\frac{u}{U_\infty} = 2\left(\frac{y}{\delta}\right) - \left(\frac{y}{\delta}\right)^2$	$0.365\frac{\mu U_\infty}{x}\sqrt{Re_x}$	$\frac{5.48x}{\sqrt{Re_x}}$	$\frac{0.73}{\sqrt{Re_x}}$	$\frac{1.46}{\sqrt{Re_L}}$
2.	$\frac{u}{U_\infty} = \frac{3}{2}\left(\frac{y}{\delta}\right) - \frac{1}{2}\left(\frac{y}{\delta}\right)^3$	$0.323\frac{\mu U_\infty}{x}\sqrt{Re_x}$	$\frac{4.64x}{\sqrt{Re_x}}$	$\frac{0.646}{\sqrt{Re_x}}$	$\frac{1.292}{\sqrt{Re_L}}$
3.	$\frac{u}{U_\infty} = 2\left(\frac{y}{\delta}\right) - 2\left(\frac{y}{\delta}\right)^3 + \left(\frac{y}{\delta}\right)^4$	$0.343\frac{\mu U_\infty}{x}\sqrt{Re_x}$	$\frac{5.84x}{\sqrt{Re_x}}$	$\frac{0.686}{\sqrt{Re_x}}$	$\frac{1.372}{\sqrt{Re_L}}$
4.	$\frac{u}{U_\infty} = \sin\left(\frac{\pi}{2}\frac{y}{\delta}\right)$	$0.327\frac{\mu U_\infty}{x}\sqrt{Re_x}$	$\frac{4.795x}{\sqrt{Re_x}}$	$\frac{0.654}{\sqrt{Re_x}}$	$\frac{1.31}{\sqrt{Re_L}}$
5.	Blasius results ($Re < 3.2 \times 10^5$)	$0.332\frac{\mu U_\infty}{x}\sqrt{Re_x}$	$\frac{5x}{\sqrt{Re_x}}$	$\frac{0.664}{\sqrt{Re_x}}$	$\frac{1.328}{\sqrt{Re_L}}$

$$\sqrt{Re_x} = \sqrt{\frac{\rho U_\infty x}{\mu}} \quad \sqrt{Re_L} = \sqrt{\frac{\rho U_\infty L}{\mu}}$$

Example 8.7 *The boundary layer thickness at a distance of 1.0 m from the leading edge of a flat plate kept at zero angle of incidence to the flow direction is 1 mm. The free-stream velocity is 20 m/s. Find the boundary layer thickness at a distance of 3m.*

Solution. Given: $x_1 = 1$ m, $\delta_1 =$ m, $\delta_1 = 1.0$ mm, $U_\infty = 20$ m/s, $x_2 = 3$ m

$$\delta \propto \sqrt{x}$$

$$\frac{\delta_2}{\delta_1} = \sqrt{\frac{x_2}{x_1}}$$

$$\delta_2 = 1.0\sqrt{\frac{3}{1}} = 1.732 \text{ mm}$$

Example 8.8 *A smooth plate 3 m long and 2 m wide is towed in oil of specific gravity 0.8 at a velocity of 2m/s along its length. Calculate the boundary layer thickness and shear stress at the trailing edge of the plate. Kinematic viscosity of oil is 10^{-4} m²/s*

Solution.

Given: $L = 3\text{m}$, $b = 2\text{m}$, $s = 0.8$, $U_\infty = 2\text{m/s}$, $\nu = 10^{-4}\ \text{m}^2/\text{s}$

$$\text{Re}_L = \frac{U_\infty L}{\gamma} = \frac{2\times 3}{10^{-4}} = 60{,}000 < 5\times 10^5$$

Therefore, the boundary layer at the trailing edge is laminar.

$$\delta_L = \frac{5L}{\sqrt{\text{Re}_L}} = \frac{5\times 3}{60{,}000} = 0.6124\,\text{m or } 61.24\,\text{mm}$$

According to Blasius, average coefficient of drag,

$$C_D^* = \frac{0.664}{\sqrt{\text{Re}_L}} = \frac{0.664}{\sqrt{60{,}000}} = 2.71\times 10^{-3}$$

Shear stress,

$$\tau_L = C_D^* \times \frac{1}{2}\rho U_\infty^2$$

$$= 2.71\times 10^{-3} \times \frac{1}{2}\times 0.8\times 10^3 \times 2^2 = 4.336\,\text{N/m}^2$$

Example 8.9 *A plate 500 mm × 200 mm has been placed longitudinally in a stream of crude oil of specific gravity 0.925 and kinetmatic viscosity of 0.9 Stoke. The free-stream velocity of oil is 5 m/s. Calculate:*

(a) The friction drag on the plate, (b) The boundary layer thickness at the trailing edge, and (c) shear stress at the trailing edge.

Solution.

Given: $L = 0.5$ m, $b = 0.2$ m, $s = 0.925$, $\nu = 0.9 \times 10^{-4}\ \text{m}^2/\text{s}$, $U_\infty = 5$ m/s

(a) $$\text{Re}_L = \frac{U_\infty L}{\nu} = \frac{5\times 0.5}{0.9\times 10^{-4}} = 27{,}778 < 5\times 10^5$$

Hence the flow over the plate is laminar.

$$C_D = \frac{1.328}{\sqrt{\text{Re}_L}} = \frac{1.328}{\sqrt{27{,}778}} = 0.007968$$

$$F_D = C_D \times \frac{1}{2}\rho A U_\infty^2 = 0.07968 \times \frac{1}{2} \times 0.925\times 10^3 \times 0.5\times 0.2\times 5^2 = 9.213\,\text{N}$$

(b) $$\delta_L = \frac{5L}{\sqrt{\text{Re}_L}} = \frac{5\times 0.5}{\sqrt{27{,}778}} = 0.015\,\text{m or 15 mm}$$

(c) $$C_D^* = \frac{0.664}{\sqrt{\text{Re}_x}} = \frac{0.664}{\sqrt{27{,}778}} = 0.03984$$

$$\tau_0 = C_D^* \times \frac{1}{2}\rho U_\infty^2 = 0.003984 \times \frac{1}{2} \times 0.925\times 10^3 \times 5^2 = 46.06\ \text{N/m}^2$$

Example 8.10 *Air is flowing over a flat plate 5m ×3m with a velocity of 4m/s. If ρ = 1.2 kg/m³ and ν = 0.15 × 10⁻⁴ m²/s, Calculate:*

(a) Length of plate over which the boundary layer thickness is laminar, and thickness of boundary layer,

(b) Shear stress at the location where boundary layer ceases to be laminar, and

(c) Total drag force on both sides of the plate for the laminar layer.

Solution.

Given: $L = 5$ m, $b = 3$ m, $U_\infty = 4$ m/s, $\rho = 1.2$ kg/m³, $\nu = 0.15 \times 10^{-4}$ m²/s

$$\text{(a)}\quad Re_L = \frac{U_\infty L}{\nu} = \frac{4\times 5}{0.15\times 10^{-4}} = 13.33\times 10^5 > 5\times 10^5$$

Hence the flow is a turbulent near the trailing edge of plate.

For laminar flow, $Re_x = 5\times 10^5$

$$\therefore \quad \frac{U_\infty x}{\nu} = \frac{4\times x}{0.15\times 10^{-4}} = 5\times 10^5$$

$$x = 1.875 \text{ m}$$

$$\delta = \frac{5x}{\sqrt{Re_x}} = \frac{5\times 1.875}{\sqrt{5\times 10^5}} = 0.01326 \text{ m or } 13.26 \text{ mm}$$

$$\text{(b)}\quad C_D^* = \frac{0.644}{\sqrt{5\times 10^5}} = 9.39\times 10^{-4}$$

$$\tau_0 = C_D^* \times \frac{1}{2}\rho U_\infty^2 = 9.39\times 10^{-4}\times \frac{1}{2}\times 1.2\times 4^2 = 0.00901 \text{ N/m}^2$$

$$\text{(c)}\quad F_D = 2\times C_D \times \frac{1}{2}\rho A U_\infty^2$$

$$C_D = \frac{1.328}{\sqrt{5\times 10^5}} = 1.878\times 10^{-3}$$

$$A = 1.875\times 3 = 5.625 \ m^2$$

$$F_D = 2\times 1.878\times 10^{-3}\times \frac{1}{2}\times 1.2\times 5.625\times 4^2 = 0.2028 \text{ N}$$

8.5 TURBULENT BOUNDARY LAYER OVER A FLAT PLATE

According to Blasius, the velocity profile for turbulent boundary layer is given by:

$$\frac{u}{U_\infty} = \left(\frac{y}{\delta}\right)^{1/7} \quad \text{for } 5\times 10^5 < Re < 10^7 \qquad \text{... (8.28)}$$

Eq. (8.28) is not applicable in the thin laminar sub-layer. In this laminar sub-layer, the velocity distribution is influenced only by viscous effects.

Wall shear stress in the turbulent boundary layer is taken as;

$$\tau_0 = 0.0225\rho U_\infty^2\left(\frac{\mu}{\rho\delta U_\infty}\right)^{1/4} \quad \text{... (8.29)}$$

1. Boundary Layer Thickness-δ

Substituting the value of $\frac{u}{U_\infty}$ in Von-Karman momentum-integral equation (8-11a), we have

$$\frac{\tau_0}{\rho U_\infty^2} = \frac{d}{dx}\left[\int_0^\delta \frac{u}{U_\infty}\left(1-\frac{u}{U_\infty}\right)dy\right]$$

$$= \frac{d}{dx}\left[\int_0^\delta\left\{\left(\frac{y}{\delta}\right)^{1/7}\left\{1-\left(\frac{y}{\delta}\right)^{1/7}\right\}dy\right]$$

$$= \frac{d}{dx}\left[\int_0^\delta\left\{\left(\frac{y}{\delta}\right)^{1/7}-\left(\frac{y}{\delta}\right)^{2/7}\right\}dy\right]$$

$$= \frac{d}{dx}\left|\frac{7y^{8/7}}{8\delta^{1/7}}-\frac{7}{9}\frac{y^{9/7}}{\delta^{2/7}}\right|_0^\delta = \frac{d}{dx}\left[\frac{7}{8}\delta-\frac{7}{9}\delta\right] = \frac{7}{72}\frac{d\delta}{dx}$$

$$\tau_0 = \frac{7}{72}\rho U_\infty^2\frac{d\delta}{dx} \quad \text{.. (8.30)}$$

Also

$$\tau_0 = 0.0225\rho U_\infty^2\left(\frac{\mu}{\rho U_\infty\delta}\right)^{1/4} \quad \text{Eq. (8.29)}$$

Equating Eqs. (8.29) and (8.30). we get

$$\frac{7}{72}\rho U_\infty^2\frac{d\delta}{dx} = 0.0225\rho U_\infty^2\left(\frac{\mu}{\rho U_\infty\delta}\right)^{1/4}$$

$$\delta^{1/4}.d\delta = 0.0225\times\frac{72}{7}\times\left(\frac{\mu}{\rho U_\infty}\right)^{1/4}dx$$

$$= 0.2314\left(\frac{\mu}{\rho U_\infty}\right)^{1/4}dx$$

Integrating both sides, we get

$$\frac{4}{5}\delta^{5/4} = 0.2314\left(\frac{\mu}{\rho U_\infty}\right)x + C$$

where C is a constant of integration.

At $x = 0$, $\delta = 0$, so that $C = 0$

$$\frac{4}{5}\delta^{5/4} = 0.2314\left(\frac{\mu}{\rho U_\infty}\right)^{1/4} x$$

$$\delta = 0.371\left(\frac{\mu}{\rho U_\infty}\right)^{1/5} \times x^{4/5} \qquad \text{... (8.31 } a)$$

$$= 0.371\left(\frac{\mu}{\rho U_\infty}\right)^{1/5} \times x^{4/5}$$

$$\therefore \qquad \delta = \frac{0.371\,x}{(\text{Re}_x)^{1/5}} \qquad \text{... (8.31 } b)$$

From Eq. (8.31 a), we observe that for for turbulent boundarylayer, δ_x $x^{4/5}$.

2. Wall shear stress, τ_0

$$\tau_0 = 0.225\rho U_\infty^2\left(\frac{\mu}{\rho U_\infty \delta}\right)^{1/4}$$

$$= 0.0225\,\rho U_\infty^2\left[\frac{\mu}{\rho U_\infty \times \dfrac{0.371\,x}{(\text{Re}_x)^{1/5}}}\right]^{1/4}$$

$$= \frac{0.0225}{(0.371)^{1/4}} \times \rho U_\infty^2\left[\frac{\mu}{\rho U_\infty x} \times (\text{Re}_x)^{1/5}\right]^{1/4}$$

$$= 0.0288\,\rho U_\infty^2\left[\frac{(\text{Re}_x)^{1/5}}{\text{Re}_x}\right]^{1/4}$$

or

$$\tau_0 = \frac{\rho U_\infty^2}{2} \times \frac{0.0576}{(\text{Re}_x)^{1/5}} \qquad \text{... (8.32)}$$

3. Local coefficient of drag, C^*_D

$$\tau_0 = \frac{\rho U_\infty^2}{2} \times \frac{0.0576}{(\text{Re}_x)^{1/5}}$$

Also

$$\tau_0 = C_D^* \times \frac{1}{2}\rho U_\infty^2$$

∴

$$C_D^* = \frac{0.0576}{(\text{Re}_x)^{1/5}} \quad \text{... (8.33)}$$

4. Drag force, F_D

Let L, b = length and width of plate respectively.

Drag froce on one side of plate,

$$F_D = \int_0^L \tau_0 \times b \times dx$$

$$= \int_0^L \frac{1}{2}\rho U_\infty^2 \times \frac{0.0576}{(\text{Re}_x)^{1/5}} \times b \times dx = \int_0^L \frac{1}{2}\rho U_\infty^2 \times \frac{0.0576}{\left(\frac{\rho U_\infty x}{\mu}\right)^{1/5}} \times b \times dx$$

$$= \frac{1}{2}\rho U_\infty^2 \times \frac{0.0576}{\left(\frac{\rho U_\infty}{\mu}\right)^{1/5}} \times b \int_0^L x^{-1/5}.dx$$

$$= \frac{1}{2}\rho U_\infty^2 \times \frac{0.576 \times b}{\left(\frac{\rho U_\infty}{\mu}\right)^{1/5}} \left| \frac{5}{4} x^{4/5} \right|_0^L$$

$$= \frac{1}{2}\rho U_\infty^2 \times \frac{0.0576 \times b}{\left(\frac{\rho U_\infty}{\mu}\right)^{1/5}} \times \frac{5}{4} \times L^{4/5}$$

$$= \frac{1}{2}\rho U_\infty^2 \times \left(0.0576 \times \frac{5}{4}\right) \times \frac{(b \times L)}{\left(\frac{\rho U_\infty L}{\mu}\right)^{1/5}}$$

$$= \frac{1}{2}\rho A U_\infty^2 \times \frac{0.072}{(\text{Re}_L)^{1/5}} \quad \text{... (8.34)}$$

5. Coefficient of drag, C_D

$$F_D = C_D \times \frac{1}{2}\rho A U_\infty^2$$

Also
$$F_D = \frac{1}{2}\rho A U_\infty^2 \times \frac{0.072}{(\mathrm{Re}_L)^{1/5}} \qquad \text{(Eq. 8.34)}$$

$$\therefore \quad C_D = \frac{0.072}{(\mathrm{Re}_L)^{1/5}} \quad \text{for} 5\times 10^5 < \mathrm{Re}_L < 10^7 \qquad \text{... (8.35)}$$

According to Prandtl and schlichting,

$$C_D = \frac{0.455}{(\log_{10} \mathrm{Re}_L)^{2.58}} \qquad \text{... (8.36)}$$

8.6 TOTAL DRAG ON A FLAT PLATE DUE TO LAMINAR AND TURBULENT BOUNDARY LAYERS

Let AB be the laminar boundary layer, B the transition point, and BC the turbulent boundary layer. The distance x_c in Fig 8.5 of the trasnsition from the leading edge can be obtained from critical Reynolds number, which is generally taken as 5×10^5.

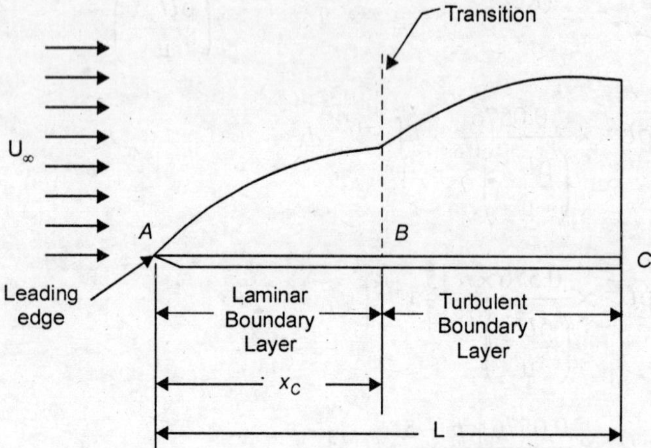

Fig. 8.5 Drag due to laminar and turbulent boundary layers on a flat plate

Drag force for the turbulent boundary layer can be estimated as follows:

$$F_{\text{turb}} = (F_{\text{turb}})_{\text{total}} - (F_{\text{turb}})_{x_c}$$

where $(F_{\text{turb}})_{\text{total}}$ = The drag force which would occur if turbulent boundary layer extends along the entire length L of the plate.

$(F_{\text{turb}})_{x_c}$ = The drag force due to fictitious turbulent boundary layer formed for the distance x_c.

If Re > 10^7, then

$$F_{\text{turb}} = \frac{0.455}{(\log_{10}\text{Re}_L)^{2.58}} \times \frac{1}{2}\rho U_\infty^2 \times L \times b - \frac{0.072}{(\text{Re}_c)^{1/5}} \times \frac{1}{2}\rho U_\infty^2 \times x_c \times b \quad \text{... (1)}$$

where L = length of the plate, b = width of the plate, U_∞ = free-stream velocity.

$$F_{\text{laminar}} = \frac{1.328}{\sqrt{\text{Re}_c}} \times \frac{1}{2}\rho U_\infty^2 \times x_c \times b = \frac{1.328\, x_c}{\sqrt{\text{Re}_c}} \times b \times \frac{1}{2}\rho U_\infty^2$$

$$F_{\text{total}} = F_{\text{laminar}} + F_{\text{turb}}$$

$$= \left[\frac{1.328\, x_c}{\sqrt{\text{Re}_c}} + \frac{0.455\, L}{(\log_{10}\text{Re}_L)^{2.58}} - \frac{0.072\, x_c}{(\text{Re}_c)^{1/5}}\right] \times b \times \frac{1}{2}\rho U_\infty^2 \quad \text{... (3)}$$

Now $\frac{\text{Re}_c}{\text{Re}_L} = \frac{x_c}{L}$ or $x_c = \frac{\text{Re}_c}{\text{Re}_L} \times L$

Eq. (3) becomes

$$F_{\text{total}} = \left[\frac{1.328\sqrt{\text{Re}_c}}{\text{Re}_L} + \frac{0.455}{(\log_{10}\text{Re}_L)^{2.58}} - \frac{0.072(\text{Re}_c)^{4/5}}{\text{Re}_L}\right]\frac{Lb\rho U_\infty^2}{2}$$

Assuming $\text{Re}_c = 5 \times 10^5$

$$F_{\text{total}} = \left[\frac{0.455}{(\log_{10}\text{Re}_L)^{2.58}} - \frac{1700}{\text{Re}_L}\right]\frac{Lb\rho U_\infty^2}{2} \quad \text{... (8.37)}$$

Also $\quad F_{\text{total}} = C_D \times \frac{1}{2}\rho A U_\infty^2 = C_D \times \frac{1}{2}\rho L b U_\infty^2$

$$\therefore \quad C_D = \frac{0.455}{(\log_{10}\text{Re}_L)^{2.58}} - \frac{1700}{\text{Re}_L} \quad \text{... (8.38)}$$

Example 8.11 *A submarine can be assumed to have cylindrical shape with rounded nose. Its length is 60 m and diameter 6 m. Calculate the total power required to overcome boundary friction if it cruises at 10 m/s velocity in sea water having density = 1030 kg/m³ and kinematic viscosity = 1×10⁻⁶ m²/s.*

Solution. Given: L = 60 m, D = 6 m, U_∞ = 10 m/s, ρ = 1030 kg/m³. $\nu = 1\times 10^{-6}$ m²/s.

$$\text{Re}_L = \frac{U_\infty L}{\nu} = \frac{10 \times 60}{1 \times 10^{-6}} = 6 \times 10^8$$

Now $\frac{U_\infty x_c}{\nu} = 5 \times 10^5$, $x_c = \frac{5 \times 10^5 \times 1 \times 10^{-6}}{10}$ 0.05 m

$x_c \ll L$, hence contribution to total drag from laminar boundary layer is negligible.

$$C_D = \frac{0.455}{(\log_{10} \text{Re}_L)^{2.58}} = \frac{0.455}{[\log_{10}(6\times10^8)]^{2.58}} = 0.001675$$

$$A = \pi DL = \pi\times6\times60 = 1130.97 \text{ m}^2$$

$$F_D = C_D\times\frac{1}{2}\rho AU_\infty^2 = 0.001675\times\frac{1}{2}\times1030\times1130.97\times10^2 = 97{,}560.3 \text{ N}$$

$$\text{Total power required, } P = \frac{F_D\times U_\infty}{10^3} = \frac{97560.3\times10}{10^3} = 975.603 \text{ kW}$$

Example 8.12 *Calculate the ratio of friction drag on the front half and rear half of the flat plate in a stream of uniform velocity, if the boundary layer is turbulent over the whole plate.*

Solution. Average coefficient of drag for turbulent boundary layer,

$$C_D = \frac{0.072}{(\text{Re}_L)^{1/5}}$$

$$\text{Re}_L = \frac{U_\infty L}{\nu}$$

$$\text{For the first half of plate, } \text{Re}_x = \frac{U_\infty x}{\nu} = \frac{U_\infty L}{2\nu}$$

Drag force per unit width for the whole plate,

$$F = F_D = C_D\times\frac{1}{2}\rho U_\infty^2\times L\times1$$

$$= \frac{0.072}{\left(\frac{U_\infty L}{\nu}\right)^{1/5}}\times\frac{1}{2}\rho U_\infty^2\times L$$

Drag force per unit width for the front half portion of plate,

$$F_1 = \frac{0.072}{\left(\frac{U_\infty L}{2\nu}\right)^{1/5}}\times\frac{1}{2}\rho U_\infty^2\times\frac{L}{2}$$

$$= \frac{0.072}{\left(\frac{U_\infty L}{\nu}\right)^{1/5}}\times\frac{1}{2}\rho U_\infty^2\times\frac{L}{2}\times(2)^{1/5}$$

Drag force for the rear half portion of the plate,

$$F_2 = F - F_1 = \frac{0.072L}{\left(\frac{U_\infty L}{\nu}\right)^{1/5}} \times \frac{1}{2}\rho U_\infty^2\left[1-\frac{1}{2}(2)^{1/5}\right]$$

$$\frac{F_1}{F_2} = \frac{\frac{1}{2}(2)^{1/5}}{1-\frac{1}{2}(2)^{1/5}} = \frac{0.574}{1-0.574} = 1.347$$

Example 8.13 *Water is flowing over a thin smooth plate of length 5 m and width 2 m at a velocity of 1.25 m/s. If the boundary layer flow changes from laminar to turbulent at a Reynolds number 5×10^5, calculate; (a) The distance from leading edge upto which boundary layer is laminar, (b) the thickness of boundary layer at the transition point, and (c) the drag force on one side of the plate. Take dynamic viscosity of water = 9.81×10^{-4} Pa.s*

Solution. $L = 5$ m, $b = 2$m, $U_\infty = 1.25$ m/s, $Re_c = 5 \times 10^5$, $\mu = 9.81 \times 10^{-4}$ Pa.s

(a) $$Re_c = \frac{\rho U_\infty x_c}{\mu}$$

$$5 \times 10^5 = \frac{10^3 \times 1.25 \times x_c}{9.81\times10^{-4}}$$

$$x_c = 0.3924 \text{ m}$$

(b) According to Blasius's solution, thickness of boundary layer at the point where the boundary layer changes from laminar to turbulent is given by,

$$\delta_c = \frac{5x_c}{\sqrt{Re_c}} = \frac{5\times0.3924}{\sqrt{5\times10^5}} = 0.002775 \text{ m or } 2.775 \text{ mm}$$

(c) Drag force on one side of plate =Drag due to laminar boundary layer + Drag due to turbulent boundary layer.

Drag due to laminar boundary layer in length x_c,

$$F_{laminar} = C_D \times \frac{1}{2}\rho A U_\infty^2$$

where $$C_D = \frac{1.328}{\sqrt{Re_c}} = \frac{1.328}{\sqrt{5\times10^5}} = 0.001878$$

$$A = x_c \times b = 0.3924 \times 2 = 0.7848 \text{ m}^2$$

$$F_{\text{laminar}} = 0.001878 \times \frac{1}{2} \times 10^3 \times 0.7848 \times (1.25)^2 = 1.15\,\text{N}$$

Drag force due to turbulent boundary layer, F_{turb} in length $(L–x_c)$

= Drag force due to turbulent boundary layer for the whole length, L of plate–Drag force due to turbulent boundary layer for the whole length, L of plate – Drag force due to turbulent flow in length x_c.

$$= (F_{\text{turb}})_{\text{total}} - (F_{\text{turb}})_{x_c}$$

$$(F_{\text{turb}})_{\text{total}} = C_D \times \frac{1}{2} \rho A U_\infty^2$$

where $$C_D = \frac{0.072}{(\text{Re}_L)^{1/5}}, \text{Re}_L = \frac{\rho U_\infty L}{\mu} = \frac{10^3 \times 1.25 \times 5}{9.81 \times 10^{-4}} = 6371005$$

$$C_D = \frac{0.072}{(6371005)^{1/5}} = 0.003137$$

$$(\text{F}_{\text{turb}})_{\text{total}} = 0.003137 \times \frac{1}{2} \times 10^3 \times 5 \times 2 \times (1.25)^2 = 24.5\,\text{N}$$

$$(F_{turb})_{x_c} = C_D \times \frac{1}{2} \rho \times x_c \times b \times U_\infty^2$$

where $$C_D = \frac{0.072}{(5 \times 10^5)^{1/5}} = 0.00522$$

$$(F_{turb})_{x_c} = 0.00522 \times \frac{1}{2} \times 10^3 \times 5 \times 2 \times (1.25)^2 = 3.2\,\text{N}$$

$$F_{\text{turb}} = 24.5 - 3.2 = 21.3\,N$$

Dreg force on one side of plate = 1.15 + 21.3 = 22.45 N

Example 8.14 *A stream lined train is 200 m long with a typical cross-section having a perimeter of 8 m above the wheels. The kinmatic viscosity of air at the prevailing temperature is 1.5 ×10^{-5} m^2/s and density 1.25 kg/m^3. Calculate the friction drag of train when running at 90 km/h.*

Solution.

Given: $L = 200$ m, $P = 8$ m, $n = 1.5 \times 10^{-5}$ m^2/s, $\rho = 1.25$ kg/m^3, $U_\infty = 90$ km/h

Surface area of train, $A = L \times P = 200 \times 8 = 1600$ m^2

Free-stream velocity, $$U_\infty = \frac{90 \times 10^3}{3600} = 25\text{ m/s}$$

$$\text{Re}_L = \frac{U_\infty L}{\nu} = \frac{25\times 200}{1.5\times 10^{-5}} = 333,333,333 > 5\times 10^5$$

The boundary layer is turbulent.

Average coefficient of drag, $C_D = \dfrac{0.455}{(\log_{10}\text{Re}_L)^{2.58}} - \dfrac{1700}{\text{Re}_L}$

$$= \frac{0.455}{[\log_{10} 333,333,333]^{2.58}} - \frac{1700}{333,333,333}$$

$$= 0.001807 - 5.1 \times 10^{-6} = 0.0018.$$

Friction drag over train surface,

$$F_D = C_D \times \frac{1}{2}\rho A U_\infty^2 = 0.0018\times\frac{1}{2}\times 1.25\times 1600\times (25)^2 = 1125\text{N}$$

Example 8.15 *Air flows at 10 m/s past a smooth rectangular flat plate 0.3 wide and 3 m long. Assuming that the turbulence level in the oncoming stream is low and that transistion occurs at Reynolds number = 5× 10⁵, Calculate ratio of total drag when the flow is parallel to the length of the plate to the value when the flow is parallel to the width. Assume for air, ρ = 1.24 kg/m³ and ν = 0.15 stoke.*

Solution.

Given: $U_\infty = 10$ m/s, $L = 3$m, $\text{Re}_c = 5 \times 10^5$. $\rho = 1.24$ kg/m³, $\nu = 0.15 \times 10^{-4}$ m²/s

(a) Flow is parallel to the length of plate.

$$Re_c = \frac{\rho U_\infty x_c}{\mu} = \frac{U_\infty x_c}{\nu}$$

$$5\times 10^5 = \frac{10\times x_c}{0.15\times 10^{-4}},\ x_c = 0.75 \text{ m}$$

Drag force on one side of plate = Drag due to laminar boundary layer + Drag due to turbulent boundary layer.

1. Drag due to laminar boundary layer $AB = 0.75$ m.

$$\text{F}_{\text{lam}} = C_D \times \frac{1}{2}\rho A U_\infty^2$$

$$C_D = \frac{1.328}{\sqrt{\text{Re}_c}} = \frac{1.328}{\sqrt{5\times 10^5}} = 0.001878$$

$$A = x_c \times b = 0.75\times 0.3 = 0.225 \text{ m}^2$$

$$F_{lam} = 0.001875 \times \frac{1}{2} \times 1.24 \times 0.225 \times 10^2 = 0.0262 \text{ N}$$

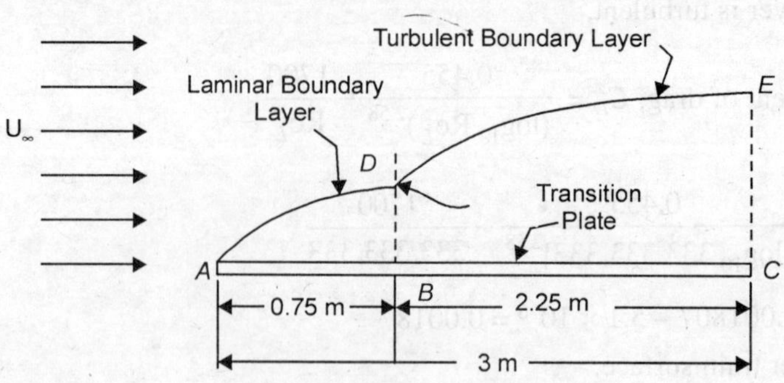

Fig. 8.6

2. Drag due to turbulent boundary layer, *BC*,

F_{turb} = Drag force due to turbulent boundary layer from *A* to *C*

— Drag force due to turbulent boundary layer from *A* to *B*

$$= (F_{turb})_{total} - (F_{turb})_{x_c}$$

$$(F_{turb})_{total} = C_D \times \frac{1}{2}\rho A U_\infty^2$$

where $$C_D = \frac{0.072}{(\text{Re}_L)^{1/5}}, \text{Re}_L = \frac{U_\infty L}{\gamma} = \frac{10 \times 3}{0.15 \times 10^{-4}} = 20 \times 10^5$$

$$C_D = \frac{0.072}{(20 \times 10^5)^{1/5}} = 0.00395$$

$$A = L \times b = 3 \times 0.3 = 0.9 \text{ m}^2$$

$$(F_{turb})_{total} = 0.00395 \times \frac{1}{2} \times 1.24 \times 0.9 \times 10^2 = 0.2204 \text{ N}$$

$$(F_{turb})_{x_c} = C_D \times \frac{1}{2}\rho A U_\infty^2$$

where $$C_D = \frac{0.072}{(\text{Re}_C)^{1/5}} = \frac{0.072}{(5 \times 10^5)^{1/5}} = 0.00522$$

$$A = x_c \times b = 0.75 \times 0.3 = 0.225 \text{m}^2$$

$$(F_{turb})_{x_c} = 0.00522 \times \frac{1}{2} \times 1.24 \times 0.225 + 10^2 = 0.0728 \text{ N}$$

$$F_{turb} = 0.22040 - 0.0728 = 0.1476 \text{ N}$$

Total drag froce, $$F_D = F_{lam} + F_{turb} = 0.0262 + 0.1476 = 0.1738 \text{ N}$$

(b) Flow is parallel to the width of plate

Now x_c (= 0.75 m) > b (= 0.3 m). Therefore, only laminar boundary layer will be formed on the plate.

$$F_D = C_D \times \frac{1}{2} \rho A U_\infty^2$$

where $$C_D = \frac{1.328}{\sqrt{Re_x}}, \ Re_x = \frac{U_\infty b}{\nu} = \frac{10 \times 0.3}{0.15 \times 10^{-4}} = 2 \times 10^5$$

$$C_D = \frac{1.328}{\sqrt{2 \times 10^5}} = 0.00297$$

$$A = 0.3 \times 3 = 0.9 \text{ m}^2$$

$$F_D = 0.00297 \times \frac{1}{2} \times 1.24 \times 0.9 \times 10^2 = 0.1657 \text{ N}$$

$$\text{Ratio of drag forces} = \frac{0.1738}{0.1657} = 1.05$$

8.7 BOUNDARY LAYER SEPARATION

The boundary layer is formed when a solid body is immeresed in a flowing fluid. The velocity in the thin layer of fluid varies from zero to free-stream velocity in a direction normal to the solid body. The thickness of the boundary layer increases along the length of the solid body. The fluid layer adjacent to the solid body surface has to do work against surface friction at the expense of its kinetic energy. This loss of kinetic energy is recovered by momentum exchange process from solid layers adjacent to the solid body surface. Thus the velocity of fluid layers goes on decreasing. At a certain point on the solid body a stage may come when the boundary layer may not be able to keep sticking to the solid body. This happens when the fluid layers can not provide kinetic energy to overcome the resistance offered by the solid body. This results in separation of boundary layer from the body surface. This phenomenon is called the body layer separation. The point on the solid body at which the boudary layer is on the verge of separation from the surface is called *point of separation.*

8.7.1 Effect of Pressure Gradient on Boundary Layer Separation

Consider the flow over a curved surface ABCDE as shown in Fig.8.7. In the region ABC of curved surface, the area of flow decreases and hence velocity increases. Therefore, the flow get accelerated in this region and the pressure decreases in the direction of flow. Hence $\frac{dp}{dx} < 0$, and the entire boundary layer moves forward. The pressure is minimum at point C. Along the region *CDE* of curved surface, the area of flow increases and hence velocity of flow decreases in the fluid. Due to decrease of velocity, $\frac{dp}{dx} > 0$. Thus in the region *CDE* the velocity of flow goes on decreasing as the kinetic energy of the layer is used to overcome the frictional resistance of the surface. The combined effect of positive pressure gradient and surface resistance reduce the momentum of the fluid. A stage comes, when the momentum of the fluid is unable to overcome the surface resistance and the boundary layer starts separating from the surface at point *D*. Downstream the point *D*, the flow takes place in the reverse direction and the velocity gradient becomes negative. Thus the positive pressure gradient helps in boundary layer separation. Large turbulent eddies are formed downstream of the point of separation. This region is called the *turbulent wake.*

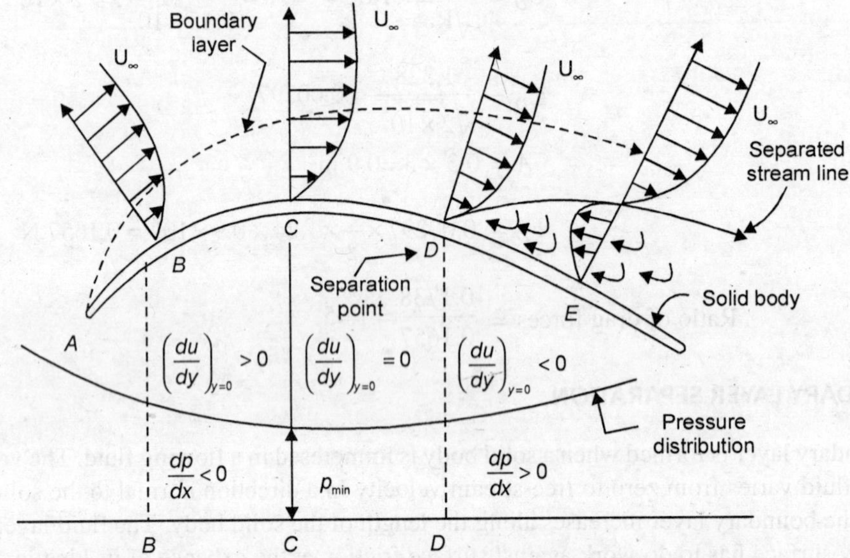

Fig. 8.7 Effect of pressure gradient on boundary layer separation.

8.7.2 Location of Sepatation Point

At the point of separation, $\left(\frac{du}{dy}\right)_{y=0} = 0.$ ... (8.39)

1. If $\left(\frac{du}{dy}\right)_{y=0} > 0$, the flow will not separate.
2. If $\left(\frac{du}{dy}\right)_{y=0} = 0$, the flow is on the verge of separation.
3. $\left(\frac{du}{dy}\right)_{y=0} < 0$, the flow has separated.

Separation of flow occurs in the following cases:

1. Diffusers 2. Pumps 3. Fans 4. Turbine blades 5. Aerofoils 6. Open channel transitions, etc.

8.7.3 Methods of Preventing separation of Boundary Layer

The following methods may be used for preventing the separationof boundary Layer:

1. Stream lining the body shape.
2. Tripping the boundary layer from laminar to turbulent by providing surface roughness.
3. Sucking the retarded flow.
4. Injecting high velocity fluid in the boundary layer
5. Providing slots near the leading edge.
6. Guiding the flow in a confined passage.
7. Providing a rotating cylinder near the leading edge.
8. Energising the flow by introducing optimum amount of swirl in the incoming flow.

8.8 DRAG AND LIFT FORCES

Consider a stationary airfoil in a real fluid, which is flowing with uniform velocity U_∞ as shown in Fig.8.8. The fluid will exert a force F_R on the airfoil perpendicular to its surface. The total force F_R can be resolved into two components, one in the direction of motion and other perpendicular to the direction of motion of fluid.

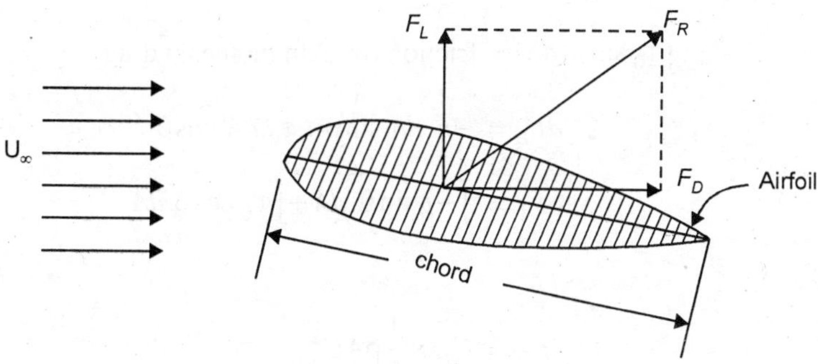

Fig. 8.8 Lift and drag on an airfoil

Drag force: The compenent of the total force in the direction of flow is called the drag force, F_D.

Lift force: The component of the total force at right angles to the direction of flow is called the lift force, F_L.

8.8.1. Expressions for Drag and Lift

Consider an arbitrary shaped solid body placed in a real fluid, which is flowing with a uniform velocity U_∞ in a horizontal direction as shown in Fig. 8.9. Consider a small elemental area dA on the surface of the body. The forces acting on the surface area dA are:

1. Pressure force equal to $p \times dA$, acting perpendicular to the surface,
2. Shear force equal to $\tau_0 \times dA$, acting along the tangential direction to the surface.

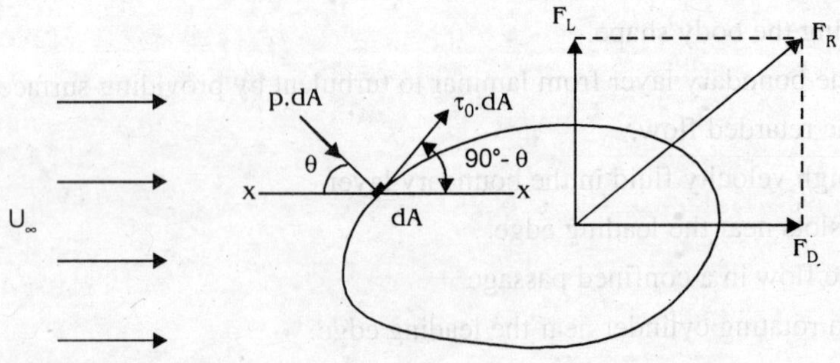

Fig. 8.9 Pressure and shear forces on an elementary area

Drag force, $$dF_D = p.dA\cdot\cos\theta + \tau_0.dA.\sin\theta$$

$$F_D = \int p\cos\theta.dA + \int \tau_0 \sin\theta.dA \quad \text{... (8.40)}$$

where $\int p\cos\theta.dA$ = pressure (or form) drag

and $\int \tau_0 \sin\theta.dA$ = friction (or skin or shear) drag.

Lift force, $$dF_L = -p.dA.\sin\theta + \tau_0.dA.\cos\theta$$

$$F_L = \int -p\sin\theta.dA + \int \tau_0 \cos\theta.dA \quad \text{... (8.41)}$$

Mathematically,

$$F_D = C_D \times \frac{1}{2}\rho A U_\infty^2 \quad \text{... (8.42)}$$

$$F_L = C_L \times \frac{1}{2}\rho A U_\infty^2 \quad \text{... (8.43)}$$

where C_D, C_L = coefficient of drag and lift respectively.

A = projected area of the body perpendicular to the direction of flow.

U_∞ = relative velocity of fluid w.r.t the body

ρ = density of fluid

Resultant force, $$F_R = \sqrt{F_D^2 + F_L^2} \quad \text{... (8.44)}$$

8.9 STREAM LINED AND BLUFF BODIES

Streamlined Body: A streamlined body is defined as that body whose surface coincides with the stream lines, when the body is placed in a flow as shown in Fig. 8.10. In this case, flow separation takes place only at the trailing edge of the body. The wake formation zone behind a streamlined body is very small so that the pressure drag will be very small. The total drag on the streamlined body will be due to friction (or shear) only. A body may be streamlined at low velocities but may not be so at higher velocities. Also when placed in a particular position in flow it may be streamlined but may not be so when placed in another position. Streamlined shapes are used for aeroplane wings and rotary axial flow machines.

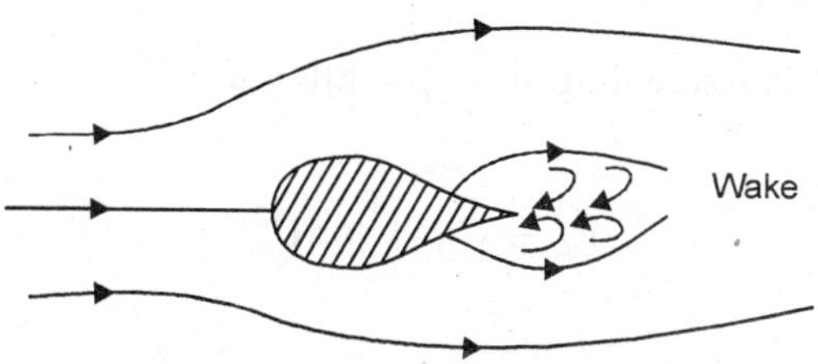

Fig. 8.10 Stream lined body

Bluff Body: A bluff body is defined as that body whose surface does not coincide with the streamlines, when placed in a flow, as shown in Fig. 8.11. In such a body there is extensive boundary layer separation accompained by a wake with large scale eddies. The pressure drag is very large as compared to the friction drag.

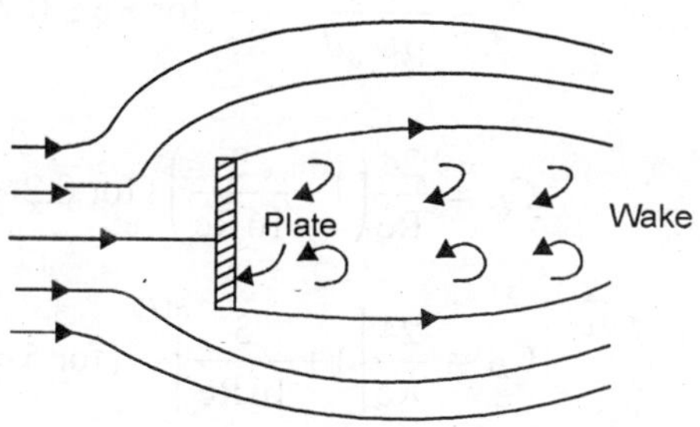

Fig. 8.11 Bluff body

8.10 DRAG ON A SPHERE

The Reynolds number is the ratio of inertia and viscous forces. For flow at very small Reynolds number (Re < 0.5), the inertia forces are negligible as compared to the viscous forces and the drag is almost proportional to the approach velocity U_∞. The flow at very small Reynolds number is known as the *creeping flow* and is largly laminar. Stokes has shown analytically that in laminar flow at very small Reynolds number, the drag experienced by the sphere referred to a the *deformation drag* is given by,

$$F_D = 3\pi\mu U_\infty d \qquad ....(8.45)$$

where d = diameter of sphere, μ = dynamic viscosity of fluid, and

U_∞ = velocity of the undisturbed fluid relative to the sphere or the terminal velocity of sphere in motion.

Of the total drag given by Eq. (8.45), two-thirds is contributed by skin-friction and the remaining one-third by pressure difference on the two-halves of the sphere. Thus,

$$\text{Skin friction drag} = \frac{2}{3}F_D = 2\pi\mu U_\infty d \quad \text{... (8.46 a)}$$

$$\text{Pressure drag} = \frac{1}{3}F_D = \pi\mu U_\infty d \quad \text{... (8.46 b)}$$

Also total drag,

$$F_D = C_D \times \frac{1}{2}\rho A U_\infty^2 \quad \text{... (8.47)}$$

where A = projected area of sphere = $\frac{\pi d^2}{4}$

Comparing Eqs. (8.45) and (8.47), we get

$$3\pi\mu U_\infty d = C_D \times \frac{1}{2}\rho \times \frac{\pi}{4}d^2 \times U_\infty^2$$

$$C_D = \frac{24\mu}{\rho U_\infty d} = \frac{24}{\text{Re}}, \text{for Re} \le 0.2 \quad \text{... (8.48 a)}$$

$$C_D = \frac{24}{\text{Re}}\left(1 + \frac{3}{16\,\text{Re}}\right), \text{for } 0.2 < \text{Re} < 5 \quad \text{... (8.48 b)}$$

$$C_D = \frac{24}{\text{Re}}\left[1 + \frac{3}{16\,\text{Re}}\right]^{1/3}, \text{for } 5 < \text{Re} < 1000 \quad \text{... (8.48 c)}$$

$$= 0.5, \text{for } 1000 < \text{Re} < 10^5 \quad \text{... (8.48 d)}$$

$$= 0.2, \text{for Re} > 10^5 \quad \text{... (8.48 e)}$$

8.10.1 Terminal Velocity of sphere

The terminal velocity is defined as the maximum constant velocity of a falling body with which the body will be travelling. The terminal velocity of a sphere falling through a liquid at rest is calculated as follows:

Let W = weight of the sphere acting downwards

F_D = drag force acting vertically upwards

F_B = buoyant force acting vertically upwards

d = diameter of sphere

γ_s = specific weight of sphere material

γ_f = specific weight of fluid

U_∞ = terminal velocity of sphere.

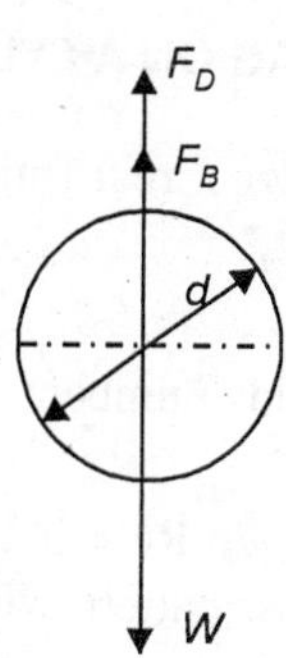

Fig 8.12 Forces acting on a sphere falling in a liquid

For equilibrium of the sphere, we have

$$W = F_D + F_B$$

or

$$F_D = W - F_B$$

$$3\pi\mu U_\infty d = \frac{\pi}{6} d^3(\gamma_s - \gamma_f)$$

$$U_\infty = \frac{d^2}{18\mu}(\gamma_s - \gamma_f) \qquad \text{... (8.49)}$$

and

$$\mu = \frac{d^2}{18U_\infty}(\gamma_s - \gamma_f) \qquad \text{... (8.50)}$$

Example 8.16 *A 5 mm metallic ball of density 7800 kg/m³ is allowed to fall in a fluid of specific gravity 0.9 and absolute viscosity 1.4 Pa.s. Calculate: (a) Drag force, (b) Skin-friction drag and pressure drag, and (c) Terminal velocity of ball in the fluid.*

Solution. Given: $d = 5$ mm, $\rho_s = 7800$ kg/m³, $s_f = 0.9$, $\mu = 1.4$ Pa.s

$\gamma_s = \rho_s g = 7800 \times 9.81 = 76{,}518$ N/m³, $\gamma_f = 0.9 \times 10^3 \times 9.81 = 8829$ N/m³

(a) Weight of ball, $W = \frac{\pi}{6} d^3 \gamma_s = \frac{\pi}{6}(5 \times 10^{-3})^3 \times 76{,}518 = 0.005$ N

Buoyant force, $F_B = \frac{\pi}{6} d^3 \gamma_f = \frac{\pi}{6}(5 \times 10^{-3}) \times 8829 = 0.000578$ N

Drag force, $F_D = W - F_B = 0.005 - 0.000\,578 = 4.422 \times 10^{-3}$ N

(b) Skin-friction drag $= \frac{2}{3} F_D = \frac{2}{3} \times 4.422 \times 10^{-3} = 2.448 \times 10^{-3}$ N

Pressure drag $= \frac{1}{3} F_D = \frac{1}{3} \times 4.22 \times 10^{-3} = 1.474 \times 10^{-3}$ N

(c) $F_D = 3\pi\mu\, d\, U_\infty$

$$U_\infty = \frac{4.422 \times 10^{-3}}{3\pi \times 1.4 \times 5 \times 10^{-3}} = 0.067 \text{ m/s}$$

$$\text{Re} = \frac{\rho U_\infty d}{\mu} = \frac{0.9 \times 10^3 \times 0.067 \times 5 \times 10^{-3}}{1.4} = 0.215 \simeq 0.2$$

8.11 DRAG ON A CYLINDER

Consider a real fluid flowing past a circular cylindr of diameter D and length L with a uniform velocity U_∞.

Reynolds number, $$\text{Re} = \frac{\rho U_\infty D}{\mu}$$

1. When Re < 0.2, the inertia force is negligible as compared to the viscous force and hence the flow pattern will be symmetrical.

   When Re < 1, $F_D \propto U_\infty$ and $C_D \propto \dfrac{1}{\text{Re}}$

2. For Re = 1 to 2000, C_D decreases and attains a minimum value of 0.95 at Re = 2000.
3. For Re = 2000 to 3×10^4, C_D increases and attains a maximum value of 1.2 at Re = 3×10^4.
4. For Re = 3×10^4 to 3×10^5, C_D decreases and becomes 0.3 at Re = 3×10^5.
5. For Re = 3×10^6, C_D increases and attains a value of 0.7 in the end.

8.12 DEVELOPEMENT OF LIFT ON A CIRCULAR CYLINDER

Consider the flow past a stationary cylinder as shown in Fig. 8.13.

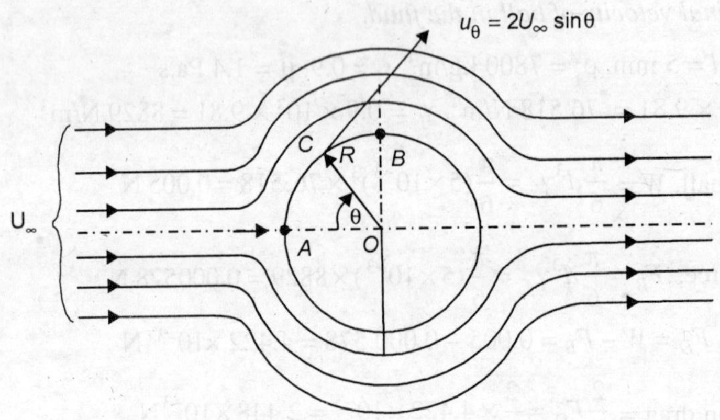

Fig. 8.13 Flow past a stationary cylinder

Let U_∞ = free-stream velocity of fluid

R = radius of cylinder.

θ = angle made by any point C on the circumference of cylinder with flow direction.

The flow pattern will be symmetrical and the velocity at point C,

$$u_\theta = 2U_\infty \sin\theta$$

At the stagnation point A, $\theta = 0°$ so that $u_A = 0$

At point B, $\theta = 90°$, $u_B = 2U_\infty$.

Since the flow pattern is symmetrical about the horizontal axis, the pressure distribution on the upper and lower halves of the cylinder are identical, and hence there is no lift acting on the cylinder.

(b) Flow past a rotating cylinder

Consider the flow of an ideal fluid past a rotating cylinder. The flow pattern shall be a combination of free vertex and uniform flow, as shown in Fig. 8.14. Due to rotation of cylinder, a constant circulation Γ is imparted to the cylinder. The peripheral velocity on the surface of the cylinder becomes,

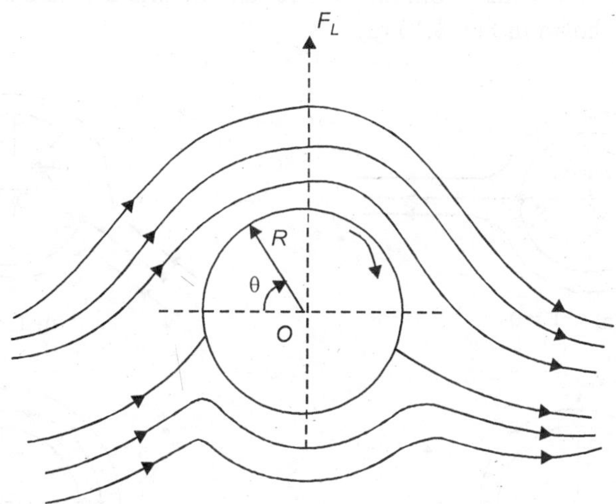

Fig. 8.14 Flow past a rotating cylinder

$$u = u_\theta + u_c$$

$$= 2U_\infty \sin\theta + \frac{\Gamma}{2\pi R} \qquad \text{... (8.52)}$$

where u_c = velocity of fluid in a free-vertex flow.

As the circulation has been taken as clockwise, the resultant velocity u around the upper half portion of the cylinder shall be higher than that arround the lower half portion. Hence on the lower half of the cylinder, pressure will be more than the pressure on the upper half portion of the cylinder. The net pressure force shall act in the upward direction and lift force is exerted on the cylinder. Since the flow is symmetrical about the vertical axis, there is no drag on the cylinder

8.12.1 Position of Stagnation Points

At the stagnation points on the surface of the cylinder, u = 0.

$$\therefore \qquad 2U_\infty \sin\theta + \frac{\Gamma}{2\pi R} = 0$$

or

$$\sin\theta = \frac{-\Gamma}{4\pi U_\infty R} \qquad \text{... (8.53)}$$

1. For $\Gamma = 0$, $\sin\theta = 0$ and $\theta = 0°$ and $180°$.
   The stagnation points are shown by S_1 and S_2 in Fig. 8.15 (a)
2. For $\tau < 4\pi RU_\infty$, $\sin\theta < -1$ and $\theta = < 90°$ and $\theta > 180°$.
   S_1 and S_2 points are shown in Fig. 8.15 (b).
3. For $\tau < 4\pi RU_\infty$, $\sin\theta = -1$ and $\theta = -90°$ and $\theta = 270°$.
   The two stagnation points coincids, as shown in Fig. 8.15 (c)
4. For $\tau < 4\pi RU_\infty$, $\sin\theta > -1$ which is not feasible. The stagnation points do not occur on the cylinder surface. They detach from the cylinder and lie into the fluid stream below the point $-90°$ and $270°$, as shown in Fig. 8.15 (d)

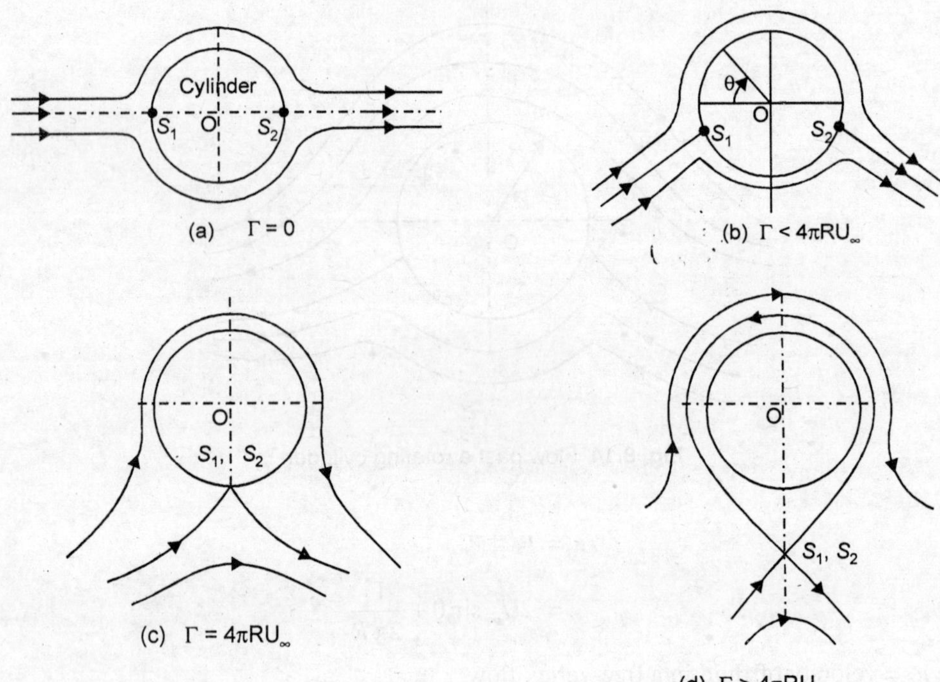

Fig 8.15 Location of stagnation points

8.12.2 Expression for Lift on Rotating cylinder

Let p_∞ = pressure of the fluid far away from the cylinder

U_∞ = velocity of the fluid far away from the cylinder

p = pressure on the surface of the cylinder

u = velocity of fluid on the surface of cylinder

$$= 2U_\infty \sin\theta + \frac{\Gamma}{2\pi R}$$

Applying Bernoulli's equation to a point far away from the cylinder and to a point on the surface of cylinder, we have

$$\frac{p_\infty}{\gamma}+\frac{U_\infty^2}{2g}=\frac{p}{\gamma}+\frac{u^2}{2g}$$

$$\frac{p}{\gamma}=\frac{p_\infty}{\gamma}+\frac{U_\infty^2}{2g}\left[1-\left(\frac{u}{U_\infty}\right)^2\right]$$

$$=\frac{p_\infty}{\gamma}+\frac{U_\infty^2}{2g}\left[1-\left(2\sin\theta+\frac{\Gamma}{2\pi RU_\infty}\right)^2\right]$$

$$p=p_\infty+\frac{1}{2}\rho U_\infty^2\left[1-\left(4\sin^2\theta+\frac{\Gamma^2}{4\pi^2R^2U_\infty^2}+\frac{2\Gamma\sin\theta}{\pi RU_\infty}\right)\right]$$

$$=p_\infty+\frac{1}{2}\rho U_\infty^2\left[1-4\sin^2\theta-\frac{\Gamma^2}{4\pi^2R^2U_\infty^2}-\frac{4\sin\theta\times\Gamma}{2\pi R\times U_\infty}\right] \quad \text{... (8.54)}$$

Consider a small element $AB = ds$ of the cylinder on its surface between θ and $\theta + d\theta$, as shown in Fig. 8.16.

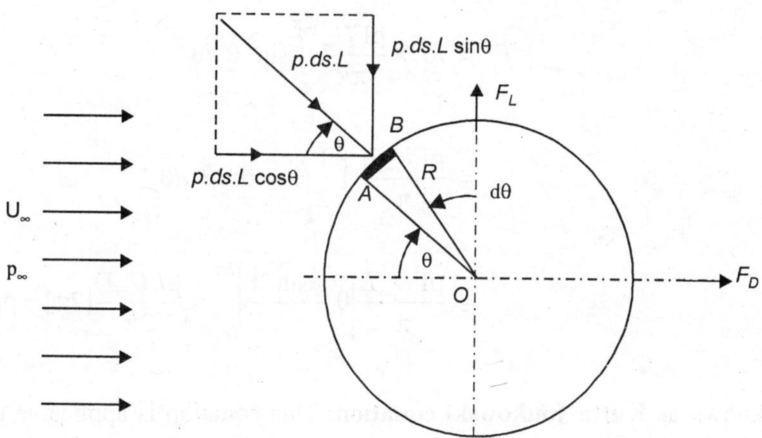

Fig. 8.16. Lift on a rotating cylinder.

Let L = length of the cylinder

R = radius of the cylinder

Then $ds = R \,.\, d\theta$

Normal force on the cylinder element

$$= p.ds.L = p.R.d\theta.L$$

Drag force on the element, $dF_D = p.Rd\theta.L\times\cos\theta$

Lift force on the element, $dF_L = -p.Rd\theta.L\times\sin\theta$

$$\text{Total drag force, } F_D = \int_0^{2\pi} pRL\cos\theta.d\theta = 0 \qquad \text{... (8.55)}$$

Because the velocity and pressure distribution is symmetrical about the *y*-axis, The concept of the zero drag force on bodies immeresed in a steady flow of an ideal fluid is called **D'Alembert's paradox.**

$$\text{Totol lift force, } F_L = -\int_0^{2\pi} pRL\sin\theta.d\theta$$

$$= -RL\left[\int_0^{2\pi} p_\infty \sin\theta.d\theta + \frac{1}{2}\rho U_\infty^2\int_0^{2\pi}\sin\theta\, d\theta - \frac{1}{2}\rho U_\infty^2\int_0^{2\pi} 4\sin^3\theta\, d\theta - \frac{1}{8}\frac{\rho\Gamma^2}{\pi^2R^2}\int_0^{2\pi}\sin\theta\, d\theta - \frac{\rho\Gamma U_\infty}{\pi R}\int_0^{2\pi}\sin^2\theta\, d\theta\right]$$

Now $\int_0^{2\pi}\sin^2\theta\, d\theta = 0$, when *n* is odd.

$$\therefore \qquad F_L = RL\times\frac{\rho\Gamma U_\infty}{\pi R}\int_0^{2\pi}\sin^2\theta\, d\theta$$

$$= \frac{\rho\Gamma U_\infty L}{\pi}\int_0^{2\pi}\frac{1}{2}(1-\cos 2\theta)\, d\theta$$

$$= \frac{\rho\Gamma U_\infty L}{\pi}\left|\theta - \frac{\sin 2\theta}{2}\right|_0^{2\pi} = \frac{\rho L U_\infty\Gamma}{2\pi}[2\pi] = \rho L U_\infty\Gamma \qquad \text{... (8.56)}$$

Eq. (8.56) is known as **Kutta-Joukowski equation**. This equation is applicable to bodies of all shapes, including airfoil, as well,.

8.12.3 Lift Coefficient for Rotating Cylinder

$$\text{Lift force,} \qquad F_L = C_L\times\frac{1}{2}\rho U_\infty^2 A$$

$$\therefore \qquad C_L = \frac{2F_L}{\rho U_\infty^2 A} = \frac{2\rho L U_\infty \Gamma}{\rho U_\infty^2 \times 2RL}$$

$$= \frac{\Gamma}{U_\infty R} \qquad \text{... (8.57)}$$

Now

$$u_c = \frac{\Gamma}{2\pi R} \text{ or } \frac{\Gamma}{R} = 2\pi u_c$$

$$\therefore \qquad C_L = \frac{2\pi u_c}{U_\infty} \qquad \text{... (8.58)}$$

where u_c = velocity of rotation of cylinder in tangential direction.

For a single stagnation point, $\theta = 270°$.

From Eq. (8.53), we have

$$\sin 270° = -\frac{\Gamma}{4\pi R U_\infty} = -1$$

or

$$\Gamma = 4\pi R U_\infty \qquad \text{... (8.59)}$$

or

$$U_\infty = \frac{\Gamma}{4\pi R}$$

$$\therefore \qquad \frac{u_c}{U_\infty} = 2 \qquad \text{... (8.60)}$$

8.12.4 Magnus Effect

When a cylinder is rotated in a uniform flow, a lift force is generated on the cylinder. This phenomenon of the lift force generated by a rotating cylinder in a uniform flow is known as Magnus effect. The Magnus effect is advantageously used in propulsion of ships, in games like table tennis, cricket, golf, etc.

Example 8.17 *A circular cylinder 2 m in diameter and 10 m long rotates at 300 rpm with its axis perpendicular to the stream of water flowing at a velocity of 20 m/s. Assuming no slip between the cylinder and circulating flow, calculate: (a) The circulation, (b) The theoretical lift, (c) The position of stagnation points, and (d) the rpm of cylinder for a single stagnation point.*

Solution.

Given: $R = 1\text{m}$, $L = 10\text{m}$, $N = 300$ rpm, $U_\infty = 20$ m/s

(a) Velocity at the surface of cylinder due to circulation a alone,

$$u_c = \frac{2\pi RN}{60} = \frac{2\pi \times 1 \times 300}{60} = 31.4 \text{ m/s}$$

Circulation, $\Gamma = 2\pi R u_c = 2\pi \times 1 \times 31.4 = 197.4 \text{ m}^2/\text{s}$

(b) Theoretical lift, $F_L = \rho L U_\infty \Gamma = 10^3 \times 10 \times 20 \times 197.4 = 39.48 \times 10^6$ N

(c) For stagnation points, $\sin\theta = -\dfrac{\Gamma}{4\pi R U_\infty}$

$$= \frac{197.4}{4\pi \times 1 \times 20} = -0.78543$$

$$\theta = -51.76°$$

$$= 231.76°, 308.24°$$

(d) For a single stagnation point, $\Gamma = 4\pi R U_\infty = 4\pi \times 1 \times 20 = 80\pi\ \text{m}^2/\text{s}$

$$u_c = \frac{\Gamma}{2\pi R} = \frac{80\pi}{2\pi \times 1} = 40 \text{ m/s}$$

$$u_c = \frac{2\pi RN}{60}$$

$$N = \frac{40 \times 60}{2\pi \times 1} = 382 \text{ rpm}$$

Example 8.18 *A cylinder whose axis is perpendicular to the stream of air having a velocity of 25 m/s, rotates at 330 rpm. The cylinder is 2 m in diameter and 10 m long. Calculate: (a) The circulation, (b) Theoretical lift force per unit length, (c) Position of stagnation points, and (d) Actual lift, drag and direction of resultant force.*

Take air density = 1.24 kg/m³, C_L = 3.4, C_D = 0.65, and $\dfrac{u_c}{U_\infty} = 1.57$ *for actual values.*

Solution.

Given: U_∞ = 25 m/s, N = 330 rpm, R = 1 m, L = 10 m, ρ = 1.24 kg/m³, C_L = 3.4, C_D = 0.65, $\dfrac{u_c}{U_\infty} = 1.57$

$$u_c = \frac{2\pi RN}{60} = \frac{2\pi \times 1 \times 330}{60} = 34.56 \text{ m/s}$$

(a) $u_c = \dfrac{\Gamma}{2\pi R}, \Gamma = 2\pi \times 1 \times 34.56 = 217.1\ \text{m}^2/\text{s}$

(b) $\Gamma_c = \rho U_\infty L \Gamma = 1.24 \times 25 \times 10 \times 217.1 = 67{,}310.7 N$

$$\frac{F_L}{L} = \frac{67{,}310.7}{10} = 6731.07 \text{ N/m}$$

(c) $\sin\theta = -\dfrac{\Gamma}{4\pi U_\infty R} = -\dfrac{217.1}{4\pi \times 25 \times 1} = -0.69105$

$\theta = -43.71°$ or $223.71°$, $316.29°$

(d) Actual lift, $F_L = C_L \times \dfrac{1}{2}\rho A U_\infty^2 = 3.4 \times \dfrac{1}{2} \times 1.24 \times 2 \times 10 \times 25^2 = 26{,}350$ N

Actual dreg, $F_D = C_D \times \dfrac{1}{2}\rho A U_\infty^2 = 0.65 \times \dfrac{1}{2} \times 1.24 \times 2 \times 10 \times 625 = 5037.5$ N

Resultant force, $F_R = \sqrt{F_L^2 + F_D^2} = \sqrt{(26{,}350)^2 + (5037.5)^2} = 26{,}827.2$ N

$$\tan\theta = \frac{F_L}{F_D} = \frac{26{,}350}{5037.5} = 5.23077$$

$$\theta = 79.177°$$

8.13 LIFT ON AN AIRFOIL

An airfoil or aerofoil is a streamlined body which may be either symmetrical or unsymmetrical, as shown in Fig. 8.17

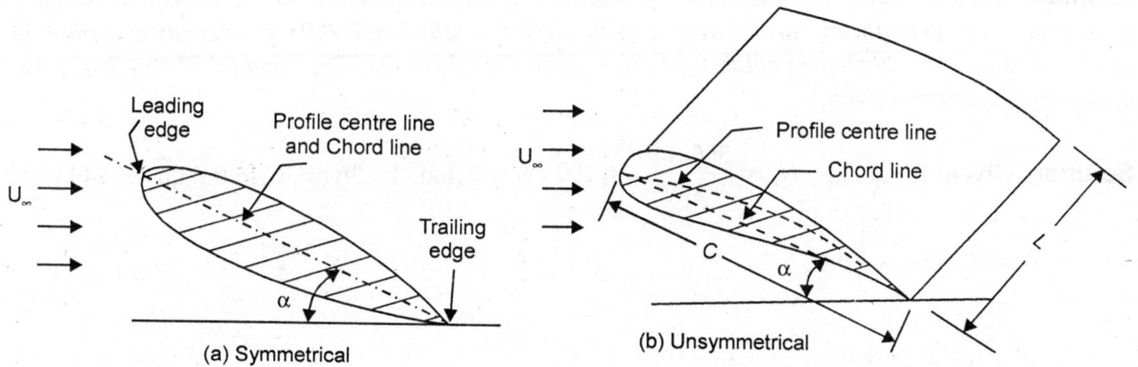

Fig. 8.17 Profiles of airfoil

Definations

Chord line: It is the line joining the leadig edge and trailing edge of the airfoil. The length of the line is called chord of airfoil (C)

Profile centreline: It is the line joining the midpoints of the profile.

Angle of attack (α) : It is the angle between the chord line and direction of fluid stream.

Camber: It is the curvature of an airfoil.

Stall: An airfoil is said to be in stall condition when the angle of attack of an airfoil is grateer than the angle of attack at maximum lift. At stall the air separates from the airfoil and eddies are formed, which increases the drag coefficient.

Aspect ratio (A.R.): It is the ratio of span of the wing (L) to its mean chord (C).

$$\text{A.R.} = \frac{L}{C} = \frac{L^2}{A} \quad \text{... (8.61)}$$

where, $A = L \times C$

Circulation developed on the airfoil, $\Gamma = \pi C U_\infty \sin\alpha$ ... (8.62)

Lift force, $F_L = \rho L U_\infty \Gamma = \rho L U_\infty \times \pi C U_\infty \sin\alpha = \pi\rho C L U_\infty^2 \sin\alpha$ ...(8.63)

Also , $F_L = C_L \times \frac{1}{2}\rho U_\infty^2 \times C \times L = \pi\rho C L U_L^2 \sin\alpha$

$\therefore$ $C_L = 2\pi \sin\alpha$

For steady state flying of an object of weight W,

$$W = F_L = C_L \times \frac{1}{2}\rho A U_\infty^2 \quad \text{... (8.65)}$$

Example 18.19 *An aeroplane weighing 40 kN is flying in a horizontal direction at 360 km/h. The span of the plane is 16 m and wing surface area is 36 m². If drag coefficient is 0.03 and density of air is 1.24 kg/m³, calculate: (a) Coefficient of lift, (b) Power required to drive the plane, and (c) Theoretical value of boudary layer circulation.*

Solution. Given: $W = 40$ kN, $U_\infty = \frac{360 \times 10^3}{3600} = 100$ m/s, $L=16$m, $A = 36$ m², $C_D = 0.03, \rho = 1.24$ kg/m³

(a) $W = F_L = C_L \times \frac{1}{2}\rho A U_\infty^2$

$$40 \times 10^3 = C_L \times \frac{1}{2} \times 1.24 \times 36 \times 100^2$$

$$C_L = 0.179$$

(b) $F_D = C_D \times \frac{1}{2}\rho A U_\infty^2 = 0.03 \times \frac{1}{2} \times 1.24 \times 36 \times 100^2 = 6696$ N

Power required, $P = F_D \times U_\infty = 6696 \times 100 \times 10^{-3} = 669.6$ kW

(c) $F_L = \rho L U_\infty \Gamma$

$$\Gamma = \frac{40 \times 10^3}{1.24 \times 16 \times 100} = 20.16 \text{ m}^2/\text{s}$$

SUMMARY

1. A thin layer adjacent to the boundary of an immersed body in a viscous fluid, where the velocity gradient is very high, is called boundary layer.
2. The viscous resistance is confined only to the boundary layer. The fluid outside the boundary layer may be considered as ideal.
3. The velocity distribution in the laminar boundary layer over a flat plate is parabolic in nature.
4. The velocity distribution in turbulent boundary layer is governed by the Prandtl's one-seventh power law.
5. In the turbulent boundary layer, a very thin layer which remains laminar, is called laminar sublayer.
6. For a laminar flow over a flat plate, Reynolds number,

$$Re_x = \frac{U_\infty x}{\nu} < 5\times10^5$$

When $Re_x > 5\times 10^5$, the flow is turbulent.

7. The boundary layer thickness is defined as that distance from the boundary at which the velocity reaches $0.99\ U_\infty$.
8. Displacement thickness, $\delta^* = \int_0^\delta \left(1-\frac{u}{U_\infty}\right)dy$
9. Momentum thickness, $\theta = \int_0^\delta \frac{u}{U_\infty}\left(1-\frac{u}{U_\infty}\right)dy$
10. Energy thickness, $\delta_e = \int_0^\delta \frac{u}{U_\infty}\left[1-\left(\frac{u}{U_\infty}\right)^2\right]dy$
11. Von-Karman momentum integral equation,

$$\frac{\tau_0}{\rho U_\infty^2} = \frac{d\theta}{dx}$$

12. Blasius results:

$$\delta = \frac{5x}{\sqrt{\text{Re}_x}},\ C_D = \frac{1.328}{\sqrt{\text{Re}_L}}$$

13. Velocity profile for turbulent boudary layer,

$$\frac{u}{U_\infty} = \left(\frac{y}{\delta}\right)^{1/7}$$

14. Shape factor, $H = \dfrac{\delta^*}{\theta}$

15. For laminar flow in a pipe,

$$\delta^* = \frac{R}{4}, \theta = \frac{R}{12}, \frac{\delta^*}{\theta} = 3$$

$$\frac{u}{U_\infty} = 1 - \frac{y^2}{R^2}$$

16. For turbulent boundary layer over a flat plate,

$$\tau_0 = 0.0225 \rho U_\infty^2 \left(\frac{\mu}{\rho U_\infty \delta} \right)^{1/4}$$

17. For $5 \times 10^5 < \text{Re} < 10^7$,

$$\delta = \frac{0.371x}{(\text{Re}_x)^{1/5}}, C_D = \frac{0.072}{(\text{Re}_L)^{1/5}}$$

For $10^7 < \text{Re} < 10^9$,

$$C_D = \frac{0.455}{(\log_{10} \text{Re}_L)^{2.58}}$$

18. Total drag on a flat plate due to laminar and turbulent layers:

$$F_{\text{total}} = \left[\frac{0.455}{(\log_{10} \text{Re}_L)^{2.58}} - \frac{1700}{\text{Re}_x} \right] \frac{Lb \rho U_\infty^2}{2}$$

$$C_D = \frac{0.455}{(\log_{10} \text{Re}_L)^{2.58}} - \frac{1700}{\text{Re}_L}$$

19. Boundary layer separation:

If $\left(\dfrac{du}{dy} \right)_{y=0} > 0$, the flow will not separate.

$\left(\dfrac{du}{dy} \right)_{y=0} = 0$, the flow is on the verge of separation.

$\left(\dfrac{du}{dy} \right)_{y=0} < 0$, the flow has separated

20. If $\frac{dp}{dx} > 0$, the boundary layer will separate.

21. Dreg force, $F_D = C_D \times \frac{1}{2}\rho A U_\infty^2$

Resultant force, $F_R = \sqrt{F_D^2 + F_L^2}$

22. Deformation drag on a sphere, $F_D = 3\pi\mu U_\infty d$

Skin-friction drag $= \frac{2}{3}F_D$

Pressure drag $= \frac{1}{3}F_D$

$$C_D = \frac{24}{\text{Re}} \; for \, \text{Re} \le 0.2$$

$$C_D = \frac{24}{\text{Re}}\left(1 + \frac{3}{16\,\text{Re}}\right) \text{ for } 0.2 < \text{Re} < 5$$

$= 0.4$ for $5 < \text{Re} < 1000$

$= 0.5$ for $1000 < \text{Re} < 10^5$

$= 0.2$ for $\text{Re} > 10^5$

Terminal velocity, $U_\infty = \frac{d^2}{18\mu}(\gamma_s - \gamma_f)$

$$W = F_D + F_B = 3\pi\mu U_\infty d + \frac{\pi}{6}d^3\gamma_f$$

23. Flow past a circular stationary cylinder:

$$u_\theta = 2U_\infty \sin\theta$$

24. Flow past a rotating cylinder:

$$u = u_\theta + u_c = 2U_\infty \sin\theta + \frac{\Gamma}{2\pi R}$$

At stagnation points, $u = 0$

$$\sin\theta = \frac{-\Gamma}{4\pi R U_\infty}$$

For a single stagnation point, the condition is

$$\Gamma = 4\pi R U_\infty$$

$$u_c = 2U_\infty$$

25. $F_L = \rho L U_\infty \Gamma$ (Kutta-Joukowski equation)

$$p = p_\infty + \frac{1}{2}\rho U_\infty^2 \left[1 - \left(2\sin\theta + \frac{\Gamma}{2\pi R U_\infty}\right)^2\right]$$

$$C_L = \frac{\Gamma}{U_\infty R} = \frac{2\pi u_c}{U_\infty}$$

26. The generation of lift by rotating cylinder in a fluid stream is called Magnus effect.

27. Flow over an airfoil.

$\Gamma = \pi c U_\infty \sin\alpha$, C = chord length, α = angle of attack

$C_L = 2\pi \sin\alpha$

$$W = F_L = C_L \times \frac{1}{2}\rho U_\infty^2 A$$

Thrust developed by the engine = F_D.

Short Answer Type Questions

1. What is boundary layer? Explain
2. What is the 'slip condition' at the boundary?
3. What is boundary layer thickness? Explain.
4. Define displacement thickness and momentum thickness.
5. Define energy thickness and shape factor.
6. What is laminar sub-layer?
7. What is the necessity of controlling the growth of boundary layer on most of the bodies?
8. What are the methods used for controlling the separation of boundary layer?
9. What is the condition for separation of boundary layer?
10. Write the Von-Karman momentum integral equation.
11. Differentiate between stream lined and bluff bodies.
12. Explain skin friction drag and pressure drag.
13. What is terminal velocity of a sphere?
14. What is D'Alembert's paradox?
15. Write Kutta-Joukowski equation.
16. What is Magnus effect?
17. What is an airfoil?
18. When an aeroplane is in steady state?

19. Define stagnation points.

20. Define the following terms for an airfoils:

(a) chord line (b) Angle attack (c) comber (d) Aspect ratio.

EXERCISES

1. The velocity distribution in the boundary layer is given by:

$$\frac{u}{U_\infty} = 2\left(\frac{y}{\delta}\right) - \left(\frac{y}{\delta}\right)^2$$

Find the shape factor.

[Ans. 2.5]

2. Air flows over a smooth flat plate with a velocity of 10 m/s. The velocity profile is given by:

$$\frac{u}{U_\infty} = 2\left(\frac{y}{\delta}\right) - \left(\frac{y}{\delta}\right)^2$$

The length of the plate is 1.2 m and width 0.9 m. If laminar boundary layer exists upto a value of Re = 2× 10^5 and kinematic viscosity of air is 0.15 Stokes, calculate: (a) The maximum distance from the leading edge upto which laminar boundary layer exists, and (b) The maximum thickness of boundary layer.

[Ans. 0.3 m, 3.67 mm]

3. Air is flowing over plate 4m × 2m with a velocity of 5m/s. If $\rho = 1.208$ kg/m^3 and $\nu = 1.47 \times 10^{-5}$ m^2/s, calculate:

(a) Length of plate over which the boundary layer is laminar and thickness of the laminar boundary layer.

(b) Shear stress at the location where boundary layer ceases to be laminar, and

(c) Total force on both sides of the portion of plate where boundary layer is laminar.

[Ans. 1.47m, 10.39 mm, 0.01418 N/m^2, 0.1662 N]

4. A 2 wide and 5 m long plate when towed through water experiences a drag of 30.38 N on both sides. Calculate the velocity of plate and the length over which the boundary layer is laminar.

[Ans. 1.0 m/s, 0.5 m]

5. A plate 0.3 m × 0.1 m is immersed in a liquid of density 998 kg/m^3 and kinematic viscosity 1× 10^{-6} m^2/s. The velocity of water is 15 m/s parallel to it. Calculate:

(a) Drag force on the portion of plate over which the boundary layer is laminar.

(b) The drag force on both sides of plate.

6. A streamlined train is 250 m long with a typical cross-section having a perimeter of 8 m above the wheels. The kinematic viscosity of air is 1.5 × 10^{-5} m^2/s and density is 1.24 kg/m^3. Calculate the friction drag of train when running at 80 km/h.

[Ans. 1088.5 N]

7. Air flows at a speed of 50 km/h on a flat plate of size 2 m long and 1 m wide. The density of air is 1.15 kg/m^3. The coefficients of lift and drag are 0.75 and 0.85 respectively. Calculate: (a) Lift force, (b) Drag force, (c) Resultant force, and (d) Power exerted by the air stream on the plate.

[Ans. 166.28 N, 33.25 N, 169.6 N at 78.69°, 461.89W]

8. A kite 0.8 m × 0.8 m, weighing 3.924 N is flying at an angle of 12° to the horizontal. The string attached to the kite makes an angle of 45° to the horizontal. The pull on the string is 24.525 N when the wind is blowing at a speed of 30 km/h. Calculate the coefficient of drag and lift. Take density of air is 1.25 kg/m³. [**Ans.** 0.624, 0.765]

9. A submarine assumed to approximate a cylinder 4 m in diameter and 20 m long is travelling submerged at 1.3 m/s in sea water. Calculate the drag exerted on it, if the drag coefficient is 0.75 for $Re > 10^5$. For water, $\rho = 1035$ kg/m³ and $\nu = 0.015$ Stokes. [**Ans.** 52.47 kN]

10. A steel ball of 3 mm diameter falls in glycerine at a terminal velocity of 0.035 m/s. The specific weights of steel and glycerine are 75 kN/m³ and 12.5 kN/m³ respectively. Calculate: (a) Dynamic viscosity of glycerine, (b) Drag force, and (c) Drag coefficient for sphere.

[**Ans.** 0.893 Pa.s, 8.837×10^{-4} N, 160.2]

11. A cylinder 1.2 m in diameter and 8 m long rotates at 90 rpm with its axis perpendicular to the air stream flowing at 30 m/s. Assuming no-slip condition between the cylinder and circulatory flow, find (a) The magnitude of circulation, (b) the lift force, and (c) the position of stagnation points.

Take specific weight of air as 12.25 N/m³. [**Ans.** 21.3 m²/s, 6.33 kN, 185.4°, 354.6°]

12. A jet plane weighing 29.4 kN and having a wing area of 20 m² flies at a velocity of 950 km/h. When the engine delivers 7350 kW, 65 percent of the power is used to overcome the drag resistance of the wing. Calculate the coefficient of lift and drag for the wing.

Take density of atmospheric air as 1.208 kg/m³. [**Ans.** 0.0349, 0.0215]

9
TURBULENT FLOW

9.1 INTRODUCTION

In a viscous fluid the particles move in random manner resulting in general mixing of the particles. Such a flow is called turbulent flow. In a turbulent flow, the fluid motion is irregular and chaotic and there is complete mixing of fluid due to collision of fluid masses with one another. The fluid masses are interchanged between adjacent layers resulting in transfer of momentum which causes additional shear stresses. The shear in turbulent flow is mainly due to momentum transfer. The contribution to fluid viscosity to total shear is small and is usually neglected. The flow in a pipe is said to be turbulent when Reynolds, number, Re > 4000.

9.2 FRICTIONAL HEAD LOSS IN PIPE FLOW

Consider a unifrm cross-section horizontal pipe, having steady flow as shown in Fig. 9.1. Consider control volume enclosed between sections 1–1 and 2–2 of the pipe.

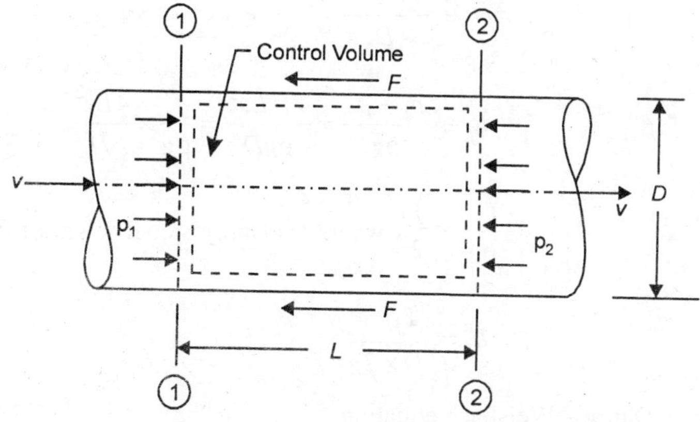

Fig.9.1 Forces acting on control volume in a pipe flow

Let p_1, p_2 = pressure at sections 1–1 and 2–2 respectively

v = velocity of flow in pipe

L = length of pipe between sections 1 – 1 and 2 – 2 respectively

D = dimeter of pipe

f' = frictional resistance per unit wetted area per unit velocity

h_f = loss of head due to friction.

F = frictional resistance force

Applying Bernoulli's equation between sections 1–1 and 2–2, we have

$$\frac{p_1}{\gamma}+\frac{v_1^2}{2g}+z_1 = \frac{p_2}{\gamma}+\frac{v_2^2}{2g}+z_2+h_f$$

Here $z_1 = z_2$ and $v_1 = v_2$

$$\therefore \quad \frac{p_1 - p_2}{\gamma} = h_f \quad \text{... (1)}$$

Frictional resistance, $F = f' \times PL \times v^2$

where P = perimeter of pipe wetted surface

$= \pi D$ ... (2)

$$F = f' \times \pi DL \times v^2$$

Resolving all forces in the horizontal direction, we have

$$(p_1 - p_2)\,A = F$$

$$(p_1 - p_2)\times\frac{\pi}{4}D^2 = f' \times \pi DL \times v^2$$

$$p_1 - p_2 = \frac{4f'Lv^2}{D}$$

or

$$h_f = \frac{p_1 - p_2}{\rho g} = \frac{4f'Lv^2}{\rho g D} = \frac{f'}{\rho g}\times\frac{4Lv^2}{D} \quad \text{... (3)}$$

Let $\frac{f'}{\rho} = \frac{f}{2}$, where f = Darcy's coefficient of friction

Then

$$h_f = \frac{4fLv^2}{D\times 2g} \quad \text{... (9.1)}$$

Eq. (9.1) is known as Darcy – Weisbach equation.

Eq. (9.1) can be written as:

$$h_f = \frac{f_1 L v^2}{D \times 2g}$$

where $f_1 = 4f$ is known as *friction factor.*

9.2.1 Coefficient of Friction in Terms of Shear Stress

Now $$(p_1 - p_2)A - F = 0$$

or $$(p_1 - p_2)A = F = \text{force due to shear stress } \tau_0$$

$$= \tau_0 \times \pi D L$$

or $$(p_1 - p_2)\frac{\pi}{4}D^2 = \tau_0 \times \pi D L$$

or $$p_1 - p_2 = \frac{4\tau_0 L}{D}$$

∴ $$h_f = \frac{p_1 - p_2}{\rho g} = \frac{4f L v^2}{D \times 2g}$$

or $$p_1 - p_2 = \frac{4fLv^2}{D \times 2g} \times \rho g = \frac{4fLv^2\rho}{2D}$$

∴ $$\frac{4\tau_0 L}{D} = \frac{4fLv^2\rho}{2D}$$

$$f = \frac{2\tau_0}{\rho v^2} \qquad \text{... (9.3)}$$

$$\tau_0 = f \times \frac{1}{2}\rho \bar{v}^2 \qquad \text{... (9.4)}$$

9.3 CHARACTERISTICS OF TURBULENT FLOW

A turbulent flow is characterised by random, irregular and haphazard movement of fluid particles, making it impossible to predict the motion of a fluid particle w.r.t. time and space. In the process of turbulent mixing the fluid with high kinetic energy is carried by the transverse fluctuations towards the boundary, while the fluid moving at low kinetic energy near the boundary is carried away from the boundary into the region of high knietic energy. This results in a more uniform velocity distribution in turbulent flow as compared to that in the laminar one. The transverse mixing of fluid tends to equalise the velocity except near the boundary where this mixing tends to vanish. Fig. 9.2 shows the velocity distributions in laminar and turbulent flows at the same Reynolds number.

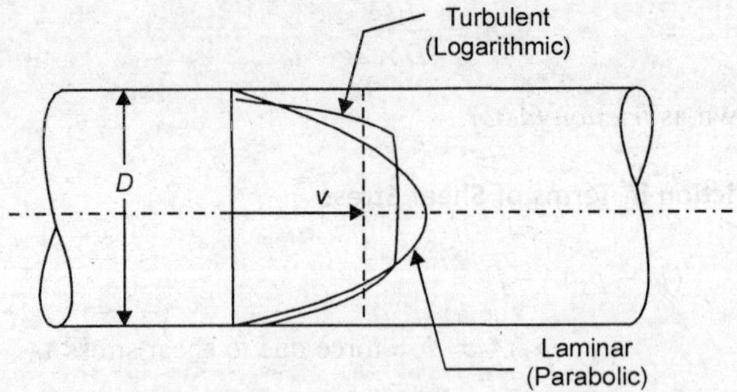

Fig. 9.2 Velocity distributions in laminar and turbulent flows in pipe at same Reynolds number

In turbulent flow, at a given point velocity and pressure are not constant with time and are random in nature. Because of this random nature, statistical methods are used to describe its charcteristics. The random nature of variation of instantaneous velocity of fluid at a point with time is shown in Fig. 9.3.

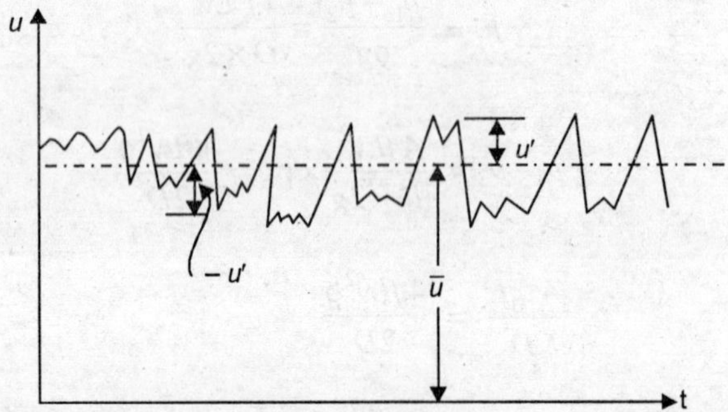

Fig. 9.3 Random velocity fluctuations at a point in turbulent flow.

The velocity in a turbulent flow may be considered as consiting of a time average portion $\overline{u}$ and a fluctuating part u'. The velocity could be written accordingly as:

$$u = \overline{u} + u'$$

where u = instaneous velocity

$\overline{u}$ = time average or temporal mean velocity

u' = velocity fluctuation.

The the temporal mean velocity components may be expressed as:

$$\overline{u} = \frac{1}{T}\int_0^T u.dt, \overline{v} = \frac{1}{2}\int_0^T v.dt, \overline{w} = \frac{1}{T}\int_0^T w.dt \quad \text{... (9.6)}$$

The temporal mean of u′ is,

$$\overline{u'} = \frac{1}{2}\int_0^T u'.dt = \frac{1}{2}\int_0^T (u-\overline{u})\,dt = \frac{1}{T}\int_0^T u.dt - \frac{1}{T}\int_0^T \overline{u}\,dt = \overline{u} - \overline{u} = 0$$

Hence $\overline{u'} = \overline{v'} = \overline{w'} = 0$ ... (9.7)

where *T* represents the interval over which the averaging is carried out. The root-mean-square value of u′, *i.e.* $\sqrt{(\overline{u'})^2}$ is a measure of the violence of turbulent fluctuations and is a quantitative measure of the *intensity of turbulence*. The mean time intervals in the sign of u′ is a measure of the *scale of turbulence.*

The *isotropic turbulence* is defined as that turbulence is which the mean-square velocity fluctuation in three coordinate directions are equal to each other. Thus

$$\overline{u'^2} = \overline{v'^2} = \overline{w'^2} \qquad \text{... (9.8)}$$

The level, degree or intensity of turbulence is defined by,

$$\frac{1}{U_\infty}\sqrt{\frac{1}{3}(\overline{u'^2} + \overline{v'^2} + \overline{w'^2})} \qquad \text{... (9.9)}$$

where U_∞ = free-stream velocity.

Average kinetic energy of turbulence per unit mass is defined as:

$$\frac{\text{Average K.E. of turbulence}}{\text{Mass}} = \frac{1}{2}(\overline{u'^2} + \overline{v'^2} + \overline{w'^2}) \qquad \text{(9.10)}$$

The turbulent energy in a boundary layer is maximum near the wall and decreases towards the free-stream.

9.4 SHEAR STRESS IN TURBULENT FLOW

The total shear stress in turbulent flow consists of a laminar part and a turbulent part, and may be expressed as:

$$\tau = \tau_l + \tau_l$$

$$= \mu\frac{du}{dy} + \eta\frac{d\overline{u}}{dy} \qquad \text{... (9.11)}$$

where μ = dyanmic viscosity of fluid.

η = eddy viscosity or dynamic coefficient of turbulence

τ_l, τ_t = shear stress due to laminar and turbulent flows respectively

$\overline{u}$ = average velocity at a distance *y* from boundary.

The ratio of η (eddy viscosity) and ρ (mass density) in known as *kinematic eddy viscosity* and is denoted by ε (epsilon). Thus

$$\varepsilon = \frac{\eta}{\rho} \qquad \text{... (9.12)}$$

9.4.1 Reynolds Expression for Turbulent Shear Stress

Reynolds considered x-component of velocity fluctuation u' as superimposed on the temporal mean velocity $\bar{u}$ while the y-component v' is considered as a cross-current or normal fluctuation (trabsverse to the flow direction).

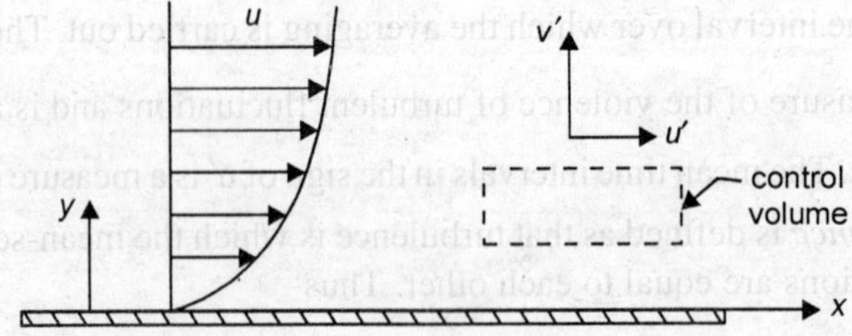

Fig. 9.4 Control volume near a wall to illustrate origin of turbulent shear stress

The function of v' is to cause an exchange of fluid particles from one layer to the other. Apply the momentum equation to the control volume within the fluid. The average normal velocity component is zero, so that the average net mass flow through each side of the control volume is zero, but it is still possible to have an average net momentum flow through each side.

Average net rate of momentum flow through the face parallel to flow per unit area is,

$$\frac{1}{T}\int_0^T \rho u v'\, dt = \frac{1}{T}\int_0^T \rho(\bar{u}+u')v' dt = \frac{\rho\bar{u}}{T}\int_0^T v' dt + \frac{\rho}{T}\int_0^T u'v'\, dt$$

$$= \frac{\rho}{T}\int_0^T u'v' dt = \rho\overline{u'v'} \qquad \text{... (9.13)}$$

From the momentum principle, Eq. (9.13) represents force per unit area and is, therefore, the *turbulent shear stress* $\bar{\tau}$. Therefore, the turbulent shear stress is caused by the momentum trasfer associated with the turbulent fluctuations.

$$\therefore \qquad \bar{\tau} = \rho\overline{u'v'} \qquad \text{... (9.14)}$$

9.4.2 Parandtl's Mixing Hypothhesis

According to Prandtl, the mixing length l is defined as the average distance between two layers in the transverse direction through which a small mass of fluid particle would move from one layer to the other adjacent layers before acquring the velocity of the new layer. He further assumed that the velocity flucuation in the x-direction u' is rleated to the mixing length

l as,

$$u' = l\frac{du}{dy}$$

and v', the fluctuation component of velocity in y-direction is of the same order of magnitude as u'. Hence

$$v' = l\frac{du}{dy}$$

$$\overline{u' \times v'} = \overline{u'v'} = \left(l\frac{du}{dy}\right) \times \left(l\frac{du}{dy}\right) = l^2\left(\frac{du}{dy}\right)^2$$

Substituting the value of $\overline{u'v'}$ in Eq. (9.14), we get

$$\bar{\tau} = \rho l^2\left(\frac{du}{dy}\right)^2 \quad \text{... (9.15)}$$

When the viscous action is also included, the total shear stress may be expressed as:

$$\bar{\tau} = \mu\frac{du}{dy} + \rho l^2\left(\frac{du}{dy}\right)^2 \quad \text{... (9.16)}$$

The viscous shear stress is negligible except near the boundary. Eq. (9.16) is used for most of turbulent fluid flow problems for determining shear stress in turbulent flow.

9.5 VELOCITY DISTRIBUTION IN TURBULENT FLOW IN PIPES

Neglecting viscous shear stress, from Eq. (9.15), we have

$$\bar{\tau} = \rho l^2\left(\frac{du}{dy}\right)^2$$

We can detemine shear stress in turbulent flow if the relation between l and y is known. Prandtl assumed that the mixing length l is a linear function of the distance y from the pipe wall. Thus

$$l \propto y$$

or $$l = ky$$

where k = a constant of proportionality, known as *Karman constant*

= 0.4

$$\therefore \quad \bar{\tau} \text{ or } \tau = \rho \times (ky)^2 \times \left(\frac{du}{dy}\right)^2 = \rho k^2 y^2\left(\frac{du}{dy}\right)^2$$

or $$\left(\frac{du}{dy}\right)^2 = \frac{\tau}{\rho k^2 y^2}$$

or $$\frac{du}{dy} = \frac{1}{ky}\sqrt{\frac{\tau}{\rho}} \quad \text{... (9.17)}$$

For small values of y, i.e. near to the pipe wall, we have

$$\tau = \tau_0 \qquad (\tau_0 = \text{boundary shear stress})$$

Eq. (9.17) becomes,

$$\frac{du}{dy} = \frac{1}{ky}\sqrt{\frac{\tau_0}{\rho}} = \frac{1}{ky} \times u_f \qquad \text{... (9.18)}$$

where $u_f = \sqrt{\dfrac{\tau_0}{\rho}}$ = shear frictin velocity or *shear velocity.*

Integrating Eq. (9.18), we get

$$\int du = u_f \int \frac{1}{ky} \cdot dy$$

$$u = \frac{u_f}{k} \ln y + C \qquad \text{... (9.19)}$$

where C = constant of integration.

At $y = R$ (radius of pipe), $u = u_{max}$

$$\therefore \qquad u_{max} = \frac{u_f}{k} \ln R + C$$

$$C = u_{max} - \frac{u_f}{k} \ln R$$

$$\therefore \qquad u = \frac{u_f}{k} \ln y + u_{max} - \frac{u_f}{k} \ln R$$

$$= u_{max} + \frac{u_f}{k} \ln \frac{y}{R}$$

For $k = 0.4$, we get

$$u = u_{max} + 2.5\, u_f \ln \frac{y}{R} \qquad \text{... (9.20)}$$

Eq. (9.20) is called Prandtl's universal velocity distribution equation for turbulent flow in pipes. This equation is applicable to smooth as well as rough pipe boundaries.

Eq. (9.20) may be written in non-dimensional form as:

$$\frac{u_{max} - u}{u_f} = 2.5 \ln \frac{R}{y}$$

$$= 2.5 \times 2.3 \log_{10}\left(\frac{R}{y}\right) = 5.75 \log_{10}\left(\frac{R}{y}\right)$$

$$\therefore \quad \frac{u_{max} - u}{u_f} = 5.75\log_{10}\left(\frac{R}{y}\right) \quad \text{... (9.21)}$$

The difference ($u_{max} - u$) is known as the *velocity defect.*

Example 9.1 *A pipe of 100 mm diameter is carrying water. If the velocities at the pipe centre and 25 mm from the pipe centre are 2m/s and 1.5 m/s respectively and flow in the pipe is turbulent, calculate the wall shearing stress.*

Solution.

Given: $R = 50$ mm, $u_{max} = 2$ m/s, $y = R - r = 50 - 25 = 25$ mm, $u = 1.5$ m/s.

$$\therefore \quad \frac{u_{max} - u}{u_f} = 5.75\log_{10}\left(\frac{R}{y}\right)$$

$$\frac{2-1.5}{u_f} = 5.75\log_{10}\left(\frac{50}{25}\right) = 1.7309$$

$$u_f = 0.289 \text{ m/s}$$

Now

$$u_f = \sqrt{\frac{\tau_0}{\rho}}$$

$$0.289 = \sqrt{\frac{\tau_0}{10^3}}$$

$$\tau_0 = 83.44 \text{ N/m}^2$$

9.5.1 Hydraulically Smooth and Rough Pipes

Let k be the average height of the irregulaties projected from the surface of a boundary as shown in Fig. 9.5. If the value of k is more than the laminar sub layer for a boundary then the boundary is called rough boundary and if the value of k is less, than the laminary sublayer then boundary is known as smooth boundary. This classification is based on boundary characteristics. For proper classification, the flow and fluid characterstics are also to be taken into account.

The flow is turbulent outside the laminar sub-layer. The eddies present in turbulent flow try to pentrate the laminar sub-layer and reach the surface of the boundary. But due to large thickness of laminar sublayer the eddies are unable to reach the surface irregularties and hence the boundary behaves as a smooth boundary. Such a boundary is called hydordynamically smooth boundary.

If the Reynolds number of flow is increased then the thickness of laminar sublayer will decrease. If the thickness of laminar sub-layer becomes much smaller than the average height of irregularities of the surface as shown in Fig. 9.5 (b), the boundary will act as rough boundary. In this case, the irreularities of the surface are above the laminar sub-layer and the eddies present in turbulent zone will come in contact with the irregularities of the sunface and lot of energy shall be lost. Such a boundary is called hydrodyanmically rough boundary.

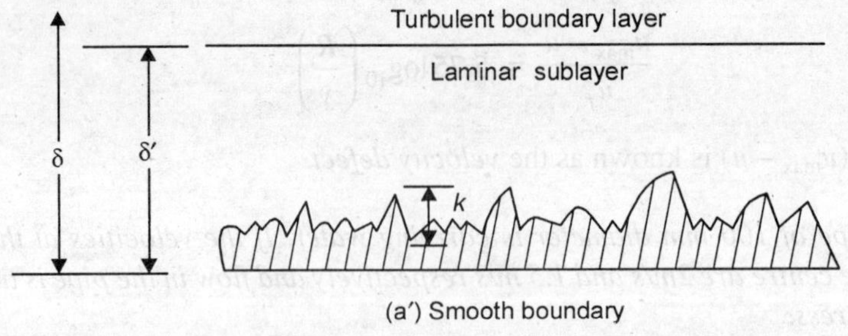

(a′) Smooth boundary

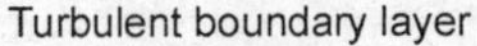

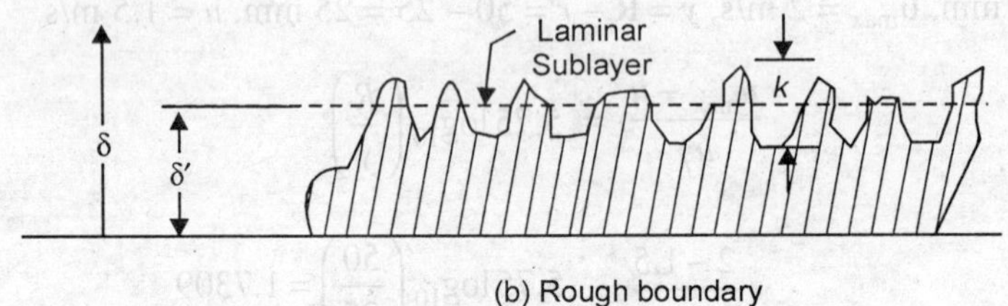

(b) Rough boundary

Fig. 9.5 Smooth and rough boundaries

According to Nikuradse's the boundary behaves as:

$$\left.\begin{array}{l} \text{1. Hydrodyanmically smooth boundary, when } \left(\dfrac{k}{\delta'}\right) < 0.25 \\ \text{2. Hydrodyanmically rough boundary, when } \left(\dfrac{k}{\delta'}\right) < 6.0 \\ \text{3. The boundary is in transition, when } 0.25 < \left(\dfrac{k}{\delta'}\right) < 6.0 \end{array}\right\} \quad \text{... (9.22)}$$

In terms of roughness Reynold's number $\dfrac{u_f\, k}{\nu}$:

$$\left.\begin{array}{l} \text{1. For smooth boundary, } \dfrac{u_f\, k}{\nu} < 4 \\ \text{2. For rough boundary, } \dfrac{u_f\, k}{\nu} > 100 \\ \text{3. For boundary in transiation stage, } 4 < \dfrac{u_f k}{\nu} < 100 \end{array}\right\} \quad \text{... (9.23)}$$

9.5.2 Velocity Distribution for Turbulent Flow in Smooth pipes

The vlocity distribution for turbulent flow in pipes is given by Eq. (9.19) as:

$$u = \frac{u_f}{k} \ln y + C$$

At $y = 0$, $u = -\infty$. This means that u is positive at some distance for away from the pipe wall. Hence at some finite distance from the wall, the velocity will be equal to zero. Let this distance from pipe wall be y'. Thus, at

$$y = y', \quad u = 0$$

$$0 = \frac{u_f}{k}\ln y' + C \quad \text{or } C = -\frac{u_f}{k}\ln y'$$

$$\therefore \quad u = \frac{u_f}{k}\ln\left(\frac{y}{y'}\right)$$

For $\quad k = 0.4$

$$u = 2.5\, u_f \ln\left(\frac{y}{y'}\right)$$

or
$$\frac{u}{u_f} = 5.75 \log_{10}\left(\frac{y}{y'}\right) \quad \text{... (9.24)}$$

According to Nikuradse for turbulent flow in smooth pipes,

$$\frac{u_f\,\delta'}{\nu} = 11.6 \quad \text{or} \quad \delta' = \frac{11.6\nu}{u_f} \quad \text{... (9.25)}$$

and
$$\frac{u_f\, y'}{\nu} = 0.108$$

or
$$y' = \frac{0.108\,\nu}{u_f} = \frac{0.108\,\nu}{11.6\,\nu} \times \delta' = \frac{\delta'}{107} \quad \text{... (9.26)}$$

Substituting the value of y' from Eq. (9.26) in Eq. (9.24), we get

$$\frac{u}{u_f} = 5.75\log_{10}\left[\frac{y \times u_f}{0.108\,\nu}\right]$$

$$= 5.75\log_{10}\left[\frac{y u_f}{\nu}\right] - 5.75\log_{10}(0.108)$$

$$= 5.75\log_{10}\left(\frac{u_f\, y}{\nu}\right) + 5.5 \quad \text{... (9.27)}$$

9.5.3 Velocity Distribution for Turbulent flow in Rough Pipes

In rough pipes, the surface irregularities are larger than the laminar sub-layer. Thus the laminar sub-layer is completely destroyed. Thus y' can be considered proportional to the height of protrusions k. According to Nikuradse,

$$y' = \frac{k}{30}$$

Eq. (9.24) becomes,

$$\frac{u}{u_f} = 5.75\log_{10}\left[\frac{30y}{k}\right]$$

$$= 5.75\log_{10}\left(\frac{y}{k}\right) + 5.75\log_{10} 30$$

$$= 5.75\log_{10}\left(\frac{y}{k}\right) + 8.5 \qquad \text{... (9.28)}$$

Example 9.2 *The velocity of flow in a rough pipe 100 mm diameter is found to increase by 25 percent from a point 10 mm from the wall to 25 mm from the wall. Calculate the height of roughness elements.*

Solution. In a rough pipe, we have

$$\frac{u}{u_f} = 5.75\log_{10}\left(\frac{y}{k}\right) + 8.5$$

where u = shear friction velocity

k = average height of roughness elements

$$\frac{u}{u_f} = 5.75\log_{10}\left(\frac{10}{k}\right) + 8.5$$

$$\frac{1.25\mu}{u_f} = 5.75\log_{10}\left(\frac{25}{k}\right) + 8.5$$

$$\frac{1}{1.25} = \frac{5.75\log_{10}\left(\frac{10}{k}\right) + 8.5}{5.75\log_{10}\left(\frac{25}{k}\right) + 8.5}$$

$$5.75\log_{10}\left(\frac{25}{k}\right) + 8.5 = 7.1875\log_{10}\left(\frac{10}{k}\right) + 10.625$$

$$5.75\log_{10}^{25} - 5.75_{\log}^{k} + 8.5 = 7.1875\log_{10}^{10} - 7.1875\log_{10}^{k} + 10.625$$

$$8.038 - 5.75\log_{10}^{k} + 8.5 = 7.1875 - 7.1875\log_{10}^{k} + 10.625$$

$$1.4375\log_{10}^{k} = 1.2745$$

$$\log_{10}^{k} = 0.8866$$

$$k = 7.702 \text{ mm}$$

Example 9.3 *A pipeline carrying water has surface protrusions of average height of 0.2 mm. The shear stress developed is 10 N/m². Find whether the pipe surface behaves as smooth, rough or in transition. For water, mass density = 10³ kg/m³ and kinematic viscosity = 0.0092 Stokes.*

Solution.

Given: $k = 0.2$ mm, $\tau = 10$ N/m², $\rho = 10^3$ kg/m³, $\nu = 0.0092 \times 10^{-4}$ m²/s

Shear frictin velocity, $$u_f = \sqrt{\frac{\tau_0}{\rho}} = \sqrt{\frac{10}{10^3}} = 0.1 \text{ m/s}$$

Roughness Reynolds number $$= \frac{u_f k}{\nu} = \frac{0.1 \times 0.2 \times 10^{-3}}{0.0092 \times 10^{-4}} = 21.44$$

$4 < \frac{u_f k}{\nu} < 100$. Hence the pipe surface behaves as in transition.

Example 9.4 *A pipe of diameter 120 mm is carrying water. The velocities at the pipe centre and 40 mm from the pipe centre are found to be 3m/s and 2.5 m/s respectively. Calculate the shearing stress.*

Solution.

Given: $k = 60$ mm, $u_{max} = 3$ m/s, $u = 2.5$ m/s at $y = 60 - 40 = 20$ mm

$$\frac{u_{max} - u}{u_f} = 5.75 \log_{10}\left(\frac{R}{y}\right)$$

$$\frac{3 - 2.5}{u_f} = 5.75 \log_{10}\left(\frac{60}{20}\right) = 2.743$$

$$u_f = 0.18225 \text{ m/s}$$

$$u_f = \sqrt{\frac{\tau_0}{\rho}}$$

$$0.18225 = \sqrt{\frac{\tau_0}{10^3}}$$

$$\tau_0 = 33.2 \text{ N/m}^2$$

Example 9.5 *A smooth pipe of 100 mm diameter and 1 km long is carrying water at the rate of 100 lit/s. The kinematic viscosity of water is 0.015 Stokes and coefficient of friction,* $f = \frac{0.0791}{(\text{Re})^{1/4}}$. *Calculate:*

(a) Loss of head, (b) Wall shearing stress, (c) Centre-line velocity, (d) Velocity and shear stress at 30 mm from the pipe wall, and (e) Thickness of laminar sub-layer.

Solution. Given: $D = 0.1$ m, $L = 100$ m, $Q = 0.01$ m³/s, $\nu = 0.015 \times 10^{-4}$ m²/s, $y = 0.03$ m

$$v = \frac{4Q}{\pi D^2} = \frac{4 \times 0.01}{g\pi \times (0.1)^2} = 1.273 \text{ m/s}$$

$$\text{Re} = \frac{vd}{\nu} = \frac{1.273 \times 0.1}{0.015 \times 10^{-4}} = 84,867$$

$$f = \frac{0.0791}{(84,867)^{1/4}} = 0.004637$$

(a) $h_f = \dfrac{4fLv^2}{D \times 2g} = \dfrac{4 \times 0.004637 \times 1000 \times (1.273)^2}{0.1 \times 2 \times 9.81} = 15.3 \text{ m}$

(b) $\tau_0 = \dfrac{f v^2 \rho}{2} = \dfrac{0.004634 \times (1.273)^2 \times 10^3}{2} = 3.755 \text{ N/m}^2$

(c) $u_f = \sqrt{\dfrac{\tau_0}{\rho}} = \sqrt{\dfrac{3.755}{10^3}} = 0.0613 \text{ m/s}$

$$\frac{u}{u_f} = 5.75 \log_{10}\left(\frac{u_f y}{\nu}\right) + 5.5$$

At $y = R$, $u = u_{\max}$

$$\frac{u_{\max}}{0.0613} = 5.75 \log_{10}\left(\frac{0.0613 \times 0.05}{0.015 \times 10^{-4}}\right) + 5.5 = 24.53$$

$$u_{\max} = 1.5 \text{ m/s}$$

(d) $\dfrac{\tau}{\tau_0} = \dfrac{r}{R}$, $r = 50 - 30 = 20\text{mm}$, R=50 mm

$$\tau = \tau_0 \times \frac{r}{R} = 3.755 \times \frac{20}{50} = 1.502 \text{ N/m}^2$$

$$\frac{u}{u_f} = 5.72 \log_{10}\left(\frac{u_f . y}{\nu}\right) + 5.5$$

$$\frac{u}{0.0613} = 5.75 \log_{10}\left(\frac{0.0613 \times 0.03}{0.015 \times 10^{-4}}\right) + 5.5 = 23.255$$

$$u = 1.425 \text{ m/s}$$

(e) $\delta' = \dfrac{11.6\nu}{u_f} = \dfrac{11.6 \times 0.015 \times 10^{-4}}{0.0613} = 2.8385 \times 10^{-4}\, m \simeq 0.284 \text{ mm}$

9.5.4 Velocity Distribution for Turbulent Flow in Terms of Average Velocity

Consider an elementary circular ring of radius r and thickness dr as shown in Fig. 9.6. The distance of the ring from pipe wall is $y = (R - r)$, where R = radius of pipe. Discharge through the ring.

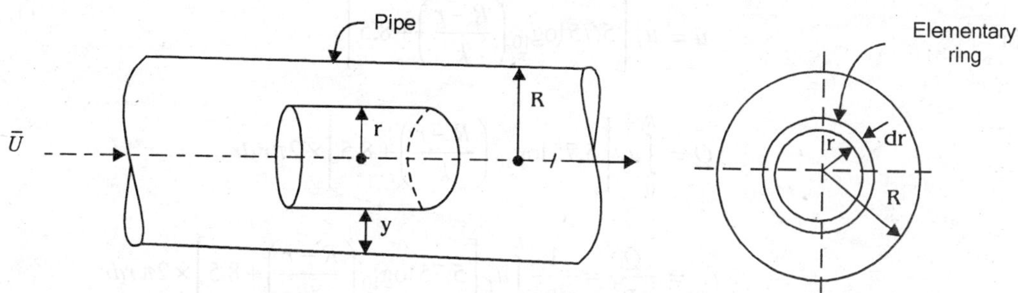

Fig. 9.6 Average velocity for turbulent flow

$$dQ = 2\pi r.dr \times u$$

Total discharge,

$$Q = \int dQ = \int_0^R u \times 2\pi r.dr$$

(a) **Smooth pipe:** For a smooth pipe, the velocity distribution is given by,

$$\frac{u}{u_f} = 5.75 \log_{10}\left(\frac{u_f \cdot y}{\nu}\right) + 5.5$$

$$u = u_f\left[5.75 \log_{10}\left\{\frac{u_f(R-r)}{\nu}\right\} + 5.5\right] \times u_f$$

$\therefore$

$$Q = \int_0^R \left[5.75 \log_{10}\left\{\frac{u_f(R-r)}{\nu}\right\} + 5.5\right] \times u_f \times 2\pi.dr$$

Average velocity,

$$\bar{U} = \frac{Q}{A} = \frac{Q}{\pi R^2}$$

$$\frac{\bar{U}}{u_f} = \frac{2}{R^2}\int_0^R \left[5.75 \log_{10}\left\{\frac{u_f(R-r)}{\nu}\right\} + 5.5\right] r\,dh$$

$$= 5.75 \log_{10}\frac{u_f R}{\nu} + 1.75 \qquad \text{... (9.29)}$$

(b) **Range Pipe:** For a rough pipe, we have

$$\frac{u}{u_f} = 5.75\log_{10}\left(\frac{y}{k}\right)+8.5 = 5.75\log_{10}\left(\frac{R-r}{k}\right)+8.5$$

or
$$u = u_f\left[5.75\log_{10}\left(\frac{R-r}{k}\right)+8.5\right]$$

$$Q = \int_0^R u_f\left[5.75\log_{10}\left(\frac{R-r}{k}\right)+8.5\right]\times 2\pi r\, dr$$

$$\bar{U} = \frac{Q}{\pi R^2} = \frac{1}{\pi R^2}\int_0^R u_f\left[5.75\log_{10}\left(\frac{R-r}{k}\right)+8.5\right]\times 2\pi r dr$$

$$\frac{\bar{U}}{u_f} = 5.75\log_{10}\left(\frac{R}{k}\right)+4.75 \qquad \text{... (9.30)}$$

Also from Eq. (9.27), we have

$$\frac{u}{u_f} = 5.75\log_{10}\left(\frac{u_f y}{\gamma}\right)+5.5$$

Combining with Eq. (9.29), we get

$$\frac{u}{u_f}-\frac{\bar{U}}{u_f} = \left[5.75\log_{10}\left(\frac{u_f y}{\gamma}\right)+5.71\right]-\left[5.75\log_{10}\frac{u_f R}{v}+1.75\right]$$

$$\frac{u-\bar{U}}{u_f} = 3.75-5.75\log_{10}\left(\frac{R}{y}\right)$$

or
$$\frac{u-\bar{U}}{u_f} = 5.75\log_{10}\left(\frac{y}{R}\right)+3.75 \qquad \text{... (1)}$$

Similarly from Eq. (9.28) and (9.30), we get

$$\frac{u}{u_f}-\frac{\bar{U}}{u_f} = \left[5.75\log_{10}\left(\frac{y}{k}\right)+8.51\right]-\left[5.75\log_{10}\left(\frac{R}{k}\right)+4.75\right]$$

$$\frac{u-\bar{U}}{u_f} = 5.75\log_{10}\left(\frac{y}{R}\right)+3.75 \qquad \text{... (2)}$$

Eqs. (1) and (2) are identical, the velocity distribution in both types of pipes is the same.

$$\therefore \quad \frac{u-\bar{U}}{u_f} = 5.75\log_{10}\left(\frac{y}{R}\right)+3.75 \qquad \text{... (9.31)}$$

Example 9.6 *Find the distance from the pipe wall at which the local velocity is equal to the average velocity for turbulent flow.*

Solution.

$$\frac{u-\bar{U}}{u_f} = 5.75\log_{10}\left(\frac{y}{R}\right)+3.75$$

for $u = \bar{U}$,

$$5.75\log_{10}\left(\frac{y}{R}\right)+3.75=0$$

$$\log_{10}\left(\frac{y}{R}\right) = -\frac{3.75}{5.75}=-0.652$$

$$\frac{y}{R} = 0.2228$$

or

$$y = 0.2228\ \text{R}$$

Example 9.7 *For turbulent flow in a pipe of 300 mm diameter, determine the discharge when the centre-line velocity is 3m/s and the velocity at point 100 mm from the centre is 1.5 m/s.*

Solution. $R = 150$ mm, $u_{max} = 3$ m/s, $y = R - h = 150 - 100 = 50$ mm

At $y = 50$ mm, $u = 1.5$ m/s

$$\frac{u_{max}-\bar{u}}{u_f} = 5.75\log_{10}\left(\frac{R}{y}\right)$$

$$\frac{3-1.5}{u_f} = 5.75\log_{10}\left(\frac{150}{50}\right)=2.743$$

$$u_f = 0.547 \text{ m/s}$$

$$\frac{u-\bar{U}}{u_f} = 5.75\log_{10}\left(\frac{y}{R}\right)+3.75$$

At $y = R$, $u = u_{max}$

$$\frac{u_{max}-\bar{U}}{u_f} = 3.75$$

$$\bar{U} = 0.95 \text{ m/s}$$

$$Q = \frac{\pi}{4}D^2\times\bar{U} = \frac{\pi}{4}(0.3)^2\times 0.95 = 0.067 \text{ m}^3/\text{s}$$

9.5.5 Velocity Distribution For Turbulent Flow in Smooth Pipes by Power Law

According to Nikuradse, the veocity distribution law for smooth pipes is given by,

$$\frac{u}{u_{max}} = \left(\frac{y}{R}\right)^{\frac{1}{n}}$$

where the exponent $\frac{1}{n}$ depends on Reynolds number.

$$\frac{1}{n} = \frac{1}{6} \text{ for Re} = 4000$$

$$= \frac{1}{7} \text{ for Re} = 1.1 \times 10^5$$

$$= \frac{1}{10} \text{ for Re} \geq 2 \times 10^6$$

For $n = \frac{1}{7}$, $$\frac{u}{u_{max}} = \left(\frac{y}{R}\right)^{1/7}$$

Eq. (9.33) is known as the $\frac{1}{7}$ th power law of velocity distribution for smooth pipes.

9.6 FRICTION COEFFICIENTS FOR SMOOTH AND ROUGH PIPES

The loss of head due to friction in pipes is given by Darcy-Weisbach equation,

$$h_f = \frac{4\,fLv^2}{D \times 2g}$$

On the basis of dimensional anlaysis, it can be shown that the coefficient of friction f depends on the Reynolds number $\left(\frac{\rho v D}{\mu}\right)$ and the ratio $\left(\frac{k}{D}\right)$.

Thus $$f = \phi\left[\left(\frac{\rho v D}{\mu}\right), \frac{k}{D}\right]$$

where k = average height of pipe wall roughness protrusions. The term $\frac{k}{D}$ is called relative roughness.

1. Laminar flow

The coefficient of friction 'f' for laminar flow in pipes is given by,

$$f = \frac{16}{\text{Re}} \quad \text{... (9.34)}$$

2. Turbulent flow

(a) **Smooth pipes:** The Blasisus empirical relation is,

$$f = \frac{0.0791}{(\text{Re})^{1/4}} \text{ for } 4000 < \text{Re} < 10^5 \quad \text{... (9.35)}$$

For Re > 10^5, Eq. (9.29) is,

$$\frac{\bar{U}}{u_f} = 5.75\log_{10}\frac{u_f R}{\nu} + 1.75 \qquad \text{... (1)}$$

and
$$f = \frac{2\tau_0}{\rho v^2} = \frac{2\tau_0}{\rho \bar{U}^2} = \frac{2}{\bar{U}^2}\times u_f^2 \qquad \left[\because v = \bar{U} \text{ and } u_f = \sqrt{\frac{\tau_0}{\rho}}\right]$$

or
$$u_f^2 = \frac{f\bar{U}^2}{2}$$

or
$$u_f = \bar{U}\sqrt{\frac{f}{2}}$$

Eq. (1) becomes,

$$\frac{\bar{U}}{\bar{U}\sqrt{\frac{f}{2}}} = 5.75\log_{10}\left(\frac{\bar{U}}{\nu}\sqrt{\frac{f}{2}}\right)R + 1.75$$

$$\frac{1}{\sqrt{\frac{f}{2}}} = 5.75\log_{10}\left(\frac{\bar{U}R}{\nu}\sqrt{\frac{f}{2}}\right) + 1.75$$

Substituting $R = \frac{D}{2}$, and simplifying, we get

$$\frac{1}{\sqrt{4f}} = 2.03\log_{10}\left(\frac{\bar{U}D}{\nu}\sqrt{4f}\right) - 0.91$$

But
$$\frac{\bar{U}D}{\nu} = \text{Re}$$

∴
$$\frac{1}{\sqrt{4f}} = 2.03\log_{10}(\text{Re}\sqrt{4f}) - 0.91 \qquad \text{... (9.36)}$$

Eq. (9.36) is valid upto Re = 4× 10^6

Von-Karman-Prandtl resistance equation

For turbulent flow in smooth pipes, this equation is given by

$$\frac{1}{\sqrt{4f}} = 2.0\log_{10}(\text{Re}\sqrt{4f}) - 0.8 \text{ for } 5\times 10^4 < \text{Re} < 4\times 10^7 \qquad \text{...(9.37)}$$

Nikuradse empirical relation

$$f = 0.0008\frac{0.05525}{(\text{Re})^{0.237}} \quad \text{... (9.38)}$$

(b) Rough pipes

For turbulent flow in rough pipes, the mean velocity $\bar{U}$ of flow is given by,

$$\frac{\bar{U}}{u_f} = 5.75\log_{10}\left(\frac{R}{k}\right) + 4.75$$

But $$u_f = \bar{U}\sqrt{\frac{f}{2}}$$

$$\therefore \quad \frac{\bar{U}}{\bar{U}\sqrt{\frac{f}{2}}} = 5.75\log_{10}\left(\frac{R}{k}\right) + 4.75$$

Simplying, we get

$$\frac{1}{\sqrt{4f}} = 2.03\log_{10}\left(\frac{R}{k}\right) + 1.68 \quad \text{... (9.39)}$$

According to Nikuradse,

$$\frac{1}{\sqrt{4f}} = 2.0\log_{10}\left(\frac{R}{k}\right) + 1.74 \quad \text{... (9.40)}$$

(c) Commerical pipes

The empirical equation developed by Colebrook and White is given by,

$$\frac{1}{\sqrt{f_1}} - 2.0\log_{10}\left(\frac{R}{k}\right) = 1.74 - 2.0\log_{10}\left[1 + 18.7\frac{(R/k)}{\text{Re}\sqrt{f_1}}\right] \quad \text{... (9.41)}$$

where f_1 (friction factor) = $4f$ (friction coefficient)

Example 9.8 *Water is flowing at the rate of 0.8 m³/s in a rough pipe of diameter 0.75 m and length 5 km. If the average height of rougness is 0.5 mm, calculate the power required to maintain this flow.*

Solution. Given: $Q = 0.8$ m³/s, $D = 0.75$ m, $L = 5$ km, $k = 0.5$ mm

$$h_f = \frac{4fLv^2}{D \times 2g}$$

For a rough pipe, $$\frac{1}{\sqrt{4f}} = 2.0\log_{10}\left(\frac{R}{k}\right)+1.74$$

$$= 2.0\log_{10}\left(\frac{0.375}{0.5\times10^{-5}}\right)+1.74 = 7.49$$

$$4f = 0.01782$$

$$v = \frac{4Q}{\pi D^2} = \frac{4\times0.8}{\pi(0.75)^2} = 1.811 \text{ m/s}$$

$$h_f = \frac{0.0178202\times5000\times(1.811)^2}{0.75\times2\times9.81} = 19.859 \text{ m}$$

Power required, $P = \gamma Q\, h_f = 9.81 \times 0.8 \times 19.859 = 155.85$ kW

Example 9.9 *A smooth pipeline of 100 mm diameter carries 3m³/min of water at 20°C with kinematic viscosity of 0.0098 Stokes. Calculate: (a) Friction factor, (b) Maximum velocity, and (c) Shear stress at the boundary.*

Solution. Given $D = 0.1$ m, $Q = 0.05$ m³/s, $\nu = 0.0098 \times 10^{-4}$ m²/s

Average velocity, $$\bar{U} = \frac{4Q}{\pi D^2} = \frac{4\times0.05}{\pi(0.1)^2} = 6.366 \text{ m/s}$$

Reynolds number, $$Re = \frac{\bar{U}D}{\nu} = \frac{6.366\times0.1}{0.0098\times10^{-4}} = 649{,}612 > 10^5$$

For smooth pipe, according to Von-Karman-Prandtl equation,

$$\frac{1}{\sqrt{4f}} = 2.0\log_{10}(\text{Re}\sqrt{4f}) - 0.8 \text{ for } 5\times10^4 < \text{Re} < 4\times10^7$$

$$= 2.0\log_{10}(649{,}612\sqrt{4f}) - 0.8$$

$$= 11.6253 + \log_{10}(4f) - 0.8$$

$$\frac{1}{\sqrt{4f}} - \log_{10}(4f) = 10.8253$$

(a) Friction factor, $f_1 = 4 \times$ coefficient of friction $f = 4f$

$$\therefore \quad \frac{1}{\sqrt{f_1}} - \log_{10} f_1 = 10.8253$$

Solving by hit and trial, we have

f_1	$\frac{1}{\sqrt{f_1}} - \log_{10} f_1$
0.01	12
0.009	12.587
0.011	11.493
0.012	11.049
0.013	10.657
0.0125	10.847

we take $f_1 = 0.0125$

(b) $f = \frac{f_1}{4} = \frac{0.0125}{4} = 0.00312$

$$u_f = \frac{\bar{U}\sqrt{f}}{2} = \frac{6.366 \times \sqrt{0.00312}}{2} = 0.178 \text{ m/s}$$

$$u = u_f\left[5.75\log_{10}\left(\frac{u_f \times y}{\nu}\right) + 5.5\right]$$

At $y = R$, $u = u_{max}$

$$u_{max} = 0.178\left[5.75\log_{10}\left(\frac{0.178 \times 0.05}{0.0098 \times 10^{-4}}\right) + 5.5\right]$$

$$= 5.03 \text{ m/s}$$

(c) Boundary shear stress.

$$u_f = \sqrt{\frac{\tau_0}{\rho}}$$

$$\tau_0 = \rho u_f^2 = 10^3 \times (0.178)^2 = 31.684 \text{N/m}^2$$

Example 9.10 *A rough plastic pipe 0.6 m in diameter and 300 m long carries water with a velocity of 3m/s. Its absolute roughness is 0.25 mm and kinematic viscosity of water is 0.9 centistokes. Determine:*

(a) Whether the flow is laminar or turbulent? (b) Head lost in friction. For laminar flow, $f = \frac{64}{\text{Re}}$, *and for turbulent flow,* $\frac{1}{\sqrt{f}} = 2\log_{10}\frac{R}{k} + 1.74$ *where* f = *frictional factor.*

Solution.

Given : D = 0.6 m, L = 300 m, v = 3 m/s, k = 0.25 mm, ν = 0.9 cst

(a) $\text{Re} = \dfrac{vD}{\nu} = \dfrac{3 \times 0.6}{0.9 \times 10^{-6}} = 2 \times 10^6 > 2000$

The flow is turbulent.

(b) $\dfrac{1}{\sqrt{f}} = 2 \log_{10}\left(\dfrac{0.3}{0.25 \times 10^{-3}}\right) + 1.74 = 7.898$

$f = 0.016$

$$h_f = \frac{fLv^2}{D \times 2g} = \frac{0.0160 \times 300 \times 3^2}{0.6 \times 2 \times 9.81} = 3.67 \text{ m}$$

Example 9.11 *A hydrodynamically smooth pipe carries water at the rate of 300 lit/s with a head loss* of 3 m in 100 m length of pipe. Find the pipe diameter. Take r = 1000 kg/m^3, n = 10^{-6} *m*^{2}*/s, Re =* $\dfrac{\rho v D}{\gamma}, h_f = \dfrac{fLv^2}{D \times 2g}$, *and* $f = 0.0032 + \dfrac{0.221}{(\text{Re})^{0.237}}$

Solution.

Given $Q = 0.3\text{m}^3/\text{s}$, $h_f = 3\text{m}$, $L = 100$ m

$$h_f = \frac{fLv^2}{D \times 2g}$$

$$3 = \frac{f \times 100 \times v^2}{D \times 2 \times 9.81}$$

or

$$f = \frac{0.5886\,D}{v^2}$$

$$Q = \frac{\pi}{4} D^2 v = 0.3$$

$$v = \frac{0.382}{D^2}$$

$$f = 0.0032 + \frac{0.221}{(\text{Re})^{0.237}}$$

$$\text{Re} = \frac{1000 \times v \times D}{10^{-6}} = 10^9 \times \frac{0.382}{D^2} \times D = \frac{382 \times 10^6}{D}$$

$$\frac{0.5886\,D}{\left(\dfrac{0.382}{D^2}\right)^2} = 0.0032 + \frac{0.221}{\left(\dfrac{382 \times 10^6}{D}\right)^{0.237}}$$

$4.034\, D^4 - 0.105\, D^{0.237} - 0.0032 = 0$

Solving by hit and trial, we get

$$\text{D} = 0.308 \text{ m}$$

SUMMARY

1. The turbulent flow is characterised by random, irregular and haphazard movement of fuid particles.
2. The flow in a pipe is turbulent when Re > 4000.
3. The shear is turbulent flow is mainly due to momentum transfer.
4. Frictional head loss in a pipe flow is given by:

$$h_f = \frac{4fLv^2}{D \times 2g}, \text{ where } f = \text{friction coefficient}$$

$$= \frac{f_1 L v^2}{D \times 2g}, \text{ where } f_1 = 4f = \text{friction factor}$$

5. Coefficient of friction, $f = \dfrac{2\tau_0}{\rho v^2}$
6. Charactersistics of turbulent flow:

Velocity in turbulent flow, $u = \bar{u} + u'$

where u = instantaneous velocity

$\bar{u}$ = temporal mean velocity

u' = velocity fluctuations

$$\bar{u} = \frac{1}{T}\int_0^T u.dt, \bar{v} = \frac{1}{T}\int_0^T v.dt, \bar{w} = \frac{1}{T}\int_0^T w\,dt$$

Temporal mean of $\overline{u'} = \dfrac{1}{T}\displaystyle\int_0^T u'.dt = 0$

$$\overline{u'} = \overline{v'} = \overline{w'} = 0$$

$$\text{Intensity of turbulence} = \frac{1}{U_\infty}\sqrt{\frac{1}{3}[(\overline{u'})^2 + (\overline{v'})^2 + (\overline{w'})^2]}$$

Isotropic turbulence is defined as that turbulence in which the mean-square velocity fluctuations in three coordinate direction are equal to each other.

$$\overline{u'} = \overline{v'} = \overline{w'}$$

7. Total shear stress in turbulent flow is given by

$$\tau = \tau_l + \tau_t = u\frac{du}{dy} + \eta\frac{d\bar{u}}{dy}$$

where η = eddy viscosity

8. Kinematic eddy viscosity, $\varepsilon = \frac{\eta}{\rho}$

9. Reynolds expression for turbulent shear stress,

$$\bar{\tau} = \overline{\rho u' v'}$$

10. According to Prandtl's mixing length hypothesis

$$\bar{\tau} = \mu \frac{du}{dy} + \rho l^2 \left(\frac{du}{dy} \right)^2$$

where l = average distance between two layers in transverse direction.

11. Prandtl's universal velocity distribution equation:

$$u = u_{max} + 2.5 u_f \ln \left(\frac{y}{R} \right)$$

where $u_f = \sqrt{\frac{\tau_0}{\rho}}$

12. Velocity defect = $u_{max} - u = 5.75\, u_f \log_{10} \left(\frac{R}{y} \right)$

13. If k = average height of irregularities of surface boundary.

δ' = thickness of laminar sub-layer

when $\frac{k}{\delta'} < 0.25$, the boundary is smooth

$\frac{k}{\delta'} > 6.0$, the boundary is rough

$0.25 < \frac{k}{\delta'} < 6.0$, the boundary is in transition.

14. Rougness Reynolds number $\frac{u_f k}{\nu}$

when $\frac{u_f k}{\nu} < 4$, the boundary is smooth

$\frac{u_f k}{\nu} > 100$, the boundary is rough

$4 < \frac{u_f k}{\nu} < 100$, the boundary is in transition.

15. Velocity distribution in turbulent flow:

$$\frac{u}{u_f} = 5.75 \log_{10} \left(\frac{u_f y}{\nu} \right) + 5.5 \text{ for smooth pipe}$$

$$= 5.75 \log_{10} \left(\frac{y}{k} \right) + 8.5 \text{ for rough pipe}$$

16. Velocity distribution in terms of average velocity:

$$\frac{\bar{U}}{u_f} = 3.75\log_{10}\left(\frac{u_f R}{\nu}\right) + 1.75 \quad \text{for smooth pipe}$$

$$\frac{\bar{U}}{u_f} = 5.75\log_{10}\left(\frac{R}{k}\right) + 4.75 \quad \text{for rough pipe}$$

$$\frac{u - \bar{U}}{u_f} = 5.75\log_{10}\left(\frac{y}{R}\right) + 3.75 \quad \text{for both smooth and rough pipes}$$

18. Coefficient of friction:

$$f = \frac{16}{\text{Re}} \quad \text{for laminar flow}$$

$$= \frac{0.0791}{(\text{Re})^{1/4}} \quad \text{for turbulent flow in smooth pipes (for } 400 < \text{Re} < 10^5)$$

$$= 0.008 + \frac{0.5525}{(\text{Re})^{0.237}} \quad \text{for } 10^5 < \text{Re} < 4\times 10^7$$

$$\frac{1}{\sqrt{4f}} = 2.0\log_{10}(\text{Re}\sqrt{4f}) - 0.8 \text{ for } 5\times 10^4 < \text{Re} < 4\times 10^7$$

$$\frac{1}{\sqrt{4f}} = 2.0\log_{10}\left(\frac{R}{k}\right) + 1.74 \text{ for rough pipes}$$

Short Answer Type Questions

1. Define turbulance and turbulent flow
2. Differentiate between coefficient of friction and friction factor.
3. Write the expression for frictional head loss in a pipe.
4. Write the expression for coefficient of friction in terms of wall shear stress in a pipe.
5. What are the characteristics of a turbulent flow?
6. Draw the velocity distribution in a pipe for laminar and turbulent flows.
7. Define shear velocity for turbulent flow in a circular pipe.
8. What is eddy velocity?
9. What is the condition for temporal mean of velocity fluctuations?
10. Define isotropic turbulence.
11. Define degree of turbulence.
12. Write theReynolds expression for turbulent shear stress.
13. Write the expression for shear stress on the basis of Prandatl mixing length theory.
14. Write the expression for Prandtl's velocity distribution in pipes for turbulent flow.

15. Define velocity defect.

16. How do you differentiate between smooth and rough pipes?

17. Write the expressions for velocity distribution for turbulent flow in smooth and rough pipes.

18. Write expression for velocity distribution in terms of average velocity for smooth and rough pipes.

19. What is the power law for velocity distribution in pipes for turbulent flows?

20. What are the parameters on which coefficients of friction depend for flow in smooth and rough pipes.

EXERCISES

1. Water flows in a pipe line having average height of protrusions on the pipe surface as 0.15 mm. The shear stress developed is 4.9 N/m^2 and kinematic viscosity of water is 0.01×10^{-4} m^2/s. Find the type of boundary of Pipe. [**Ans.** in transition]

2. A rough pipe is of diameter 0.1 m. The velocity at a point 30 mm from the pipe wall is 25 % more than the velocity at a point 10 mm from the pipe wall. Calculate the average height of roughness. [**Ans.** 3.714 mm]

3. Calculate the wall shear stress in a pipe of diameter 0.1 m carrying water. The velocity of water at the pipe centre is 2m/s and at 30 mm from the pipe centre is 1.5 m/s. Assume turbulent flow. [**Ans.** 47.67 N/m^2]

4. For turbulent flow in a pipe of diameter 0.3 m, calculate the discharge when the centre line velocity is 2m/s and 1.6 m/s at a point 0.1 m from the centre [**Ans.** 0.1027 m^3/s]

5. Water is flowing through a rough pipe of diameter 0.5 m and length 4 km at the rate of 0.5 m^3/s. Calculate the power required to maintain this flow. The average height of roughness may be taken as 0.4 mm. [**Ans.** 241.13 kW]

6. A rough pipe of diameter 0.4 m and length 1.0 km carries water at the rate of 0.4 m^3/s. The pipe wall roughness is 0.012 mm. Calculate the coefficient of friction, wall shear stress, maximum velocity and velocity at a distance of 0.15 m from the pipe wall.

[**Ans.** 0.0024, 3.18 m/s, 12. N/m2, 3.6 m/s 3.52 m/s]

7. Water is flowing through a rough pipe of diameter 0.6 m at the rate of 0.6 m^3/s. The wall roughness is 3 mm. Calculate the power lost per km length of pipe

8. A smooth pipe of diameter 0.3 m and length 600 m carries water at the rate of 0.04 m^3/s. Calculate the head lost due to friction, wall shear stress, maximum velocity, and thickness of laminar sub-layer. Take kinematic viscosity of water as 0.018 Stokes.[**Ans.** 0.589 m, 0.722 N/m^2, 0.665 m/s, 0.78 mm]

10

FLUID MEASURMENTS

10.1 INTRODUCTION

Fluid measurments is an important task to be performed in the study of fluid mechanics. In this chapter, we shall study the methods for the measurment of pressure, velocity, discharge, and viscosity of fluids.

10.2 MEASUREMENT OF PRESSURE

The pressure of a fluid may be measured by the following devices:

(a) Manometers, and *(b)* Elastic elements

***(a)* Manometers:** A manometer may be defined as a device used for measuring the pressure at a point in a fluid by balancing the column of fluid by the same or another column of liquid. The pressure at a depth h below the free-surface in a liquid of specific weight γ is given by, $p = \gamma h$. This hydrostatic law of pressure variation is utilised in measuring pressure. The process of measuring pressure consists of installing a glass tube at or near the point of interest. The height of the liquid that rises in a glass tube (commonly known as a piezometer) measured above the point, is a measure of the pressure at the point. The piezometers when used for pressure measurements are called manometers.

The manometers may be classified as:

(i) Simple manometers

(ii) Differential manometers

(iii) Micro-Manometers

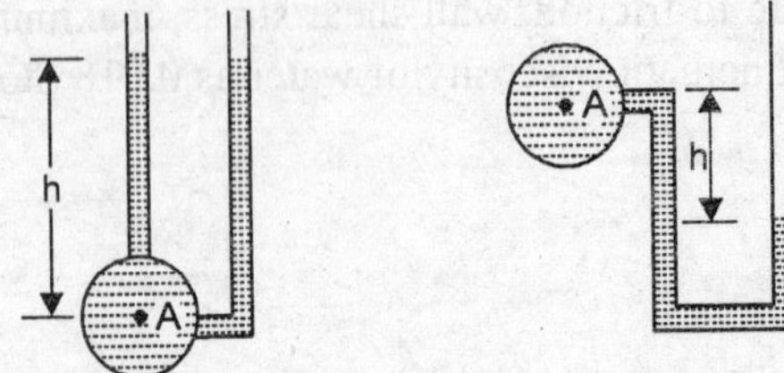

(a) For positive pressure at A (b) For negative pressure at A

Fig. 10.1 Simple Manometers

1. Piezometers Fig. 10.1 *(a)* shows a simple manometer for measuring positive pressure and Fig. 10.1 *(b)* for measuring negative pressure at A . $p_A = \gamma h$, where γ = specific weight of the fluid. This can be used for measuring small and moderate pressure because as the pressure gets higher the length of piezometer becomes correspondingly larger.

A piezometer tube is not suitable for measuring negative pressure, because air will enter in the pipe through the tabe.

2. U-tube Manometer A piezometer cannot be used when large gas pressure is to be measured, because a gas forms no free atmospheric surface. In such a case a U-tube manometer may be used. It consists of a glass tube bent in U-shape, one point of which is connected to a point at which pressure is to be measured and other end remains open to the atmosphere as shown in fig 10.2.

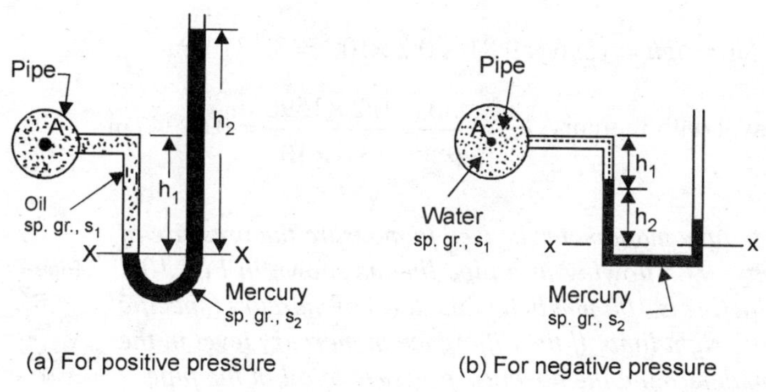

Fig. 10.2 U-tube manometers.

The U-tube manometer for measuring positive pressure is shown in fig. 10.2 *(a)*. Let A be the point in the pipe at which pressure is to be measured and X-X is the datum line as shown in Fig. 10.2 *(a)*.

Let h_1 = height of the light liquid like oil, in the left limb of U-tube, above the datum line.

h_2 = height of the heavy liquid like mercury, in the right limb of U-tube, above the datum line.

p_A = pressure in the pipe

γ_1 = specific weight of light liquid.

s_1, s_2 = specific gravity of light and heavy liquid respecvely.

For the equlibrium of pressure head in the two limbs at X-X, we have

$$\frac{p_A}{\gamma_1} + h_1 s_1 = h_2 s_2$$

$$\therefore \quad \frac{p_A}{\gamma_1} = h_2 s_2 - h_1 s_1 \quad \text{... (10.1a)}$$

For measuring negative pressure, as shown in Fig. 10.2 *(b)*, we have

$$\frac{p_A}{\gamma_1} + h_1 s_1 + h_2 s_2 = 0$$

$$\therefore \quad \frac{p_A}{\gamma_1} = -(h_1 s_1 + h_2 s_2) \quad \text{... (10.1 } b\text{)}$$

Example 10.1 *A manometer is used to measure the pressure drop for flow of water through a pipe. The difference in level was found to be 20 cm. If the manometer fluid is of density 1596 kg/m³, find the pressure drop in S.I. units. If the manometeric fluid is changed to mercury of specific gravity 13.6, what will be the difference in level?*

Solution.

Pressure drop, $\Delta h = \rho g h = 1596 \times 9.81 \times 0.2 \times 10^{-3} = 3.131$ kPa

Difference in level with mercury, $h_m = \frac{h\rho}{\rho m} = \frac{0.2 \times 1596}{13.6 \times 10^3} = 0.02347\,\text{m}$

Example 10.2 *A U-tube manometer is used to measure the pressure of oil of specific gravity 0.85 flowing in a pipe line as shown in Fig. 10.3 The centre of the pipeline is 100 mm below the level of mercury (specific gravity = 13.6) in the right limb. If the difference of mercury level in the two limb is 150 mm, determine the absolute pressure of oil in the pipe.*

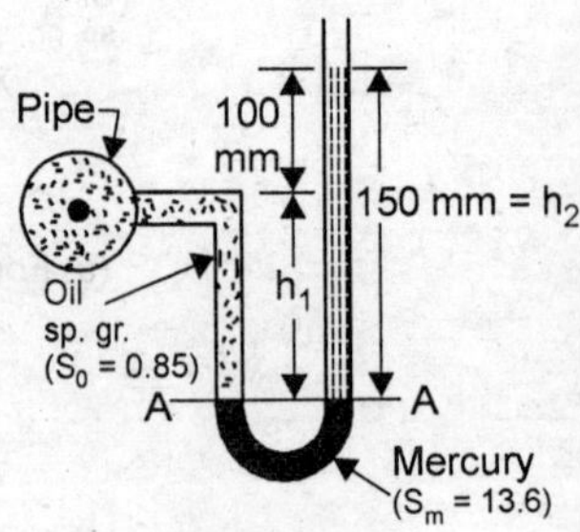

Fig. 10.3

Solution Let p_g = gauge pressure in pipe.

$$\frac{p_g}{\gamma_w} + s_o h_1 - s_m h_2 = 0$$

$$p_g = \gamma_w(-s_o h_1 + s_m h_2)$$

$$= 10^3 \times 9.81 (-0.85 \times 50 + 13.6 \times 150) \times 10^{-3}$$

$$= 19.595 \text{ kN/m}^2$$

$$p_a = p_{atm} + p_g = 1.01325 \times 10^5 + 19.595 \times 10^3 = 1.2092 \times 10^5 \text{ N/m}^2$$

Example 10.3. *A U-tube manometer containing mercury was used to find the negative pressure in a pipe containing water as shown in Fig. 10.4. The right limb was open to the atmosphere. Find the vaccum pressure in the pipe, if the difference of mercury level in the two limbs was 100 mm and height of water in the left limb from the centre of pipe was 40 mm below, Also calculate the absolute pressure in the pipe.*

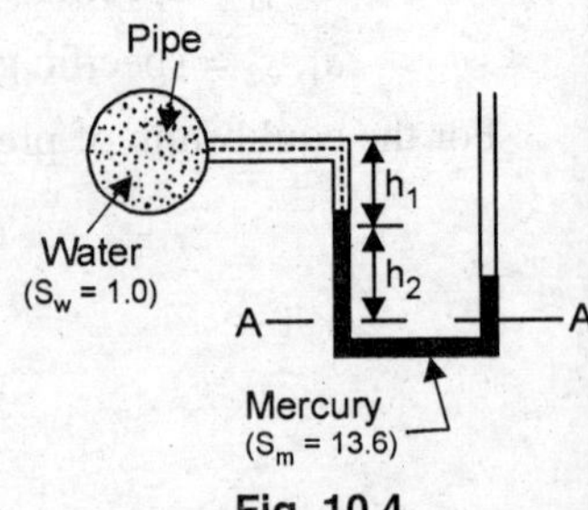

Fig. 10.4

Solution.

Let $\quad p_g$ = gauge pressure in pipe

$h_1 = 40$ mm, $h_2 = 100$ mm

$$\frac{p_g}{\gamma_w} + s_w h_1 + s_m h_2 = 0$$

$$p_g = -\gamma_w (s_w h_1 + s_m h_2)$$
$$= -10^3 \times 9.81\ (1.0 \times 40 + 13.6 \times 100) \times 10^{-3}$$
$$= -13734 \text{ N/m}^2 \text{ or } 13.734 \text{ kPa (vacuum)}$$
$$p_a = p_{atm} + p_g = 1.01325 \times 10^5 - 0.13734 \times 10^5$$
$$= -0.875951 \times 10^5 \text{ N/m}^2 \text{ or } 87.5791 \text{ kPa}$$

Example 10.4 *Fig. 10.5 shows a single column manometer connected to a pipe containing liquid of specific gravity 0.85. The ratio of area of reservoir to that of the limb is 100. Find the pressure in the pipe.*

Solution

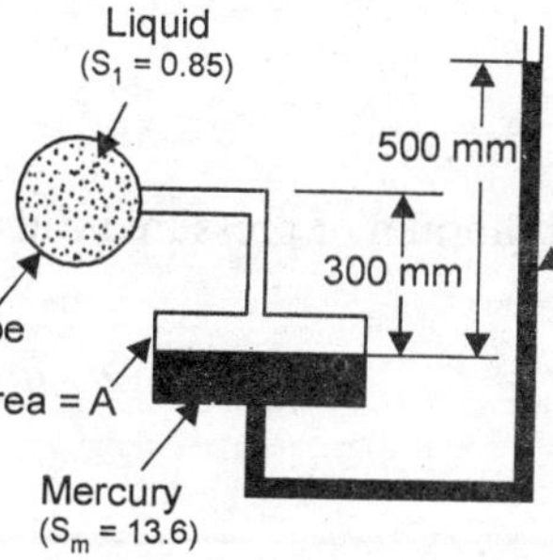

Given: $s_1 = 0.85$, $s_2 = 13.6$, $A/a = 100$,

$h_1 = 300$ mm, $h_2 = 500$ mm

Let Δh = fall of liquid level in reservoir.

Then $A \times \Delta h = a \times h_2$

$$\Delta h = \frac{a}{A} h_2$$

Now, $h + (h_1 + \Delta h)s_1 = (h_2 + \Delta h)s_2$

$$h = \Delta h(s_2 - s_1) + h_2 s_2 - h_1 s_1$$

$$= -\frac{a}{A} h_2 (s_2 - s_1) + h_2 s_2 - h_1 s_1$$

$$= \frac{1}{100} \times 500\ (13.6 - 0.85) + 500 \times 13.6 - 300 \times 0.85$$

$= 6608.75$ mm of water of 6.60875 m of water

$$p = \gamma_w h = 9.81 \times 10^3 \times 6.60875 = 64.83 \text{ kPa}$$

Fig. 10.5

Example 10.5 *Fig. 10.6 (a) shows a conical vessel having its outlet at A to which a U-tube manometer is connected. The reading of the manometer shown in the figure is when the vessel is empty. Find reading of the manometer when the vessel is completely filled with water.*

Solution.

Empty vessel (Fig. 10.6 *a*)

Let h_1 = height of water above *x-x*.

Specific gravity of water, $s_1 = 1.0$

Specific gravity of mercury, $s_2 = 13.6$

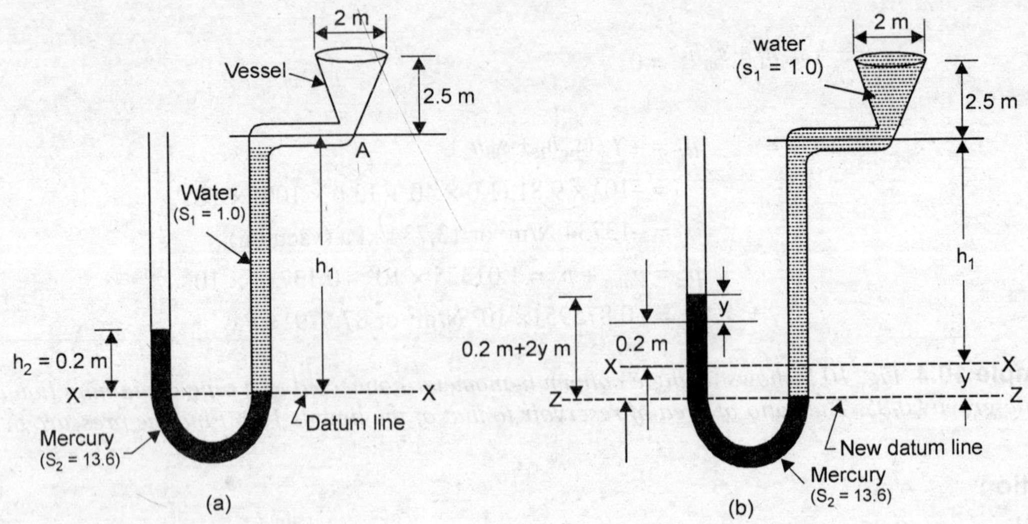

Fig. 10.6

For equilibrium of pressure heads at *x*-*x* datum line, we get

$$h_1 s_1 = h_2 s_2$$

$$h_1 \times 1.0 = h_2 \times 13.6 = 0.2 \times 13.6 = 2.72$$

$$h_1 = 2.72 \text{ m}$$

Vessel full of water (Fig. 10.6 *b*)

Let y = height by which mercury level go down from *x*-*x* in right limb, *m*

= height by which mercury level go up from *x*-*x* is left limb, *m*

Now the new datum line is $z - z$.

For equilibrium of pressure heads at *z*–*z*, we have

$$(0.2 + 2y) \times 13.6 = (h_1 + y + 2.5) \times 1.0$$

$$2.72 + 27.2\,y = 2.72 + y + 2.5$$

$$y = 0.0954 \text{ m} \quad \text{or} \quad 95.4 \text{ mm}$$

Thus the reading of manometer = 95. 4 mm

(ii) Differential manometer

A differential manometer is used to measure the difference in pressure at two points. The differential manometer shown in Fig. 10.7 consists of a glass U-tube connected to the points between which the pressure differenc is to be determined. A heavier liquid which is immiscible with the other two liquids. occupies the lower part of the U-tube. The petcocks at *C* and *D* permit the expulsion of air. Due to difference in pressure at points *A* and *B* there exists a difference in level *h* in the manometer liquid.

At plane x–x, pressure is same.

$$\frac{p_A}{\gamma_w} - s_1 z_1 = \frac{p_B}{\gamma} - s_2 z_2 + sh$$

where γ_w = specific weight of water

$$\frac{p_A - p_B}{\gamma_w} = sh + s_1 z_1 - s_2 z_2 \quad \text{... (10.2 a)}$$

For $s_1 = s_2 = 1.0$ and $s = 13.6$

$$\frac{p_A - p_B}{\gamma_w} = 12.6\, h \quad \text{... (10.2 b)}$$

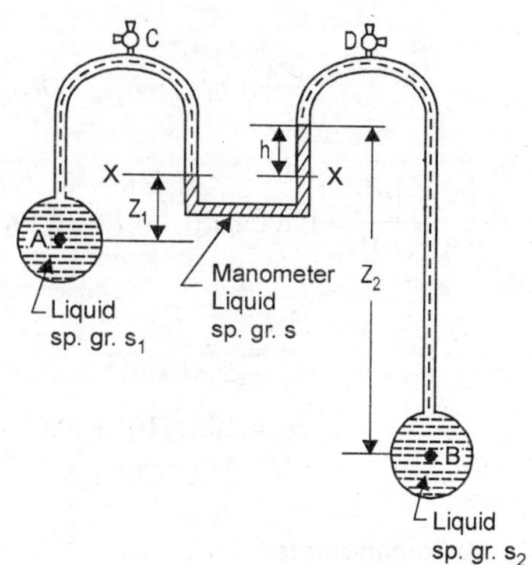

Fig 10.7 Differential manometer

Example 10.6 *A differential manometer connected at two points A and B in a pipe containing an oil of specific gravity 0.9 shows a difference in mercury levels of 150 mm. Find the difference in pressure at the two points.*

Solution.

Given: $h = 150$ mm

The differential manometer is shown in Fig. 10.8.

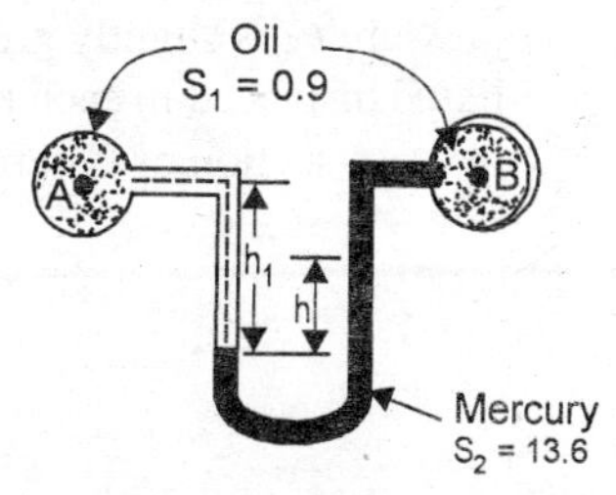

Fig 10.8

$$\frac{p_A}{\gamma_w} + s_1 h_1 - s_2 h - s_1 (h_1 - h) = \frac{p_B}{\gamma_w}$$

$$\frac{p_A - p_B}{\gamma_w} = s_2 h + s_1 (h_1 - h) - s_1 h_1$$

$$= s_2 h - s_1 h$$

$$= (s_2 - s_1)\, h$$

$$p_A - p_B = \gamma_w (s_2 - s_1) h$$

$$= 10^3 \times 9.81 \times (13.6 - 0.9) \times 150 \times 10^{-3}$$

$$= 18.688 \text{ kPa}$$

Example 10.7 Fig. 10.9 shows a differential manometer connected at two points A and B. At A air pressure is 100 kPa. Find the abosolute pressure at *B*.

Solution:

Given : $h_1 = 650$ mm, $h_2 = 150$ mm, $h_3 = 250$ mm

$s_w = 1$, $s_0 = 0.85$, $s_m = 13.6$, $p_A = 100$ kPa

$$\frac{p_A}{\gamma_w} + h_1 s_w - h_2 s_m - h_3 s_0 = \frac{p_B}{\gamma_w}$$

$$\frac{100 \times 10^3}{9.81 \times 10^3} + 0.65 \times 1.0 - 0.15 \times 13.6 - 0.25 \times 0.85$$

$$= \frac{p_B}{9.81 \times 10^3}$$

$$p_B = 100 \times 10^3 + (0.65 - 2.04 - 0.2125) \times 9.81 \times 10^3$$

$$= 84.279 \text{ kPa}$$

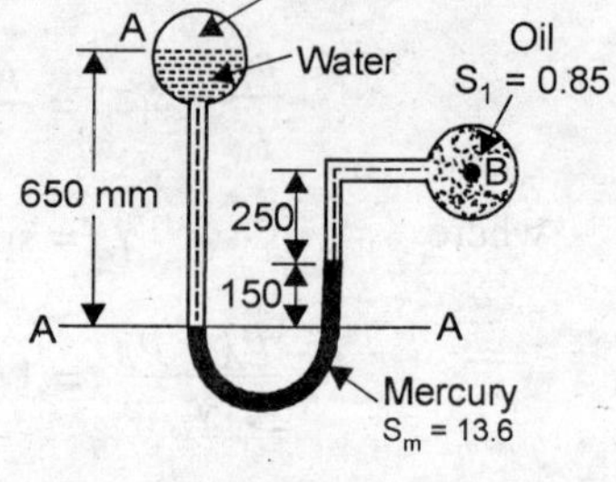

Fig. 10.9

(iii) Micromanometer

(a) The micromanometers are used for measuring small difference of pressure. The micromanometer shown in Fig. 10.10 utilises two manometer liquids which are immisicibel with each other and also with the fluid whose pressure difference is to be measued. The heavier liquid fills the lower part of the U-tube upto *O–O* and then the lighter liquid is added on both sides, filling the tanks *C* and *D* upto the level X-*X*. The fluid whose pressure difference is to be determined fills the space above *X–X*. When the pressure p_A is slightly greater than p_B, the liquid levels will be as shown in Fig. 10.10. The volume of liquid displaced in each tank is equal to the volume of liquid displaced in the U-tube. If '*a*' is the area of cross-section of U-tube and *A* that of the tank, then

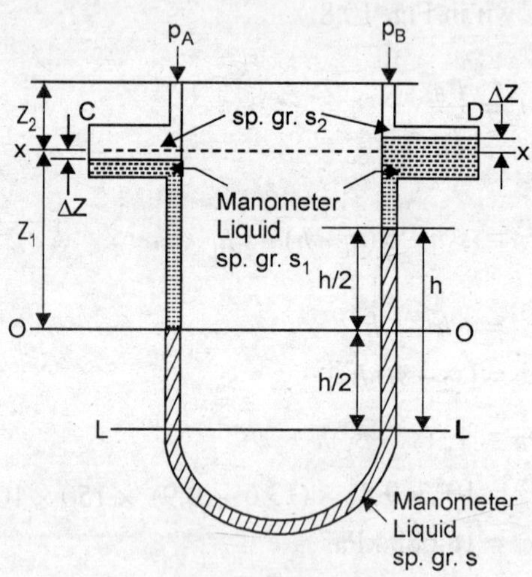

Fig 10.10 Micromanometer

$$A \times \Delta z = \frac{h}{2} \times a \quad \text{or} \quad \Delta z = \frac{ha}{2A}$$

Equatting the pressure along *L-L* in U-tube,

$$\frac{p_A}{\gamma_w} + s_2(z_2 + \Delta z) + s_1\left(z_1 - \Delta z + \frac{h}{2}\right) = \frac{p_B}{\gamma_w} + s_2(z_2 + \Delta z) + s_1\left(z_1 - \frac{h}{2} + \Delta z\right) + sh$$

$$\frac{p_A - p_B}{\gamma_w} = h\left[s - s_1\left(1 - \frac{a}{A}\right) - s_2\frac{a}{A}\right] \quad \text{... (10.3 a)}$$

$$= KH$$

where K = quantity within the large brackets

or $$\frac{p_A - p_B}{\gamma_w} \propto h \quad \text{... (10.3 b)}$$

(b) Inclined Manometer

An inclined manometer as shown in Fig. 10.11 is used to measure small pressure precisely in low velocity gas flow. When the opening of the tank C is exposed to the atmosphere, the inclined glass tube has the liquid at level O. But when a positive small pressure is applied to the tank liquid, the level of the liquid in the inclined tube rises and is measured by the inclined distance l. The change in the level of the tank is negligible if the cross-section is made very large as compared to that of the tube. The applied pressure will then be equal to,

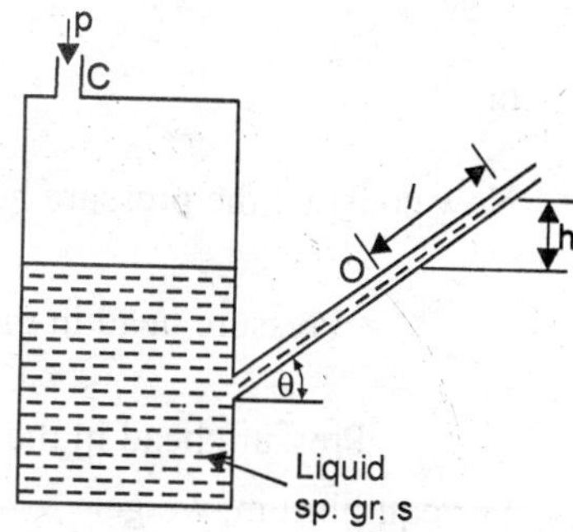

Fig. 10.11 Inclined Manometer

$$p = s\gamma_w l \sin\theta = \gamma l \sin\theta \quad \text{... (10.4)}$$

where γ = specific weight of manometer liquid,

This type of manometer, when calibrated directly in cm of water, is called *draft gauge.*

(c) Single column Manometer

In a single column manometer a shallow reservoir having a large cross-sectional area as compared to the area of the tube is connected to one limb of the manometer, as shown in Fig. 10.12. For any variation in pressure, the change in the liquid level in the reservoir will be so small that it may be neglected, and the pressure is indicated by the height of the liquid in the other limb. As such only one reading in the narrow limb of the manometer need to be taken for all pressure measurements.

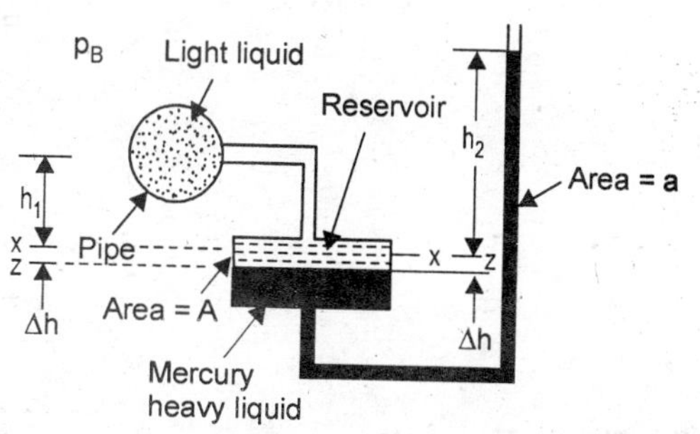

Fig. 10.12 Single column manometer

Let X-X be the datum line in the reservoir when the single column manometer is not connected to the pipe.

When the manometer is connected to a pipe containing light liquid under high pressure, the pressure in the pipe will force the light liquid to push the heavy liquid in the reservoir downwards. As the area of reservoir is very large, the fall of the heavy liquid level will be very small. However the rise of level of heavy liquid in right limb will be quite considerable.

Let h_1 = height of centre of pipe above X-X

h_2 = rise of heavy liquid level in right limb from X-X.

Δh = fall of heavy liquid level in reservoir.

p = pressure in pipe

A = cross-sectional area of reservoir

a = cross-sectional area of right limb.

s_1, s_2 = specific gravity of light and heavy liquids respectively.

Then
$$A \times \Delta h = a \times h_2$$

or
$$\Delta h = \frac{a \times h_2}{A}$$

Now consider the pressure heads above the datum line $z - z$.

$$\text{Pressure head in the left limb} = \frac{p}{\gamma_1} + (h_1 + \Delta h)s_1$$

$$\text{Pressure head in the right limb} = (h_2 + \Delta h)\,s_2$$

For equilibrium, we get

$$\frac{p}{\gamma_1} + (h_1 + \Delta h)s_1 = (h_2 + \Delta h)s_2$$

or
$$\frac{p}{\gamma_1} = (h_2 + \Delta h)s_2 - (h_1 + \Delta h)s_1$$

$$= \Delta h(s_2 - s_1) + h_2 s_2 - h_1 s_1$$

$$= \frac{a \times h_2}{A}(s_2 - s_1) + h_2 s_2 - h_1 s_1 \quad \text{... (10.5 a)}$$

$$\simeq h_2 s_2 - h_1 s_1, \text{ when } \frac{a}{A} \to 0$$

$$\therefore \quad h = \frac{p}{\gamma_1} = h_2 s_2 - h_1 s_1 \quad \text{... (10.5 b)}$$

(b) Elastic Elements

Elastic elements get deformed when subjected to pressure. The deformation gives an indication of pressure. The elastic element may be of the following forms as shown in Fig. 10.13.

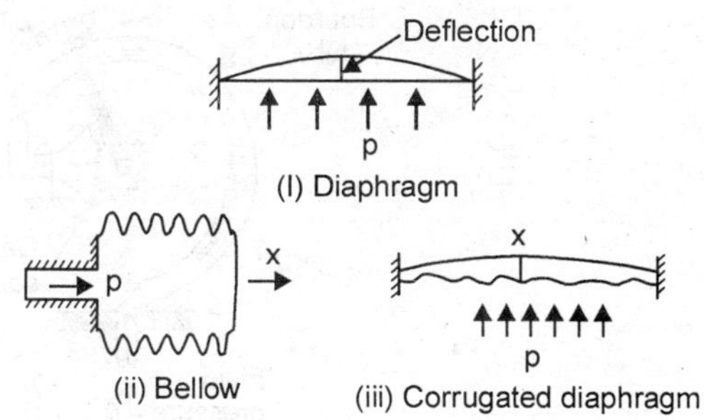

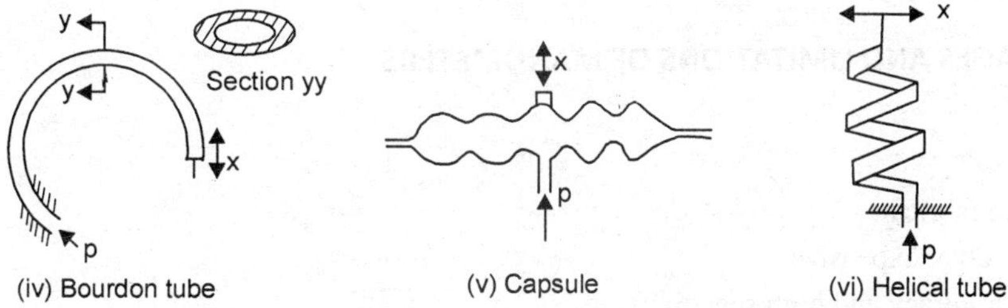

Fig. 10.13

(i) Diaphragm *(ii)* Bellows *(iii)* Corrugated diaphragm
(iv) Bourdon tube *(v)* Capsule *(vi)* Helical tube

The deformation may be measured by mechanical or electrical means. Usually, an electromechanical transducer is used along with the elastic element, as shown in Fig. 10.14. The output voltage can be indicated and recorded by a suitable instrument like an oscilloscope or a recorder.

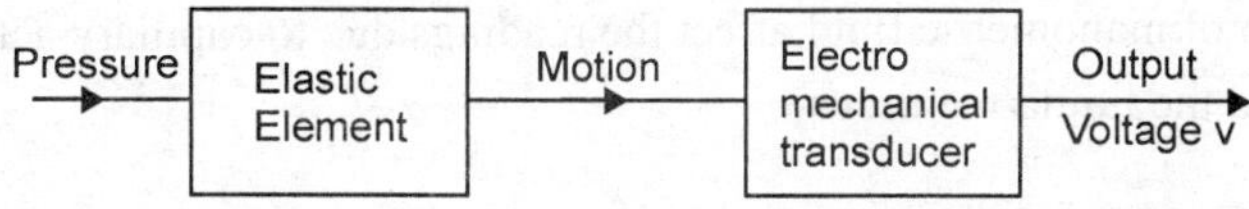

Fig 10.14 Pressure measurement using an elastic element and an electromechanical transducer

(a) Bourdon Gauge

A Bourdon gauge shown in Fig. 10.15 is commonly used for measuring pressure. This uses an elastic Bourden tube. The cross-section of the tube tends to round out under pressure. The tube uncoils since the inner and outer arc lengths remain almost equal to their original lengths. The motion of the end of the tube is amplified and indicated by a pointer moving on a calibrated scale.

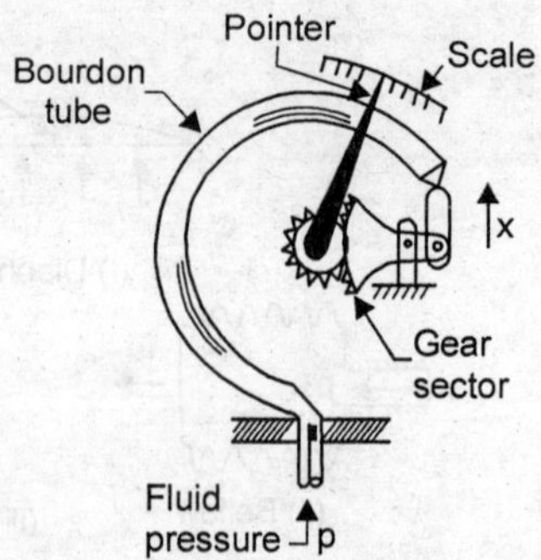

Fig. 10.15 Bourdon pressure gauge

10.3 ADVANTAGES AND LIMITATIONS OF MANOMETERS

Advantages

1. Easy to fabricate.
2. Relatively inexpensive.
3. Good accuracy and high sensitivity.
4. High sensitivity.
5. No effect of vibrations.
6. Specially suitable for low pressure and low differential pressures.
7. Sensitivity can be easily changed by changing the quantity of manometric liquid.

Limitations

1. Bulky and large in size.
2. Fragile and get easily broken.
3. Affected by temperature, altitude and gravity.
4. Surface tension of manometric fluid affect the readings due to capillary action.
5. Difficult to read the liquid meniscus.

10.4 FLOW VELOCITY MEASUREMENT

In many experimental studies of fluid flow phenomena, it is necessary to determine the magnitude and/or direction of flow-velocity at a point in a fluid and how this varies from point to point. Usually velocity probes are inserted into the fluid flow to obtain accurate flow velocity. Such probes are:

1. Pitot static tube
2. Hot-wire anemometer for gases
3. Hot-film anemometer for liquids
4. Laser-doppler anemometer.

1. Pitot-Static Tube

A pitot-static tube is used to measure both the static as well as the stagnation pressure. It consists of two circular concentric tubes one inside the other with a circular space in between, as shown in Fig. 10.16

The static pressure is measured through two or more holes drilled through the outer tube into an annular space. It is assumed that the fluid flows along the outside of the tube in such a way that the true static pressure is obtained. The stagnation pressure is caused at the tip as shown in Fig. 10.16.

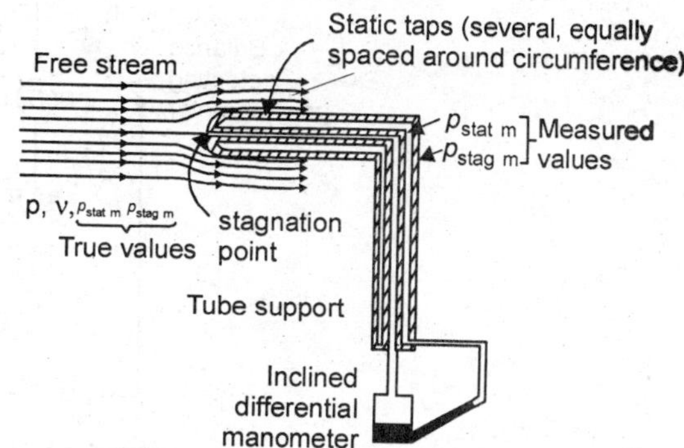

Fig. 10.16 Pitot-static tube

Assuming steady one-dimensional flow of an incompressible frictionless fluid, the flow velocity v is given by,

$$v = \sqrt{\frac{2(p_{stag} - p_{stat})}{\rho}} \qquad \text{... (10.6)}$$

where

ρ = fluid mass density

p_{stag} = stagnation or total pressure, free stream.

P_{stat} = static pressure, free stream.

We can also write,

$$v = C\sqrt{2gh} \qquad \text{... (10.7)}$$

where

C = correction coefficient for the Pitot tube

$$h = h_{stag} - h_{stat} = \frac{p_{stag} - p_{stat}}{\gamma}$$

2. Hot-wire Anemometer

It is a commonly used device for measuring the mean and fluctuating velocities in fluid flow. The flow sensing element is a short length of 5 μm diameter platinum-tungsten wire welded between two prongs of the probe and heated electrically as a part of Wheatstone bridge, as shown in Fig. 10.17. *(a)*. When the probe is introduced in the fluid stream, it tends to get cooled by the instantaneous velocity and consequently there is a decrease in its resistance. The rate of cooling of the wire is a direct measure of the flow velocity.

The hot wire anemometer may be used in two ways to measure the velocity.

***(i)* Constant current method:** It uses an electric circuit adjusted to feed a constant current to the hot wire. The knowledge of this current and the resistance of the wire defines the power being fed to the wire which is a function of flow velocity. The wire attains an equilibrium temperature when the I^2R heat generated in it is just balanced by the convective heat loss from its surface.

***(ii)* Constant temperature method:** In this method, the current through the wire is adjusted to keep the wire temperature (as measured by its resistance) constant. The current required to do this then becomes a measure of flow velocity.

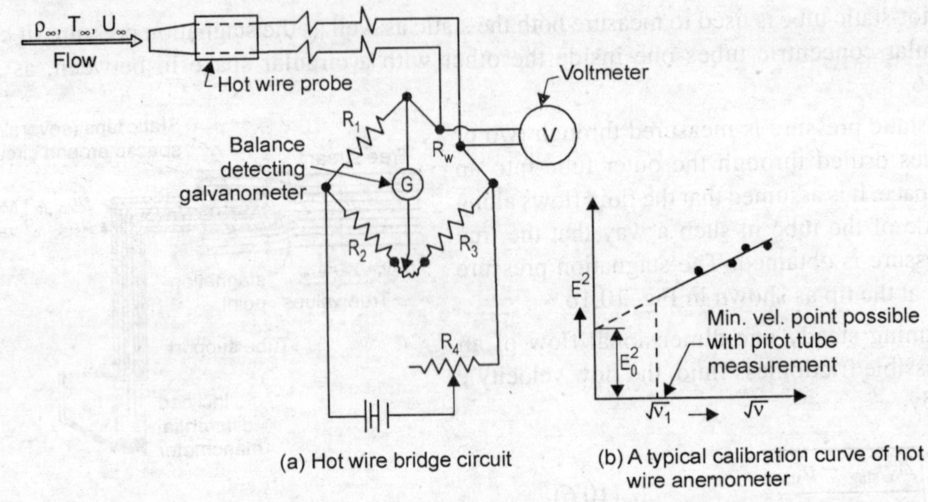

(a) Hot wire bridge circuit

(b) A typical calibration curve of hot wire anemometer

Fig. 10.17 Schematic diagram of hot wire operation

For equilibrium conditions the. energy balance for a hot wire is:

$$I^2R_w = hA\ (Tw - Tf) \qquad \text{... (10.8)}$$

where
I = wire current
R_w = wire resistance
T_w = wire temperature
T_f = temperature of flowing fluid
h = film coefficient of heat transfer
A = heat transfer area.

Now h is a function of flow velocity for a given fluid density. For a range of velocities, this has the general form,

$$h = C_0 + C_1\sqrt{v} \qquad \text{... (10.9)}$$

Induced voltage,
$$E^2 = A_0 + A_1\ \sqrt{v} \qquad \text{... (10.10)}$$

For the measurement of average (steady) velocities, constant-temperature method may be used. For accurate work, a given hot-wire probe must be calibrated in the fluid in which it is to be used. That is, it is exposed to known velocities and its output is recorded over a range of velocities. The calibration plot between E^2 and $\sqrt{v}$ is obtained. The curve is extrapolated in the lower range to give the value of E_0 for zero velocity, as shown in Fig. 10.17 (b). The calibration equation is generally written as:

$$E^2 = E_0^2 + B\sqrt{v} \qquad \text{(10. 11)}$$

3. Hot-film Anemometer

Here the resistance element is a thin film of platinum deposited on a glass base. The film takes the place of hot wire. The required circuitry is basically similar to that used in constant temperature hot-wire anemometer. The film transducers have great mechanical strength and may be used at very' high temperatures by constructing them with internal cooling-water passages.

Fig. 10.18 shows two configurations of sensors for hot-film anemometer. It is used to measure the flow velocity of liquids.

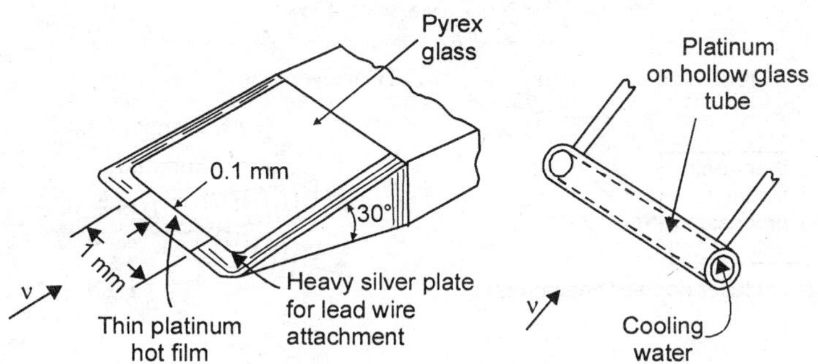

Fig. 10.18 Hot-film sensors

4. Laser Doppler Anemometer

The operating principle of this device is based on the Doppler shift in frequency of the light scattered by an object moving relative to the radiating source. The technique basically consists of focussing laser beams at the point in the fluid where velocity is to be measured. At this focal point, the laser light scattered from the fluid or liquid particles entrained in the fluid is sensed by a photo-detector. Signal processing of the photo-detector output yields the magnitude of Doppler frequency shift which is directly proportional to the instantaneous flow velocity.

The dual optical arrangement for the laser Doppler anemometer is shown in Fig.10.19 *(a)*. Helium-Neon laser of 5-15 mW power is used when the measured flow is at a distance of 10 -20 cm. The laser beam is split into two equal parts by means of a beam splitter. The focussing lens is put in front of the beam splitter to focus the two beams at a point where the flow velocity of fluid is to be measured. At the focal point, the two split laser beams cross each other and form an interference fringe pattern. If any microscopic dust or dirt particle pass through the region of high intensity, they would scatter light and cause a Doppler shift in the frequency of scattered light. Thus, the light received by the photo-detector will show a varying electrical signal, the frequency of which is proportional to the rate at which the particles cross the interference fringes.

Doppler shift in frequency, $\Delta f = 2nv \sin\left(\frac{\theta}{2}\right)\lambda_0$

where n = index of refraction of fluid

λ_0 = wave length of laser beam in vacuum

θ = angle between two converging beams.

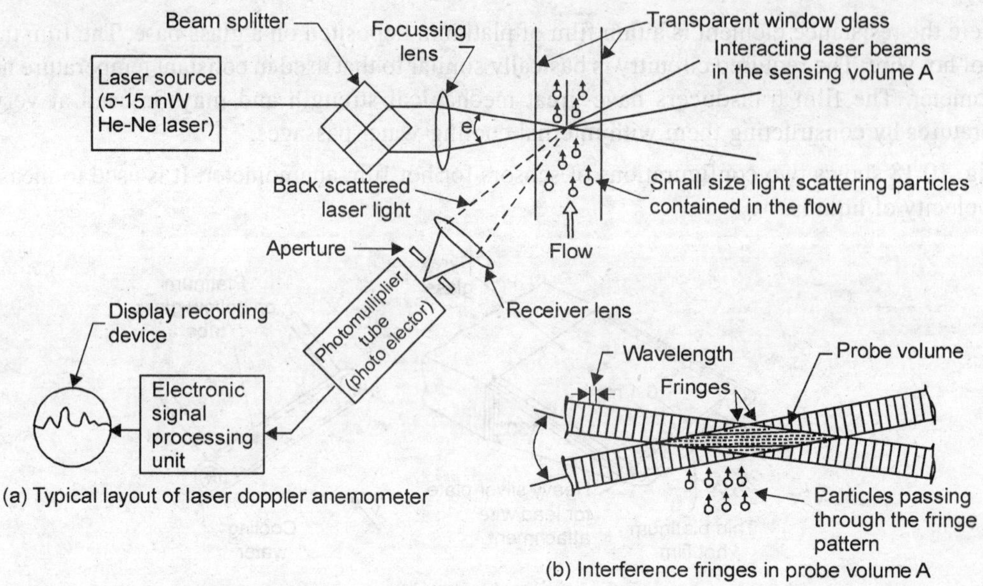

Fig. 10.19 Details of laser doppler anemometer in dual beam or fringe mode

10.5 FLOW MEASUREMENTS

Fluid flow measurements can be made by the following methods:

1. Positive displacement meters.
   (*i*) Nutating disc meter
   (*ii*) Sliding vane type meter
   (*iii*) Lobed impeller flow meter

2. Variable head meters
   (*i*) Venturimeter
   (*ii*) Orifice meter
   (*iii*) Nozzle meter
   (*iv*) Pitot tube

3. Variable area meters

(*i*) Rotameter
(*ii*) Pitot static tube
(*iii*) Drag force flow meter
(*iv*) Turbine flow meter

10.5.1 Positive Displacement Meters

In these devices the metered fluid is repeatedly filled and emptied from a space of known volume The principle of measurement is that the liquid flows through a meter and moves the measuring element that seals the measuring chamber into a series of measuring compartments each holding a definite volume.

Each element is successively filled from the flow at the inlet and emptied at the outlet of the meter. These devices are widely used in applications where a high accuracy and repeatability under steady flow conditions are required.

1. Nutating disc Meter

It consists of an eccentrically mounted disc which nutates (or wobblers) in the metering chamber with spherical sides as shown in Fig. 10.20 .The wobbling disc provides the rotational motion along with sweeping motion. This motion is used to drive the indicating/ recording mechanism which can be directly calibrated in terms of fluid discharge. It is used as a domestic water meter.

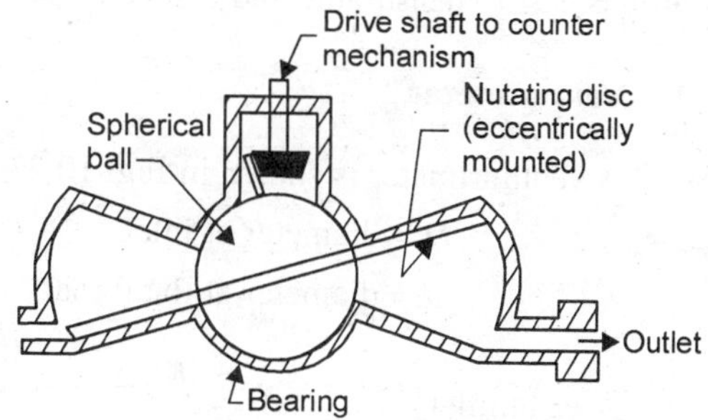

Fig. 10.20 Nutating disc meter

2. Sliding Vane Type Meter

This type of meter shown in Fig. 10.21, has an accurately machined body having rotor with four evenly spaced slots which form the guides for the vanes. The metered liquid entering the inlet revolves the rotor and the vanes around a cam causing the vanes to move radially. The vane nearest to the inlet part begins to move outwards and becomes fully extended at point *A*. The vane ahead at point *B* is already fully extended and thus a measuring chamber of known volume is formed between the two vanes. A continuous series of chambers at the rate of four per revolutions are formed which delivers the flow at the outlet.

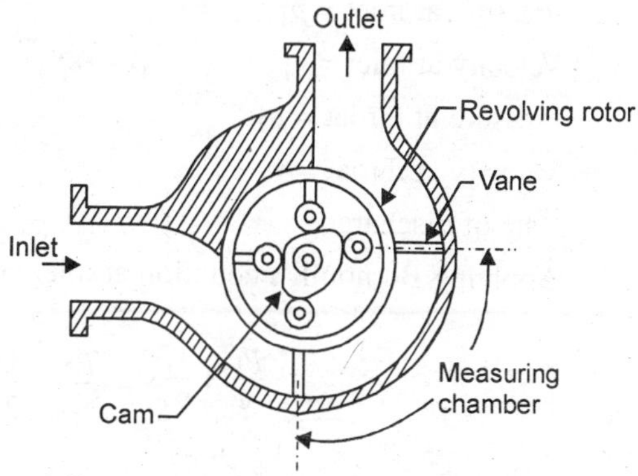

Fig. 10.21 Sliding vane type meter

3. Lobed-Impeller Meter

Fig.10.22, consists of two rotors mounted on two parallel shafts. The rotors are lobed and revolve in opposite directions in a close fitting chamber. The rotor lobes have cycloidal or involute shape. The incoming liquid is trapped between the two rotors and is conveyed to the outlet as a result of the rotation of the rotors. For every rotation of each rotors, a volume corresponding to twice the area *A* is passed through the meter. Thus the number of revolutions of the rotor gives an indication of the volumetric flow. It is used for metering gas flows.

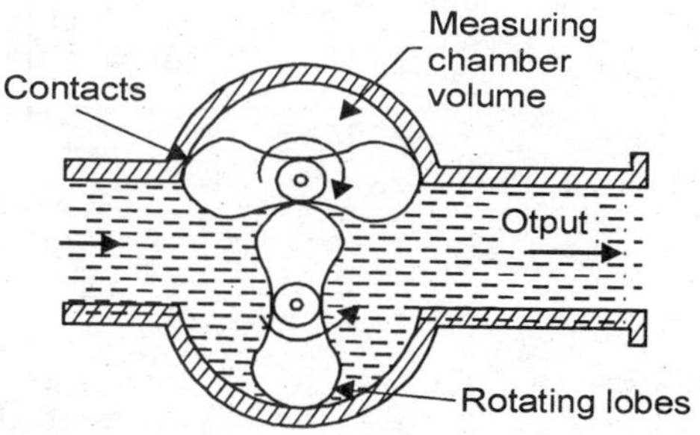

Fig. 10.22 Lobed impeller flow meter

10.5.2 Variable Head Meters

These are obstruction type of flow meters which introduce constriction in the flow passage. This results in the acceleration of the fluid which causes a reduction in the pressure. The magnitude of pressure drop is easily measurable and can be related to the rate of flow.

1. **Venturimeter**

A ventunrimeter is shown in Fig. 10.23

D = diameter at inlet

d = diameter at the throat

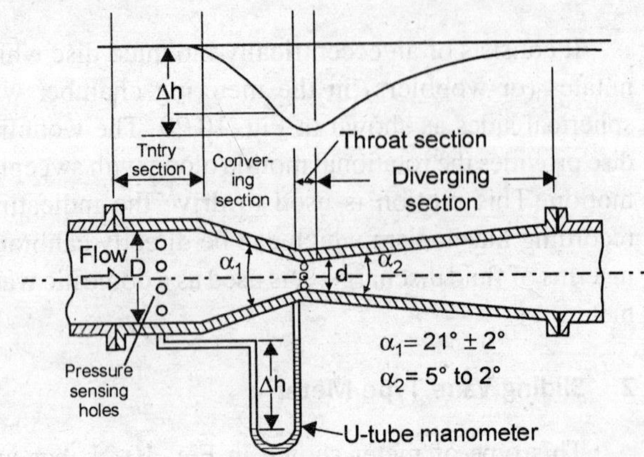

Fig. 10.23 Venturimeter

Area at inlet, $$A_1 = \frac{\pi}{4}D^2$$

Throat area, $$A_2 = \frac{\pi}{4}d^2$$

Pressure at inlet = p_1

Velocity at inlet = v_1 $\beta = 45°$

Pressure at throat = p_2

Velocity at throat = v_2

Rate of discharge, $Q = A_1 v_1 = A_2 v_2$

Applying Bernoulli's equation at inlet and throat, we have

$$\frac{p_1}{\gamma} + \frac{v_1^2}{2g} = \frac{p_2}{\gamma} + \frac{v_2^2}{2g}$$

$$\frac{p_1 - p_2}{\gamma} = \frac{v_2^2 - v_1^2}{2g}$$

or $$\Delta h = \frac{v_2^2}{2g}\left[1 - \left(\frac{A_2}{A_1}\right)^2\right]$$

$$v_2^2 = 2g\,\Delta h \frac{A_1^2}{(A_1^2 - A_2^2)}$$

$$v_2 = 2g\,\Delta h \frac{A_1^2}{(A_1^2 - A_2^2)}$$

$$v_2 = \frac{A_1}{\sqrt{A_1^2 - A_2^2}}.\sqrt{2g\Delta h}$$

$$Q = A_2 v_2 = \frac{A_1 A_2}{\sqrt{A_1^2 - A_2^2}} . \sqrt{2g\Delta h} = k\sqrt{\Delta h} \quad \text{... (10.12)}$$

where $$k = \frac{A_1 A_2}{\sqrt{A_1^2 - A_2^2}} \times \sqrt{2g} = \text{venturimeter constant}$$

If C_d = coefficient of discharge = $\dfrac{\text{actual discharge}}{\text{ideal discharge}}$

Then $$Q_{\text{actual}} = C_d k \sqrt{\Delta h}$$

2. Orifice meter

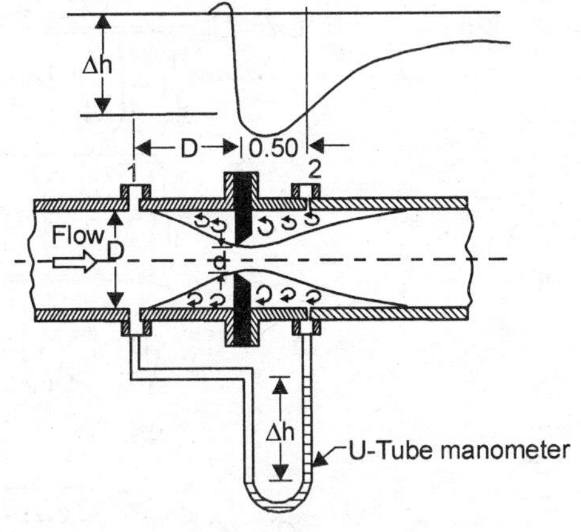

Fig. 10.24 Orifice meter

The orificemeter is shown in Fig. 10.24.

Let A_0 = area of orifice

A_c = area of vena-contracta

Coefficient of contraction, $C_c = \dfrac{A_c}{A_0}$, $Q = A_1 v_1 = A_c v_c$

Applying Bernoulii's equation between inlet and vena-contracta, we have

$$\frac{p_1}{\gamma} + \frac{v_1^2}{2g} + z_1 = \frac{p_c}{\gamma} + \frac{v_c^2}{2g} + z_2; \quad \Delta h = \frac{p_1 - p_c}{\gamma} + (z_1 - z_2) = \frac{v_c^2 - v_1^2}{2g}$$

$$v_c = \sqrt{2g.\Delta h + v_1^2} = \sqrt{2g\,\Delta h + v_c^2 \left(\frac{A_c}{A_1}\right)^2} = \sqrt{2g.\Delta h + v_c^2 C_2^2 \left(\frac{A_0}{A_1}\right)^2}$$

$$v_c^2 = 2g.\Delta h + \left(\frac{A_0}{A_1}\right)^2 C_c^2 v_c^2$$

or $$v_c^2 \left[1 - \left(\frac{A_0}{A_1}\right)^2 C_c^2\right] = 2g.\Delta h$$

$$v_c = \frac{\sqrt{2g.\Delta h}}{\sqrt{1 - \left(\dfrac{A_0}{A_1}\right)^2 C_c^2}}$$

$$Q = A_0 v_c = A_c C_c v_c = A_0 C_c \frac{\sqrt{2g\,\Delta h}}{\sqrt{1-\left(\dfrac{A_0}{A_1}\right)^2 C_c^2}}$$

Let
$$C_d = C_e \frac{\sqrt{1-\left(\dfrac{A_0}{A_1}\right)^2}}{\sqrt{1-\left(\dfrac{A_0}{A_1}\right)^2 C_2^2}}$$

or
$$C_c = C_d \frac{\sqrt{1-\left(\dfrac{A_0}{A_1}\right)^2 C_c^2}}{\sqrt{1-\left(\dfrac{A_0}{A_1}\right)^2}}$$

Fig. 10.25 Nozzle meter

$$\therefore \qquad Q = C_d \frac{A_0 A_1 \sqrt{2g.\Delta h}}{\sqrt{A_1^2 - A_0^2}} \qquad \text{... (10.13)}$$

It is quite cheap, easy to install and takes almost no space. However, it is susceptible to accuracies resulting from erosion, corrosion, clogging etc, due to flow of dirty water.

3. Nozzle meter

It is a constriction with well rounded entrance placed in the pipe line (Fig.10.25). It can be installed between the flanges of the pipeline. The discharge equation is same as for the venturimeter. The coefficient of contraction, $C_c = 1$ for the nozzle. The nozzlemeter occupies less space. However, it is difficult to install and is further limited to moderate pipe sizes.

4. Pitot-Tube

It measures the stagnation pressure at any point in the flow. The static pressure may be measured by a piezometer. The pitot-tube consists of a tube having a 90° bend of shorter length, opened at its both ends. The bent leg is directed upstream, so that a stagnation point is created immediately ahead of its inlet as shown in Fig.10.26 On account of the stagnation pressure so created, the liquid rises in the vertical limb. A piezometer installed on the boundary indicates the static pressure p_0.

Stagnation pressure, $\dfrac{p_s}{\gamma} = \dfrac{p_0}{\gamma} + \dfrac{v_0^2}{2g}$

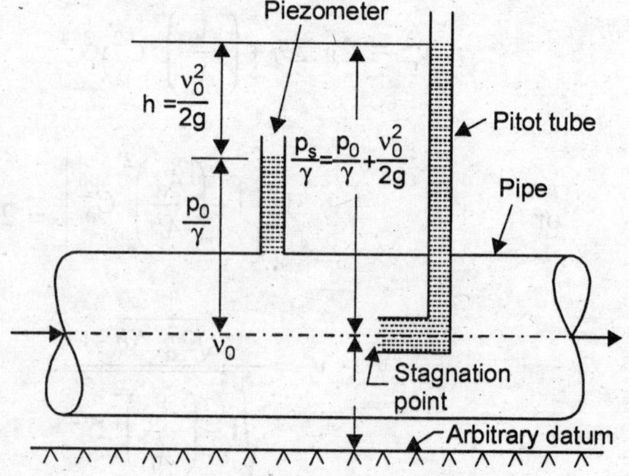

Fig. 10.26 Pitot tube

or $$p_s = p_0 + \rho \frac{v_0^2}{2}$$

$$v_0 = \sqrt{\frac{2(p_s - p_0)}{\rho}} = \sqrt{2gh}$$

where h = dynamic pressure head

$$Q = Av_0 \quad \text{... (10.14)}$$

where A = pipe area across flow.

10.5.3 Variable area Meters

In the variable area meters, the area of the restriction is altered to maintain a steady pressure difference.

1. Rotameter

It is a commonly used device to measure discharge. In this device, shown in Fig. 10.27 the flow enters the bottom of a vertically placed tapered tube and causes the bob or float to move upwards. The float will rise to a point in the tube where the drag force and buoyant force is balanced by the weight of the float. The position of the float in the tube is taken as an indication of the flow rate.

The force balance equation of the float is:

$$F_{drag} + F_{buoyancy} = F_{weight}$$

$$(p_d - p_u) + \rho_f g V_b = \rho_b g V_b$$

where V_b = volume of bob

A_b = area of bob

ρ_f, ρ_b = density of flowing fluid and bob respectively

p_d, p_u = pressure at the downward and upward faces of bob respectively

$$p_d - p_u = \frac{V_b}{A_b} g(\rho_b - \rho_f)$$

$$Q_{actual} = C_d \frac{A_t(A_t - A_b)}{\sqrt{A_1^2 - (A_t - A_b)^2}} \sqrt{2g} \sqrt{\frac{p_d - p_u}{\rho_f g}}$$

where A_t = area of tube at the float level

$A_t - A_b$ = minimum annular area between tube and bob

C_d = coefficient of discharge.

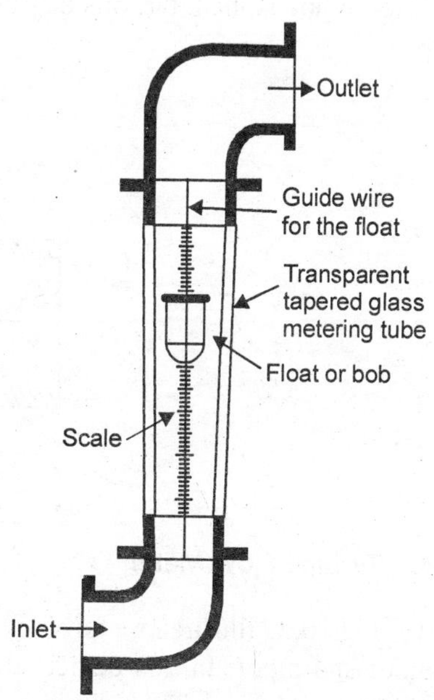

Fig. 10.27 Rotameter

$$Q_{actual} = \frac{C_d(A_t - A_b)}{\sqrt{1-\left[\frac{A_t - A_b}{A_t}\right]^2}}.\sqrt{2g}\sqrt{\frac{V_b}{A_b} \times \frac{(\rho b - \rho_f)}{\rho_f}} \quad ...(10.15)$$

If $\frac{A_t - A_b}{A_t} << 1$ and variation in C_d with float position is small, then

$$Q_{\text{actual}} = K(A_t - A_b) = K\sqrt{\frac{\rho_b - \rho_f}{\rho_f}} \quad ...(10.16)$$

where K = rotameter constant

2. Pitot Static Tube

Refer to art. 10.4.

3. Drag Force Flow Meter

For sufficiently high Reynolds number, the drag coefficient is reasonably constant. Therefore, for a given density, drag force F_d is proportional to square of volume flow rate V^2. The drag force of the body can be measured by attaching the drag body to a force measuring device. A cantilever beam type drag force flow meter using bonded strain gauges is shown in Fig.10.28. It gives very good dynamic response. It is specially suitable to measure flows in which differential pressure measurements before and after the obstruction are not conveniently obtainable. Common examples of such flow are the highly viscous flows of hot asphalt, tar, oils or slurries at high pressure.

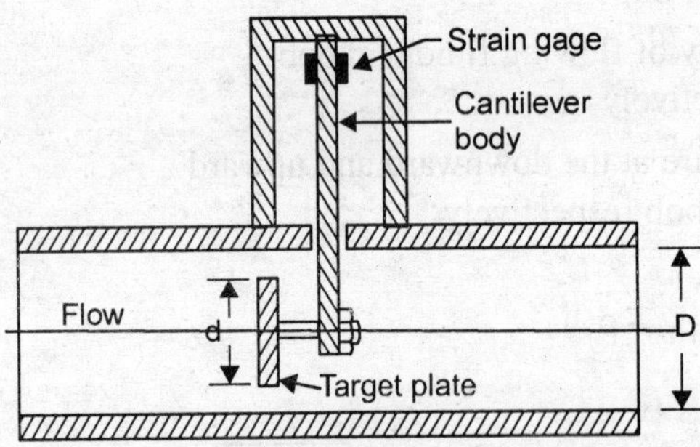

Fig. 10.28 Drag force flow meter

4. Turbine Flow Meter

Fig.10.29 illustrates a turbine flow meter with an electrical output suitable for measuring the flow in tubes and pipes. In this device, the rotor motion, proportional to flow rate, is sensed by means of a reluctance pick-up. The voltage pulses generated are counted by means of an electronic counter to give the total flow. It is simultaneously possible to obtain the flow and rate of flow.

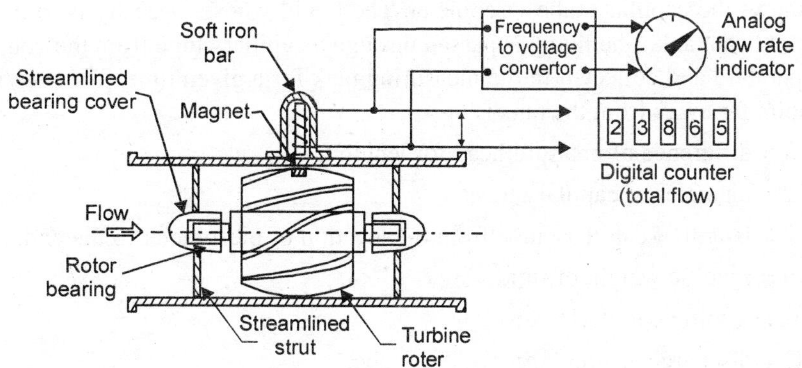

Fig. 10.29 Turbine flow meter

Volume flow rate, $Q = k\omega$

where k = constant for the meter

ω = angular velocity of rotor, rad/s

This is used by weather stations, measuring flows in rivers, streams, channels etc.

10.6 COEFFICIENT OF VISCOSITY MEASURMENTS

The coeffficient of viscosity of a liquid can be determined by the following methods:

1. Capillary tube method
2. Falling sphere resistance method.
3. Rotating cylinder method.
4. Orifice type viscometer.

1. Capillary Tube Method

In the capillary tube method, the viscosity of a liquid is determined by measuring the pressure difference for a given length of capillary tube. The Hagen-Poiseuille law is used for calculating viscosity.

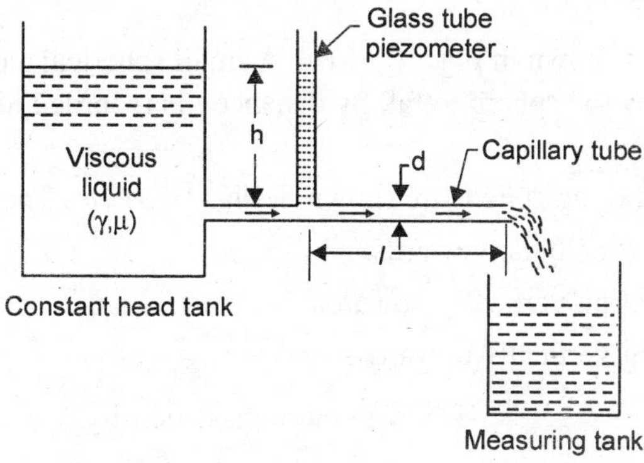

Fig. 10.30 Capillary tube viscometer.

Fig 10.30 shows the capillary tube viscometer. The liquid whose viscosity is to be determined is filled in a constant head tank. The liquid is passed through a capillary tube from the constant head tank at constant temperature and collected in the measuring tank for a given time. The pressure head '*h*' is measured at a point far away from the tank.

Let h = difference of pressure head for length l.

d = diameter of capillary tube

l = length of capillary tube from the location of piezometer to the end.

γ = specific weight of fluid

μ = coefficient of viscosity

Q = discharge + thrughthe copillary tube.

Applying Hagen– poiscuille's equation, we have

$$h = \frac{32\mu\bar{u}l}{\gamma d^2}$$

But

$$\bar{u} = \frac{4Q}{\pi d^2}$$

$$h = \frac{32\mu \times \frac{4Q}{\pi d^2} \times l}{\gamma d^2} = \frac{128\mu Ql}{\pi\gamma d^4}$$

$$\mu = \frac{\pi\gamma h d^4}{128 Ql} \qquad \text{... (10.17)}$$

2. Falling sphere Resistance Method

The falling sphere method of measuring viscosity of a liquid is based on Stoke's law, according to which the drag force, F_D on a small sphere of diameter, d, moving with a constant velocity, v through a viscous fluid of dynamic viscosity, μ, is given by:

$$F_D = 3\pi\mu v d$$

The experimental set-up is shown in Fig. 10.31 *(a)*. A small spherical ball is released into the liquid to be tested. The time taken by the sphere to fall by distance l is recorded to calculate the velocity v of sphere.

The forces acting on the falling sphere are shown in Fig. 10.31 *(b)*. These forces are:

1. Weight, W: acting vertically downwards.
2. Buoyant force, F_B: acting vertically upwards
3. Drag force, F_D: acting vertically upwards.

Now, W = Volume × mass density × g

$$= \frac{\pi}{6} d^3 \times \rho_s \times g$$

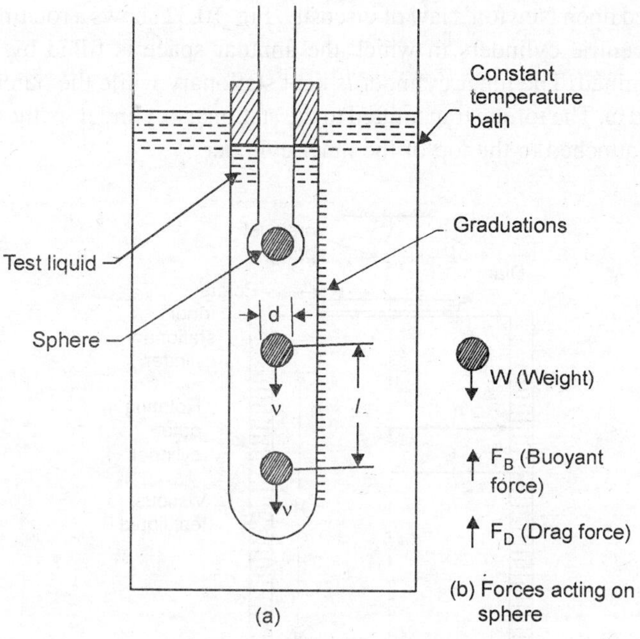

Fig. 10.31 Falling sphere viscometer.

$$F_B = \text{Weight of liquid displaced}$$

$$= \frac{\pi}{6} d^3 \times \rho_l \times g$$

where ρ_s, γ_l = density of sphere and test liquid respectively.

$$F_D = 3\pi \mu v d$$

For equilibrium of sphere,

$$F_B + F_D = W$$

or

$$F_D = W - F_B$$

$$3\pi\mu v d = \frac{\pi}{6} d^3 \rho_s g - \frac{\pi}{6} d^3 \rho_l g$$

$$= \frac{\pi}{6} d^3 g (\rho_s - \rho_l)$$

$$\mu = \frac{g d^2}{18 v} (\rho_s - \rho_l) \qquad \text{... (10.18)}$$

3. Rotating Cylinder Method

This method is based upon Newton's law of viscosity. Fig. 10.32 shows a rotating cylinder viscometer. It consists of two concentric cylinders in which the annular space is filled by the test liquid whose viscosity is to be determined. The inner cylinder is kept stationary while the outer cylinder is rotated at a constant angular speed ω. The torque transmitted to the stationary cylinder by the test liquid is measured by the torsional spring attached to the top of the inner cylinder.

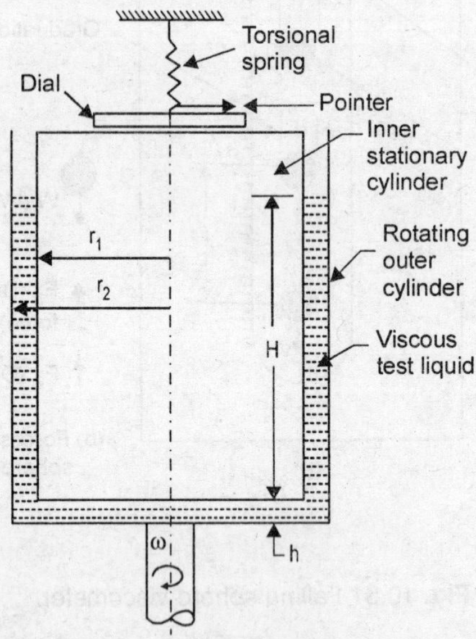

Fig 10.32 Rotating cylinder viscometer

According to the Newton's law of viscosity,

Shear stress, $\tau = \mu \times \text{velocity gradient}, \dfrac{du}{dy}$

$$\therefore \quad \frac{du}{dy} = \frac{\omega r_2 - 0}{(r_2 - r_1)} = \frac{\omega r_2}{(r_2 - r_1)}$$

$$\therefore \quad \tau = \frac{\mu \omega r_2}{(r_2 - r_1)}$$

Shear force, $F = \tau \times \text{surface area}$

$$= \tau \times 2\pi r_1 H$$

$$= \frac{\mu \omega r_2}{(r_2 - r_1)} \times 2\pi r_1 H$$

where H = height of the liquid in annular space.

Torque on the inner cylinder, $T_1 = F \times r_1$

$$= \frac{\mu \omega r_2}{(r_2 - r_1)} \times 2\pi r_1 H \times r_1$$

$$= \frac{2\pi \mu \omega^2 r_1 r_2 H}{(r_2 - r_1)}$$

Let h = gap between the bottom of the two cylinders

Then torque T_2 applied on the inner cylinder is given by Eq. (6.48) as,

$$T_2 = \frac{\mu}{60h} \pi^2 N r_1^4$$

where $\omega = \frac{2\pi N}{60}$ or $N = \frac{30\omega}{\pi}$ rpm

$$T_2 = \frac{\pi \mu}{2h} \times \omega r_1^4$$

Total torque acting on the inner cylinder, $T = T_1 + T_2$

$$= \frac{2\pi \mu \omega r_1^2 r_2 M}{(r_2 - r_1)} + \frac{\pi \mu \omega r_1^4}{2h}$$

$$= 2\pi \mu \, \omega r_1^2 \left[\frac{r_2 H}{(r_2 - r_1)} + \frac{r_1^2}{4h} \right]$$

$$\mu = \frac{2(r_2 - r_1) h T}{\pi r_1^2 \omega \, [4Hhr_2 + r_1^2 (r_2 - r_1)]} \qquad \text{... (10.19)}$$

4. Orifice Type Viscometer

In this method, the time taken by a certain quantity of a liquid whose viscosity is to be determined, to flow through a short capillary tube is recorded. The coefficient of viscosity is then determined by comparing with the coefficient of viscosity of a liquid whose viscosity is known or by the use of conversion factors. The operation of Saybolt, Redwood and Engler is based on this principle.

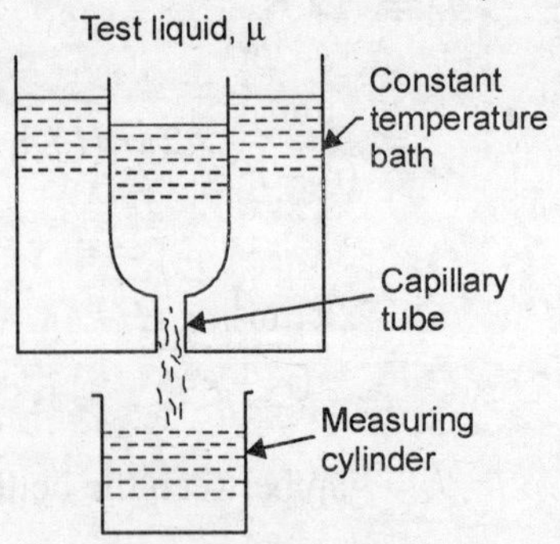

Fig. 10.33 Saybolt viscometer

Fig 10.33 Shows a Saybolt viscometer. It consists of a tank in the bottom of which is short capillary tube. The tank filled with a test liquid is surrounded by a constant temperature bath. The viscosity is determined by measuring the time required for 60 cm^3 of liquid to flow out of the tank through the capillary tube. The initial height of liquid in the tank is previously adjusted to a standard height. From the time measurment, the kinematic viscosity of liquid is determined from the relation,

$$\nu = At - \frac{B}{t} \qquad \text{... (10.20)}$$

where A = 0.24, B = 190, t = time noted in seconds, and
ν = kinematic viscosity in stokes.

Example 10.8 *The viscosity of an oil of specific gravity 0.9 is measured by a capillarly tube of diameter 40 mm. The diference of pressure head between two points 1.5 m apart is 0.4 m of water. The mass of oil collected in a measuring tank is 50 kg in 80 seconds. Calculated the viscosity of oil.*

Solution.

Given: s = 0.9, d = 40 mm l = 1.5 m, h = 0.4 m of water. m = 50 kg, t = 80 s.

Mass flow rate of oil, $$\dot{m} = \frac{m}{t} = \frac{50}{80} = 0.625 \text{ kg/s}$$

Density of oil, $$\rho = s \times \rho_w = 0.9 \times 10^3 = 900 \text{ kg/m}^3$$

Discharge, $$Q = \frac{\dot{m}}{\rho} = \frac{0.625}{900} = 0.000694 \text{ m}^3/\text{s}$$

Viscosity, $$\mu = \frac{\pi \rho g h d^4}{128 Q l} = \frac{\pi \times 900 \times 9.81 \times 0.4 \times (0.04)^4}{128 \times 6.944 \times 10^{-4} \times 1.5} = 0.213 \text{ Pa.s}$$

Example 10.9 *A sphere of diameter 2 mm falls 160 mm in 20 s in a viscous liquid. The density of sphere is 7600 kg/m^3 and that of liquid is 900 kg/m^3. Calculate the coefficient of viscosity of the liquid.*

Solution

Given: $d = 2$ mm, $l = 160$ mm, $t = 20$ s, $\rho_s = 7600$ kg/m^3, $\rho_l = 900$ kg/m^3.

Velocity of sphere,
$$v = \frac{l}{t} = \frac{0.160}{20} = 0.008\,\text{m/s}$$

Viscosity of liquid,
$$\mu = \frac{gd^2}{18v}(\rho_s - \rho_l)$$
$$= \frac{9.81\times(2\times10^{-3})^2}{18\times0.008}(7600-900)$$
$$= 1.826 \text{ Pa.s}$$

Example 10.10 *The viscosity of a liquid is determined by rotating cylinder method. The inner stationary cylinder is of 200 mm diameter and the outer cylinder is of 205 mm diameter. The annulus space carries a liquid, whose viscosity is to be determined, upto a height of 300 mm. The clearance at the bottom of two cylinders is 5mm. The outer cylinder is rotaled at 375* rpm. *The forque mmeaswed on the torsion meter attached to the inner cylinder is 5.9 N.m. Calculate the viscosity of liquid.*

Solution.

Given: $r_1 = 100$ mm, $r_2 = 102.5$ mm, $H = 300 - h = 300 - 5 = 295$ mm, $h = 5$ mm, $N = 375$ rpm, $T = 5.9$ N.m

$$\omega = \frac{2\pi N}{60} = \frac{2\pi\times375}{60} = 39.27 \text{ rad/s}$$

$$\mu = \frac{2(r_2 - r_1)\times h\times T}{\pi r_1^2\omega[4Hhr_2 + r_1^2(r_2 - r_1)]}$$

$$= \frac{2(102.5-100)\times5\times10^{-6}\times5.9}{\pi\times(0.1)^2\times39.27[4\times0.295\times0.005\times0.1025+(0.1)^2(102.5-100)\times10^{-3}]}$$

$$= 01898 \text{ Pa.s}$$

Example 10.11 A single column monometer is connected to a pipe carrying a liquid of specific gravity 0.9 as shown in Fig. 10.34. Find the pressure in the pipe if the area of the reservoir is 100 times the area of the tube. The specific gravity of mercury is 13.6.

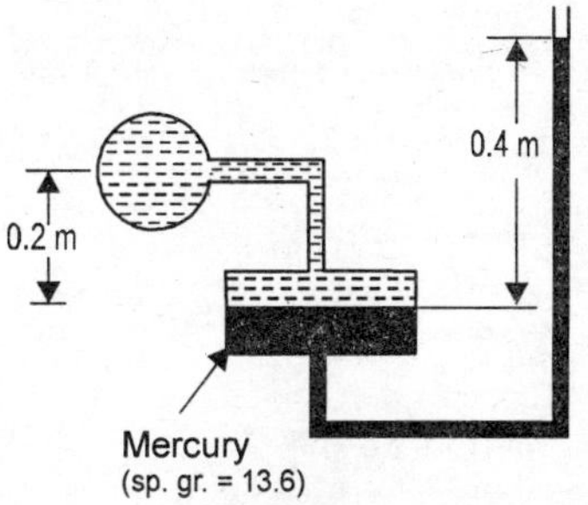

Fig 10.34

Solution

Given: $s_1 = 0.9$, $s_2 = 13.6$, $\frac{A}{a} = 100$, $h_1 = 0.2$ m, $h_2 = 0.4$ m

$$\frac{p_A}{\gamma_1} = \frac{a}{A}h_2(s_2 - s_1) + h_2 s_2 - h_1 s_1$$

$$= \frac{0.4}{100}(13.6-0.9)+0.4\times 13.6-0.2\times 0.9$$

$$= 5.3108$$

$$p_A = 0.9 \times 10^3 \times 5.3108 \times 9.81 = 46889 \text{ N/m}^2$$

Example 10.12 *A pitot-static tube placed in the centre of a 300 mm diameter pipe has one orifice pointing upstream and the other perpendicular to it. The mean velocity in the pipe is 0.8 times the central velocity. Calculate the discharge through the pipe if the pressure difference between the two orifices is 60 mm of water. Take the velocity coefficient of pitot tube as, $C_v = 0.98$.*

Solution.

Given: $d = 0.3$ *m*, $v_m = 0.8\, v_{max}$, $h = 0.06$ *m*, $C_v = 0.98$

$$v_{\max} = C_v\sqrt{2gh} = 0.98\sqrt{2\times 9.81\times 0.06} = 1.063 \text{ m/s}$$

$$v_m = 0.8 \times 1.063 = 0.8504 \text{ m/s}$$

$$Q = \frac{\pi}{4}d^2 v_m = \frac{\pi}{4}(0.3)^2 \times 0.8504 = 0.06\,\text{m}^3/s$$

Example 10.13 *Determine the throat diameter of a venturimeter in an installation of 100 mm diameter pipe carrying water. The maximum range available in water–Hg differential pressure gauge is 50 cm of Hg deflection. Find the maximum throat diameter which will induce full gauge defelection when the flow rate is 20 l/s. Assume $C_d = 0.984$.* **(IES, 2002)**

Solution $d_1 = 100$ mm, $Q = 20$ *lps* $= 0.02$ m³/s, $C_d = 0.984$

$$Q = \left[\frac{a_1 a_2}{(a_1^2 - a_2^2)^{0.5}}\right] C_d (2gh)^{0.5}$$

$$h = x\left(\frac{S_{\text{mercury}}}{S_{\text{water}}} - 1\right) = 0.5\left[\frac{13.6}{1} - 1\right] = 6.3\,\text{m}$$

$$0.02 = \left[\frac{\left\{\left(\frac{\pi}{4}\right)0.2^2 \times \left(\frac{\pi}{4}\right) d_2^2\right\}}{\left\{\left(\frac{\pi}{4}\right)(0.1^4 - d_2^4)^{0.5}\right\}}\right] \times 0.984 \times [2\times 9.81\times 6.3]^{0.5}$$

$$0.02 = \frac{0.085922\, d_2^2}{(0.1^4 - d_2^4)^{0.5}}$$

$$0.1 - d_2^4 = 18.456\, d_2^4$$

$$d_2 = 47.61 \text{ mm}$$

SUMMARY

1. The pressure of a fluid may be measured by manometers and elastic elements.
2. A manometer measures the pressure at a point in a fluid by balancing the column of fluid by the same or another column of liquid.
3. The manometers may be classified as simple, differential, and micro-manometers.
4. Simple manometers are: piezometer, and U-tube manometer.
5. A differential manometer is used to measure the difference in pressure at two points.
6. The micro–manometers are used for measuring small difference of pressures.
7. The inclined manometer is used to measure small pressure precisely in low velocity gas flow.
8. A single column manometer require only one reading to be taken in narrow limb of the manometer.
9. Flow velocity can be measured by: Pitot static tube, hot-wire anemometer, hot-film anemometer, and laser-doppler anemometer.
10. Pitot-static tube:

$$v = C\sqrt{2gh}$$

where C = correction coefficient for the Pitot tube

$$h = \frac{p_{stag} - p_{stat}}{\gamma}$$

11. Discharge can be measured by: Rotameter, venturimeter, orificemeter, Pitot tube, Pitot static tube, turbine flow meter, etc.
12. Pitot tube:

Stagnation pressure, $p_s = p_0 + \frac{1}{2}\rho v_0^2$

where p_0, v_0 = static pressure and velocity respectively.

$$v_0 = \sqrt{\frac{2(p_s - p_0)}{\rho}} = \sqrt{2gh}$$

Dynamic pressure, $$h = \frac{v_0^2}{2g}$$

$$Q = Av_0$$

13. Venturimeter :

$$Q = C_d \frac{A_1 A_2}{\sqrt{A_1^2 - A_2^2}} \cdot \sqrt{2g \cdot \Delta h}$$

14. Orificemeter :

$$Q = C_d \frac{A_0 A_1}{\sqrt{A_1^2 - A_0^2}} . \sqrt{2g . \Delta h}$$

15. Rotameter :

$$Q = C_d \frac{(A_t - A_b)}{\sqrt{1 - \left(\frac{A_t - A_b}{A_t}\right)^2}} . \sqrt{2g} \sqrt{\frac{V_b}{A_b} \times \left(\frac{\rho_b - \rho_f}{\rho_f}\right)}$$

where v_b, A_b = volume and area of bob respectively

ρ_f, ρ_b = density of flowing fluid and bob respectively.

A_t = area of tube

$$p_d - p_u = \frac{V_b}{A_b} g(\rho_b - \rho_f)$$

p_d, p_u = pressure at the downward and upward faces of bob respectively.

16. Dynamic viscosity measurements :

(i) Capillary tube method.

$$\mu = \frac{\pi \gamma h d^4}{128 Q l}$$

(ii) Falling sphere method

$$\mu = \frac{g d^2}{18 v} (\rho_s - \rho_l)$$

(iii) Rotating cylinder method.

$$\mu = \frac{2(r_2 - r_1) h T}{\pi r_1^2 \omega [4 H h r_2 + r_1^2 (r_2 - r_1)]}$$

(iv) Saybolt viscometer.

$$\nu = \frac{\mu}{\rho} = 0.24 t - \frac{190}{t}$$

where t = time in seconds.

Short Answer Type Questions

1. What is a manometer?
2. How manometer are classified?
3. What is a piezometer?
4. What is a differential manometer?
5. Explain the working of U-tube manometer.
6. What is the main advantage of a single leg manometer?
7. What are micro-manometers?
8. Explain the working of a Bourdon gauge.
9. What are the methods for measuring flow velocity?
10. What are the uses of Pitot-static tube?
11. Where do we use hot-wire anemometer?
12. Name the variable head meters for flow measurements.
13. Name the variable area meters for flow measurements.
14. List the various methods for the measurement of dynamic viscosity.

EXERCISES

1. The pressure of water in a pipe line was measured by means of a simple manometer as shown in Fig. 10.35. Calculate the pressure of water. **[Ans.** 65,727 Pa]

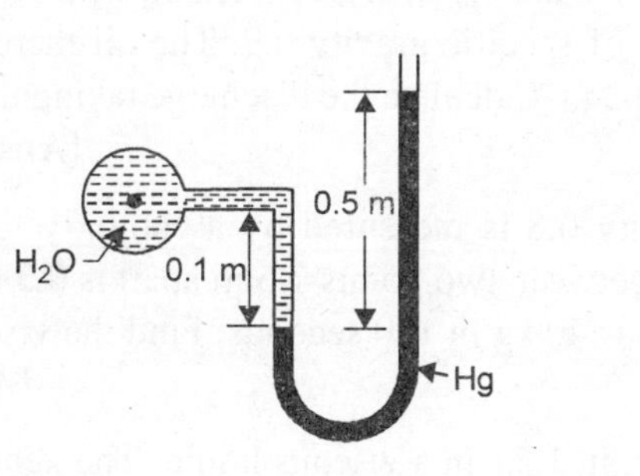

Fig. 10.35

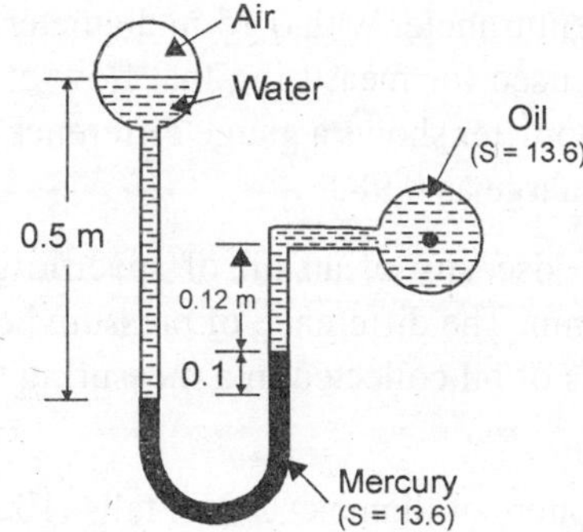

Fig. 10.36

2. Fig. 10.36 shows a differential manometer. If air pressure in A is 80 kPa, find the absolute pressure at B. **[Ans.** 72.497 kPa]

3. A micrometer, having a ratio of reservoir to limb area as 40, was used to determine the pressure in a pipe containing water. Determine the pressure in the pipe for manometer readings as shown in Fig. 10.37 **[Ans.** 1.0632 m]

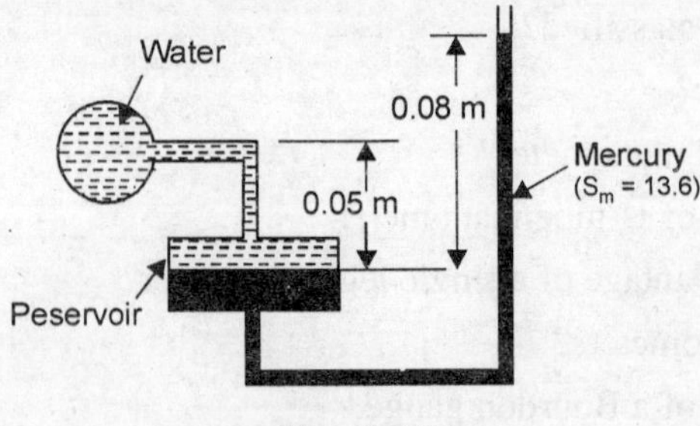

Fig. 10.37

4. Calculate the pressure at point A in Fig. 10.38. The slope of the tube is 4 horizontal to 1 vertical. The ratio of areas of reservoir to tube is 100. The fluid in the pipe is air and that in the manometer is kerosene having specific gravity = 0.8. **[Ans.** d]

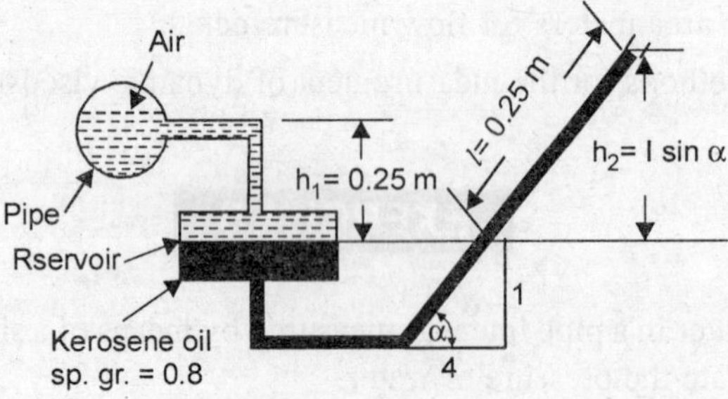

Fig. 10.38

5. A venturimeter with 0.15 m diameter at inlet and 0.1 m at throat is laid with its axis horizontal. It is used for measuring that flow of oil of specific gravity 0.9. The oil-mercury differential manometer shows a gauge difference of 0.2 m. Calculate the discharge taking the coefficient of discharge as 0.98. **[Ans.** 0.06393 m^3/s]

6. The viscosity of an oil of specific gravity 0.8 is measured by a capillary tube of diameter 40 mm. The difference of pressure head between two points 1.5 m apart is 0.3 m of water. The mass of oil collected in a measuring tank is 40 kg in 120 seconds. Find the viscosity of oil. **[Ans.** 0.236 Pa.s]

7. A sphere of diameter 3 mm falls 100 mm in 1.5 s in a viscous liquid. The density of sphere is 7000 kg/m^3 and of liquid is 800 kg/m^3. Find the dynamic viscosity of the liquid. **[Ans.** 4.561 Pa.s]

8. The viscosity of a liquid is determined by rotating cylinder method, in which case the inner cylinder 0.250 m is stationary. The outer cylinder of diameter 0.255 m contains the liquid upto a height of 0.4 m. The clearance at the bottom of the two cylinders is 6 mm. The outer cylinder is rotated at 300 rpm. The torque registered on the torsion metre attached to the inner cylinder is 4.905 N.m Calculate the viscosity of liquid **[Ans.** 0.077 Pa.s]

11

DIMENSIONAL AND MODEL ANALYSIS

PART-I: DIMENSIONAL ANALYSIS

11.1 INTRODUCTION

Dimensional analysis is a mathematical technique for solving engineering problems. It makes use of the dimensions of variables on which the problem depends. A physical phenomenon can be expressed by an equation giving relationship between different quantities. Such quantities are dimensional and non-dimensional. Dimensional analysis helps in determining a systematic arrangement of the variables in the physical relationship, combining dimensional variables to form non-dimensional parameters. It has become an important tool for analysing fluid flow problems.

11.2 SYSTEMS OF DIMENSIONS

The various physical quantities can be expressed in terms of *fundamental quatities*. The fundamental (or primary) quantities are: mass (M), length (L), time (T), and temperature (θ). The quantities which are expressed in terms of the fundamental quantities are called *derived* (or secondary) *quantities,* e.g. velocity, area, acceleration etc. The expression for a derived quantity in terms of the primary quantities is called the *dimension* of the physical quantity.

The two common system of dimensioning a physical quantity are; M–L–T and F–L–T system of units, where F = force. The dimensions of various quantities used in both the systems are given in Table 11.1.

11.3 DIMENSIONAL HOMOGENEITY

Dimensional homogeneity states that every term in an equation when reduced to fundamental dimensions must contain identical powers of each dimension. A dimensionally homogenous equation is independent of the fundamental units of measurment if the units there in are consistent. Let us consider the velocity equation,

$$v = \sqrt{2gh}$$

$$[LT^{-1}] = [2 \times LT^{-2} \times L]^{1/2} = [L^2\,T^{-2}]^{1/2} = [LT^{-1}]$$

Similarly,

$$h_f = \frac{4flv^2}{2gd}$$

$$[L] = \frac{[L][LT^{-1}]^2}{[LT^{-2}][L]} = [L]$$

Dimensional homogenity is based on the *Fourier's principle of homogeneity.*

Table 11.1 Dimensions of various physical quantities.

Physical quantiy	Symbol	Dimensions	
		M–L–T system	F–L–T system
1. Fundamental Quantities			
Mass	M	M	$FL^{-1}T^2$
Length	L	*L*	L
Time	T	T	T
Force	F	MLT^{-2}	F
2. Geometric Quantities			
Area	A	L^2	L^2
Volume	V	L^3	L^3
Moment of inertia	I	L^4	L^4
3. Kinematic Quantities			
Linear velocity	*u, v, w, U, V*	LT^{-1}	LT^{-1}
Angular velocity	ω	T^{-1}	T^{-1}
Acceleration	*a* or *f*	LT^{-2}	LT^{-2}
Angular acceleration	∝	T^{-2}	T^{-2}
Discharge	*q, Q*	L^3T^{-1}	L^3T^{-1}
Gravity	*g*	LT^{-2}	LT^{-2}
Kinematic viscosity	ν	L^2T^{-1}	L^2T^{-2}
Stream function, Circulation	ψ, Γ	L^2T^{-1}	L^2T^{-1}
Vorticity	ξ	T^{-1}	T^{-1}
4. Dynamic Quantities			
Force, Thrust	F	MLT^{-2}	F
Density	ρ	ML^{-3}	$FL^{-4}T^2$
Specific weight	γ	$ML^{-2}T^{-2}$	FL^{-3}
Dynamic viscosity	μ	$ML^{-1}T^{-1}$	$FL^{-2}T$
Surface tension	σ	MT^{-2}	FL^{-1}
Pressure, shear stress	*p*, τ	$ML^{-1}T^{-2}$	FL^{-2}
Modulus of elasticity, Bulk modulus	E, K	$ML^{-1}T^{-2}$	FL^{-2}
Momentum	–	MLT^{-1}	FT
Angular momentum	–	ML^2T^{-1}	FLT
Work	W	ML^2T^{-2}	FL
Energy	E	ML^2T^{-2}	FL
Torque	T	ML^2T^{-2}	FL
Power	P	ML^2T^{-3}	FLT^{-1}

11.4 METHODS OF DIMENSIONAL ANALYSIS

The two commonly used methods for dimensional analysis are:

1. Rayleigh's methods, and 2. Buckingham's π-method.

11.4.1 Rayleigh's Method

The Reyleigh's method gives a special form of relationship among the dimensionless groups. Its inherent drawback is that it does not provide any information regrading the number of dimensionless groups to be obtained as a result of dimensional analysis. It is used for determining the expression for a variable which depends upon maximum of three or four variables only. In case the number of independent variables becomes more, then it is very difficult to find the expression for the dependent variable.

The method involves the following steps:

(i) List all the independent variables which are likely to influence the value of dependent variables.

(ii) Write the functional relationship between dependent and independent variables.

$$y = f(x_1, x_2, x_3, ...) \quad ...(11.1)$$

(iii) Write Eq. (11.1) in the form:

$$y = K(x_1^a, x_2^b, x_c^c, ...) \quad ...(11.2)$$

where K = a dimensionaless coefficient to be determined.

(iv) Express each of the qunatities on both sides of Eq. (11.2) in the fundemental units of M,L and T.

(v) Apply the principle of dimensional homogeneity to obtain a set of simultaneous equations involving the exponents a, b, c, ...

(vi) Solve the simultaneous equations to obtain the values of exponents a, b, c,..... If the number of exponents involved is more than 3, then the exponents of D, v and ρ are evaluated in terms of other exponents.

(vii) Substite the values of exponents in the main equation and form the non-dimensional parameters by grouping the variables with like exponents.

Example 11.1 *Find an expression for the drag froce on a smooth sphere of diameter D, moving with a uniform velocity v in a fluid of density* ρ *and dynamic viscosity* μ.

Solution. The drag force F_D is a function of, D, v ,ρ and μ.

$$\therefore \quad F_D = f(D, v, \rho, \mu)$$

or

$$F_D = \phi(D^a, v^b, \rho^c, \mu^d)$$

where ϕ is a non-dimensional constant.

Using M–L–T system, we have

$$MLT^{-2} = \phi\,[L^a]\,[LT^{-1}]^b\,[ML^{-3}]^c\,[ML^{-1}T^{-1}]^d$$

$$[M]^1\,[L]^1\,[T]^{-2} = \phi\,[L]^{a+b-3c-d}\,[M]^{c+d}\,[T]^{-b-d}$$

For M: $$c + d = 1 \quad ...(1)$$

For L: $$a + b - 3c - d = 1 \quad ...(2)$$

For T: $$-a - b = -2 \quad ...(3)$$

There are four unknowns (a, b, c, d) and three equations. Therefore, it is not possible to find the values of a, b, c and d explicitly. Therefore, three of them can be expressed in terms of the fourth variable, which is most important. The role of viscosity μ is very important for drag force and hence a, b, c are expressed in terms of d.

$$\therefore \quad c = 1 - d$$

$$b = 2 - d$$

$$a = 1 - (2 - d) + 3(1 - d) + d = 2 - d$$

For

$$F_D = \phi[D]^{2-d}[v]^{2-d}[\rho]^{1-d}[\mu]^d$$

$$= \phi[D^2 v^2 \rho (D^{-d} v^{-d} \rho^{-d} \mu^d)]$$

$$= \phi\left[\rho D^2 v^2 \left(\frac{\mu}{\rho v D}\right)^d\right]$$

$$= \rho D^2 v^2 \phi\left(\frac{\mu}{\rho v D}\right)$$

Example 11.2 *The resistance force R of a supersonic plane during flight depends upon the length of aircraft l, velocity v, air viscosity μ, air density ρ, and bulk modulus of air K. Determine the functional relationship between these variables and the resisting force.*

Solution:

$$R = f_1(l, v, \mu, \rho, K)$$

or

$$R = f_2(l^a, v^b, \mu^c, \rho^d, K^e)$$

where f_2 is a non-dimensional constant.

Using the M–L–T system, we have

$$[M][L][T]^{-2} = f_2 [L]^a [T^{-1}]^b [ML^{-1}T^{-1}]^c [ML^{-3}]^d [ML^{-1} T^{-2}]^e$$

$$= f_2 [M]^{c+d+e} [L]^{a+b-c-3d-e} [T]^{-b-c-2e}$$

For dimensional homogeneity, we have

For M: $\quad c + d + e = 1$

For L: $\quad a + b - c - 3d - e = 1$

For T: $\quad -b - c - 2e = -2$

There are five variables and three equations. For a supersonic plane, μ and K are very important variables. Therefore, we express a, b, and d in terms of c and e.

$$d = 1 - c - e$$

$$b = 2 - c - 2e$$

$$a = 1 - (2 - c - 2e) + c + 3(1 - c - e) + e$$

$$= 1 - 2 + c + 2e + c + 3 - 3c - 3e + e$$

$$= 2 - c$$

$$\therefore \quad R = f_2[l]^{2-c}[v]^{2-c-2e}(\mu)^c[\rho]^{1-c-e}[K]^e$$

$$= f_2 l^2 v^2 \rho \left(\frac{\mu}{l v \rho}\right)^c \left(\frac{K}{v^2 \rho}\right)^e$$

$$= l^2 v^2 \rho \phi \left[\left(\frac{\mu}{l v \rho}\right), \left(\frac{K}{v^2 \rho}\right)\right]$$

Example 11.3 *The pressure drop Δp in a pipe of diameter D and length l depends on the density ρ and viscosity μ of the flowing fluid, mean velocity v of flow and average height of protuberance t. Show that the pressure drop can be expressed in the form:*

$$\Delta p = \rho v^2 \phi \left(\frac{l}{D}, \frac{\mu}{v D \rho}, \frac{t}{D}\right)$$

Solution

$$\Delta p = f_1 (D, l, \rho, \mu, v, t)$$

$$= f_2 [D^a, l^b, \rho^c, \mu^d, v^e, t^f]$$

where f_2 is a non-dimensional constant.

Using M–L–T system, we have

$$[M][L]^{-1} [T]^{-2} = f^2 [L]^a [L]^b [ML^{-3}]^c [ML^{-1} T^{-1}]^d [LT^{-1}]^e [L]^f$$

$$= f_2 [M]^{c+d} [L]^{a+b-3c-d+e+f} [T]^{-d-e}$$

For dimensional homogeneity, we have

For M: $c + d = 1$

For L: $a + b - 3c - d + e + f = -1$

For T: $-d - e = -2$

There are six unknowns and three equations. For pressure drop in pipe, l, μ and t are very important variables. Therefore, we determine a, c and e in terms of b, d and f.

$$c = 1 - d$$

$$e = 2 - d$$

$$a = -1 - b + 3c + d - e - f$$

$$= -1 - b + 3(1 - d) + d - (2 - d) - f$$

$$= -1 - b + 3 - 3d + d - 2 + d - f$$

$$= -b - d - f = -(b + d + f)$$

$$\Delta p = f_2 [D^{-(b+d+f)} . l^b \cdot \rho^{2-d} \cdot \mu^d \cdot v^{2-d} \cdot t^f]$$

$$= f_2 \left[\rho v^2 \left(\frac{l}{D}\right)^b \left(\frac{\mu}{v D \rho}\right)^d \left(\frac{t}{D}\right)^f\right]$$

$$= \rho v^2 \phi \left[\frac{l}{D}, \frac{\mu}{v D \rho}, \frac{t}{D}\right]$$

This is usually written in the form,

$$h_f = \frac{4flv^2}{2gD}$$

where f = coefficient of friction, $\frac{vD\rho}{\mu}$ = Re, and $\frac{t}{D}$ = surface finish of pipe.

Example 11.4 *A partially submerged body is towed in water. The resistance R to its motion depends on the density ρ, the viscosity μ of water, length l of the body, velocity v of the body, and acceleration due to gravity g. Show that the resistance to motion can be expressed in the form:*

$$R = \rho l^2 v^2 \phi\left[\left(\frac{\mu}{\rho l v}\right), \left(\frac{lg}{v^2}\right)\right]$$

Solution.

$$R = f(\rho, \mu, l, v, g)$$

or

$$R = C(\rho^a, \mu^b, l^c, v^d, g^e)$$

Where C = a non-dimensional constant.

Using M–L–T system, we have

$$[M][L][T]^{-2} = C[ML^{-3}]^a . [ML^{-1}T^{-1}]^b \cdot [L]^c [LT^{-1}]^d \cdot [LT^{-2}]^e$$

$$= C[M]^{a+b} \cdot [L]^{-3a-b+c+d+e} \cdot [T]^{-b-d-2e}$$

For dimensional homogeneity the exponents of each dimension on both sides of the equation must be identical. Thus,

For M: $a + b = 1$

For L: $-3a - b + c + d + e = 1$

For T: $-b - d - 2e = -2$

There are five unknown variables and three equations. The μ and g variables are more important for resistance to motion of a partially submerged body. Therefore, we evaluate a, c, d in terms of b and e.

$$a = 1 - b$$

$$d = 2 - b - 2e$$

$$c = 1 + 3a + b - d - e$$

$$= 1 + 3(1-b) + b - (2 - b - 2e) - e$$

$$= 1 + 3 - 3b + b - 2 + b + 2e - e$$

$$= 2 - b + e$$

$$\therefore \quad R = C[\rho^{1-b} \cdot \mu^b \cdot l^{2-b+e} \cdot v^{2-b-2e} \cdot g^e]$$

$$= C[\rho l^2 v^2 (\rho^{-b}\mu^b l^{-b} v^{-b})(l^e v^{-2e} g^e)]$$

$$= C\left[\rho l^2 v^2 \left(\frac{\mu}{\rho l v}\right)^b \cdot \left(\frac{lg}{v^2}\right)^e\right]$$

$$= \rho l^2 v^2 \phi\left[\left(\frac{\mu}{\rho l v}\right), \left(\frac{lg}{v^2}\right)\right]$$

Example 11.5 *The efficiency η of a fan depends on density ρ, the dynamic viscosity μ of the fluid, the angular velocity ω, diameter D of rotor and the discharge Q. Express η in terms of dimensionless parameters.*

Solution.

$$\eta = f(\rho, \mu, \omega, D, Q)$$

or

$$\eta = C(\rho^a, \mu^b, \omega^c, D^d, Q^e)$$

where C = non-dimensional constant.

Using M–L–T system, the corresponding equation for dimensions is:

$$[M]^0 [L]^0 [T]^0 = C [ML^{-3}]^2 \ [ML^{-1}T^{-1}]^b \ [T^{-1}]^c \ [L]^d \ [L^3T^{-1}]^e$$

$$= C[M]^{a+b}[L]^{-3a-b+d+3e}[T]^{-b-c-e}$$

For dimensional homogeneity, the exponent of each dimesnion on both sides of the equation must be identical. Thus,

For M: $a + b = 0$

For L: $-3a - b + d + 3e = 0$

For T: $-b - c - e = 0$

There are five unknown variables and three equations. The efficiency of the fan is most affected by μ and Q. Therefore, we evaluate a,c,d in terms of b and e.

$$a = -b$$

$$c = -b - e$$

$$d = 3a + b - 3e$$

$$= -3b + b - 3e = -2b - 3e$$

∴

$$\eta = C[\rho^{-b} \cdot \mu^b \cdot \omega^{-(b+e)} \cdot D^{-(2b+3e)} \cdot Q^e]$$

$$= C\left[\left(\frac{\mu}{\rho\omega D^2}\right)^d, \left(\frac{Q}{\omega D^3}\right)^e\right]$$

$$= \phi\left[\left(\frac{\mu}{\rho\omega D^2}\right), \left(\frac{Q}{\omega D^3}\right)\right]$$

Example 11.6 *A circular cylinder of a given Length/Diameter ratio is kept in steady rotation at N rev./sec. in a uniform stream of fluid of velocity v. Assuming that the power required to maintain the motion depends on density ρ, kinematic viscosity ν of the fluid, and the diameter D of the cylinder. Using Rayleigh's method show that,*

$$P = \left(\frac{\rho v^3}{D}\right) f\left(\frac{vD}{\nu}, \frac{ND^2}{\nu}\right) \qquad \textit{[IES. 2003]}$$

Solution.

Let $P = f(\rho, v, d, \nu, N)$

or $P = C[\rho^a, v^b, D^c, \nu^d, N^e]$

where C = non-dimensional constant

Using M–L–T system, the corresponding equation for dimensions is:

$$M L^2 - T^{-3} = C[ML^{-3}]^a [LT^{-1}]^b [L]^c [L^2T^{-1}] [T^{-1}]^e$$

$$= C[M]^a [L]^{-3a+b+c+2d} [T]^{-b-d-e}$$

For dimensional homogeneity the exponents of each dimension on both sides of the equation must be identical Thus,

For M: $a = 1$

For L: $-3a + b + c + 2d = 2$

For T: $-b - d - e = -3$

There are five unknown variables and three equations. The fluid velocity v *and* rotational speed N of cylinder are important variables. Therefore, we express a, c, d in terms of b and e.

$$a = 1$$

$$d = 3 - b - e$$

$$c = 2 + 3a - b - 2d$$

$$= 2 + 3 - b - 6 + 2b + 2e$$

$$= -1 + b + 2e$$

$$\therefore \quad P = C[(\rho)^1 \cdot (v)^b \cdot (D)^{-1+b+2e} \cdot (\nu)^{3-b-e} \cdot (N)^e]$$

$$= C\left[\left(\frac{\rho v^3}{D}\right), \left(\frac{vD}{\nu}\right)^b, \left(\frac{ND^2}{\nu}\right)^e\right]$$

$$= \left(\frac{\rho v^3}{D}\right) \phi\left[\left(\frac{vD}{\nu}\right), \left(\frac{ND^2}{\nu}\right)\right]$$

11.4.2 Buckingham's π–Theorem method

The Rayleigh's method of dimensional analysis becomes laborious and cumbersome when the number of pysical variables are large. Buckingham's π-method has the advantage that the number of dimensionless groups, called *pi* (Greek letter π) to be formed are known in advance.

Buckingham's π–Theorem

This theorem states that if there are n number of variables (both dependent and independent) in a dimensionally homogeneous equation and these variables contain m fundmental dimensions then the number of dimensionless groups, which can be formed, shall be $(n - m)$. These dimensionless group are called π-terms or groups.

Consider a phenomenon involving n variables x_i, $i = 1$ to n, such that a variable say x_1 depends upon independent variables $x_2, x_3, \dots, x_n$ and upon no other, the general functional relationship between the dependent variable and independent variables may be expressed as:

$$x_1 = f(x_2, x_3, \dots, x_n) \quad \dots (11.3)$$

On account of the mathematical equilibrium between dependent and independent variables, Eq. (11.3) can be written as,

$$\phi(x_1, x_2, \dots, x_n) = 0 \quad \dots (11.4)$$

It is a dimensionally homogeneous equation and contains n varibles. If there are m fundamental dimensions, then according to Buckingham's π-theorem Eq. (11.4) can be written in terms of $(n-m)$ number of π –terms. Hence Eq. (11.4) becomes as,

$$\phi(\pi_1, \pi_2, \dots, \pi_{n-m}) = 0 \quad \dots (11.4)$$

in which each π-term may contain maximum of $(m + 1)$ variables, only one of which is required to be changed from term to term. These m variables which appear repeatedly in each of π-terms are called *repeating variables*. The repeating variables are chosen from among the variables such that they together involve all the fundamental dimensions and they themselves do not form a dimensionless parameter. If each of the fundamental dimensions (M, L, T) are involved, then $m = 3$ and the result will be an equation with $(n - 3)$ dimensionless terms.

Guidelines for selecting repeating variables

The following points should be kept in view while selecting repeating variables:

1. Select the first repeating variable from those describing the geometry of flow e.g.size and shape of fluid passage or of the moving body such as diameter, length, etc.
2. Select the second repeating variable from those representing the fluid properties such as density, viscosity, surface tension, elasticity and vapour pressure.
3. Select the third repeating variable from those characterising the fluid motion such as velocity, acceleration, discharge, pressure, force, power.

Some dont's while selecting repeating variables

1. The repeating variables must not form the non-dimensional parameters among themselves.
2. Dependent variable should not be selected as repeating variable.
3. No two repeating variables should have the same dimensions.

Choice of repeating variables

1. l, v, ρ 2. l, v, μ 3. d, v, ρ 4. d, v, μ.

Example 11.7 *Solve example 11.4 by using Buckingham's π-theorem.*

Solution:

1. $$R = f(v, l, \mu, \rho, g)$$

or $$\phi(R, v, l, \mu, \rho, g) = 0$$

Here $n = 6$ and $m = 3$

Number of dimensionless π – terms = $n - m = 6 - 3 = 3$

2. Selection of repeating variables

R: is a dependent variable and should not be selected as a repeating variable.

Geometric property variable : l having dimension L

Fluid property variable : ρ having dimension ML^{-3}

Fluid motion variable : v having dimension LT^{-1}

l, ρ, v do not form a dimensionless group.

Hence l, ρ, v are the repeating variables.

3. Number of variables forming a π-term = $m + 1 = 3 + 1 = 4$

$$\pi_1 = l^{a_1},\ v^{b_1},\ \rho^{c_1},\ R$$

$$\pi_2 = l^{a_2},\ v^{b_2},\ \rho^{c_2},\ \mu$$

$$\pi_3 = l^{a_3},\ v^{b_3},\ \rho^{c_3},\ g$$

4. π – terms solutions:

(*i*) π_1 – term

$$[M]^0[L]^0[T]^0 = [L]^{a_1}\cdot[LT^{-1}]^{b_1}\cdot[ML^{-3}]^{c_1}\cdot[MLT^{-2}]$$

$$= [M]^{c_1+1}\cdot[L]^{a_1+b_1-3c_1+1}\cdot[T]^{-b_1-2}$$

For M: $c_1 + 1 = 0,\ c_1 = -1$

For L: $a_1 + b_1 - 3c_1 + 1 = 0,\ a_1 + b_1 = -4$

For T: $-b_1 - 2 = 0,\quad b_1 = -2$

Then $a_1 = -4 - b_1 = -4 + 2 = -2$

$$\pi_1 = l^{-2}v^{-2}\rho^{-1}R = \frac{R}{l^2v^2\rho}$$

(*ii*) π_2 – term

$$[M]^0[L]^0[T]^0 = [L]^{a_2}[LT^{-1}]^{b_2}[ML^{-3}]^{c_2}[ML^{-1}T^{-1}]$$

$$= [M]^{c_2+1}\cdot[L]^{a_2+b_2-3c_2-1}\cdot[T]^{-b_2-1}$$

For M: $c_2 + 1 = 0,\ c_2 = -1$

For L: $a_2 + b_2 - 3c_2 - 1 = 0, \quad a_2 + b_2 = -2$

For T: $-b_2 - 1 = 0, \quad b_2 = -1$, then $a_2 = -1$

$$\pi_2 = l^{-1}v^{-1}\rho^{-1}\mu = \frac{\mu}{lv\rho}$$

(iii) π_s – term

$$[M]^0[L]^0[T]^0 = [L]^{a_3}[LT^{-1}]^{b_3}[ML^{-3}]^{c_3}[LT^{-2}]$$

$$= [M]^{c_3}[L]^{a_3+b_3-3c_3+1}[T]^{-b_3-2}$$

For M: $c_3 = 0$

For L: $a_3 + b_3 - 3c_3 + 1 = 0,\ a_3 + b_3 + 1 = 0$

For T: $-b_3 - 2 = 0,\ b_3 - 2$, then $a_3 - 2 + 1 = 0,\ a_3 = 1$

$$\pi_3 = l^1 v^{-2}\rho^0 g = \frac{lg}{v^2}$$

5. $\phi\left[\frac{R}{l^2v^2\rho}, \frac{\mu}{lvp}, \frac{lg}{v^2}\right]$

or

$$\frac{R}{l^2v^2\rho} = \phi_1\left[\frac{\rho lv}{\mu}, \frac{v}{\sqrt{lg}}\right]$$

$$R = l^2v^2\rho\phi_1\left[\frac{\rho vl}{\mu}, \frac{v}{\sqrt{lg}}\right]$$

$$= l^2v^2\rho\phi_1(\text{Re}, \text{Fr})$$

Example 11.8 *Using Buckingham's π-theorem, show that the velocity through a circular orifice is given by:*

$$v = \sqrt{2gH}\,\phi\left[\frac{D}{H}, \frac{\mu}{\rho vH}\right]$$

where H = head causing flow, D = diameter of orifice, μ = dynamic viscosity, ρ = mass density, and g = acceleration due to gravity.

Solution.

$$v = f(H, D, \mu, \rho, g)$$

or $$f_1(v, H, D, \mu, \rho, g) = 0$$

Here $n = 6$, $m = 3$

Number of π-terms $= n - m = 6 - 3 = 3$

$$\therefore \quad f_1(\pi_1\, \pi_2, \pi_3) = 0$$

2. Number of variables in each π–term $= m + 1 = 3 + 1 = 4$

Number of repeating variables $= m = 3$

Property	**Repeating Variable**
Geometric	H
Fluid	ρ
Motion	g

v being dependent variable is not being chosen as repeating variable.

3. $\pi_1 = H^{a_1} g^{b_1} \rho^{c_1}\, v$

$\pi_2 = H^{a_2} g^{b_2} \rho^{c_2}\, D$

$\pi_3 = H^{a_3} g^{b_3} \rho^{c_3}\, \mu$

4. π – term solutions:

(i) π_1 – term

$$[M]^0[L]^0[T]^0 = [L]^{a_1}[LT^{-2}]^{b_1}[ML^{-3}]^{c_1}[LT^{-1}]$$

$$= [M]^{c_1}[L]^{a_1+b_1-3c_1+1}[T]^{-2b_1-1}$$

For M: $c_1 = 0$

For L: $a_1 + b_1 - 3c_1 + 1 = 0,\ a_1 + b_1 + 1 = 0$

For T: $-2b_1 - 1 = 0,\ b_1 = -\frac{1}{2}$, then $a_1 = -1 + \frac{1}{2} = -\frac{1}{2}$

$$\pi_1 = [H]^{-\frac{1}{2}}[g]^{-1/2}[\rho]^0[v] = \frac{v}{\sqrt{gH}}$$

(ii) π_2-term:

$$[M]^0[L]^0[T]^0 = [L]^{a_2}[LT^{-2}]^{b_2}[ML^{-3}]^{c_2}[L]^1$$

$$= [M]^{c_2}[L]^{a_2+b_2-3c_2+1}[T]^{-2b_2}$$

For M: $c_2 = 0$

For L: $a_2 + b_2 - 3c_2 + 1 = 0,\ a_2 + b_2 + 1 = 0$

For T: $-2b_2 = 0, b_2 = 0,$ then $a_2 = -1$

$$\pi_2 = [H]^{-1}[g]^0[\rho]^0[D] = \frac{D}{H}$$

(iii) π_3 – term:

$$[M]^0[L]^0[T]^0 = [L]^{a_3}[LT^{-2}]^{b_3}[ML^{-3}]^{c_3}[ML^{-1}T^{-1}]$$

$$= [M]^{c_3+1}[L]^{a_3+b_3-3c_3-1}[T]^{-2b_3-1}$$

For M: $c_3 + 1 = 0, c_3 = -1$

For L: $a_3 + b_3 - 3c_3 - 1 = 0, \quad a_3 + b_3 + 2 = 0$

For T: $-2b_3 - 1 = 0, \quad b_3 = \frac{1}{2},$ then $a_3 = -2 + \frac{1}{2} = -\frac{3}{2}$

$$\pi_3 = [H]^{-3/2}[g]^{-1/2}[\rho]^{-1}[\mu]$$

$$= \frac{\mu}{H^{3/2}\rho\sqrt{g}} = \frac{\mu}{H\rho\sqrt{gH}} = \frac{\mu}{H\rho v}\cdot\pi_1$$

$$\therefore \quad f_1\left(\frac{v}{\sqrt{gH}}, \frac{D}{H}, \frac{\mu}{H\rho v}\cdot\pi_1\right) = 0$$

or
$$\frac{v}{\sqrt{gH}} = \phi\left[\frac{D}{H}, \frac{\mu}{H\rho v}\cdot\pi_1\right]$$

or
$$v = \sqrt{2gH}\,\phi\left[\frac{D}{H}, \frac{\mu}{\rho v H}\right]$$

Hence proved.

Example 11.9 *State the Buckingham's π-theorem and hence obtain an expression for the thrust developed by a propeller which depends upon the angular velocity ω, apporach velocity V, dynamic viscosity μ, density ρ, propeller diameter D and the compressibility of the medium measured by the local velocity of sound C.* ***[IES, 2005]***

Solution. Buckingham's Pi-theorem. If there are n variables in a dimensionally homogeneous equation and if these variables contain m primary dimensions, then the variables can be grouped into $(n - m)$ non-dimensional parameters. The non-dimensional groups are called π-terms.

For the propeller, we have

$$F(T, \omega, V, D, \mu, \rho, C) = 0$$

Here $n = 7$, $m = 3$

Therefore, number of π-terms = $7 - 3 = 4$

Choose V,D,ρ as the repeating variables.

(i)

$$\pi_1 = v^{a_1} d^{b_1} \rho^{c_1} T$$

$$M^0 L^0 T^0 = [LT^{-1}]^{a_1} [L]^{b_1} [ML^{-3}]^{d} MLT^{-2}$$

$$= M^{c_1+1} L^{a_1+b_1-3c_1+1} T^{a_1-2}$$

$$0 = c_1 + 1; 0 = a_1 + b_1 - 3c_1 + 1; 0 = -a_1 - 2$$

$$a_1 = -2,\ b_1 = -2,\ c_1 = -1$$

$$\pi_1 = V^{-2} D^{-2} \rho^{-1} T = \frac{T}{V^2 D^2 \rho}$$

(ii)

$$\pi_2 = V^{a_2} D^{b_2} \rho^{c_2} \mu$$

$$M^0 L^0 T^0 = [LT^{-1}]^{a2} [L]^{b2} [ML^{-3}]^{c_2} ML^{-1}T^{-1}$$

$$= M^{c_2+1} L^{a_2+b_2-3c_2-1} T^{a_2-1}$$

$$0 = c_2 + 1; 0 = a_2 + b_2 - 3c_2 - 1; 0 = -a_2 - 1$$

$$a_2 = -1,\ b_2 = -1,\ c_2 = -1$$

$$\pi_2 = V^{-1} D^{-1} \rho^{-1} \mu = \frac{\mu}{VD\rho}$$

(iii)

$$\pi_3 = V^{a3} D^{b3} \rho^{c3} \omega$$

$$M^0 L^0 T^0 = [LT^{-1}]^{a_3} [L]^{b_3} [ML^{-3}]^{c_3} T^{-1}$$

$$= M^{c3} L^{a_3+b_3-3c_3-1} T^{a_3-1}$$

$$0 = c_3; 0 = a_3 + b_3 - 3c_3; 0 = -a_3 - 1$$

$$a_3 = -1,\ b_3 = c_3 = 0$$

$$\pi_3 = V^{-1} D^{1} \rho^{0} \omega = \frac{D\omega}{V}$$

(iv)

$$\pi_4 = V^{a4} D^{b4} \rho^{c4} C$$

$$M^0 L^0 T^0 = [LT^{-1}]^{a4} [L]^{b4} [ML]^{c04} LT^{-1}$$

$$= M^{c_4} L^{a_4+b_4+3c_4+1} T^{-a_4-1}$$

$$0 = c_4; 0 = a_4 + b_4 - 3c_4 + 1; 0 = -a_4 - 1$$

$$a_4 = -1,\ b_4 = 0,\ c_4 = 0$$

$$\pi_4 = V^{-}D^0\rho^0 C = \frac{C}{V}$$

$$\phi = \left[\frac{\mu}{VD\rho}, \frac{D\omega}{V}, \frac{C}{V}\right]$$

$$\frac{T}{V^2D^2\rho} = \phi\left[\frac{\mu}{VD\rho}, \frac{D\omega}{V}, \frac{C}{V}\right]$$

$$T = (V^2D^2\rho)\phi\left[\frac{\mu}{VD\rho}, \frac{D\omega}{V}, \frac{C}{V}\right]$$

$$= (V^2D^2\rho)\phi\left[\frac{1}{\text{Re}}, \frac{D\omega}{V}, \frac{1}{M}\right]$$

Example 11.10 *In a compressor, the frictional torque is T in the impeller. If the diameter is D, speed N, fluid viscosity μ and density ρ, prove using Buckingham's π-theorem of dimensional analysis that*

$$T = \rho . N^2 . D^5 . f\left[\frac{\mu}{\rho N D^2}\right]$$ *[IES, 2007]*

Solution: Frictional torque in a compressor,

$$T = f(D, N, \mu, \rho)$$

Here $n = 5, m = 3$

Number of dimensionless parameters $= n - m = 5 - 3 = 2$

The dimensions of variables are

$D : L$

$N : T^{-1}$

$\mu : ML^{-1}T^{-1}$

$\rho : ML^{-3}$

$T : ML^{-2}T^{-2}$

Let D, ρ, N be the repeating variables

$$\pi_1 = D^{a_1}\rho^{b_1}, N^{c_1}\mu$$

$$M^0 L^0 T^0 = [L]^{a_1}\cdot[ML^{-3}]^{b_1}\cdot[T^{-1}]^{c_1} ML^{-1}T^{-1}$$

$$= L^{a_1-3b_1-1}M^{b_1+1}T^{-c_1-1}$$

$$b_1+1 = 0, b_1 = -1$$

$$a_1 - 3b_1 - 1 = 0, a_1 = -2$$

$$-c_1 - 1 = 0, c_1 = -1$$

$$\pi_1 = D^{-2}\rho^{-1}N^{-1}\mu = \frac{\mu}{\rho ND^2}$$

$$\pi_2 = D^{a_2}\rho^{b_2}N^{c_2}T$$

$$M^0 L^0 T^0 = [L]^{a_2}\cdot[ML^{-3}]^{b_2}[T^{-1}]^{c_2} ML^2T^{-2}$$

$$= M^{b_2+1}L^{a_2-3b_2+2}T^{-c_2-2}$$

$$b_2 + 1 = 0, b_2 = -1$$

$$a_2 - 3b_2 + 2 = 0, a_2 = -5$$

$$-c_2 - 2 = 0, c_2 = -2$$

$$\pi_2 = D^{-5}\rho^{-1}N^{-2}T = \frac{T}{\rho N^2 D^5}$$

Now,
$$f(\pi_1, \pi_2) = 0$$

$$\pi_2 = f(\pi_1)$$

$$\frac{T}{pN^2D^5} = f\left(\frac{\mu}{\rho ND^2}\right)$$

$$T = \rho N^2 D^5 f\left(\frac{\mu}{\rho ND^2}\right)$$

Example 11.11 *The rate of flow Q of a liquid having mass density ρ and viscosity μ is to be measured by means of an orificemeter. The diameter of the orificementer is d which is installed in a pipeline of diameter D. The pressure drop across the orifice plate is Δp. Obtain the expression for Q by using the method of dimensional analysis.* ***[IES, 2006]***

Solution
$$Q = f(d, D, \rho, \mu, \Delta p)$$

Number of variables, $n = 6$

Number of fundamental dimensions, $m = 3$

Number of dimensionless group (π-terms) = $n - m = 6 - 3 = 3$

Let us select ρ, Q, d as repeating variables.

$$\pi_1 = [\rho]^{a_1}[Q]^{b_1}[d]^{c_1}[D]^1$$

$$M^0 L^0 T^0 = [ML^{-3}]^{a_1}[L^3T^{-1}]^{b_1}[L]^{c_1}[L]^1$$

$$= [M]^{a_1}[L]^{-3a_1+3b_1+c_1+1}[T]^{-b_1}$$

For dimensional homogeneity, we have

$$a_1 = 0$$

$$-3a_1 + 3b_1 + c_1 + 1 = 0$$

$$-b_1 = 0$$

Thus $$a_1 = 0,\ b_1 = 0,\ c_1 = 1$$

$$\pi_1 = \left[\frac{D}{d}\right]$$

$$\pi_2 = [\rho]^{a_2}[Q]^{b_2}[d]^{c_2}[\mu]^1$$

$$M^0 L^0 T^0 = \left[ML^{-3}\right]^{a_2}\left[L^3 T^{-1}\right]^{b_2}[L]^{c_2}\left[ML^{-1}T^{-1}\right]^1$$

$$= [M]^{a_2+1}[L]^{3a_2+3b_3+c_2-1}[T]^{-b_2-2}$$

For dimensional homogeneity, we have

$$a_2 + 1 = 0$$

$$-3a_2 + 3b_2 + c_2 - 1 = 0$$

$$-b_2 - 1 = 0$$

Thus $$a_2 = -1,\ b_2 = -1,\ c_2 = 1$$

$$\pi_2 = \left[\frac{\mu d}{\rho Q}\right]$$

$$\pi_3 = [\rho]^{a_3}[Q]^{b_3}[d]^{c_3}[\Delta p]^1$$

$$M^0L^0T^0 = [ML^{-3}]^{a_3}[L^3T^{-1}]^{b_3}[L]^{c_3}[ML^{-1}T^{-2}]^1$$

$$= [M]^{a_3+1}[L]^{3a_3+3b_3+c_3-1}[T]^{-b_3-2}$$

For dimensional homogeneity, we have

$$a_3 + 1 = 0$$

$$-3a_3 + 3b_3 + c_3 - 1 = 0$$

$$-b_3 - 2 = 0$$

Thus, $a_3 = -1$, $b_3 = -2$, $c_3 = 4$

$$\pi_3 = \left[\frac{d^4 \Delta p}{\rho Q^2}\right]$$

We may write that,

$$\phi(\pi_1, \pi_2, \pi_3) = 0$$

or, $$\phi\left(\frac{D}{d},\frac{\mu d}{\rho Q},\frac{d^4\Delta p}{Q^2\rho}\right) = 0$$

or, $$\phi\left(\frac{d}{D},\frac{\rho Q}{\mu d},\frac{Q^2\rho}{d^4\Delta p}\right) = 0$$

or, $$\frac{Q^2\rho}{d^4\Delta p} = f_1\left(\frac{pQ}{\mu d},\frac{d}{D}\right)$$

or, $$\frac{Q}{\left[d^2\left(\frac{\Delta p}{\rho}\right)^{1/2}\right]} = f_2\left(\frac{\rho Q}{\mu D},\frac{d}{D}\right)$$

or, $$Q = Cd^2\left[\frac{\Delta p}{\rho}\right]^{1/2}$$

where C is dependent on $\frac{\rho Q}{\mu d}$ and $\frac{d}{D}$.

Example 11.12. *The frictional torque T of a disc of diameter D rotating at a speed N in a fluid of viscosity μ and density ρ in a turbulent flow is given by*

$$T = D^5N^2\rho Q\left[\frac{\mu}{D^2N\rho}\right]$$ ***[IES, 2008]***

Prove this by the method of dimensions.

Solution: $$T = f(D, N, \mu, \rho) \quad (1)$$

or $$f_1(T,D,N,\mu,\rho) = 0 \quad (2)$$

Total number of variables, $n = 5$

Writting dimensions of each variable, we have

$$T = [ML^2T^{-2}]$$
$$D = [L]$$
$$N = [T^{-1}]$$
$$\mu = [ML^{-1}T^{-1}]$$
$$\rho = [ML^{-3}]$$

Number of fundamental dimensions, $m = 3$

Number of π-terms = $n - m = 5 - 3 = 2$

Eqn. (2) can be written as: $f_1 (\pi_1, \pi_2) = 0$

Choosing D, N, ρ as repeating variables, we have

$$\pi_1 = D^{a_1} N^{b_1} \rho^{a_1} T$$

$$\pi_2 = D^{a_2} N^{b_2} \rho^{c_2} \mu$$

π_1-terms:

$$\pi_1 = D^{a_1} N^{b_1} \rho^{c_1} T$$

$$M^0 L^0 T^0 = [L]^{a_1} [T^{-1}]^{b_1} [ML^{-3}]^{c_1} [ML^2 T^{-2}]$$

$$= [M]^{c_1+1} [L]^{a_1 - 3c_1 + 2} [T]^{-b_1 - 2}$$

$$c_1 + 1 = 0$$

$$a_1 - 3c_1 + 2 = 0$$

$$-b_1 - 2 = 0$$

$\therefore$

$$b_1 = -2$$

$$c_1 = -1$$

$$a_1 = 3c_1 - 2 = 3 \times (-1) - 2 = -5$$

$$\pi_1 = D^{-5} N^{-2} \rho^{-1} T = \frac{T}{D^5 N^2 \rho}.$$

π_2-term:

$$\pi_2 = D^{a_2} N^{b_2} \rho^{c_2} \mu$$

$$M^0 L^0 T^0 = [L]^{a_2} [T^{-1}]^{b_2} [ML^{-3}]^{c_2} [ML^{-1} T^{-1}]$$

$$= [M]^{c_2+1} [L]^{a_2 - 3c_2 - 1} [T]^{-b_2 - 1}$$

$$c_2 + 1 = 0$$

$$a_2 - 3c_2 - 1 = 0$$

$$-b_2 - 1 = 0$$

$\therefore$

$$b_2 = -1$$

$$c_2 = -1$$

$$a_2 = 3c_2 + 1 = 3(-1) + 1 = -2$$

$$\pi_2 = D^{-2} N^{-1} \rho^{1} \mu = \frac{\mu}{D^2 N \rho}$$

Eqn. (3) becomes,

$$f_1\left(\frac{T}{D^5 N^2 \rho}, \frac{\mu}{D^2 N \rho}\right) = 0$$

$$\frac{T}{D^5N^2\rho} = \phi\left(\frac{\mu}{D^2N\rho}\right)$$

or
$$T = D^5N^2\rho\phi\left(\frac{\mu}{D^2N\rho}\right)$$

Hence proved.

Example 11.13 *The discharge Q over a V-shaped notch is known to depend on the angle θ of the notch, the head H of the water surface, the velocity of approach V_0 and the gravity g. Find the dimensionless form of the discharge equation.* ***[IES, 2008]***

Solution.
$$Q = f(\theta, H, V_0, g) \quad \text{... (1)}$$

or
$$f_1(Q, \theta, H, V_0, g) = 0 \quad \text{... (2)}$$

Total number of variables, $n = 5$

Number of fundamental dimensions, $m = 2$ [L and T]

Number of π-terms = $n - m = 5 - 2 = 3$

Writing dimensions of each variable, we have

$$Q = [L^3 T^{-1}]$$
$$\theta = [M^0 L^0 T^0]$$
$$H = [L]$$
$$V_0 = [LT^{-1}]$$
$$g = [LT^{-2}]$$

Eqn. (2) can be written as: $f_1(\pi_1, \pi_2, \pi_3) = 0$ ... (3)

Choosing H and g as repeating variables, we have

$$\pi_1 = H^{a_1} g^{b_1} Q$$
$$\pi_2 = H^{a_2} g^{b_2} V_0$$
$$\pi_3 = H^{a_3} g^{b_3} \theta$$

π_1-terms:

$$\pi_1 = H^{a_1} g^{b_1} Q$$
$$M^0L^0T^0 = [L]^{a_1}[LT^{-2}]^{b_1}[L^3T^{-1}]$$
$$= [L]^{a_1+b_1+3}[T]^{-2b_1-1}$$
$$a_1 + b_1 + 3 = 0$$
$$-2b_1 - 1 = 0$$
$$b_1 = -\frac{1}{2}$$

$$a_1 = -3+\frac{1}{2}=-\frac{5}{2}$$

$$\pi_1 = H^{-5/2}g^{-1/2}Q=\frac{Q}{g^{1/2}H^{5/2}}$$

π_2-term:

$$\pi_2 = H^{a_2}g^{b_2}V_0$$

$$[M^0L^0T^0] = [L]^{a_2}[LT^{-2}]^{b_2}[LT^{-1}]=[L]^{a_2+b_2+1}[T]^{-2b_2-1}$$

$$a_2 + b_2 + 1 = 0$$

$$-2b_2 - 1 = 0$$

$$b_2 = -\frac{1}{2}$$

$$a_2 = -1+\frac{1}{2}=-\frac{1}{2}$$

$$\pi_2 = H^{-1/2}g^{-1/2}V_0\frac{V_0}{\sqrt{gH}}$$

π_3-terms:

$$\pi_3 = H^{a_3}g^{b_3}\theta$$

$$[M^0L^0T^0] = [L]^{a_3}[LT^{-2}]^{b_3}[M^0L^0T^0]$$

$$= [L]^{a_3+b_3}[T]^{-2b_3}$$

$$a_3 + b_3 = 0$$

$$-2b_3 = 0$$

$$b_3 = 0$$

$$a_3 = 0$$

$$\pi_3 = H^0 g^0 \theta = 0$$

Substituting the values of π_1, π_2 and π_3 in Eqn. (3), we get,

$$f_1\left(\frac{Q}{H^{5/2}g^{1/2}},\frac{V_0}{\sqrt{gh}},\theta\right) = 0$$

Example 11.14 *The pressure difference Δp in a pipe of diameter D and length l due to turbulent flow depends on the velocity v, viscosity μ, density ρ and roughness k. Using Buckingham's π-theorem, obtain an expression for Δp.*

Solution.

$$\Delta p = f(D,l,v,\mu,\rho,k)$$

or $$f_1(\Delta p, D, l, v, \mu, \rho, k) = 0$$

$n = 7$, $m = 3$

$$\text{Number of } \pi\text{-terms} = n - m = 7 - 3 = 4$$

$$f_1(\pi_1, \pi_2, \pi_3, \pi_4) = 0$$

$$\text{Number of repeating variables} = m = 3$$

$$\text{Number of variables in each } \pi\text{-term} = m + 1 = 4$$

Choose D, v, ρ as repeating variables. The π-terms are:

$$\pi_1 = D^{a_1} \cdot v^{b_1} \cdot \rho^{c_1} \cdot \Delta p$$

$$\pi_2 = D^{a_2} \cdot v^{b_2} \cdot \rho^{c_2} \cdot l$$

$$\pi_3 = D^{a_3} \cdot v^{b_3} \cdot \rho^{c_3} \cdot \mu$$

$$\pi_4 = D^{a_4} \cdot v^{b_4} \cdot \rho^{c_4} \cdot k$$

π_1-term:

$$[M]^0\,[L]^0\,[T]^0 = [L]^{a_1} \cdot [LT^{-1}]^{b_1} \cdot [ML^{-3}]^{c_1} \cdot [ML^{-1}T^{-2}]$$

$$= [M]^{c_1+1} \cdot [L]^{a_1+b_1-3c_1-1} \cdot [T]^{-b_1-2}$$

For M: $c_1 + 1 = 0,\ c_1 = -1$

For L: $a_1 + b_1 - 3c_1 - 1 = 0, \quad a_1 + b_1 + 2 = 0$

For T: $-b_1 - 2 = 0,\ b_1 = -2$, then $a_1 = -2 + 2 = 0$

$$\pi_1 = D^0 \cdot v^{-2} \cdot \rho^{-1} \cdot \Delta p = \frac{\Delta p}{\rho v^2}$$

π_2-term:

$$[M]^0\,[L]^0\,[T]^0 = [L]^{a_2} \cdot [LT^{-1}]^{b_2} \cdot [ML^{-3}]^{c_2} \cdot [L]^1$$

$$= [M]^{c_2} \cdot [L]^{a_2+b_2-3c_2+1} \cdot [T]^{-b_2}$$

For M: $c_2 = 0$

For L: $a_2 + b_2 - 3c_2 + 1 = 0,\ a_2 + b_2 + 1 = 0$

For T: $-b_2 = 0$, then $a_2 = -1$

$$\pi^2 = D^{-1} \cdot v^0 \rho^{\circ} . l = \frac{l}{D}$$

π_3-term

$$[M]^0\,[L]^0\,[T]^0 = [L]^{a_3}[LT^{-1}]^{b_3}[ML^{-3}]^{c_3}[ML^{-1}T^{-1}]$$

$$= [M]^{c_3+1}.[L]^{a_3+b_3-3c_3-1}[T]^{-b_3-1}$$

For M: $c_3 + 1 = 0,\ c_3 = -1$

For L: $a_3 + b_3 - 3c_3 - 1 = 0,\ a_3 + b_3 + 2 = 0$

For T: $-b_3 - 1 = 0,\ b_3 = -1$, then $a_3 = -1$

$$\pi_3 = D^{-1}.V^{-1}\rho^{-1}\mu = \frac{\mu}{Dv\rho}$$

π_4-term

$$[M]^0\,[L]^0\,[T]^0 = [L]^{a_4}.[LT^{-1}]^{b_4}[ML^{-3}]^{c_4}[L]^1$$

$$= [M]^{c4}.[L]^{a_4+b_4-3c_4+1}[T]^{-b_4}$$

For M: $c_4 = 0$

For L: $a_4 + b_4 - 3\,c_4 + 1 = 0,\ a_4 + b_4 + 1 = 1$

Fot T: $-b_4 = 0$, then $a_4 = -1$

$$\pi_4 = D^{-1}v^0\rho^0 k = \frac{k}{D}$$

$$\therefore \quad f_1\left(\frac{\Delta p}{\rho v^2}, \frac{l}{D}, \frac{\mu}{Dv\rho}, \frac{k}{D}\right) = 0$$

or
$$\frac{\Delta p}{\rho v^2} = \phi\left[\frac{l}{D}, \frac{\mu}{Dv\rho}, \frac{k}{D}\right]$$

$$= \frac{l}{D}\phi\left[\frac{\mu}{Dv\rho}, \frac{k}{D}\right]$$

or
$$\frac{\Delta p}{\rho g} = \frac{v^2}{g} \cdot \frac{l}{D} \cdot \phi\left[\frac{\mu}{Dv\rho}, \frac{k}{D}\right]$$

$$= \frac{v^2}{g} \cdot \frac{l}{D}\phi\left[\frac{1}{\text{Re}}, \frac{k}{D}\right]$$

Now
$$f = \phi\left[\frac{1}{\text{Re}}, \frac{k}{D}\right] = \text{coefficient of friction}$$

Then
$$\frac{\Delta p}{pg} = \frac{4f}{2} \cdot \frac{v^2 l}{gD}$$

$$\therefore \quad \frac{\Delta p}{\rho g} = h_f = \frac{4flv^2}{D \times 2g}$$

PART-II: MODEL ANALYSIS

11.5 INTRODUCTION

A *model* is the small scale replica of the actual structure of machine called the prototype. The models are made to study the performance of actual structures and machines to save cost. The main objectives achieved through model analysis and subsequent model testing are:

1. To obtain a relationship amongst variables inflluencing a problem in term of dimensionless parameters.
2. To obtain experimental data from model tests for predicting the performance of its prototype.
3. To check the validity of assumptions made in the design of a structure or machine.

11.6 SIMILITUDE

To achieve similarity between the flow in the prototype and its model, every dimensionless parameter referrring to the conditions in the model must have the same numerical value as the corresponding parameter referring to the prototype. For complete similarity to exist between the model and its prototype, it is necessary that they must be geometrically, kinematically and dynamically similar.

11.6.1 Geometric similarity

For geometric similarity to exist between the model and the prototype, the ratios of corresponding lengths in the model and the prototype must be same. The geometric similarity is thus the similarity of shape.

Let L_m, B_m, H_m, D_m, A_m, V_m = length, width, height, diameter, area and volume respectively of the model.

L_p, B_p, H_p, D_p, A_p, V_p = corresponding parmeters for the prototype.

where m and p are subscripts for model and prototype.

Then for geometric similarity, we have

$$\frac{L_m}{L_p} = \frac{B_m}{B_p} = \frac{H_m}{H_p} = \frac{D_m}{D_p} = l_r \qquad \text{... (11.6)}$$

where l_r = scale ratio or scale factor.

Similarity $$\frac{A_m}{A_p} = l_r^2 \qquad \text{... (11.7)}$$

and $$\frac{V_m}{V_p} = l_r^3 \qquad \text{... (11.8)}$$

11.6.2 Kinematic Similarity

The kinematic similarity is the similarity of motion. For kinematic similarity to exist the streamline pattern in the model must be same as that in the prototype. It follows that flow characteristics such as velocity, acceleration and discharge must be same at all corresponding points. Thus for kinematic similarity, we have

$$\frac{v_p}{v_m} = v_r = \text{velocity ratio} \qquad \text{... (11.9)}$$

$$\frac{a_p}{a_m} = a_r = \text{acceleration ratio} \qquad \text{... (11.10)}$$

$$\frac{Q_p}{Q_m} = Q_r = \text{discharge ratio} \qquad \text{... (11.11)}$$

11.6.3 Dynamic Similarity

The dynamic similarity is the similarity of forces. The flows in the model and prototype are dynamically similar if at all the corresponding points, identical types of forces are parallel and bear the same ratio.

For dymaic similarity, we have

$$\frac{(F_i)_m}{(F_i)_p} = \frac{(F_v)_m}{(F_v)_p} = \frac{(F_g)_m}{(F_g)_p} = F_r = \text{force ratio} \qquad \text{... (11.12)}$$

where i, v, g = subscripts for inertia, viscous and gravity respectively.

11.7 DIMENSIONLESS NUMBER AND THEIR SIGNIFICANCE

1. Reynolds Number, Re

It is defined as the ratio of the inertia force to the viscous force.

Inertia force, F_i = mass × accelearatin

$$= \rho Q v = \rho A v^2$$

Viscous. force, F_v = shear stress × area

$$= \tau \times A$$

$$= \mu \frac{du}{dy} \times A = \mu \frac{v}{L} \times A$$

Reynolds number,

$$\text{Re} = \frac{F_i}{F_v} = \frac{\rho A v^2}{\mu \frac{v}{L} A} = \frac{\rho v L}{\mu} = \frac{vL}{\nu} \qquad \text{... (11.13 a)}$$

$$= \frac{vd}{\nu} \qquad \text{... (11.13 b)}$$

where d = pipe diameter

$$\nu = \frac{\mu}{\rho} = \text{kinematic viscosity}$$

(i) Reynolds number signifies the relative predominace of inertia force to viscous force occuring in flow systems.

(ii) It is taken as the criterion of dynamic similarity in flow where viscous forces predominate, *e.g.* motion of submarine, flow around automobiles and aeroplanes at low velocities, and flow through pipes of smaller size.

2. Froude's Number, Fr

It is defined as the square root of the ratio of inertia force and gravity force.

$$Fr = \sqrt{\frac{F_i}{F_g}}$$

$$= \sqrt{\frac{\rho A v^2}{\rho A l g}} = \frac{v}{\sqrt{l g}} \quad \text{... (11.14)}$$

The Froude number governs the dynamic similarity of the flow where gravitational force is predominant over all other foces, *e.g.* flow over notches and weirs, flow in open channels, motion of ship in turbulent sea, etc.

3. Euler's Number, Eu

It is defined as the square root of the ratio of inertia force to pressure force.

$$Eu = \sqrt{\frac{F_i}{F_p}}$$

$$= \sqrt{\frac{\rho A v^2}{p A}} = \sqrt{\frac{v^2}{(p/\rho)}} = \frac{v}{\sqrt{p/\rho}} \quad \text{... (11.15)}$$

The Euler number is important in fluid flow problems in which pressure gradient exists, e.g. discharge through orifices and mouthpieces, pressure rise due to sudden closure of valves, and flow through pipes.

4. Weber's Number, We

It is defined as the square root of the ratio of inertia force to the surface teusion.

$$\text{We} = \sqrt{\frac{F_i}{F_\sigma}}$$

$$= \sqrt{\frac{\rho A v^2}{\sigma l}} = \sqrt{\frac{\rho l^2 v^2}{\sigma l}} = \frac{v}{\sqrt{\left(\frac{\sigma}{\rho l}\right)}} \qquad \text{... (11.16)}$$

This number is important for capillary flow of water in soil, flow of blood in veins and arteries, etc.

5. Mach Number, M

It is defined as the square root of the ratio of inertia force to elastic force.

$$M = \sqrt{\frac{F_i}{F_e}}$$

$$= \sqrt{\frac{\rho A v^2}{\sigma_e \times A}} = \sqrt{\frac{\rho l^2 v^2}{\sigma_e l^2}} = \frac{v}{\sqrt{\frac{\sigma_e}{\rho}}} = \frac{v}{v_s} \qquad \text{... (11.17)}$$

where σ_e = elastic stress

$v_s = \sqrt{\frac{\sigma_e}{\rho}}$ = velocity of sound in fluid

The Mach number is important in compressible fluid flow problems, high velocity in pipes and motion of projectiles and missiles.

11.8 MODEL LAWS

1. Reynolds Model Law

$$(\text{Re})_m = (\text{Re})_p$$

$$\left[\frac{\rho v l}{\mu}\right]_m = \left[\frac{\rho v l}{\mu}\right]_p \qquad \text{... (11.18)}$$

or $$\frac{\rho_p}{\rho_m} \times \frac{v_p}{v_m} \times \frac{l_p}{l_m} \times \frac{\mu_m}{\mu_p} = 1$$

Let $\frac{\rho_p}{\rho_m} = \rho_r, \frac{v_p}{v_m} = v_r, \frac{l_p}{l_m} = l_r, \frac{\mu_p}{\mu_r} = \mu_r$ be the scale ratios.

Then $$\frac{\rho_r\, v_r\, l_r}{\mu_r} = 1 \qquad \text{... (11.19)}$$

For a pipe, $$\left(\frac{\rho\, v\, d}{\mu}\right)_m = \left(\frac{\rho\, v\, d}{\mu}\right)_p$$

For the same fluid,

$$v_m\, d_m = v_p\, d_p \qquad \text{... (11.21 } a\text{)}$$

and $$v_m\, l_m = v_p\, l_p \qquad \text{... (11.21 } b\text{)}$$

2. Froude Model Law

$$(\text{Fr})_m = (\text{Fr})_p \qquad \text{... (11.22)}$$

$$\left[\frac{v}{\sqrt{gl}}\right]_m = \left[\frac{v}{\sqrt{gl}}\right]_p$$

For $$g_m = g_p$$

$$\frac{v_m}{\sqrt{l_m}} = \frac{v_p}{\sqrt{l_p}} \qquad \text{... (11.23)}$$

or $$\frac{v_p}{v_m} = \sqrt{\frac{l_p}{l_m}} = \sqrt{l_r}$$

3. Euler Model Law

$$(Eu)_m = (Eu)_p$$

$$\left[\frac{v}{(p/\rho)}\right]_m = \left[\frac{v}{(p/\rho)}\right]_p \qquad \text{... (11.24)}$$

For $\rho_m = \rho_p$

$$\frac{v_m}{v_p} = \sqrt{\frac{p_m}{p_p}} \qquad \text{... (11.25)}$$

4. Weber Model Law

$$(We)m = (We)_p$$

$$\left[\frac{v}{\left(\dfrac{\sigma}{\rho l}\right)}\right]_m = \left[\frac{v}{\left(\dfrac{\sigma}{\rho l}\right)}\right]_p \quad \text{... (11.26)}$$

For $\rho_m = \rho_p$

$$\frac{v_m}{v_p} = \frac{\sigma_m}{\sigma_p} \times \frac{l_p}{l_m} \quad \text{... (11.27)}$$

5. Mach Model Law

$$M_m = M_p$$

$$\frac{v_m}{\sqrt{\left(\dfrac{\sigma_e}{\rho}\right)_m}} = \frac{v_p}{\sqrt{\left(\dfrac{\sigma_e}{\rho}\right)_p}} \quad \text{... (11.28)}$$

For $\rho_m = \rho_p$

$$\frac{v_m}{v_p} = \sqrt{\frac{(\sigma_e)_m}{(\sigma_e)_p}} = \sqrt{\frac{(v_s)_m}{(v_s)_p}} \quad \text{... (11.29)}$$

The scale ratios are given in Table 11.2.

Table 11.2 Scale ratio.

Model Law	Time scale ratio T_r	Acceleration scale ratio a_r	Force scale ratio F_r	Discharge scale ratio Q_r	Pressure scale ratio p_r	Energy or work scale ratio E_r	Momentum or impuse scale ratio M_r	Torque scale ratio T_r	Power scale ratio P_r
1. Reynolds Law	$\frac{l_r}{v_r}$	$\frac{v_r}{T_r}$	$\rho_r l_r^2 v_r a_r$	$\rho_r v_r l_r^2$	–	–	–	–	–
2. Froude Law	$\sqrt{l_r}$	1	l_r^3	$l_r^{2.5}$	l_r	l_r^4	$l_r^{3.5}$	l_r^4	$l_r^{3.5}$

Example 11.15 *The force required to tow a 1:30 scale model of a motor boat in a lake at a speed of 2 m/s is 0.5 N. Calculate the corresponding velcoity of the prototype of 50 m for dynamically similar conditions. What would be the force required to propel the prtotype at that velocity in the same lake?*

Solution:

Given: $\frac{l_p}{l_m} = 30,\ v_m = 2 \text{ m/s},\ P_m = 0.5 \text{ N}$

Froude number, $(Fr)_m = (Fr)_p$

$$\left(\frac{v}{\sqrt{gl}}\right)_m = \left(\frac{v}{\sqrt{gl}}\right)_p$$

$$v_p = v_m\sqrt{\frac{l_p}{l_m}} = 2\sqrt{30} = 10.95\,\text{m/s}$$

Force,
$$P = \rho l^2 v^2$$

$$\frac{P_p}{P_m} = \frac{\rho_p}{\rho_m}\times\left(\frac{l_p}{l_m}\right)^2\times\left(\frac{v_p}{v_m}\right)^2$$

$$= \left(\frac{l_p}{l_m}\right)^3 \qquad \left[\because \frac{\rho_p}{\rho_m} = \frac{l_p}{l_m} \text{ and } \frac{v_p}{v_m} = \sqrt{\frac{l_p}{l_m}}\right]$$

$$= 30^3 = 27{,}000$$

$$P_p = 0.5\times 27{,}000 = 13{,}500\text{ N}$$

Example 11.16 *The performance of a spillway is to be studied by means of a model constructed to a scale of 1:10. Neglecting the viscous and surface tension effects, determine (a) rate of flow in the model for a prototype discharge of 2000 m^3/s, and (b) the disssipation of energy in the prototype hydraulic jump, if the jump in the model dissipates 0.35 kW.*

Solution.
$$(Fr)_m = (Fr)_p$$

$$\left(\frac{v}{\sqrt{gl}}\right)_m = \left(\frac{v}{\sqrt{gl}}\right)_p$$

$$\frac{v_p}{v_m} = \sqrt{\frac{l_p}{l_m}}$$

and
$$\frac{Q_p}{Q_m} = \left(\frac{l_p}{l_m}\right)^2\frac{v_p}{v_m} = \left(\frac{l_p}{l_m}\right)^{5/2}$$

$$Q_m = 2000\times\left(\frac{1}{10}\right)^{5/2} = 6.32\text{ m}^3\text{/s}$$

Power,
$$P = \gamma QH$$

$$\frac{P_p}{P_m} = \frac{\gamma_p}{\gamma_m}\times\frac{Q_p}{Q_m}\times\frac{H_p}{H_m} = \left(\frac{l_p}{l_m}\right)^{5/2}\times\frac{l_p}{l_m} = \left(\frac{l_p}{l_m}\right)^{7/2}$$

Example 11.17 *Show that for a flow governed only by gravity, inertia and pressure forces, the ratio of rates of flow in two dynamically similar systems equals to 5/2 power of the length ratio.* ***[IES, 2006]***

Solution Gravity force, $F_g = M \times g = \rho l^3 \times g$

Inertia force, $F_i = m \times a = \rho l^2 v^2$

Pressure force, $F_p = \Delta p \times l^2$

$$\frac{F_i}{F_g} = \frac{\rho l^2 \times v^2}{\rho l^3 \times g} = \frac{v^2}{l g}$$

Froude number $Fr = \frac{v}{\sqrt{l g}}$

Euler number, $E = \frac{F_i}{F_p} = \frac{\rho l^2 v^2}{\Delta p \times l^2} = \frac{\rho v^2}{\Delta p}$

The discharge depends on velocity and area of cross-section.

From Froude number, for dynamic similarity, we have

Velocity, ratio $V_r = \frac{V_p}{V_m} = \sqrt{\frac{l_p}{l_m}} = \sqrt{l_r}$ ... (1)

where $l_r = \frac{l_p}{l_m}$ = length ratio

Time ratio, $t_r = \frac{t_p}{t_m} = \left(\frac{l_p}{l_m}\right) \times \left(\frac{V_m}{V_p}\right) = \frac{l_r}{\sqrt{l_r}} = \sqrt{l_r}$ ... (2)

$$Q = AV = \frac{l^2 \times l}{t} = \frac{l^3}{t}$$

Discharge ratio, $Q_r = \frac{l_r^3}{t_r} = \frac{l^3}{\sqrt{l_r}} = l_r^{5/2}$ ... (3)

Hence proved.

Example. 11.18 *(a) Define Froude number, Reynolds number, Mach number, Euler number and Weber number.*

(b) Considering Froude number as the criterion of dynamic similarity for a certain flow situation, work out the scale factors for velocity, time, discharge, acceleration, force, work and power in terms of the scale factor for length. ***[IES, 2003]***

Solution.

(*a*) Froude number, $F_r = \left[\frac{\text{Inertia force}}{\text{Gravity force}}\right]^{0.5} = \frac{v}{[gL]^{0.5}}$

Reynolds number, $Re = \frac{\text{Inertia force}}{\text{Gravity force}} = \frac{vL}{\nu}$

Mach number, $\text{M} = \left[\frac{\text{Inertia force}}{\text{Surface tension force}}\right]^{0.5} = \frac{v}{v_s}$

Euler number, $Eu = \left[\frac{\text{Inertia force}}{\textit{Compressibility}}\right]^{0.5} = \frac{v}{[\Delta p/\rho]^{0.5}}$

Weber number $\text{We} = \left[\frac{\text{Inertia force}}{\textit{Surface tension force}}\right]^{0.5} = \frac{\rho v^2 L}{\sigma}$

(b) Froude number, $Fr = \frac{v}{[gL]^{0.5}}$

$$\frac{v_m}{[gL_m]^{0.5}} = \frac{v_p}{[gL_p]^{0.5}}$$

$$v_r = \frac{v_p}{v_m} = \left[\frac{L_p}{L_m}\right]^{0.5} = (L_r)^{0.5}$$

Time, $T_r = \frac{L_r}{v_r} = \frac{L_r}{(L_r)^{0.5}} = (L_r)^{0.5}$

Discharge, $Q_r = v_r L_r^2 = L_r^{5/2}$

Acceleration, $a_r = \frac{v_r}{T_r} = 1$

Force, $F_r = \rho_r v_r^2 L_r^2 = \rho_r L_r^3$

Work, $W_r = \text{Force} \times \text{Distance} = \rho_r L_r^3 L_r = \rho_r L_r^4$

Power $P_r = \text{Force} \times \text{Velocity} = \rho_r (L_r)^{7/2}$

Example 11.19. *An Orificemeter to carry water is calibrated with air in a geometrically similar model at 1/5 prototype scale. Dynamically similar flow will be obtained when the discharge ratio (air to water) is:*

(i) 0.4 (i) 2.5 (iii) 62.5

Assume the ratio of kinematic viscosity of air to water as 12.5.

Solution.

Given:
$$l_r = \frac{1}{5} = \frac{l_m}{l_p}, \frac{\nu_a}{\nu_a} = \frac{\nu_m}{\nu_p} = 12.5$$

$$(\text{Re})_m = (\text{Re})_p$$

$$\frac{v_m}{v_p} = \frac{l_p}{l_m} \times \frac{\nu_m}{\nu_p} = 5 \times 12.5 = 62.5$$

$$\frac{Q_m}{Q_p} = \left(\frac{l_m}{l_p}\right)^2 \times \frac{v_m}{v_p} = \left(\frac{1}{5}\right)^2 \times 62.5 = 2.5 \text{ i.e. (ii).}$$

Example 11.20 *A model of submarine is scaled down to* $\frac{1}{20}$ *of prototype and is to be tested in a wind tunnel where free stream pressure is 2 MPa (abs.) and temperature is 50°C. The speed of prototype is 8 m/s Determine the free stream velocity of air and the ratio of the drags between model and prototype. Assume kinematic viscosity of sea water as* 1.4×10^{-6} m^2/s*, viscosity of air as 0.0184 cP, and density of sea water as 1020* kg/m^3*.*

Solution.

Given: $l_r = \frac{l_m}{l_p} = \frac{1}{20}$, $p = 2$ MPa, $t = 50°C$, $v_p = 8$ m/s, $\nu_p = 1.4 \times 10^{-6}$ m^2/s,

$\mu_a = 0.0184 \times 10^{-3}$ Pa.s Pa.s, $\rho_p = 1020$ kg/m^3

$$\rho_a = \frac{p}{RT} = \frac{2 \times 10^6}{287 \times 323} = 21.57 \text{ kg/m}^3$$

$$\nu_m = \nu_a = \frac{\mu_a}{\rho_a} = \frac{0.0184 \times 10^{-3}}{21.57} = 8.53 \times 10^{-7} \text{ m}^2/\text{s}$$

For dynamic similarity,

$$(\text{Re})_p = (\text{Re})_m$$

$$\frac{v_p l_p}{\nu_p} = \frac{v_m l_m}{\nu_m}$$

$$v_m = \frac{l_p}{l_m} \times \frac{\nu_m}{\nu_p} \times v_p = 20 \times \frac{8.53 \times 10^{-7}}{1.4 \times 10^{-6}} \times 8 = 97.486 \text{ m/s}$$

$$\frac{(F_D)_m}{(F_D)_p} = \frac{\rho_m}{\rho_p} \times \left(\frac{l_m}{l_p}\right)^2 \times \left(\frac{v_m}{v_p}\right)^2$$

$$= \frac{21.57}{1020} \times \left(\frac{1}{2}\right)^2 \times \left(\frac{97.486}{8}\right)^2$$

$$= 0.00785$$

Example 11.21 *A geometrically similar model of an air duct is built to* $\frac{1}{25}$ *scale and tested with water which is 50 times more viscous and 800 times denser than air. When tested under dynamically similar conditions, the pressure drop is 2 bar in model. Calculate corresponding pressure drop in prototype and express in water column.*

Solution.

Given: $\frac{l_p}{l_m} = 25, \frac{\mu_p}{\mu_m} = \frac{1}{50}, \frac{\rho_p}{\rho_m} = \frac{1}{800}, (\Delta p)_m = 2 \text{ bar}$

For dynamic similarity, $(\text{Re})_m = (\text{Re})_p$

$$\left(\frac{\rho v l}{\mu}\right)_m = \left(\frac{\rho v l}{\mu}\right)_p$$

$$\frac{v_p}{v_m} = \frac{\mu_p}{\mu_m} \times \frac{\rho_m}{\rho_p} \times \frac{l_m}{l_p} = \frac{1}{50} \times 800 \times \frac{1}{25} = 0.64$$

Pressure, $$p = \frac{F}{A} = \frac{\rho l^2 v^2}{l^2} = \rho v^2$$

$$\frac{(\Delta p)_p}{(\Delta p)_m} = \frac{\rho_p}{\rho_m} \times \left(\frac{v_p}{v_m}\right)^2 = \frac{1}{800} \times (0.64)^2 = 5.12 \times 10^{-4}$$

$$(\Delta p)_p = 2 \times 10^5 \times 5.12 \times 10^{-4} = 102.4 \text{ Pa}$$

$$h_w = \frac{(\Delta p)_p}{\gamma_w} = \frac{102.4}{9810} = 0.010438 \text{ m of water}$$

Example 11.22 *In the model test of a spillway the discharge and velocity of flow over the model were 2.5 m*^{3}*/s and 1.5 m/s respectively. Calculate the velocity and discharge over the prototype which is 25 times the model size.*

Solution.

Given: $Q_m = 2.5$ m³/s, $v_m = 1.5$ m/s, $l_r = 25$

For $$(\text{Fr})_m = (\text{Fr})_p$$

$$\frac{v_p}{v_m} = \sqrt{l_r} = \sqrt{25} = 5$$

$$v_p = 1.5 \times 5 = 7.5 \text{ m/s}$$

$$\frac{Q_p}{Q_m} = l_r^{2.5} = 25^{2.5} = 3125$$

$$Q_p = 2.5 \times 3125 = 7812.5 \text{ m}^3/\text{s}$$

Example 11.23 *A small flow meter is designed to measure gas flow in 10 mm gas pipeline. For the flow rate of 0.004 m³/s the pressure drop across the meter is 4.5 kPa. An enlarged geometrically similar model of the prototype is tested in a 300 mm diameter water pipe. Calculate the scale factors for velocity and discharge and the pressure drop in the model.*

Take $\rho_g = 12$ *kg/m³*, $\mu_g = 18 \times 10^{-6}$ *Pa.s*

$\rho_w = 1000$ *kg/m³*, $\mu_w = 11.8 \times 10^{-4}$ *Pa.s*

Solution.

Given: $d_p = 10$ mm, $Q_p = 0.004$ m³/s, $(\Delta p)_p = 4.5$ kPa. $d_m = 300$ mm, $\rho_p = 12$ kg/m³, $\mu_p = 18 \times 10^{-6}$ Pa.s, $\rho_m = 1000$ kg/m³, $\mu_m = 11.8 \times 10^{-4}$ Pa.s

(*i*) For $(\text{Re})_m = (\text{Re})_p$

$$\left(\frac{v d\rho}{\mu}\right)_m = \left(\frac{v d\rho}{\mu}\right)_p$$

$$\frac{v_m}{v_p} = \frac{d_p}{d_m} \times \frac{\rho_p}{\rho_m} \times \frac{\mu_m}{\mu_p} = \frac{10}{300} \times \frac{12}{1000} \times \frac{11.8 \times 10^{-4}}{18 \times 10^{-6}} = 0.0251$$

$$\frac{Q_m}{Q_p} = \frac{v_m}{v_p} \cdot \left(\frac{dm}{d_p}\right)^2 = 0.0251 \times \left(\frac{300}{10}\right)^2 = 22.6$$

(*ii*) For $(Eu)_m = (Eu)_p$

$$\left(\frac{\rho v^2}{\Delta p}\right)_m = \left(\frac{\rho v^2}{\Delta p}\right)_p$$

$$(\Delta_p)_m = \frac{\rho_m}{\rho_p} \times \left(\frac{v_m}{v_p}\right)^2 \times (\Delta p)_p$$

$$= \frac{1000}{12} \times (0.0251)^2 \times 4.5$$

$$= 0.236 \text{ kPa}$$

11.9 USES OF DIMENSIONAL ANALYSIS

1. To check the dimensional homogeneity of any equation of fluid motion.
2. To derive a rational formula for a flow phenomenon.
3. To derive an equation expressed in terms of non-dimensional parameters.
4. To analyse a complex fluid flow phenomenon.

11.10 ADVANTAGES OF MODEL TESTING

1. Model tests are economical and convenient.
2. The performance of hydraulic structures can be predicted in advance.
3. The safety and reliability of such hydraulic structures for which analytical methods are not possible can be ascertained by model testing.
4. Defects in existing structures can be detected and rectified.

11.11 DISTORTED MODELS

A distorted model is one which is not geometrically similar to its prototype. In such a model different scale ratios for the linear dimensions are adopted.

The following distortions may be introduced in a model:

1. **Geometric distortion:** This distortion is due to using different scales for vertical and horizontal dimensions.
2. **Configuration distoration:** The model is geometrically similar but its bed slope is increased by keeping it in a tilted position compared to the position of prototype.
3. **Hydraulic distortion:** This distortion occurs due to a change in some hydraulic quantity such as discharge.
4. **Material distortion:** This distortion involves the use of different materials for the model and prototype.

Distorted models are used for rivers, dams across very wide rivers, and harbours.

The distorted models are used for the following reasons:

1. Maintaining accuracy in vertical dimensions.
2. Maintaining turbulent flow.
3. Accommodating available facilities.
4. Obtaining suitable roughness condition.
5. Obtaining suitable bed material and its adequate movement.

11.11.1 Merits and Demerits of Distorted Models

Merits

1. Precise measurements are possible due to increase in the depth of fluid.
2. Operation of model is simplified due to reduced size.
3. Reduction in model cost due to reduced size.
4. The effects of surface tension can be minimised.
5. Better results due to increased Reynolds number.

Demerits

1. Slopes, bends and cuts cannot be truly reproduced.
2. Depth distortion gives rise to different wave patterns.
3. Due to different horizontal and vertical scales, pressure and velocity disributions are not truly reproduced.
4. Results are difficult to extrapolate and interpolate.
5. The observer expereinces an unfavourable psychological effect.

SUMMARY

1. Dimensional analysis is a mathematical technique which uses dimensions to solve engineering problems.
2. Dimensional homogeneity states that every term in an equation when reduced to fundamental dimensions must contain identical powers of each dimension.
3. Rayleigh's method and Buckingham's π-method are commonly used for dimensional analysis.
4. Rayleigh's method is used when number of variables is small.
5. Buckingham's π-theorem states that if there are n number of variables in a dimensionally homogeneous equation and these variables contain m fundamental dimensions then number of dimensionless groups, which can be formed shall be $(n - m)$.
6. The dimensionless groups are called π-terms.
7. The repeating variables can be selected from thos describing the geometry of flow, fluid properties and fluid motion.
8. Model analysis is an experimental method of finding solutions of complex flow problems.
9. A model is a small scale replica of the prototype. It must have geometric, kinematic and dynamic similarities with the prototype.
10. Reynolds number, $$\text{Re} = \frac{\text{Inertia force}}{\text{Viscous force}} = \frac{vL}{\nu}$$

    Reynolds number is used for ascertaining the Laminar flow. It is used for dynamic similarity in model testing.

11. Mach number, $M = \left[\dfrac{\text{Inertia force}}{\text{Compressibility force}}\right]^{0.5} = \dfrac{v}{v_s}$

where v_s = Velocity of sound.

It is used in model testing of aricraft and missiles.

12. Euler number, $Eu = \left[\dfrac{\text{Inertia force}}{\text{Pressure force}}\right]^{0.5} = \dfrac{v}{\left(\dfrac{\Delta p}{\rho}\right)^{0.5}}$

It is used in cavitation studied where the pressure force is important in addition to the viscous force, and the dynamic similarity is obtained when the Euler and Reynolds Number are kept the same for the model and the prototype.

13. Weber number, $We = \left[\dfrac{\text{Inertia force}}{\text{Surface tension force}}\right]^{0.5} = \sqrt{\dfrac{\rho v^2 l}{\sigma}}$

It is used in river and harbour modeling.

14. Froude number, $Fr = \left[\dfrac{\text{Inertia force}}{\text{Gravity force}}\right]^{0.5} = \dfrac{v}{(gL)^{0.5}}$

It is used in modeling of spillways, weirs, stilling basins, channel transistion, sluices, etc.

15. For dynamic similarity,

$$(\text{Re})_p = (\text{Re})_m$$
$$(\text{Fr})_p = (\text{Fr})_m$$
$$M_p = M_m$$
$$(\text{Eu})_p = (\text{Eu})_m$$
$$(\text{We})_p = (\text{We})_m$$

Short Answer Type Questions

1. Define dimensional analysis.
2. Define dimensional homogeneity.
3. What are the methods commonly used for dimesional analysis?
5. What are repeating variables? How are these selected?
6. Define Buckingham's π-theorem.
7. What is model analysis? What are its objectives?
8. Explain geometeric, kinematic and dynamic similarity.
9. What are dimensionless numbers?
10. Define the following: Mach number, Froude number, and Mach number.
11. Define Euler number and Weber number.

13. What are distorted models? When they are used?

14. How distortions can be introduced in models?

EXERCISES

1. Prove by dimensional analysis that the power P developed by hydralic turbine is given by:

$$P = \rho N^3 D^5 \phi\left[\frac{N^2 D^2}{gH}\right]$$

where ρ = mass density of liquid, N = rotational speed,
D = runner diameter, H = working head, and
g = acceleration due to gravity.

2. Show that velocity through an orifice can be expressed as:

$$v = \sqrt{2gH}\ \phi\left[\frac{D}{H}, \frac{\mu}{\rho v H}, \frac{\sigma}{\rho v^2 H}\right]$$

where H = head causing flow, D = orifice diameter,
μ = dyanamic viscosity, ρ = mass density, and
σ = surface tension

3. Prove that the shear stress, τ in a fluid flowing through a pipe can be expressed as:

$$\tau = \rho v^2 \phi\left[\frac{\mu}{\rho D v}, \frac{k}{D}\right]$$

where D = pipe diameter, ρ = mass, density,
v = velocity, μ = dynamic viscosity, and
k = height of roghness projection.

4. The resistance R to the motion of a completely submerged body depends upon the length of the body L, velocity of flow v, mass density of fluid ρ, and kinematic viscosity of fluid ν. Using dimensional analysis, prove that

$$R = \rho v^2 L^2 \phi\left[\frac{vL}{\nu}\right]$$

5. Show from first principles that where gravity is predominant over all other forces, prototype ratio of discharge is equal to $l_r^{5/2}$, where l_r is the length ratio of model to prototype.

6. In a prototype, both gravity and viscous forces predominate. A model study is conducted in a oil of viscosity 0.05 Stokes. If the model-prototype ratio is 1:3, how many times more viscoous should the prototype oil be? **[Ans.** 5.2 times]

7. The propeller of a ship operates in close proximaty to the free surface. The propeller of an ordinary cargo ship 190 m long is about 2.5 m in diameter and turns at 100 r.p.m. If a 1:100 scale model of the ship and propeller is to be tested in a towing tank,
   (a) What should be the speed of a model propeller?
   (b) What is the ratio of power required by the model and that of the ship? [**Ans.** 1000rpm, 10^{-7}]
8. A seaplane is to take off at 140 km/h. If the maximum speed vailable for conducting the model tests is 6m/s, what is the largest model scale to which the model can be constructed?[**Ans.** 1:42]
9. A surface vessel of 120 m length is to be tested by a model 3 m long. If the vessel travels at 10 m/s, at what speed must model be towed for dynamic similarity between model and prototype? If the drag of the model is 9.37 N, what prototype drag is to be expected?

   [**Ans.** 1.58 m/s, 596. 4 kN]
10. In an aeroplane model of size $\frac{1}{50}$ of its prototype the pressure drop is 4 bar. The model is tested in water. Find the corresponding pressure drop in the prototype. Take specific weight of air = 0.0124 kN/m^3. The viscosity of water is 0.01 Poise and that of air is 0.00018 Poise.

    [**Ans.** 0.00042 bar]
11. In order to predict the pressure drop in a large air duct a model is constructed with linear dimension $\frac{1}{10}$th that of the prototype, and water is used as the test fluid. If water is 1000 times denser than air and has 100 times the viscosity of air, determine the pressure drop in the prototype for the conditions corresponding to a pressure drop of 70 kPa in the model. [**Ans.** 0.07 kPa]
12. In 1:30 model of a spillway, the velocity and discharge are 1.5 m/s and 2m^3/s. Find the corresponding velocity and discharge in the prototype. [**Ans.** 8.2 m/s, 9859 m^3/s]

12
NOTCHES AND WEIRS

12.1 DEFINITIONS

Notch: A notch may be defined as an opening in the side of a tank or a small channel in such a way that the liquid surface in the tank or channel is below the top edge of the opening. It is used for measuring the rate of flow of a liquid through a small channel or a tank. It is generally made of a metallic plate.

Weir: A weir may be defined as any regular obstruction in an open stream over which the flow takes place. It is made of masonary or concrete. It is generally in the form of a vertical wall, with a sharp edged at the top, running throughout across the open channel. It is used for meauring the rate of flow of water in streams or rivers.

Nappe or vein: The sheet of water flowing through a notch or over a weir is called nappe or vein.

Crest or sill: The bottom edge of a notch or top or the weir over which water flows is known as the crest or sill.

12.2 CLASSIFICATION OF NOTCHES AND WEIRS

12.2.1 Types of Notches

The notches may be classfied as follows:

(a) According to the shape of the opening:

1. Rectangular notch
2. Triangular notch
3. Trapezoidal notch
4. Stepped notch.

(b) According to the effect of the sides on the nappe:

1. Notch with end contraction.
2. Notch without end contraction or suppressed notch.

12.2.2 Types of Weirs

The weirs may be classified as follows:

(a) According to the shape of opening:

1. Rectangular weir
2. Trapezoidal (or cipolletti) weir
3. Triangular weir.

(b) According to the shape of the crest:

1. Sharp-crested weir
2. Broad–crested weir
3. Narrow-crested weir
4. Qgee-shaped weir.

(c) According to the effect of sides on the emerging nappe:

1. Weir with end contraction.
2. Weir without end contraction.

(d) According to the nature of discharge:

1. Ordinary weir
2. Submerged or drowned weir.

12.3 Discharge Over a Rectangular Notch or Weir

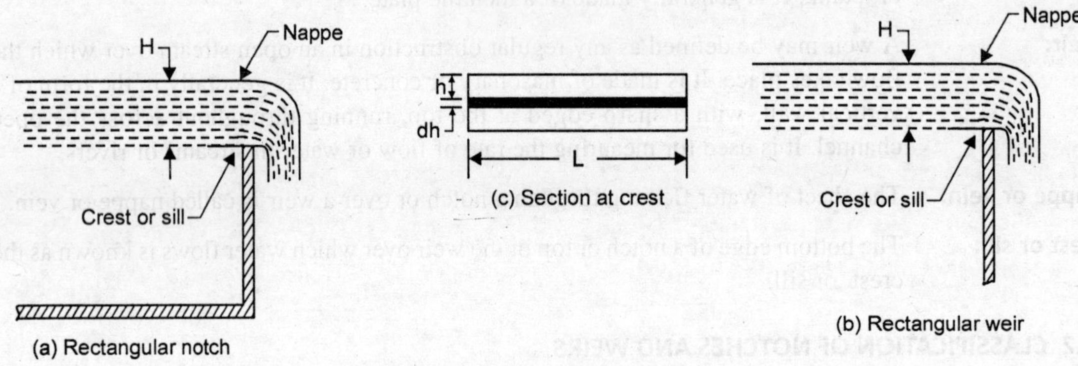

Fig. 12.1 Rectangular notch and weir

Consider a rectangur notch or weir provided in a channel carrying water as shown in Fig. 12.1 *(a)* and *(b)* respectively.

From Fig 12.2 *(b)*, we have

Let H = head of water over the crest

L = length of notch or weir

C_d = coeffcient of discharge

Consider an elementary horizontal strip of water of thickness dh and length L at a depth h from the free surface of water, as shown in Fig. 12.1 (c).

Area of strip, $$dA = L \times dh$$

Theoretical velocity of water flowing through the strip, $v = \sqrt{2gh}$

Discharge through the strip, $$dQ = C_d \times dA \times v$$

$$= C_d \times L \times dh \times \sqrt{2gh}$$

Total discharge over the whole notch or weir,

$$Q = C_d \times L \times \sqrt{2g} \int_0^H \sqrt{h}\, dh$$

$$= C_d L \times \sqrt{2g} \left| \frac{h^{3/2}}{(3/2)} \right|_0^H$$

$$= \frac{2}{3} C_d L \sqrt{2g} \cdot H^{3/2} \qquad \text{... (12.1)}$$

12.4 DISCHARGE OVER A TRIANGULAR NOTCH OR WEIR

Consider a triangular notch or weir as shown in Fig. 12.2.

Let H = head of water above the apex of notch

θ = angle of notch

C_d = coefficientof discharge.

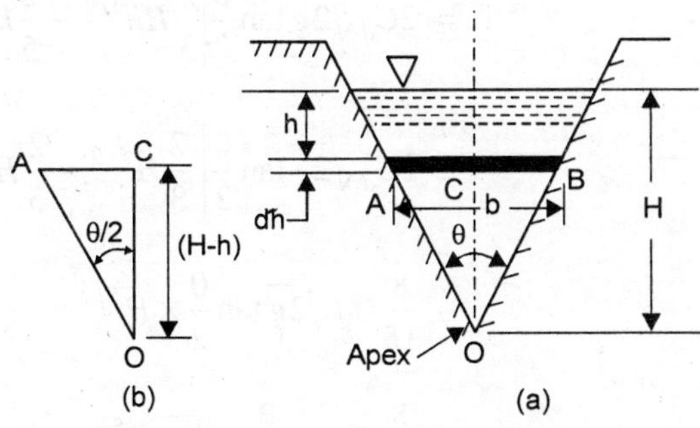

Fig. 12.2 Triangular notch or weir

Consider a horizontal strip of water of thickness dh at a depth h from the free surface of water as shown in Fig. 12.2 (a).

$$\tan\frac{\theta}{2} = \frac{AC}{OC} = \frac{AC}{(H-h)}$$

$$AC = (H-h)\tan\frac{\theta}{2}$$

Width of strip, $$b = AB = 2AC = 2(H-h)\tan\frac{\theta}{2}$$

Area of strip, $$dA = b \times dh = 2(H-h)\tan\frac{\theta}{2} \times dh$$

Theoretical velocity of water through strip, $v = \sqrt{2gh}$

Discharge through the strip, $$dQ = C_d \times dA \times v$$

$$= C_d \times 2(H-h)\tan\frac{\theta}{2} \times dh \times \sqrt{2gh}$$

$$= 2C_d (H-h)\tan\frac{\theta}{2} \times \sqrt{2g} \times \sqrt{h} \times dh$$

Total discharge, $$Q = 2C_d\sqrt{2g}\,\tan\frac{\theta}{2}\int_0^H (H-h)\sqrt{h}\cdot dh$$

$$= 2C_d\sqrt{2g}\,\tan\frac{\theta}{2}\int_0^H (H\sqrt{h} - h^{3/2})\,dh$$

$$= 2C_d\sqrt{2g}\,\tan\frac{\theta}{2}\left|\frac{2}{3}Hh^{3/2} - \frac{2}{5}h^{5/2}\right|_0^H$$

$$= 2C_d\sqrt{2g}\,\tan\frac{\theta}{2}\left[\frac{2}{3}H^{5/2} - \frac{2}{5}H^{5/2}\right]$$

$$= \frac{8}{15}C_d\sqrt{2g}\,\tan\frac{\theta}{2} \times H^{5/2}$$

$$= \frac{8}{15}C_d\,\tan\frac{\theta}{2} \times \sqrt{2g}H^{5/2} \qquad \text{... (12.2)}$$

For $C_d = 0.6$ and $\theta = 90°$, $\tan\frac{\theta}{2} = 1$, and

$$Q = 1.417\ H^{5/2} \qquad \text{... (12.3)}$$

Advantages of triangular notch over rectangular notch are:

1. The expression for discharge for a right-angled V-notch or weir is very simple.
2. A triangular notch give accurate results than a rectangular notch for measuring low discharge.
3. Discharge calculations required only one value of H only.
5. No ventilation is required.

12.5 DISCHARGE OVER A TRAPEZOIDAL NOTCH OR WEIR

Consider a trapezoidal notch or weir as shown in Fig. 12.3. It is a combination of a rectangular notch and a triangular notch or weir. Thus the total discharge will be equal to the sum of discharge through a rectangular notch or weir and a triangular notch or weir.

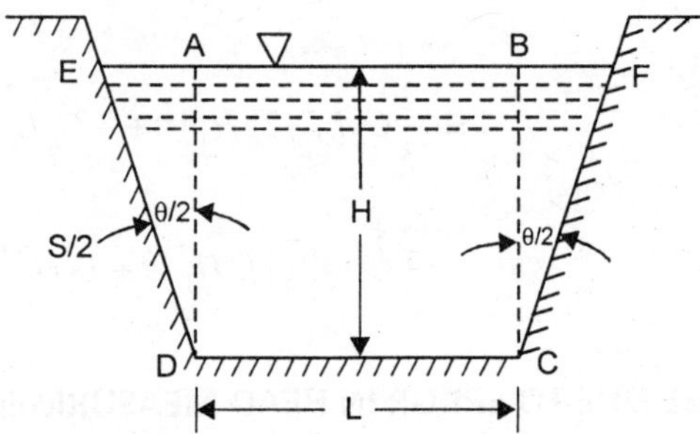

Fig 12.3 Trapezoidal notch

Let C_{d_1} = coefficient of discharge for rectangular portion ABCD

C_{d_2} = coefficient of discharge for triangular portions ADE and BCF.

Total discharge, $Q = Q_1 + Q_2$

$$= \frac{2}{3} C_{d_1} L \sqrt{2g} H^{3/2} + \frac{8}{15} C_{d_2} \tan\frac{\theta}{2} \sqrt{2g} H^{5/2} \quad \text{... (12.4)}$$

12.6 DISCHARGE OVER A STEPPED NOTCH

A stepped notch shown in Fig. 12.4 is a combination of rectangular notches. The discharge through the stepped notch is equal to the sum of the discharges through the different rectangular notches.

Let H_1, H_2, H_3 = heads of water above the crests of notch 1, 2 and 3 respectively.

L_1, L_2, L_3 = length of notches 1, 2 and 3 respectively.

C_d = coefficient of discharge for all notches

Total discharge,

$$Q = Q_1 + Q_2 + Q_3$$

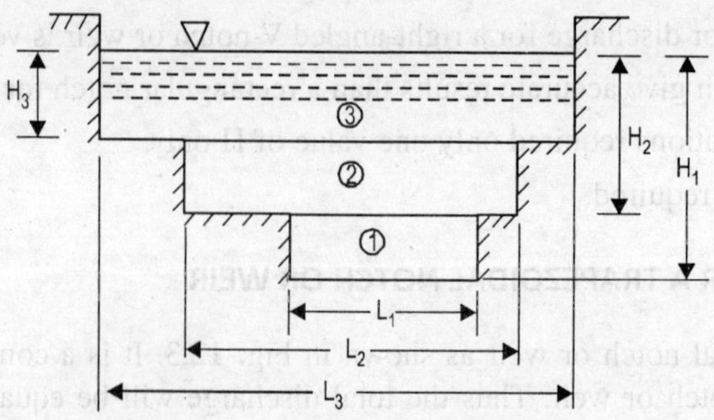

Fig. 12.4

$$= \frac{2}{3}C_d L_1\sqrt{2g}H_1^{3/2} + \frac{2}{3}C_d L_2\sqrt{2g}H_2^{3/2} + \frac{2}{3}C_d\sqrt{2g}H_3^{3/2}$$

$$= \frac{2}{3}C_d\sqrt{2g}[L_1H_1^{3/2} + L_2H_2^{3/2} + L_3H_3^{3/2}] \qquad \text{... (12.5)}$$

12.7 EFFECT ON DISCHARGE DUE TO ERROR IN HEAD MEASURMENT

1. Rectangular Notch or Weir

$$Q = \frac{2}{3}C_d L\sqrt{2g}H^{3/2}$$

$$= KH^{3/2} \qquad \text{... (1)}$$

where $K = \frac{2}{3}C_d L\sqrt{2g}$

Differentiating Eq. (1) w.r.t H, we get

$$dQ = \frac{3}{2}KH^{1/2}.dH$$

$$\frac{dQ}{Q} = \frac{K \times \frac{3}{2}H^{1/2} \cdot dH}{KH^{3/2}} = \frac{3}{2}\frac{dH}{H} \qquad \text{... (12.6)}$$

2. Triangular Notch or Weir

$$Q = \frac{8}{15}C_d \tan\frac{\theta}{2} \times \sqrt{2g}\, H^{5/2}$$

$$= KH^{5/2} \qquad \text{... (1)}$$

where $$K = \frac{8}{15} C_d \tan\frac{\theta}{2} \times \sqrt{2g}$$

Differentiating Eq. (1) w.r.t H, we get

$$dQ = \frac{5}{2} K H^{3/2} . dH$$

$$\frac{dQ}{Q} = \frac{K \times \frac{5}{2} H^{3/2} \cdot dH}{K H^{5/2}} = \frac{5}{2} \frac{dH}{H} \quad \text{... (12.7)}$$

12.8 TIME REQUIRED FOR EMPTYING A RESERVOIR OR TANK

1. Retangular Notch or Weier

Consider a reservoir or tank of uniform cross-sectional area, A on the one side of which a rectangular notch or weir is provided.

Let C_d = Coefficient of discharge

L = length of crest of of notch of weir

H_1, H_2 = initial and final head of liquid above the crest of notch or weir respectively.

T = time required to lower the head of liquid from H_1 to H_2.

Let at any instant the head of liquid surface above the crest of notch or weir be h and in a small interval of time dT the liquid surface falls by dh. Then

$$-A \cdot dh = Q \times dT$$

–ve sign is taken because h decreases with time T.

Now $$Q = \frac{2}{3} C_d L \sqrt{2g} h^{3/2}$$

$$-A.dh = \frac{2}{3} C_d L \sqrt{2g} h^{3/2} \times dT$$

or $$dT = -\frac{A}{\frac{2}{3} C_d L \sqrt{2g}} \cdot \frac{dh}{h^{3/2}}$$

Total time, $$\int_0^T dT = -\frac{3A}{2 C_d L \sqrt{2g}} \int_{H_1}^{H_2} h^{-3/2} . dh$$

$$= -\frac{3A}{2C_d L\sqrt{2g}} \left| \frac{h^{-1/2}}{-\frac{1}{2}} \right|_{H_1}^{H_2}$$

$$T = \frac{3A}{C_d L\sqrt{2g}}[H_2^{-1/2} - H_1^{-1/2}]$$

$$= \frac{3A}{C_d L\sqrt{2g}}\left[\frac{1}{\sqrt{H_2}} - \frac{1}{\sqrt{H_1}}\right] \quad \text{... (12.8)}$$

2. Triangular Notch or Weir

Proceeding as above, we have

$$-A.dh = Q \times dT$$

$$= \frac{8}{15} C_d \tan\frac{\theta}{2} \times \sqrt{2g} \times h^{5/2} \times dT$$

$$\int_0^T dT = \frac{-15A}{8C_d \tan\frac{\theta}{2} \times \sqrt{2g}} \int_{H_1}^{H_2} h^{-5/2} dh$$

$$T = \frac{-15A}{8C_d \tan\frac{\theta}{2} \times \sqrt{2g}} \left| \frac{h^{-3/2}}{-\frac{3}{2}} \right|_{H_1}^{H_2}$$

$$= \frac{5A}{4C_d \tan\frac{\theta}{2} \times \sqrt{2g}}\left[\frac{1}{H_2^{3/2}} - \frac{1}{H_1^{3/2}}\right] \quad \text{... (12.9)}$$

12.9 VELOCITY OF APPROACH

The velocity of approach is defined as the velocity with which water approaches or reaches the notch or weir before it flows over it.

Let v_a = *velocity of approach,*

Then additional head due to velocity of approach,

Let $H_a = \dfrac{v_a^2}{2g}$

∴ Initial head of water = $H + H_a$

Final head of water = H_a

The velocity of approach is determined by finding the discharge over the notch or weir by neglecting the velocity of approach.

$$v_a = \frac{Q}{A}$$

where A = cross-sectional area of channel on the upstream side of notch or weir.

The velocity of approach is used to find the additional head H_a. Again the discharge is calculated and above process is repeated for obtaining more accurate discharge.

Then corrected discharge over rectangular notch on weir is,

$$Q_{\text{correct}} = \frac{2}{3} C_d L \sqrt{2g}\,[(H + H_a)^{3/2} - H_a^{3/2}] \qquad \text{... (12.10)}$$

Corrected discharge for triangular notch or weir,

$$Q_{\text{corrected}} = \frac{8}{15} C_d \tan\frac{\theta}{2} \times \sqrt{2g}\,[(H + H_a)^{5/2} - H_a^{5/2}] \qquad \text{... (12.11)}$$

12.10 EMPIRICAL FORMULAE FOR DISCHARGE OVER RECTANGULAR WEIR

1. Francis's Formula

When the length of crest is equal to width of channel then the weir is called *suppressed* weir. But if the weir is not suppressed, the effect of end contraction is taken into account. Francis on the basis of his experiments established that end contraction decreases the effective length of the crest of weir and hence decreases the discharge. Each end contraction reduces the crest length by $0.1 \times H$, where H = Head over the weir.

For a rectangular weir there are two end contractions and hence effective length,

$$L_{\text{eff}} = L - 0.2\,H$$

as shown in Fig. 12.5

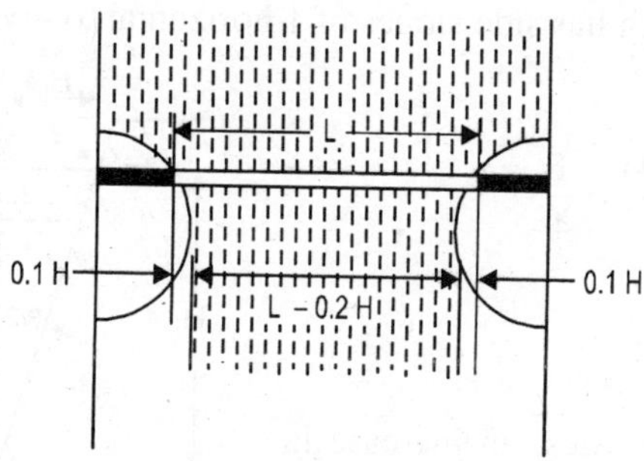

Fig. 12.5

$$Q = \frac{2}{3}C_d[L - 0.2H] \times \sqrt{2g}\, H^{3/2} \quad \text{... (12.12)}$$

If $C_d = 0.623$ and $g = 9.81$ m/s², then

$$Q = 1.84\text{L } [\text{L}-0.2\ H]H^{3/2} \quad \text{... (12.13)}$$

If end contractions are suppressed, then

$$Q = 1.84 \text{ L } H^{3/2} \quad \text{... (12.14)}$$

If velocity of approach is considered, then

$$Q = 1.84 \text{ L } [(H + H_a)^{3/2} - H_a^{3/2}] \quad \text{... (12.15)}$$

2. Bazin's Formula

According to Bazin's formula,

$$Q = m \times L \times \sqrt{2g} \times H^{3/2} \quad \text{... (12.16)}$$

where $m = \frac{2}{3}C_d = 0.405 + \frac{0.003}{H}$

H = head over the weir

If velocity of appoach is considered, then

$$Q = m_1 \times L \times 2\sqrt{2g}\left[(H + H_a)^{3/2}\right] \quad \text{... (12.17)}$$

where $m_1 = 0.405 + \frac{0.003}{(H + H_a)}$

12.11 CIPOLLETTI WEIR OR NOTCH

It is a trapezoidal weir, which has side slopes of 1 horizontal to 4 vertical, as shown in Fig. 12.6.

In ΔADE,

$$\tan\frac{\theta}{2} = \frac{AE}{AD} = \frac{(H/4)}{H} = \frac{1}{4}$$

$$\therefore \quad \frac{\theta}{2} = 14°\text{–}2'$$

By giving this slope to the sides, an increase in discharge through the triangular portions ADE and BCF of the weir is obtained.

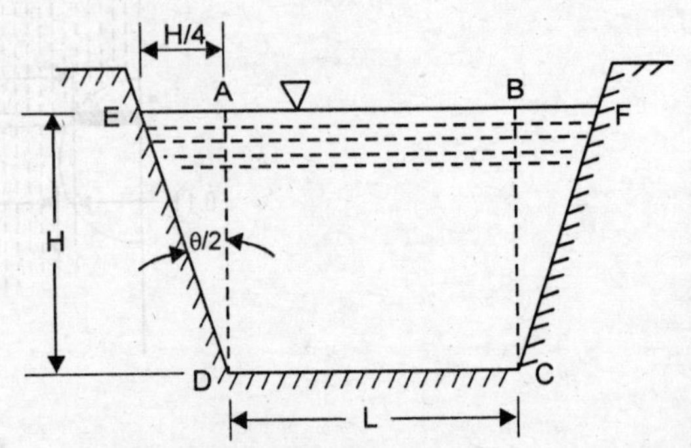

Fig. 12.6 Cippolletti weir

For a rectangular weir with end contractions,

$$Q = \frac{2}{3}C_d \times (L - 0.2H)\sqrt{2g}H^{3/2}$$

$$= \frac{2}{3}C_d L\sqrt{2g}H^{3/2} - \frac{2}{15}C_d\sqrt{2g}H^{5/2}$$

We observe that due to end contractions the discharge decreases by $\frac{2}{15}C_d\sqrt{2g}H^{5/2}$. This decrease in discharge can be compensated by giving a suitable slope to the sides so that the discharge through the trianglular portion is equal to $\frac{2}{15}C_d\sqrt{2g}H^{5/2}$.

The discharge through triangular notch of angle θ is given by,

$$Q = \frac{8}{15}\ C_d\sqrt{2g}\tan\frac{\theta}{2} \times H^{5/2}$$

$$\therefore \qquad \frac{8}{15}C_d \times \sqrt{2g} \times \tan\frac{\theta}{2} \times H^{5/2} = \frac{2}{15}C_d \times \sqrt{2g} \times H^{5/2}$$

$$\tan\frac{\theta}{2} = \frac{2}{15} \times \frac{15}{8} = \frac{1}{4}$$

or

$$\frac{\theta}{2} = 14°\text{–}2'$$

Thus discharge through cippolletti weir is, ... (12.18)

$$Q = \frac{2}{3}C_d \times L \times \sqrt{2g} \times H^{3/2}$$

If velocity of approach is considered, then

$$Q = \frac{2}{3}C_d \times L\sqrt{2g}[(H + H_a)^{3/2} - H_a^{3/2}] \qquad \text{... (12.19)}$$

12.12 DISCHARGE OVER A BROAD-CRESTED WEIR

A weir having a wide crest is known as broad-crested weir, as shown in Fig. 12.7.

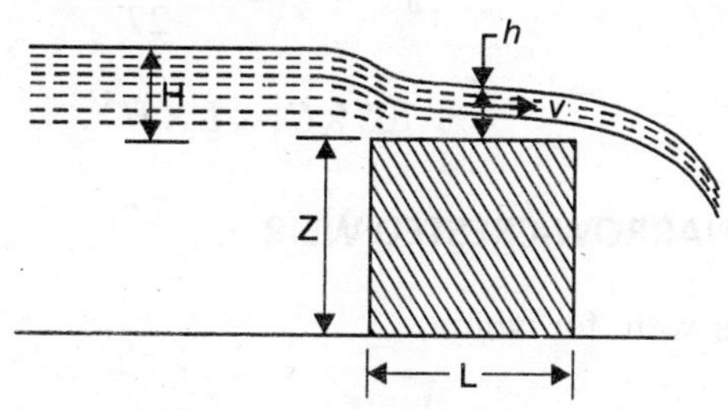

Fig. 12.7 Broad-crested weir.

Let H = head of water above the crest

L = length of crest

(i) If $2L > H$, the weir is called broad–crested weir.

(ii) If 2L < H, the weir is called a narrow crested weir.

Let h = constant head of water at the middle of weir

v = velocity of flow over the weir.

Applying Bunoulli's theorem to the still water surface on the upstream side and running water at the end of weir, we have

$$0 + 0 + H = 0 + \frac{v^2}{2g} + h$$

$$\therefore \quad \frac{v^2}{2g} = H - h$$

or

$$v = \sqrt{2g(H-h)}$$

Dicharge over weir,

$$Q = C_d \times L \times h\sqrt{2g(H-h)}$$

$$= C_d \times L \times \sqrt{2g(Hh^2 - h^3)} \quad \text{... (12.20)}$$

The discharge will be maximum, if $(Hh^2 - h^3)$ in maximum.

or

$$\frac{d}{dh}(Hh^2 - h^3) = 0$$

$$2Hh - 3h^2 = 0$$

$$h = \frac{2}{3}H$$

$$Q_{max} = C_d \times L \times \sqrt{2g\left(H \times \frac{4}{9}H^2 - \frac{8}{27}H^3\right)}$$

$$= C_d \times L \times \sqrt{2g \times \frac{4}{27}H^3}$$

$$= 1.705 \times C_d \times L \times H^{3/2} \quad \text{... (12.21)}$$

12.13 DISCHARGE OVER A NARROW-CRESTED WEIR

It is similar to a rectangular weir, for which,

$$Q = \frac{2}{3} \times C_d \times L \times \sqrt{2g} \times H^{3/2} \quad \text{... (12.22)}$$

12.14 DISCHA4GE OVER AN OGEE WEIR

In an Ogee weir, the crest of weir rises upto maximum height of 0.115 H, where H = height of water above inlet of the weir and then falls as shown in Fig. 12.8. The discharge for an Ogee weir is the same as that of a rectangular weir.

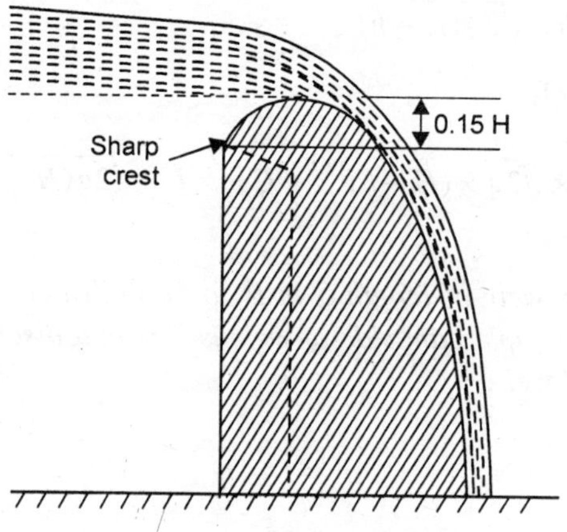

Fig. 12.8 An Ogee weir

∴ $$Q = \frac{2}{3} \times C_d \times L \times \sqrt{2g} \times H^{3/2} \qquad \text{... (12.23)}$$

12.15 DISCHARGE OVER SUBMERGED OR DROWNED WEIR

In a submerged weir, the water on the downstream side of the weir is above the crest of the weir, then the weir is called to be a submerged weir, as shown Fig. 12.9. The total discharge over the weir is obtained by dividing the weir into two parts. The portion between upstream and downstrream water surface may be treated as free weir and portion between down stream water surface and crest of weir as a drowned weir.

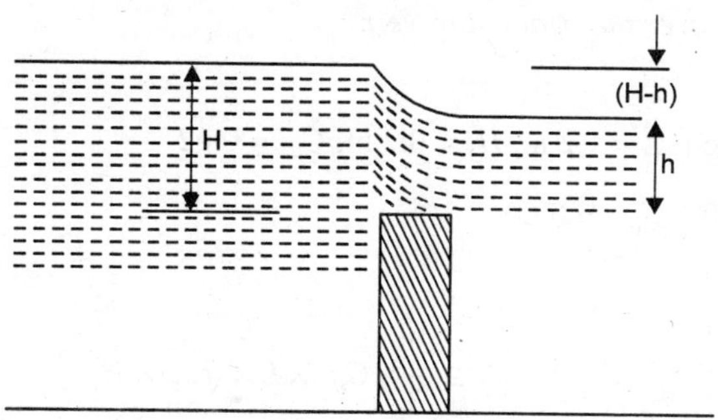

Fig. 12.9 Submerged weir

Let H = height of water on the upstream side of weir

h = height of water on the downstream side of weir

Then $Q_1 = \frac{2}{3} C_{d_1} \times L \times \sqrt{2g}\,(H-h)^{3/2}$

$Q_2 = C_{d_2} \times L \times h \times \sqrt{2g(H-h)}$

Total discharge, $Q = Q_1 + Q_2$

$$= \frac{2}{3} C_{d_1} \times L \times \sqrt{2g} \times (H-h)^{3/2} + C_{d_2} \times L \times \sqrt{2g(H-h)} \quad \text{... (12.24)}$$

Example 12.1 *A rectangular weir of length 5 m is to be built across a reciangular channel. The maximum depth of water on the upstream side of weir is 2 m and discharge is 2m³/s. Neglecting end contractions, find the height of weir.*

Take $C_d = 0.65$

solution.

Given: $L = 5$ m, $H_1 = 2$m, $Q = 2\text{m}^3/\text{s}$, $C_d = 0.65$

For a rectangular weir,

$$Q = \frac{2}{3} \times C_d \times L \times \sqrt{2g} \times H^{3/2}$$

$$2 = \frac{2}{3} \times 0.65 \times 5 \times \sqrt{02 \times 9.81} \times H^{3/2}$$

$$H = 0.3515 \text{ m}$$

Height of weir, $H_2 = H_1 - H = 2 - 0.351 = 1.6485$ m

Example 12.2 *Water flows over a rectangular weir 1.2 m wide at a depth of 0.2 m afterwards passes through a triangular right-angled weir. Taking C_d for rectangular and triangular weir as 0.62 and 0.60 respectively, find the depth over the triangular weir.*

solution.

Given: Rectangular weir: $L = 1.2$ m, $H_1 = 0.2$ m, $C_{d1} = 0.62$

Triangular weir: $\theta = 90°$, $C_{d_2} = 0.60$

Rectangular weir:

$$Q = \frac{2}{3} \times C_{d_1} \times L \times \sqrt{2g} \times H_1^{3/2}$$

$$= \frac{2}{3} \times 0.62 \times 1.2 \times \sqrt{2 \times 9.81} \times (0.2)^{3/2} = 0.1965 m^3 / s$$

Triangalar weir:

$$Q = \frac{8}{15} \times C_{d_2} \times \tan\frac{\theta}{2} \times \sqrt{2g} \times H_2^{5/2}$$

$$0.1965 = \frac{8}{15} \times 0.60 \times \tan 45° \times \sqrt{2 \times 9.81} \times H_2^{5/2}$$

$$H_2 = 0.4537 \text{ m}$$

Example 12.3 *Find the discharge through a trapezoidal notch which is 1 wide at the top and 0.4 m at the bottom and is 0.3 m in height. The head of water on the notch is 0.2 m. Assume C_d for the rectancular portion = 0.62 and C_d = 0.60 for the triangular portion.*

solution. Refer to Fig. 12.3

$$\tan\frac{\theta}{2} = \frac{AE}{AD} = \frac{\left(\frac{EF - DC}{2}\right)}{AD} = \frac{(1.0 - 0.4)}{2 \times 0.3} = 1$$

$$Q = \frac{2}{3} \times C_{d_1} \times L \times \sqrt{2g} \times H^{3/2} + \frac{8}{15} \times C_{d_2} \times \tan\frac{\theta}{2} \times \sqrt{2g} \times H^{5/2}$$

$$= \frac{2}{3} \times 0.62 \times 0.4 \times \sqrt{2 \times 9.81} \times (0.2)^{3/2} + \frac{8}{15} \times 0.60 \times 1 \times \sqrt{2 \times 9.81} \times (0.2)^{5/2}$$

$$= 0.06550 + 0.02536 = 0.09086 \text{ m}^3/\text{s}$$

Example 12.4 *A rectangular notch 0.4 m long is used for measuring a discharge of 0.03 m³/s. An error of 2 mm was made while measuring the head over the notch. Calculate the percentage error in the discharge. Take C_d = 0.62.*

Solution.

Given: $L = 0.4$, $Q = 0.03$ m³/s, $dH = 0.002$ m, $C_d = 0.62$

For rectangular notch,

$$Q = \frac{2}{3} \times C_d \times L \times \sqrt{2g} \times H^{3/2}$$

$$0.03 = \frac{2}{3} \times 0.62 \times 0.4 \times \sqrt{2 \times 9.81} \times H^{3/2}$$

$$H = 0.11887 \text{ m}$$

$$\frac{dQ}{Q} = \frac{3}{2} \times \frac{dH}{H} = \frac{3}{2} \times \frac{0.002}{0.11887} = 0.0252 \text{ or } 2.52\%$$

Example 12.6 *The head of water over a triangular notch of 60° is 0.6 m and coefficient of dicharge is 0.62. The flow measured by it is to be within an accuracy of 1.5 percent up or down. Calculate limiting values of head.*

Solution.

Given: $\theta = 60°$, $H = 0.6$ m, $C_d = 0.62$, $\frac{dQ}{Q} = \pm 1.5\% = \pm 0.015$

For triangular notch,

$$Q = \frac{8}{15} \times C_d \times \sqrt{2g} \times \tan\frac{\theta}{2} \times H^{5/2}$$

$$= \frac{8}{15} \times 0.62 \times \sqrt{2 \times 9.81} \times \tan 30° \times (0.6)^{5/2} = 0.149487 \text{ m}^3/\text{s}$$

$$\frac{dQ}{Q} = \frac{5}{2} \times \frac{dH}{H}$$

$$\pm 0.015 = \frac{5}{2} \times \frac{dH}{H}$$

$$\frac{dH}{H} = \pm 0.006$$

$$dH = \pm 0.006 \times 0.6 = \pm 0.0036 \text{ m}$$

$$H \pm dH = 0.6 \pm 0.0036$$

$$= 0.6036 \text{ m and } 0.5964 \text{ m}$$

Example 12.7 *The quantity of water flowing through a channel is measured by a rectangular notch weir. The water head over the notch is thrice the breadth of the rectangular notch. If the rectangular notch is replaced by a V-notch, calculate the angle of V-notch, other conditions remaining the same. The co-effficient of discharge for V-notch is 0.6 and for a rectangular notch is 0.64.* **(IES, 2005)**

Solution.

Rectangular notch,

$$Q_1 = \frac{2}{3}.C_{d1} L (2g)^{0.5} H^{3/2}$$

$$L = \frac{H}{3}$$

$$Q_1 = \frac{2}{9} \cdot C_{d1} (2g)^{0.5} H^{5/2}$$

V-notch,

$$Q_2 = \frac{8}{15}.C_{d_2} \tan\theta (2g)^{0.5} H^{5/2}$$

For $Q_1 = Q_2$, we get

$$\frac{2}{9}\times 0.64 = \frac{8}{15}\times 0.6\times \tan\theta$$

$$\tan\theta = 0.444$$

$$\theta = 23.96^\circ = 24^\circ$$

Example 12.8 *While conducting flow measurment using a rectangular notch, an error of 2% in head over the notch and an error of –3% in length of the notch occurred. Determine the percentage error in discharge.*

Solution.

For a rectangular notch.

$$Q = \frac{2}{3}C_d\sqrt{2g}LH^{3/2} = KLH^{3/2}$$

$$\frac{\partial Q}{\partial L} = KH^{3/2}, \frac{\partial Q}{\partial H} = \frac{3}{2}KLH^{1/2}$$

$$dQ = \frac{\partial Q}{\partial L}dL + \frac{\partial Q}{\partial H}dH$$

$$= KH^{3/2}dL + \frac{3}{2}KLH^{1/2}dH$$

$$\frac{dQ}{Q} = \frac{dL}{L} + \frac{3}{2}\frac{dH}{H}$$

$$= -0.03 + \frac{3}{2}\times 0.02$$

$$= -0.03 + 0.03 = 0$$

Example 12.9 *A reservoir of dimensions 40 m ×50 m contains water. Its level is to be lowered from 2 m to 1 m by a rectangular notch of length 1 m. Calculate the time required. Take $C_d = 0.62$.*

Solution.

Given: $H_1 = 2$ *m*, $H_2 = 1$ *m*, $A = 40 \times 50 = 2000$ *m*^{2}, $L = 1m$, $C_d = 0.62$.

Time required by rectangular notch,

$$T = \frac{3A}{C_d\times L\times\sqrt{2g}}\left[\frac{1}{\sqrt{H_2}} - \frac{1}{\sqrt{H_1}}\right]$$

$$= \frac{3\times 2000}{0.62\times 1\times\sqrt{2\times 9.81}}\left[\frac{1}{\sqrt{1}} - \frac{1}{\sqrt{2}}\right]$$

$$= 639.91 \text{ s or } 10 \text{ min } 40 \text{ s}$$

Example 12.10 *A right-angled V-notch is inserted in the side of a tank of length 3 m and width 2m. Initial height of water above the apex of notch is 0.4 m. Calculate the height of water above the apex if the time required to lower the head in tank from 0.4 m to final height is 2 minutes. Take $C_d = 0.60$.*

Solution.

Given: $\theta = 90°$, $A = 3\times 2 = 6\ m^2$, $H_1 = 0.4\ m$, $T = 2\times 60 = 120\ s$, $C_d = 0.60$

Let H_2 = final height of water above the apex of notch

$$T = \frac{5A}{4\times C_d \times \tan\frac{\theta}{2}\times\sqrt{2g}}\left[\frac{1}{H_2^{3/2}} - \frac{1}{H_1^{3/2}}\right]$$

$$120 = \frac{5\times 6}{4\times 0.60\times \tan 45°\times\sqrt{2\times 9.81}}\left[\frac{1}{H_2^{3/2}} - \frac{1}{(0.4)^{3/2}}\right]$$

$$42.523 = \frac{1}{H_2^{3/2}} - 3.953$$

$$H_2 = 0.0773 \text{ m or } 77.3 \text{ mm}$$

Example 12.11 *The velocity of approach for flow of water over a rectangular weir of length 50 m is 0.5 m/s. The head of water over the weir is 1.25 m. Calculate the discharge if $C_d = 0.60$.*

Solution.

Given: $L = 50$ m, $v_a = 0.5$ m/s, $H = 1.25$ m, $C_d = 0.60$

Additional head due to velocity of approach,

$$H_a = \frac{v_a^2}{2g} = \frac{(0.5)^2}{2\times 9.81} = 0.0127\,\text{m}$$

$$Q = \frac{2}{3}\times C_d \times L\times\sqrt{2g}\,[(H + H_a)^{3/2} - H_a^{3/2}]$$

$$= \frac{2}{3}\times 0.60.\times 50\times\sqrt{2\times 9.81}\,[(1.25 + 0.0127)^{3/2} - (0.0127)^{3/2}]$$

$$= 125.57\ \text{m}^3/\text{s}$$

Example 12.12 *A suppressed rectangular weir is constructed across a channel of 0.8 m width with a head of 0.5 m and crest 0.7 m above the bed of channel. Considering velocity of approach, estimate the discharge. Take $C_d = 0.62$.*

Solution.

Given: $b = 0.8$ m, $H = 0.5$ m, depth of channel = 0.5 + 0.7 = 1.2 m, $C_d = 0.62$

For a suppressed weir, width of channel = width of weir = 0.8 m

Area of channel, $A = 0.8 \times 1.2 = 0.96\ m^2$

Dicharge over a rectangular weir without velocity of appoach,

$$Q = \frac{2}{3} \times C_d \times b \times \sqrt{2g} \times H^{3/2}$$

$$= \frac{2}{3} \times 0.62 \times 0.8 \times \sqrt{2 \times 9.81} \times (0.5)^{3/2}$$

$$= 0.5178\ m^3/s$$

Velocity of approach ,

$$v_a = \frac{Q}{\text{Area of channel}} = \frac{0.5178}{0.96} = 0.5394\ m/s$$

Additional head,

$$H_a = \frac{v_a^2}{2g} = \frac{(0.5394)^2}{2 \times 9.81} = 0.0148\ m$$

$$Q = \frac{2}{3} \times C_d \times b \times \sqrt{2g}\,[(H + H_a)^{3/2} - H_a^{3/2}]$$

$$= 0.5384\ m^3/s$$

Example 12.13 *A weir 30 m long is divided into 10 equal bays by vertical posts, each 0.5 wide. Calculate the discharge over the weir if the head over the crest is 1.5 m and velocity of approach is 2m/s.*

Solution

Given: $L_1 = 30$ m, $N = 10$, $b = 0.5$ m, $H = 1.5$ m, $v_a = 2$m/s

For 10 bays, number of vertical posts = $N–1 = 10 – 1 = 9$

Effective length of weir, $L = L_1 – 9 \times b = 30 – 9 \times 0.5 = 25.5$ m

Additional head,

$$H_a = \frac{v_a^2}{2g} = \frac{2^2}{2 \times 9.81} = 0.2039\ m$$

Number of end contractions, $n = 2 \times N = 2 \times 10 = 20$

Using Francis formula, we have

$$Q = 1.84\,[L – 0.1 \times n\,(H + H_a)]\,[(H + H_a)^{3/2} – H_a^{3/2}]$$

$$= 1.84\,[25.5 – 0.1 \times 20\,(1.5 + 0.2039)]\,[(1.5 + 0.2039)^{3/2} – (0.2039)^{3/2}]$$

$$= 86.668\ m^3/s$$

Example 12.14 *A cipolletti weir of crest length 0.75 m discharges water with a head of 0.4 m. Calculate the discharge over the weir if the channel is 1 m wide and 0.5 m deep. Take $C_d = 0.60$.*

Solution.

Given: $L = 0.75m$, $H = 0.4\ m$, $A = 1 \times 0.5 = 0.5\ m^3$, $C_d = 0.60$

$$Q = \frac{2}{3} \times C_d \times L \times \sqrt{2g} \times H^{3/2}$$

$$= \frac{2}{3} \times 0.60 \times 0.75\sqrt{2 \times 9.81} \times (0.4)^{3/2} = 0.3362\,\text{m}^3/\text{s}$$

$$v_a = \frac{Q}{A} = \frac{0.3362}{0.5} = 0.672\,\text{m/s}$$

$$v_a = \frac{Q}{A} = \frac{0.3362}{0.5} = 0.672\,\text{m/s}$$

$$\text{H}_a = \frac{v_a^2}{2g} = \frac{(0.672)^2}{2 \times 9.81} = 0.023\,\text{m}$$

$$\text{Q} = \frac{2}{3} \times C_d \times L \times \sqrt{2g}\,[(H + H_a)^{3/2} - H_a^{3/2}]$$

$$= \frac{2}{3} \times 0.60 \times 0.75 \times \sqrt{2 \times 9.81}\;[(0.4 + 0.23)^{3/2} - (0.023)^{3/2}]$$

$$= 0.3609\ \text{m}^3/\text{s}$$

SUMMARY

1. A notch is a device used for measuring the flow rate of a liquid through a small channel. It is generally made out of metallic plate.
2. A weir is a concrete or masonary structure placed in an open channel to measure flow rate.
3. The main difference between a notch and weir is, that the notch is of small size and the weir is of a large size.
4. Discharge through a rectangular notch or weir:

   (i) Neglecting velocity of approach and sides contraction.

$$Q = \frac{2}{3} \times C_d \times L \times \sqrt{2g} \times H^{3/2}$$

   (ii) Considering velocity of approach.

$$v_a = \frac{Q}{\text{Area of channel}}$$

$$H_a = \frac{v_a^2}{2g}$$

$$Q = \frac{2}{3} \times C_d \times L \times \sqrt{2g}\,[(H + H_a)^{3/2} - H_a^{3/2}]$$

(iii) Considering sides contraction.

Effective length, L_{eff} = L–0.2 H for a single notch or weir

= L–0.1 *n*H for *n* notches or weirs

$$Q = \frac{2}{3} \times C_d \times L_{eff} \times \sqrt{2g} \times H^{3/2}$$

5. Discharge through a triangular notch or weir:

$$Q = \frac{8}{15} \times C_d \times \sqrt{2g} \times \tan\frac{\theta}{2} \times H^{5/2}$$

6. Discharge through a trapezoidal notch or weir:

$$Q = \frac{2}{3} \times C_{d_1} \times L \times \sqrt{2g} \times H^{3/2} + \frac{8}{15} \times C_{d2} \times \sqrt{2g} \times \tan\frac{\theta}{2} \times H^{5/2}$$

7. Error in discharge due to error in measurment of head over a notch or weir:

$$\frac{dQ}{Q} = \frac{3}{2}\frac{dH}{H} \text{ for a rectangular notch}$$

$$= \frac{5}{2}\frac{dH}{H} \text{ for a triangular notch}$$

8. Time required to empty a reservoir or trank by a notch:

$$T = \frac{3A}{C_d L\sqrt{2g}}\left[\frac{1}{\sqrt{H_2}} - \frac{1}{\sqrt{H_1}}\right] \text{ by a rectangular notch}$$

$$= \frac{5A}{4C_d \tan\left(\frac{\theta}{2}\right)\sqrt{2g}}\left[\frac{1}{H_2^{3/2}} - \frac{1}{H_1^{3/2}}\right] \text{ by a triangular notch}$$

9. Framcis's formula for a rectangular weir:

$Q = 1.84\,[L - 0.2\,H]H^{3/2}$, for two end contractions.

$= 1.84\,L\,H^{3/2}$, if end contractions are suppressed.

$= 1.84L[(H + H_a)^{3/2} - H_a^{3/2}]$, if velocity of approach is considered.

10. Bazin's formula for a rectangular weir:

$Q = mL\sqrt{2g} \times H^{3/2}$, without velocity of approach

$= mL\sqrt{2g} \times (H + H_a)^{3/2}$, considering velocity of approach where $m = 0.405 + \frac{0.003}{H} = \frac{2}{3}C_d$, without velocity of approach

$$= 0.405 + \frac{0.003}{H + H_a}, \text{ with velocity of approach}$$

11. A cipolletti weir is a trapezoidal weir having sides of 1 horizontal to 4 vertical.

$$Q = \frac{2}{3} \times C_d \times L \times \sqrt{2g} \times H^{3/2}, \text{ without velocity of approach.}$$

$$= \frac{2}{3} \times C_d \times L \times \sqrt{2g} \times [(H + H_a)^{3/2} - H_a^{3/2}], \text{ with velocity of approach.}$$

12. Discharge over a broad-crested weir:

$$Q = C_d \times L \times \sqrt{2g} \times \sqrt{(Hh^2 - h^3)}$$

Condition for maximum discharge, $h = \frac{2}{3}H$

$$Q_{max} = 1.705\ C_d \times L \times H^{3/2}$$

13. Discharge over submerged or drowned weir:

$$Q = \frac{2}{3} \times C_{d1} \times L \times \sqrt{2g} \times (H - h)^{3/2} + C_{d_2} \times L \times h \times \sqrt{2g} \times \sqrt{(H - h)}$$

Short Answer Type Questions

1. Define the terms: Notch, Weir, Nappe and Crest.
2. What is the main difference between a notch and weir?
3. How do you classify notches and weirs?
4. What are the advantages of a triangular notch over a rectangular notch?
5. What is velocity of approach?
6. Define end contraction of a weir. How this is accounted for?
7. What is a cipolletti weir? What is its advantage?
8. Differentiate between broad-crested and narrow-crested weirs.
9. What is the condition for maximum discharge over a broad-crested weir?
10. What is a suppressed weir?
11. What is a drowned weir?
12. What is an Ogee weir?

EXERCISES

1. A rectangular sharp-crested weir is 1.2 m long. How high should it be placed in a channel to maintain an upstream depth of 1.5 m for a discharge of 0.46 m^3/s? Take $C_d = 0.64$

   [Ans. 1.16 m]

2. The head upstream of a rectangular weir 2 m long is 0.1 m. Cross-sectional area of up stream flow is 0.3m^2. Estimate to a first approximation the discharge allowing for velocity of approach. Take $C_d = 0.60$. **[Ans.** 0.1242 m^3/s]

3. A reservoir has an area of 9×10^4 m^2 and is provided with a Francis weir, 5 m long with end contraction. If initial level is 1m above the sill of the weir how long will it take for the water level to fall to 0.5 m above the sill? **[Ans.** $2h$ – 16 min]

4. Find the discharge over a triangular notch of angle 60° when the head over the notch is 0.20 m. Take $C_d = 0.62$. **[Ans.** 0.0169 m^3/s]

5. A right-angled V-notch was used for measuring a discharge of 0.03 m^3/s. An error of 2 mm was made in measuring the head over the notch. Calculate the percentage error in the discharge. Take $C_d = 0.62$. **[Ans.** 2.37 %]

6. Calculate the time required to lower the water level from 3 m to 1.5 m in a reservoir of dimensions 80 m × 80 m, by a rectangular notch of length 2 m. Take $C_d = 0.60$. **[Ans.** 14 min–23s]

7. Water is flowing in a rectangular channel 1.2 m wide and 0.8 m deep. Calculate the discharge over a rectangular weir of crest length 0.7 m if the head of water over the crest of weir is 0.25 m. Take $C_d = 0.60$. **[Ans.** 0.1557 m3/s]

8. A discharge of 1500 m^3/s is to pass over a rectangular weir. The weir is divided into a number of openings each of 7.5 m span. If the velocity of approach is 3m/s, find the number of openings required so that head of water over the crest of weir does not exceed 1.8 m. **[Ans.** 38]

9. A broad-crested weir 40 m long has 0.4 m height of water above its crest. Calculate the maximum discharge. Take $C_d = 0.60$ and neglect velocity of approach. **[Ans.** 10.35 m3/s]

10. In a submerged weir of 3 m length the height of water on the upstream and downstream sides are 0.2 m and 0.1 m respectively. Calculate discharge over the weir if the discharge coefficients for free and drowned portion are 0.6 and 0.8 respectively.

13

FLOW THROUGH OPEN CHANNELS

13.1 INTRODUCTION

An open channel may be defined as a passage in which liquid flows with its upper surface exposed to atmosphere. The flow in open channels is characterised by the existence of a free surface. The free-surface may be defined as the surface of contact between the liquid and the overlying gaseous fluid. The interface is subjected to a constant pressure throughout its length and breadth. Some examples of open channel flow are: pipe flow when flowing water has free surface, a sewer under ordinary conditions of flow, flow of water in a canal, etc.

13.2 CLASSIFICATION OF FLOW

The open channel flow may be classified as shown in Fig. 13.1.

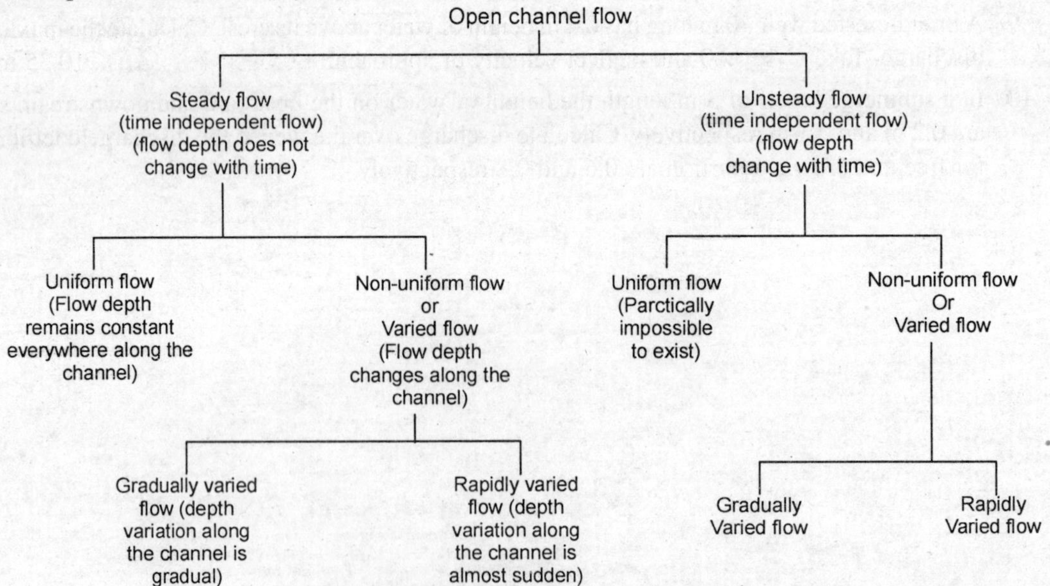

Fig. 13.1 Classification of open channel flow

The open channel flow can be classified based on the change in flow depth with respect to time and space.

1. Classification with respect to time.
   (i) **Steady flow:** The flow is steady if the depth of flow at a section does not change during the time interval under consideration.
   (ii) **Unsteady flow:** If the flow depth at a section changes with time, the flow is called unsteady.
2. Classification with respect to sapce.
   (i) **Spatially varied flow:** If the water flows into or out of the channel at different points along its length then the flow is known as spatially varied or discountinuous flow.
   (ii) **Uniform flow:** When the depth of flow remains same at every section along the channel then the flow is said to be uniform.
   (iii) **Non-uniform or varied flow:** The flow is said to be non-uniform or varried if the depth of flow changes along the length of the channel.

The non-uniform flow may be further classified as follows:

(a) Gradually varied flow–in which changes in depth are gradual.

(b) Rapidly varied flow–in which the depth changes abruptly within a short distance.

The above classification of open channel flow is shown in Fig. 13.1

13.2.1 Laminar and Turbulent Flow

In open channel flow, Reynolds number is defined as:

Reynolds number,
$$\text{Re} = \frac{\rho v R}{\mu} \qquad \text{...(13.1)}$$

When ρ = mass density of fluid

v = mean velocity of flow

R = hydraulic radius

$$= \frac{\text{area of flow}, A}{\text{wetted perimeter}, P}$$

μ = dynamic viscosity of fluid.

When *(i)* Re < 500, the flow is laminar

(ii) 500 < Re < 2000, the flow is transitional

(iii) Re > 2000, the flow is turbulent.

13.2.2 Subcritical, Critical and Supercritical Flow

The gravitational force is a predominant force in open channel flow. Therefore, Froude number is an important parameter for analysing flow.

Froud number, $$Fr = \frac{v}{\sqrt{gD}} \quad \text{...(13.2)}$$

where v = mean velocity of flow

D = hydraulic depth = $\frac{\text{area of flow, } A}{\text{width at the free surface, } T}$

g = acceleration due to gravity

The magnitude of Froud number determines the type of flow carried by the channel.

(i) If Fr < 1, the flow is said to be sub-critical or tranquil or streaming

(ii) If Fr = 1, the flow is known as critical

(iii) If Fr > 1, the flow is called supercritical.

13.3 DEFINITIONS

The important terms related to flow through open channel are shown in Fig. 13.2

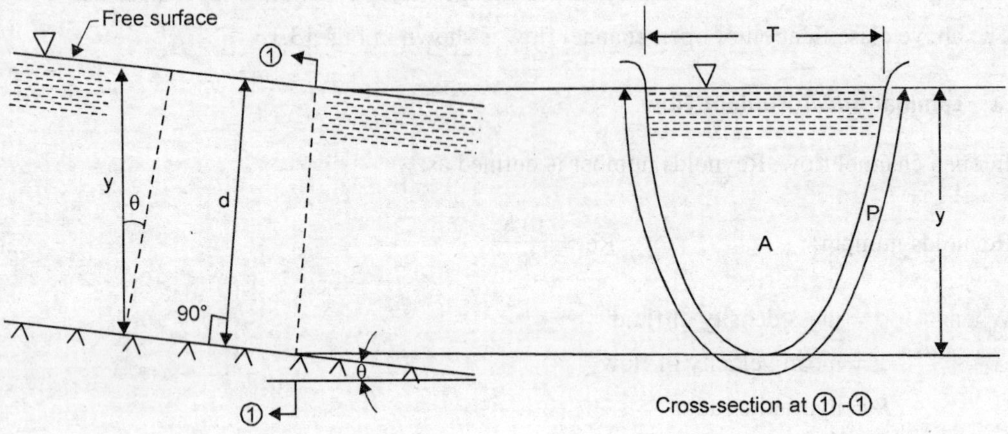

Fig 13.2 Terms related to flow through open channel.

Depth of flow (y): It is the vertical distance of the lowest point (or bed) of the channel section from the free surface.

Depth of flow sectin (d): It is the depth of flow normal to the bed of a channel.

$$d = y \cos \theta \quad \text{...(13.3)}$$

θ = angle which the channel bed makes with the horizontal.

Since θ is very small, therefore, $\cos\theta \simeq 1$ and $d = y$

Top width (T): It is the width of the channel section at the free surface.

Wetted area (A): It is the cross-sectional area of the flow section of the channel.

Wetted perimeter (P): It is the length of the channel boundary in contact with the flowing water at any section.

Hydraulic radius (R): It is the ratio of the cross-sectional area A of flow to wetted perimeter P. It is also called hydraulic mean depth.

$$R = \frac{A}{P} \qquad ...(13.4)$$

Hydraulic depth (D): It is the ratio of wetted area A to top width of free surface T.

$$D = \frac{A}{T} \qquad ...(13.5)$$

(i) Rectangular section (Fig. 13.3*a*)

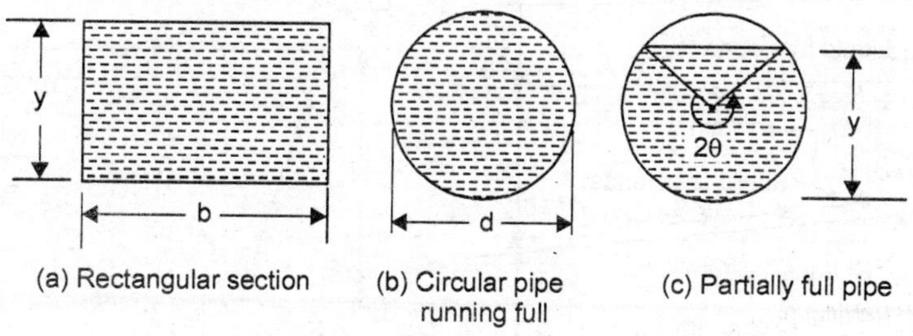

Fig. 13.3 Channel cross-sections

$$A = b \times y$$

$$P = b + 2y$$

$$R = \frac{A}{P} = \frac{by}{b+2y} \simeq y \quad \text{since } b >> y \qquad ...(13.6)$$

(ii) Pipe running full (Fig. 13.3 *b*):

$$R = \frac{A}{P} = \frac{\frac{\pi}{4}d^2}{\pi d} = \frac{d}{4} \qquad ...(13.7)$$

(iii) Partially full pipe (Fig. 13.3 c)

$$R = \frac{A}{P} = \frac{\frac{1}{2}r^2(2\theta - \sin 2\theta)}{2r\theta} \qquad ...(13.8)$$

13.4 FORMULAE FOR UNIFORM FLOW

When flow takes place in an open channel, two types of forces are developed. The first one is the force which causes the flow, and is due to the slope provided to the channel. The second one is the force of resistance which opposes the motion. The force causing the flow is equal to the component of the weight of water in the direction of flow. The uniform flow will be obtained when the resistance force equals the force causing the flow.

The following formulae have been developed by Chezy and Manning based on their experimental results:

1. Chezy's formula 2. Manning's formula

13.4.1 Chezy's formula

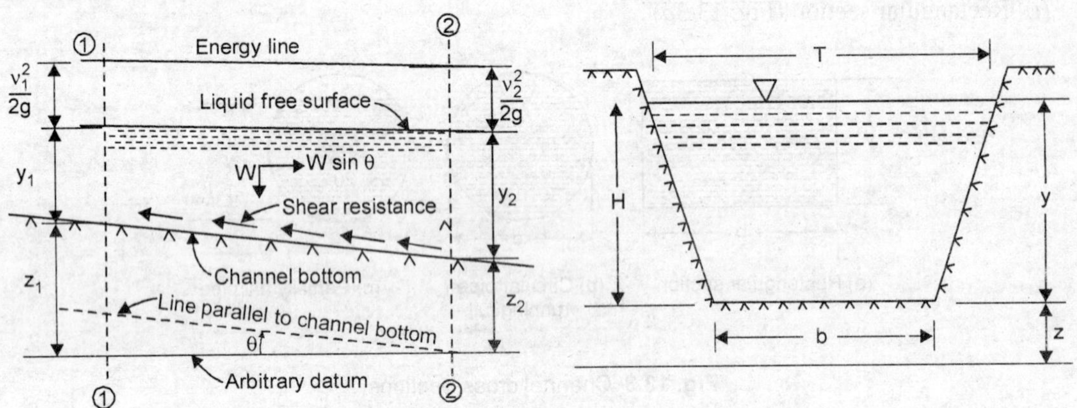

Fig. 13.4 Uniform flow in open channel.

Consider a longitudinal section of an open channel in which flow is steady and uniform, as shown in Fig. 13.4 *(a)*. The Chezy's formula is based on the following assumptions:

1. The force resisting the flow pei unit of wetted area is proportional to the square of the velocity.
2. The force causing the flow must be equal to the force of resistance.

Consider the body of water contained in between sections 1–1 and 2–2 distance L apart.

Let v = velocity of flow

P = wetted perimeter

Then according to first assumption,

Resistance force per unit wetted area $\propto v^2 = K_0 v^2$

where K_0 = a constant

Wetted area, $A = P \times L$

$$\therefore \quad \text{Resisting force} = K_0 L p v^2 \quad \text{...(1)}$$

The force causing flow is equal to the component of the weight of water in the direction of flow.

Let W = weight of water contained in between sections 1– 2.

$= \gamma AL$

θ = angle of inclination of channel bottom with the horizontal.

Then force causing flow = W sin $\theta = \gamma AL \sin\theta = \gamma ALS$ ... (2)

where $S = \sin\theta$

It may be noted that in uniform flow, we have

Channel slope = Water surface slope = Energy line slope.

Using the second assumption and equating Eqs. (1) and (2), we get

$$\gamma ALS = K_0 PLv^2$$

or

$$v = \sqrt{\left(\frac{\gamma}{K_0}\right)\left(\frac{A}{P}\right)S} = C\sqrt{RS} \quad \text{... (13.9)}$$

where

$$R = \frac{A}{P} = \text{hydraulic radius}$$

$$C = \sqrt{\frac{\gamma}{K_0}} = \text{flow resistance factor, known as Chezy's constant}$$

Discharge,

$$Q = Av = AC\sqrt{RS} \quad \text{... (13.10)}$$

13.4.2 Empirical Formulae for Chezy's Constant

The dimension of C is

$$= \frac{v}{\sqrt{RS}} = \frac{LT^{-1}}{\sqrt{L}} = \frac{\sqrt{L}}{T} = L^{1/2}T^{-1}$$

1. Bezin's Formula (in MKS units)

$$C = \frac{157.6}{1.81 + \frac{K}{\sqrt{R}}} \quad \text{... (13.11)}$$

where K = Bazin's constant, which depends on roughness of channel surface (see Table 13.1)

R = hydraulic radius

2. Ganguillet-Kutter Formula (in MKS units)

$$C = \frac{23 + \frac{0.00155}{S} + \frac{1}{N}}{1 + \left(23 + \frac{0.00155}{S}\right)\frac{N}{\sqrt{R}}} \quad \text{... (13.12)}$$

where N = roughness coefficient and is known as Kutter's constant (see Table 13.2)

3. Mannings's Formula

$$C = \frac{1}{n} R^{1/6}$$

where n = coefficient of roughness (or rugosity coefficient) called Manning's n. ...(13.3)
(See Table 13.3)

Table 13.1 Values of Bazin's constant K

S.No.	Surface of channel	Bazin's constant (K)
1.	Smooth cement plaster or planed wood	0.11
2.	Concrete, brick, or unplaned wood	0.21
3.	Smooth rubble masonry or poor brickwork	0.83
4.	Earth channels in very good condition	1.54
5.	Earth channels in rough condition	3.17
6.	Dredged earth channels, average condition	2.36

Table 13.2 Values of Kutter's Constant N

S.No.	Surface of channel	N (Kutter's/Manning's constant)
1.	Smooth cement plaster or planed wood	0.010
2.	Very smooth concrete and planed timber	0.011
3.	Smooth concrete	0.012
4.	Glazed brickwork	0.013
5.	Vitrified clay	0.014
6.	Brick surface lined with cement mortar	0.015
7.	Earth channels in best condition	0.17
8.	Straight unlined earth channels in good condition	0.20
9.	Rivers and earth channels in fair condition	0.025
10.	Canal and river of rough surface with weeds	0.030

13.4.3 Average Shear Stress

The resistance force may be expressed in terms of the average shear stress acting on the wetted parameter of the channel.

Let τ_0 = average shear stress on the wetted perimeter. Then

Force of resistance = $\tau_0 PL$ ... (13.14)

For uniform flow,

$$\tau_0 PL = \gamma ALS$$

$$\tau_0 = \gamma \times \frac{A}{P} \times S = \gamma RS \quad \text{... (13.15)}$$

Example 13.1 *Water flows uniformly at a depth of 1.5 m in rectangular canal 4 m wide laid on a slope of 1:1000. What is the mean shear stress on the sides and bottom of the canal. Also determine the type of flow in the canal. Take Chezy's constant = 70 and kinematic viscosity = 10^{-6} m²/s.*

Solution.

Given: $y = 1.5$ m, $b = 4$m, $S = \frac{1}{1000}$, $C = 70$, $\nu = 10^{-6}$ m²/s

Hydraulic radius,
$$R = \frac{A}{P} = \frac{b \times y}{b + 2y} = \frac{4 \times 1.5}{4 + 2 \times 1.5} = \frac{6}{7}$$

Mean shear stress,
$$\tau_0 = \gamma RS = 9810 \times \frac{6}{7} \times \frac{1}{1000} = 8.4 \text{ N/m}^2$$

Average velocity,
$$v = C\sqrt{RS} = 70\sqrt{\frac{6}{7} \times \frac{1}{1000}} = 2.04949 \text{ m/s}$$

Reynolds number,
$$\text{Re} = \frac{vR}{\nu} = \frac{2.0494 \times \frac{6}{7}}{10^{-6}} = 1.75622 \times 10^6 > 2000$$

Froude number,
$$\text{Fr} = \frac{v}{\sqrt{gD}}$$

$$\text{D} = \frac{A}{T} = \frac{by}{b} = y = 1.5\text{m}$$

$$\text{Fr} = \frac{2.0494}{\sqrt{9.81 \times 1.5}} = 0.534 < 1$$

The flow is subcritical.

Example 13.2 *Calculate the discharge through a trapezoidal channel of width 5 m and side slope of 1 horizontal to 2.5 vertical. The depth of flow of water is 2.5 m and Chezy's constant is 50. The slope of bed of channel is 1 in 3000.*

Solution.

Given: $b = 5$ m, $y = 2.5$ m, $C = 50$, $S = \frac{1}{3000}$

Refer to Fig. 13.5.

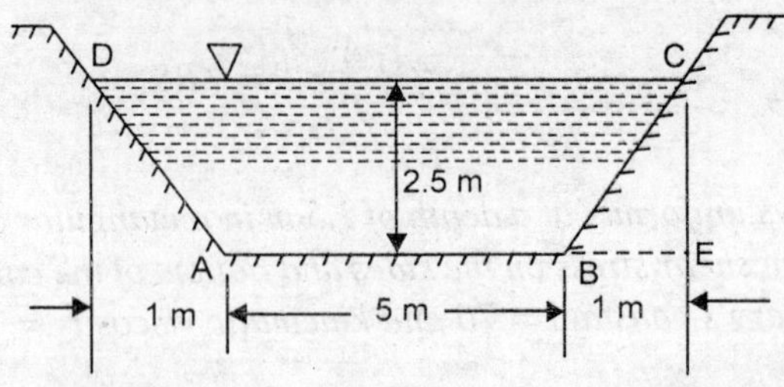

Fig. 13.5

$$BE = \frac{2.5}{2.5} = 1\,\text{m}$$

$$T = CD = 5 + 2 \times 1 = 7\ \text{m}$$

Area of trapezium, $$A = \left(\frac{AB + CD}{2}\right) \times CE$$

$$= \left(\frac{5+7}{2}\right) \times 2.5 = 15\,\text{m}^2$$

Wetted perimeter, $$P = 5 + 2\sqrt{1 + (2.5)^2} = 10.385\,\text{m}$$

Hydraulic radius, $$R = \frac{A}{P} = \frac{15}{10.385} = 1.444$$

$$Q = AC\sqrt{RS} = 15 \times 50\sqrt{1.444 \times \frac{1}{3000}} = 16.456\ \text{m}^3/\text{s}$$

Example 13.3 *A triangular gutter with included angle 60° conveys water at a uniform depth of 0.3 m. If the discharge is 0.042 m³/s, determine the gradient of trough. Take Chezy's constant = 55.*

Solution.

Given: $2\theta = 60°$, $y = 0.3$ m, $Q = 0.042$ m³/s, $C = 55$

Refer to Fig. 13.6.

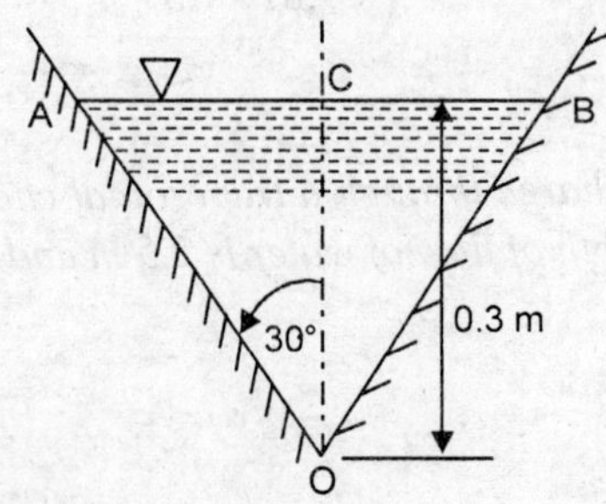

Fig. 13.6

$$\frac{AC}{OC} = \tan 30°, AC = 0.3 \tan 30° = 0.1732 \text{ m}$$

$$\frac{OC}{OA} = \cos 30°, OA = \frac{0.3}{\cos 30°} = 0.346 \text{ m}$$

$$AB = 2AC = 2 \times 0.1732 = 0.3464 \text{ m}$$

Area across flow, $A = \frac{1}{2} \times AB \times OC$

$$= \frac{1}{2} \times 0.3464 \times 0.3 = 0.05196 \text{ m}^2$$

Wetted perimeter, $P = 2 \times OA = 2 \times 0.346 = 0.692 \text{ m}$

Hydraulic radius, $R = \frac{A}{P} = \frac{0.05196}{0.692} = 0.075 \text{ m}$

$$Q = AC\sqrt{RS}$$

$$0.042 = 0.05196 \times 55 \sqrt{0.075 \times 5}$$

$$S = 2.38 \times 10^{-3} = \frac{1}{347}$$

Gradient of trough = 1:347

13.4.4 Manning's Formula

According to Manning's the mean velocity of flow in the open channel is given by,

$$v = \frac{1}{n} R^{2/3} S^{1/2} \text{ m/s} \quad \text{... (13.16)}$$

where n = coefficient of roughness, known as Manning's n. (see tabel 13.3)

Table 13.3 Mannings'sn for various types of channel surfaces (for straight and regular channels)

Type of channel surface	Manning's n
Smooth Cement, Planed	0.010—0.014
Canvas, Unplaned wood	0.011—0.015
Finished concrete	0.011—0.013
Unfinished concrete	0.013—0.016
Brick masonry	0.012—0.020
Cast iron	0.013—0.017
Riveted steel	0.017—0.020
Rubble masonry	0.020—0.030
Earth	0.020—0.030
Gravel	0.022—0.035
Earth with vegetation	0.025—0.040

Example 13.4 *Determine the rate of flow through a 3 m wide concrete lined rectangular channel having a slope of 1 in 10,000 when the depth of flow is 1 m.*

Taake n = 0.012 in Manning's formula.

Solution.

Given: $b = 3\text{m}, y = 1 \text{ m}, S = \dfrac{1}{10^4}, n = 0.012$

$$A = by = 3 \times 1 = 3\text{m}^2$$

$$P = b + 2y = 3 + 2 = 5\text{m}$$

$$R = \frac{A}{P} = \frac{3}{5} = 0.6$$

$$Q = \frac{1}{n} AR^{2./3} S^{1/2} = \frac{1}{0.012} \times 3(0.6)^{2/3} \times (10^{-4})^{1/2} = 1.778 \text{m}^3/\text{s}$$

13.5 HYDRAULICALLY EFFICIENT CHANNEL CROSS-SECTION

The most efficient cross-section of an open channel, from hydraulic point of view, is that which for a given slope, roughness coefficient and area of flow, carries the maximum discharge.

$$Q = AC\sqrt{RS} \qquad \text{... Clezy's formula}$$

or

$$Q = \frac{1}{n} AR^{2/3} s\frac{1}{2} \qquad \text{... Manning's formula}$$

From above equations, it is evident that for given channel slope, roughness coefficient and area of flow, the discharge will be maximum, when the hydraulic radius R is maximum. Now $R = \dfrac{A}{P}$. Therefore for Q to be maximum, P has to be minimum. Such a channel will not only be hydraulically efficient, but will also be economical

13.5.1 Hydraulically Efficient Rectangular Channel

Consider the rectangular channel shown in Fig. 13.7.

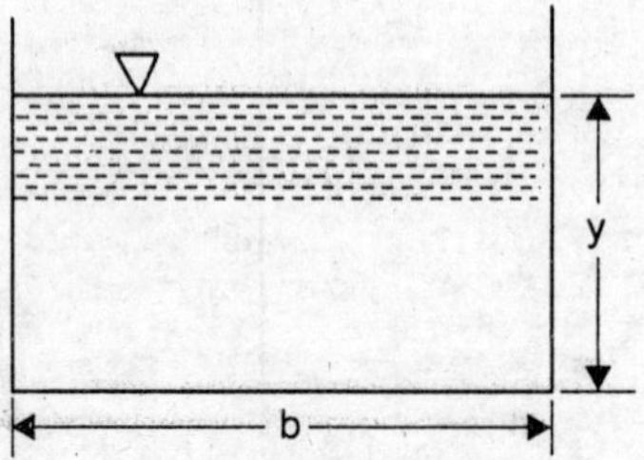

Fig. 13.7. Rectangular channel

$$A = by$$

$$P = b + 2y = \frac{A}{y} + 2y$$

For P to be minimum, $\frac{dP}{dy} = 0$ and $\frac{d^2P}{dy^2}$ should be positive.

$$-\frac{A}{y^2} + 2 = 0$$

or

$$A = 2y^2$$

$$b = \frac{A}{y} = 2y \qquad \text{... (13.17 } a)$$

or

$$y = \frac{b}{2}$$

Also $\frac{d^2P}{dy^2} = \frac{2A}{y^3}$ which is positive ... (13.17 *b*)

Thus for a rectanular channel to be hydraulically most efficient channel, its width should be twice the depth of flow.

13.5.2 Hydraulically Efficient Trapezoidal Channel

Consider a trapezoidal channel shown in Fig. 13.8

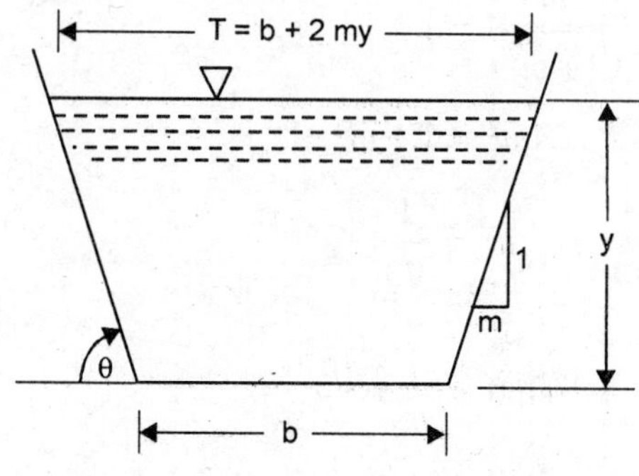

Fig. 13.8 Trapezoidal channel.

$$A = \frac{1}{2}(b + T)y = \frac{1}{2}(b + b + 2my)\,y$$

$$= (b + my)y$$

$$P = b + 2y\sqrt{1+m^2} = \left(\frac{A}{y} - my\right) + 2y\sqrt{1+m^2}$$

For P to be minimum,

$$\frac{dP}{dy} = \left(-\frac{A}{y^2} - m\right) + 2\sqrt{1+m^2} = 0$$

$$A = y^2(2\sqrt{1+m^2} - m)$$

$$= y(b + my)$$

$$b = 2y(\sqrt{1+m^2} - m)$$

$$\frac{T}{2} = \frac{b+2my}{2} = y\sqrt{1+m^2} \qquad \text{... (13.18)}$$

∴ Half the top width = sloping side of cross-section.

$\frac{d^2P}{dy^2} = \frac{2A}{y^3}$, which is positive and the condition obtained is for minimum P.

Now differentiate P w.r.t. m.

$$\frac{dP}{dm} = -y + \frac{2my}{\sqrt{1+m^2}} = 0$$

$$\frac{2m}{\sqrt{1+m^2}} = 1$$

or $$4m^2 = 1 + m^2$$

$$m = \frac{1}{\sqrt{3}}$$

$$\tan\theta = \frac{1}{m} = \sqrt{3}$$

∴ $$\theta = 60° \qquad \text{...(13.19)}$$

i.e. Best side slope is. 60° to the horizontal.

Therefore, P is minimum when $\theta = 60°$

Maximum value of hydraulic radius,

$$R_{max} = \frac{A}{P_{min}} = \frac{(b+my)\,y}{b+2y\sqrt{1+m^2}}$$

$$= \frac{(b+my)\,y}{b+b+2my} = \frac{y}{2} \qquad \text{... (13.20)}$$

A semi-cicle drawn from O with radius equal to depth of flow will touch the three sides of the trapezoidal channel, as shown in Fig. 13.9.

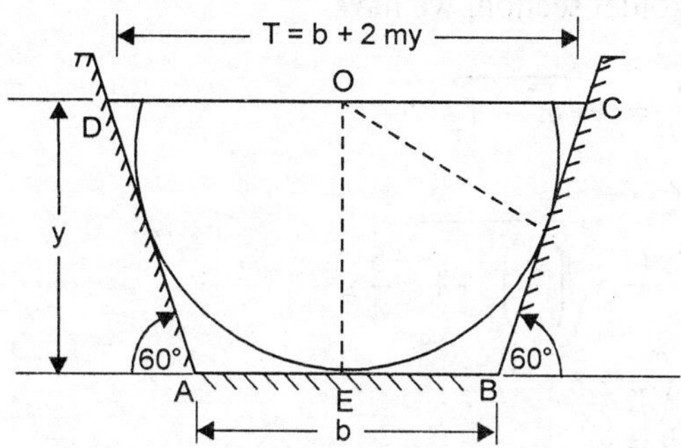

Fig.13.9 Most economical section of trapezoidal channel.

Substituting $m = \frac{1}{\sqrt{3}}$ in Eq. (13.18), we get

$$\frac{b+\frac{2}{\sqrt{3}}y}{2} = y\sqrt{1+\frac{1}{3}} = \frac{2y}{\sqrt{3}}$$

$$\frac{b}{2}+\frac{y}{\sqrt{3}} = \frac{2y}{\sqrt{3}} \qquad \text{... (13.21)}$$

$$b = \frac{2y}{\sqrt{3}}$$

and

$$P = b+2y\sqrt{1+\frac{1}{3}}$$

$$= b+\frac{4y}{\sqrt{3}} = \frac{2y}{\sqrt{3}}+\frac{4y}{\sqrt{3}} = \frac{6y}{\sqrt{3}} = 3b \qquad \text{... (13.22)}$$

Thus for a side of 60°, the length of sloping side is equal to the base width of the trapezoidal section.

Example 13.5 *A trapezoidal channel has side slopes of 3 horizontal to 4 vertical and the slope of its bed is 1 in 2500. Determine the optimum dimensions of the channel, if it is to carry water at 0.5 m³/s. Take Chezy's constant = 80.*

Solution:

Given: $m = \frac{3}{4}$, $S = \frac{1}{2500}$, $Q = 0.5\text{m}^3/\text{s}$, $C = 80$

Refer to Fig. 13.10.

For most economical trapezoidal section, we have

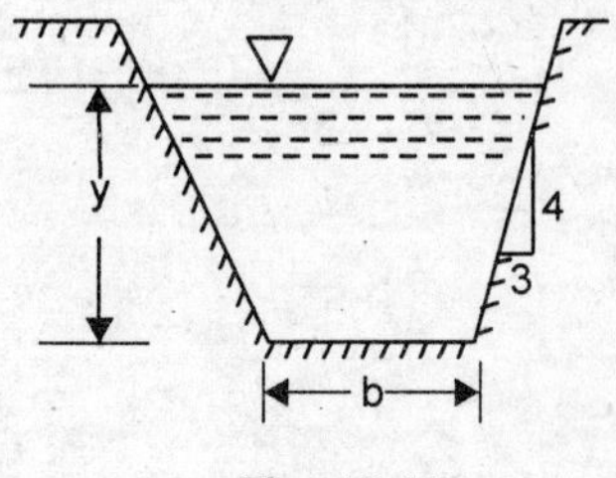

Fig. 13.10

$$\frac{b+2my}{2} = y\sqrt{m^2+1}$$

$$\frac{b+2\times\frac{3}{4}y}{2} = y\sqrt{\left(\frac{3}{4}\right)^2+1} = \frac{5}{4}y$$

$$b + 1.5\, y = 2.5\, y$$

$$b = y$$

$$Q = AC\sqrt{RS}$$

For most economical section, $R = \frac{y}{2}$

$$A = (b+my)\times y = \left(y+\frac{3}{4}y\right)y = 1.75\,y^2$$

$$0.5 = 1.75y^2 \times 80\sqrt{\frac{y}{2}\times\frac{1}{2500}}$$

$$= 1.98\, y^{5/2}$$

$$y = 0.577 \text{ m}$$

$$b = y = 0.577 \text{ m}$$

Example 13.6 *A trapezoidal channel with side slopes of 1:1 has to be designed to carry 10 m³/s of water at a velocity of 2m/s, so that the amount of concrete lining for the bed and sides is minimum.*

(a) Calculate the area of lining required for 1 m length of canal.

(b) If the rugosity coefficient n = 0.015, calculate the bed slope of the canal for unifrom flow.

Solution

Given: $m = 1$, $Q = 10\text{m}^3/\text{s}$, $v = 2\text{m/s}$, $n = 0.015$

Area across flow, $A = \frac{Q}{v} = \frac{10}{2} = 5\text{m}^2$

(*a*) For minimum amount of concrete lining, the wetted perimeter must be minimum. For a trapezoidal channel,

$$\frac{b+2my}{2} = y\sqrt{m^2+1} \text{ and } R = \frac{y}{2}$$

$$\frac{b+2\times 1\times y}{2} = y\sqrt{1+1} = \sqrt{2}\,y$$

$$b + 2y = 2\sqrt{2}y$$

$$b = 2y(\sqrt{2}-1) = 0.828\,y$$

Area of flow, $A = (b + my)\,y$

$$5 = (0.828\,y + 1 \times y)\,y = 1.828\,y^2$$

$$y = 1.654 \text{ m}$$

$$b = 0.828 \times 1.654 = 1.369 \text{ m}$$

Area of lining per metre length of channel

$$= P\times 1 = (b + 2y\sqrt{m^2+1})\times 1$$

$$= 1.369 + 2\times 1.654\sqrt{2}$$

$$= 6.047 \text{ m}$$

(*b*) According to Manning's formula,

$$Q = \frac{A}{n}R^{2/3}S^{1/2}$$

$$10 = \frac{5}{0.015}\times\left(\frac{1.654}{2}\right)^{2/3}\times S^{1/2}$$

$$S = 1 \text{ in } 862$$

Example 13.7 *A trapezoidal channel having cross-sectional areal A_1, wetted perimeter P_1, Manning's coefficient n, and laid to a slope of S carries a discharge Q, at a depth of flow equal to y. To increase the discharge, the base width of the channel is widened by x, keeping other parameters same. Prove. that:*

$$\left(\frac{Q_2}{Q_1}\right)^3 \times \left(1+\frac{x}{P_1}\right)^2 = \frac{(1+xy)^5}{A_1}$$

Solution According to Manning's formula, rate of flow,

Where Q_2 = new discharge in the channel.

$$Q = \frac{A}{n} \times R^{2/3} S^{1/2} = \text{KAR}^{2/3} = KA\left(\frac{A}{P}\right)^{2/3} = K\frac{A^{5/3}}{P^{2/3}}$$

where $$K = \frac{S^{1/2}}{n}$$

$$A_2 = A_1 + xy, \; P_2 = P_1 + x$$

$$Q_1 = K\frac{A_1}{P_1^{2/3}} \text{ and } Q_2 = K\frac{A_2^{5/3}}{P_2^{2/3}}$$

$$\frac{Q_2}{Q_1} = \left(\frac{A_1 + xy}{A_1}\right)^{5/3} \times \left(\frac{P_1}{P_1 + x}\right)^{2/3}$$

$$= \left(1 + \frac{xy}{A_1}\right)^{5/3} \times \left(\frac{1}{1 + \frac{x}{P_1}}\right)^{2/3}$$

Taking cube of both sides, we get

$$\left(\frac{Q_2}{Q_1}\right)^3 = \left(1 + \frac{xy}{A_1}\right)^5 \times \left(\frac{1}{1 + \frac{x}{P_1}}\right)^2$$

$$\left(\frac{Q_2}{Q_1}\right)^3 \times \left(1 + \frac{x}{P_1}\right)^2 = \left(1 + \frac{xy}{A_1}\right)^5$$

Hence Proved.

Example 13.8 *A trapezoidal canal is to carry 50 m³/s of water with a mean velocity of 0.6 m/s. One side of the canal is vertical and the other has a slope of 2 horizontal to 1 vertical. Find the minimum hydraulic slope, if Manning's n = 0.013.*

Solution.

Given: $Q = 50$ m³/s, $v = 0.6$ m/s, $m = 2$, $n = 0.013$

Refer to Fig. 13.11.

Area of flow, $$A = \left(\frac{b + b + my}{2}\right) \times y$$

$$= \left(\frac{2b + 2y}{2}\right) y$$

$$= (b + y)y$$

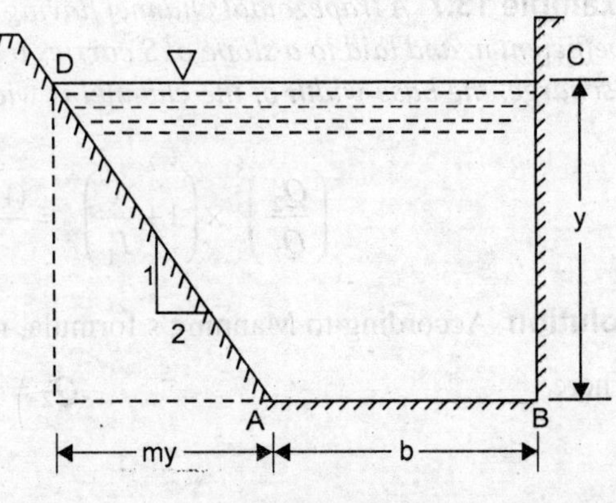

Fig. 13.11

$$= \frac{Q}{v} = \frac{50}{0.6} = 83.333$$

$$b = \frac{83.333}{y} - y$$

Wetted perimeter,
$$P = b + y + \sqrt{m^2+1} \times y = b + y(1+\sqrt{m^2+1})$$

$$= b + y(1+\sqrt{5}) = b + 3.236\,y$$

Hydraulic radius,
$$R = \frac{A}{P} = \frac{83.333}{b+3.236\,y} = \frac{83.333}{\frac{83.333}{y} - y + 3.236\,y}$$

$$= \frac{83.333}{\frac{83.333}{y} + 2.236\,y}$$

$$Q = \frac{A}{n} R^{2/3} S^{1/2}$$

or
$$S^{1/2} = \frac{Qn}{AR^{2/3}} = \frac{50 \times 0.013}{83.333 \times R^{2/3}} = \frac{7.8 \times 10^{-3}}{R^{2/3}}$$

For S to be minimum, R has to to be maximum.

or $\frac{83.333}{\frac{83.333}{y} + 2.236\,y}$ is to be maximum

or $\frac{83.333}{y} + 2.236y$ has to be minimum.

$$\therefore \quad \frac{d}{dy}\left[\frac{83.333}{y} + 2.236y\right] = 0$$

$$-\frac{83.333}{y^2} + 2.236 = 0$$

$$y^2 = 37.268$$

$$y = 6.10 \text{ m}$$

$$R = \frac{83.333}{\frac{83.333}{6.10} + 2.236 \times 6.1} = 3.05\,\text{m}$$

$$S_{\min}^{1/2} = \frac{7.8\times10^{-3}}{(3.05)^{2/3}} = 3.708\times10^{-3}$$

$$S_{\min} = 1.375\times 10^{-5}$$

13.5.3 Hydraulically Efficient Triangular Channel

Consider a triangular channel as shown in Fig. 13.12.

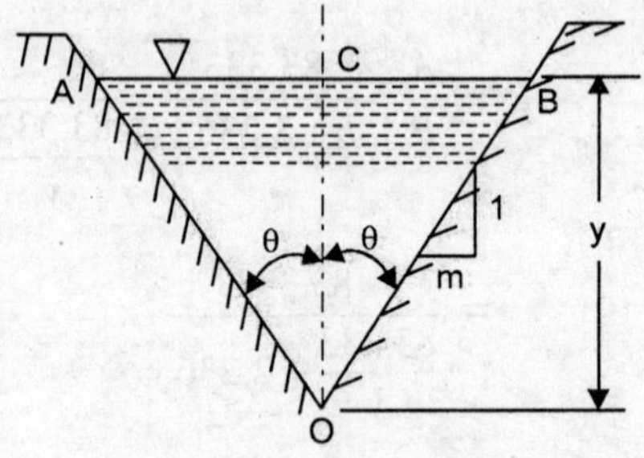

Fig. 13.12 Triangular channel

Let y = depth of flow

θ = angle made by the sides with the vertical.

In ΔOBC,
$$\tan\theta = \frac{BC}{OC} = \frac{BC}{y}$$

or
$$BC = y\tan\theta \qquad \text{... (1)}$$

Also
$$\cos\theta = \frac{OC}{OB} = \frac{y}{OB}$$

or
$$OB = y\sec\theta \qquad \text{... (2)}$$

Area of flow,
$$A = \frac{1}{2}\times AB\times OC = BC\times OC$$

$$= y^2\tan\theta \qquad \text{... (3)}$$

Perimieter,
$$P = 2\times OB = 2y\sec\theta \qquad \text{... (4)}$$

From Eq. (3),
$$y = \sqrt{\frac{A}{\tan\theta}}$$

Eq. (4) becomes,
$$P = 2\sqrt{\frac{A}{\tan\theta}}\times\sec\theta \qquad \text{... (5)}$$

For minimum parameter, $$\frac{dP}{d\theta} = 0$$

or $$\frac{d}{d\theta}\left[2\sqrt{A}\times\frac{\sec\theta}{\sqrt{\tan\theta}}\right] = 0$$

$$2\sqrt{A}\left[\frac{\sqrt{\tan\theta}\times\sec\theta.\tan\theta-\sin\theta\times\frac{1}{2}(\tan\theta)^{-1/2}.\sec^2\theta}{\tan\theta}\right] = 0$$

$$\sec\theta\times(\tan\theta)^{3/2}-\frac{1}{2}\frac{\sec^3\theta}{(\tan\theta)^{1/2}} = 0$$

$$2\sec\theta.\tan^2\theta-\sec^3\theta = 0$$

or $$2\tan^2-\sec^2\theta = 0$$

or $$\sqrt{2}\tan\theta = \sec\theta$$

or $$\sin\theta = \frac{1}{\sqrt{2}}$$

or $$\theta = 45^\circ \text{ and } m = 1 \qquad \text{... (13.23)}$$

Hydraulic radius, $$R = \frac{A}{P} = \frac{y^2\tan\theta}{2y\sec\theta}$$

$$= \frac{y\tan 45^\circ}{2\sec 45^\circ} = \frac{y}{2\sqrt{2}} \qquad \text{... (13.24)}$$

Hence a triangular channel section will be most econonical when each of its sloping sides makes angle of 45° with the vertical.

Example 13.9 *Water flows in a channel of the shape of isosceles triangle of breadth 'a' and sides making an angle of 45° with the bed. Determine the relation between depth of flow d and breadth 'a' for maximum velocity and for maximum discharge condition. Use Manning's fomula and note that 'd' is less than 0.5 a.*

Solution.

Refer to Fig. 13.13.

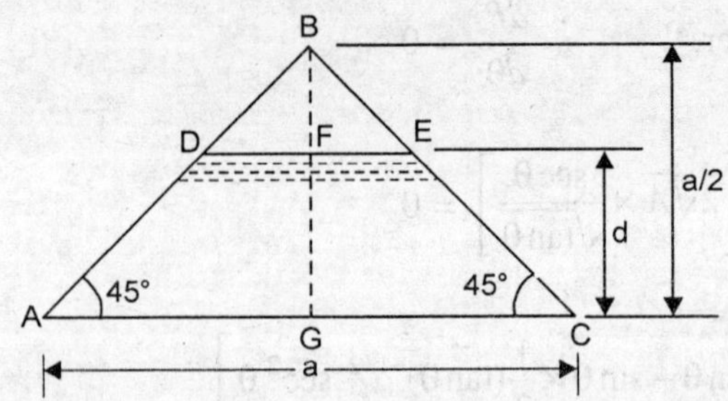

Fig. 13.13

According to Manning's formula,

$$Q = \frac{A}{n} R^{2/3} S^{1/2}$$

and

$$v = \frac{1}{n} R^{2/3} S^{1/2}$$

Area of flow,

$$A = \frac{1}{2}(AC + DE) \times d$$

ΔS BDE and ABC are similar. Thus

$$\frac{DE}{AC} = \frac{BF}{BG}$$

$$DE = AC \times \frac{BF}{BG} = \frac{a \times \left(\frac{a}{2} - d\right)}{(a/2)} = a - 2d$$

$$A = \frac{1}{2}(a + a - 2d)d = (a - d)d$$

Wetted perimeter,

$$P = AC + 2AD = a + 2 \times \frac{d}{\sin 45^\circ} = a + 2\sqrt{2}\, d$$

Hydraulic radius,

$$R = \frac{A}{P} = \frac{(a - d)d}{(a + 2\sqrt{2})d}$$

$$v = \frac{1}{n}\left[\frac{(a - d)d}{(a + 2\sqrt{2}d)}\right]^{2/3} S^{1/2}$$

For maximum velocity, $\frac{dv}{dd} = 0$

$$\frac{d}{dd}\left[\frac{S^{1/2}}{n}\left\{\frac{(a-d)d}{(a+2\sqrt{2}\,d)}\right\}^{2/3}\right] = 0$$

$$\left(\frac{S^{1/2}}{n}\right)\left[\frac{2}{3}\left\{\frac{(a-d)}{(a+2\sqrt{2}d)}\right\}^{-1/3}\left\{\frac{a+2\sqrt{2}d\,(a-2d)-(a-d)d\times\sqrt{2}}{(a+2\sqrt{2}\,d)^2}\right\}\right] = 0$$

On simplification, we get

$$a^2 - 2ad - 2\sqrt{2}\,d^2 = 0$$

$$a = \frac{1}{2}[2d \pm \sqrt{4d^2 + 8\sqrt{2}\,d^2}\,]$$

$$= 2.955\,d$$

or $\frac{d}{a} = 0.3382$

For maximum discharge, $\frac{dQ}{dd} = 0$

$$Q = \frac{(a-d)d}{n}\left[\frac{(a-d)d}{(a+2\sqrt{2}\,d)}\right]^{2/3} S^{1/2}$$

$$\frac{d}{dd}\left[\frac{S^{1/2}}{n}\{(a-d)d\}^{5/3}\times\frac{1}{(a+2\sqrt{2}d)^{2/3}}\right] = 0$$

$$\left(\frac{S^{1/2}}{n}\right)\left[\frac{(a+2\sqrt{2}d)^{2/3}\times\frac{5}{3}\{(a-d)d\}^{2/3}(a-2d)-\{(a-d)d\}^{5/3}\times\frac{2}{3}(a+2\sqrt{2}d)^{-1/3}\times 2\sqrt{2}}{(a+2\sqrt{2}d)^{4/3}}\right] = 0$$

On simiplification, we get

$$5a^2 - 1.5147\,ad - 22.6274\,d^2 = 0$$

or $a^2 - 0.3029\,ad - 4.52553\,d^2 = 0$

$$a = \frac{1}{2}[0.3029d \pm \sqrt{(-0.3029\,d)^2 + 4\times 4.5255d^2}\,]$$

$$= \frac{1}{2}[0.3029d \pm 4.2654d]$$

$$= 2.284\ d$$

$$\frac{d}{a} = 0.4378$$

13.5.4 Hydraulically Efficient Circular Channel

Closed conduits such as sewage pipes and drains run partly full most of the time (except during floods) and therefore, have a free surface. Flow in such pipes is governed by the same principles as appled to open channel flow.

Consider the circular channel shown in Fig. 13.14 shown with depth of water y.

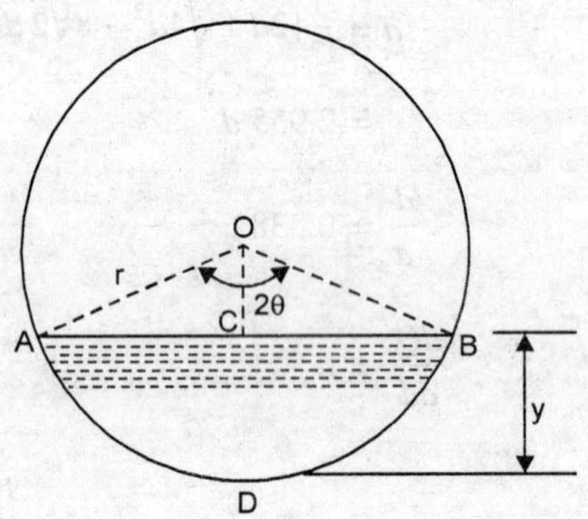

Fig. 13.14 Circular channel

Area of flow,

$$A = \text{area ADBA}$$

$$= \text{area of sector } OADBO - \text{area of } \Delta OAB$$

$$= \frac{\pi r^2}{2\pi} \times 2\theta - \frac{1}{2} \times AB \times OC$$

$$= r^2\theta - \frac{1}{2} \times 2r^2 \sin\theta \times r\cos\theta$$

$$= r^2\theta - \frac{1}{2} r^2 \sin 2\theta$$

$$= r^2\left(\theta - \frac{\sin 2\theta}{2}\right)$$

Wetted perimeter, $$P = 2r\theta$$

Hydraulic radius, $$R = \frac{A}{P} = \frac{r^2\left(\theta - \frac{\sin 2\theta}{2}\right)}{2r\theta} = \frac{r}{2\theta}\left(\theta - \frac{\sin 2\theta}{2}\right) = \frac{r}{2}\left(1 - \frac{\sin 2\theta}{2\theta}\right)$$

(a) Conditions for maximum velocity

Velocity of flow according to Chezy's formula,

$$v = C\sqrt{RS} = C\sqrt{\frac{A}{P}S}$$

Velocity will be maximum when $\frac{A}{P}$ is maximum.

or $$\frac{d}{d\theta}\left(\frac{A}{P}\right) = 0$$

$$\frac{d}{d\theta}\left[\frac{r}{2}\left(1 - \frac{\sin 2\theta}{2\theta}\right)\right] = 0$$

$$\frac{r}{2}\left[0 - \frac{1}{2}\left\{\frac{\theta.2\cos 2\theta - \sin 2\theta \times 1}{\theta^2}\right\}\right] = 0$$

or $$2\theta\cos 2\theta = \sin 2\theta = 0$$

or $$\tan 2\theta = 2\theta$$

By hit and trial, we find that,

$$2\theta = 257.5° \text{ or } \theta = 128.75°$$

Depth of flow, $$y = OD - OC = r - r\cos\theta$$

$$= r\,(1 - \cos 128.75°)$$

$$= 1.62\,r = 0.81\text{ d} \qquad \text{... (13.25)}$$

where d = diameter of channel.

Thus maximuim velocity occurs when the depth of flow is 0.81 times the diameter of circular channel.

Hydraulic radius, $$R = \frac{r}{2}\left(1 - \frac{\sin 257.5°}{2 \times \frac{\pi}{180} \times 128.75}\right)$$

$$= 0.6086\,r = 0.3043\,d \qquad \text{... (13.26)}$$

(b) **Condition for maximum discharge.**

$$Q = Av = AC\sqrt{RS} = AC\sqrt{\frac{A}{P}S} = C\sqrt{\frac{A^3}{P}}\sqrt{S}$$

For discharge to be maximum, $\left(\frac{A^3}{P}\right)$ is to be maximum.

$$\therefore \qquad \frac{d}{d\theta}\left(\frac{A^3}{p}\right) = 0$$

$$\frac{d}{d\theta}\left[\frac{r^6\left(\theta - \frac{\sin 2\theta}{2}\right)^3}{2r\theta}\right] = 0$$

$$\left(\frac{d}{d\theta}\left[\frac{(\theta - \frac{\sin 2\theta)^3}{2}}{\theta}\right]\right) = 0$$

$$\frac{\theta \times 3\left(\theta - \frac{\sin 2\theta}{2}\right)^2 (1 - \cos 2\theta) - \left(\theta - \frac{\sin 2\theta}{2}\right)^2 \times 1}{\theta^2} = 0$$

$$\left(\theta - \frac{\sin 2\theta}{2}\right)^2 \left[3\theta(1 - \cos 2\theta) - \left(\theta - \frac{\sin 2\theta}{2}\right)\right] = 0$$

Either $$\theta - \frac{\sin 2\theta}{2} = 0 \qquad \text{... (1)}$$

$$\therefore \qquad \sin 2\theta = 2\theta$$

or $$3\theta(1 - \cos 2\theta) - \left(\theta - \frac{\sin 2\theta}{2}\right) = 0$$

$$3\theta - 3\theta\cos 2\theta - \theta + \frac{\sin 2\theta}{2} = 0$$

$$2\theta - 3\theta\cos 2\theta + \frac{\sin 2\theta}{2} = 0$$

$$4\theta - 6\theta\cos 2\theta + \sin 2\theta = 0 \quad \text{... (2)}$$

By hit and trial, we find that,

$$2\theta = 308° \quad \text{or } \theta = 154°$$

$$y = r(1 - \cos\theta)$$

$$= r(1 - \cos 154°)$$

$$= 1.9\, r = 0.95\, d \quad \text{... (13.27)}$$

$$R = \frac{r}{2\theta}\left(\theta - \frac{\sin 2\theta}{2}\right)$$

$$\theta = \frac{\pi}{180}\times 154 = 2.687 \text{ rad}$$

$$R = \frac{r}{2\times 2.687}\left(2.687 - \frac{\sin 308°}{2}\right)$$

$$= 0.573\, r = 0.2865\, r \simeq 0.29\, r \quad \text{... (13.28)}$$

Example 13.10 *A concrete lined cicular channel of 4 m diameter has a bed slope of 1 in 500. Calculate the velocity and flow rate for the conditions of:*

(a) maximum velocity, and (b) maximum flow rate.

Take Chezy's constant = 52

Solution.

Given: $d = 4\text{m}$, $S = \frac{1}{500}$, $C = 52$

(*a*) For maximum velocity condition,

$$2\theta = 257.5° = 257.5\times\frac{\pi}{180} = 4.49 \text{ rad}$$

Depth, fof flow, $y = 0.81\, d = 0.81 \times 4 = 3.24$ m

Area of flow, $A = \frac{r^2}{2}(2\theta - \sin 2\theta)$

$$= \frac{4}{2}(4.49 - \sin 257.5°)$$

$$= 10.93 \text{ m}^2$$

Wetted perimeter, $P = 2r\theta = 4.49 \times 2 = 8.98$ m

Hydraulic radius, $$R = \frac{A}{P} = \frac{10.93}{8.98} = 1.217$$

Velocity of flow, $$v = C\sqrt{RS} = 52\sqrt{1.217 \times \frac{1}{500}} = 2.565 \text{ m/s}$$

Flow rate, $Q = Av = 10.93 \times 2.565 = 28.035 \text{ m}^3/\text{s}$

(b) For maximum flow rate condition,

$$2\theta = 308° = 308 \times \frac{\pi}{180} = 5.375 \text{ rad}$$

Depth of flow, $$y = 0.95\, d = 0.95 \times 4 = 3.80 \text{ m}$$

Area of flow, $$A = \frac{r^2}{2}(2\theta - \sin 2\theta)$$

$$= \frac{4}{2}(5.375 - \sin 308°)$$

$$= 12.326 \text{ m}^2$$

wetted perimeter, $$P = 2r\theta = 5.375 \times 2 = 10.75 \text{ m}$$

Hydraulic radius, $$R = \frac{A}{P} = \frac{12.326}{10.75} = 1.1466 \text{ m}$$

Flow velocity, $$v = C\sqrt{RS} = 52\sqrt{1.1466 \times \frac{1}{500}} = 2.49 \text{ m/s}$$

Flow rate, $$Q = AV = 12.326 \times 2.49 = 30.69 \text{ m}^3/\text{s}$$

13.6 VELOCITY DISTRIBUTION IN OPEN CHANNELS

As in pipes, the velocities in open channels are retarded near the solid boundaries. An additional factor influencing the velocity profile in open channels is introduced on account of the free surface. The surface tension of the liquid-air interface (*i.e.*, the free surface) produces a resistance to flow and causes maximum velocity to occur at some distance below the free-surface. A strong wind when blowing in upstream or downstream direction also affects the flow and accordingly modifies the velocity profile. In the vertical direction, the velocity is minimum at the channel bottom and it attains a maximum value at a distance from the free surface which varies from 0.05 to 0.25 times the flow depth, the higher value is applicable to section closer to the bank. At the free-surface, the velocity is somewhat less than the maximum value due to surface tension effects and possible air resistance. The sketches of Fig. 13.15 *(a)* show the pattern of velocity distributions in the horizontal and vertical directions for a rectangular channel. From the inspection of these velocity distribution curves the following observation can be made :

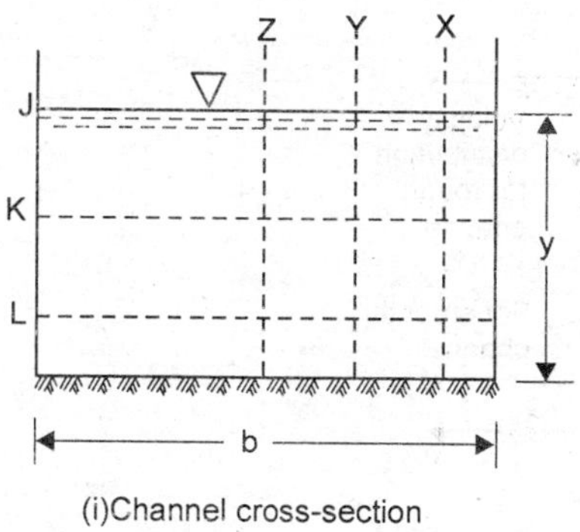

(i)Channel cross-section

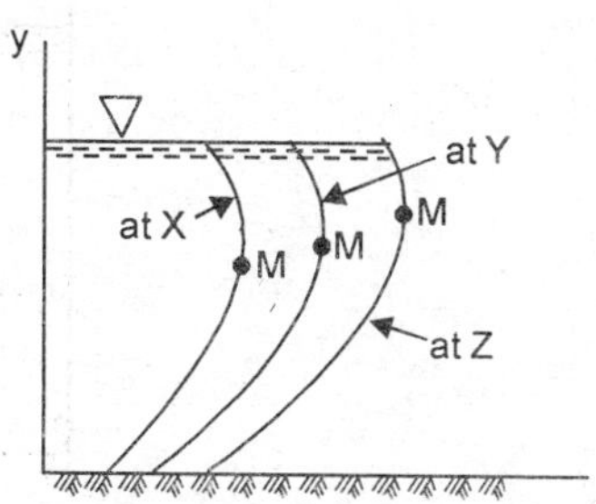

(ii) Velocity distribution in the vertical plane.
Note. *M* represents the point of maximum velocity

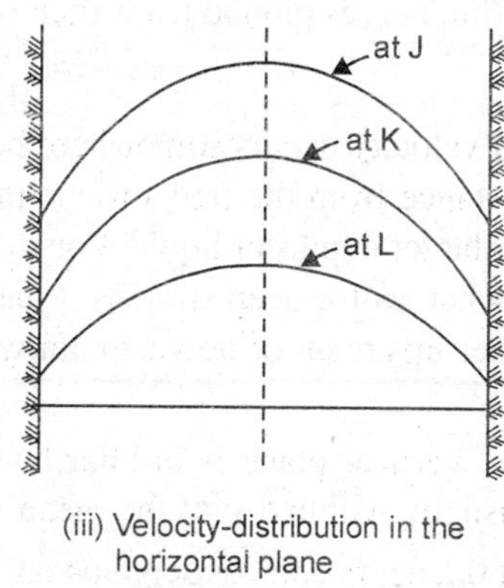

(iii) Velocity-distribution in the horizontal plane

Fig. 13.15 (a) Velocity distribution in a rectangular channel.

(1) As we go away from the bank (evident from the locations of vertical sections *X*, *Y* and Z) the point of maximum velocity shifts upwards and at the central section Z, the maximum velocity occurs almost at the free surface.

The velocity distribution in channels depends upon the shape of the channel section, the channel roughness and the presence of bends in its alignment. The effect of roughness is to increase the curvature of the velocity distribution curve. Fig. 13.15 (b) shows the comparison between the velocity distribution curves for smooth and rough channels. The presence of bends also modifies the velocity distribution. If bend is a sharp one, the centrifugal acceleration which acts in the radially outward direction has a higher value, causing higher velocities near the outer bank of the channel.

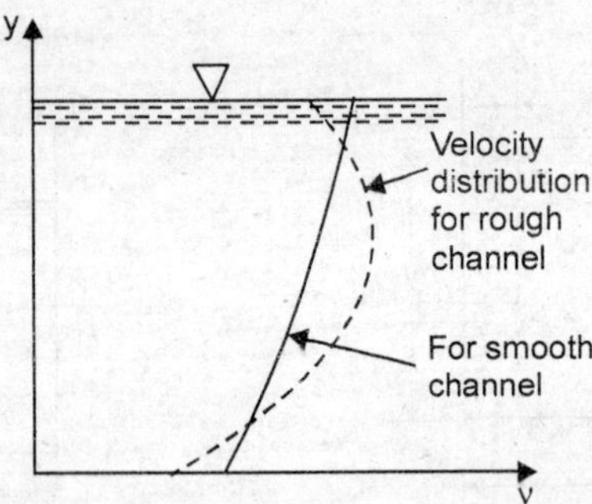

Fig. 13.15 (b) Velocity Distribution in rough and smooth channel.

From measurement of velocities in natural streams (*i.e.*, channels characterized by irregular cross-section) and. the velocity distribution curves plotted from such observations, the following findings can be made :

(2) In general, the maximum velocity occurs somewhere between the liquid surface and one-third of the flow depth, the distance from the free-surface to the point of maximum velocity being proportionally greater for larger depths of liquid. For shallow streams, the maximum velocity is very near to the surface whereas for deep streams it may lie at about one-third of the depth. A strong wind blowing either upstream or downstream will affect distribution of velocity in the vertical plane.

(3) The mean velocity in the vertical plane is ordinarily at 0.55 to 0.65 of the flow depth. The velocity at 0.6 depth is usually within 5% of the mean velocity.

(4) The mean of velocities at 0.2 depth and 0.8 depth usually gives the mean velocity in the vertical within 2%.

(5) The mean velocity in the vertical is ordinarily 0.80 to 0.95 of the surface velocity. The smaller value applies to the shallower streams.

13.7 HYDRAULIC JUMP

In an open channel, when flow changes from super-critical state to subcritical state, the depth of flow is increased. This increase in depth occurs in a short length of the channel, and is known as the hydraulic jump. A hydraulic jump is, therefore, an abrupt rise in the depth of flow signifying a change of flow from super-critical state to sub-critical state. It involves violent mixing due to high turbulence and air entrainment.

Hydraulic jump generally occurs at such places as the foot of spillway and downstream of the sluice gate where the velocities are very high. The Hydraulic jump converts kinetic energy of rapidly flowing stream into potential energy. It is responsible for high loss of energy and is frequently employed as an effective energgy dissipating device. The hydraulic jump is also known as a *standing wave*.

13.7.1 Analysis of Hydraulic Jump

The folloiwng assumptions are made in the analysis of hydraulic jump:

1. The flow is uniform and the pressure distribution is hydrostatic before and after the jump.
2. The channel is horizontal or of very small slope.
3. Loss of head due to friction in the channel is negligible.
4. The momentum correction factor is unity.

The hydraulic jump may be analysed by applying the continuity equation, momentum principle and the energy equation. For simplicity in mathematical treatment, we consider a horizontal channel of rectangular section.

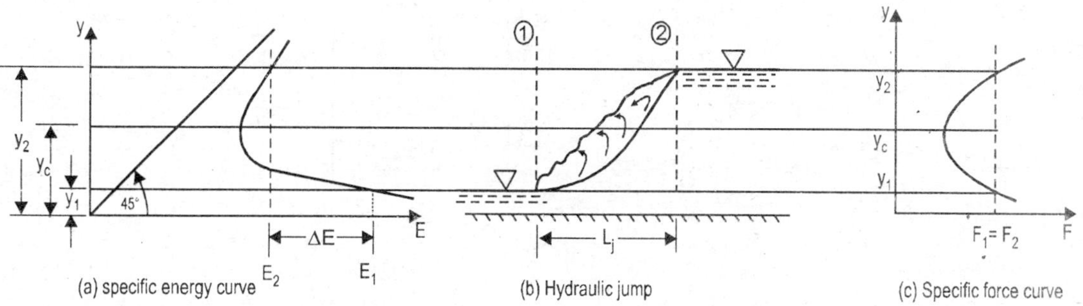

Fig. 13.16 Hydraulic jump in a horizontal rectangular channel.

Applying momentum equation to the body of water enclosed between sections (1) and (2) of the hydraulic jump shown in Fig. 13.16 *(b)*, we have

$$\frac{\gamma Q}{g}(v_2 - v_1) = P_1 - P_2 \quad \text{...(1)}$$

where

Hydrostatic force at section 1, $P_1 = \gamma Z_1 A_1 = \frac{\gamma y_1}{2} \times b\, y_1 = \frac{1}{2}\gamma b\, y_1^2$

Hydrostatic force at section 2. $P_2 = \frac{1}{2}\gamma\, b\, y_2^2$

b = width of rectangular channel

From continuity equation, we have

$$Q = b_1 y_1 v_1 = b_2 y_2 v_2 \quad \text{...(2)}$$

If $\quad q$ = discharge per unit width.

Then $\quad Q = bq$

Eq. (1) becomes,

$$\frac{\gamma(bq)}{g}\left(\frac{q}{y_2}-\frac{q}{y_1}\right) = \frac{1}{2}\gamma b y_1^2 - \frac{1}{2}\gamma b y_2^2$$

which simplifies to,

$$\frac{2q^2}{g} = y_1 y_2 (y_1 + y_2) \qquad \text{... (3)}$$

or $$y_2^2 + y_1 y_2 - \frac{2q^2}{g y_1} = 0$$

$$y_2 = \frac{y_1}{2}\left[-1 \pm \sqrt{1+\frac{8q^2}{g y_1^3}}\right]$$

$$= \frac{y_1}{2}\left[-1 \pm \sqrt{1+\frac{8v_1^2}{g\, y_1}}\right]$$

or $$\frac{y_2}{y_1} = \frac{1}{2}[\sqrt{1+8F_1^2}-1] \qquad \text{... (13.28)}$$

where $F_1 = \frac{v_1}{\sqrt{g y_1}}$, is the Froude number of approaching fluid.

and $$\frac{y_1}{y_2} = \frac{1}{2}[\sqrt{1+8F_2^2}-1] \qquad \text{...(13.29)}$$

where $F_2 = \frac{v_2}{\sqrt{g y_2}}$, is the Froude number downstream of the jump.

(i) For $F_1 < 1, \frac{y_2}{y_1} < 1$, This is the condition for a falling liquid surface and the flow is sub-critical.

(ii) For $F_1 = 1, \frac{y_2}{y_1} = 1$, and the flow is critical.

(iii) For $F_1 > 1, \frac{y_2}{y_1} > 1$, It is essential for the formation of hydraulic jump.

Relation between Pre-Jump and Post–Jump Froude Numbers

$$F_1 = \frac{v_1}{\sqrt{g y_1}} \quad \text{and } F_2 = \frac{v_2}{\sqrt{g y_2}}$$

From the continuity equation, we have

$$g = v_1 y_1 = v_2 y_2$$

or

$$F_1\sqrt{gy_1}\cdot y_1 = F_2\sqrt{g\,y_2}\cdot y_2$$

$$F_1 y_1^{3/2} = F_2 y_2^{3/2}$$

or

$$\frac{y_2}{y_1} = \left(\frac{F_1}{F_2}\right)^{2/3}$$

Substituting in Eq. (13.28), we get

$$\left(\frac{F_1}{F_2}\right)^{2/3} = \frac{1}{2}[\sqrt{1+8F_1^2}-1]$$

or

$$F_2^2 = \frac{8F_1^2}{[\sqrt{1+8F_1^2}-1]^3}$$

$$F_2 = \frac{2\sqrt{2}\,F_1}{[\sqrt{1+8F_1^2}-1]^{3/2}} \quad \text{... (13.30)}$$

Height of hydraulic jump, $H_j = y_2 - y_1$ ... (13.31)

Length of hydraulic jump, $L_j = (5 \text{ to } 7)\, H_j$ ... (13.32)

Loss of energy due to hydraulic jump

$$\Delta E = E_1 - E_2$$

$$= \left(y_1 + \frac{v_1^2}{2g}\right) - \left(y_2 + \frac{v_2^2}{2g}\right)$$

$$= \frac{(v_1^2 - v_2^2)}{2g} - (y_2 - y_1)$$

$$= \frac{q^2}{2g}\left[\frac{1}{y_1^2} - \frac{1}{y_2^2}\right] - (y_2 - y_1)$$

$$= \frac{q^2}{2g}\left[\frac{y_2^2 - y_1^2}{y_1^2\, y_2^2}\right] - (y_2 - y_1)$$

From Eq (3), we have

$$\frac{2q^2}{g} = y_1 y_2 (y_1 + y_2)$$

$$\therefore \quad \Delta E = \frac{y_1 y_2 (y_1 + y_2)}{4}\left[\frac{y_2^2 - y_1^2}{y_1^2\, y_2^2}\right] - (y_2 - y_1)$$

$$= \frac{(y_2 + y_1)(y_2^2 - y_1^2)}{4\, y_1\, y_2} - (y_2 - y_1)$$

$$= (y_2 - y_1)\left[\frac{(y_2 + y_1)^2}{4\, y_1\, y_2} - 1\right]$$

$$= (y_2 - y_1)\left[\frac{(y_2 - y_1)^2}{4 y_1\, y_2}\right]$$

$$= \frac{(y_2 - y_1)^3}{4 y_1\, y_2} \quad \text{... (13.33)}$$

Example 13.11 *A sluice gate discharges water into horizontal rectangular channel with a velocity of 10 m/s and depth of flow of 1.2 m. Calculate the depth of flow of water after the jump and loss in total head.*

Solution.

Given: $v_1 = 10$ m/s, $y_1 = 1.2$ m

$q = v_1\, y_1 = 10 \times 1.2 = 12$ m^3/s per m

$$y_2 = -\frac{y_1}{2} + \sqrt{\frac{y_1^2}{4} + \frac{2q^2}{g y_1}}$$

$$= -\frac{1.2}{2} + \sqrt{\frac{(1.2)^2}{4} + \frac{2 \times 12^2}{9.81 \times 1.2}}$$

$$= -0.6 + 4.98 = 4.382 \text{ m}$$

Loss in total head, $$\Delta E = \frac{(y_2 - y_1)^3}{4 y_1\, y_2} = \frac{(4.382 - 1.2)^3}{4 \times 1.2 \times 4.382} = 1.53\,\text{m}$$

Example. 13.12 *In a rectangular channel of width 0.6 m, a hydraulic jump occurs at a point where depth of water flow is 0.15 m and Froude number is 2.5. Calculate: (a) Specific energy (b) Critical and subsequent depth (c) Loss of head, and (d) Energy dissipated.*

Solution.

Given: $b = 0.6$ m, $y_1 = 0.15$, $F_1 = 2.5$

$$F_1 = \frac{v_1}{\sqrt{gy_1}}$$

$$2.5 = \frac{v_1}{\sqrt{9.81 \times 0.5}}$$

$$v_1 = 3.03 \text{ m/s}$$

$$q = v_1 y_1 = 3.03 \times 0.15 = 0.4545 \text{ m}^3\text{/s per m.}$$

(*a*) Specific energy:

$$E_1 = y_1 + \frac{v_1^2}{2g} = 0.15 + \frac{(3.03)^2}{2 \times 9.81} = 0.618 \text{ m}$$

(*b*) Critical depth:

$$y_e = \left(\frac{q^2}{g}\right)^{1/3} = \left[\frac{0.4545}{9.81}\right]^{1/3} = 0.276 \text{ m}$$

$$y_2 = \frac{y_1}{2}[\sqrt{1 + 8F_1^2} - 1]$$

$$= \frac{0.15}{2}[\sqrt{1 + 8(2.4)^2} - 1]$$

$$= 0.461 \text{ m}$$

(*c*) Loss of head,

$$\Delta E = \frac{(y_2 - y_1)^3}{4 y_1 y_2}$$

$$= \frac{(0.461 - 0.15)^3}{4 \times 0.15 \times 0.461} = 0.108 \text{ m}$$

(*e*) Energy dissipated:

$$P = \gamma Q \cdot \Delta E$$

$$Q = A_1 v_1 = b y_1 v_1 = 0.6 \times 0.15 \times 0.03 = 0.2727 \text{ m}^3\text{/s}$$

$$P = 9810 \times 0.2727 \times 0.108 = 288.92 \text{ W}$$

SUMMARY

1. An open channel may be defined as a passage in which liquid flows with its upper surface exposed to atmosphere.
2. The flow in a channel is said to be uniform, if the depth of flow, slope of channel, cross-section and velocity of flow remain constant over a given length of channel.
3. Type of flow:

   (i) Re < 500 : laminar

   500 < Re < 2000: transitional

   Re > 2000 : turbulent

   (ii) Fr < 1.0 : Sub-critical

   Fr = 1.0 : Sub-critical

   Fr > 1.0 : supercritcal
4. Velocity of flow and flow rate:

   (i) Chezy's formula.

$$v = C\sqrt{RS}, Q = AC\sqrt{RS}$$

where $R = \dfrac{A}{P}$ = hydraulic radius.

A = area of flow

P = wetted perimeter

S = slope of bed

C = Chezy's constant

$$= \frac{157.6}{1.81 \dfrac{K}{\sqrt{R}}}, \text{ Bezin's formula}$$

$$= \frac{23 + \dfrac{0.00155}{S} + \dfrac{1}{N}}{1 + \left(23 + \dfrac{0.00155}{S}\right)\dfrac{N}{\sqrt{R}}}, \text{ Ganguillet-Kutter formula}$$

N = roughness coefficient

(ii) Manning' s formula.

$$Q = \frac{A}{n} R^{2/S} S^{1/2}$$

where n = coefficient of roughness or Kutter's constant.

5. Hydraulicaly efficient (or economical) section of the channel is one which gives the maximum discharge for a given amount of excavation.

6. Condition for maximum discharge:

(a) Rectangular section.

(i) $b = 2y$ *(ii)* $R = \frac{y}{2}$

(b) Trapezoidal section.

(i) Half top width = sloping side

or $$\frac{b+2my}{2} = y\sqrt{m^2+1}$$

(ii) $R = \frac{y}{2}$

(iii) Best side slope, $\theta = 60°$ or $m = \frac{1}{\sqrt{3}} = \frac{1}{\tan\theta}$

(c) Triangular section.

(i) Each sloping side makes and angle of 45° with the vertical.

(ii) $R = \frac{y}{2\sqrt{2}}$

(d) Circular section.

(i) Condition for maximum velocity

$$y = 0.81\,d$$
$$R = 0.305\,d$$

(ii) Condition for maximum discharge.

$$y = 0.95\,d$$
$$R = 0.29\,d$$

(iii) $A = r^2\left(\theta - \frac{\sin 2\theta}{2}\right)$

$P = 2r\theta$

7. Specific energy of flow per unit weight,

$$E = y + \frac{v^2}{2g}$$

8. The depth of flow at which specific energy is minimum is called the critical depth.

$y_c = \left(\frac{q^2}{g}\right)^{1/3}$, where q = discharge per unit width

9. Critical velocity of flow, $v_c = \sqrt{g \times y_c}$

10. Hydraulic jump is an abrupt rise in the depth of flow signifying a change of flow from super-crtical state to sub-critical state.

$$\frac{y_2}{y_1} = \frac{1}{2}(\sqrt{1+8F_1^2} - 1)$$

$$y_2 = -\frac{y_1}{2} + \sqrt{\frac{y_1^2}{4} + \frac{2v_1^2\, y_1}{g}}$$

$$y_2 = -\frac{y_1}{2} + \sqrt{\frac{y_1^2}{4} + \frac{2q^2}{2gy_1}}.$$

Height of hydraulic jump, $H_j = y_2 - y_1$

Length of hydraulic jump, $L_j = (5 \text{ to } 7)\, H_j$

Loss of energy, $\Delta E = \dfrac{(y_2 - y_1)^3}{4y_1\, y_2}$

$$F_2 = \frac{2\sqrt{2}\, F_1}{[\sqrt{1+8F_1^2} - 1]^{3/2}}$$

$$\frac{y_2}{y_1} = \left(\frac{F_1}{F_2}\right)^{2/3}$$

Short Answer Type Questions

1. Define an open channel.
2. What are the various types of channels?
3. State the conditions for uniform flow in open channels.
4. Define wetted perimeter and hydraulic radius.
5. Write Chezy's formula and explain all the terms.
6. Write Manning's formula and explain all the terms.
7. What is hydraulically efficient channel?
8. Define specific energy and critical depth.
9. What is a hydraulic jump?
10. What is the condition for hydraulic jump to form?

EXERCISES

1. A rectangular channel is 6 m wide and 1.6 m deep. The bed slope of the channel is 1 in 1000. Calculate the rate of flow for uniform flow. Take Chezy's constant equal to 50.

   [**Ans.** 15.5 m^3/s]

2. Calculate the discharge through a trapezoidal channel of width 8 m and side slopes of 1 horizontal to 3 vertical. The depth of flow of water is 2.4 m and the slope of bed is 1 in 4500. Take Chezy's constant C = 60

   [**Ans.** 24 m^3/s]

3. A triangular gutter, whose sides include an angle of 60°, conveys water at a uniform depth of 3m. If the slope of bed is 1 in 1000, calculate the rate of flow of water. Take Chezy's constant C = 52.

   [**Ans.** 7.4 m^3/s]

4. A channel has semicircular bottom of 3 m diameter and two sides as vertical when the depth of flow is 2.7 m. Calculate the discharge if slope of bed is 1 in 2000 and Chezy's constant C = 60.

   [**Ans.** 9.58 m^3/s]

5. A channel of rectangular cross-section is to be dug in a rocky portion of soil. The channel is to convey 6m^3/s of water with an average velocity of 2m/s. Find the most economical cross-section of the channel. Take Chezy's constant C = 62. [**Ans.** $b = 2.45$ m, $y = 1.225$ m, $S = \frac{1}{589}$]

6. Determine the dimensions of a most economical trapezoidal earth-lined channel to carry 15m^3/s at a slope of 1 in 3000. Take Manning's $n = 0.02$. [**Ans.** $b = 3.19$ m, $y = 2.76$ m]

7. A concrete lined circular channel of 3m diameter has a bed slope of 1 in 600. Determine the velocity and the flow rate for the condition of (a) maximum velocity, and *(b)* maximum discharge. Take Chezy's constant, $C = 50$. [**Ans.** *(a)* 1.95 m/s, 12 m^3/s *(b)* 1.346 m/s, 9.33 m^3/s]

8. A control sluice spanning the entry of a 40 m wide rectangular channel having a mild slope admits water at the rate of 16m^3/s and velocity of 3m/s. Find whether a hydraulic jump is expected in the channel downstream from the sluice. [**Ans.** No]

9. A hydraulic jump occurs in a rectangular channel and the depths of flow before and after the jump are 0.5 m and 2 m respectively. Calculate the critical depth of flow. [**Ans.** 1.077 m]

10. A rectangular channel carrying a super -critical flow is to be provided with a hydraulic jump type of energy dissipator. Energy loss required in the hydraulic jump is 6 m and inlet Froude number is 10. Determine the subsequent depths. [**Ans.** 0.162 m, 2.21 m]

14 POTENTIAL FLOW

14.1 INTRODUCTION

An incompressible and inviscid (or non-viscous) fluid may be termed as an ideal or potential fluid. For such a fluid density is constant and viscosity is zero. Also the shear stress will be zero in the case of ideal or potential flow. The velocity of flow will be uniform. Under certain conditions, certain real fluid's like water and air may be treated as ideal fluids. In this chapter, we shall study the two-dimensional flow fields of ideal fluids in the horizontal plane.

14.2 RECTILINEAR UNIFORM FLOW

In analysing the flow about solid bodies immersed in a fluid stream, the approaching flow is frequently of infinite extend and possesses straight and parallel streamlines and uniform velocity distribution. Such a flow with velocity U in horizontal direction (along x-axis) is shown in Fig.14.1(a).

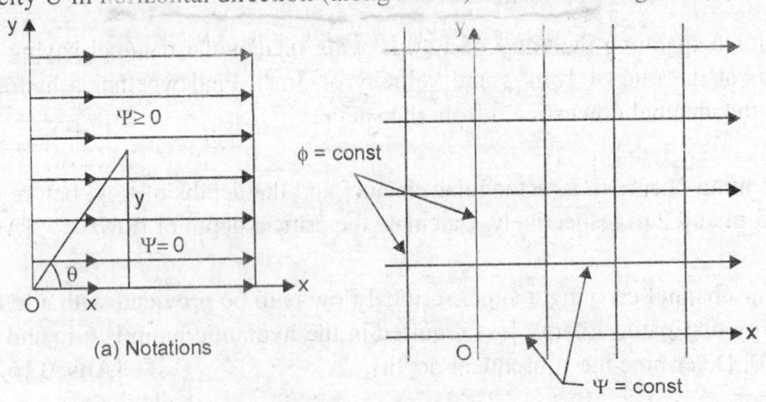

Fig. 14.1 Uniform flow

For such a flow, $u = U$ and $v = 0$.

Stream function, $\psi = \int \frac{\partial \psi}{\partial y} \cdot dy + \int \left(-\frac{\partial \psi}{\partial x}\right) dx + C$

$= \int U\, dy + \int 0\, dx + C$

$= Uy + C$

If at $y = 0$, $\psi = 0$, then $C = 0$

Thus $$\psi = Uy \qquad ...(14.1)$$

Eq. (14.1) represents that streamlines for which ψ = const, are lines parallel to x-axis.

In polar coordinates, we have

$$\psi = U.r \sin\theta \qquad ...(14.2)$$

Now $$u = -\frac{\partial \phi}{\partial x} = U, v = -\frac{\partial \phi}{\partial y} = 0$$

$$\phi = \int \frac{\partial \phi}{\partial x} dx + \int \frac{\partial \phi}{\partial y} dy + C_1$$

$$= \int (-U) dx + 0 + C_1$$

$$= -Ux + C_1$$

when $x = 0$, let $\phi = 0$, so then $C_1 = 0$

$$\therefore \phi = -Ux = -Ur \cos\theta \qquad ... (14.3)$$

The equipotential lines, ϕ = const, are parallel to the y-axis. Fig. 14.1*(b)* shows a flownet of streamlines and equipotential lines which intersect orthogonally. The flownet consist of perfect squares throughout the flow. Both the velocity and pressure remain contant throughout the flow field.

14.3 SOURCE AND SINK

Consider a symmetrical flow field consisting of radial streamlines directed outward from a common point O, as shown in Fig. 14.2. Such a flow is called the source flow. According to continuity principle the velocities will decrease as the streamlines spread. The symmetry will require that all velocities will be same at the same radial distance from the origin. Accross all circles, the same discharge will pass, and thus the velocity at any radius r in the flow field may be written as,

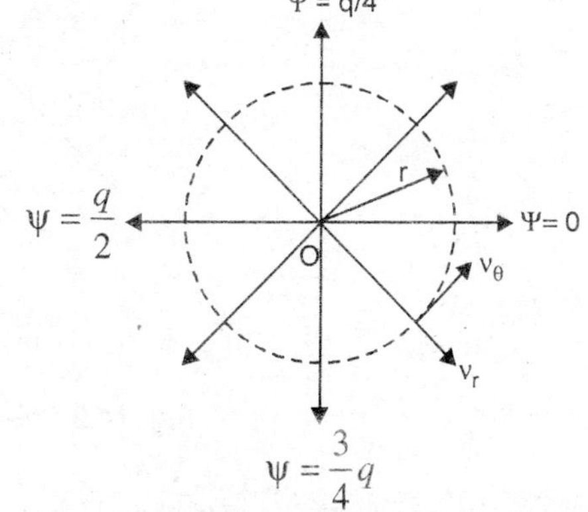

Fig. 14.2 Source Flow

$$v_r = \frac{q}{2\pi r} \text{ and } v_\theta = 0$$

where q = strength of the source, and is defined as the rate of flow passing through any circle enclosing the source.

$$\psi = \int \frac{1}{r} \frac{\partial \psi}{\partial \theta} \cdot r d\theta + \int \frac{\partial \psi}{\partial r} \cdot dr + C$$

$$= \int \left(\frac{q}{2\pi r} \right) \cdot r d\theta + \int (0) \cdot dr + C$$

$$= \frac{q\theta}{2\pi} + C$$

Let for $\theta = 0°$, $\psi = 0$. Then C = 0

$$\therefore \qquad \psi = \frac{q}{2\pi}\theta$$

$$\text{Velocity potential, } \phi = \int \left(\frac{\partial \phi}{r\partial \theta} \right) r d\theta + \int \left(\frac{\partial \phi}{\partial r} \right) dr + C \qquad \text{... (14.4)}$$

$$= \int (-v_\theta) r d\theta + \int \left(-\frac{q}{2\pi r} \right) dr + C$$

Since $v_\theta = -\dfrac{\partial \phi}{r\partial \theta} = 0$ and $v_r = -\dfrac{\partial \phi}{\partial r} = \dfrac{q}{2\pi r}$

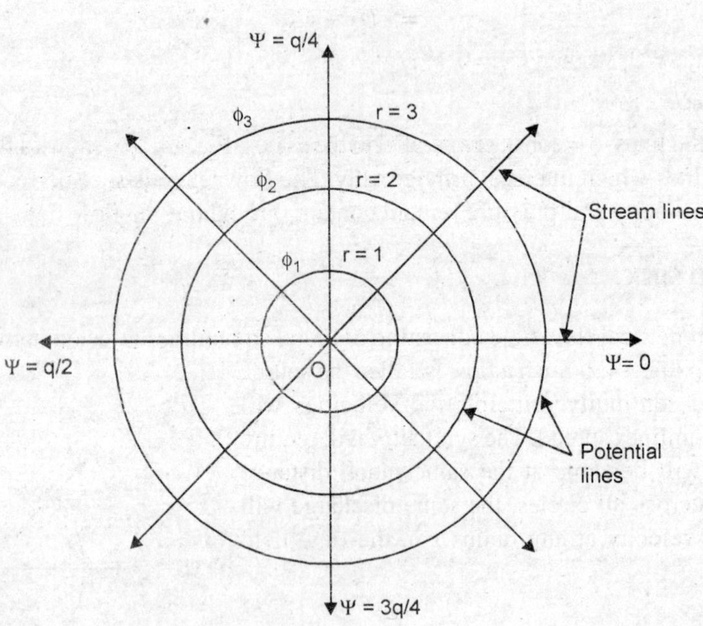

Fig. 14.3 Stream lines and potential lines for a source

Let when $r = 1$, $\phi = 0$, then C = 0

$$\therefore \quad \phi = -\frac{q}{2\pi}\ln r \qquad ...(14.5)$$

The inward flow to a point is called the *sink* flow. Its stream function may be written as,

$$\psi = -\frac{q}{2\pi}\theta \qquad ...(14.6)$$

and the velocity potential,

$$\phi = \frac{q}{2\pi}\ln r \qquad ...(14.7)$$

The streamlines and potential lines for a source are shown in Fig. 14.3. Eq. (14.7) shows that the velocity potential function is a function of r. For a given radius r, ϕ will be constant. Hence it will be a circle with origin at the source.

Pressure distribution in a plane source flow

Consider two points 1 and 2 in the plane source flow which is horizontal. In that case the datum head will be same for the two points of flow.

Let p_1 = pressure at point 1 which is at a radius r from the source.

v_{r1} = velocity at point 1

p_2 = pressure at point 2, which is at a large distance away from the source.

v_{r2} = velocity at point 2 = 0 $\qquad \left(\because V_r = \frac{q}{2\pi r} \simeq \theta \text{ for large } r\right)$

Applying Bernoulli's equation, we get

$$\frac{p_1}{\rho g} + \frac{v_{r1}^2}{2g} = \frac{p_2}{\rho g} + 0$$

$$p_1 - p_2 = -\frac{\rho v_{r1}^2}{2}$$

$$= -\frac{\rho}{2}\left(\frac{q}{2\pi r}\right)^2 = -\frac{\rho q^2}{8\pi^2 r^2} \qquad ...(14.8)$$

Eq. (14.8) shows that $p \propto \frac{1}{r^2}$

Example 14.1 *Two discs are placed in a horizontal plane, one above the other. The water enters at the centre of lower disc and flows radially outwards from a source of strength 0.65m²/s. The pressure at a radius 60mm is 200kPa. Calculate: (a) the pressure at a radius of 480mm, and (b) stream function at an angle of 60° if $\psi = 0$ at $\theta = 0°$.*

Solution.

Given: *(a)* $q = 0.65 m^2/s$, $r_1 = 60 mm$, $p_1 = 200$ kPa, $r_2 = 480$ mm, $\theta = 60°$

Radial velocity, $$v_{r1} = \frac{q}{2\pi r_1} = \frac{0.65}{2\pi \times 0.06} = 1.724 m/s$$

$$v_{r2} = \frac{q}{2\pi r_2} = \frac{0.65}{2\pi \times 0.48} = 0.215 m/s$$

Applying Bernoulli's equation at $r_1 = 0.06$m and $r_2 = 0.48$m, we have

$$\frac{p_1}{\rho g} + \frac{v_{r1}^2}{2g} = \frac{p_2}{\rho g} + \frac{v_{r2}^2}{2g}$$

or $$\frac{p_1}{\rho} + \frac{v_{r1}}{2} = \frac{p_2}{\rho} + \frac{v_{r2}^2}{2}$$

$$\frac{200 \times 10^3}{10^3} + \frac{(1.724)}{2} = \frac{p_2}{10^3} + \frac{(0.215)^2}{2}$$

$$p_2 = 201.46 kPa$$

(b) $$\psi = \frac{q}{2\pi}\theta$$

$$= \frac{0.65}{2\pi} \times \frac{\pi}{180} \times 60 = 0.1083 m^2/s$$

14.4 FREE VORTEX FLOW

Free Vortex flow is a circulatory flow of a fluid such that its streamlines are concentric circles. For a free vortex flow, we have

$$v_\theta \times r = \text{const} = C \text{ (say)}$$

Also circulation around a stream line of an irrotational vortex is,

$$\Gamma = 2\pi r\, v_\theta = 2\pi \times C$$

where v_θ = tangential velocity at any radius r from the centre.

$$= \frac{\Gamma}{2\pi r}$$

For a free vortex flow, the velocity components are:

$$v_\theta = \frac{\Gamma}{2\pi r} \text{ and } v_r = 0$$

Equation of Stream Function

By definition, we have

$$v_\theta = -\frac{\partial\psi}{\partial r} \text{ and } v_r = \frac{1}{r}\frac{\partial\psi}{\partial\theta}$$

For a free Vortex, $v_r = 0$

$$v_\theta = \frac{\Gamma}{2\pi r} = \frac{-\partial\psi}{\partial r}$$

or
$$d\psi = -\frac{\Gamma}{2\pi r}dr$$

Integrating, we get

$$\int d\psi = \int -\left(\frac{\Gamma}{2\pi}\right)\frac{dr}{r}$$

$$\psi = -\frac{\Gamma}{2\pi}\ln r \qquad \text{...(14.9)}$$

Eq. (14.9) shows that stream function is a function of radius. For a given value of r, the stream function is constant. Hence, stream lines are concentric circles. The stream lines are shown in Fig.14.4.

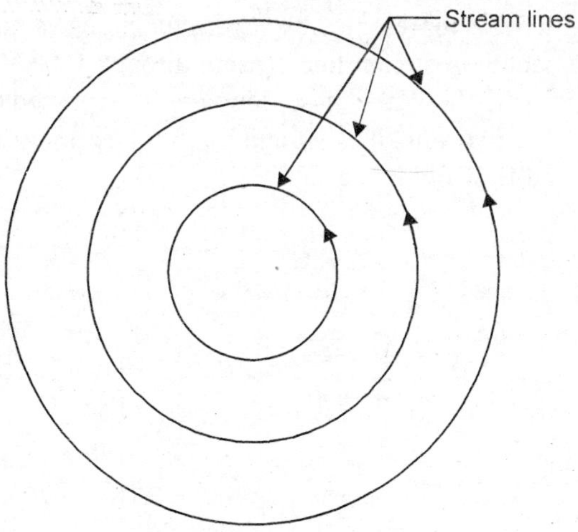

Fig. 14.4 Steam lines in free vortex

Equation of potential function

By definition, we have

$$v_r = \frac{\partial \phi}{\partial r}, v_\theta = \frac{1}{r}\frac{\partial \phi}{\partial \theta}$$

For a free vortex flow, $v_r = 0$ and $v_\theta = \frac{\Gamma}{2\pi r}$

$$\therefore \quad \frac{1}{r}\frac{\partial \phi}{\partial \theta} = \frac{\Gamma}{2\pi r}$$

or $$d\phi = \frac{\Gamma}{2\pi} d_\theta$$

Integrating, we get

$$\int d\phi = \int \frac{\Gamma}{2\pi} d\theta$$

$$\phi = \frac{\Gamma}{2\pi} \cdot \theta \qquad ...(14.10)$$

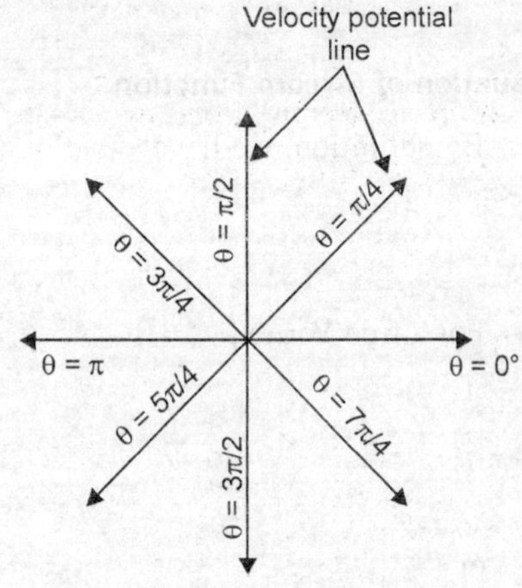

Fig. 14.5 Potential lines in free vortex

Eq. (14.10) shows that velocity potential function is a function of θ. For a given value of θ, ϕ = const. Hence equipotential lines are radial lines as shown in Fig. 14.5.

14.5 SUPERPOSITION OF FLOW FIELDS

Important flow fields may be developed by suitably combining uniform rectilinear, source sink and vortex flows. Consider two flow fields represented by stream functions ψ_A and ψ_B. Let at a point P in a flow field ψ_A, the velocity tangential to stream line passing through P be V_A; at the same point in the flow field ψ_B, the similar velocity be V_B. When these flows are superpossed the resulting velocity at the point must be the vector sum of the two velocities V_A and V_B. At other points the same vector addition of velocities occurs and a combined flow field ψ results.

From Fig. 14.6, we have

$$u = u_A \nrightarrow u_B$$

or $$\frac{\partial \psi}{\partial y} = \frac{\partial \psi_A}{\partial y} \nrightarrow \frac{\partial \psi_B}{\partial y} = \frac{\partial}{\partial y}(\psi_A \nrightarrow \psi_B)$$

or $$\psi = \psi_A \nrightarrow \psi_B \qquad ...(14.10)$$

where the symbol $\nrightarrow$ represents algebraic addition.

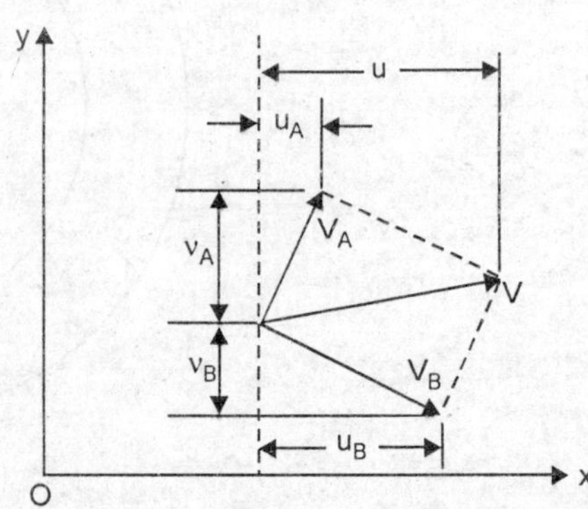

Fig. 14.6 Notations for combining two fields

14.5.1 Source in a Rectilinear Flow

Consider a source at the origin of coordinates superposed with the uniform flow as shown in Fig.14.7 The stream function for the combined flow fields,

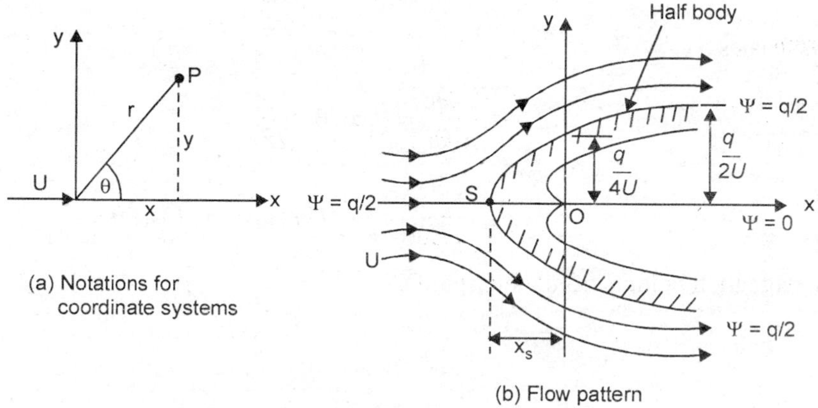

Fig. 14.7 Source placed in a rectilinear flow

$$\psi = U \cdot r \sin\theta + \frac{q}{2\pi}\cdot\theta = U\cdot y + \frac{q}{2\pi}\cdot\theta \quad ...(14.11)$$

This fields a flow field which will represent the streamline picture about the nose of a solid body of special shape placed in a free stream, as shown in Fig. 14.7 *(b)*. The velocity potential for the body may be written as:

$$\left.\begin{aligned}\phi &= -U.r\cos\theta - \frac{q}{2\pi}\ln r \\ &= -U.x - \frac{q}{2\pi}\ln(x^2+y^2)^{1/2}\end{aligned}\right\} \quad ...(14.12)$$

The velocity components in Cartesian Coordinates are:

$$u = -\frac{\partial\phi}{\partial x} = U + \frac{\partial}{\partial x}\left[\frac{q}{2\pi}\ln(x^2+y^2)^{1/2}\right]$$

$$= U + \frac{q}{2\pi}\times\frac{1}{(x^2+y^2)^{1/2}}\times\frac{1}{2}(x^2+y^2)^{-1/2}\times 2x$$

$$= U + \frac{q}{2\pi}\times\frac{x}{(x^2+y^2)}$$

$$= U + \frac{qx}{2\pi r^2} = U + \frac{q\cos\theta}{2\pi r} \qquad ...(14.13)$$

$$v = -\frac{\partial \phi}{\partial y} = \frac{qy}{2\pi r^2} = \frac{q}{2\pi r}\sin\theta \qquad ...(14.14)$$

In polar coordinates,

$$v_r = -\frac{\partial \phi}{\partial r} = U\cos\theta + \frac{q}{2\pi r} \qquad ...(4.15)$$

$$v_\theta = -\frac{1}{r}\frac{\partial \phi}{\partial \theta} = -\frac{1}{r}[U.r\sin\theta] = -U\sin\theta \qquad ...(14.16)$$

Distance of stagnation point S from the origin O,

$$x_s = \frac{q}{2\pi U} \qquad ...(14.17)$$

Equation of streamline passing through the stagnation point is,

$$\psi_s = U.\left(\frac{q}{2\pi U}\right)\sin\pi + \frac{q\pi}{2\pi} = \frac{q}{2} \qquad ...(14.18)\backslash$$

The equation of stream line passing through S, $\psi = \psi_s$ and hence that of body contour is given by:

$$Ur\sin\theta + \frac{q\theta}{2\pi} = \frac{q}{2}$$

or

$$y = r\sin\theta = \frac{q}{2U}\left(1 - \frac{\theta}{\pi}\right) \qquad ...(14.19)$$

Eq. (14.19) represents the streamline passing through the stagnation point S and divides the flow into two regions. This dividing streamline can be replaced by a solid wall. At $\theta = 0$ ($x \to \infty$), Eq.(14.19) becomes

$$y = \frac{q}{2U} \qquad ...(14.20)$$

and gives the maximum value of the height of the solid wall. Eq.(14.20) shows that the body contour of the solid wall approaches this line asymptotically. The body contour formed by a source in a rectangular flow is unique in the sense that it has a head or nose but no tail and is known as *'Rankine half body'*.

At

$$\left.\begin{aligned} \theta = \frac{\pi}{2}, \quad & y = \frac{q}{4U} \\ \theta = \pi, \quad & y = 0 \\ \theta = \frac{3\pi}{2}, \quad & y = \frac{-q}{4U} \end{aligned}\right\} \qquad ...(14.21)$$

Pressure at any point in the flow field

Let p_∞ = pressure at infinity where velocity is U.

p = pressure at any point P in the flow field where velocity is V.

Applying Bernoulli's theorem at a point at infinity and point P, we have

$$\frac{p_\infty}{\rho g} + \frac{U^2}{2g} = \frac{p}{\rho g} + \frac{U^2}{2g}$$

or

$$\frac{U^2 - V^2}{2} = \frac{p - p_\infty}{\rho}$$

Pressure coefficient,

$$C_p = \frac{p - p_\infty}{\frac{1}{2}\rho U^2} = \frac{U^2 - V^2}{2} \times \frac{2}{U^2} = 1 - \left(\frac{V}{U}\right)^2 \quad ...(14.22)$$

Example 14.2 *A uniform flow with a velocity of 2.5m/s is flowing over a plane source of stength 25m²/s. The uniform flow and source flow are in the same plane. A point P situated in the flow field is at a distance of 0.5m from the source and at an angle of 45° to the uniform flow. Determine: (a) stream function at point P, (b) resultant velocity of flow at P, and (c) location of stagnation point from the source.*

Solution:

Given: U = 2.5 m/s, q = 25m²/s, r = 0.5m, θ = 45°

(a)
$$\psi = U.r \sin\theta + \frac{q}{2\pi}\theta$$

$$= 2.5 \times 0.5 \sin 45° + \frac{25}{2\pi} \times \frac{\pi}{180} \times 45 = 4m^2/s.$$

(b)
$$v_r = U.\cos\theta + \frac{q}{2\pi r}$$

$$= 2.5\cos 45° + \frac{25}{2\pi \times 0.5} = 9.72m/s$$

$$v_\theta = -U\sin\theta = -2.5\sin 45° = -1.768m/s$$

Resultant velocity,
$$v = \sqrt{v_r^2 + v_\theta^2}$$

$$= \sqrt{(9.72)^2 + (-1.768)^2} = 9.88m/s$$

(c)
$$x_s = \frac{q}{2\pi U} = \frac{25}{2\pi \times 2.5} = 1.59m$$

Example 14.3 *A uniform flow with a velocity of 2m/s is flowing over a source placed at the origin. The stagnation point occurs at (–0.4, 0). Determine: (a) Strength of the source, (b) Maximum width of Rankine half body, and (c) other principal dimensions of the Rankine half body.*

Solution.

Given: U = 2m/s, $x_s = 0.4$m

(a) $$x_s = \frac{q}{2\pi U}$$

$$q = 2\pi U x_s = 2\pi \times 2 \times 0.4 = 5.026 \text{m}^2/\text{s}$$

(b) $$y = \frac{q}{2U}\left(1 - \frac{\theta}{\pi}\right)$$

For $$\theta = 0° \; y = y_{max} = \frac{q}{2U} = \frac{5.026}{2 \times 2} = 1.257\text{m}$$

Maximum width of Rankine body = $2y_{max}$ = 2.514m

(c) At $\theta = \frac{\pi}{2}, y = \frac{q}{4U} = \frac{1.257}{2} = 0.628m$

Width of body at origin = 2 × 0.628 = 1.256 m

At $\theta = \pi$, y = 0

14.5.2 DOUBLET

A doublet is a limiting combination of a source and sink of equal strength, which are allowed to approach each other in such a manner that the product of their strength and the distance between them always remains a constant. The axis of the doublet is from the sink towards the source.

Consider a doublet comprising of source of strength q at $(a, 0)$ and a sink of equal strength at $(-a, 0)$ as shown in Fig. 14.8.

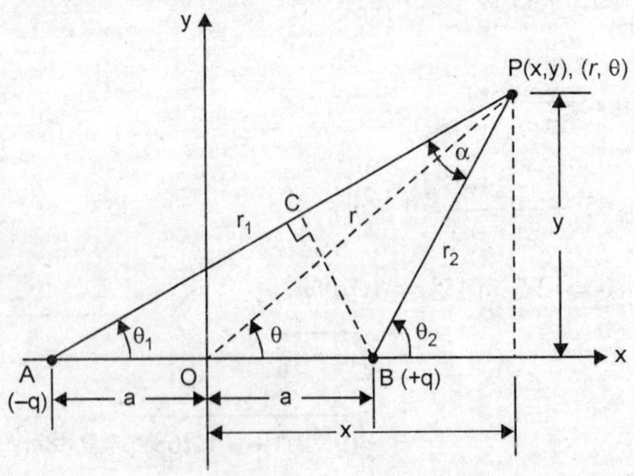

Fig. 14.8 Notations for a 2 – D doublet

Let x, y = Cartesian coordinates of point P.

r, θ = polar coordinates of point P.

r_1, θ_1 = position of point P w.r.t. sink placed at A.

r_2, θ_2 = position of point P w.r.t. source placed at B.

α = angle subtended at P by the join of sink and source, i.e. ∠APB

$= \theta_2 - \theta_1$

Stream function of the combined flow field,

$$\psi = \psi_2 + \psi_1 = \frac{q}{2\pi}(\theta_2 - \theta_1) = \frac{q\alpha}{2x} \qquad \text{... (1)}$$

Drop BC perpendicular on AP and consider the common side of ∆s ABC and BCP

$$\frac{BC}{AB} = \sin\theta_1,\ BC = 2a\sin\theta_1$$

$$\frac{BC}{BP} = \sin\alpha,\ BC = r_2\sin\alpha$$

$$\therefore \quad r_2 \sin\alpha = 2a\sin\theta \qquad \text{... (2)}$$

As a → 0, $\sin\alpha \to \alpha$, $r_2 \to r_1 \to r$

and $\theta_2 \to \theta_1 \to \theta$

$$\therefore r\alpha = 2a\sin\theta$$

or

$$\alpha = \frac{2a}{r}\sin\theta \qquad \text{...(3)}$$

Eq.(1) becomes,

$$\psi = \frac{q}{2\pi}\left(\frac{2a}{r}\sin\theta\right) \qquad \text{...(14.23)}$$

Now consider a condition under which as $a \to 0$, q is increased such that $2qa$ remains constant and equal to m.

$$\therefore \quad m = 2qa \qquad \text{...(14.24)}$$

where m as known as the "*strength of the doublet*".

$$\therefore \quad \psi = \frac{m}{2\pi r}\sin\theta \qquad \text{...(14.25)}$$

The velocity potential for the doublet,

$$\phi = \phi_1 + \phi_2 = +\frac{q}{2\pi}\ln r_1 - \frac{q}{2\pi}\ln r_2 = \frac{q}{2\pi}(\ln r_1 - \ln r_2)$$

$$= \frac{q}{2\pi}\ln\frac{r_1}{r_2}$$

To take the limit as $a \rightarrow 0$ and q increased in such a manner that $2qa = m$, the form of the above expression for ϕ must be modified as under:

$$r_2^2 = r^2 + a^2 - 2ar\cos\theta = r^2\left[1+\left(\frac{a}{r}\right)^2 - 2\left(\frac{a}{r}\right)\cos\theta\right]$$

$$r_1^2 = r^2 + a^2 + 2ar\cos\theta = r^2\left[1+\left(\frac{a}{r}\right)^2 + 2\left(\frac{a}{r}\right)\cos\theta\right]$$

$$\phi = \frac{q}{4\pi}(\ln r_1^2 - \ln r_2^2)$$

$$= \frac{q}{4\pi}\left[\ln r^2 + \ln\{1+\left(\frac{a}{r}\right)^2 + 2\left(\frac{a}{r}\right)\cos\theta\} - \ln r^2 - \ln\{1+\left(\frac{a}{r}\right)^2 - 2\left(\frac{a}{r}\right)\cos\theta\}\right]$$

Now $\ln(1+x) = x - \frac{x^2}{2} + \frac{x^3}{3} - \frac{x^4}{4} + \ldots.$

$$\phi = \frac{q}{4\pi}\left[\left(\frac{a}{r}\right)^2 + 2\left(\frac{a}{r}\right)\cos\theta - \frac{1}{2}\left\{\left(\frac{a}{r}\right)^2 + 2\left(\frac{a}{r}\right)\cos\theta\right\}^2\right.$$

$$+\frac{1}{3}\left\{\left(\frac{a}{r}\right)^2 + 2\left(\frac{a}{r}\right)\cos\theta\right\}^3 - \left\{\left(\frac{a}{r}\right)^2 - 2\left(\frac{a}{r}\right)\cos\theta\right\}$$

$$\left. + \frac{1}{2}\left\{\left(\frac{a}{r}\right)^2 - 2\left(\frac{a}{r}\right)\cos\theta\right\}^2 - \frac{1}{3}\left\{\left(\frac{a}{r}\right)^2 + 2\left(\frac{a}{r}\right)\cos\theta\right\}^3\right]$$

$$= \frac{qa}{\pi}\left[\frac{\cos\theta}{r} - \left(\frac{a}{r}\right)^2\frac{\cos\theta}{r} + \left(\frac{a}{r}\right)^4\frac{\cos\theta}{r} + \ldots\right]$$

When $a \rightarrow 0$

$$\phi = \frac{m}{2\pi}\frac{\cos\theta}{r} \qquad \ldots(14.26)$$

If the positions of source and sink are interchanged so that source is placed upstream, then the strength of doublet becomes $(-m)$ and Eq. (14.25) becomes,

$$\psi = -\frac{m \sin\theta}{2\pi r} \qquad \text{...(14.27a)}$$

and
$$\phi = -\frac{m \cos\theta}{2\pi r} \qquad \text{...(14.27b)}$$

For any constant value of ψ, Eq. (14.24) can be written as:

$$r = r_0 \sin\theta$$

where
$$r_0 = \frac{m}{2\pi\psi_c} \qquad \text{...(14.28)}$$

ψ_c = arbitrary constant for ψ

Eq. (14.28) represents a circle which passes through the origin with centre on the y–axis and a diameter of r_0.

Similarly for any arbitrary constant value of ϕ, Eq. (14.26) can be expressed as:

$$r = R_0 \cos\theta \qquad \text{...(14.29)}$$

where
$$R_0 = \frac{m}{2\pi\phi_c}$$

ϕ_c = arbitrary constant for ϕ

Eq. (14.29) represents a circle which passes through the origin with centre on x-axis and a diameter of R_0. The streamlines and equipotential lines for the doublet are shown in Fig.14.9.

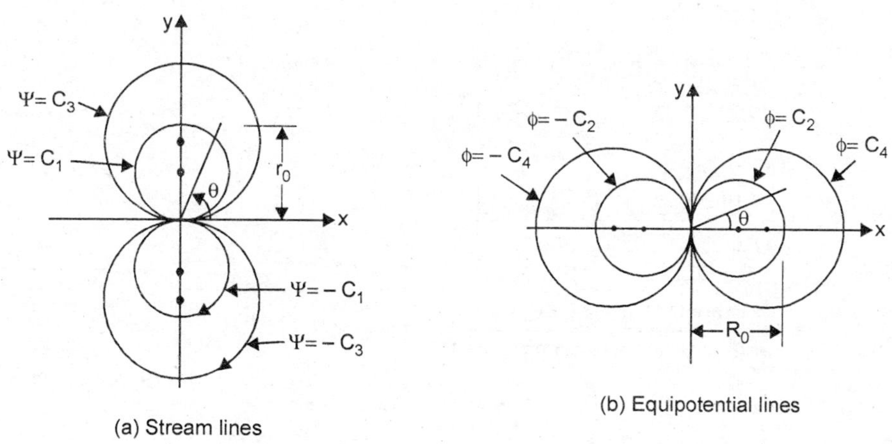

Fig. 14.9 Streamlines and equipotential lines for doublet flow

Example 14.4 A point P (0.5, 1) is situated in the flow field of a doublet of strength $5 m^2/s$. Calculate the velocity at this point and the value of stream function.

Solution.

Given: $x = 0.5$, $y = 1$, $m = 5\text{m}^2/\text{s}$

$$r = \sqrt{x^2 + y^2} = \sqrt{(0.5)^2 + 1} = 1.118$$

$$\sin\theta = \frac{y}{r} = \frac{1}{1.118} = 0.894$$

Stream function,

$$\psi = -\frac{m\sin\theta}{2\pi r} = -\frac{5\times 0.894}{2\pi\times 1.118} = -0.636$$

$$u = \frac{\partial\psi}{\partial y} = \frac{\partial}{\partial y}\left[-\frac{5y}{2\pi r^2}\right]$$

$$= \frac{\partial}{\partial y}\left[-\frac{5}{2\pi}\times\frac{y}{x^2+y^2}\right]$$

$$= -0.796\left[\frac{(x^2+y^2) - y\times 2y}{(x^2+y^2)^2}\right]$$

$$= -0.796\left[\frac{x^2-y^2}{(x^2+y^2)^2}\right] = -0.796\left[\frac{0.25-1}{(1.25)^2}\right] = 0.382 \text{ m/s}$$

$$v = -\frac{\partial\psi}{\partial x} = -\frac{\partial}{\partial x}\left[-\frac{m}{2\pi}\times\frac{y}{(x^2+y^2)}\right]$$

$$= \frac{m}{2\pi}\left[\frac{(x^2+y^2)\times 0 - y\times 2x}{(x^2+y^2)^2}\right]$$

$$= \frac{-mxy}{\pi(x^2+y^2)^2} = -\frac{5\times 0.5\times 1}{\pi(0.25+1)^2} = -0.509 \text{ m/s}$$

Resultant velocity,

$$V = \sqrt{u^2+v^2} = \sqrt{(0.382)^2 + (-0.509)^2}$$

$$= 0.636 \text{ m/s}$$

14.5.3 Doublet in Rectilinear Flow

The stream function for the combined uniform rectilinear flow and doublet is given by

$$\psi = Ur\sin\theta - \frac{m\sin\theta}{2\pi r} \qquad ...(14.30)$$

For $\psi = 0$, we get

$$\left(U - \frac{m}{2\pi r^2}\right)\sin\theta = 0 \qquad ...(14.31)$$

Its solution is:

$$\left.\begin{aligned} \theta &= 0\, or\, \pi \\ and\, r &= \left(\frac{m}{2\pi U}\right)^{1/2} = R \end{aligned}\right\} \qquad ...(14.32)$$

Hence $\psi = 0$ represents the streamline which is a circle of radius R joined by x-axis on both sides. A cylinder of radius R may be taken as a solid boundary. The stream function and velocity potential for uniform flow around a circular cylinder of radius R, by substituting for m from Eq. (14.32) are:

$$\psi = U\left(r - \frac{R^2}{r}\right)\sin\theta \qquad ... (14.33)$$

$$\phi = -U\left(r + \frac{R^2}{r}\right)\cos\theta \qquad ... (14.34)$$

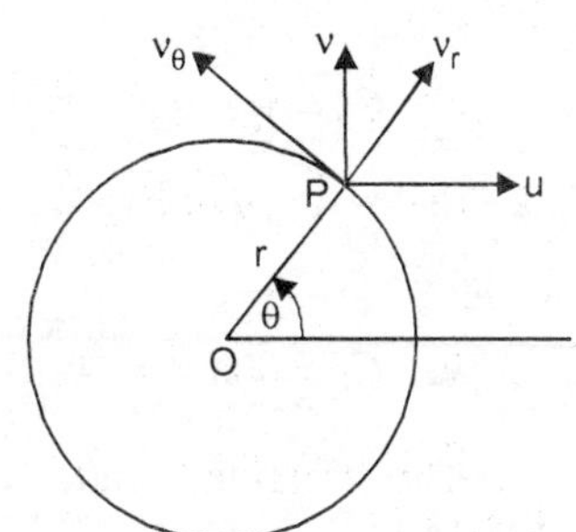

Fig. 14.10 Velocity components on the surface of a cylinder

The velocities are:

$$v_r = -\frac{\partial\phi}{\partial r} = U\left(1 - \frac{R^2}{r^2}\right)\cos\theta \qquad ...(14.35)$$

$$v_\theta = -\frac{\partial\phi}{r\partial\theta} = -U\left(1 + \frac{R^2}{r^2}\right)\sin\theta \qquad ...(14.36)$$

From Fig. 14.10, we have

$$u = v_r\cos\theta - v_\theta\sin\theta = U\cos^2\theta + U\sin^2\theta = U$$

$$v = v_r\sin\theta + v_\theta\cos\theta = U(\sin\theta\cos\theta - \cos\theta\sin\theta) = 0$$

Validity of boundary conditions:

(i) At $r = R$, $v_r = 0$

(ii) As $r \to \infty$, $u = U$ and $v = 0$.

From Eq. (14.36), and $r = R$

$$v_\theta = -2U\sin\theta \qquad ...(14.37)$$

–ve sign in Eq.(14.37) indicates that v_θ is in clockwise direction. Eq.(14.37) also reveals that the stagnation points are located at $\theta = 0$ and π.

At $\theta = \frac{\pi}{2}$, v_θ attains a maximum value. ...(14.38)

$$|(v_\theta)_{max}| = 2U$$

The pressure at any point on the surface of cylinder can be determined by using the Bernoulli's equation between a point far away in the free stream, where pressure is p_∞ and velocity U, and a point on the cylinder, where pressure is p and velocity v_θ.

$$p + \frac{\rho v_\theta^2}{2} = p_\infty + \frac{\rho U^2}{2}$$

Substituting for v_θ from Eq. (14.37), we get

$$p - p_\infty = \frac{\rho}{2}(U^2 - v_\theta^2) = \frac{1}{2}\rho U^2(1 - 4\sin^2\theta) \quad ...(14.39)$$

Pressure coefficient,

$$C_p = \frac{p - p_\infty}{\frac{1}{2}\rho U^2} = 1 - 4\sin^2\theta \quad ...(14.40)$$

(i) $C_p = 1$ at $\theta = 0$ and π ...(14.41*a*)

(ii) $C_p = -3$ at $\theta = \frac{\pi}{2}$ *and* $\frac{3\pi}{2}$ ...(14.41*b*)

(iii) $C_p = 0$ at $\theta = \pm\frac{\pi}{6}$ *and* $\pm\frac{5\pi}{6}$ ...(14.41*c*)

The streamline pattern for flow due to double placed in a uniform rectilinear flow is shown in Fig.14.11.

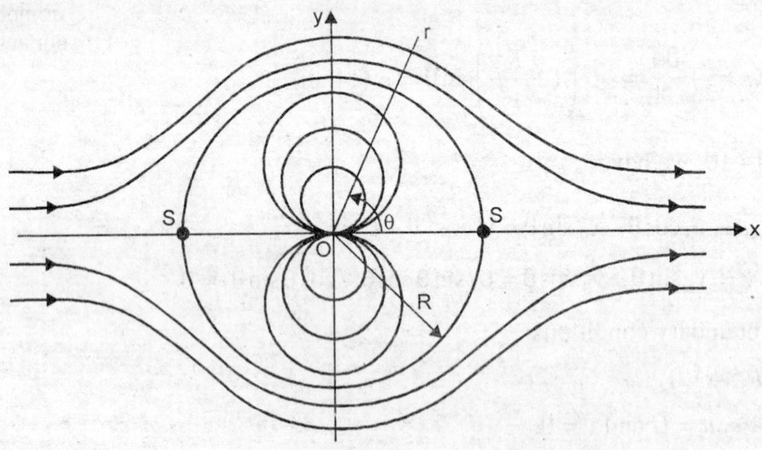

Fig. 14.11 Streamline pattern in a combined flow due to doublet placed in a uniform rectilinear flow

14.5.4 Drag and Lift Forces on cylinder

Consider an elementary area $R.d\theta$ of the cylinder surface as shown in Fig.14.12. The x– and y–components of the pressure force representing the drag and lift forces per unit length of the cylinder are:

Drag force,

$$F_x = F_D = -\int_0^{2\pi} (p - p_\infty)Rd\theta.\cos\theta$$

$$= -\frac{1}{2}\rho U^2 \int_0^{2\pi} (1 - 4\sin^2\theta)\cos\theta \cdot d\theta = 0 \quad \text{... (14.42)}$$

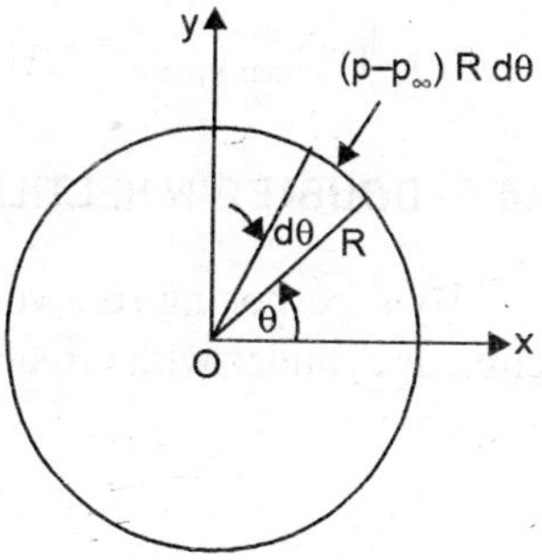

Lift force,

$$F_y = F_L = -\int_0^{2\pi} (p - p_\infty)Rd\theta.\sin\theta$$

$$= -\frac{1}{2}\rho U^2 \int_0^{2x} (1 - 4\sin^2\theta)\sin\theta\, d\theta = 0 \quad \text{... (14.43)}$$

Fig. 14.12

This theoretical result of non-existence of drag on a body placed in a uniform stream is called "*d'Alembert's paradox*".

Example 14.5 A uniform flow of 10 m/s is flowing over a doublet of strength 20 m²/s. The doublet is in the line of uniform flow. Determine:

(a) Shape of the Rankine body

(b) Radius of the Rankine circle

(c) Value of streamline function at Rankine circle.

(d) Resultant velocity at a point on the Rankine Circle at an angle of 30° from x-axis.

(e) Value of maximum veloeity on the Rankine circle and location of the point where velocity is maximum.

Solution:

Given: U = 10 m/s, m = 20 m²/s.

(a) When a uniform rectilinear flow is flowing over a doublet and doublet and uniform flow are in line, then the shape of the Rankine body is a circle of radius,

$$R = \sqrt{\frac{m}{2\pi U}}$$

(b)

$$R = \sqrt{\frac{20}{2\pi \times 10}} = 0.564m$$

(c) At the Rankine circle, $\psi = 0$.

(d) $v_r = 0$, $v_\theta = -2U\sin\theta = -2 \times 10 \times \sin 30° = -10$ m/s.

$$V = \sqrt{v_r^2 + v_\theta^2} = -10\, m/s$$

(e) $\left|(v_\theta)_{max}\right|_{\theta=90°} = 2U = 2 \times 10 = 20$ m/s.

14.6 DOUBLET IN RECTILINEAR FLOW WITH CIRCULATION

By superposing free vortex, doublet and uniform rectilinear flow fields results in a flow around a circular cylinder with circulation. The stream function for clockwise vortex is,

$$\psi = U\left(r - \frac{R^2}{r}\right)\sin\theta + \frac{\Gamma}{2\pi}\ln r \qquad ...(14.44)$$

Velocity potential for the combined flow field is given by,

$$\phi = -U\left(r + \frac{R^2}{r}\right)\cos\theta + \frac{\Gamma\theta}{2\pi} \qquad ...(14.45)$$

The velocity components are:

$$v_r = -\frac{\partial\psi}{r\partial\theta} = U\left(1 - \frac{R^2}{r^2}\right)\cos\theta \qquad ...(14.46)$$

$$v_\theta = -\frac{\partial\psi}{\partial r} = -U\left(1 + \frac{R^2}{r^2}\right)\sin\theta - \frac{\Gamma}{2\pi r} \qquad ...(14.47)$$

On the surface of the cylinder, $r = R$, and

$$v_r = 0$$

$$v_\theta = -2U\sin\theta - \frac{\Gamma}{2\pi R} \qquad ...(14.48)$$

At the stagnation points, $v_\theta = 0$. Let $\theta = -\alpha$, then

$$\sin\alpha = \frac{\Gamma}{4\pi RU} \qquad ...(14.49)$$

(i) For $\Gamma < 4\pi RU$, there are two symmetrically located stagnation points as shown in Fig.14.13.

(ii) For $\alpha = \frac{\pi}{2}$ $\sin\alpha = 1$ and $\Gamma = 4\pi RU$. The two stagnation points merge into one, as shown in Fig.14.14.

(iii) For $\Gamma > 4\pi$ RU, the stagnation point will be found to be located at some distance below the circle, on the vertical axis of symmetry as shown in Fig. 14.15.

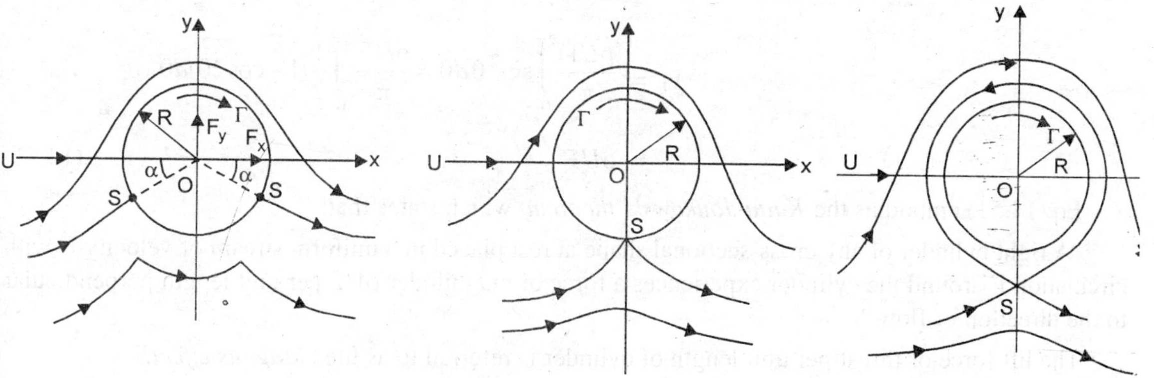

Fig. 14.13 $\Gamma < 4\pi RU$ **Fig. 14.14** $\Gamma = 4\pi RU$ **Fig. 14.15** $\Gamma > 4\pi RU$

The pressure at any point in the flow field may be found by using Bernoulli's equation. Applying this equation between a point in the free stream and another point on the surface of cylinder, we have

$$p + \frac{\rho}{2} v_\theta^2 = p_\infty + \frac{\rho}{2} U^2$$

$$p + \frac{\rho}{2}\left(2U\sin\theta + \frac{\Gamma}{2\pi R}\right)^2 = p_\infty + \frac{\rho}{2} U^2$$

$$p - p_\infty = \frac{1}{2}\rho U^2 - \frac{1}{2}\rho\left(2U\sin\theta + \frac{\Gamma}{2\pi R}\right)^2$$

$$F_D = F_x = -\int_0^{2\pi} (p - p_\infty) R d\theta . \cos\theta = 0 \qquad \text{...(14.50)}$$

$$F_L = F_y = -\int_0^{2\pi} (p - p_\infty) R d\theta . \sin\theta$$

$$= -\int_0^{2\pi} \left[\frac{1}{2}\rho U^2 - \frac{1}{2}\rho\left(2U\sin\theta + \frac{\Gamma}{2\pi R}\right)^2\right] R\sin\theta . d\theta$$

Now $$-\int_0^{2\pi} \sin\theta . d\theta = -\int_0^{2\pi} \sin^3\theta \, d\theta = 0$$

$\therefore$ $$F_L = \frac{\rho U\Gamma}{\pi}\int_0^{2\pi} \sin^2\theta \, d\theta = \frac{\rho U\Gamma}{\pi}\int_0^{2\pi} \frac{1}{2}(1-\cos 2\theta)d\theta$$

$$= \rho U\Gamma \quad ...(14.51)$$

Eq.(14.51) embodies the *Kutta-Joukowski theorem*, which states that:

"A right cylinder of any cross-sectional shape at rest placed in a uniform stream of velocity U with circulation Γ around the cylinder experiences a force of magnituder ρUΓ per unit length perpendicular to the direction of flow."

The lift force or thrust per unit length of cylinder is referred to as the "*Magnus effect*."

SUMMARY

1. The flow of an incompressible, non-viscous fluid is called potential flow.
2. Rectilinear uniform flow has straight and parallel streamlines and uniform velocity distribution.
3. For a uniform rectilinear flow,

   $$\psi = Uy, \quad \phi = -Ux$$

4. A source is a point from which the fluid is emitted uniformly in all directions.
5. The strength of a source is defined as the volume flow rate per unit depth per unit time. Its units are m^2/s.

   $$v_r = \frac{q}{2\pi r}, \quad v_\theta = 0$$

   $$\psi = \frac{q\theta}{2\pi}, \quad \phi = -\frac{q}{2\pi}\ln r, \quad q = \text{strength of the source.}$$

6. In a sink flow the fluid flows radially inwards towards the origin.

   $$\psi = -\frac{q\theta}{2\pi}, \phi = \frac{q}{2\pi}\ln r$$

7. In a plane source flow,

   $$p_1 - p_2 = -\frac{\rho q^2}{8\pi^2 r^2}$$

8. Free vertex flow is a circulatory flow of a fluid such that its stream lines are concentric circles.

   $$v_r = 0, \quad v_\theta = \frac{\Gamma}{2\pi r}$$

$$\phi = -\frac{\Gamma}{2\pi}\ln r$$

$$\phi = \frac{\Gamma}{2\pi}\theta$$

9. For superpositon of two flows,

$$\psi = \psi_1 + \psi_2$$

$$\phi = \phi_1 + \phi_2$$

10. Source + Rectilear flow:

$$\psi = Ur\sin\theta + \frac{q}{2\pi}\theta = Uy + \frac{q}{2\pi}\theta$$

$$\phi = -Ur\cos\theta - \frac{q}{2\pi}\ln r = -Ux - \frac{q}{2\pi}(x^2 + y^2)^{1/2}$$

$$u = U + \frac{qx}{2\pi r^2} = U + \frac{q\cos\theta}{2\pi r}$$

$$v = \frac{qy}{2\pi r^2} = \frac{q\sin\theta}{2\pi r}$$

Distance of stagnation point from origin, $x_s = \dfrac{q}{2\pi U}$

Rankine half body contour is given by,

$$y = \frac{q}{2U}\left(1 - \frac{\theta}{\pi}\right)$$

Pressure coefficient, $C_p = \dfrac{p - p_\infty}{\frac{1}{2}\rho U^2} = 1 - \left(\dfrac{U}{V}\right)^2$

11. A doublet is a limiting combination of a source and sink of equal strength, which are allowd to approach each other in such a manner that the product of their strength and the distance between them always remains constant.

Strength of the doublet, $m = 2qa$

$$\psi = -\frac{m\sin\theta}{2\pi r}, \ \phi = \frac{-m\cos\theta}{2\pi r}$$

12. Doublet in rectilinear flow.

Radius of cylinder, $R = \left(\dfrac{m}{2\pi U}\right)^{1/2}$

$$v_r = U\left(1-\frac{R^2}{r^2}\right)\cos\theta$$

$$v_\theta = -U\left(1+\frac{R^2}{r^2}\right)\sin\theta$$

$$|(v_\theta)_{max}| = 2U \text{ at } r = R \text{ and } \theta = \frac{\pi}{2}$$

$$C_p = 1 - 4\sin^2\theta$$

$$F_D = F_L = 0$$

13. Doblet in rectilinear flow with circulation

$$v_r = U\left(1-\frac{R^2}{r^2}\right)\cos\theta$$

$$v_\theta = -U\left(1+\frac{R^2}{r^2}\right)\sin\theta - \frac{\Gamma}{2\pi r}$$

At the stagnation point, $\sin\alpha = \dfrac{\Gamma}{4\pi RU}$, $\theta = -\alpha$

$$F_D = 0,\ F_L = \rho U\Gamma$$

14. Magnus effect is the lift force per unit length of the cylinder.

Short Answer Type Questions

1. Define a potential flow.
2. What is a rectilinear uniform flow?
3. Define a source and sink.
4. Define strength of the source.
5. What is a free vortex flow?
6. How two or more flow fields are superposed?
7. Define pressure coefficient.
8. Define a doublet.
9. What is strength of the doublet?
10. What is the equivalent flow of doublet superposed on rectilinear flow?
11. What is a Rankine half body?

12. What is d'Alembert's paradox?

13. State Kutta-Joukowski theorem.

14. What is Magnus effect?

EXERCISES

1. A uniform flow field defined by $\psi = y$ is superposed on $\vec{u} = x\hat{i} - y\hat{j}$. Determine the stream function and velocity potential for the combined flow [**Ans.** $\psi = xy + y$, $\phi = \frac{x^2}{2} + x - \frac{y^2}{2}$]

2. A source of strength 1.8 m^2/s is superposed by uniform velocity 0.6 m/s. Calculate the fluid velocity at (0.75m, 120°). [**Ans.** 0.526 m/s, 81.03°]

3. A uniform rectilinear flow at 5 m/s is superposed on a source placed at the origin. If the stagnation point occurs at (–0.32, 0), find the strength of source, maximum width of Rankine half body and flow velocity at $\left(0.5, \frac{\pi}{2}\right)$. [**Ans.** 10$m^2/s$, 1m, 5.94 m/s]

4. A flow field is prescribed by: $\phi = Ux + \frac{q}{2\pi}\ln r$.

Check whether this field is physically possible or not. If so, what does the flow pattern represent? Find the main dimensions of the body profile for $U = 2$ m/s and $q = 5m^2/s$.

[**Ans.** Yes, $y_{max} = 1.25$ m at $\theta = 0°$, $y = 0.625$ m, at $\theta = \frac{\pi}{2}, x_s = 0.398\,\text{m}$]

5. For a doublet of strength 20 m^2/s, calculate the velocity at point P(1, 2) and the value of stream function passing through it. [**Ans.** 0.636 m/s, –1.274 m^2/s]

6. A uniform flow with a velocity of 20m/s is flowing over a source of strength 10m^2/s. The uniform flow and source flow are in the same plane. Obtain the equation of dividing stream line and sketch the flow pattern. [**Ans.** $y = 0.25 - \frac{\theta}{4\pi}$]

7. A uniform flow of 10m/s is flowing over a doublet of strength 15 m^2/s. The doublet is in the line of uniform flow. The polar coordinates of a point P in the flow field are 0.9 m and 30°. Find: (a) Streamline function, and *(b)* the resultant velocity at the point.

[**Ans.** 3.177 m^2/s, 8.89 m/s]

MULTIPLE CHOICE QUESTIONS

For Fluid Mechanics

Multiple Choice Questions

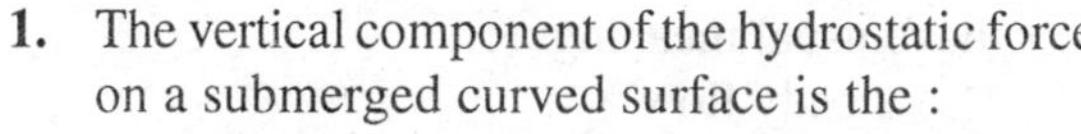

1. The vertical component of the hydrostatic force on a submerged curved surface is the :
   (a) mass of liquid vertically above it
   (b) weight of the liquid vertically above it
   (c) force on a vertical projection of the surface
   (d) product of pressure at the centroid and the surface area

2. If a hydraulic press has a ram of 12.5 cm diameter and plunger of 1.25 cm diameter, what force would be required on the plunger to raise a mass of 1 tonne on the ram?
   (a) 981 N (b) 98.1 N
   (c) 9.81 N (d) 0.98 N

3. A liquid mass readjusts itself and undergoes a rigid body type of motion when it is subjected to a :
   (a) constant angular velocity
   (b) constant angular acceleration
   (c) linearly varying velocity
   (d) linearly varying acceleration

4. In a two-dimensional flow, where u is the x-component and v is the y-component of velocity, the equation of streamline is given by :
   (a) $udx - vdy = 0$
   (b) $vdx - udy = 0$
   (c) $uv\, dx\, dy = 0$
   (d) $udx + vdy = 0$

5. The continuity equation $\frac{\partial u}{\partial x} + \frac{\partial v}{\partial y} + \frac{\partial w}{\partial z} = 0$ is valid only for :
   (a) ideal fluid flow
   (b) incompressible fluid whether the flow is steady or not
   (c) steady flow, whether it is compressible or not
   (d) steady flow and compressible fluids

6. From a reservoir, water is drained through two pipes of 10 cm and 20 cm diameter respectively. If the frictional head loss in both the pipes is same, then the ratio of discharge through the larger pipe to that through the smaller pipe will be :
   (a) $\sqrt{2}$ (b) $2\sqrt{2}$
   (c) 4 (d) $4\sqrt{2}$

7. While water **passes** through a given pipe at a mean velocity 'V', the flow is found to change from laminar to turbulent. If another fluid of specific gravity 0.8 and coefficient of viscosity 20% of that of water, is passed through the same pipe, the transition of flow from laminar to turbulent is expected if the flow velocity is :
   (a) 2V (b) V (c) V/2 (d) V/4

8. A hydraulic jump is formed in a 5 m wide rectangular channel with subsequent depths of 0.2 m and 0.8 m. The discharge in the channel, in m³/s, is:

(a) 2.43 (b) 3.45
(c) 4.43 (d) 5.00

9. If the velocity distribution in a turbulent boundary layer is given by $\frac{u}{u_\infty} = \left(\frac{y}{\delta}\right)^{1/9}$ then the ratio of displacement thickness to nominal boundary layer thickness will be :

(a) 1.0 (b) 0.6
(c) 0.3 (d) 0.1

10. The barometric pressure at the base of mountain is 750 mm Hg and at the top 600 mm Hg. If the average air density is 1 kg/m³, the height of the mountain is approximately :

(a) 2000 m (b) 3000 m
(c) 4000 m (d) 5000 m

11. At the interface of a liquid and a gas at rest, the pressure is :

(a) higher on the concave side compared to that on the convex side
(b) higher on the convex side compared to that on the concave side
(c) equal on both sides
(d) equal to surface tension divided by radius of curvature on both sides

12. A rectangular tank of base size 3 m × 3 m contains oil of specific gravity 0.8 upto a height of 8 m. When it is accelerated at 2.45 m/s² vertically upwards, the force on the base of the tank will be :

(a) 29400 N (b) 38240 N
(c) 78400 N (d) 49050 N

13. Match List-I with List-II and select the correct answer using the codes given below the lists :

List-I (Device)	List-II (Use)
A. Barometer	1. Gauge pressure
B. Hydrometer	2. Local atmospheric pressure
C. U-tube manometer	3. Relative density
D. Bourdon gauge	4. Pressure differential

Codes :

	A	B	C	D
(a)	2	3	1	4
(b)	3	2	1	4
(c)	3	2	4	1
(d)	2	3	4	1

14. The continuity equation for 3-dimensional flow $\frac{\partial u}{\partial x} + \frac{\partial v}{\partial y} + \frac{\partial w}{\partial z} = 0$ is applicable to (symbols have usual meanings) :

(a) steady flow
(b) uniform flow
(c) ideal fluid flow
(d) ideal as well as viscous fluid flow

15. Bernoulli's equation $\frac{p}{\rho} + \frac{v^2}{2} + gh =$ constant, is applicable for :

(a) steady, frictionless and incompressible flow along a streamline
(b) uniform and frictionless flow along a streamline when ρ is a function p
(c) steady and frictionless flow along a stremline when ρ is a function of p
(d) steady, uniform and incomperssible flow along a streamline.

16. Two identical pipes of length 'L', diameter 'd' and friction factor 'f' are connected in parallel between two points For the same total volume flow rate with pipe of same diameter 'd' and same friction factor 'f' the single length of the pipe will be:

(a) $\frac{L}{2}$ (b) $\frac{L}{\sqrt{2}}$

(c) $\sqrt{2}L$ (d) $\frac{L}{4}$

17. Two venturimeters of different area ratios are connected at different locations of a pipeline to measure discharge. Similar manometers are used across the two venturimeters to register the head differences. The first venturimeter of area ratio 2 registers a head difference 'h' while the second venturimeter registers '$5h$'. The area ratio for the second venturimeter is :

(a) 3 (b) 4 (c) 5 (d) 6

18. The thickness of laminar boundary layer at a distance 'X' from the leading edge over a flat plate varies as :

(a) X (b) $X^{1/2}$
(c) $X^{1/5}$ (d) $X^{4/5}$

19. Consider the following coefficients :
(Re = Reynold number)

1. 1.328 $\text{Re}^{-(0.5)}$ for laminar flow
2. 0.72 $\text{Re}^{-(0.2)}$ for turbulent flow
3. 0.072 $\text{Re}^{-(0.2)}$ for turbulent flow
4. 1.028 $\text{Re}^{-(0.5)}$ for laminar flow.

The Coefficients for drag for a flat plate would include :

(a) 1 and 2 (b) 2 and 4
(c) 1 and 3 (d) 3 and 4.

20. Consider the following statements :

1. Gases are considered incompressible when Mach number is less than 0.2
2. A Newtonian fluid is incompressible and non-viscous
3. An ideal fluid has negligible surface tension

Which of these statements (s) is/are correct?

(a) 2 and 3 (b) 2 alone
(c) 1 alone (d) 1 and 3

21. The standard sea level atmospheric pressure is equivalent to :

(a) 10.2 m of fresh water of $\rho = 998$ kg/m^3
(b) 10.1 m of salt water of $\rho = 1025$ kg/m^3
(c) 25.5 m of kerosene of $\rho = 800$ kg/m^3
(d) 6.4m of carbon tetrachloride of $\rho = 1590$ kg/m^3

22. Which one of the following represents equilibrium of a static fluid? (Symbols have the usual meaning)

(a) $\dfrac{dp}{dz} = -\dfrac{\rho}{g}$ (b) $\dfrac{dp}{g} = -\dfrac{dz}{\rho}$

(c) $\rho\, dp = -\dfrac{dz}{g}$ (d) $\dfrac{dp}{\rho} = -g dz$

23. A tank with four equal vertical faces of width/ and depth h is filled up with a liquid. If the force on any vertical side is equal to the force at the bottom, then value of $\dfrac{h}{l}$ will be :

(a) 2 (b) $\sqrt{2}$ (c) 1 (d) $\dfrac{1}{2}$

24. Match List-I (Physical properties of fluid) with List-II (Dimension/Definitions) and select the correct answer using codes given below the Lists :

List-I	List-II
A. Absolute viscosity	1. $\dfrac{du}{dy}$ is constant
B. Kinematic viscosity	2. Newton per metre
C. Newtonian fluid	3. Poise
D. Surface tension	4. $\dfrac{\text{Stress}}{\text{Strain}}$ is constant
	5. Stokes

Codes :

	A	B	C	D
(a)	5	3	1	2
(b)	3	5	2	4
(c)	5	3	4	2
(d)	3	5	1	2

25. A float valve of the 'ball-cock' type is required to close an opening of a supply pipe feeding a cistern as shown in the given Fig. 1

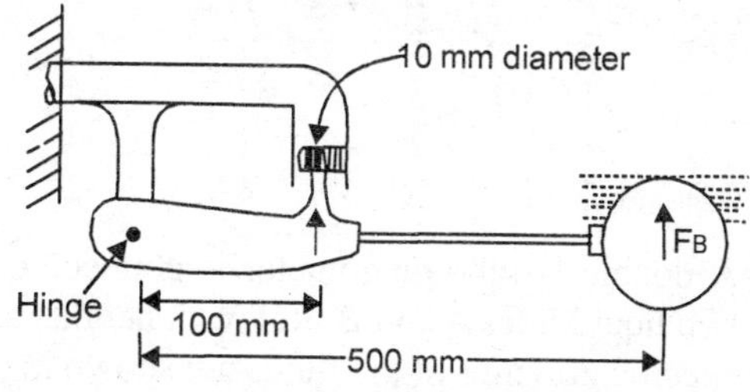

Fig. 1

The buoyant force F_B required to be exerted by the float to keep the valve closed against a pressure of 0. 28 N/mm² is :

(a) 4.4 N (b) 5.6 N
(c) 7.5 N (d) 9.2 N

26. Which one of the following equations represents the continuity equation for steady compressible fluid flow ?

(a) $\nabla.\rho\bar{V} + \frac{\partial \rho}{\partial t} = 0$

(b) $\nabla.\rho\bar{V} + \frac{\partial \rho}{\partial t} = 0$

(c) $\nabla.\bar{V} = 0$

(d) $\nabla.\rho\bar{V} = 0$

27. A liquid of specific gravity 0.82 flows with a velocity 4.4 m/s. Its velocity head is :

(a) 0.82 m (b) 4.43 m (c) 1 m (d) 19.6 m

28. Match List-I with List-II and select the correct answer using the codes given below the Lists :

List-I	List-II
A. Orifice meter	1. Measurement of flow in a channel
B. Broad creasted weir	2. Measurement of velocity in a pipe/ channel
C. Pitol tube	3. Measurement of flow in a pipe of any inclination
D. Rotameter	4. Measurement of upward flow in a vertical pipe

Codes :

	A	B	C	D
(a)	3	1	4	2
(b)	1	3	2	4
(c)	3	1	2	4
(d)	1	3	4	2

29. A double U-tube manometer is connected to two liquid lines *A* and B. Relevant heights and specific gravities of the fluids are shown in the given Fig. 2. The pressure difference, in head of water, between fluids at *A* and *B* is :

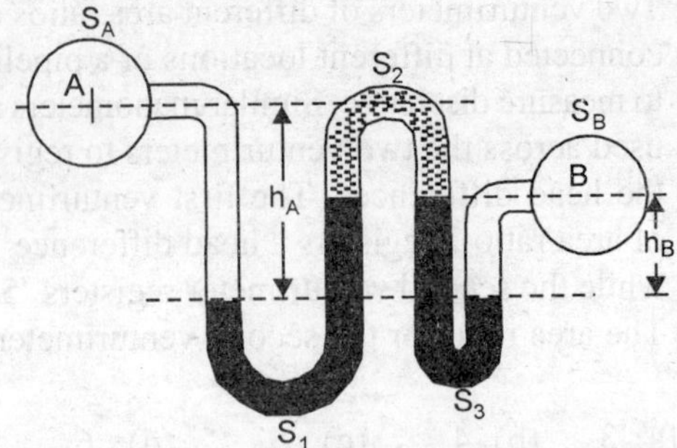

Fig. 2

(a) $S_A h_A + S_1 h_B - S_3 h_B + S_B h_B$

(b) $S_A h_A - S_1 h_B - S_2(h_A - h_B) + S_3 h_B - S_B h_B$

(c) $S_A h_A + S_1 h_B + S_2(h_A - h_B) - S_3 h_B + S_B h_B$

(d) $S_A h_A - S_1 h_B + S_3 h_B - S_B h_B$

30. Consider the following statements :

(a) Streak line indicates instantaneous position of particles of fluid passing through a point.

(b) Streamlines are paths traced by a fluid particle with constant velocity.

(c) Fluid particles cannot cross streamlines irrespective of the type of flow.

(d) Streamlines converge as the fluid is accelerated, and diverge when retarded.

Which of these statements are correct?

(a) 1 and 4 (b) 1, 3 and 4
(c) 1, 2, and 4 (d) 2 and 3

31. If the diameter of a capillary tube is doubled, the capillary-rise will become :

(a) $\sqrt{2}$ times less

(b) double

(c) half

(d) $\sqrt{2}$ times more

32. A stepped pipeline with four different cross-sections discharges water at the rate of 2 litres per second. Match List-I (Areas of pipe is sq. cm) with List-II (Velocities of water in cm/s) and select the correct answer using the codes given below the Lists :

List-I	List-II
A. 500	1. 4
B. 100	2. 5
C. 400	3. 10
D. 200	4. 15
	5. 20

Codes :

	A	B	C	D
(a)	5	1	2	3
(b)	1	4	2	3
(c)	1	4	3	4
(d)	3	2	5	1

33. Fluid flow rate Q, can be measured easily with the help of a venturi tube, in which the difference of two pressures, Δp, measured at an upstream point and at the smallest cross-section of the tube, is used. If a relation $\Delta p \propto Q^n$ exists, then n is equal to :

(a) $\frac{1}{3}$ (b) $\frac{1}{2}$

(c) 1 (d) 2

34. The drag force D on a certain object in a certain flow is a function of the coefficient of viscosity μ, the flow speed, V and the body dimension L (for geometrically similar objects); then D is proportional to :

(a) $L \mu V$ (b) $\frac{\mu^2 V^2}{L^2}$

(c) $\mu^2 V^2 L^2$ (d) $\frac{\mu L}{V}$

35. A metallic piece weighs 80 N in air and 60 N in water. The relative density of the metallic piece is about :

(a) 8 (b) 6

(c) 4 (d) 2

36. Match List-I (Nature of equilibrium of floating body) with List-II (Conditions for equilibrium) and select the correct answer using the codes given below the lists :

List-I (Nature of equilibrium of floating body)	List-II (Conditions for equilibrium)
A. Unstable equilibrium	1. $\overline{MG} = 0$
B. Netutral equilibrium	2. M is above G
C. Stable equilibrium	M is below G
	4. $\overline{BG} = 0$

(where M, G and B are metacentre, centre of gravity and centre of buoyancy respectively.

Codes :

	A	B	C
(a)	1	3	2
(b)	3	1	2
(c)	1	3	4
(d)	4	2	3

37. The velocity potential function in a two dimensional flow fluid is given by $\phi = x^2 - y^2$. The magnitude of velocity at the point (1, 1) is :

(a) 2 (b) 4

(c) $2\sqrt{2}$ (d) $4\sqrt{2}$

38. The relations between shear stress (τ) and velocity gradient for ideal fluids, Newtonian fluids and non-Newtonian fluids are given below. Select the correct combination :

(a) $\tau = 0; \tau = \mu\left(\frac{du}{dy}\right)^2; \tau = \mu\left(\frac{du}{dy}\right)^3$

(b) $\tau = 0; \tau = \mu\left(\frac{du}{dy}\right)^2; \tau = \mu\left(\frac{du}{dy}\right)^2$

(c) $\tau = \mu\left(\frac{du}{dy}\right); \tau = \mu\left(\frac{du}{dy}\right)^2; \tau = \mu\left(\frac{du}{dy}\right)^3$

(d) $\tau = \mu\left(\frac{du}{dy}\right); \tau = \mu\left(\frac{du}{dy}\right)^2; \tau = 0$

39. Consider the following statements :

1. For stream function to exist, the flow should be irrotational.

2. Potential functions are possible even though continuity is not satisfied.
3. Stremlines diverge where the flow is accelerated.
4. Bernoulli's equation will be satisfied for flow across a cross-section.

Which of the above statements is/are correct?

(a) 1, 2, 3 and 4
(b) 1, 3 and 4
(c) 3 and 4
(d) 2 only

40. Match List-I (Device) with List-II (Use) and select the correct answer using the codes given below the lists :

List-I (Device)	List-II (Use)
A. Pitot tube	1. Boundary shear stress
B. Preston tube	2. Turbulent veloicty fluctuations
C. Flow Nozzle	3. The total head
D. Hot wire anemo-meter	4. Flow rate

Codes :

	A	B	C	D
(a)	4	2	3	1
(b)	3	1	4	2
(c)	4	1	3	2
(d)	3	2	4	1

41. Flow separation is caused by :

(a) thinning of boundary layer thickness to zero
(b) a negative pressure gradient
(c) a positive pressure gradient
(d) reduction of pressure to local vapour pressure

42. A laminar flow is taking place in a pipe. Match List-I (Term) with List-II (Expression) and select the correct answer using the codes given below the lists :

List-I (Term)	List-II (Expression)
A. Discharge, Q	1. $\dfrac{16\mu}{\rho VD}$
B. Pressure drop, $\dfrac{\Delta P}{L}$	2. $\dfrac{\pi d^2 \Delta p}{128\mu L}$
C. Friction factor, f	3. $\dfrac{32\mu V}{D^2}$
	4. $\dfrac{\pi d^4 \Delta p}{128\mu L}$

Codes :

	A	B	C
(a)	2	3	4
(b)	4	3	1
(c)	4	1	3
(d)	1	4	2

43. The Magnus effect is defined as :

(a) the generation of lift per unit drage force
(b) the circulation induced in an aircraft wing
(c) the separation of boundary layer near the trailing edge of a slender body
(d) the generation of lift on a rotating cylinder in a uniform flow.

44. A $\dfrac{1}{25}$ model of a ship is to be tested for estimating the wave drag. If the speed of the ship is 1 m/s, then the speed at which the model must be tested is :

(a) 0.04 m/s (b) 0.2 m/s
(c) 5.0 m/s (d) 25.0 m/s

45. Consider the following statements for a two-dimensional potential flow :

1. Laplace equation for stream function must be satisfied.
2. Laplace equation for velocity potential must be satisfied.
3. Stremlines and equipotential lines are mutually perpendicular.
4. Streamlines can intersect each other in very high speed flows.

Which of the above statements are correct?

(a) 1 and 4 (b) 2 and 4

(c) 1, 2 and 3 (d) 2, 3 and 4

46.

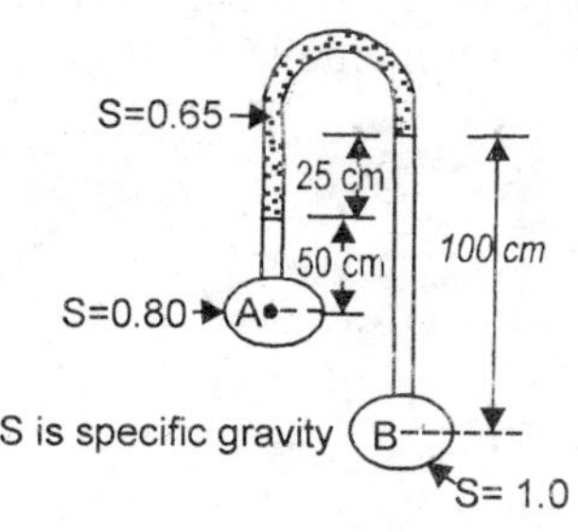

Fig. 3

The pressure difference between point B and A (as shown in the above Fig. 3) in centimetres of water is :

(a) – 44 (b) 44

(c) – 76 (d) 76

47. A solid P floats with half of its volume immersed in water and solid Q floats with two-thirds of the volume immersed in water. The densities of solider P and Q are in ratio :

(a) 1 : 2 (b) 1:3

(c) 2:3 (d) 3:4

48. The Bernoulli's equation refers to conservation of :

(a) mass (b) linear momentum

(c) angular momentum(d) energy

49. The continuity equation in a differential form is :

(a) $\frac{dA}{A} + \frac{dV}{V} + \frac{d\rho}{\rho} = \text{constant}$

(b) $\frac{d}{dA} + \frac{V}{dV} + \frac{\rho}{d\rho} = \text{constant}$

(c) $\frac{dA}{A} + \frac{dV}{V} + \frac{d\rho}{\rho} = 0$

(d) $A\,dA + V\,dV + \rho\,d\rho = 0$

50. Velocity defect in boundary layer theory is defined as :

(a) the error in the measurement of velocity at any point in the boundary layer

(b) the difference between the velocity at a point within the boundary layer and the free stream velocity

(c) the difference between the velocity at any point within the boundary layer and the velocity nearer the boundary

(d) the ratio between the velocity at a point in the boundary layer and the free stream velocity.

51. The drage coefficient for laminar flow varies with Reynolds number (Re) as:

(a) $Re^{1/2}$ (b) Re

(c) Re^{-1} (d) $Re^{-1/2}$

52. Match List-I (Basic Ideal Flow) with List-II (Example) and select the correct answer using the codes given below the lists :

List-I (Basic Ideal Flow)	**List-II (Example)**
A. Superposition of a uniform flow over a doublet	1. Flow over a half Rankine body
B. Superposition of a uniform flow over a source and sink	2. Flow over a Rankine oval
C. Superposition of a uniform flow over a source	3. Flow over a rotating body
D. Superposition of a free vortex flow along with a uniform flow over a doublet	4. Flow over a stationary body

Codes :

	A	B	C	D
(a)	4	1	2	3
(b)	3	2	1	4
(c)	4	2	1	3
(d)	3	1	2	4

53. The square root of the ratio of inertia force to gravity force is called :

(a) Reynolds number (b) Froude number

(c) Mach number (d) Euler number

54. The thickness of turbulent boundary layer at a distance x from the leading edge over a flat plate varies as:

(a) $x^{4/5}$ (b) $x^{1/2}$ (c) $x^{1/5}$ (d) $x^{3/5}$

55. For two-dimensionaly fluid element in x-y plane the rotational component is given by:

(a) $\omega_z = \frac{1}{2}\left(\frac{\partial u}{\partial x} - \frac{\partial v}{\partial y}\right)$

(b) $\omega_z = \frac{1}{2}\left(\frac{\partial u}{\partial x} + \frac{\partial v}{\partial y}\right)$

(c) $\omega_z = \frac{1}{2}\left(\frac{\partial v}{\partial x} + \frac{\partial u}{\partial y}\right)$

(d) $\omega_z = \frac{1}{2}\left(\frac{\partial v}{\partial x} - \frac{\partial u}{\partial y}\right)$

56. Which of the following relations must hold for an irrotational two-dimensional flow in the $x - y$ plane.

(a) $\frac{\partial v}{\partial y} - \frac{\partial u}{\partial x} = 0$ (b) $\frac{\partial u}{\partial z} - \frac{\partial v}{\partial x} = 0$

(c) $\frac{\partial w}{\partial y} - \frac{\partial v}{\partial z} = 0$ (d) $\frac{\partial v}{\partial x} - \frac{\partial u}{\partial y} = 0$

57. The coefficient of friction f in terms of shear stress 'τ_0' is given by:

(a) $f = \frac{\rho v^2}{2\tau_0}$ (b) $f = \frac{\tau_0}{\rho v^2}$

(c) $f = \frac{2\tau_0}{\rho v^2}$ (d) $\frac{2\rho v_0}{\tau_0}$

58. Velocity distribution in a turbulent boundary layer follows:

(a) Logarithmic law (b) Parabolic law
(c) Linear law (d) Cubic law

59. What is the discharge for laminar flow through a pipe of diameter 40 mm having centre line velocity of 1.5 m/s?

(a) $\frac{3\pi}{59}$ m^3/s (b) $\frac{3\pi}{2,500}$ m^3/s

(c) $\frac{3\pi}{5,000}$ m^3/s (d) $\frac{3\pi}{10,000}$ m^3/s

60. Which one of the following is the expression for momentum thickness θ of a boundary layer?

(a) $\theta = \int_0^\delta \left[1 - \frac{U}{U_0}\right]^2 dy$

(b) $\theta = \int_0^\delta \left[1 - \frac{U}{U_0}\right]^2 dy$

(c) $\theta = \int_0^\delta \frac{U}{U_0}\left[1 - \frac{U}{U_0}\right] dy$

(d) $\theta = \int_0^\delta \frac{U}{U_0}\left[1 - \left(\frac{U}{U_0}\right)^2\right] dy$

61. The displacement thickness at a section, for an air stream (ρ = 1.2 kg/m^3) moving with a velocity of 10 m/s over a flat plate is 0.5 mm. What is the loss of mass rate of flow of air due to boundary layer formation in kg per metre width of plate per second?

(a) 6×10^{-3} (b) 6×10^{-5}
(c) 3×10^{-3} (d) 2×10^{-3}

62. In a laminar boundary layer over a flat plate, what would be the ratio of wall shear stresses τ_1 and τ_2 at the two sections which lie at distances x_1 = 30 cm and x_2 = 90 cm from the leading edge of the plate?

(a) $\frac{\tau_1}{\tau_2} = 3.0$ (b) $\frac{\tau_1}{\tau_2} = \frac{1}{3}$

(c) $\frac{\tau_1}{\tau_2} = (3.0)^{1/2}$ (d) $\frac{\tau_1}{\tau_2} = (3.0)^{1/3}$

63. Which of the following is caused by the occurrence of a normal shock in the diverging section of a convergent-divergent nozzle?

1. Velocity jump 2. Pressure jump
3. Velocity drop 4. Pressure drop.

Select the correct answer using the codes given below:

(a) 1 only (b) 1 and 2
(c) 2 and 3 (d) 1 and 4

64. A circular plate of 1.5 m diameter is submerged in water with its greatest and least depths below the water surface being 2 m and 0.75 m respectively. What is the approximate magnitude of the total thrust on one face of the plate?
(a) 24 kN (b) 28 kN
(c) 12 kN (d) 16 kN

65. Which one of the following statements is correct?
A steady flow of diverging straight stream lines:
(a) is a uiform flow with local acceleration
(b) has convecfive normal acceleration
(c) has convective tangential acceleration
(d) has both convective normal and tangential accelerations.

66. Which one of the following is the expression of the rotational component for a two-dimensional fluid element in x–y plane?

(a) $\omega_z = \frac{1}{2}\left(\frac{\partial v}{\partial x} - \frac{\partial u}{dy}\right)$

(b) $\omega_z = \frac{1}{2}\left(\frac{\partial v}{\partial x} + \frac{\partial u}{dy}\right)$

(c) $\omega_z = \frac{1}{2}\left(\frac{\partial v}{\partial x} - \frac{\partial v}{dy}\right)$

(d) $\omega_z = \frac{1}{2}\left(\frac{\partial u}{\partial x} + \frac{\partial v}{dy}\right)$

67. What is the percentage error in the estimation of the discharge due to an eıror of 2% in the measurement of the reading of a differential manometer connected to an orifice meter?
(a) 4 (b) 3 (c) 2 (d) 1

68. Which one of the following is the characteristics of a fully developed laminar flow :
(a) The pressure drop in the flow direction is zero
(b) The velocity profile changes uniformly in the flow direction
(c) The velocity profile does not change in the flow direction
(d) The Reynolds number for the flow is critical.

69. Which one of the following is the correct relationship between the boundary layer thickness δ, displacement thickness δ* and the momentum thickness θ?
(a) $\delta < \delta^* > \theta$ (b) $\delta^* > \theta > \delta$
(c) $\theta > \delta > \delta^*$ (d) $\theta > \delta^* > \delta$

70. The thickness of turbulent boundary layer at a distance x from the leading edge on a flat plate varies as:
(a) $x^{4/5}$ (b) $x^{3/5}$
(c) $x^{1/2}$ (d) $x^{1/5}$

71. When a falt plate of 0.1 m^2 area is pulled at a constant velocity of 30 cm/s parallel to another stationary plate pocaled at a distance 0.01 cm from it and the space in between filled with a fluid of dynamic viscosity = 0.01 Ns/m^2, the force required to be applied is :
(a) 0.3 N (b) 3 N
(c) 10 N (d) 16 N

72. Which one of the following is the continuity equation in differential form? (The symbols have usual meanings)

(a) $\frac{dA}{A} + \frac{dV}{V} + \frac{d\rho}{\rho} = \text{constant}$

(b) $\frac{dA}{A} + \frac{dV}{V} + \frac{d\rho}{p} = 0$

(c) $\frac{A}{dA} = \frac{V}{dV} + \frac{\rho}{d\rho} = \text{constant}$

(d) $AdA + VdA + \rho d\rho = 0$

73. What will be the maximum efficiency of the pipeline if one-third of the available head in flow through the pipeline is consumed by friction?
(a) 33.33% (b) 50.00%
(c) 66.66% (d) 75.00%

74. Which one of the following is satisfied if the flow is irrotational for a two-dimensional fluid element in the x–y plane?

(a) $\frac{\partial v}{\partial x} = \frac{\partial u}{\partial y}$ (b) $\frac{\partial v}{\partial x} = -\frac{\partial u}{\partial y}$

(c) $\frac{\partial u}{\partial x} = \frac{\partial v}{\partial y}$ (d) $\frac{\partial u}{\partial x} = -\frac{\partial v}{\partial y}$

75. The velocity distributions in laminar boundary layer is given by the relation $u/u_\infty = y/\delta$. What is the momentum thickness for the boundary layer?
(a) $\delta/2$ (b) $\delta/3$
(c) $\delta/4$ (d) $\delta/6$

76. How is the Von-Karman momentum integral equation expressed (θ is momentum thickness)?

(a) $\frac{\tau_0}{\frac{1}{2}\rho U_\infty^2} = \frac{\partial \theta}{\partial x}$ (b) $\frac{\tau_0}{2\rho U_\infty^2} = \frac{\partial \theta}{\partial x}$

(c) $\frac{\tau_0}{\rho U_\infty^2} = \frac{\partial \theta}{\partial x}$ (d) $\frac{\tau_0}{\frac{1}{3}\rho U_\infty^2} = \frac{\partial \theta}{\partial x}$

77. Air (kinematic viscosity = 15×10^{-6} m^2/s) with a free stream velocity of 10 m/s flows over a smooth two-dimensional flat palte. If the critical Reynolds number is 5×10^5, what is the maximum distance from the leading edge up to which laminar boundry layer exists?
(a) 30 cm (b) 75 cm
(c) 150 cm (d) 300 cm

78. Using the Prandtl's mixing length concept, how is the turbulent shear stress expressed?

(a) $\rho l \frac{\overline{du}}{dy}$ (b) $\rho l^2 \frac{\overline{du}}{dy}$

(c) $\rho l \left(\frac{\overline{du}}{dy}\right)^2$ (d) $\rho l^2 \left(\frac{\overline{du}}{dy}\right)^2$

79. Consider the following statements :

The state of stress in a fluid consists of normal pressure only if the fluid:
1. is at rest
2. is in uniform motion
3. has non-uniform velcoity profile
4. has zero viscosity.

Which of the statements given above are correct?
(a) 1, 2 and 3 (b) 1, 2 and 4
(c) 1, 3 and 4 (d) 2, 3 and 4

80. Darcy-Weisback equation for the head loss in a flow through a pipe is given by $h_l = 4\,f l v^2$ (2gd) (the symbols have the usual meaning). For the laminar flow through a circular pipe, how does the friction factor f vary with the Reynolds number (Re):
(a) f = 8/Re (b) f = 16/Re
(c) f = 32/Re (d) f = 64/Re

81. The velocity distribution in laminar boundary layer is given by the relation $u/u_x = y/\delta$.

What is the displacement thickness for the boundary layer?
(a) δ (b) $\delta/2$
(c) $\delta/3$ (d) $\delta/4$

82. Which one of the following equations gives the velocity distribution across a circular pipe having a viscous flow?
(a) $u = u_{max}\,[1-(r/R)^2]$
(b) $u = u_{max}\,[R^2 - r^2]$
(c) $u = u_{max}\,[1-(r/R)]^2$
(d) $u = u_{max}\,[1+(r/R)]^2$

83. In a two-dimensional fluid flow, $u = 6x + xy$. Which one of the following gives the component of the velocity to satisfy the continuity equation?
(a) $6x + xy$ (b) $6 + xy$
(c) $-(6y + xy)$ (d) $-\left(6y + \frac{1}{2}y^2\right)$

84. If $u = ax$ and $v = -ay$ give the velocity distribution for a two-dimensional flow, what is the equation of a stream line passing through the point (3, 1)?
(a) $xy = 3$ (b) $x + y = 4$
(c) $x + 3y = 6$ (d) $x^2y = 9$

85. Consider the following statements :
1. Boundary-layer thickness in laminar flow is greater than that of turbulent flow.
2. Boundary-layer thickness of turbulent flow is greater than that of laminar flow.

3. Velocity distributes uniformly in a turbulent boundary-layer.
4. Velocity has a gradul variation in a laminar boundary layer.

Which of the statements given above are correct?

(a) 1, 3 and 4 only (b) 1, 2, 3 and 4
(c) 1 and 2 only (d) 2 3 and 4, only

87. Which one of the following is the correct statement?

The velocity profiles for fully developed laminar and turbulent flow, respectively, in a pipe are :

(a) parabolic and parabolic
(b) parabolic and elliptic
(c) linear and 1/7 power law
(d) parabolic and 1/7 power law

88. Which one of the following is the correct statement?

For the case of laminar flow between two fixed parallel plates, the shear stress is :

(a) constant across the passage
(b) maximum at the centre and zero at the bouundary
(c) zero all through the passage
(d) maximum at the boundary and zero at the centre

89. Consider the following statements :

The coefficient of discharge C_d of a venturimeter takes into account

1. the effect of roughness of the surface
2. non-uniform velocity distributions at inlet and throat section.
3. Reynolds number of flow
4. discharge
5. length of throat
6. diameter of throat
7. diameter ratio.

Which of the statements given above are correct?

(a) 1, 2, 4 and 5 (b) 1, 4, 5 and 6
(c) 1, 2, 3 and 7 (d) 2, 6 and 7

90. Which one of the following expresses the error in discharge due to error in the measurement of head over a triangular notch?

(a) $\frac{dQ}{Q} = \frac{5dH}{2H}$ (b) $\frac{dQ}{Q} = \frac{3dH}{2H}$

(c) $\frac{dQ}{Q} = \frac{7dH}{2H}$ (d) $\frac{dQ}{Q} = \frac{1dH}{2H}$

91. Which one of the following is the correct statement?

A differential manometer connected to a Pitot-static tube used for measuring fluid velocity gives :

(a) static pressure
(b) total pressure
(c) dynamic pressure
(d) difference between total pressure and dynamic pressure.

92. Which one of the following correctly represents the shear stress distribution across a section of a circular pipe having a viscous flow?

(a) $\tau = \frac{\partial pr^2}{\partial x}$

(b) $\tau = \frac{\partial p(r/2)}{\partial x}$

(c) $\tau = -\frac{\partial p(r/2)}{\partial x}$

(d) $\tau = \frac{\partial p(r)}{\partial x}$

93. A pipe of diameter (D) conveying a discharge Q is to be replaced by parallel pipes of smaller diameter (d) to discharge the same quantity. What will be the ratio of $\frac{D}{d}$? (f is same for all pipes)

(a) $\frac{D}{d} = 2$ (b) $\frac{D}{d} = \sqrt{2}$

(c) $\frac{D}{d} = 4^{1/5}$ (d) $\frac{D}{d} = 4^{1/3}$

94. Match List-I (Forms of Bernoulli's Equation) with List-II (Units of these forms) and select the correct answer using codes given below the lists:

List-I (Forms of Bernoulli's Equation)	List-II (Units of these forms)
A. $p+wz+\frac{\rho V^2}{2}$	1. total energy per unit volume
B. $p+wz+\frac{\rho V^2}{2}$	2. total energy per unit mass
C. $\frac{p}{w}+z+\frac{V^2}{2g}$	3. total energy per unit weight

Codes :

	A	B	C
(a)	1	2	3
(b)	1	3	2
(c)	2	1	3
(d)	2	3	1

95. Which one of the following is the correct statement?

A frictionaless, incompressible fluid flows steadily through a convergent nozzle, then its

(a) energy must decrease
(b) velocity must decrease
(c) pressure must decrease
(d) momentum must decrease

96. Which one of the following statements is correct? For stability of a floating body,

(a) M should lie between G and B (in that order)
(b) M should lie above B and G (in that order)
(c) M should lie below B and G (in that order)
(d) M should coincide with B and G

97. Consider the following statements :

A rectangular block of wood of size $L \times B \times H$ will float in water in such a way that:

(a) the longest dimension is vertical
(b) the longest dimension is horizontal
(c) the metacentre is above centre of gravity
(d) the centre of buoyancy is above the centre of gravity

Which of the statements given above is/are correct?

(a) 1 only (b) 2 only 3, only
(c) 2, 3 and 4 (d) 1, 3 and 4

98. Bernoulli's equation is derived by making which one of the following assumptions?

(a) The flow is steady only
(b) The flow is uniform and incompressible
(c) The flow is non-viscous, uniform and steady
(d) The flow is steady, non-viscous, incompressible and irrotational.

99. A skater weighing 1000 N skates at a speed of 20 m/s on ice maintained at 0°C The average skating area supporting the skater is 0.001 m^2 and the coefficient of friction between the skates and ice is 0.02. What will be the average thickness of a film of water existing at the interface between the skater and ice? (Take dynamic viscosity of water as 0.001 Ns/m^2)

(a) 10^{-5}m (b) 10^{-6} m
(c) 10^{-7} m
(d) Not possible to estimate since there cannot be a possibility of formation of a thin film of water at interface

100. Flow separation is likely to take place when the pressure gradient in the direction of flow is :

(a) zero
(b) adverse
(c) slightly favourable
(d) strongly favourable

101. Hydrodynamic and thermal boundary layer thickness are equal for Prandtl number:

(a) equal to zero (b) less than 1
(c) equal to 1 (d) more than 1

102. In a rotameter as the flow rate increases, the float :

(a) rotates at higher speed
(b) rotates at lower speed
(c) rises in the tube
(d) drops in the tube

103. A two-dimensional fluid flow is described by the velocity components $u = 5x^3$, $v = -15x^2y$. The stream function at point will be :

(a) 2 m^2/s (b) 2 m^3/s
(c) 10 m^3/s (b) 20 m^3/s

104. Match the following:

List-I	List-II
A. Circular sewer maximum discharge	1. y = 0.938 D
B. Maximum velocity in circular sewer	2. y = 0.81 D
C. Triangular Channel	3. $y_c = \frac{4}{5}E, \frac{v_c^2}{2g} = \frac{y_c}{4}$
D. Bourdon gauge	4. $y_c = \frac{2}{3}E, \frac{v_c^2}{2g} = \frac{y_c}{4}$

Codes :

	A	B	C	D
(a)	4	4	2	1
(b)	3	4	1	2
(c)	2	3	1	4
(d)	1	4	2	3

105. Aging of pipe implies :
(a) pipe becoming smoother with time
(b) relative roughness decreasing periodically
(c) increase in absolute roughness peroidically with time
(d) increase in absolute roughness linearly with time

106. A nozzle has velocity head at outlet of 10m. If it is kept vertical the height reached by the stream is :
(a) 100 m (b) 10 m
(c) $\sqrt{10}$ m (d) $\frac{1}{\sqrt{10}}$

107. Laminar sub-layer may develop during flow over a flat-plate. It exists in :
(a) laminar zone (b) transition zone
(c) turbulent zone
(d) laminar and transition zone

108. The velocity components for a two dimensional incompressible flow of a fluid are $u = x - 4y$ and $v = -y - 4x$. It can be concluded that :
(a) the flow does not satisfy the continuity equation
(b) the flow is rotational
(c) the flow is irrotational
(d) none of the above.

109. A $\frac{1}{25}$ model of a ship it to be tested for estimating the wave drag. If the speed of the prototype is 1.0 m/s, then the speed at which the model must be tested is :
(a) 0.04 m/s (b) 0.2 m/s
(c) 5.0 m/s (d) 25.0 m/s

110. In which of the following cases frictional drag is predominating?
1. Tennis ball 2. Parachute
3. Arrow 4. Cyclist
(a) 1 and 2 only (b) 2 and 3
(c) 2, 3 and 4 only (d) 1, 2 and 3 only

111. Match the following :

List-I	List-II
A. Wave drag of a ship	1. Cavitation in pumps and turbines
B. Pressure coefficient	2. ρr. L^3 r
C. Thoma numbers	3. Re ≈ 0.1
D. Stokes Law	4. $\frac{\Delta p}{\rho V^2/2}$

Codes :

	A	B	C	D
(a)	1	2	4	3
(b)	4	3	1	2
(c)	1	3	4	2
(d)	2	4	1	3

112. The terminal velocity of small sphere falling in a viscous fluid is :
(a) proportinal to the diameter of the sphere
(b) inversely proportional to the viscosity of the fluid
(c) inversely proportional to the diameter of the sphere
(d) proportional to the density of the fluid.

113. Which of the following statement is ture?
(a) For an ideal fluid μ = 0. ρ = constant, $K = 0$

(b) *A* floating body is in stable, unstable or neutral equilibrium according to as the metacentric height is zero, positive or negative.
(c) The exact analysis of viscous flow problems can be made by Euler's equation
(d) The most economical diameter of a pipe is the one for which the annual fixed cost and annual power cost (to overcome friction) are minimum.

114. Which of the following statements is false?
(a) The lift force (per unit length) on a body depends on the density of the fluid
(b) For laminar flow through a pipe, the loss of head is directly proportional to velocity as well as viscosity of fluid
(c) A hydraulic jump occurs when a super-critical flow comes across sub critical flow
(d) At the stall point for the airfoil, the lift is minimum and the drag decreases sharply beyond it.

115. Which of the following is not a dimensionless group?
(a) $\frac{\Delta p}{\rho N^2 D^2}$ (b) $\frac{gH}{N^2 D^2}$
(c) $\frac{\rho \omega D^2}{\mu}$ (d) $\frac{\Delta p}{\rho V^3}$

116. In the Fig. 4 below, Stoke's law is valid for region,
(a) AB (b) BC
(c) CD (d) BD

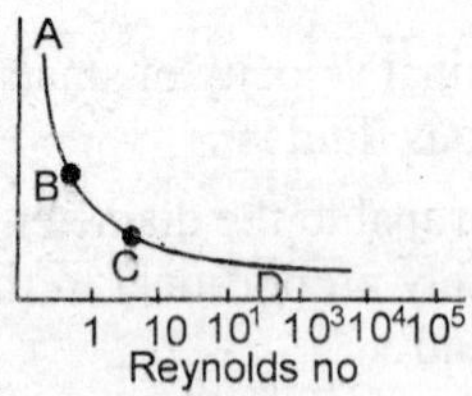

Fig. 4

117. For a real fluid moving with uniform velocity, the pressure :
(a) depends upon depth and orientation
(b) is independent of depth but depends upon orientation
(c) is independent of orientation but depends upon depth
(d) is independent of both depth and orientation

118. The vertical component of force on a curved surface submerged in a static liquid is equal to the :
(a) mass of the liquid above the curved surface
(b) weight of the liquid above curved surface
(c) product of pressure at *C.G.* multiplied by the area of the curved surface
(d) product of pressure at C.G. multiplied by the projected area of the curved surface

119. In the situation shown in the given Fig. 5 the length EC is 3 m and M is the mid-point of BC. The hydrostatic force on BC measured per unit width (width being perpendicular to the plane of the paper) with '*g*' being the acceleration due to gravity, will be :

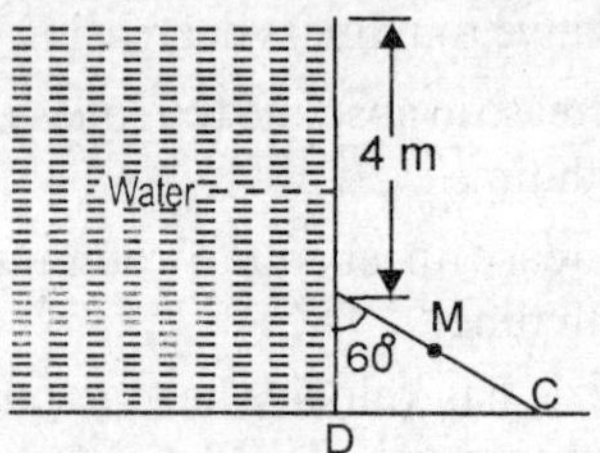

Fig. 5

(a) 16500 g N/m passing through M
(b) 16500 g N/m passing through a point between M and C
(c) 14250 g N/m passing through M
(d) 14250 g N/m passing through a point between M and C

120. Two pipe lines at different pressure, p_A and P_B, each carrying the same liquid of specific gravity S_1 are connected to a U-tube with a liquid of specific gravity S_2 resulting in the differences h_1, h_2 and h_3 as shown in the Fig. 6 The difference in pressure head between points *A* and *B* in terms of head of water is :

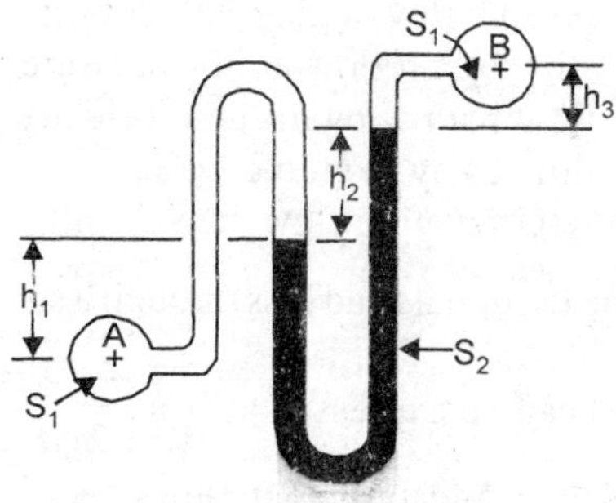

Fig. 6

(a) $h_1S_2 + h_2S_1 + h_3S_1$ (b) $h_1S_1 + h_2S_2 - h_3S_1$
(c) $h_1S_1 - h_2S_2 - h_3S_1$ (d) $h_1S_1 + h_2S_2 + h_3S_1$

121. If the governing equation for a flow field is given by $\nabla^2\phi = 0$ and the velocity is given by $\vec{V} = \nabla\phi$, the :

(a) $\nabla \times \vec{V} = 0$ (b) $\nabla \times \vec{V} = 1$

(c) $\nabla^2 \times \vec{V} = 1$

(d) $(\vec{V} \cdot \nabla)\vec{V} = \frac{\partial \vec{V}}{\partial t}$

122. Decrease in temperature, in general, results in :

(a) an increase in viscosities of both gases and liquids
(b) a decrease in the viscosity of liquids and gases
(c) an increase in the viscosity of liquids and an decrease in that of gases
(d) a decrease in the viscosity of liquids and an increase in that of gases

123. Which of the following equations are forms of continuity equations ?

($\vec{V}$ is the velocity and $\forall$ is volume)

1. $A_1\vec{V}_1 = A_2\vec{V}_2$

2. $\frac{du}{\partial x} + \frac{dv}{\partial y} = 0$

3. $\int_S \rho\vec{V}.dA + \frac{\partial}{\partial t}\int_\forall \rho d\forall = 0$

4. $\frac{1}{r}\frac{\partial(rv_r)}{\partial r} + \frac{\partial v_z}{\partial z} = 0$

Select the correct answer using the codes below :

Codes :

(a) 1, 2, 3 and 4 (b) 1, and 2
(c) 3 and 4 (d) 2, 3 and 4

124. Consider the following assumptions :

1. Steady flow
2. Inviscid flow
3. Flow along a stream line
4. Conservative force field

For an ideal fluid, which of the statements are correct?

(a) 1 and 2 (b) 1, 2 and 4
(c) 2, 3 and 4 (d) 1, 3 and 4

125. A pipe friction test shows that, over the range of speeds used for the test, the non-dimensional friction factor '*f*' varies inversely with Reynolds Number. From this, one can conclude that the :

(a) fluid must be compressible
(b) fluid must be ideal
(c) pipe must be smooth
(d) flow must be laminar

126. Chezy's formula is given by (*m, i, C* and *V* are, respectively, the hydraulic mean depth, slope of the channel, Chezy's constant and average velocity of flow) :

(a) $V = i\sqrt{mC}$ (b) $V = C\sqrt{im}$

(c) $V = m\sqrt{iC}$ (d) $V = \sqrt{miC}$

127. In rough turbulent flow in a pipe, the friction factor would depend upon :

(a) velocity of flow
(b) pipe diameter
(c) type of fluid flowing
(d) pipe condition and pipe diameter

128. The differential manometer connected to Pitot static tube used for measuring fluid velocity gives :

(a) static pressure
(b) total pressure
(c) dynamic pressure
(d) difference between total pressure and dynamic pressure

129. The speed of the air emerging from the blades of a running table fan is intended to be measured as a function of time. The point of measurement is very close to the blade and the time period of the speed fluctuation is four times the time taken by the blade to complete one revolution. The appropriate method of measurement would involve the use of :

(a) a Pitot tube
(b) a hot wire anemometer
(c) high speed photography
(d) a Schlieren system.

130. Match List-I with List-II and select the correct answer using the codes given below the lists:

List-I (Discharge measuring device)	List-II (Characteristic feature)
A. Rotameter	1. Vena contracta
B. Venturimeter	2. End contraction
C. Orificemeter	3. Tapering tube
D. Flow nozzle	4. Convergent-divergent
	5. Bell mouth entry

Codes :

	A	B	C	D
(a)	1	2	3	4
(b)	3	4	1	5
(c)	5	4	2	1
(d)	3	5	1	2

131. In turbulent flow over an impervious solid wall :

(a) viscous stress is zero at the wall
(b) viscous stress is of the same order of magnitude as the Reynold's stress
(c) the Reynold's stress is zero at the wall
(d) viscous stress is much smaller than Reynold's stress.

132. The 'velocity defect law' is so named because it governs a :

(a) reverse flow region near a wall
(b) slip-stream flow at low pressure
(c) flow with a logarithmic velocity profile a little away from the wall
(d) re-circulating flow near a wall

133. The frictional head loss through a straight pipe (h_f) can be expressed as $h_f = \dfrac{4flv^2}{2gd}$ for both laminar and turbulent flows. For a laminar flow, 'f' is given by (Re is the Reynolds Number based on pipe diameter)

(a) 24/Re (b) 32/Re
(c) 64/Re (d) 128/Re

134. The laminar boudary layer thickness in zero-pressure-gradient flow over a flat plate along the x-direction varies as (x is the distance from the leading edge) :

(a) $x^{-1/2}$ (b) $x^{1/7}$
(c) $x^{1/2}$ (d) x

135. In the region of the boundary layer nearest to the wall where vorticity is not equal to zero, the viscous forces are :

(a) of the same order of magnitude as the inertial forces
(b) more than inertial forces
(c) less than inertial forces
(d) negligible

136. Drag on cylinders and spheres decrease when the Reynolds number is in the region of 2 × 10 since :

(a) flow separation occurs due to transition to turbulence
(b) flow separation is delayed due to onset of turbulence
(c) flow separation is advanced due to transition to turbulence
(d) flow reattachment occurs

137. Match the common observations in List-I with the explanations in List-II and select the correct answer using the codes given below the lists :

List-I	List-II
A. Singing of telephone wires	1. Vortex flow
B. Velocity profile in a pipe is initially parabolic and then flattens	2. Drag
C. Formation of cyclones	3. Vortex shedding
D. Shape of rotameter tube	4. Turbulence
	5. Compressibility

Codes :

	A	B	C	D
(a)	5	2	1	4
(b)	3	4	5	2
(c)	3	4	1	2
(d)	5	2	1	4

138. In turbomachinery, the relevant parameters are volume flow rate, density, viscosity bulk modulus, pressure difference, power consumption, rotational speed and a characteristic dimension. According to Buckingham *pi* (π) theorem, the number of independent non-dimensional groups for this case is :
(a) 3 (b) 4 (c) 5 (d) 6

139. List-I gives 4 dimensionless number and List-II given the types of force which are one of the constitutents describing the number. Match List-I with List-II and select the correct answer using the codes given below the lists :

List-I	List-II
A. Euler number	1. Pressure force
B. Froude number	2. Gravity force
C. Mach number	3. Viscous force
D. Webber number	4. Surface tension
	5. Elastic force

Codes :

	A	B	C	D
(a)	2	3	4	5
(b)	3	2	4	5
(c)	2	1	3	4
(d)	1	2	5	4

140. The realisation velocity potential on fluid flow indicates that the :
(a) flow must be irrotational
(b) circulation around any closed curve must have a finite value
(c) flow is rotational and satisfies the continuity equation
(d) vorticity must be non-zero.

141. Both free vortex and forced vortex can be expressed mathematically as function of tangential velocity V and the corresponding radius r. Free vortex and forced vortex are definable through V and r as :

	Free Vortex	Forced vortex
(a)	$V = r \times$ const.	$Vr =$ const.
(b)	$V \times r =$ const.	$V^2 = r \times$ const.
(c)	$V \times r =$ const.	$V = r \times$ const.
(d)	$V^2 \times r =$ const.	$V = r \times$ const.

142. The hydrostatic force on the curved surface AB shown in the given Fig. 7 acts

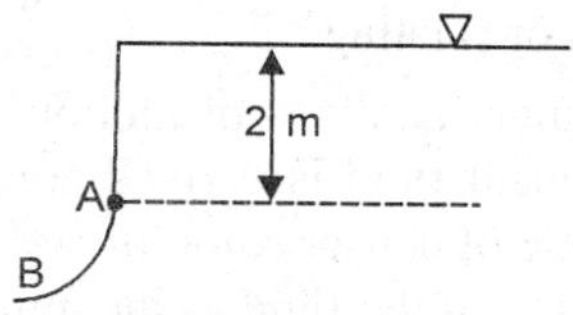

Fig. 7

(a) vertically downwards
(b) vertically upwards
(c) downawards, but at an angle with the vertical plane
(d) upward, but at an angle with the vertical plane

143. For stable equilibrium of a floating body,
(a) the body should be lighter in weight.
(b) the body should be made up of material whose specific gravity is less than that of the liquid in which it is floating.
(c) the metacentre should be above the centre of gravity of the body.
(d) the metacentre should be above the centre of buoyancy.

144. Consider the following statements :
For a two-dimensional potential flow.
1. Laplace equation for stream function must be satisfied.
2. Laplace equation for velocity potential must be satisfied.

3. Streamlines and equipotential lines are mutually perpendicular.
4. Stream function and potential function are not interchangeable

Of these statements :
(a) 1 and 4 are correct
(b) 2 and 4 are correct
(c) 1, 2 and 3 are correct
(d) 2, 3 and 4 are correct.

145. The pressure distribution for an ideal flow over a circular cylinder is such that :
(a) the maximum positive pressure intensity occurs at either of the stagnation points
(b) the maximum negative pressure is less than the maximum positive pressure
(c) the net force on the cylinder is in the direction of flow
(d) there is no lift whether the cylinder is at rest or rotating.

146. The Euler equations of motion for the flow of an ideal fluid is derived considering the principle of conservation of :
(a) mass and the fluid as incompressible and inviscid
(b) momentum and the fluid as incompressible and viscous
(c) momentum and the fluid as incompressible and inviscid
(d) energy and the fluid as incompressible and inviscid.

147. The continuity equation for a steady flow states that :
(a) velocity field is continuous at all points in flow field
(b) the velocity is tangential to the streamlines
(c) the stream function exists for steady flows
(d) the efflux rate of mass through the control surfaces is zero.

148. Bernoulli's equation represents the :
(a) forces at any point in the field and is obtained by integrating the momentum equation for viscous flows
(b) energies at any point in the flow field and is obtained by integrating the Euler equations
(c) momentum at any point in the flow field and is obtained by integrating the equation of continuity
(d) moment of momentum and is obtained by integrating the energy equation.

149. A fully developed laminar viscous flow through a circular tube has the ratio of maximum velocity to average velocity as :
(a) 3.0 (b) 2.5
(c) 2.0 (d) 1.5.

150. In a laminar flow through a pipe of diameter D, the total discharge Q, is expressed as (μ is the dynamic viscosity of the fluid and $\left(-\frac{dp}{dx}\right)$ is the pressure gradient :

(a) $-\frac{\pi D^4}{128\mu}\cdot\frac{dp}{dx}$ (b) $-\frac{\pi D^4}{64\mu}\cdot\frac{dp}{dx}$

(b) $-\frac{\pi D^4}{32\mu}\cdot\frac{dp}{dx}$ (d) $-\frac{\pi D^4}{16\mu}\cdot\left(\frac{dp}{dx}\right)$.

151. A trapezoidal open channel has the cross section as shown in the given Fig. 8 In order to have maximum hydraulic efficiency, the hydraulic radius, R and the length of the side, l should be :

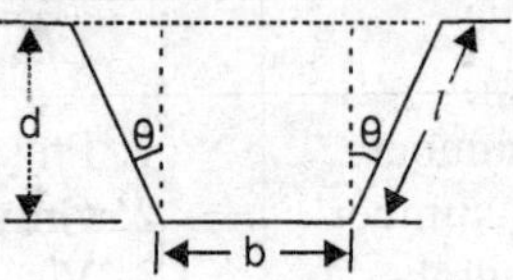

Fig. 8

(a) $\frac{d}{4}$ and $\frac{2}{\sqrt{3}}d$ respectively

(b) $\frac{d}{4}$ and $\frac{\sqrt{2}}{3}d$ respectively

(c) $\frac{d}{2}$ and $\frac{2}{\sqrt{3}}d$ respectively

(d) $\frac{d}{2}$ and $\frac{\sqrt{2}}{3}d$ respectively.

152. The universal velocity distribution for turbulent flow in a channel is given (u is the friction velocity and $\eta = u/y$. The kinematic viscosity is η and y is the distance from the wall.

(a) $\frac{u}{u^*} = 2.5 \text{ In } \eta + 5.5$

(b) $\frac{u}{u^*} = \eta$

(c) $\frac{u}{u_{max}} = 5.75 \ln \eta + 5.5$

(d) $\frac{u}{u^*} = 5.75 \ln \eta + 5.5$

153. Flow separation at a solid surface takes place due to :

(a) friction at the solid surface
(b) decrease in pressure along the flow direction
(c) increase in pressure along the flow direction
(d) positive pressure gradient along the flow direction and friction at the boundary.

154. Reynolds stresses may be defined as the :

(a) stresses (normal and tangential) due to viscosity of the fluid.
(b) additional normal stresses due to fluctuating velocity components in a turbulent flow
(c) additional shear stresses due to fluctuating velocity components in a turbulent flow
(d) additional normal and shear stresses due to fluctuating velocity components in the flow.

155. When we consider the momentum exchanges between two adjacent layers in a turbulent flow, can it be postulated that if at an instant there is an increase in u in the x direction it will be followed by a change in v in the y direction?.

(a) yes, in such a manner that $\overline{\vec{u}\vec{v}} = 0$
(b) yes, in such a manner that $\overline{\vec{u}\vec{v}}$ = non-zero and positive
(c) yes, in such a manner that $\overline{\vec{u}\vec{v}}$ = non-zero and negative
(d) no, as u and v are not dependent on each other.

156. According to Ptandtl's mixing length theory, the turbulent kinematic viscosity, v_t is expressed in terms of the mixing length, l and the velocity gradient, $\frac{\partial \vec{u}}{\partial y}$ as :

(a) $v_t = l\left(\frac{\partial \vec{u}}{\partial y}\right)$ (b) $v_t = l^2\left|\frac{\partial \vec{u}}{\partial y}\right|$

(c) $v_t = l^2\left[\frac{\partial \vec{u}}{\partial y}\right]$ (d) $v_t = l\left|\frac{\partial \vec{u}}{\partial y}\right|$

157. In the case of laminar boundary layer on a flat plate, the local skin friction coefficient is given by (symbols have the usual meaning) :

(a) $C_f = \frac{4.91x}{\sqrt{\text{Re}_x}}$ (b) $C_f = \frac{0.664}{\sqrt{\text{Re}_x}}$

(c) $C_f = \frac{1.328}{\sqrt{\text{Re}_x}}$ (d) $C_f = 0.332\sqrt{\text{Re}_x}$

158. The energy thickness for a laminar boundary layer flow depends on local and free stream velocities within and outside the boundary layer δ. The expression for the energy thickness is given by (symbols have the meaning).

(a) $\int_0^\delta \left(1 - \frac{u}{U_\infty}\right) dy$

(b) $\int_0^\delta \frac{u}{U_\infty}\left(1 - \frac{u}{U_\infty}\right) dy$

(c) $\int_0^\delta \left(1 - \frac{u}{U_\infty}\right)^2 dy$

(d) $\int_0^\delta \frac{u}{U_\infty}\left(1 - \frac{u^2}{U_\infty^2}\right) dy$.

159. For laminar flow over a flat plate, the thickness of the boundary layer at a distance from the leading edge is found to be 5 mm. Thickness of the boundary layer at a downstream section, which is at twice the distance of previous section from the leading edge will be :

(a) 10 mm (b) $5\sqrt{2}$ mm

(c) $5/\sqrt{2}$ mm (d) 2.5mm.

160. Consider the following two statements on the flow over immersed bodies :

Statement 1 : Drag can be created without lift

Statement 1 : Lift cannot be created without drag.

Of these statements:

(a) 1 is correct and 2 is incorrect
(b) 1 is incorrect and 2 correct
(c) both 1 and 2 are correct
(d) both 1 and 2 are incorrect.

161. In the design of tall chimney for a factory, in addition to strength and rigidity considerations care should be taken to check its safety against wind, because :

(a) the wind creates a large bending moment on the chimney
(b) the wind can have any direction
(c) additional forces are exerted on the chimney due to mixing of the effluent gases with the surrounding air
(d) of the effect of vortex shedding on the chimney.

162. In the case of falling sphere in a fluid for the Stoke's law to be valid, the :

(a) sphere must be metallic
(b) fluid density should be constant
(c) flow around the sphere should be turbulent
(d) the Reynolds number based on sphere diameter should be less than unity.

163. Match List-I with List-II and select the correct answer using the codes given below the Lists:

List-I (Forces)	List-II (Dimensionless groups)
A. Viscous force	1. Reynolds number
B. Elastic force	2. Froude number
C. Surface tension	3. Weber number
D. Gravity	4. Mach number.

Codes :

	A	B	C	D
(a)	1	4	2	3
(b)	1	2	4	3
(c)	3	4	1	2
(d)	1	4	3	2

164. If two flows are specified as :

(i) $u = y, v = \frac{3}{2}x$ and

(ii) $u = xy^2, v = x^2 y,$

then it can be concluded that :

(a) both the flows are rotational
(b) both the flows are irrotationl
(c) flow (*i*) is rotational while flow (*ii*) is irrotational
(d) flow (*i*) is irrotational while flow (*ii*) is rotational.

165. The streamlines and the lines of constant velocity potential in an inviscid rotational flow field form:

(a) parallel grid lines placed in accordance with their magnitude
(b) intersecting grid net with arbitrary orientation
(c) an orthogonal grid system
(d) none of the above.

166. The Magnus effect is defined as :

(a) the generation of lift per unit drag force
(b) the circulation induced in an aircraft wing
(c) the separation of boundary layer near the trailing edge of a slender body
(d) the generation of lift on a rotating cylinder in a uniform flow.

167. The transition Reynolds number for flow over a flat plate is 5×10^5. What is the distance from the leading edge at which transition will occur for flow of water with a uniform velocity of 1 m/s? [For water, the kinematic viscosity, v – 0.858×10^{-6} m^2 /s.].

(a) 1 m (b) 0.43
(c) 43 m (d) 103 m.

168. The depth of a fluid is measured in vertical z-direction; x and y are the other two directions and are mutually perpendicular. The static pressure variation in the fluid is given by (symbols have the usual meaning).
(a) $dp/dz = g$ (b) $dp/dz = \rho$
(c) $dp/dz = \rho g$ (d) $dp/dz = (-)\rho g$.

169. In a two-dimensional flow, the velocity components in x and y directions in terms of stream function (ψ) are :

(a) $u = \partial\psi/\partial x, v = \partial\psi/\partial y$

(b) $u = \partial\psi/\partial y, v = -\partial\psi/\partial x$

(c) $u = (-)\partial\psi/\partial y, v = \partial\psi/\partial x$

(d) $u = \partial\psi/\partial x, v = -\partial\psi/\partial y$

170. If H is the total head at inlet and h_1 is the head lost due to friction, efficiency of. power transmission through a straight pipe is given by :
(a) $(H - h_1)/H$ (b) $H/(H + h_1)$
(c) $(H - h_1)/(H + h_1)$ (d) $H/(H - h_1)$

171. A dimensionless group formed with the variables ρ (density), ω (angular velocity) μ (dynamic viscosity) and D (charcteristic diameter) is:
(a) $\rho\omega\mu/D^2$ (b) $\rho\omega D^2/\mu$
(c) $\rho\omega\mu D^2$ (d) $\rho\omega\mu D$

172. Flow over a half body is studied by utilising a free stream velocity of 5 m/s superimposed on a source at the origin. The body has a maximum width of 2m. The co-ordinates of the stagnation point are:
(a) $x = 0.32\ m, y = 0$
(b) $x = 0, y = 0$
(c) $x = (-)\ 0.32$ m, $y = 0$
(d) $x = 3m, y = 2m$

173. The following terms relate to floating bodies :

Centre of gravity ...G, Meta Centre ...M, Weight of floating body ...W, Buoyant force ...F_B

Match List-I with List-II and select the correct answer.

List-I (Condition)	List-II (Result)
A. G is above M	1. Stable equilibrium
B. G and M coincide	2. Unstable equilibrium
C. G is below M	3. Floating body
D. $F_B > W$	4. Neutral equilibrium.

Codes :

	A	B	C	D
(a)	1	3	2	4
(b)	3	1	4	2
(c)	2	3	4	1
(d)	2	4	I	3

174. For fully developed laminar flow through a pipe the volumetric flow is given by (symbols have the usual meaning) :

(a) $\frac{\pi}{8\mu}R^4\left(-\frac{dp}{dz}\right)$ (b) $\frac{\pi}{4\mu}\left(-\frac{dp}{dz}\right)$

(c) $\frac{\pi}{32\mu}R^4\left(-\frac{dp}{dz}\right)$

(d) $\frac{\pi}{16\mu}R^4\left(1-\frac{dp}{dz}\right)$

175. Prandtl Pitot tube was used to measure the velocity of a fluid of specific gravity S_1, The differential manometer, with a fluid of specific gravity S_2, connected to the Pitot tube recorded a level difference as h. The velocity V is given by the expression (Fig. 9) :

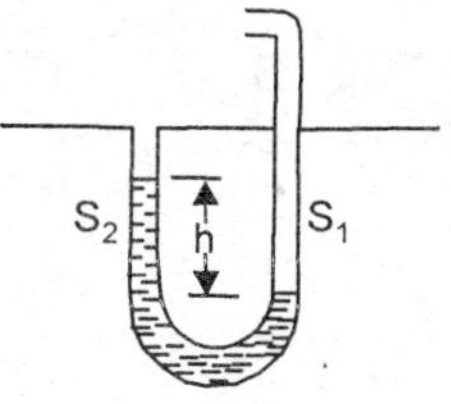

Fig. 9

(a) $\sqrt{2gh(S_1/S_2-1)}$ (b) $\sqrt{2ghS_2/S_1-1)}$

(c) $\sqrt{2gh(S_1-S_2)}$ (d) $\sqrt{2gh(S_2-S_1)}$

176. The expression $(p + \rho gz + \rho v^2/2)$ commonly used to express Bernoulli's equation, has units of :

(a) total energy per unit mass
(b) total energy per unit weight
(c) total energy per unit volume
(d) total energy per unit cross-sectional area of flow.

177. The frictional head loss in a turbulent flow through a pipe varies :

(a) directly as the average velocity
(b) directly as the square of the average velocity
(c) inversely as the square of the average velocity
(d) inversely as the square of the internal diameter of the pipe.

178. Match List-I with List-II and select the correct answers.

List-I (Properties of fluids)	List-II (Definition/ Results)
A. Ideal fluid	1. Viscosity does not changes with rate of deformation.
B. Newtonian fluid	2. Fluid of zero viscosity.
C. μ/p	3. Dynamic viscosity.
D. Mercury in glass	4. Capillary depression.
	5. Kinematic viscosity.
	6. Capillary rise.

Codes :

	A	B	C	D
(a)	1	2	4	6
(b)	1	2	3	4
(c)	2	1	3	6
(d)	2	1	5	4

179. List-I gives the different items related to a boundary layer while List-II gives the mathematical expression. Match List-I with List-II and select the correct answer using the codes given below the lists: (symbols have their usually meaning).

List-I	List-II
A. Boundary layer thickness	1. $y = \delta,\ u = 0.99\ U_x$
B. Displacement thickness	2. $\int_0^\delta \left(1 - \frac{u}{U_\infty}\right) dy$
C. Momentum thickness	3. $\int_0^\delta \frac{u}{U_\infty}\left(1 - \frac{u}{U_\infty}\right) dy$
D. Energy thickness	4. $\int_0^\delta \frac{u}{U_\infty}\left(1 - \frac{u^2}{U_\infty^2}\right) dy$

Codes :

	A	B	C	D
(a)	1	2	3	4
(b)	2	1	4	3
(c)	2	1	3	4
(d)	1	2	4	3

180. The differential form of continuity equation for two-dimensional flow of fluid may be written in the following form: $\rho\frac{\partial u}{\partial x} + \rho\frac{\partial v}{\partial y} = 0$ in which u and v are velocities in the x and y direction and ρ is the density. This is valid for :

(a) compressible, steady flow
(b) compressible, unsteady flow
(c) incompressible, unsteady flow
(d) incompressible, steady flow.

181. Water hammer in pipe lines takes place when :

(a) fluid is flowing with high velocity
(b) fluid is flowing with high pressure
(c) flowing fluid is suddenly brought to rest by closing a valve
(d) flowing fluid is brought to rest by gradually closing a valve.

182. A laminar boundary layer occurs over a flat plate at zero incidence of she flow. The thickness of boundary layer at a section 2 m from the leading edge is 2 mm The thickness of boundary layer at a section 4 m from the leading edge will be :

(a) $2 \times (2)^2$ mm (b) $2 \times (2)^{1/2}$ mm
(c) $2 \times (2)^{4/5}$ mn (d) $2 \times (2)^{1/5}$ mm

183. The model of a propeller, 3 m in diameter, cruising at 10 m/s in air, is tested in a wind tunnel on a 1 :10 scale model. If a thrust of 50 N is measured on the model at 5 m/s wind speed, then the thrust on the prototype will be :
(a) 20,000 N (b) 2,000 N
(c) 500 N (d) 200 N

184. For solid spheres falling vertically downwards under gravity in a viscous fluid, the terminal velocity, V_1 varies with diameter 'D' of the sphere as:
(a) $V_1 \propto D^{1/2}$ for all diameters
(b) $V_1 \propto D^2$ for all diameters
(c) $V_1 \propto D^{1/2}$ for large D and $V_1 \propto D^2$ for small D
(d) $V_1 \propto D^2$ for large D and $V_1 \propto D^{1/2}$ for small D.

185. In the region of the boundary layer nearest to the wall where velocity in not equal to zero, the viscous forces are :
(a) of the same order of magnitude as the inertial forces
(b) more than inertial forces
(c) less than inertial forces
(d) negligible.

186. Given that:

u = velocity in the x-direcrion

v = velocity in the y-direction.

A two-dimensional flow in x–y plane is irrotational if :

(a) $\frac{\partial u}{\partial x} = \frac{\partial v}{\partial y}$ (b) $\frac{\partial u}{\partial x} = \frac{\partial u}{\partial y}$

(c) $\frac{\partial v}{\partial x} = \frac{\partial u}{\partial y}$ (d) $\frac{\partial v}{\partial x} = \frac{\partial v}{\partial y}$

187. Of the possible irrotiational flow functions given below, the incorrect relation is (where ψ = stream function and ϕ = velocity potential) :

(a) $\psi = xy$ (b) $\psi = A(x^2 - y^2)$

(c) $\phi = ur \cos\theta + \frac{u}{r} \cos\theta$

(d) $\phi = \left(r - \frac{2}{r}\right) \sin\theta$

188. In which one of the following cases must separation of boundary layer occur?
(a) $dp/dx < 0$ (b) $dp/dx = 0$
(c) $dp/dx > 0$
(d) $dp/dx > 0$ and the velocity profile has a point of inflection.

189. If z is vertically upwards, ρ is the density and g gravitational acceleration (see Fig. 10) then the pressure $\partial p/\partial z$ in fluid at rest due to gravity is given by :

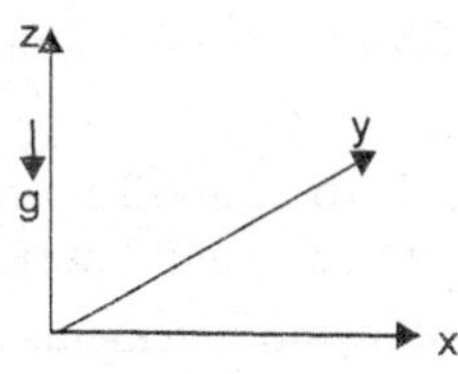

Fig. 10

(a) $\rho g z^2/2$ (b) $-\rho g$
(c) $-\rho g z$ (d) $\rho\, g/z$

190. A rectangular water tank, full to the brim, has its length, breadth and height in the ratio of 2 : 1 : 2. The ratio of hydrostatic forces at the bottom to that at any larger vertical surface is :
(a) 1/2 (b) 1 (c) 2 (d) 4

191. The manometer shown in the given Fig. 11 cannects two pipes, carrying oil and water: respectively. From the figure one :

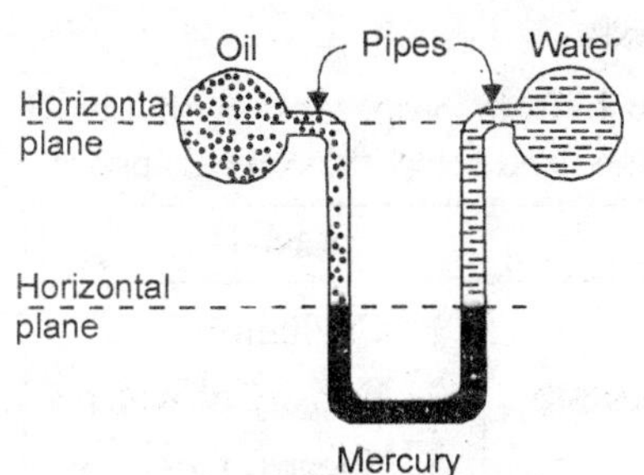

Fig. 11

(a) can conclude that the pressures in the pipes are equal.
(b) can conclude that the pressure in the oil pipe is higher.
(c) can conclude that the pressure in the water pipe is higher
(d) cannot compare the pressure in the two pipes for want of sufficient data.

192. Consider the following statements :

The metacentric height of a floating body depends
(a) directly on the shape of its water-line area.
(b) on the volume of liquid displaced by the body.
(c) on the distance between the metacentre and the centre of gravity.
(d) on the second moment of water-line area.

Of these statements :
(a) 1 and 2 are correct (b) 2 and 3 are correct
(c) 3 an 4 are correct (d) 1 and 4 are correct

193. Which one of the following statements is true to two-dimennsional flow of ideal fluids?
(a) Potential function exists if stream function exists.
(b) Stream function may or may not exist.
(c) Both potential function and stream function must exist for every flow.
(d) Stream function will exist but potential function may or may not exist.

194. The curl of a given velocity field $(\nabla \times \vec{V})$ indicates the rate of
(a) increase or decrease of flow at a point
(b) twisting of the lines of flow
(c) deformation
(d) translation.

195. Match List-I (fluid properties) with List-II (related terms) and select the correct answer :

List-I	List-II
A. Capillarity	1. Cavitation
B. Vapour pressure	2. Density of water
C. Viscosity	3. Shear forces
D. Specific gravity	4. Surface tension

Codes :

	A	B	C	D
(a)	1	4	2	3
(b)	1	4	3	2
(c)	4	1	2	3
(d)	4	1	3	2

196. The general form of expression for the continuity equation in a Cartesian coordinate system for incompressible or compressible flow is given by :

(a) $\frac{\partial u}{\partial x}+\frac{\partial v}{\partial y}+\frac{\partial w}{\partial z}=0$

(b) $\frac{\partial(\rho u)}{\partial x}+\frac{\partial(\rho u)}{\partial y}+\frac{\partial(\rho w)}{\partial z}=0$

(c) $\frac{\partial p}{\partial t}+\frac{\partial(\rho u)}{\partial x}+\frac{\partial(\rho v)}{\partial y}+\frac{\partial(\rho w)}{\partial z}=1$

(d) $\frac{\partial p}{\partial t}+\frac{\partial(\rho u)}{\partial x}+\frac{\partial(\rho v)}{\partial y}+\frac{\partial(\rho w)}{\partial z}=1$

197. In a two-dimensional flow in x-y plane, if $\frac{\partial u}{\partial y}=\frac{\partial v}{\partial x}$, then the fluid element will undergo :
(a) translation only
(b) translation and rotation
(c) translation and deformation
(d) rotation and deformation.

198. Water flow through a pipline having four different diameters at 4 stations is shown in the given Fig. 12.

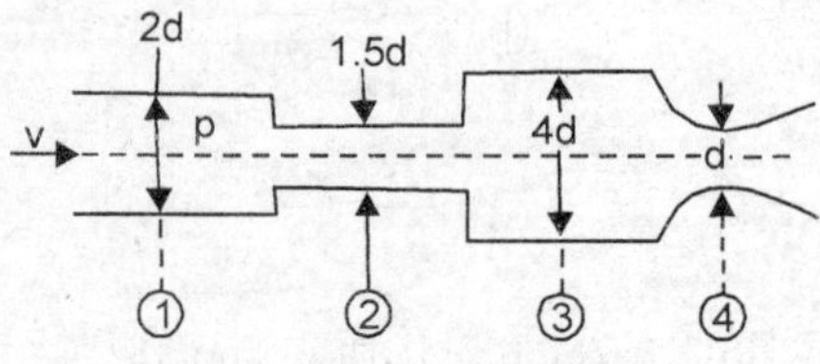

Fig. 12

The correct sequence of station numbers in the decreasing order of pressures is:

(a) 3, 1, 4, 2 (b) 1, 3, 2, 4
(c) 1, 3, 4, 2 (d) 3, 1, 2, 4

During the measurement of viscosity of air flowing through a pipe, we use the relation.

$\rho = \frac{\pi d^4}{128Q} \cdot \left(-\frac{dp}{dx}\right)$ under the condition that in the measuring section:

a) there is a viscous zone near the wall and an inviscid core persists at the centre.
b) the entire cross-section is viscous.
c) the flow can be assumed as potential flow.
d) the flow is irrotational.

If energy grade and hydraulic grade lines are drawn for flow through an inclined pipeline the following 4 quantities can be directly observed :

1. Static head 2. Friction head
3. Datum head 4. Velocity head.

Starting from the arbitrary datum line, the above types of heads will be in the sequence :

a) 3, 2, 1, 4 (b) 3, 4, 1, 2
c) 3, 4, 1, 2 (d) 3, 1, 4, 2

If calibration chart is prepared for a hot-wire anemometer for measuring the mean velocities, the highest level of accuracy can be :

a) equal to accuracy of a Pitot tube
b) equal to accuracy of a rotameter
c) equal to accuracy of a venturimeter
d) more than that of all the three instruments mentioned above.

At the point of boundary layer separation :

a) shear stress is maximum
b) shear stress is zero
c) velocity is negative
d) density variation is maximum.

All experiments thus far indicated that there can be a laminar flow in a pipe if the Reynolds number is below :

a) 2300 (b) 4000
c) 2000 (d) 40000

If $\frac{\partial u}{\partial x} + \frac{\partial v}{\partial y} = 0$ for a turbulent flow then it signifies that :

(a) bulk momentum transport is conserved
(b) $u'v'$ is non-zero and positive
(c) turbulence is anisotropic
(d) none of the above is true.

205. The predominant forces acting on a element of fluid in the boundary layer over a flat plate placed in a uniform stream include :

(a) inertia and pressure forces
(b) viscous and pressure forces
(c) viscous and body forces
(d) viscous and inertial forces.

206. Which one of the following velocity distributions of u/u_∞ satifies the boundary conditions for laminar flow on a flat plate? (Here u_∞ is the free stream velocity, u is velocity at any normal distance y from the flat plate, $\eta = y/\delta$ and δ is boundary layer thickness) :

(a) $\eta - \eta^2$ (b) $1.5\eta - 0.5\eta^3$
(c) $3\eta - \eta^2$ (d) $\cos(\pi\eta/2)$

207. The turbulent boundary-layer thickness varies as :

(a) $x^{4/5}$ (b) $x^{1/5}$
(c) $x^{1/2}$ (d) $x^{1/7}$

208. During the flow over a circular cylinder, the drag coefficient drops significantly at a critical Reyonlds Number of 2×10^5. This is due to :

(a) excessive momentum loss in the boundary layer
(b) Separation point travelling upstream
(c) reduction in skin-friction drag
(d) the delay in separation due to transition to turbulence.

209. Match List-I with List-II and select the correct answer :

List-I (Predominant force)	List-II Dimensionless numbers)
A. Compressibility force	1. Euler number
B. Gravity force	2. Prandtl Number
C. Surface tension force	3. Mach number
D. Viscous force	4. Reynolds numbers
	5. Weber number.

Codes :

	A	B	C	D
(a)	1	2	3	4
(b)	3	2	5	4
(c)	3	1	4	2
(d)	2	3	5	1

210. Kinematic similarity between model and prototype is the similarity of :
(a) shape (b) discharge
(c) stream (d) forces

211. A cylindrical gate is holding water on one side as shown in the given Fig. 13 The resultant vertical component of force of water per meter width of gate will be :

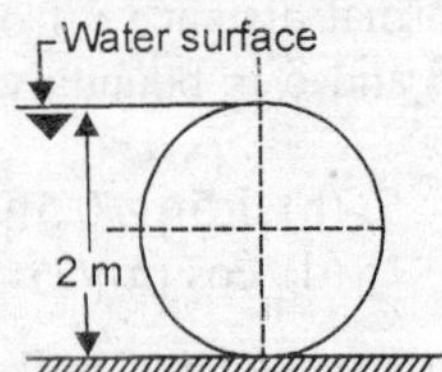

Fig. 13

(a) zero (b) 7700.8 N/m
(c) 15401.7 N/m (d) 30803.4 N/m

212. A differential manometer is used to measure the difference in pressure at points A and B in terms of specific weight of water, γ. The specific gravities of the; liquids X,Y and Z are respectively s_1, s_2 and s_3. The correct difference $\left(\frac{P_A}{\gamma} - \frac{P_B}{\gamma}\right)$ is given by (Fig. 14)

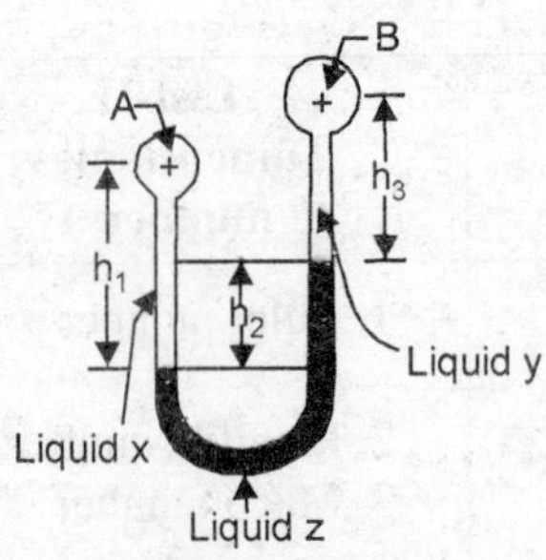

Fig. 14

(a) $h_3 s_2 - h_1 s_1 + h_2 s_3$
(b) $h_1 s_1 + h_2 s_3 - h_3 s_2$
(c) $h_3 s_1 - h_1 s_2 + h_2 s_3$
(d) $h_1 s_1 + h_2 s_2 - h_3 s_3$

213. A large metacentric height in a vessel :
(a) improves stability and makes periodic time to oscillation longer
(b) impairs stability and makes periodic time of oscillation shorter
(c) has no effect on stability or the periodic time of oscillation
(d) improves stability and makes the periodic time of osciallation shorter

214. The parameters for ideal fluid flow around a rotating circular cylinder can be obtained by superposition of some elementary flows. Which one of the following sets would describe the flow around a rotating circular cylinder?
(a) Doublet, vortex and uniform flow
(b) Source, vortex and uniform flow
(c) Sink, vortex and uniform flow
(d) Vortex and uniform flow

215. For an irrotational flow, the velocity potential lines and the streamlines are always :
(a) parallel to each other
(b) coplanar
(c) orthogonal to each other
(d) inclinded to the horizontal

216. The dimensions of surface tension is :
(a) N/m^2 (b) N/m
(c) J/m^2 (d) W/m

217. Which one of following is the bulk modulus K of a fluid? (Symbols have the usual meaning) :
(a) $\rho \frac{dp}{d\rho}$ (b) $\frac{dp}{\rho d\rho}$
(c) $\frac{\rho d\rho}{dp}$ (d) $\frac{d\rho}{\rho dp}$

218. A hydraulic jump occurs in a channel :
(a) wherenever the flow is supercritical
(b) if the flow is controlled by a sluice gate

(c) if the bed slope changes from mild to steep
(d) if the bed slope changes from steep

219. Which one of the following statements is true for fully developed through pipes?
(a) The flow is parallel, has no inertia effects, the pressure gradient is of constant value and the pressure force is balanced by the viscous force
(b) The flow is parallel, the pressure gradient is proportional to the intertia force and there is no viscous effect
(c) The flow is parallel, the pressure gradient is negligible and the inertia force is balanced by the viscous force
(d) The flow is not parallel, the core region accelerates and the viscous drag is far too less than the inertia force

220. Given,

H = height of liquid
b = width of notch
a = cross-sectional area
a_1 = area at inlet,
a_2 = area at the throat, and
C_d = coefficient of drag,

Match List-I with List-II and select the correct answer using the codes given below the Lists:

List-I	List-II
A. Discharge through venturimeter	1. $\frac{2}{3}C_d b\sqrt{2g}H^{3/2}$
B. Discharge through an external mouthpiece	2. $\frac{8}{15}C_d\sqrt{2g}H^{5/2}$
C. Discharge over rectangular notch	3. $\frac{C_d a_1 a_2}{\sqrt{a_1^2 - a_2^2}}\sqrt{2gH}$
D. Discharge over right-angled notch.	4. $0.855a\sqrt{2gH}$

Codes :

	A	B	C	D
(a)	1	2	3	4
(b)	3	4	1	2
(c)	2	1	3	4
(d)	2	3	1	4

221. Flow separation is caused by :
(a) reduction of pressure to local vapour pressure
(b) a negative pressure gradient
(c) a positive pressure gradient
(d) thinning of boundary layer thickness to zero.

222. In a turbulent flow, $\bar{u}, \bar{v}$, and $\bar{w}$ are time average velocity components. The fluctuating components are u', v', and w' respectively. The turbulence is said to be isotropic if :
(a) $\bar{u} = \bar{v} = \bar{w}$
(b) $\bar{u} + u' = \bar{v} + v' = \bar{w} + w'$
(c) $(\vec{u'})^2 = (\vec{v'})^2 = (\vec{w'})^2$
(d) none of the situations

223. Shear stress in a turbulent flow is due to :
(a) the viscous property of the fluid
(b) the fluid density
(c) fluctuation of velocity in the direction of flow
(d) fluctuation of velocity in the direction of flow as well as tranverse to it.

224. In a turbulent flow, 'l' is the Prandtl's mixing length and $\partial\bar{u}/\partial y$ is the gradient of the average velocity in the direction normal to flow. The final expression for the turbulent velocity v_t is given by :
(a) $v_t = l\left(\frac{\partial \bar{u}}{\partial y}\right)$
(b) $v_t = \left|\frac{\partial \bar{u}}{\partial y}\right|$
(c) $v_t = l^2\left(\frac{\partial \bar{u}}{\partial y}\right)$
(d) $v_t = \left(\frac{\partial \bar{u}}{\partial y}\right)$

225. During the growth of turbulent boundary layer over a flat plate for a moderately high Reynolds number, the boundary layer thickness, δ varies as :

(a) $x^{1/3}$ (b) $x^{1/2}$
(c) $x^{4/5}$ (d) $x^{7/8}$

226. Given that :

δ = boundary layer thickness,

δ^* = displacement thickness

δ_e = energy thickness

θ = momentum thickness

The shape factor H of a boundary layer is given by :

(a) $H = \frac{\delta_e}{\delta}$ (b) $H = \frac{\delta *}{\theta}$

(c) $H = \frac{\delta}{\theta}$ (d) $H = \frac{\delta}{\delta *}$

227. If U_∞ = free stream velocity, u = velocity at y, and δ = boundary layer thickness, then in a boundary layer flow, the momentum thickness δ is given by :

(a) $\theta = \int_0^\delta \frac{u}{U_\infty}\left(1 - \frac{u}{U_\infty}\right)dy$

(b) $\theta = \int_0^\delta \frac{u}{U_\infty}\left(1 - \frac{u^2}{U_\infty^2}\right)dy$

(c) $\theta = \int_0^\delta \frac{u^2}{U_\infty^2}\left(1 - \frac{u}{U_\infty}\right)dy$

(d) $\theta = \int_0^\delta \left(1 - \frac{u}{U_\infty}\right)dy$

228. Telephone wires often snap due to crossflow of wind past the wires. The main reason for this is :

(a) The force exerted by the wind on the wires is large in magnitude
(b) Poor quality of the work executed
(c) Wide variation of wind velocity in magnitude and direction
(d) Vortex shedding

229. The variables controlling the motion of a floacting vessel through water are the drag force F, the speed v, the length l, the density ρ dynamic viscosity μ of water, gravitational constant g. If the dimensional groups are Reynolds nu (*Re*), Weber number (*We*) Prandtl numbe and Froude number (*Fr*), the expressio F is given by :

(a) $\frac{F}{\rho v^2 l^2} = f(\text{Re})$

(b) $\frac{F}{\rho v^2 l^2} = f(\text{Re}, \text{Pr})$

(c) $\frac{F}{\rho v^2 l^2} = f(Fr, Wr)$

(b) $\frac{F}{\rho v^2 l^2} = f(\text{Re}, Fr)$

230. Euler number is defined as the ratio of i force to :

(a) viscous force (b) elastic force
(c) pressure force (d) gravity force

231. An inviscid irrotational flow field of vortex motion has a circulation constat Ω tangential velocity at any point in the field is given by Ω/r, where, r, is the distance from the centre. At the centre, is a mathematical singularity which c phiysically substituted by a forced v motion ($r = r_c$), the angular velocity ω is by,

(a) $\Omega/(r_c)^2$(b) Ω/r_c (c) Ωr_c (d)

232. The normal stress is the same in direc at a point in a fluid only when :

(a) The fluid is frictional
(b) The fluid is frictionless and incompre
(c) The fluid has zero viscosity and is a
(d) One fluid layer has no motion relat an adjacent layer.

233. Which of the following forces act on a at rest?

1. Gravity force
2. Hydrostatic force
3. Surface tension
4. Viscous force

Select the correct answer using the given below :

(a) 1, 2, 3 and 4 (b) 1, 2 and 3
(c) 2, and 4 (d) 1, 3 and 4

234. A stepped cylindrical container is filled with a liquid as shown in the given Fig. 15 The container with its axis vertical, is first placed with its larger diameter downward and then upward. The ratio of the forces at the bottom in the two cases will be :

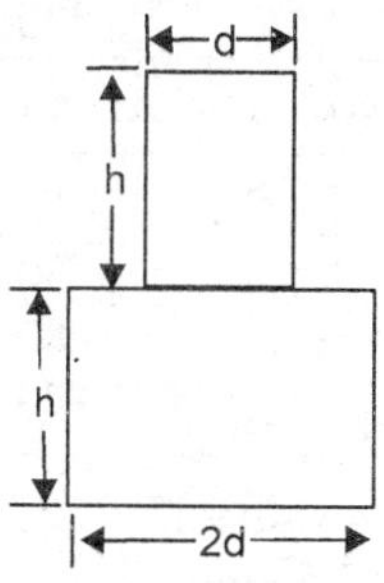

Fig. 15

(a) $\frac{1}{2}$ (b) 1
(c) 2 (d) 4

235. A circular annular plate haviang outer and inner diameters of 1.4 m and 0.6 respectiveiy is immersed in water with its plane making an angle of 60° with the horizontal The centre of the circular annular plate is 1.85 m below the free surface. The hydrostatic thrus on one side of the plate is :

(a) 1975 N (b) 19.75 N
(c) 11.4 N (d) 22.8 N

236. A house-top water tank is made of flat plates and is full to the brim. Its height its twice that of any side. The ratio of force on the bottom of the tank to that on any side will be :

(a) 4 (b) 2
(c) 1 (d) 1/2

237. A right-circular cylinder open at the top is filled with liquid of relative density 1.2. It is rotated about its vertical axis at such a speed that half the liquid spills out. The pressure at the centre of the bottom will be :

(a) Zero
(b) One-fourth of the value when the cylinder was full
(c) Half of the value when the cylinder was full
(d) Not determinable from the given data.

238. In the Fig.16 shown below, air is contained in the pipe and water is the manometer liquid. The pressure at A' is approximately

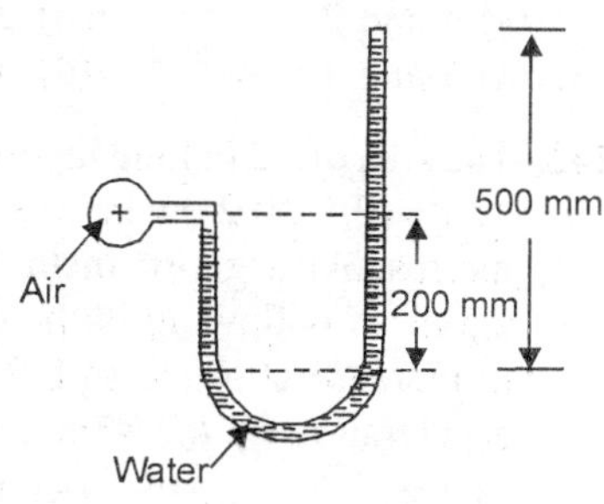

Fig. 16

(a) 10.14 m of water absolute
(b) 0.2 m of water
(c) 1.2 of water vacuum
(d) 4901 Pa

239. Consider the following statements:

Filling up a part of the empty hood of a ship with ballasts will :

1. Reduce the metacentric height
2. Lower the position of the centre of gravity
3. Elevate the position of centre of gravity.
4. Elevate the position of centre of buoyancy.

Of these statements :

(a) 1, 3 and 4 are correct
(b) 1 and 2 are correct
(c) 3 and 4 are correct
(d) 2 and 4 are correct

240. A cylindrical piece of cork weighing 'W floats with its axis in horizontal position in a liquid of relative density 4. By anchoring the bottom, the corkpiece is made to float at neutal equilibrium position with its axis vertical. The vertically downward force exerted by anchoring would be :

(a) 0.5 W (b) W
(c) 2 W (d) 4 W

241. Consider the following assumptions :

1. The fluid is compressible.
2. The fluid is inviscid.
3. The fluid is incompressible and homogeneous.
4. The fluid is viscous and compressible.

The Euler's equation of motion requires as assumptions indicated in :

(a) 1 and 2 (b) 2 and 3
(c) 1 and 4 (d) 3 and 4

242. The area of a 2 m long tapered duct decreases as A = (0.5 – 0.2x) where 'x' is the distance in metres. At a given instant a discharge of 0.5 m^3 /s is flowing in the duct and is found to increase at a rate of 0.2 m^3 /s. The local acceleration (in m^2 /s) at x = 0 will be :

(a) 1.4 (b) 1.0
(c) 0.4 (d) 0.667

243. Surface tension is due to:

(a) Viscous forces
(b) Cohesion
(c) Adhesion
(d) The difference between adhesive and cohesive forces.

244. Newton's law of viscosity depends upon the :

(a) Stress and strain in a fluid
(b) Shear stress, pressure and velocity
(c) Shear stress and rate of strain
(d) Viscosity and shear stress

245. Irrotational flow occurs when :

(a) Flow takes place in a duct of uniform cross-section at constant mass flow rate
(b) Streamlines are curved
(c) There is no net rotation of the fluid element about its mass centre.
(d) Fluid element does not undergo any change in size or shape.

246. A pipe flow system with flow direction is shown in the given Fig. 17. The following table gives the velocities and the corresponding areas :

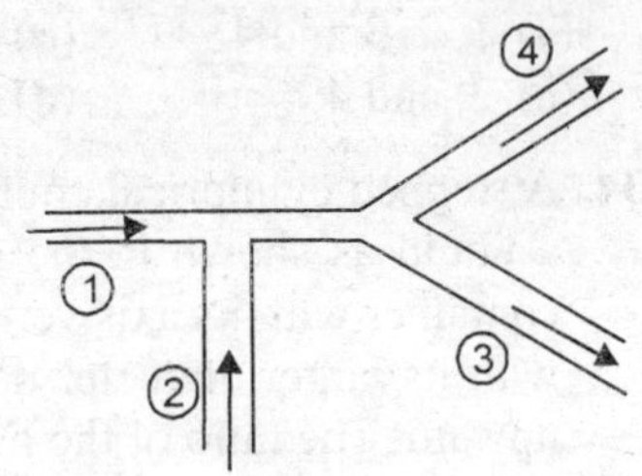

Fig. 17

Pipe No.	Area (cm)	Velocity (cm/s)
1.	50	10
2.	50	V_2
3.	80	5
4.	70	5

The value of v_2 is :

(a) 2.5 cm/s (b) 5.0 cm/s
(c) 7.5 cm/s (d) 10.0 cm/s.

247. A liquid flow downard through a tapered vertical portion of pipe. At the entrance and exit of the pipe, the static pressures are equal. If for a vertical heigh 'h' the velocity becomes four times, then the ratio of 'h' to the velocity head at entrance will be (Fig.18)

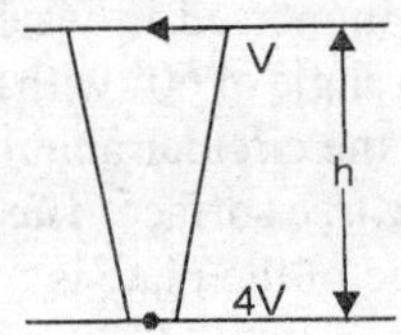

Fig. 18

(a) 3 (b) 8 (c) 15 (d) 24

248. The equivalent length of the stepped pipeline shown in the Fig.19 can be expressed in terms of the diameter 'D' as

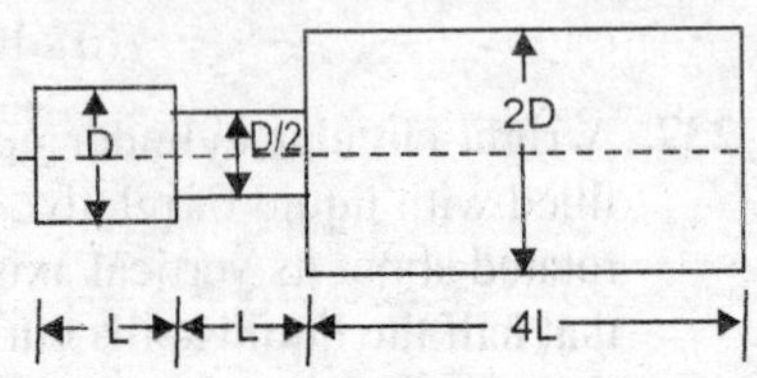

Fig. 19

(a) 5.25 L (b) 9.5 L

(c) $33\frac{1}{32}L$ (d) 33 L

249. A horizontal pipe of cross-sectional area 5 cm^2 is connected to a venturimeter of throat area 3 cm^2 as shown in the given Fig. 20. The manometer reading is equivalent to 5 cm of water. The discharge in cm^3/s nearly:

(a) 0.45 (b) 5.5

(c) 21.0 (d) 370

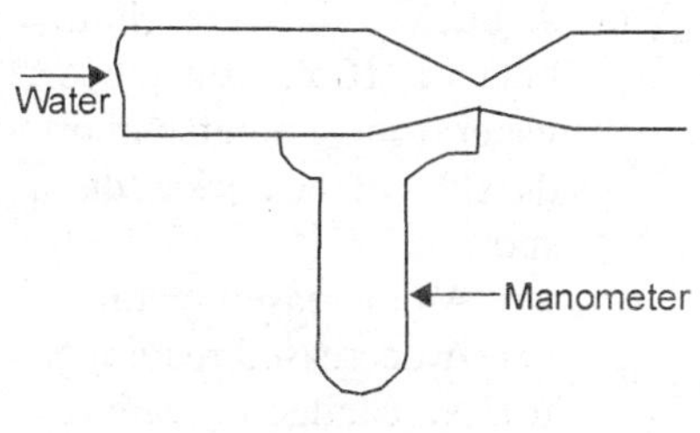

Fig. 20

250. In a fully trubulent flow through a rough pipe, the friction factor 'f' is (Re is the Reynolds number and ξ_s/D is relative roughness)

(a) A function of Re

(b) A function of Re and ξ_s/D

(c) A function of ξ_s/D

(d) independent of Re and ξ_s/D

251. In a boundary layer developed along the flow, the pressure decreases in the downstream direction. The boundary layer thickness would :

(a) Tend to decrease (b) Remain constant

(c) Increase rapidly

(d) Increase gradulally

252. Which one of the following statements is true of flow around a submerged body?

(a) For subsonic, non-viscous flow, the drag is zero

(b) For supersonic flow the drag coefficient is dependent equally on Mach number and Reynolds number

(c) The lift and drag coefficient of an aerofoil is independent of Reynolds number

(d) For incompressible flow around an aerofoil, the profile drag is the sum of form drag and skin friction drag.

253. If 'n' variables in a physical phenomenon contained 'm' fundamental dimensions, then the variables can be arranged into:

(a) n dimensionless terms

(b) m dimensionless terms

(c) $(n - m)$ dimensionless terms

(d) $(n + m)$ dimensionless terms

254. Given power 'P' of a pump, the head 'H' and the discharge 'Q' and the specific weight 'w' of the liquid, dimensional analysis would lead to the result that 'P is proportional to :

(a) $H^{1/2} Q^2 w$ (b) $H^{1/2} Q w$

(c) $H Q^{1/2} w$ (d) HQw

255. A 1 : 20 model of a spillway dissipates 0.25 hp. The corresponding prototype horsepower dissipated will be :

(a) 0.25 (b) 5.00 (c) 447.20 (d) 8944.30

256. If the stream function is given by $\psi = 3xy$, then the velocity at a point (2, 3) will be :

(a) 7.21 unit (b) 10.82 unit

(c) 18 unit (d) 54 unit

257. A triangular dam of height h and base width b is filled to its top with water as shown in the given Fig. 21. The condition of stability is :

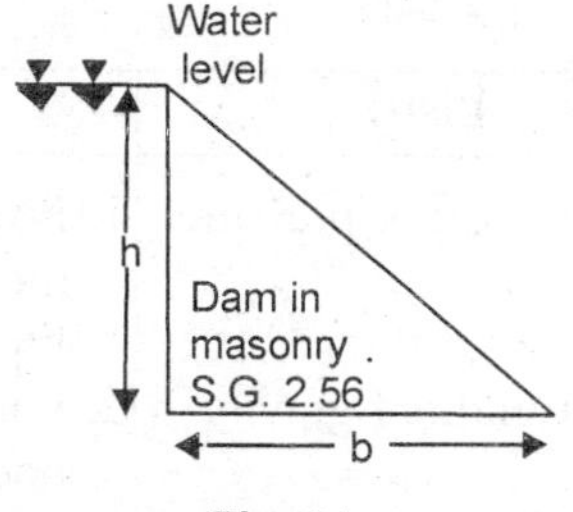

Fig. 21

(a) $b = h$ (b) $b = 2.6\,h$

(c) $b = \sqrt{3}\,h$ (d) $b = 0.625\,h$

258. Stability of a freely floating object is assured if its centre of :

(a) Buoyancy lies below its centre of gravity

(b) Gravity coincides with its centre of buoyancy

(c) Gravity lies below its metacenter

(d) Buoyancy lies below its metacenter

259. A vertical sluice gate, 2.5 m wide and weighing 500 kg is held in position due to horizontal force of water on one side and associated friction force, When the water level drops down to 2 m above the bottom of the gate, the gate just starts sliding down. The coefficient of friction between the gate and the supporting structure is :
(a) 0.20 (b) 0.10 (c) 0.05 (d) 0.02

260. The reading of gauge 'A' shown in the given Fig. 22 is :

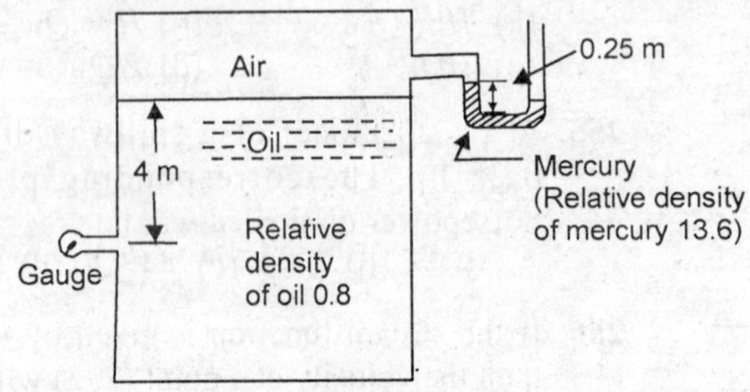

Fig. 22

(a) – 31.392 kPa (b) – 1.962 kPa
(c) 31.392 kPa (d) + 19.62 kPa

261. Match List-I with List-II regarding a body party submerged in a liquid and select the correct answer using the codes given below the lists :

List-I	List-II
A. Centre of pressure	1. Point of application of the weight of displaced liquid
B. Centre of gravity	2. Point about which the body starts oscillating when tilted by a small angle
C. Centre of buoyancy	3. Point of application of hydrostatic pressure
D. Metacentre	4. Point of application of the weight of the body

Codes :

	A	B	C	D
(a)	4	3	1	2
(b)	4	3	2	3
(c)	3	4	1	2
(d)	3	4	2	1

262. If a piece of metal having a specific gr of 13.6 is placed in mercury of specific gr 13.6, then :
(a) The metal piece will sink to the bo
(b) The metal piece will simply float the mercury with no immersion
(c) The metal piece will be immerse mercury by half
(d) The whole of the metal piece wi immersed with its top surface ju mercury level

263. A bucket of water hangs with a sp balance. If an iron piece is supended water from another support without tou the sides of the bucket, the spring balanc show :
(a) An increased reading
(b) A decreased reading
(c) No change in reading
(d) Increased or decreased rea depending on the depth of immers

264. The least radius of gyration of a ship i and the metacentric height is 750 mm time period of oscillation of the ship is
(a) 42.41 s (b) 75.4 *s*
(c) 20.85 s (d) 85 s

165. If the surface tension of water-air interf 0.073 N/m, the gauge pressure inside drop of 1 mm diameter will be :
(a) 0.146 N/m² (b) 0.73 N/m²
(c) 146 N/m² (d) 292 N/m²

266. The elbow nozzle assembly, shown i given Fig. 23 is in horizontal plane velocity of jet issuing from the nozzle

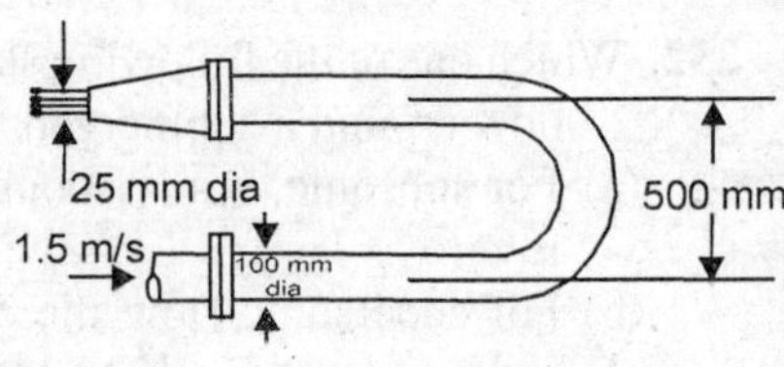

Fig. 23

(a) 4 m/s (b) 16 m/s
(c) 24 m/s (d) 30 m/s

267. The pipe cross-sections and fluid flow are shown in the given Fig. 24. The ve in the pipe labelled as *A* is :

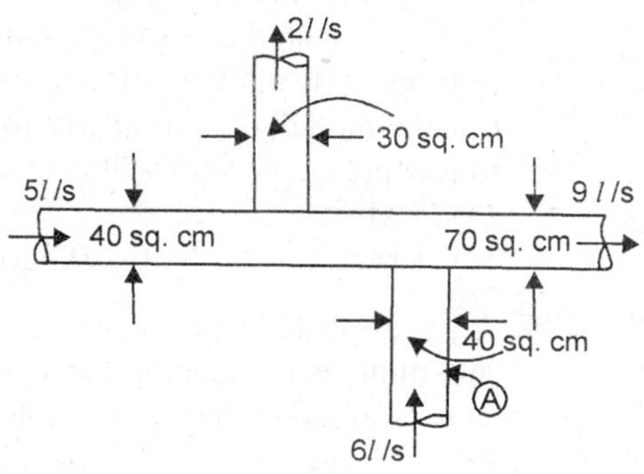

Fig. 24

(a) 1.5 m/s (b) 3 m/s
(c) 15 m/s (d) 30 m/s

268. Point A of head 'H_A' is at a higher elevation than point B of head 'H_B'. The head loss between these points is H_L. The flow will take place (see Fig. 25)

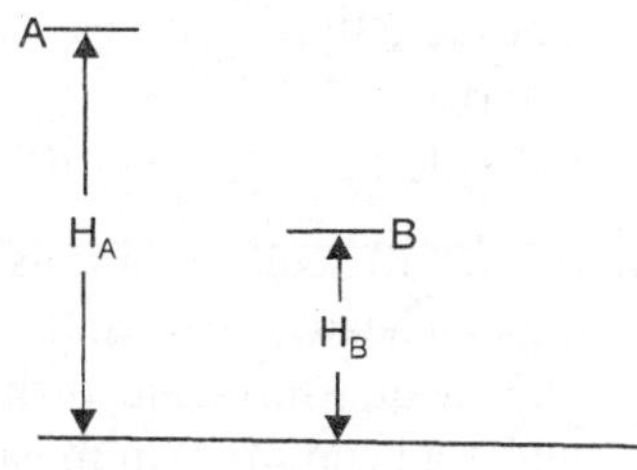

Fig. 25

(a) Always from A to B
(b) From A to B if $H_A + H_L = H_B$
(c) From B to if $H_A + H_L = H_B$
(d) From B to A if $H_B + H_L = H_A$

269. Consider the following statements regarding hydraulic jump :

1. There occurs a transformation of super critical flow to sub-critical flow.
2. The flow is unifrom and distribution is due to hydrostatic force before and after the jump.
3. There occurs a loss of energy due to eddy formation and turbulence

Which of these statements are correct?

(a) 1, 2 and 3 (b) 1 and 2
(c) 2 and 3 (d) 1 and 3

270. Match List-I (Pipe flow) with List-II (Types of acceleration) and select the correct answer using the code given below the list:

List-I	List-II
A. Flow at constant rate passing through a bend	1. Zero acceleration
B. Flow at constant rate passing through a straight uniform diameter pipe	2. Local and convective acceleration
C. Gradually changing flow through a bend	3. Convective acceleration
D. Gradualy changing flow through a straight pipe	4. Local acceleration

Codes :

	A	B	C	D
(a)	3	1	2	4
(b)	3	1	4	2
(c)	1	3	2	4
(d)	1	3	4	2

271. The value of friction factor is misjudged by +25% in using Darcy-Weisbach equation The resulting error in the discharge will be:

(a) + 25% (b) – 18.25 %
(c) – 12.5% (d) +12.5%

272. For turbulent boundary layer flow, the thickness of laminar sublayer 'δ_1' is given by:

(a) $\frac{v}{u*}$ (b) $\frac{5v}{u*}$

(c) $575 \log \frac{v}{u*}$ (d) $2300 \frac{v}{u*}$

273. The correct sequence in ascending order of the magnitude of the given parameters is :

(a) Boundary layer thickness, momentum thickness, displacement thickness
(b) Displacement thickness, boundary layer thickness, momentum thickness
(c) Momentum thickness, displacement thickness, boundary laver thickness
(d) Momentum thickness, boundary layer thickness, displacement thickness

274. Consider the following statement :

1. The cause of stalling of an aerofoil is the boundary layer separation and formation of increased zone of wake.
2. An aerofoil should have a rounded nose in supersonic flow to prevent formation of new shock.
3. When an aerofoil operates at an angle of incidence greater than of stalling, the lift decreasses and drage increases.
4. A rough ball when at certain speed can attain longer range due to reduction of lift as the roughness induces early separation.

Which of the these statements are correct?

(a) 3 and 4 (b) 1 and 2
(c) 2 and 4 (d) 1 and 3

275. A parachutist has a mass of 90 kg and a projected frontai area of 0.30 m^2 in free fall. The drag coefficient based on frontal area is found to be 0.75. If the air density is 1.28 kg/m^3 the terminal velocity of the parachutist will be:

(a) 104.4 m/s (b) 78.3 m/s
(c) 25 m/s (d) 18.5 m/s

276. If the number of fundamental dimensions equal 'm' then the repeating variables shall be equal to :

(a) *m* and none of repeat variables shall represent the dependent variable
(b) *m* + 1 and one of the repeating variable shall represent the dependent variable
(c) *m* + 1 and none of the repeating variables shall represent the dependent variable
(d) *m* and one of the repeating variable shall represent the dependent variable

277. A sphere is moving in water with a velocity of 1.6 m/s. Another sphere of twice the diameter is placed in a wind tunnel and tested with air which is 750 times less dense and 60 times less viscous than water. The velocity of air that will give dynamically similar conditions is:

(a) 5 m/s (b) 10 m/s
(c) 20 m/s (d) 40 m/s

278. A ship model 1/60 scale with negligible friction is tested in a towing tank at a speed of 0.6 m/s. If a force of 0.5 kg is required to tow the model, the propulsive force required to tow prototype ship will be:

(a) 5 MN (b) 3 MN
(c) 1 MN (d) 0.5 MN

279. A 1 : 256 scale model of a reservoir is drained in 4 minutes by opening the sluice gate. The time required to empty the protoype will be:

(a) 128 min. (b) 64 min.
(c) 32 min. (d) 25.4 min.

280. Air at 2 bar 60°C enters a constant area pipe of 60 mm diameter with a velocity of 40 m/s. During the flow through the pipe, heat is added to the air stream. Frictional effects are negligible and the values of C_p and C_v are that of standard air. The Mach number of the flow corresponding to the maximum entropy will be corresponding to the maximum entropy will be :

(a) 0.845 (b) 1
(c) 0.1212 (d) 1.183

281. If the full-scale turbine is required to work under a head of 30 *m* and to run at 428 *r.p.m.*, then a quarter-scale turbine model tested under a head of 10 m must run at :

(a) 143 r.p.m. (b) 341 r.p.m.
(c) 428 r.p.m. (d) 988 r.p.m.

282. The dimensionless group formed by wavelength. λ, density of fluid ρ acceleration due to gravity *g*, and surface tension σ, is:

(a) $\frac{\sigma}{\lambda^2 g\rho}$ (b) $\frac{\sigma}{\lambda g^2\rho}$
(c) $\frac{\sigma g}{\lambda^2 \rho}$ (d) $\frac{\rho}{\lambda^2 g\sigma}$

283. Which one of the following sets of standard flows is superimposed to represent the flow around a rotating cylinder?

(a) Doublet, vortex and uniform flow
(b) Source, vortex and uniform flow
(c) Sink, vortex and uniform flow
(d) Vortex and uniform flow.

284. A float of cubical shape has sides of 10 cm. The float valve seat to have a flow area of 0.5 cm^2 as shown in the given Fig. 26. If the pressure of water in the pipeline is 1 bar, the rise of water level h in the tank to just stop the water flow will be :

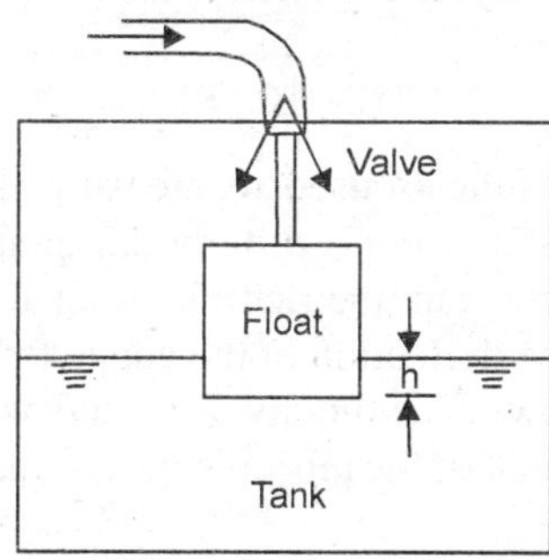

Fig. 26

(a) 7.5 cm (b) 5.0 cm
(c) 25 cm (d) 0.5 cm

285. A U-tube manometer is connected to a pipeline conveying water as shown in the given Fig. 27. The pressure head of water in the pipeline is :

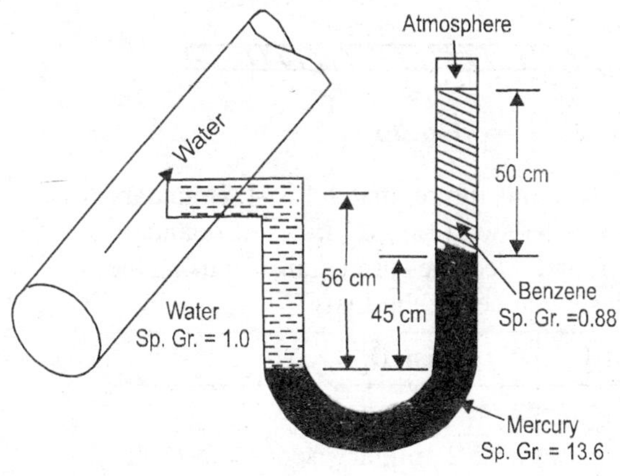

Fig. 27

(a) 7.12 m (b) 6.56 m
(c) 6.0 m (d) 5.12 m

286. The eye of a tornado has radius of 40 m. If the maximum wind velocity is 50 m/s the velocity at a distance of 80 m radius is :
(a) 100 m/s (b) 2500 m/s
(c) 31.25 m/s (d) 25 m/s

287. If a vessel containing liquid moves downward with constant acceleration g, then :
(a) The pressure throughout the liquid mass is atmospheric
(b) The pressure in the liquid mass is greater than the hydrostatic pressure
(c) There will be vacuum in the liquid
(d) The pressure throughout the liquid mass is greater than atmospheric.

288. Improved streamlining produces 25% reduction in the drage coefficient of a torpedo. When it is travelling fully submerged and assuming the driving power to remain the same, the increase in speed will be :
(a) 10 % (b) 20%
(c) 25% (d) 30%

289. If a bullet is fired in standard air at 15°C at the Mach angle of 30°, the velocity of the bullet would be :
(a) 513.5 m/s (b) 585.5 m/s
(c) 6455 m/s (d) 6805 m/s

290. A stream function is given by $(x^2 - y^2)$. The potential function of the flow will be :
(a) $2xy + f(x)$
(b) $2xy$ + constant
(c) $2(x^2 - y^2)$ (d) $2xy + f(x)$

291. The height of a cylindrical container is twice that of its diameter. The ratio of the horizontal forces on the wall of the cylinder when it is completely filled to that when it is half filled with the same liquid, is :
(a) 2 (b) 3
(c) 3,5 (d) 4

292. The velocities and corresponding flow areas of the branches labelled (1), (2), (3), (4) and (5) for a pipe system shown in given Fig. 28 are given in the following table :

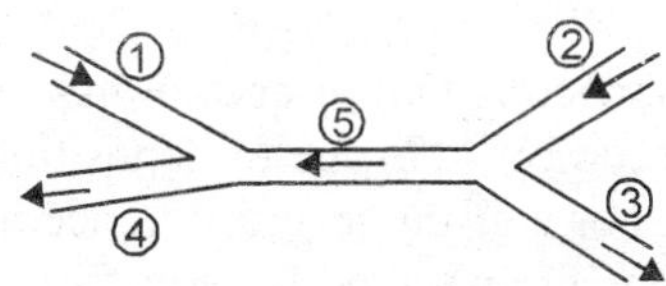

Fig. 28

Pipe Label	Velocity	Area
(1)	5 cm/s	4 sq cm
(2)	6 cm/s	5 sq cm
(3)	v_3 cm/s	2 sq cm
(4)	4 cm/s	10 sq cm
(5)	v_5 cm/s	8 sq cm

The velocity v_5 would be :

(a) 2.5 cm/s (b) 5 cm/s
(c) 7.5 cm/s (d) 10 cm/s

293. A pipe is connected in series to another pipe whose diameter is twice and length is 32 times that of the first pipe. The ratio of frictional head losses for the first pipe to those for the second pipe is (both the pipes have the same frictional constant) :

(a) 8 (a) 4 (c) 2 (d) 1

294. Which one of the following statements is correct?

(a) Hydraulic grade line and energy grade line are the same in fluid flow problems.
(b) Energy grade line lies above the hydraulic grade line and is always parallel to it.
(c) Energy grade line lies above the hydraulic grade line and they are separated from each other by a vertical distance equal to the velocity head.
(d) The hydraulic grade line slopes upwards meeting the energy grade line only at the exit of flow.

295. If laminar flow takes place in two pipes, having realtive roughnesses of 0.002 and 0.003, at a Reynolds number of 1815, then :

(a) the pipe of relative roughness of 0.003 has higher friction factor.
(b) the pipe of relative roughness of 0.003 has a lower friction factor.
(c) both pipes have the same friction factor.
(d) no comparison is possible due to inadequate data.

296. A pipeline connecting two reservoirs has its diameter reduced by 20% due to deposition of chemicals. For a given head difference in the reservoirs with unaltered friction factor, this would cause a reduction in discharge of :

(a) 42.8% (b) 20%
(c) 17.8% (d) 10.6%

297. A tank containing water has two orifices of the same size at depths of 20 cm and 90 cm below the free surface of water. The ratio of discharges through these orifices is :

(a) 1 : 1 (b) 2 : 3
(c) 4 : 9 (d) 16 : 81

298. A Pitot static tube is used to measure the velocity of water using a differential gauge which contains a manometric fluid of relative density 1.4. The deflection of the gauge fluid when water flows at a volocity of 1.2 m/s will be (the coefficient of the tube maybe assumed to be 1) :

(a) 183.5 mm (c) 52.4 mm
(c) 5.24 m (d) 73.4 mm

299. The development of the boundary layer zones labelled *P, Q, R* and S over a flat plate is shown in the given Fig. 29.

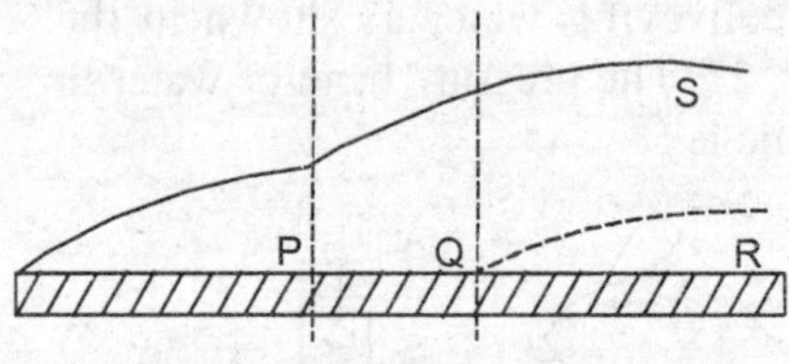

Fig. 29

Based on this figure, match List-I (Boundary layer zones) with List-II (Types of boundary layer) and select the correct' answer using the codes given below the Lists :

List-I	List-II
A. P	1. Transitional
B. Q	2. Laminar viscous sub-layer
C. R	3. Laminar
D. S	4. Turbulent

Codes :

	A	B	C	D
(a)	3	1	2	4,
(b)	3	2	1	4
(c)	4	2	1	3
(d)	4	1	2	3

A pipe of 20 cm diameter and 30 km length transports oil from a tanker to the shore with a velocity of 0.318 m/s. The flow is laminar. If $\mu = 0.1$ N-m/s^2, the power required for the flow would be :
(a) 9.25 kW (b) 8.36 kW
(c) 7.63 kW (d) 10.13 kW

In a turbulent boundary layer over the entire length of a plate, the boundary layer thickness increases with its distance X from the leading edge as :
(a) $X^{1/2}$ (b) $X^{1/5}$
(c) $X^{2/5}$ (d) $X^{4/5}$

Separation of fluid flow is caused by :
(a) reduction of pressure in the direction of flow
(b) reduction of the boundary layer thickness
(c) presence of adverse pressure gradient
(d) presence of favourable pressure gradient

When pressure drag over a body is large as compared to the friction drag, then the shape of the body is that of :
a) an aerofoil
b) a streamilined body
c) a two-dimensional body
d) a bluff body

A circular cylinder of 400 mm diameter is rotated about its axis in stream of water having uniform velocity of 4 m/s. When both the stagnation points concide, the lift force experienced by the cylinder is :
a) 160 kN/m (b) 10.05 kN/m
c) 80 kN/m (d) 40.2 kN/m

An automobile moving at a velocity of 0 km/hr is experiencing a wind resistance of 2 kN. If the automobile is moving at a velocity of 50 km/hr. the power required to overcome the wind resistance is :
) 43.4 kW (b) 3.125 kW
) 2.5 kW (d) 27.776 kW

When a cylinder is placed in an ideal fluid and the flow is uniform, the pressure coefficient C_p is equal to :
) $1 - \sin^2 \theta$ (b) $1 - 2 \sin^2 \theta$
) $1 - 4 \sin^2 \theta$ (d) $1 - 8 \sin^2 \theta$

307. If the upstream Mach number of a normal shock occurring in air ($k = 1.4$) is 1.68, than the Mach numbr after the shock is :
(a) 0.84 (b) 0.646
(c) 0.336 (d) 0.564

308. A rectangular tank of square cross-section is having its height equal to twice the length of any side at the base. If the tank is filled up with a liquid, the ratio of the total hydrostatic force on any vertical wall to that at the bottom is :
(a) 2.0 (b) 1.5 (c) 1.0 (d) 0.5

309. Differential pressure head measured by mercury oil differential manometer (specific gravity of oil is 0.9) equivalent to a 600 mm difference of mercury levels will nearly be :
(a) 7.62 m of oil (b) 76.2 m of oil
(c) 7.34 m of oil (d) 8.47 m of oil.

310. A block of aluminium having mass of 12 kg is suspended by a wire and lowered until submerged into a tank containing oil of relative density 0.8. Taking the relative density of aluminium as 2.4 the tension in the wire will be (take $g = 10$ m/s^2) :
(a) 12000 N (b) 800 N
(c) 120 N (d) 80 N.

311. A barge 30 m long and 10 m wide has a draft of 3 m when floating with its sides in vertical position. If its centre of gravity is 25 m above the bottom, the nearest value of metacentric height is :
(a) 3.28 m (b) 2.78 m
(c) 1.78 m (d) zero.

312. A cylindrical vessel having its height equal to its diameter is filled with liquid and moved horizontally at an acceleration equal to acceleration due to gravity. The ratio of the liquid left in the vessel to the liquid at static equilibrium condition is :
(a) 0.2 (b) 0.4 (c) 0.5 (d) 0.75.

313. The shear stress developed in a lubricating oil, of viscosity 9.81 poise, filled between two parallel plates 1 cm parat and moving with relative velocity of 2 m/s is :

(a) 20 N/m (b) 19.62 N/m^2
(c) 29.62 N/m^2 (d) 40 N/m^2.

314. The convective acceleration of fluid in the x-direction is given by :

(a) $u\frac{\partial u}{\partial x}+v\frac{\partial u}{\partial t}+w\frac{\partial w}{\partial z}$

(b) $\frac{\partial u}{\partial t}+v\frac{\partial v}{\partial t}+w\frac{\partial u}{\partial t}$

(c) $u\frac{\partial u}{\partial x}+v\frac{\partial v}{\partial y}+w\frac{\partial w}{\partial z}$

(d) $u\frac{\partial u}{\partial x}+v\frac{\partial u}{\partial y}+w\frac{\partial u}{\partial z}$.

315. Match List-I (Types of flow) with List-II (Basic ideal flows) and select the correct answer using the codes given below :

List-I (Types of flow)	List-II (Basic ideal flows)
A. Flow over a stationary cylinder	1. source + sink + uniform flow
B. Flow over a half Rankine body	2. doublet + uniform flow
C. Flow over a rotating body	3. source + uniform flow
D. Flow over a Rankine oval	4. doublet + free vortex + uniform flow

Codes:

	A	B	C	D
(a)	1	4	3	2
(b)	2	4	3	1
(c)	1	3	4	2
(d)	2	3	4	1

316. A glass tube with a 90° bend is open at both the ends. It is inserted into a flowing stream of oil, S = 0.90, so that one opening is directed upstream and the other is directed upward. Oil inside the tube is 50 mm higher than the surface of flowing oil. The velocity measured by the tube is, nearly.

(a) 0.89 m/s (b) 0.99 m/s
(c) 1.40 m/s (d) 1.90 m/s.

317. At location-I of a horizontal line, the fluid pressure head is 32 cm and velocity head is 4 cm. The reduction in area at location II is such that the pressure head drops down to zero. The ratio of velocities at location-II to that at location-I is (Fig. 30) :

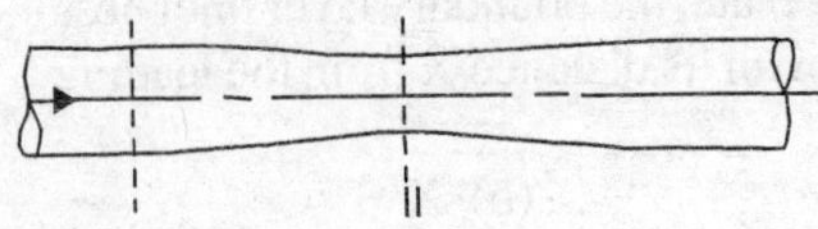

Fig. 30

(a) 3 (b) 2.5
(c) 2 (d) 1.5 .

318. For maximum transmission of power through a pipe line with total H, the head lost due to friction h_f is given by :

(a) 0.1 H (b) $\frac{H}{3}$

(c) $\frac{H}{2}$ (d) $\frac{2H}{3}$

319. Two pipelines of equal length and with diameters of 15 cm and 10 cm are in parallel and connect two reservoirs. The difference in water levels in the reservoirs is 3 m. If the friction is assumed to be equal, the ratio of the discharges due to the larger dia pipe to that of the smaller dia pipe is, nearly :

(a) 3.375 (b) 2.756
(c) 2.25 (d) 1.5.

320. The critical depth of a rectangular channel of width 4.0 m for a discharge of 12 m^3/s is, nearly :

(a) 300 mm (b) 30 mm
(c) 0.972 (d) 0.674 m.

321. An open channel flow encounters a hydraulic jump as shown in the Fig. 31. The following fluid conditions are observed between *A* and *B* :

1. Critical depth
2. Steady non-uniform flow

3. Unsteady non-uniform flow
4. Steady uniform flow.

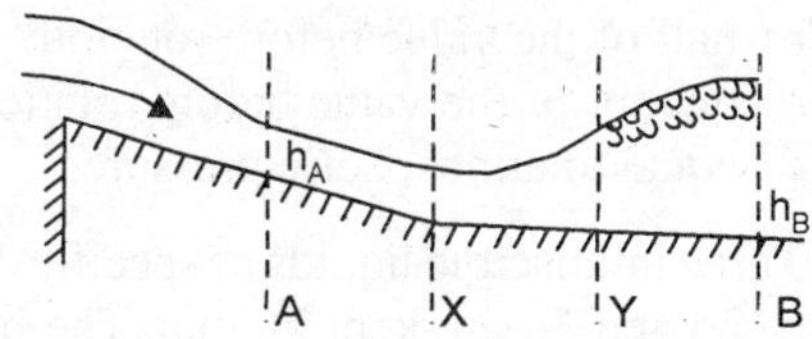

Fig. 31

The correct sequence of the flow conditions in the direction of flow is :

(a) 1, 2, 3, 4 (b) 1, 4, 2, 3
(c) 2, 1, 4, 3 (d) 4, 2, 3, 1.

322. Laminar developed flow at an average velocity of 5 m/s occurs in a pipe of 10 cm radius. The velocity at 5 cm radius is :

(a) 7.5 m/s (b) 10 m/s
(c) 2.5 m/s (d) 5 m/s.

323. In a fully-developed turbulent pipe flow, assuming $\frac{1}{7}$*th* power law, the ratio of time mean velocity at the centre of the pipe to the average velocity of the flow is :

(a) 2.0 (b) 1.5
(c) 1.22 (d) 0.817.

324. The pressure drop in a 100 mm diameter horizontal pipe is 50 kPa over a length of 10m. The shear stress at the pipe wall is :

(a) 0.25 kPa (b) 0.125 kPa
(c) 0.50 kPa (d) 25.0 kPa.

325. The velocity distribution in the boundary layer is given as $\frac{u}{U_\infty} = \frac{y}{\delta}$, where u is the velocity at a distance y from the boundary, U_∞ is the free stream velocity and δ is the boundary layer thickness at a certain distance from the leading edge of plate. The ratio of displacement to momentum thicknesses is :

(a) 5 (b) 4
(c) 3 (d) 2

326. For the velocity profile $\frac{u}{U_\infty} = \eta$, the momentum thickness of a laminar boundary layer on a flat plate at a distance of 1 m from leading edge for air (kinematic viscosity = 2×10^3 m^2/s flowing at a free stream velocity of 2 m/s is given by :

(a) 3.16 mm (b) 2.1 mm
(c) 3.16 mm (d) 21 mm.

327. According to Blasius law, the local skin friction coefficent in the boundary layer over a flat plate is given by :

(a) $\frac{0.332}{\sqrt{R_e}}$ (b) $\frac{0.664}{\sqrt{R_e}}$
(c) $\frac{0.647}{\sqrt{R_e}}$ (d) $\frac{1.328}{\sqrt{R_e}}$.

328. Match List-I with List-II and select the correct answer using the codes given below the Lists:

List-I	List-II
A. Stokes' law	1. Strouhal number
B. Bluff body	2. Creeping moton
C. Streamline body	3. Pressure drag
D. Karman Vortex Street	4. Skin friction drag

Codes :

	A	B	C	D
(a)	2	3	1	4
(b)	3	2	4	1
(c)	2	3	4	1
(d)	3	2	1	4

329. Match List-I (Dimensionless numbers) with List-II (Definition as the ratio of) and select the correct answer from the codes given below the Lists :

List-I (Dimensionless numbers)	List-II (Definition as the ratio of)
A. Reynolds number	1. Inertia force and elastic force
B. Froude Number	2. Inertia force and surface tension force
C. Weber number	3. Intertia force and gravity force
D. Mach number	4. Inertia force and viscous force

Codes :

	A	B	C	D
(a)	1	2	3	4
(b)	4	3	2	1
(c)	1	3	2	4
(d)	4	2	3	1

330. The stream function in a 2-dimensional flow field is given by $\psi = xy$. The potential function is :

(a) $\frac{(x^2 + y^2)}{2}$ (b) $\frac{(y^2 - x^2)}{2}$

(c) xy (d) $x^2y + y^2x$.

331. A ship with hull length of 100 m is to run with a speed of 10 m/s. For dynamic similarity, the velocity for a 1 : 25 model of the ship in a towing tank should be :

(a) 2 m/s (b) 10 m/s

(c) 20 m/s (d) 25 m/s.

332. A standard 90° V-notch weir is used to measure discharge. The discharge is Q_1 for a height H_1 above the sill and Q_2 is the discharge at H_2. If $\frac{H_2}{H_1}$ is 4, then $\frac{Q_2}{Q_1}$ is :

(a) 32 (b) $16\sqrt{2}$

(c) 16 (d) 8.

333. A right circular cylinder is filled with liquid upto its top level. It is rotated about is vertical axis at such a speed that half the liquid spills out, then the pressure at the point of intersection of the axis and bottom surface is:

(a) same as before rotation
(b) half of the value before rotation
(c) quarter of the value before rotation
(d) equal to atmospheric pressure.

334. Three immiscible liquids of specific densities ρ, 2ρ and 3ρ are kept in a jar. The height of the liquids in the jar and at the piezometer fitted to the bottom of the jar are as shown in the given Fig. 32. The ratio H/h is :

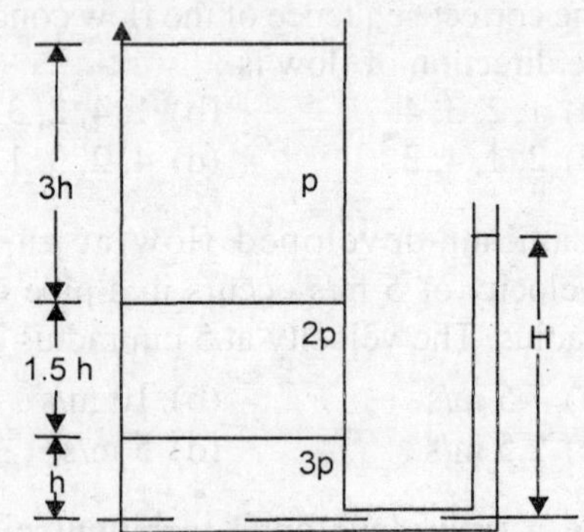

Fig. 32

(a) 4 (b) 3.5

(c) 3 (d) 2.5.

335. The hydraulic mean depth (where A = area and P = wetted perimeter) is given by :

(a) $\frac{P}{A}$ (b) $\frac{P^2}{A}$

(c) $\frac{A}{P}$ (d) $\sqrt{\frac{A}{P}}$.

336. Which of the following is/are related to measure the discharge by a rectangular notch?

1. $\frac{3}{2} C_d \cdot b\sqrt{2g} \cdot H^2$

2. $\frac{2}{3} C_d . b\sqrt{2g} \cdot H^{3/2}$

3. $\frac{2}{3} C_d \cdot b\sqrt{2g} \cdot H^{5/2}$

4. $\frac{2}{3} C_d \cdot b\sqrt{2g} \cdot H^{1/2}$

Select the correct answer from the codes given below :

Codes :

(a) 1 and 3 (b) 2 and 3
(c) 2 alone (d) 4 alone.

337. The critical value of Reynolds number for transition from laminar to turbulent boundary layer in external flows is taken as :

(a) 2300 (b) 4000
(c) 5×10^5 (d) 3×10^6.

338. The boundary layer flow separates from the surface if :

(a) $\frac{du}{dy} = 0$ and $\frac{dp}{dx} = 0$

(b) $\frac{du}{dy} = 0$ and $\frac{dp}{dx} > 0$

(c) $\frac{du}{dy} = 0$ and $\frac{dp}{dx} < 0$

(d) the boundary layer thickness is zero.

339. The laminar boundary layer thickness, δ at any point x for flow over a flat plate is given by $\frac{\delta}{x} =$:

(a) $\frac{0.664}{\sqrt{\text{Re}_x}}$ (b) $\frac{1.328}{\sqrt{\text{Re}_x}}$

(c) $\frac{1.75}{\sqrt{\text{Re}_x}}$ (d) $\frac{5.0}{\sqrt{\text{Re}_x}}$.

340. Volumetric flow rate Q, acceleration due to gravity g and head H form a dimensioness group, which is given by :

(a) $\sqrt{\frac{gH^5}{Q}}$ (b) $\frac{Q}{gH}$

(c) $\frac{Q}{\sqrt{g^3 H}}$ (d) $\frac{Q}{\sqrt{g^2 H}}$.

341. A model test is to be conducted in a water processing a 1 : 20 model of a submarine, which is to travel at a speed to 12 krn/h deep where sea surface. The water temperature in the tunnel is maintained, so that its kinematre vrscosity is half that of sea water. At what speed is the model test to be conducted to produce useful data for the prototype?

(a) 12 km/h (b) 240 km/h
(c) 26 km/h (d) 120 km/h.

342. A model test is to be conducted for an under water structure, which is likely to be exposed to strong water currents. The significant forces are known to be dependent on structure geometry, fluid velocity, fluid density and viscosity, fluid depth and acceleration due to gravity. Choose from the codes given below which of the following numbers must match for the model with that of the prototype :

1. Mach number
2. Weber number
3. Froude number
4. Reynolds number.

Codes :

(a) 3 alone (b) 1, 2, 3 and 4
(c) 1 and 2 (d) 3 and 4.

343. During subsonic, adiabatic flow of gases in pipes with friction, the flow friction, the flow properties go through particular mode of changes. Match List-I (Flow properties) with List-II (Mode of changes) and select the correct answer from the codes given below the Lists :

List-I . (Flow Properties)	List-II (Mode of changes)
A. Pressure	1. Increases in flow direction
B. Density	2. Deceases with flow direction
C. Temperature	3. Relative density
D. Velocity	4. Pressure differential

Codes :

	A	B	C	D
(a)	1	1	2	2
(b)	2	2	2	1
(c)	2	2	1	2
(d)	2	1	1	2

344. Hydrostatic law of pressure is given as :

(a) $\frac{\partial p}{\partial z} = \rho g$ (b) $\frac{\partial p}{\partial z} = 0$

(c) $\frac{\partial p}{\partial z} = z$ (d) $\frac{\partial p}{\partial z} = \text{constant}$.

345. In a pipe-flow, pressure is to be measured at a particular cross-section using the most appropriate instrument. Match List-I (Expected pressure range) with List-II (appropriate measuring device) and select the correct answer using the codes given below the Lists :

List-I (Expected)	List-II (Appropriate)
A. Steady flow with small positive gauge pressure	1. Bourdon pressure gauge
B. Steady flow with small negative and positive gauge pressure	2. Pressure transducer
C. Steady flow	3. Simple piezometer
D. Unsteady flow with fluctuating pressure	4. U-tube manometer

Codes :

	A	B	C	D
(a)	3	2	1	4
(b)	1	4	3	2
(c)	3	4	1	2
(d)	1	2	3	4

346. The capillary rise at 20°C in clean glass tube of 1 mm diameter containing water is approximately :

(a) 15 mm (b) 50 mm
(c) 20 mm (d) 30 mm.

347. Pressure drop of water flowing through a pipe (density 1000 kg/m^3) between two points is measured by using a vertical U–tube manometer. Manometer uses a liquid with density 2000 kg/m^3. The difference in height of manometric liquid in the two limbs of the manometer is obsereved to be 10 cm. The pressure drop between the two points is :

(a) 98.1 N/m^2 (b) 981 N/m^2
(c) 1962 N/m^2 (d) 19620 N/m^2

348. Match List-I (Stability) with List-II (Conditions) and select the correct answer using the codes given below the lists :

List-I (Stability)	List-II (Conditions)
A. Stable equilibrium of a floating body	1. Centre of buoyancy below the centre of gravity
B. Stable equilibrium of a submerged body	2. Metacentre above the centre of gravity
C. Unstable equilibrium of a floating body	3. Centre of buoyancy above the centre of gravity
D. Unstable equilibrium of a submerged body	4. Metacentre below the centre of gravity

Codes :

	A	B	C	D
(a)	4	3	2	1
(b)	2	3	4	1
(c)	4	1	2	3
(d)	2	1	4	3

349. A dam is having a curved surface as shown in the Fig. 33.

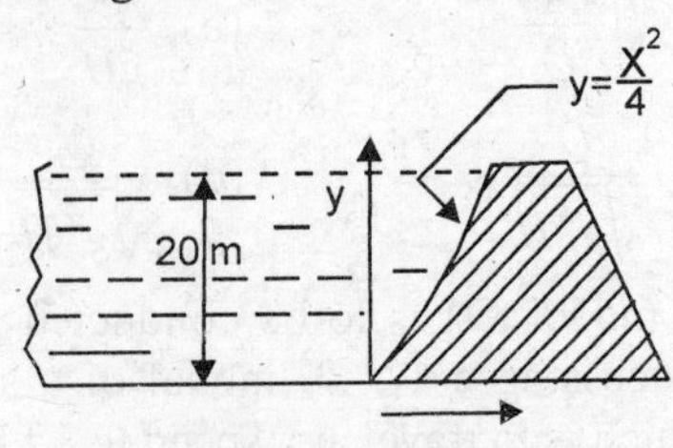

Fig. 33

The height of the water retained by the dam is 20 m, density of water is 1000 kg/m^3 Assuming g as 9.81 m/s^2, the horizontal force acting on the dam per unit length is :

(a) 1.962×10^2 N (b) 2×10^5 N
(c) 1.962×10^6 N (d) 3.924×10^6 N.

350. The velocity potential of a velocity field is given by $\phi = x^2 - y^2$ + const. Its stream function will be given by :

(a) $-2xy$ + constant (b) $+2xy$ + constant
(c) $-2xy + f(x)$ (d) $-2xy + f(y)$.

351. A sternline is a line :

(a) which is along the path of the particle
(b) which is always parallel to the main direction of flow
(c) along which there is no flow
(d) on which tangent drawn at any point gives the direction of velocity.

352. Match List-I (Example) with List-II (Type of flow) and select the correct answer using the codes given below :

List-I (Example)	List-II (Type of flow)
A. Flow in a straight long pipe with varying fow rate	1. Uniform, steady
B. Flow of gas through the nozzle of a jet engine	2. Uniform, unsteady
C. Flow of water through the hose of a fire fighting pump	3. Uniform, unsteady
D. Flow in a river during tidal wave	4. Non-uniform, unsteady

Codes :

	A	B	C	D
(a)	1	4	3	2
(b)	3	2	1	4
(c)	1	2	3	4
(d)	3	4	1	2

353. Match List-I (Type of fluid) with List-II (Variation of shear stress) and select the correct answer using the codes given below:

List-I (Type of fluid)	List-II (Variation of shear stress)
A. Ideal fluid	1. Shear stress varies linearly with the rate of strain
B. Newtonian fluid	2. Shear stress does not vary linearly with the rate of strain
C. Non-Newtonian fluid	3. Fluid behaves like a solid until a minimum yield stress beyond which it exhibits a linear relatioship between shear stress and the rate of strain
D. Bingham plastic	4. Shear stress is zero

Codes :

	A	B	C	D
(a)	3	1	2	4
(b)	4	2	1	3
(c)	3	2	1	4
(d)	4	1	2	3

354. The equation of a velocity distribution over a plate is given by $u = 2y - y^2$ where u is the velocity in m/s at a point y meter from the plate measured perpendicularly. Assuming $\mu = 8.60$ poise, the shear stress at point 15 cm from the boundary is :

(a) 1.72 N/m^2 (b) 1.46 N/m^2
(c) 14.62 N/m^2 (d) 17.20 N/m^2.

355. Match List-I (Fluid parameters) with List-II (Basic dimensions) and select the correct answers using the codes given below the lists:

List-I (Fluid parameters)	List-II (Basic dimensions)
A. Dynamic viscosity	1. $\frac{M}{T^2}$
B. Chezy's roungness coefficient	2. $\frac{M}{LT^2}$
C. Bulk modulus of elasticity	3. $\frac{M}{LT}$
D. Surface tension (σ)	4. $\frac{\sqrt{L}}{T}$

Codes :

	A	B	C	D
(a)	3	2	4	1
(b)	1	4	2	3
(c)	3	4	2	1
(d)	1	2	4	3

356. The force of impingement of a jet on a vane increases if :
(a) the vane angle is increased
(b) the vane angle is decreased
(c) the pressure is reduced
(d) the vane is moved against the jet.

357. Which of the following assumptions are made for deriving Bernoulli's equation ?
1. Flow is steady and incompressible
2. Flow is unsteady and compressible
3. Effect of friction is neglected and flow is along a stream line
4. Effect of friction is taken into consideration and flow is along a stream line.

Select the correct answer using the codes given below :

Codes :
(a) 1 and 3 (b) 2 and 3
(c) 1 and 4 (d) 2 and 4.

358. While measuring the velocity of air (ρ = 1.2 kg/m^3), the difference in the stagnation and static pressures of a pitot static tube was found to be 380 Pa. the velocity at the location in m/s is :
(a) 24.03 (b) 4.02
(c) 17.8 (d) 25.17.

359. The drag force exerted by a fluid on a body immersed in the fluid is due to :
(a) pressure and viscous forces
(b) pressure and gravity forces
(c) pressure and surface tension forces
(d) viscous and gravity forces.

360. Which phenomenon will occur when the valve at the discharge end of a pipe connected to a reservoir is suddenly closed ?
(a) Cavitation (b) Erosion
(c) Hammering (d) Surging.

361. When can piezometer be not used for pressure measurement is pipes ?
(a) The pressure difference is low
(b) The velocity is high
(c) The fluid in the pipe is a gas
(d) The fluid in the pipe is highly viscous.

362. Which one of the following causes lift on an immersed body in a fluid stream ?
(a) Buoyant forces
(b) Resultant fluid force on the body
(c) Dynamic fluid force component exerted on the body parallel to the approach velocity
(d) Dynamic fluid force component exerted on the body normal to the approach velocity.

363. If for a flow, a stream function ψ exists and satisfies the Laplace equation, then which one of the following is the correct statement ?
(a) The continuity equation is satisfied and the flow is irrotational
(b) The continuity equation is satisfied and the flow is rotational
(c) The flow is irrotional but does not satisfy the continuity equation
(d) The flow is rotational.

364. When a vertical cylindrical vessel containing water is rotated about its axis, then the free surface of water becomes :
(a) a cycloid of revolution
(b) an ellipsoid of revolution
(c) a hyperboloid of revolution
(d) a paraboloid of revolution.

365. Which one of the following statements is correct? The pressure centre is :
(a) the centroid of the pressure prism
(b) a point on the line of action of the resultant force
(c) at the centroid of the submerged area
(d) always above the centroid of the area.

366. Consider the following properties of a fluid :
1. Viscosity
2. Surface tension
3. Capillarity
4. Vapour pressure.

Which of the above properties can be attributed to the flow of jet of oil in an unbroken stream?
(a) 1 only (b) 2 only
(c) 1 and 3 (d) 2 and 4

367. A U-tube open at both ends and made of 8 mm diameter glass tube has mercury up to a height of 10 cm in both the limbs. If 19 cm of water is added to one of the limbs, what is the difference in mercury levels in the two limbs at equilibrium?
(a) 4.5 mm (b) 1.0 cm
(c) 2.8 cm (d) 3.2 cm.

368. Which one of the dimensionless number identifies the compressibility effect of a fluid?
(a) Euler number (b) Froude number
(c) Mach number (d) Weber number.

369. What is the shape factor of a hemispherical body placed on a flat surface with respect to itself?
(a) Zero (b) 0.25 (c) 0.5 (d) 1.0

370. For linear distributon of velocity in the boundary layer on a flat plate, what is the ratio of displacement thickness (δ^*) the boundary layer thickness (δ)?

(a) $\frac{1}{4}$ (b) $\frac{1}{3}$
(c) $\frac{1}{2}$ (d) $\frac{1}{5}$.

371. Consider the following statements :
1. For achieving dynamic similarity in model studies on ships, Froude numbers are equated
2. Reynolds number should be equated for studies on aerofoil for dynamic similarity
3. In model studies on a spillway, the ratio of width to height is equated for kinematic similarity.

Which of the statements given above are correct?
(a) 1, 2 and 3 (b) 1 and 2
(c) 2 and 3 (d) 1 and 3.

372. How is the best hydraulic channel cross-section defined?
(a) The section with minimum roughness coefficient
(b) The section that has a maximum area for a given flow
(c) The section that has minimum wetted perimeter
(d) The section that has maximum wetted area.

373. Which one of the following is the condition for equilibrium for a floating body?
(a) The metacentre coincides with the centre of gravity
(b) The metacentre is below the centre of gravity
(c) The metacentre is above the centre of gravity
(d) The centre of buoyancy is below the crentre of gravity.

374. Which of the following assumptions/ conditions are true in the case of Rayleigh flow?
1. Perfect gas
2. Constant area duct
3. Steady one-dimensional real flow
4. Heattransfer during the flow.

Select the correct answer using the code given below :
(a) 1, 2 and 3 (b) 2, 3 and 4
(c) 1, 3 and 4 (d) 1, 2 and 4.

375. If the cross-section of a nozzle is increasing in the direction of flow in supersonic flow, then in the downstream direction :
(a) both pressure and velocity will increase
(b) both pressure and velocity will decrease
(c) pressure will increase but velocity will decrease
(d) pressure will decrease but velocity will increase.

376. A 12 cm diameter straight pipe is laid at a uniform downgrade and flow rate is maintained uch that velocity head in the pipe is 0.5 m. If the pressure in the pipe is observed to be inform along the length when the down slope of the pipe is 1 in 10, what is the friction actor for the pipe?
(a) 0.012 (b) 0.024
(c) 0.042 (d) 0.050.

377. The energy grade line (EGL) for steady flow in a uniform diameter pipe is shown below in Fig. 34. Which of the following items is contained in the box ?

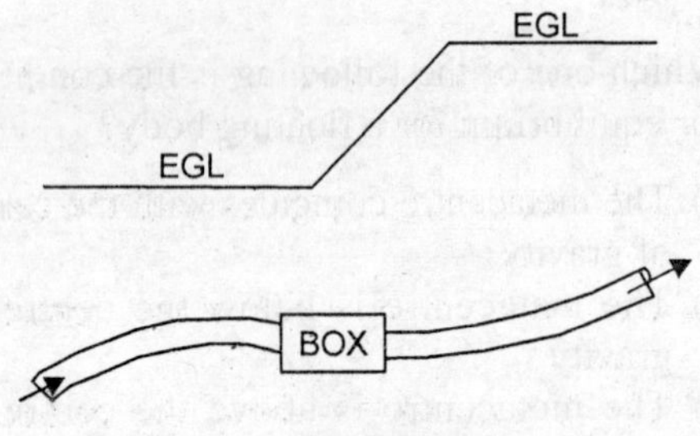

Fig. 34

(a) A pump (b) A turbine
(c) A partially closed valve
(d) An abrupt expansion.

378. A compound pipeline consists of two pieces of idential pipes. The equivalent length of same diameter and same friction factor, for the compound pipeline is L_1 when pipes are connected in series, and is L_2 when connected in parallel. What is the ratio of equivalent length L_1/L_2?
(a) 32 : 1 (b) 8 : 1
(c) 2 : 1 (d) $\sqrt{2}:1$.

379. For irrotational and incompressible flow, the velocity potential and steam function are given by ϕ and ψ, respectively. Which one of the following sets is correct ?
(a) $\nabla^2\phi =, \nabla^2\psi \neq 0$
(b) $\nabla^2\phi \neq 0, \nabla^2\psi = 0$
(c) $\nabla^2\phi = 0, \nabla^2\psi \neq 0$
(d) $\nabla^2\phi \neq 0, \nabla^2\psi \neq 0$.

380. In a two-dimensional incompressible steady flow, the velocity component $u = Ae^x$ is obtained. What is the other component v of velocity ?
(a) $v = Ae^{xy}$ (b) $v = Ae^{xy}$
(c) $v = -Ae^x + f(x)$ (d) $v = -Ae^y x + f(y)$.

381. Consider the following statements about hydrostatic force on a submerged surface:
1. It remains the same even when the surface is turned
2. It acts vertically even when the surface is turned.

Which of these is/are correct?
(a) Only 1 (b) Only 2
(c) Both 1 and 2 (d) Neither 1 nor 2.

382. The depth of centre of pressure for a rectangular lamina immersed vertically in water up to height 'h' is given by :
(a) $h/2$ (b) $h/4$ (c) $2h/3$ (d) $3h/2$.

383. The vertical component of force on a curved surface submerged in a static liquid is equal to the :
(a) Weight of liquid column above the C.G. of the curved surface
(b) Weight of liquid above the curved surface
(c) Product of pressure at C.G., multiplied by the area of the curved surface
(d) Product of pressure at C.G., mulripled by the projected area of the curved surface.

384. The point of application of a horizontal force on a curved surface submerged in liquid is:

(a) $\frac{I_G}{A\bar{h}} = \bar{h}$ (b) $\frac{I_G + A\bar{h}^2}{A\bar{h}}$

(c) $\frac{A\bar{h}}{I_G} + \bar{h}$ (d) $\frac{I_G}{\bar{h}} + A\bar{h}$.

where A = area of the immersed surface

$\bar{h}$ = depth of centre of surface immersed

I_G = moment of inertia about centre of gravity

385. To measure the pressure head of the fluid of specific gravity S flowing through a pipeline, a simple micro-manometer containing a fluid of specific gravity S_1 is connected to it. The readings are as indicated in the Fig. 35. The pressure head in the pipeline is:

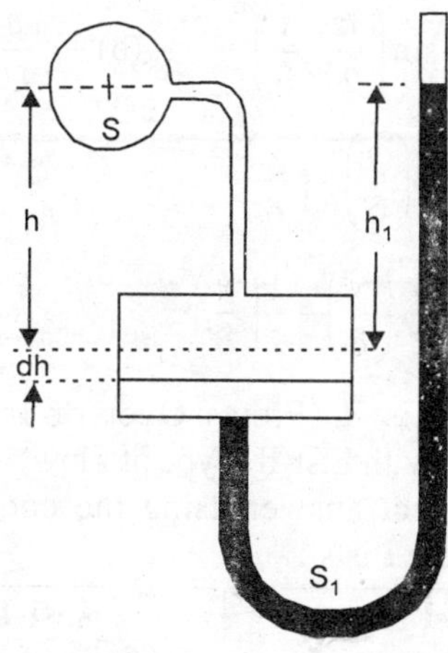

Fig. 35

(a) $h_1S_1 - hS - \Delta h\ (S_1 - S)$
(b) $h_1S_1 - hS + \Delta h\ (S_1 - S)$
(c) $hS - h_1S_1 - \Delta h\ (S_1 - S)$
(d) $hS - h_1S_1 + \Delta h\ (S_1 - S)$.

386. Match List-I (Rheological Equation) with List-II (Types of Fluids) and select the correct answer using the codes given below the Lists:

List-I (Rheological Equation)	List-II (Types of Fluids)
A. $\tau = \mu\left(\frac{du}{dy}\right)^n$, $n = 1$	1. Bingham plastic
B. $\tau = \mu\left(\frac{du}{dy}\right)^n$, $n < 1$	2. Dilatant fluid
C. $\tau = \mu\left(\frac{du}{dy}\right)^n$, $n > 1$	3. Newtonian fluid
D. $\tau = \tau_0 + \mu\left(\frac{du}{dy}\right)^n$, $n = 1$	4. Pseudo-plastic fluid

Codes :

	A	B	C	D
(a)	3	2	4	1
(b)	4	1	2	3
(c)	3	4	2	1
(d)	4	2	1	3

387. Which one of the following streamfunctions is a possible irrotational flow field?

(a) $\psi = x^3y$ (b) $\psi = 2xy$
(c) $\psi = Ax^3y^2$ (d) $\psi = Ax + By^2$.

388. The expression $\frac{\partial\phi}{\partial t} + \int\frac{\partial p}{\rho} + \frac{1}{2}|\nabla\phi|^2 + gz =$ constant represents :

(a) steady flow energy equation
(b) unsteady irrotational Bernoulli's equation
(c) steady rotational Bernoulli's equation
(d) unsteady rotational Bernoulli's equation.

389. For a circular channel, the wetted parameter (where R = radius of circular channel θ = half the angle subtended by the water surface at the centre) is given by :

(a) $R\theta/2$ (b) $3R\theta$
(c) $2R\theta$ (d) $R\theta$.

390. Consider the following statements :
A hydraulic jump occurs in an open channel
1. when the Froude number is equal to or less than one
2. at the toe of a spillway
3. downstream of a sluice gate in a canai
4. when the bed slope suddenly changes.

Which of these are correct ?
(a) 1, 2, 3 and 4 (b) 1, 2 and 3
(c) 2, 3 and 4 (d) 1 and 4.

391. An orifice meter with $C_d = 0.61$ is substituted by venturimeter with C, = 0.98 in a pipeline carrying crude oil having the same throat diameter as that of the orifice. For the same flow rate, the ratio of the pressure drops for the venturimeter and the orifice meter is:
(a) 0.61/0.98 (b) $(0.61)^2/(0.98)^2$
(c) 0.98/0.61 (d) $(0.98 - 0.61)^2$.

392. The instrument preferred in the measurement of highly fluctuating velocities in air flow is :
(a) Pitot-static tube
(b) Propeller type anemomenter
(c) Three cup anemometer
(d) Hot wire anemometer.

393. The friction factor for fully developed turbulent flow through a rough circular pipe depends on :
(a) Reynolds number
(b) Relative roughness
(c) Reynolds number and relative roughness
(d) Reynolds number, relative roughness and diameter to length ratio.

394. In a steady flow of an oil in a pipe in the fully developed laminar regime, the shear stress is:
(a) Constant across the pipe
(b) Maximum at the centre and decreases parabolically towards the pipe wall boundary
(c) Zero at the boundary and increases linearly towards the centre
(d) Zero at the centre and increases towards pipe wall.

395. Velocity of pressure waves due to pressure distrubances imposed in a liquid is equal to:
(a) $\left(\frac{E}{\rho}\right)^{1/2}$ (b) $(E\rho)^{1/2}$
(c) $\left(\frac{\rho}{E}\right)^{1/2}$ (d) $\left(\frac{1}{\rho E}\right)^{1/2}$

396. Consider the following statements pertaining to boundary layer :
1. Boundary layer is a thin layer adjacent to the boundary where maximum viscous energy dissipation takes place
2. Boundary layer thickness is a thickness by which the ideal flow is shifted
3. Separation of boundary layer is caused by presence of adverse pressure gradient.

Which of these statements are correct ?
(a) 1, 2 and 3 (b) 1 and 2
(c) 1 and 3 (d) 2 and 3.

397. The velocity profile for turbulent layer over a flat plate is :
(a) $\frac{u}{U} = \sin\left(\frac{\pi}{2} - \frac{y}{\delta}\right)$ (b) $\frac{u}{U} = \left(\frac{y}{\delta}\right)^{\frac{1}{7}}$
(c) $\frac{u}{U} = 2\left(\frac{y}{\delta}\right) - \left(\frac{y}{\delta}\right)^2$
(d) $\frac{u}{U} = \frac{3}{2}\left(\frac{y}{\delta}\right) - \frac{1}{2}\left(\frac{y}{\delta}\right)^3$.

398. Match List-I (Flows Over or Inside the system) with List-II (Type of Flow) and select the correct answer using the codes given below the Lists :

List-I (Flows Over or Inside the Systems)	**List-II (Type of Flow)**
A. Flow over a sphere	1. Two dimensional flow
B. Flow over a long circular cylinder	2. One dimensional flow
C. Flow in a pipe bend	3. Axisymmetric flow
D. Fully developed flow in a pipe at constant flow rate	4. Three dimensional flow

Codes :

	A	B	C	D
(a)	3	1	2	4
(b)	1	4	3	2
(c)	3	1	4	2
(d)	1	4	2	3

399. Consider the following statements :

1. Dimensional analysis is used to determine the number of variables involved in a certain phenomenon
2. The group of repeating variables in dimensional analysis should include all the fundamental units
3. Buckingham's π-theorem stipulates the number of dimensionless groups for a givp-phenomenon
4. The coefficient in Chezy's equation has no dimension.

Which of these are correct?

(a) 1, 2, 3 and 4 (b) 2, 3 and 4
(c) 1 and 4 (d) 2 and 3.

400. Match List-I (Flow/Wave) with List-II (Dimensionless Number) and select the correct answer using the codes given below the Lists :

List-I (Flow/Wave)	List-II (Dimension-less Number)
A. Capillary waves in channel	1. Reynolds number
B. Testing of aerofoils	2. Froud number
C. Flow around bridge piers	3. Weber number
D. Turbulent flow through pipes	4. Euler number
	5. Mach number

Codes:

	A	B	C	D
(a)	5	4	3	2
(b)	3	5	4	1
(c)	5	4	2	1
(d)	3	5	2	1

401. Consider the following statements:

1. Complete similarity between model and prototype envisages geometric and dynamic similarities only
2. Distorted models are necessary where geometric similarity is not possible due to practical reasons
3. In testing of model of a ship, the surface tension forces are generally neglected
4. The scale effect takes care of the effect of dissimilarity between model and prototype.

Which of these statements are correct?

(a) 1 and 3 (b) 1, 2 and 4
(c) 2 and 3 (d) 2 and 4.

402. If coefficient of contraction at the vena contracta is equal to 0.62, then what will be the dynamic loss coefficient in sudden contraction in air-conditioning duct?

(a) 0.25 (b) 0.375
(c) 0.55 (d) 0.65.

403. An oil specific gravity 0.9 has viscosity of 0.28 Stokes at 38°C. What will be its viscosity in Ns/m^2?

(a 0.2520 (b) 0.0311
(c) 0.0252 (d) 0.0206.

404. A mercury-water manometer has a gauge difference of 500 mm (difference in elevation of meniscil). What will be difference in pressure?

(a) 0.5 m (b) 6.3 m
(c) 6.8 m (d) 7.3 m.

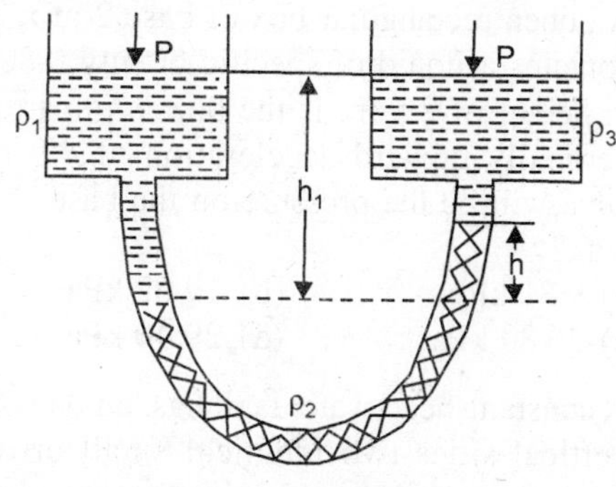

Fig. 36

The balancing column shown in the above Fig. 36 contains 3 liquids of different densities, ρ_1, ρ_2 and ρ_3. The liquid level of one limb is h_1 below the top level and there is a difference of h relative to that in the other limb. What will be the expression for h?

(a) $\frac{\rho_2 - \rho_3}{\rho_1 - \rho_3} h_1$ (b) $\frac{\rho_2 - \rho_3}{\rho_1 - \rho_3} h_1$

(c) $\frac{\rho_1 - \rho_3}{\rho_2 - \rho_3} h_1$ (d) $\frac{\rho_1 - \rho_2}{\rho_2 - \rho_3} h_1$.

406. A circular annular plate bounded by two concentric circles of diameter 1.2 m and 0.8 m is immersed in water with its plane making an angle of 45° with the horizontal. The centre of the circles is 1.625 m below the free surface. What will be the total pressure force on the face of the plate?

(a) 7.07 kN (b) 10.00 kN
(c) 14.14 kN (d) 18.00 kN.

407. A plate of rectangular shape having the dimensions of 0.4 m × 0.6 m is immersed in water with its longer side vertical. The total hydrostatic thrust on one side of the plate is estimated as 18.3 kN. All other conditions remaining the same, the plate is turned through 90° such that its longer side rematins vertical. What would be the total force on one face of the plate?

(a) 9.14 kN (b) 18.3 kN
(c) 36.6 kN (d) 12.2 kN.

408. An open rectangular box of base 2 m × 2 m contains a liquid of specific gravity 0.80 up to height of 2.5 m. If the box is imparted a vertically upward acceleration of 4.9 m/s, what will be the pressure on the base of the tank?

(a) 9.81 kPa (b) 19.62 kPa
(c) 36.80 kPa (d) 29.40 kPa.

409. A constant-head water tank has, on one of its vertical sides two identical small orifices issuing two horizontal jets in the same vertical plane. The vertical distance between the centres of orifices is 1.5 m and the jet trajectories intersect at a point 0.5 m below the lower orifice. What is the approximate height of water level in the tank above the point of intersection of trajectories?

(a) 1.0 m (b) 2.5 m
(c) 0.5 m (d) 2.0 m.

410. The components of velocity in two-demensional frictionless incompressible flow are $u = t^2 + 3y$ and $v = 3t + 3x$. What is the approximate resultant total acceleration at the point (3, 2) and $t = 2$?

(a) 5 (b) 49
(c) 59 (d) 54.

411. The stream function $y = x^3 - y^3$ is observed for a two dimensional flow field. What is the magnitude of the velocity at point (1, – 1)?

(a) 4.24 (b) 2.83
(c) 0 (d) – 2.83.

412. Which one of the following statements is correct?

Irrotational flow is characterized as the one in which :

(a) the fluid flows along a straight line
(b) the fluid does not rotate as it moves along
(c) the net rotation of fluid particles about their mass centres remains zero
(d) the streamlines of flow are curved and closely spaced.

413. A 20 cm diameter 500 m long water pipe with friction factor $c_f = 0.025$, leads from a constant-head reservoir and terminates at the delivery end into a nozzle discharging into air. (Neglect all energy losses other than those due to pipe friction). What is the approximate diameter of the jet for maximum power?

(a) 6.67 mm (b) 5.98 mm
(c) 66.7 (d) 59.8 mm.

414. Which of the following functions represent the velocity potential in a two-dimensional flow of an ideal fluid?

1. $2x + 3y$ 2. $4x^2 - 3y^2$
3. $\cos(x - y)$ 4. $\tan^{-1}\left(\frac{x}{y}\right)$.

Select the correct answer using the codes given below :

(a) 1 and 3 (b) 1 and 4
(c) 2 and 3 (d) 2 and 4.

415. The pressure drop for a relatively low Reynolds number flow in a 600 mm, 30 m long pipeline is 70 kPa. What is the wall shear stress?

(a) 0 Pa (b) 350 Pa
(c) 700 Pa (d) 1400 Pa.

416. A pitot-static tube ($C = 1$) is used to measure air flow. With water in the differential manometer and a gauge difference of 75 mm, what is the value of air speed if $\rho = 1.16$ kg/m^3?

(a) 1.21 m/s (b) 16.2 m/s
(c) 35.6 m/s (d) 71.2 m/s.

417. Match List-I (Measuring devices) with List-II (Measured Parameter) and select the correct answer using the codes given below the Lists:

List-I (Measuring Devices)	List-II (Measured Parameter)
A. Pitot tube	1. Flow static pressure
B. Micro-manometer	2. Rate of flow (indirect)
C. Pipe bend meter	3. Differential pressure
D. Wall pressure tap	4. Flow stagnation pressure

Codes:

	A	B	C	D
(a)	1	3	2	4
(b)	4	3	2	1
(c)	1	2	3	4
(d)	4	2	3	1

418. Match List-I (Variable in Laminar Boundary Layer Flow over a flat Plate Set Parallel to the Stream) with List-II (Related Expression with usual notations) and select the correct answer using the codes given below the Lists:

List-I (Variable is Laminar Boundary Layer Flow over a Flat Plate Set Parallel to the Stream)	List-II (Related Expression with usual notations)
A. Boundary layer thickness	1. $1.729\sqrt{Ux/v}$
B. Average skin-friction coefficient	2. $0.332rU2/\sqrt{Ux/v}$
C. Shear stress at boundary	3. $5/\sqrt{v/xU}$
D. Displacement thickness	4. $0.664\sqrt{UL/x}$
	5. $1.328\sqrt{UL/v}$

Codes :

	A	B	C	D
(a)	3	5	4	2
(b)	2	4	1	3
(c)	3	5	2	1
(d)	5	4	1	2

419. A flat plate, 2m × 0.4 m is set parallel to a uniform stream of air (density 1.2 kg/m^3 and viscosity 16 centistokes) with its shorter edges along the flow. The air velocity is 30 km/ h. What is the approximate estimated thickness of boundary layer at the downstream end of the plate?

(a) 1.96 mm (b) 4.38 m
(c) 13.12 mm (d) 9.15 mm.

420. What is the ratio of displacement thickness to momentum thickness for linear velocity distribution over a flat plate?

(a) 1.5 (b) 2.0
(c) 2.5 (d) 3.0.

421. Consider the following statement regarding laminar sublayer of boundary layer flow:

1. The laminar sublayer exists only in a region that occurs before the formation of laminar boundary layer

2. The laminar sublayer is a region next to the wall where the viscous force is predominant while the rest of the flow is turbulent
3. The laminar sublayer occurs only in turbulent flow past a smooth plate.

Which of the statements given above is/are correct?

(a) 1, 2 and 3 (b) 1 and 2
(c) Only 2 (d) 1 and 3.

422. Match List-I (Flow Depth) with List-II (Basic Hydraulic Condition Associated therewith) and select the correct answer using the codes given below the Lists:

List-I (Flow Depth)	List-II (Basic Hydraulic Condition Associated therewith)
A. Conjugate depth	1. Uniform flow
B. Critical depth	2. Same specific energy
C. Alternate depth	3. Minimum specific energy
D. Normal depth	4. Same specific force
	5. Same bed slope

Codes :

	A	B	C	D
(a)	3	5	4	2
(b)	2	4	1	3
(c)	4	3	2	1
(d)	5	4	1	2

423. A 40 mm diameter 2 m long straight uniform pipe carries a steady flow of water (viscosity 1.02 centipoise) at the rate of 2.0 litres per minute. What is the approximate value of the stress on the internal wall of the pipe?

(a) 0.0166 dyne/cm^2 (b) 0.0812 dyne/cm^2
(c) 8.12 dyne/cm^2 (d) 0.9932 dyne/cm^2.

424. Which one of the following is the correct expression for the area of flow for a circular channel? (Where θ = half the angle subtended by water surface at the centre and R = radius of the circular channel):

(a) $R^2\left(2\theta - \dfrac{\sin 2\theta}{2}\right)$ (b) $R^2\left(\theta - \dfrac{\sin 2\theta}{2}\right)$

(c) $R^2(2\theta - \sin 2\theta)$ (d) $2R^2(\theta - \sin 2\theta)$.

425. A 1.0 m long model of a ship is towed at a speed of 81 cm/s in a towing tank. To what speed of the ship, 64 m long does this correspond to?

(a) 7.20 m/s (b) 6.48 m/s
(c) 5.76 m/s (d) 3.60 m/s.

426. In a flow condition where both viscous and gravity forces dominate and both the Froude number and the Reynolds number are the same in model and prototype; and the ratio of kinematic viscosity of model to that of the prototype is 0.0894. What is the model scale?

(a) 1 : 3.3 (b) 3.3 : 1
(c) 5 : 1 (d) 1 : 5.

427. A ship whose full length is 100 m is to travel at 10 m/s. For dynamic similarity, with what velocity should a 1 : 25 model of the ship be towed?

(a) 2 m/s (b) 10 m/s
(c) 25 m/s (d) 250 m/s.

428. Match List-I (Type of Model) with List-II (Transference Ratio for Velocity) and select the correct answer using the codes given below the Lists:

List-I (Type of Model)	List-II (Transference Ratio for Velocity)
A. Reynolds model	1. $\sqrt{R_r / \rho_r}$
B. Froude model	2. $\sqrt{\sigma_r / (\rho_r l_r)}$
C. Weber model	3. $\mu_r / (r_r l_r)$
D. Mach model	4. $\sqrt{g_r l_r}$.

(Where symbols g, μ, σ and *K* have their usual meanings and subscript *r* refers to the ratio):

Codes :

	A	B	C	D
(a)	3	1	3	4
(b)	3	4	2	1
(c)	2	1	3	4
(d)	2	4	3	1

429. For a steady two-dimensional flow, the scalar components of the velocity field are $v_x = 2x$, $v = 2y$, $v_{zx} = 0$. What are the components of aceleration?

(a) $a_x = 0, a_y = 0$ (b) $a_x = 4x, a_y = 0$
(c) $a_x = 0, a_y = 4y$ (d) $a_r = 4x, a_y = 4y$

430. Consider the following statements regarding a path line in fluid flow :

1. A path line is a line traced by a single particle over a time interval
2. A path line shows the positions of the same particle at successive time instants
3. A path line shows the instantaneous positions of a number of particles, passing through a common point, at some previous time instants.

Which of the statements given above are correct?

(a) Only 1 and 3 (b) Only 1 and 2
(c) Only 2 and 3 (d) 1, 2 and 3.

431. A beaker of water is falling freely under the influence of gravity. Point *B* is on the surface and point C is vertically below B near the bottom of the beaker. If p_B is the pressure at point *B* and p_c the pressure at point C, then which one of the following is correct?

(a) $pB = pc$ (b) $pB < Pc$
(c) $pB > pc$.
(d) Insufficient to determine any of the above.

432. A homogeneous solid of any arbitary shape floats upright in a homogeneous liquid with immersed volume *V* and is in stable equilibrium. If the solid is overturned and made to float upside down in a different homogeneous liquid with exactly same volume *V* above the liquid surface, then the equilibrium.

(a) would be stable
(b) would be neutral
(c) would be unstable
(d) may or may not be stable.

433. A 25 cm long prismatic homogeneous solid floats in water with its axis vertical and 10 cm projecting above water surface. If the same solid floats in some oil with axis vertical and 5 cm projecting above the liquid surface, what is the specific gravity of the oil?

(a) 0.60 (b) 0.70 (c) 0.75 (d) 0.80.

434. A semi-circular plane area of diameter 1 m, is subjected to a uniform gas pressure of 420 kN/m^2. What is the moment of thrust (aporoximately) on the area about its straight edge?

(a) 35 kNm (b) 41 kNm
(c) 55 kNm (d) 82 kNm.

435. A horizontal oil tank is in the shape of a cylinder with hemispherical ends. If it is exactly half full, what is the ratio of magnitude of the vertical component of resultant hydrostatic thrust on one hemispherical end to that of the horizontal component?

(a) 2π (b) $\pi/2$
(c) $4/(3\pi)$ (d) $2\pi/(4)$.

436. The standard atmospheric pressure is 762 mm of Hg. At this place, what does an absolute pressure of 380 mm of Hg correspond to?

(a) 320 mm of Hg vacuum
(b) 382 mm of Hg vacuum
(c) 62 mm of Hg vacuum
(d) 62 mm of Hg gauge.

437. A manometer is made of a tube of uniform bore of 0.5 cm^2 cross-sectional area, with one limb vertical and the other limb inclined at 30° to the horizontal. Both of its limbs are open to atmosphere and, initially, it is partly filled with a manometer liquid of specific gravity 1.25. If then an additional volume of 7.5 cm^3 of water is poured in the inclined tube, what is the rise of the meniscus in the vetical tube?

(a) 4 cm (b) 7.5 cm
(c) 12 cm (d) 15 cm.

438. A vertical clean glass tube of uniform bore is used as a piezometer to measure the pressure of liquid at a point. The liquid has a specific weight of 15 kN/m^3 and a surface tension of 0.06 N/m in contact with air. If for the liquid, the angle of contact with glass is zero and the capillary rise in the tube is not to exceed 2 mm, what is the required minimum diameter of the tube?
(a) 6 mm (b) 8 mm
(c) 10 mm (d) 12 mm.

439. When the pressure on a given mass of liquid is increased from 3.0 MPa to 35 MPa, the density of the liquid increases from 500 kg/m^3 to 501 kg/m^3. What is the average value of bulk modulus of the liquid over the given pressure range?
(a) 700 MPa (b) 600 MPa
(c) 500 MPa (d) 250 MPa.

440. Which one the following is measured by a rotameter ?
(a) Velocity of fluids
(b) Discharge of fluids
(c) Viscosity of fluids
(d) Rotational speed of solid shafts.

441. A venturimeter in an oil (sp. gr. 0.8) pipe is connected to a differential manometer in which the gauge liquid is mercury (sp. gr. 13.6). For a flow rate of 0.16 m^3/s, the manometer registers a gauge differential of 20 cm. The oil-mercury manoneter being unavailable, an air oil differential manometer is connected to the same venturimeter. Neglecting variation of discharge coefficient for the venturimeter, what is the new gauge differential for a flow rate of 0.08 m^3/s?
(a) 64 cm (b) 68 cm
(c) 80 cm (d) 85 cm.

442. How are the velocity coefficient C_v· the discharge coefficient C_d, and the contraction coefficient C_c of an orifice related?
(a) $C_v = C_C C_d$ (b) $C_c = C_v C_d$
(c) $C_d = C_C C_v$ (d) $C_c C_v C_d = 1$.

443. Which one of the following statements is correct while using boundary layer equations, Bernoulli's equation :
(a) can be used anywhere
(b) can be used only outside the boundary layer
(c) can be used only inside the boundary layer
(d) cannot be used either inside or outside the boundary layer.

444. The velocity distribution in the boundary layer over a flat plate, set parallel to the direction of in incompressible free stream, is given by one-sixth power law of the form $u/U = (y/6)^{1/6}$. What is the ratio of displacement thickness to boundary layer thickness?
(a) 1:8 (b) 1 : 7
(c) 7 : 72 (d) 3 : 8.

445. An open channel of symmetric right-angled traingular cross-section is conveying a discharge Q. Taking g as the acceleration due to gravity, what is the critical depth?
(a) $\left(\frac{Q^2}{g}\right)^{1/3}$ (b) $\left(\frac{2Q^2}{g}\right)^{1/3}$
(c) $\left(\frac{Q^2}{g}\right)^{1/5}$ (d) $\left(\frac{2Q^2}{g}\right)^{1/5}$

446. In M-L-T system, what is the dimension of specific speed for a rotodynamic pump?
(a) $L^{-3/4}\,T^{3/2}$ (b) $M^{1/2}L^{1/4}T^{-5/2}$
(c) $L^{3/4}T^{-3/2}$ (d) $L^{3/4}T^{3/2}$.

446. In the boundary layer, the flow is:
(a) viscous and rotational
(b) inviscid and irrotational
(c) inviscid and rotational
(d) viscous and irrotational.

448. It is observed in a flow problem that pressure, inertia and gravity forces are important. Then, similarity requires that :
(a) Reynolds and Weber number be equal
(b) Mach and Froude numbers be equal
(c) Euler and Froude numbers be equal
(d) Reynolds and Mach numbers be equal.

449. The pressure drop in a pipe flow is directly proportional to the mean velocity. It can be deduced that the :

(a) flow is laminar
(b) flow is turbulent
(c) pipe is smooth (d) pipe is rough.

450. A hydrometer weighs 0.03 N and has a stem at the upper end which is cylindrical and 3 mm in diameter. It will float deeper in oil of specific gravity 0.75 than in alcohol of specific gravity 0.8 by how much amount?
(a) 10.7 mm (b) 43.3 mm
(c) 33 mm (d) 36 mm.

451. A wooden rectangular block of length l is made to float in water with its axis vertical. The centre of gravity of the floating body is 0.15 l above the centre of buoyancy. What is the specific gravity of the wooden block?
(a) 0.6 (b) 0.65
(c) 0.7 (d) 0.75.

452. If B is the centre of buoyancy, G is the centre of gravity and M is the metacentre of a floating body, the body will be in stable equilibrium if:
(a) $MG = 0$ (b) M is below G
(c) $BC = 0$ (d) M is above G.

453. The metacentric height of a passenger ship is kept lower than that of a naval or a cargo ship because:
(a) Apparent weight will increase
(b) Otherwise it will be in neutral equilibrium
(c) It will decrease the frequency of rolling
(d) Otherwise it will sink and be totally immersed.

454. Which one of the following is the correct statement ?

Stream line, path line and streak line are identical when the
(a) flow is steady
(b) flow is uniform
(c) flow velocities do not change steadily with time
(d) flow is neither steady nor uniform.

455. In cylindrical vortex motion about a vertical axis, r_1 and r_2 are the radical distances of two points on the horiontal plane ($r_2 > r_1$). If for a given tangential fluid velocity at r_1, the pressure difference between the points in free vortex is one-half of that when the vorte is a forced one, then what is the value of the ratio (r_2/r_1)?

(a) $\sqrt{\frac{3}{2}}$ (b) $\sqrt{2}$

(c) $\frac{3}{2}$ (d) $\sqrt{3}$.

456. The relation $\frac{\partial^2\phi}{\partial x^2}+\frac{\partial^2\phi}{\partial y^2}=0$ for an irrotational flow is known as which one of the following?
(a) Navier-Stokes equation
(b) Laplace equation
(c) Reynolds equation
(d) Euler's equation.

457. Which one of the following stream functions ψ is a possible irrotational flow field?
(a) $\psi = y^2 - x^2$ (b) $\psi = A \sin (xy)$
(c) $\psi = A\, x^2y^2$ (d) $\psi = Ax + By^2$.

458. Match List I with List-II and select the correct answer using the code given below the lists:

List-I (Condition)	**List-II (Regulating fact)**
A. Existence of stream function	1. Irrotationality of flow
B. Existence of velocity potential	2. Continuity of flow
C. Absence of temportal variations	3. Uniform flow
D. Constant velocity vector	4. Steady flow

Codes :

	A	B	C	D
(a)	4	3	2	1
(b)	2	1	4	3
(c)	4	1	2	3
(d)	2	3	4	1

459.

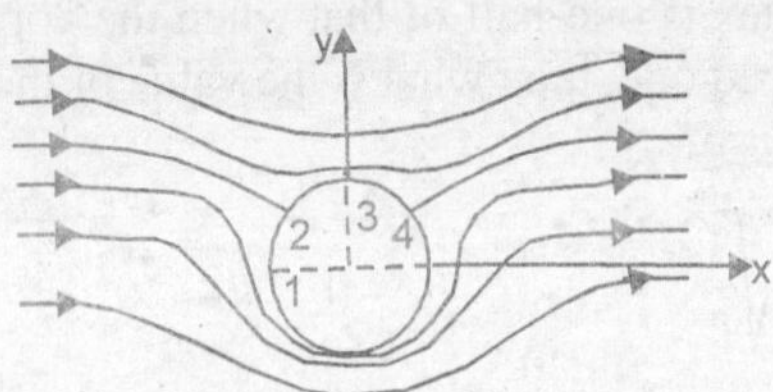

Fig. 37

A cylindrical object is rotated with constant angular velocity about its symmetry axis in -a uniform flow field of a ideal fluid producing streamlines as shown in the Fig. 37 given above. At which points(s), is the pressure on the cylinder surface maximum?

(a) Only at point 3 (b) Only at point 2
(c) At points 1 and 3 (d) At points 2 and 4.

460. Which one of the following non-dimensional numbers is used for transition from laminar to turbulent flow in free convection?

(a) Reynolds number
(b) Grashof number
(c) Peclet number
(d) Rayleigh number.

461. Velocity for flow through a pipe, measure at the centre is found to be 2 m/s. Reynolds number is around 800. What is the average velocity in the pipe?

(a) 2 m/s (b) 1.7 m/s
(c) 1 m/s (d) 0.5 m/s.

462. Match List-I with List-II and select the correct answer using the code given below the lists:

List-I (Measuring instrument)	List-II (Variable to be measured)
A. Hot-wire anemometer	1. Discharge
B. Pitot-tube	2. Rotational speed
C. V-notch weir	3. Velocity fluctuations
D. Techometer	4. Stagnation pressure.

Codes :

	A	B	C	D
(a)	4	3	2	1
(b)	3	4	2	1
(c)	4	3	2	1
(d)	3	4	1	2

463. In orifice meter, having an orifice of diameter d is fitted in a pipe of diameter D. For this orifice meter, what is the co-efficient of discharge C_d?

(a) A function of Reynolds number only
(b) A function of d/D only
(c) A function of d/D and Reynolds number
(d) Independent of/D and Reynolds number.

464. The velocity of a water stream is being measured by L-shaped Pitot-tube and the reading is 20 cm. Then, what is the approximate value of velocity?

(a) 19.6 m/s (b) 2.0 m/s
(c) 9.8 m/s (d) 20 cm/s.

465. Which one of the following conditions will linearlize the Navier-Stokes equations to make it amenable for analytical solutions?

(a) Low Reynolds number (Re < < 1)
(b) High Reynolds number (Re >> 1)
(c) Low Mach number (M << 1)
(d) High Mach number (M >> 1).

466. The head loss in turbulent flow in pipe varies:

(a) Directly as the velocity
(b) Inversely as the square of the velocity
(c) Inversely as the square of the diameter
(d) Approximately as the square of the velocity.

467. A ship's model, with scale 1 : 100, has a wave resistance of 10 N at its design speed. What is the corresponding prototype wave resistance in kN?

(a) 100 (b) 1000
(c) 10000
(d) Cannot be determined because of insufficient data.

468. In a fluid machine, the relevant parameters are volume flow rate, density, viscosity, bulk modulus, pressure difference, power

consumption, rotational speed and characteristic dimension. Using the Buckingham *pi* (π) theorem, what would be the number of independent non-dimensional groups?
(a) 3 (b) 4
(c) 5
(d) None of the above.

469. Consider the following statements:
1. The phenomenon of lift produced by imposing circulation over a doublet in a uniform flow is known as Magnus effect.
2. The path-deviation of a cricket ball from its original trajectory is due to the Magnus effect.

Which of the statement given above is/are correct?
(a) 1 only (b) 2 only
(c) Both 1 and 2 (d) Neither 1 nor 2.

470. If the relationship between the shear τ and the rate of shear strain (du/dy) is expressed as :

$\tau = \mu\,(du/dy)^n$,

then the fluid with exponent n > 1 is known as which one of the following?
(a) Bingham plastic
(b) Dilatant fluid
(c) Newtonian fiuid
(d) Pseudo plastic fluid.

471. What are the dimensions of kinematic viscosity of a fluid?
(a) LT^{-2} (b) L^2T^{-1}
(c) $ML^{-1}T^{-1}$ (d) $ML^{-2}T^{-2}$.

472. The pressure difference of two very light gases in two rigid vessels is being measured by a vertical U-tube water filled manometer. The reading is found to be 10 cm. What is the pressure difference?
(a) 9.81 kPa (b) 0.0981 bar
(c) 98.1 Pa (d) 981 N/m^2.

473. A circular plate 1.5 m diameter is submerged in water with its greatest and least depths below the surface being 2 m and 0.75 m respectively. What is the total pressure (approximately) on one face of the plate?
(a) 12 kN (b) 16 kN
(c) 24 kN
(d) None of the above.

474. Whenever a plate is submerged at an angle with the direction of flow of liquid, it is subjected to some pressure. What is the component of this pressure in the direction of flow of liquid known as?
(a) Stagnation pressure (b) Lift
(c) Drag (d) Bulk modulus.

475. Resultant pressure of the liquid in case of an immersed body acts through which one of the following?
(a) Centre of gravity (b) Centre of pressure
(c) Metacentre
(d) Centre of buoyancy.

476. In an experiment, the following shear stress-time rate of shear strain values are obtained for a fluid:

Time rate of shear strain (1/s) :	0	2	3	4
Shear stress (kPa) :	0	1.4	206	4

How can the fluid be classified?
(a) Newtonian fluid (b) Bingham plastic
(c) Pseudo plastic (d) Dilatant.

477. A velocity field is given by, $u = 3xy$ and $v = \frac{3}{2}(x^2 - y^2)$. What is the relevant equation of a streamline?

(a) $\frac{dx}{dy} = \frac{(x^2 - y^2)}{xy}$ (b) $\frac{dx}{dy} = \frac{3xy}{(x^2 - y^2)}$

(c) $\frac{dx}{dy} = \frac{2xy}{(x^2 - y^2)}$ (d) $\frac{dx}{dy} = \frac{(x^2 - y^2)}{2xy}$.

478. For steady incompressible flow, if the u-component of velocity is $u = Ae^x$, then what is the v-component of velocity?
(a) Ae^y (b) Ae^xy
(c) $-Ae^xy$ (d) $-Ae^x$.

479. An incompressible fluid flows radially outward from a line source in a steady manner. How does the velocity in any radial direction vary?

(a) r (b) r^2
(c) $1/r^2$ (d) $1/r$.

480. Consider the following statements :

1. The friction factor in laminar flow through pipes is independent of roughness
2. The friction factor for laminar flow through pipes is directly proportional to Reynolds number
3. In fully turbulent flow, through pipes, friction factor is independent of Reynolds number.

Which of the statements given above are correct?

(a) 1, 2 and 3 (b) 1 and 3 only
(c) 2 and 3 only (d) 1 and 2 only.

481. What is hydraulic diameter used in place of diameter for non-circular ducts equal to?

(a) A/m (b) $4\,A/m$
(c) $A/(4m)$ (d) $4\,m/A$.

where A = area of flow and m = perimeter,

482. The velocity distribution in a turbulent boundary layer is given by $u/U = (y/\delta)^{1/7}$. What is the displacement thickness δ?

(a) δ (b) $\delta/7$
(c) $(7/8)\,\delta$ (d) $\delta/8$.

483. For hydraulically most efficient symmetric trapezoidal section of an open channel, which one of the following is the false characterization?

(a) Wetted perimeter is minimum for a given area of flow section
(b) Hydraulic radius is half the flow depth
(c) Width at top liquid is twice the hydraulic depth
(d) Discharge is maximum for given area of flow, bed slope and roughness.

484. The power consumed per unit length in laminar flow for the same discharge, varies directly as D^n, where D is the diameter of the pipe. What is the value of 'n' ?

(a) 1/2 (b) –1/2 (c) – 2 (d) – 4.

485. The velocity profile in a laminar boundary layer is given by $u/U = y/\delta$. The ratio of momentum thickness to displacement thickness for the boundary is given by which one of the following?

(a) 2 : 3 (b) 1:2
(c) 1 : 6 (d) 1 : 3.

486. Match List-I with List-II and select the correct answer using the code given below the lists:

List-I (Variable)	List-II (Dimensional expression)
A. Dynamic viscosity	1. $M L^2 T^{-3}$
B. Moment of momentum	2. $M L^{-1} T^{-2}$
C. Power	3. $M L^{-1} T^{-1}$
D. Volume modulus of elasticity	4. $M L^2 T^{-2}$
	5. $M L^2 T^{-1}$

Codes :

	A	B	C	D
(a)	1	4	2	3
(b)	3	5	1	2
(c)	1	5	2	3
(d)	3	4	1	2

487. Match List-I with List-II and select the correct answer using the code given below the list :

List-I (Dimensionless number)	List-II (Nature of force involved)
A. Euler number	1. Surface tension
B. Weber number	2. Gravity
C. Mach number	3. Pressure
D. Froude number	4. Elastic

Codes :

	A	B	C	D
(a)	3	1	4	2
(b)	3	4	1	2
(c)	4	1	2	3
(d)	4	2	1	3

488. The stream function in a flow field is given by $\psi = 2xy$. In the same flow field, what is the velocity at a point (2, 1)?

(a) 4 unit (b) 5.4 unit
(c) 1.73 unit (d) 4.47 unit.

489. A rectangular plate 0.75 m × 2.4 m is immersed in a liquid of relative density 0.85 with its 0.75 m side horizontal and just at the water surface. If the plane of plate makes an angle of 60° with the horizontal, what is the aproximate pressure force on one side of the plate?

(a) 7.80 kN (b) 15.60 kN
(c) 18.00 kN (d) 24.00 kN.

490. In an open U tube containing mercury, kerosene of specific gravity 0.8 is poured into one of its limbs so that the length of column of kerosene is about 40 cm. The level of mercury column in the limb is lowered approximately by how much?

(a) 2.4 cm (b) 1.2 cm
(c) 3.6 cm (d) 0.6 cm.

491. Which one of the following is correct?

The capillary rise on depression in a small diameter tube is

(a) Directly proportional to the specific weight of the fluid
(b) Inversely proportional to the surface tension
(c) Inversely proportional to the diameter
(d) Directly proportional to the surface area.

492. What is pressure inside a soap bubble, over the atmospheric pressure if its diameter is 2 cm and the surface tension is 0.1 N/m?

(a) 0.4 N/m^2 (b) 4.0 N/m^2
(c) 40.0 N/m^2 (d) 400.0 N/m^2.

493. Which one of the following is correct?

In a fully developed region of the pipe flow,

(a) The velocity profile continuously changes from linear to parabolic shape
(b) The pressure gradient remains constant in the downstream direction
(c) The pressure gradient continuously changes exceeding the wall shess in the downstream direction
(d) The pipe is not running full.

494. Which one of the following is correct?

In the flow past bluff bodies :

(a) Pressure drag is smaler than friction drag
(b) Friction drag occupies the major part of total drag
(c) Pressure drag occupies the major part of total drag
(d) Pressure drag is less than that of streamlined body.

495. Two flows are specified as:

(A) $\mu = y$, $v = -(3/2)\,x$ (B) $u = xy^2$, $v = xy^2$
Which one of the following can be concluded?

(a) Both flows are rotational
(b) Both flows are irrotational
(c) Flow *A* is rotational while *B* is irrotational
(d) Flow *A* is irrotational while *B* is rotational.

496. Which one of the following is correct? For flow of an ideal fluid over a cylinder, from the front stagnation point,

(a) Pressure first decreases then increases
(b) Velocity first decreases then increases
(c) Pressure remains the same
(d) Velocity remains the same.

497. A steady, incompressible flow is given by:

$u = 2x^2 + y^2$ and $v = -4xy$

What is the convective acceleration along *x*-direction at point (1, 2)?

(a) $a_x = 6$ unit (b) $a_s = 24$ unit
(c) $a_x = -8$ unit (d) $a_x = -24$ unit.

498. Which one of the following is correct?

During sensible cooling of moist air, its relative humidity

(a) Increases
(b) Does not change
(c) Decreases
(d) Affects specific humidity.

499. Consider the following statements related to the fluid properties:

1. Vapour pressure of water at 373 K is 101.5 $\times 10^3$ N/m^2
2. Capillary height in cm for water in contact with glass tube and air is (tube dia)/0.268
3. Blood is a Newtonian fluid.

Which of the statements given above is/are correct?

(a) 1 only (b) 1 and 3
(c) 1 and 2 (d) 2 only.

500. What is unit of dynamic viscosity of a fluid termed 'poise' equivalent to?

(a) dyne/cm^2 (b) gms/cm
(c) dyne s/cm^2 (d) gm-cm/s.

501. What is the pressure difference between inside and outside of a droplet of water?

(a) $2\,\sigma/d$ (b) $4\,\sigma/\delta$
(c) $8\,\sigma/d$ (d) $12\,\sigma/d$.

where 'σ' is the surface tension and '*d*' is the diameter of the droplet.

502. The lower portion of a *U*-tube of uniform bore, having both limbs vertical and open to atmosphere, is initially filled with a liquid of specific gravity 3*S*. A lighter liquid of specific gravity *S* is then poured into one of the limbs such that the length of column of lighter liquid is *X*. What is the resulting movement of the meniscus of the heavier liquid in the other limb?

(a) *X* (b) *X*/2 (c) *X*/3 (d) *X*/6.

503. How is the difference of pressure head, "*h*" measured by a mercury-oil differential manometer expressed?

(a) $h = x\left[1 - \dfrac{S_g}{S_0}\right]$ (b) $h = x[S_g - S_0]$

(c) $h = x[S_0 - S_g]$ (d) $h = x\left[\dfrac{S_g}{S_0} - 1\right]$.

where x = manometer rading; S_g and S_0 are the specific gravities of mercury and oil, respectively.

504. What is the vertical component of pressure force on submerged curved surface equal to?

(a) Its horizontal component
(b) The force on a vertical projection of the curved surface
(c) The product of the pressure at centroid and surface area
(d) The gravity force of liquid vertically above the curved surface up to the free surface.

505. What is the depth of centre of pressure of a vertical immersed surface from free surface of liquid equal to?

(a) $\dfrac{I_G}{A\bar{h}} + \bar{h}$ (b) $\dfrac{I_G A}{\bar{h}} + \bar{h}$

(c) $\dfrac{I_G \bar{h}}{A} + \bar{h}$ (d) $\dfrac{A\bar{h}}{I_G} + \bar{h}$.

(Symbols have their usual meaning)

506. What is buoyant force?

(a) Lateral force acting on a submerged body
(b) Resultant force acting on a submerged body
(c) Resultant force due to water on a body
(d) Resultant hydrostatic force on a body due to fluid surrounding it.

507. How is the metacentric height, *GM* expressed?

(a) $GM = BG - (I/V)$ (b) $GM = (V/t) - BG$
(c) $GM = (I/V) - BG$ (d) $GM = BG - (V/I)$.

where I = Moment of inertia of the plane of the floating body at the water surface,

V = Volume of the body submerged in water,

BG = Distance between the centre of gravity (G) and the centre of buoyancy (B)

508. Stability of a floating body can be improved by which of the following:

1. Making its width large
2. Making the draft small
3. Keeping the centre of mass low
4. Reducing its density.

Select the correct answer using the code given below:

(a) 1, 2, 3 and 4 (b) 1, 2 and 3 only
(c) 1 and 2 only (d) 3 and 4 only.

509. The distance from the centre of buoyancy to the metacentre is given by I/V_d, where V_d is the volume of fluid displaced. What does I represented?
(a) Moment of inertia of a horizontal section of the body taken at the surface of the fluid
(b) Moment of inertia about about its vertical centroidal axis
(c) Polar moment of inertia
(d) Moment of inertia about its horizontal centroidal axis.

510. What acceleration would cause the free surface of a liquid contained in an open tank moving in a horizontal track to dip by 45°?
(a) $g/2$ (b) $2g$
(c) g (d) $(3/2)g$.

511. A liquid flows out of an orifice of an enclosed tank filled to a depth 'h' that contains air above the liquid at a pressure 'p'. If the speed of the fluid flowing out is to be reduced:
(a) the tank should be open to the atmosphere
(b) the cross-sectional area of the orifice is increased
(c) the air pressure above the liquid should be increased
(d) the amount of liquid in the tank should be increased.

512. A 2-cm-dia pipe transports water at 20 m/s. It exits out of 100 small 2-mm-dia holes. Exit in velocity, in m/s will be:
(a) 120 (b) 80
(c) 40 (d) 20.

513. What flow rate, in m^3 /s is needed using a 20 : 1 scale model of a dam over which 4 m^3 / s of water flows?
(a) 0.010 (b) 0.0068
(c) 0.0047 (d) 0.0022.

514. In forced vorted flow, velocity v and radial distance r are related as:
(a) $v \propto r$ (b) $v \propto \frac{1}{r}$
(c) $v \propto \frac{1}{r^2}$ (d) $v \propto r^2$.

515. In 1 : 20 model of a stilling basin, the height of the hydraulic jump in the model is observed to be 0.20 meter. The height of the hydraulic jump in the prototype will be:
(a) 1 m (b) 2 m
(c) 4 m (d) 8 m.

516. The continuity equation:

$$\frac{\partial u}{x} + \frac{\partial v}{y} + \frac{\partial w}{z} = 0$$

represents conservation of :
(a) mass
(b) mass and energy
(c) ass, energy and velocity
(d) mass, momentum, energy and velocity.

517. Which of the following flow is rotational?
(a) $u = y;\ v = \frac{3}{2}\ x$
(b) $u = xy^2;\ v = x^2y$
(c) Both (a) and (b) above
(d) none of the above.

ANSWERS

1. (b)	**2.** (b)	**3.** (a)	**4.** (b)	**5.** (a)	**6.** (d)	**7.** (d)	**8.** (c)	**9.** (d)	**10** (a)
11. (d)	**12.** (c)	**13.** (d)	**14.** (c)	**15.** (a)	**16.** (a)	**17.** (b)	**18.** (b)	**19.** (c)	**20.** (b)
21. (b)	**22.** (d)	**23.** (a)	**24.** (d)	**25.** (a)	**26.** (a)	**27.** (c)	**28.** (c)	**29.** (d)	**30.** (c)
31. (c)	**32.** (b)	**33.** (d)	**34.** (a)	**35.** (d)	**36.** (b)	**37.** (c)	**38.** (b)	**39.** (b)	**40.** (b)
41. (c)	**42.** (b)	**43.** (d)	**44.** (b)	**45.** (c)	**46.** (b)	**47.** (d)	**48.** (d)	**49.** (c)	**50.** (c)

51. (d) **52.** (c) **53.** (b) **54.** (a) **55.** (d) **56.** (d) **57.** (c) **58.** (d) **59.** (c) **60.** (c)

61. (a) **62.** (c) **63.** (d) **64.** (a) **65.** (d) **66.** (a) **67.** (d) **68.** (b) **69.** (a) **70.** (a)

71. (b) **72.** (b) **73.** (c) **74.** (a) **75.** (d) **76.** (c) **77.** (b) **78.** (d) **79.** (b) **80.** (d)

81. (b) **82.** (a) **83.** (d) **84.** (a) **85.** (a) **86.** (d) **87.** (d) **88.** (d) **89.** (a) **90.** (a)

91. (c) **92.** (c) **93.** (c) **94.** (a) **95.** (c) **96.** (b) **97.** (b) **98.** (d) **99.** () **100.** (c)

101. (c) **102.** (c) **103.** (c) **104.** (d) **105.** (b) **106.** (b) **107.** (c) **108.** (c) **109.** (b) **110.** (d)

111. (d) **112.** (b) **113.** (d) **114.** (d) **115.** (d) **116.** (a) **117.** (d) **118.** (b) **119.** (d) **120.** (d)

121. (a) **122.** (c) **123.** (b) **124.** (a) **125.** (c) **126.** (b) **127.** (d) **128.** (c) **129.** (d) **130.** (b)

131. (d) **132.** (c) **133.** (c) **134.** (c) **135.** (c) **136.** (d) **137.** (c) **138.** (c) **139.** (d) **140.** (a)

141. (b) **142.** (d) **143.** (c) **144.** (c) **145.** (a) **146.** (a) **147.** (c) **148.** (b) **149.** (c) **150.** (a)

151. (c) **152.** (d) **153.** (c) **154.** (c) **155.** (a) **156.** (b) **157.** (b) **158.** (d) **159.** (b) **160.** (c)

161. (d) **162.** (d) **163.** (d) **164.** (c) **165.** (c) **166.** (d) **167.** (b) **168.** (d) **169.** (b) **170.** (a)

171. (b) **172.** (c) **173.** (d) **174.** (a) **175.** (d) **176.** (c) **177.** (b) **178.** (c) **179.** (a) **180.** (d)

181. (c) **182.** (b) **183.** (a) **184.** (b) **185.** (d) **186.** (c) **187.** (d) **188.** (c) **189.** (b) **190.** (b)

191. (b) **192.** (b) **193.** (c) **194.** (a) **195.** (d) **196.** (a) **197.** (d) **198.** (d) **199.** (a) **200.** (d)

201. (d) **202.** (b) **203.** (a) **204.** (a) **205.** (b) **206.** (b) **207.** (a) **208.** (a) **209.** (b) **210.** (c)

211. (d) **212.** (a) **213.** (d) **214.** (a) **215.** (c) **216.** (b) **217.** (d) **218.** (b) **219.** (c) **220.** (b)

221. (c) **222.** (c) **223.** (a) **224.** (c) **225.** (c) **226.** (b) **227.** (a) **228.** (d) **229.** (d) **230.** (c)

231. (a) **232.** (b) **233.** (b) **234.** (d) **235.** (d) **236.** (c) **237.** (a) **238.** (d) **239.** (d) **240.** (d)

241. (b) **242.** (c) **243.** (b) **244.** (c) **245.** (c) **246.** (b) **247.** (c) **248.** (d) **249.** (d) **250.** (b)

251. (d) **252.** (b) **253.** (c) **254.** (d) **255.** (d) **256.** (b) **257.** (d) **258.** (c) **259.** (b) **260.** (b)

261. (c) **262.** (d) **263.** (c) **264.** (c) **265.** (d) **266.** (c) **267.** (a) **268.** (c) **269.** (d) **270.** (a)

271. (c) **272.** (b) **273.** (c) **274.** (d) **275.** (b) **276.** (c) **277.** (b) **278.** (c) **279.** (b) **280.** (c)

281. (d) **282.** (a) **283.** (a) **284.** (b) **285.** (c) **286.** (d) **287.** (a) **288.** (a) **289.** (d) **290.** (b)

291. (d) **292.** (a) **293.** (b) **294.** (c) **295.** (c) **296.** (a) **297.** (b) **298.** (d) **299.** (a) **300.** (c)

301. (d) **302.** (d) **303.** (d) **304.** (d) **305.** (a) **306.** (c) **307.** (b) **308.** (c)) **309.** (d) **310.** (d)

311. (c) **312.** (d) **313.** (b) **314.** (d) **315.** (d) **316.** (b) **317.** (a) **318.** (b) **319.** (b) **320.** (c)

321. (a) **322.** (a) **323.** (c) **324.** (b) **325.** (c) **326.** (b) **327** (b) **328.** (a) **329.** ([illegible]) **330.** (b)

331. (a) **332.** (d) **333.** (d) **334.** (c) **335.** (c) **336.** (c) **337.** (c) **338.** (b) **339.** (d) **340.** (a)

341. (d) **342.** (d) **343.** (c) **344.** (a) **345.** (b) **346.** (d) **347.** (c) **348.** (d) **349.** (c) **350.** (a)

351. (d) **352.** (b) **353.** (d) **354.** (c) **355.** (c) **356.** (a) **357.** (a) **358.** (d) **359.** (a) **360.** (c)

361. (d) **362.** (d) **363.** (a) **364.** (d) **365.** (a) **366.** (a) **367.** (c) **368.** (a) **369.** (c) **370.** (c)

371. (a) **372.** (c) **373.** (c) **374.** (c) **375.** (d) **376.** (b) **377.** (a) **378.** (b) **379.** (a) **380.** (c)

381. (a) **382.** (c) **383.** (b) **384.** (b) **385.** (b) **386.** (c) **387.** (b) **388.** (b) **389.** (c) **390.** (c)

391. (b) **392.** (d) **393.** (b) **394.** (d) **395.** (a) **396.** (b) **397.** (b) **398.** (c) **399.** (b) **400.** (d)

401. (c) **402.** (b) **402.** (c) **404.** (b) **405.** (c) **406.** (b) **407.** (b) **408.** (d) **409.** (b) **410.** (c)

411. (a) **412.** (c) **413.** (d) **414.** (b) **415.** (b) **416.** (c) **417.** (b) **418.** (c) **419.** (b) **420.** (d)

421. (c) **422.** (c) **423.** (b) **424.** (b) **425.** (b) **426.** (c) **427.** (a) **428.** (b) **429.** (d) **430.** (b)

431. (c) **432.** (a) **433.** (c) **434.** (a) **435.** (b) **436.** (a) **437.** (a) **438.** (b) **439.** (d) **440.** (b)

441. (c) **442.** (c) **443.** (b) **444.** (b) **445.** (a) **446.** (c) **447.** (c) **448.** (c) **449.** (a) **450.** (d)

451. (c) **452.** (d) **453.** (c) **454.** (a) **455.** (b) **456.** (b) **457.** (a) **458.** (b) **459.** (d) **460.** (d)

461. (c) **462.** (d) **463.** (c) **464.** (b) **465.** (a) **466.** (d) **467.** (c) **468.** (c) **469.** (c) **470.** (b)

471. (b) **472.** (d) **473.** (c) **474.** (c) **475.** (b) **476.** (d) **477.** (c) **478.** (c) **479.** (d) **480.** (b)

481. (b) **482.** (d) **483.** (c) **484.** (d) **485.** (d) **486.** (b) **487.** (a) **488.** (d) **489.** (b) **490.** (a)

491. (c) **492.** (c) **493.** (a) **494.** (c) **495.** (c) **496.** (a) **497.** (c) **498.** (a) **499.** (a) **500.** (c)

501. (b) **502.** (c) **503.** (d) **504.** (d) **505.** (a) **506.** (d) **507.** (c) **508.** (a) **509.** (a) **510.** (c)

511. (a) **512.** (d) **513.** (d) **514.** (a) **515.** (c) **516.** (a) **517.** (a)

EXPLANATORY NOTES

2. $$\frac{F}{\left(\frac{\pi}{4}\times(1.25)^2\right)} = \frac{1000\times 9.81}{\frac{\pi}{4}\times(12.5)^2}$$

$$F = \left(\frac{1.25}{12.5}\right)^2 \times 1000\times 9.81$$

$$= 98.1 \text{ N}$$

Ans. (b)

4. Equation of a stream line is :

$$\frac{dx}{u} = \frac{dy}{v}$$

or $$vdx - u\,dy = 0$$ **Ans.** (b)

5. Continuity equation is applicable for incompressible fluids and steady motion.
Ans. (a)

6. Frictional head loss in a pipe

$$h_f = \frac{4flv^2}{2gd}$$

$$Q = \frac{\pi}{4}d^2v \quad \text{or} \quad v = \frac{4Q}{\pi d^2}$$

$$\therefore \quad h_f = \frac{4fl}{2gd}\times\frac{16Q^2}{\pi^2 d^4} = \left(\frac{32fl}{\pi^2 g}\right)\times\frac{Q^2}{d^5} \propto \frac{Q^2}{d^5}$$

For $h_{f1} = h_{f2}$

$$\frac{Q_1^2}{d_1^5} = \frac{Q_2^2}{d_2^5}$$

$$\frac{Q_2}{Q_1} = \left(\frac{d_2}{d_1}\right)^{5/2} = \left(\frac{20}{10}\right)^{5/2} \; 2^{5/2} = \sqrt{32} = 4\sqrt{2}$$

Ans. (d)

7. $\text{Re} = \dfrac{vd}{\nu}$

$$\nu = \frac{\mu}{\rho}, \nu_2 = \frac{0.2\mu}{0.8\rho} = \frac{\mu}{4\rho} = \frac{\nu}{4}$$

$$\frac{vd}{\nu} = \frac{4v_2 d}{\nu}$$

$$v_2 = \frac{v}{4}$$

Ans. (d)

8. $y_1 = 0.2\text{m}, y_2 = 0.8$ m,

$$\frac{y_2}{y_1} = \frac{1}{2}[\sqrt{1+8F_1^2} - 1]$$

$$\frac{0.8}{y_1} = \frac{1}{2}[\sqrt{1+8F_1^2} - 1]$$

$$8 = \sqrt{1+8F_1^2} - 1$$

$$1+8F_1^2 = 9^2 = 81$$

$$8F_1^2 = 80$$

$$F_1^2 = 10$$

$$F_1 = 3.16$$

$$F_1^2 = \frac{v_1^2}{g\,y_1}$$

$$10 = \frac{v_1^2}{9.81\times 0.2}, v_1^2 = 19.62$$

$$v_1 = 4.429 \text{ m/s}$$

$$Q = by_1v_1 = 5\times 0.2\times 4.429 \simeq 4.43 \text{ m}^3/\text{s}$$

Ans. (c)

9. Displacement thickness,

$$\delta^* = \int_0^\delta \left(1 - \frac{u}{U_x}\right)dy$$

$$= \int_0^\delta \left[1 - \left(\frac{y}{\delta}\right)^{1/9}\right]dy$$

$$= \left| y - \frac{1}{\delta^{1/9}} \cdot y^{10/9} \times \frac{9}{10} \right|_0^{\delta}$$

$$= \delta - \frac{9}{10}\delta = \frac{\delta}{10}$$

$$\frac{\delta'}{\delta} = \frac{1}{10} = 0.1$$ **Ans.** (d)

10. Height of mountain, $h = \dfrac{p_0 - P}{\gamma_a}$

$$p_0 = 750 \times 10^{-3} \times 13.6 \times 9.81 \times 10^3$$
$$= 100062 \text{ N/m}^2$$
$$p = 600 \times 10^{-3} \times 13.6 \times 9.81 \times 10^3$$
$$= 80049.6 \text{ N/m}^2$$
$$\gamma_a = 9.81 \times 1 = 9.81 \text{ N/m}^3$$
$$h = \frac{100062 - 80049.6}{9.81}$$
$$= 2040\text{m} \simeq 2000 \text{ m}$$ **Ans.** (a)

12. Hydrostatic force $= 0.8 \times 10^3 \times 9.80 \times 8$

$= 62720 \text{ N} \downarrow$

Accelerating force $= 3 \times 3 \times 8 \times 0.8 \times 10^3 \times 2.45$

$= 141120 \text{ N} \uparrow$

Force on the base of the tank $= 141120 - 62720$

$= 78400 \text{ N} \uparrow$ **Ans.** (d)

13. $A - 2, B - 3, C - 4, D - 1$ **Ans.** (d)

16. $Q = Q_1 + Q_2$ **Ans.** (a)

$$v_1 = \frac{4Q_1}{\pi d_1^2}, v_2 = \frac{4Q_2}{\pi d_2^2}$$

$$\frac{Q_1}{Q_2} = \frac{A_1 v_1}{A_2 v_2} = \frac{\frac{\pi}{4} d_1^2 v_1}{\frac{\pi}{4} d_2^2 v_2} = \left(\frac{d_1}{d_2}\right)^2 \left(\frac{v_1}{v_2}\right)$$

For $d_1 = d_2 = d$

$$\frac{Q_1}{Q_2} = \frac{v_1}{v_2}$$

Frictional head loss, $h_f = \dfrac{4 f l v^2}{2gd}$

$$Q_1 = \frac{\pi}{4} d^2 v, v = \frac{4Q}{\pi d^2} \text{ or } v \propto Q$$

$$\frac{4f L_e v^2}{2gd} = \frac{4f L v_1^2}{2gd} + \frac{4fL v_2^2}{2gd}$$

$$L_e v^2 = L v_1^2 + L v_2^2$$

$$L_e Q^2 = L Q_1^2 + L Q_2^2$$

$$Q_1 = Q_2 = \frac{Q}{2} \text{ for same } d$$

$$L_e Q^2 = \frac{LQ^2}{4} + L\frac{Q^2}{4} = \frac{LQ^2}{2}$$

$$L_e = \frac{L}{2}$$

17. For venturimeter,

$$Q = C_d \left[\frac{A_1 A_2}{\sqrt{A_1^2 - A_2^2}}\right] \times \left[2g\left(\frac{s_2}{s_1} - 1\right)h\right]^{1/2}$$

$$\alpha \propto \frac{A_1}{\sqrt{\left(\frac{A_1}{A_2}\right)^2 - 1}} \sqrt{h}$$

For $\dfrac{A_1}{A_2} = 2$

$$Q_1 \propto \sqrt{\frac{h}{3}}$$

$$Q_1 \propto \frac{\sqrt{5h}}{\sqrt{\left(\frac{A_1}{A_2}\right)^2 - 1}}$$

For $Q_1 = Q_2$

$$\sqrt{\frac{h}{3}} = \frac{\sqrt{5h}}{\left[\left(\frac{A_1}{A_2}\right)^2 - 1\right]^{1/2}}$$

$$\sqrt{\left(\frac{A_1}{A_2}\right)^2 - 1} = \sqrt{5 \times 3}$$

$$\left(\frac{A_1}{A_2}\right)^2 - 1 = 15$$

$$\left(\frac{A_1}{A_2}\right)^2 = 16$$

$$\frac{A_1}{A_2} = 4$$ **Ans.** (b)

18. Thickness of laminar boundary layer over a flat plate,

$$\delta = \frac{5x}{\sqrt{\text{Re}_x}} = \frac{5x}{\sqrt{\frac{vx}{\nu}}} \propto x^{1/2}$$ **Ans.** (b)

21. At sea level, h = 760 mm of Hg

(a) $\dfrac{760 \times 10^{-3} \times 13.6 \times 9.81 \times 10^3}{9.81 \times 998} = 10.35\,\text{m}$

(b) $\dfrac{760 \times 10^{-3} \times 13.6 \times 10^3}{1025} = 10.08 \simeq 10.1\,\text{m}$

(c) $760 \times 10^{-3} \times 13.6/0.8 = 12.92$ m

(d) $760 \times 10^{-3} \times 13.6/1.590 = 6.5$ m **Ans.**(b)

22. Bernoulli's equation is :

$$\frac{v^2}{2g} + \frac{p}{\rho g} + z = \text{const}$$

For $v = 0$, $\dfrac{p}{\rho g} + z = \text{const}$

$$\frac{p}{\rho g} + gz = \text{const.}$$

$$dp/\rho + g\,dz = 0 \text{ or } \frac{dp}{\rho} = -g\,dz$$ **Ans.** (d)

23. Force on the bottom, $F_b = h\gamma l^2$

Force on the sides $F_s = 0.5\,\gamma\, l\, h^2$

For $F_s = F_b$

$$0.5\,\gamma\, lh^2 = \gamma\, hl^2$$

$$\therefore \quad \frac{h}{l} = 2$$ **Ans.** (a)

24. $A - 3, B - 5, C - 1, D - 2$ **Ans.** (d)

25. $F_B \times 500 = \frac{\pi}{4}(10)^2 \times 0.28 \times 100$ **Ans.** (a)

$$F_B = 4.398 \simeq 4.4\text{N}$$

27. Velocity head $= \dfrac{v^2}{2g} = \dfrac{(4.4)^2}{2 \times 9.81} = 0.987 \simeq 1\text{m}$

Ans. (c)

28. $A - 3, B - 1, C - 2, D - 4$ **Ans.** (c)

29. $\dfrac{p_A}{\gamma} + S_A\,h_A - S_1\,h_B + S_3 h_B - S_B\,h_B = \dfrac{p_B}{\gamma}$

$$\frac{p_B - p_A}{\gamma} = S_A h_A - S_1 h_B + S_3 h_B - S_B\, h_B$$

Ans. (d)

31. Capillary rise, $h \propto \dfrac{1}{d}$ **Ans.** (c)

32. $Q = Av = 2000\ \text{cm}^3/\text{s}$ **Ans.** (b)

$A - 1, B - 5, C - 2, D - 3$

33. $Q \propto \sqrt{h}$ or $h \propto Q^2$ **Ans.** (d)

34. $L\mu V = \dfrac{m \times kg}{m \times s} \times \dfrac{m}{s} = \text{kg.m/s}^2 = \text{N}$ **Ans.** (a)

35. Relative density $= \dfrac{\text{weight in air}}{\text{weight is water}}$

$$= \frac{80}{60} = 1.33 \simeq 2$$ **Ans.** (d)

36. $A-3, B-1, C-2$ **Ans.** (b)

37. $\phi = x^2 - y^2$

$$u = -\frac{\partial\phi}{\partial x} = -2x$$

$$v = -\frac{\partial\phi}{\partial y} = +2y$$

$$V = \sqrt{u^2+v^2} = \sqrt{4x^2+4y^2}$$

$$V_{(1,1)} = \sqrt{8} = 2\sqrt{2}$$ **Ans.** (c)

40. $A-3, B-1, C-4, D-2$ **Ans.** (b)

42. $A-4, B-1, C-1$ **Ans.** (b)

44. $v_p = v_m\sqrt{\frac{l_p}{l_m}}$

$$1 = v_m\sqrt{25} = 5v_m$$

$$v_m = \frac{1}{5} = 0.2\,\text{m/s}$$ **Ans.** (b)

46. $\frac{p_A}{\gamma} - 0.8 \times 50 - 0.65 \times 25 + 1.0 \times 100 = \frac{p_B}{\gamma}$

$$\frac{p_B - p_A}{\gamma} = -40 - 16.25 + 100$$

$$= 43.75\text{ m} \simeq 44\text{ m}$$ **Ans.** (b)

47. Weight of solid P, W_P = weight of water displaced

$$= \frac{1}{2}V_P \times 10^3 \times 9.81\text{ N}$$

Weight of solid Q, $W_Q = \frac{1}{2}\ V_Q \times 10^3 \times 9.81$ N

$$\rho_P = \frac{W_p}{9.81V_p} = \frac{1}{2}\times 10^3$$

$$\rho_Q = \frac{W_Q}{9.81V_p} = \frac{1}{2}\times 10^3$$

$$\frac{\rho_P}{\rho_Q} = \frac{1}{2}\times\frac{3}{2} = \frac{3}{4} \text{ or } 3:4$$ **Ans.** (d)

49. $Q = A\rho v = \text{const}$

Taking *ln* of both sides

$ln\,A + ln\,\rho + ln\,v = ln$ const

Taking differential, we have

$$\frac{dA}{A} + \frac{d\rho}{\rho} + \frac{dv}{v} = 0$$ **Ans.** (d)

52. $A-4, B-2, C-1, D-3$ **Ans.** (c)

53. $\frac{\delta}{x} = \frac{0.37}{(\text{Re}_x)^{1/5}} = \frac{0.37}{\left(\frac{Ux}{\nu}\right)^{1/5}} \propto x^{-1/5}$ or $\delta \propto x^{4/5}$

Ans. (a)

59. $Q = Av = \frac{\pi}{4}(0.04)^2 \times 1.5 = \frac{3\pi}{5000}\text{m}^3/\text{s}$ **Ans.** (c)

61. $\delta^* = 0.5$ mm, $\rho = 1.2$ kg/m^3, $u = 10$ m/s

Loss of mass rate of flow of air = $\delta^*\,\rho u$

$= 0.5 \times 10^{-3} \times 1.2 \times 10 = 6\times 10^{-3}$ kg/m

Ans. (a)

62. Boundary shear stress $\propto \frac{1}{\sqrt{\text{Re}_x}} \propto \frac{1}{\sqrt{x}}$

$$\frac{\tau_1}{\tau_2} = \sqrt{\frac{x_2}{x_1}} = \sqrt{\frac{90}{30}} = \sqrt{3}$$ **Ans.** (c)

64.

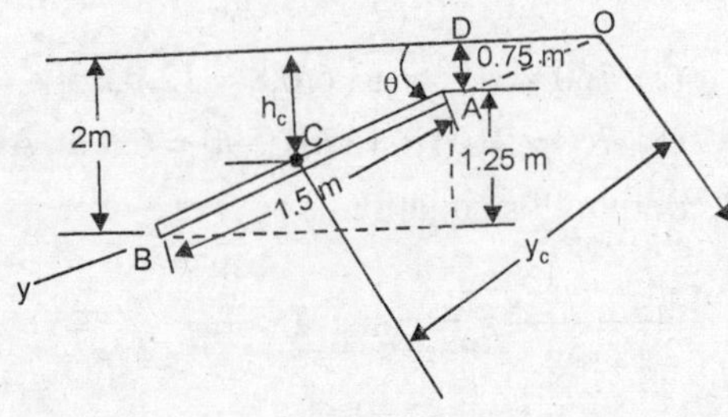

Fig. 38

$$\sin\theta = \frac{1.25}{1.5} = 0.833$$

$$\theta = 56.44°$$

$$F = \gamma A y_c \sin\theta$$

$$\frac{AD}{OA} = \sin\theta, OA = \frac{0.75}{0.833} = 0.9\,\text{m}$$

$$y_c = 0.9 + \frac{1.5}{2} = 1.65\,\text{m}$$

$$F = 9.81\times10^3 \times \frac{\pi}{4}(1.5)^2 \times 1.65 \times 0.833$$

$$= 23.837 \simeq 24\text{ kN}$$ **Ans.** (a)

67. For an orificemeter, $Q \propto \sqrt{H}$ or $Q = K\sqrt{H}$, where, K = a const. **Ans.** (d)

$$\ln Q = \ln K + \frac{1}{2}\ln H$$

$$\frac{dQ}{Q} = 0 + \frac{1}{2}\frac{dH}{H} = \frac{1}{2}\times 2 = 1\%$$

71. $A = 0.1$ m^2, $U = 30$ cm/s, $b = 0.01$ cm, $\mu = 0.01$ N.s/m^2

$$\tau = \frac{\mu U}{b} - \frac{1}{2}\frac{dp}{dx}(b-2y)$$

$$q = \frac{Ub}{2} - \frac{b^3}{12\mu}\cdot\frac{dp}{dx}$$

For $q = 0$, $\dfrac{dp}{dx} = \dfrac{6\mu U}{b^2}$

$$\tau = \frac{\mu U}{b} - \frac{1}{2}\times\frac{6\mu U}{b^2}(b-2y)$$

$$= \frac{\mu U}{b} - \frac{3\mu U}{b^2}(b-2y)$$

Average shear stress

$$\tau_{avg} = \frac{1}{b}\int_0^b\left[\frac{\mu U}{b} - \frac{3\mu U}{b^2}(b-2y)\right]dy$$

$$= \frac{1}{b}\left[\frac{\mu U y}{b} = \frac{3\mu U}{b^2}(by - y^2)\right]_0^b$$

$$= \frac{1}{b}\left[\mu U - 3\mu U(b^2 - b^2)\right]^0$$

$$= \frac{\mu U}{b}$$

$$= \frac{0.01\times0.3}{0.01\times10^{-2}} = 30\text{ N/m}^2$$

Force required = $t_{avg} \times A = 30 \times 0.1 = 3$ N

Ans. (b)

75. $\dfrac{U}{U_\infty} = \dfrac{y}{\delta}$ **Ans.** (d)

Momentum thickness,

$$\theta = \int_0^\delta \frac{u}{U_\infty}\left(1 - \frac{u}{U_\infty}\right)by$$

$$= \int_0^\delta \frac{y}{\delta}\left(1 - \frac{y}{\delta}\right)by$$

$$= \left|\frac{y^2}{2\delta} - \frac{y^3}{3\delta^2}\right|_0^\delta = \frac{\delta}{2} - \frac{\delta}{3} = \frac{\delta}{6}$$

77. $\nu = 15\times10^{-6}$ m^2/s, $U_\infty = 10$ m/s **Ans.** (b)

$Re_c = 5\times10^5$

$$Re_c = \frac{U_\infty x}{\nu}$$

$$x = \frac{5\times10^5\times15\times10^{-6}}{10} = 0.75\text{m or }75\text{cm}$$

81. $\delta^* = \displaystyle\int_0^\delta\left(1 - \frac{u}{U_\infty}\right)dy$ **Ans.** (c)

$$= \int_0^\delta\left(1 - \frac{y}{\delta}\right)dy$$

$$= \left| y - \frac{y^2}{2\delta} \right|_0^\delta = \delta - \frac{\delta}{2} = \frac{\delta}{2}$$

83. $u = 6x + xy$

Continuity equation in two-dimensions is :

$$\frac{\partial u}{\partial x} + \frac{\partial v}{\partial y} = 0$$

$$u = -\frac{\partial \phi}{\partial x}, v = -\frac{\partial \phi}{\partial y}$$

$$\frac{\partial u}{\partial x} = 6 + y$$

(a) $\frac{\partial v}{\partial y} = x$

$6 + y + x \neq 0$

(b) $\frac{\partial v}{\partial y} = x,\ 6 + y + x \neq 0$

(c) $\frac{\partial v}{\partial y} = -(6 + x) = -(6 + x)$

$6 + y - 6 - 6x \neq 0$

(d) $\frac{\partial v}{\partial y} = -(6 + y)$

$(6 + y) - (6 + y) = 0$ **Ans.** (d)

84. The equation of stream line is ;

$$\frac{dx}{u} = \frac{dy}{v}$$

$$\frac{dx}{ax} = \frac{dy}{-ay}$$

$$\frac{1}{a} lnx = -\frac{1}{a} lny + C$$

The stream line passes through point (3, 1)

$$\frac{1}{a} ln\ 3 = -\frac{1}{a} ln\ 1 + C = 0 + C$$

$$Cc = \frac{1}{a} \ln 3$$

$$\frac{1}{a} \ln x = -\frac{1}{a} \ln y + \frac{1}{a} \ln 3$$

or $\ln x = -\ln y + \ln 3$

$\ln x + \ln y = \ln 3$

$\ln xy = \ln 3$

$xy = 3$ **Ans.** (a)

85. Pressure inside a soap bubble,

$$p = \frac{8\sigma}{d}$$

$$\sigma = \frac{pd}{8} = \frac{25 \times 50 \times 10^{-3}}{8} = 0.156 \text{ N/m}$$

Ans. (a)

90. For a triangular notch, **Ans.** (a)

$$Q \,\alpha\, H^{5/2}$$

$$\ln Q = \frac{5}{2} \ln H$$

$$\frac{dQ}{Q} = \frac{5}{2} \frac{dH}{H}$$

93. $Q = Q_1 + Q_2$ **Ans.** (c)

Since $d_1 = d_2 = d$

$\therefore$ $Q_1 = Q_2$

$\therefore$ $Q = 2Q_1$

$D^2 v = 2d^2 v_1$ ...(i)

Also $\frac{fLv^2}{2gD} = \frac{flv^2}{2gd_1}$

$$\frac{v^2}{D} = \frac{v_1^2}{d^{11}}$$

$$\left(\frac{v}{v^1}\right)^2 = \frac{D}{d_1} = \frac{D}{d} \quad \text{...(ii)}$$

from eq. (i), $\frac{v}{v_1} = 2\left(\frac{d}{D}\right)^2$

$$\left(\frac{v}{v_1}\right)^2 = 4\left(\frac{d}{D}\right)^4$$

$$\therefore \quad \frac{D}{d} = 4\left(\frac{d}{D}\right)^4$$

$$\frac{D}{d} \times 4\left(\frac{d}{D}\right)^4 = 4$$

$$\left(\frac{D}{d}\right)^5 = 4$$

$$\frac{D}{d} = 4^{1/5}$$

94. $A-1, B-2, C-3$ **Ans.** (a)

103. $u = 5x^3, v = -15x^2 y$

$$u = \frac{\partial\psi}{\partial y} = 5x^3$$

$$y = 5x^3 y + c_1$$

$$v = \frac{\partial\psi}{\partial y} = -15x^2 y$$

$$\psi = 5x^3 y + c_2$$

Taking $c_1 = c_2 = 0$

$\psi = 5 \times 1 \times 2 = 10$ m³/s **Ans.** (a)

104. $A-1, B-4, C-2, D-3$ **Ans.** (d)

106. Velocity head, $\frac{v^2}{2g} = h = 10$ m **Ans.** (b)

108. $u = x - 4y, v = -y - 4x$ **Ans.** (c)

$$\frac{\partial u}{\partial x} = 1, \frac{\partial u}{\partial y} = -4$$

$$\frac{\partial v}{\partial x} = -4; \frac{\partial v}{\partial y} = -1$$

Continuity equation is : $\frac{\partial u}{\partial x} + \frac{\partial v}{\partial y} = 1 - 1 = 0$

Hence continuity equation is satisfied.

For the flow to be irrotational, $\frac{\partial u}{\partial x} = \frac{\partial v}{\partial y}$

i.e., –4 = –4. Hence the flow is irrorational

109. $\frac{v_m}{v_p} = \sqrt{\frac{l_m}{l_p}}$, $v_m = 1.0\sqrt{\frac{1}{25}} = 0.2$ m/s

Ans. (b)

111. $A-2, B-4, C-1, D-3$ **Ans.** (d)

112. Terminal velocity of a small sphere falling in a viscous fluid,

$$v = \frac{d^2((\gamma_s - \gamma_f)}{18\mu}$$ **Ans.** (b)

115. $\frac{\Delta p}{pv^3} = \frac{mL^{-1}T^{-2}}{[ML^{-3}][LT^{-1}]^3} = M^{1-1}L^{-1+3+3}T^{-2+3}$

$= L^{-1}T$, Which is non-dimensional

Ans. (d)

119. Hydrostatic force on $BC = \sqrt{F_v^2 + F_h^2}$

Vertical component, F_v = Weight of liquid above BC

$$= \frac{1}{2}(4 + 5.5) \times 3 \sin 60° \times g \times 10^3$$

$$= 12.34 \times 10^3 \times g \text{ N/m}$$

Horizontal component, F_h

= projected area of BC × depth upon centre of $BD \times g \times 10^3$

$= 3 \cos 60° \times (4 + 0.75) \times g \times 10^3$ N/m

$= 7.125 \times 10^3 \times g$ N/m

Resultant force $= 10^3 \times g\,[(12.34)^2 + (7.125)^2]^{1/2}$

$= 14250\,g$ N/m and acts at $\frac{2}{3}BD$ or between M and C. **Ans.** (d)

120. $\frac{p_A}{\gamma} - s_1h_1 - s_2h_2 - s_1h_3 = \frac{p_B}{\gamma}$

$\frac{p_A - p_B}{\gamma} = s_1h_1 + s_2h_2 + s_1h_3$ **Ans.** (d)

125. If $f \alpha \frac{1}{\text{Re}}$, then the pipe must be smooth.

Ans. (c)

130. A–3, B–4, C–1, D–5 **Ans.** (b)

137. A –3, B – 4, C – 1, D – 2 **Ans.** (c)

138. $n = 8$, $m = 3$ **Ans.** (c)

Number of independent non-dimensional groups = $n - m = 8 - 3 = 5$

139. A – 1, B – 2, C – 5, D – 2 **Ans.** (d)

159. $\frac{\delta_2}{\delta_1} = \sqrt{\frac{x_2}{x_1}}, \delta_2 = 5\sqrt{2}$ mm **Ans.** (b)

163. A – 1, B – 4, C – 3, D – 2 **Ans.** (d)

164. For irrotational flow, $\frac{\partial v}{\partial x} = \frac{\partial u}{\partial y}$ **Ans.** (c)

(i) $\frac{\partial v}{\partial x} = \frac{3}{2}, \frac{\partial u}{\partial y} = 1$, flow is rotational

(*ii*) $\frac{\partial v}{\partial x} = 2xy, \frac{\partial u}{\partial y} = 2xy$, flow is irrotational

167. $Re_x = \frac{vx}{v}$

$$5 \times 10^5 = \frac{1 \times x}{0.858 \times 10^{-6}}$$

$x = 0.429 \simeq 0.43$ m **Ans.** (b)

172. Coordinates of stagnation point are :

$$x = -\frac{q}{2\pi u} = -\frac{2v}{2\pi u} = -\frac{1}{\pi} = -0.32 \text{ m}$$

$y = 0$ **Ans.** (c)

173. A–2, B–4, C–1, D–3 **Ans.** (d)

178. A–2, B–1, C–3, D–6 **Ans.** (c)

179. A–1, B–2, C–3, D–4 **Ans.** (a)

182. $\delta_2 = \delta_1\sqrt{\frac{x_2}{x_1}} = 2\sqrt{\frac{4}{2}} = 2\sqrt{2}$ mm **Ans.** (d)

183. $F_P = F_m \left(\frac{l_p}{l_m}\right)^2 \left(\frac{v_p}{v_m}\right)^2$ **Ans.** (a)

$$= 50\,(10)^2 \times \left(\frac{10}{5}\right)^2 = 20{,}000 \text{ N}$$

187. For irrotational flow, $\frac{\partial v}{\partial x} = \frac{\partial u}{\partial y}$

(*i*) $u = \frac{\partial \psi}{\partial y} = x, \; v = \frac{-\partial \psi}{\partial x} = y$

$\frac{\partial v}{\partial x} = 0, \frac{\partial u}{\partial y} = 0$. Flow is irrotational

(*ii*) $u = -2Ay, \; v = -2Ax$

$\frac{\partial v}{\partial x} = -2A, \frac{\partial u}{\partial y} = -2A$

Flow is irrotational.

(*iii*) $v_r = \frac{\delta \phi}{\partial r} = u \cos\theta - \frac{1}{r^2} u \cos\theta$

$$= \left(1 - \frac{1}{r^2}\right) u \cos\theta$$

$$v_\theta = -\frac{1}{r}\frac{\partial \phi}{\partial \theta} = -\frac{1}{r}\left[-ur\sin\theta - \frac{u}{r}\sin\theta\right]$$

$$= u\sin\theta + \frac{u}{r^2}\sin\theta$$

$$\text{Voritcity} = \frac{v_\theta}{r} + \frac{\partial v_\theta}{\partial r} - \frac{1}{r}\frac{\partial v_r}{\partial \theta}$$

$$= \frac{u}{r}\sin\theta + \frac{u}{r^3}\sin\theta - \frac{2u}{r^3}\sin\theta$$

$$+\frac{1}{r}\left(1-\frac{1}{r^2}\right)u\sin\theta = 0$$

Hence flow is irrotational

By elimination, (*d*) is not irrotational. **Ans.** (d)

189. $p = -\rho g z$

$$\frac{\partial p}{\partial z} = -\rho g$$ **Ans.** (b)

190. L : B : H :: 2 : 1 : 2

$$F_{\text{bottom}} = \rho g h \times LH$$

$$F_{\text{face}} = \frac{1}{2}\rho g h \times LH$$

$$\frac{F_{\text{bottom}}}{F_{\text{Face}}} = \frac{2B}{H} = 2\times\frac{1}{2} = 1$$ **Ans.** (b)

191. $\frac{p_0}{\gamma_0} s_0 h - s_w h = \frac{p_w}{\gamma_w}$

$$\frac{p_0}{\gamma_0} - \frac{p_w}{\gamma_w} = (s_w - s_0)h$$

As $s_w > s_0$, therefore $p_0 > p_w$ **Ans.** (b)

195. *A*–4, *B*–1, *C*–3, *D*–2 **Ans.** (d)

198 $p \propto \frac{1}{v} \propto d^2$ Hence, the sequence is 3, 1, 2, 4

$$Q = \frac{\pi}{4}d^2\, v,\; v \propto \frac{1}{d^2}$$ **Ans.** (d)

202. At the point of boundary layer separation,

$$\left(\frac{\partial u}{\partial y}\right)_{y=0} = 0$$

Hence, $\tau = \mu\frac{\partial u}{\partial y} = 0$ **Ans.** (b)

206. $\frac{u}{u_\infty} = 1$ in case of (b) at $y = \delta$ **Ans.** (b)

209. *A* – 3, *B* – 2, *C* – 5, *D* – 4 **Ans.** (b)

211. $F = \frac{\pi}{2}\times d\times l\times h\times g\times 10^3$

$$= \frac{\pi}{2}\, 2\times 1\times 2\times 9.81\times 10^3$$

$$= 30803.4 \text{ N/m}$$ **Ans.** (d)

212. $\frac{p_A}{\gamma} + s_1h_1 - s_3h_2 - s_2h_3 = \frac{p_B}{\gamma}$ **Ans.** (d)

$$\frac{p_A - p_B}{\gamma} = h_3 s_2 + h_2 s_3 - h_1 s_1$$ **Ans.** (a)

220. *A* – 3, *B*– 4, *C*– 1, *D* – 2 **Ans.** (b)

221. For flow separation, $\frac{\partial p}{\partial x} > 0.$ **Ans.** (c)

231. At the interface of free and forced vortex, tangential velocities are equal.

$$\frac{\Gamma}{2\pi r_c} = \omega r_c$$

or $\omega = \frac{\Gamma}{2\pi}\times\frac{1}{r_c^2} = \frac{\Omega}{r_c^2}$

where $\Omega = \frac{\Gamma}{2\pi}$ **Ans.** (a)

234. $F = \rho g\, Ah$

or $F \propto Ah$

For same *h*, $F \propto$ A

$$\frac{F_2}{F_1} = \frac{A_2}{A_1} = \left(\frac{2d}{d}\right)^2 = 4$$ **Ans.** (d)

235. Hydrostatic thrust on inclined plane

$$= \rho g\, A\bar{x}$$

$$= 10^3\times 9.81\times\frac{\pi}{4}[(1.4)^2-(0.6)^2]\times 1.85$$

$$= 22.8 \text{ kN}$$ **Ans.** (d)

236. $F_{\text{bottom}} = \rho g\, LBh$

$$F_{side} = \frac{1}{2}\rho g L h^2$$

$$\frac{F_{bottom}}{F_{side}} = \frac{2B}{h} = \frac{2B}{2B} = 1$$ **Ans.** (c)

237. Let r = radius, h = height of cylinder

z = depth of paraboloid

Volume of water spilled out

= Volume of paraboloid of revolution

$0.5 \times \pi r^2 h = 0.5 \times 5\ \pi r^2 z$

$z = h$

Therefore, no water is left in the cylinder at the centre resulting in zero pressure at the centre of bottom. **Ans.** (a)

238. Neglecting static head due to air,

$p_A = \rho g h_w = 10^3 \times 9.81 \times 500 \times 10^{-3}$

$= 4905$ Pa $\simeq$ 4901 Pa **Ans.** (d)

240. Relative density of cork = 4

Therefore it will float upto $\frac{1}{4}$ th of its height in the liquid. Centre of gravity and centre of buoyancy of cork must coincide to make it float in neutral equilibrium. Thereofre, the cork must be immersed to its full height. For this purpose, downward force to be applied to cork = buoyant force on cork = 4 W

Ans. (d)

242. $A = 05 - 0.2\,x$, $v = \frac{Q}{A}$

Acceleration, $a = \frac{dv}{dt} = \frac{d}{dt}\left(\frac{Q}{A}\right)$

$= \frac{1}{A} \cdot \frac{dQ}{dt}$

At $x = 0$, $A = 0.5$ m²

$\frac{dQ}{dt} = 0.2$ m³/s

$a = \frac{1}{0.5} \times 0.2 = 0.4$ m/s² **Ans.** (c)

246. $Q_1 + Q_2 = Q_3 + Q_4$

$50 \times 10 + 50\,v_2 = 80 \times 5 + 70 \times 5$

$v_2 = \frac{250}{50} = 5$ cm/s **Ans.** (b)

247. Applying Bernoulli's equation, we have

$$\frac{v^2}{2g} + h = \frac{(4v)^2}{2g}$$

$$h = \frac{15v^2}{2g}$$

$$\frac{h}{\left[\frac{v^2}{2g}\right]} = 15$$ **Ans.** (c)

248. For equivalent pipe,

$$\frac{L_e}{D^5} = \frac{L}{D^5} + \frac{L}{(D/2)^5} + \frac{4L}{(2D)^5}$$

$$= \frac{L}{D^5}\left[1 + 32 + \frac{1}{8}\right] = \left(33\frac{1}{8}\right)\frac{L}{D^5}$$

$$L_e = 33\frac{1}{8}L$$ **Ans.** (d)

249. $$Q = \frac{A_1 A_2}{\sqrt{A_1^2 - A_2^2}}\sqrt{2gh}$$

$$= \frac{5 \times 3}{\sqrt{25 - 9}}\sqrt{2 \times 9.81 \times 5}$$

$= 370$ cm³/s **Ans.** (d)

255. Power, $P \propto (\text{model ratio})^{7/2} \propto 20^{7/2}$

$P_p = P_m \times 20^{7/2} = 0.25 \times 20^{7/2}$

$= 8944.30$ HP **Ans.** (d)

256. $\psi = 3xy$

$$u = \frac{\partial \psi}{\partial y} = 3x,\ v = -\frac{\partial \psi}{\partial x} = -3y$$

Resultant velocity,

$$V = \sqrt{u^2 + v^2}$$

$$= \sqrt{9x^2 + 9y^2}$$

$$= \sqrt{9\times4 + 9\times9} = \sqrt{117} = 10.82 \text{ unit}$$

Ans. (b)

257. Horizontal water pressure, $P_h = \frac{\gamma h^2}{2}$ and acts at $\frac{h}{3}$ from the base of dam.

Weight of dam masonry, $W = s \times \gamma \times \frac{1}{2} bh$

$= 2.56\,\gamma \times \frac{1}{2} bh = 1.28\ \gamma bh$, and acts at CG. of dam.

Let R = resultant of P_h and W

x = distance of R from toe of dam

e = eccentricity of the point of application

$$x = \frac{b}{2} - e$$

To avoid tension at the base, $e = \frac{b}{6}$

$$\therefore \quad x = \frac{b}{2} - \frac{b}{6} = \frac{b}{3}$$

Taking moments about the toe of dam, we have

$$P_h \times \frac{h}{3} + W \times \frac{2b}{3} - W \times \frac{2b}{3} = 0$$

$$P_h \times h - Wb = 0$$

$$\frac{\gamma h^3}{2} - 1.28\ \gamma b^2 h = 0$$

$$h^2 = 2.56\ b^2 \text{ or } b = 0.625\ h$$

Ans. (d)

259. Area of gate at sliding, $A = 2 \times 2.5 = 5\ \text{m}^2$

Force on upstream of gate, $P = \frac{1}{2}\rho g h A$

$$= \frac{1}{2} \times 10^3 \times 9.81 \times 2 = 49050\text{N}$$

Frictional force, $F = \mu P$

$$500 \times 9.81 = \mu \times 49050$$

$$\mu = 0.10$$

Ans. (d)

260. Air pressure from manometer reading

$= \rho g h = -\ 1000 \times 9.81 \times 0.25 \times 13.6$

$= -\ 33.354$ kPa

Pressure of oil $= s\,\rho\, gh =\ 0.8 \times 10^3 \times 9.81 \times 4$

$= 31.392$ kPa

Resultant pressure read by gauge A

$= 31.392 - 33.534 = -\ 1.962$ kPa **Ans.** (b)

261. $A - 3, B - 4, C - 1, D - 2$ **Ans.** (c)

263. Buoyant force of equal amount will act on the iron piece and thus spring balance reading will not change. **Ans.** (c)

264. Time peroid of oscillation,

$$T = 2\pi\left[\frac{(\text{Radius of gyration})^2}{\text{metacentric height} \times g}\right]^{1/2}$$

$$= 2\pi\left[\frac{9^2}{0.75\times9.81}\right]^{1/2} = 20.85s$$ **Ans.**(c)

265. $p = \frac{4\sigma}{d} = \frac{4\times0.073}{1\times10^{-3}} = 292\ \text{N/m}^2$ **Ans.** (d)

266. $A_1 v_1 = A_2 v_2$

$$v_2 = v_1\left(\frac{d_1}{d_2}\right)^2 = 1.5\left(\frac{100}{25}\right)^2 = 24 \text{ m/s}$$

Ans. (c)

267. $Q = Av$

$\text{A} \times 10^3 = 40\ v$

$v = 150$ cm/s or 1.5 m/s **Ans.** (a)

270. $A - 3, B - 1, C - 2, D - 4$ **Ans.** (a)

271. $h_f = \frac{4flv^2}{2gd} = \left[\frac{4fl}{2gd}\right]\left[\frac{16Q^2}{\pi^2 d^4}\right]$

$$\therefore \quad Q = \frac{1}{\sqrt{f}}$$

$$Q_2/Q_1 = \sqrt{\frac{f_1}{f_2}} = \sqrt{\frac{1}{1.25}} = 0.8944$$

$$\text{Error} = \frac{Q_2 - Q_1}{Q_1} = \frac{Q_2}{Q_1} - 1$$

$$= 0.8944 - 1 = -0.1056 \text{ or } -10.56\%$$

$\approx -12.5\ \%$ **Ans.** (c)

272. In turbulent flow, thickness of laminar sublayer,

$\delta_1 = \frac{5\nu}{u^*}$ **Ans.** (b)

275. Weight of parachute $= \frac{1}{2} C_D \rho U_\infty^2 A$

$$90 \times 9.81 = \frac{1}{2} \times 0.75 \times 1.28 \times U_\infty^2 \times 0.3$$

$U_x = 78.3$ m/s **Ans.** (b)

277. $Re_1 = Re_2$

$$\frac{v_1 d_1 \rho_1}{\mu_1} = \frac{v_2 d_2 \rho_2}{\mu_2}$$

$$\frac{v_2}{v_1} = \frac{\rho_1}{\rho_2} \times \frac{d_1}{d_2} \times \frac{\mu_1}{\mu_2} = 750 \times \frac{1}{2} \times \frac{1}{60} = \frac{25}{4}$$

$v_2 = \frac{1.6 \times 25}{4} = 10$ m/s **Ans.** (b)

278. For dynamic similarity, Froude law gives,

$$\left(\frac{v}{\sqrt{gL}}\right)_m = \left(\frac{v}{\sqrt{gL}}\right)_p$$

$$\frac{v_m}{v_p} = \sqrt{\frac{L_m}{L_p}} = \sqrt{60}$$

Propulsive force of prototype,

$$F_p = F_m \left(\frac{v_m}{v_p}\right)^2 \left(\frac{L_m}{L_p}\right)^2 = F_m \left(\frac{L_m}{L_p}\right)^3$$

$= 0.5 \times 9.81 \times 60^3 = 1$ MN **Ans.** (c)

279. $t_p = t_m \sqrt{\frac{l_m}{l_p}} = 4\sqrt{256} = 64$ min **Ans.** (b)

280. $v_s = \sqrt{\gamma RT} = \sqrt{1.4 \times 287 \times 333} = 365.8$ m/s

$M = \frac{v_a}{v_s} = \frac{40}{365.8} = 0.11 = 0.1212$ **Ans.** (c)

281. $N_m = N_p \left(\frac{D_p}{D_m}\right)\left(\frac{H_m}{H_p}\right)^{1/2}$

$$= 428 \times 4 \times \left(\frac{10}{30}\right)^{1/2} = 988 \text{ rpm}$$

Ans. (d)

282. $\frac{\sigma}{\lambda^2 g\rho} = \frac{N}{m} \times \frac{1}{m^2} \times \frac{s^2}{m^2} \times \frac{m^3}{kg}$ **Ans.** (a)

$$= \text{N} \times \left(\frac{s^2}{kg \times m}\right) = \frac{N}{N} = \text{dimensionless.}$$

284. Force on valve = Upward force on float due to buoyancy

$$1 \times 10^5 \times 0.5 \times 10^{-4} = 10 \times 10\, h \times 10^3 \times 9.81 \times 10^{-6}$$

$h = 5$ cm **Ans.** (b)

285. $\frac{p_w - p_a}{\gamma} = (50 \times 0.88 + 45 \times 13.6 - 56 \times 1.0) \times 10^{-2}$

$= 6$ m **Ans.** (c)

286. $v \propto \frac{1}{r}$

$$v_2 = v_1\left(\frac{r_1}{r_2}\right) = 50 \times \frac{40}{80} = 25 \text{ m/s}$$ **Ans.** (d)

287. $$\frac{dp}{dh} = \gamma\left(1+\frac{a_z}{g}\right)$$

$$dp = \gamma\left(1+\frac{a_z}{g}\right)dh$$

For $a_z = -g,\ dp = 0$ **Ans.** (a)

Therefore, pressure throughout the liquid mass is atmospheric.

288. $F_D \propto C_D v^2$ and power, $P = F_D \times v$

$$P \propto C_D v^3$$

For $$P_1 = P_2,\ \frac{v_2}{v_1} = \left(\frac{C_{D1}}{C_{D2}}\right)^{1/3} = \left(\frac{1}{0.75}\right)^{1/3}$$

$$= 1.1006$$

$$\frac{v_2 - v_1}{v_1} = \frac{1.1006-1}{1} = 0.1006 \text{ or } 10.06\% \simeq 10\%$$

Ans. (a)

289. Mach number, $M = \dfrac{1}{\sin\alpha} = \dfrac{1}{\sin 30^\circ} = 2$

Velocity of sound,

$$v_s = \sqrt{\gamma RT} = \sqrt{1.4\times 287\times 288}$$

$$= 340.17 \text{ m/s}$$

$$\frac{v}{v_s} = 2$$

$$v = 2 \times 340.17$$

$$= 680.34 \simeq 680.5 \text{ m/s}$$ **Ans.** (d)

290. $$\psi = x^2 - y^2$$

$$u = \frac{\partial\psi}{\partial y} = -\frac{\partial\phi}{\partial x} = -2y$$

or $$\frac{\partial\phi}{\partial y} = 2y$$

$$\phi = 2xy + c_1$$

$$v = -\frac{\partial\psi}{\partial x} = -\frac{\partial\phi}{\partial x} = 2x$$

or $$\frac{\partial\phi}{\partial y} = 2x,\ \phi = 2xy + c_2$$

$\therefore$ $$\phi = 2xy + \text{const}$$ **Ans.** (b)

291. Horizontal force, $F \propto h^2$

$$\frac{F_1}{F_2} = \left[\frac{h}{(h/2)}\right]^2 = 4$$ **Ans.** (d)

292. $$Q_5 = Q_4 - Q_1$$

$$8v_5 = 4\times 10 - 5\times 4 = 20$$

$$v_5 = 2.5 \text{ cm/s}$$

$$Q_2 - Q_3 = Q_5$$

$$6\times 5 - 2v_3 = 8\times 25$$

$$v_3 = \frac{10}{2} = 5 \text{ cm/s}$$ **Ans.** (a)

293. $$h_f = \frac{4flv^2}{2gd}$$

For $Q_1 = Q_2,\ A_1 v_1 = A_2 v_2$

$$\frac{v_2}{v_1} = \frac{A_1}{A_2} = \left(\frac{d_1}{d_2}\right)^2$$

$$\frac{h_{f2}}{h_{f1}} = \left(\frac{l_2}{l_1}\right)\left(\frac{d_1}{d_2}\right)^3 = 32\times\left(\frac{1}{3}\right)^3 = 4$$

Ans. (b)

295. Friction factor in pipe, $f = \dfrac{64}{\text{Re}}$

The friction factor does not depend on relative roughness of pipe. **Ans.** (c)

296. $Q = Av,\ v = \dfrac{Q}{A} \propto \dfrac{Q}{d^2}$

Frictional head loss in pipe.

$$h_f = \frac{4flv^2}{2gd} \propto \frac{v^2}{d} \propto \frac{Q^2}{d^5}$$

$$Q \propto d^{5/2}$$

$$\frac{Q_2}{Q_1} = \left(\frac{d_2}{d_1}\right)^{5/2} = (0.8)^{5/2} = 0.5724$$

$$1 - \frac{Q_2}{Q_1} = 1 - 0.5724 = 0.4276$$

$$\frac{Q_1 - Q_2}{Q_1} = 0.4276 \text{ or } 42.76\% \simeq 42.8\%$$

Ans. (a)

297. $v = \sqrt{2gh} \propto \sqrt{h}$

$$Q \propto v \propto \sqrt{h}$$

$$Q_1/Q_2 = \sqrt{h_1/h_2} = \sqrt{\frac{40}{90}} = \frac{2}{3} \text{ or } 2:3$$

Ans. (b)

298. $v_0 = C_d\sqrt{2gh}$

$$1.2 = 1.0\sqrt{2 \times 9.81 \times h}$$

$$h = 73.4 \text{ mm}$$

Ans. (d)

299. $A - 3, B - 1, C - 2, D - 4$ **Ans.** (a)

300. $R_c = \frac{vd\rho}{\mu} = \frac{0.318 \times 0.2 \times 10^3}{0.1} = 636$

$$4f = \frac{64}{R_c} = \frac{64}{636} = 0.1006$$

$$hf = \frac{4flv^2}{2gd} = \frac{0.1006 \times 30 \times 10^3 \times (0.318^2)}{2 \times 9.81 \times 0.2}$$

$$= 77.78 \text{m}$$

$$Q = \frac{\pi}{4}d^2 v = \frac{\pi}{4}(0.2)^2 \times 0.318$$

$$= 9.99 \times 10^{-3} \text{ m}^3/\text{s}$$

Power required, $P = \frac{\gamma Q h_f}{10^3}$

$$= \frac{9.81 \times 10^3 \times 9.99 \times 10^{-3} \times 77.78}{10^3}$$

$$= 7.63 \text{ kW}$$

Ans. (c)

304. $\Gamma = 4\pi U r = 4\pi \times 4 \times 0.2 = 10.05 \text{ m}^2/\text{s}$

Lift, $L = \rho U \Gamma = 10^3 \times 4 \times 10.05 = 40.2 \text{ kN/m}$

Ans. (d)

305. $F_D \propto v^2$

$$F_{D2} = F_{D1}\left(\frac{v_2}{v_1}\right) = 2\left(\frac{50}{40}\right) = \frac{25}{8} \text{ kN}$$

Power required, $P = \frac{F_{D_2} v_2}{10^3} = \frac{25 \times 5010^3}{8 \times 3600}$

$$= 43.4 \text{ kW}$$

Ans. (a)

307. $M_2^2 = \frac{2 + (k-1)M_1^2}{2k\,M_1^2 - (k-1)}$

$$= \frac{2 + (14.1 - 1)(1.68)^2}{2 \times 1.4 \times (1.68)2 - (1.4 - 1)}$$

$$= 0.417$$

$$M_2 = 0.646$$

Ans. (b)

308. $F_{\text{base}} = \gamma h A = \gamma \times 2a \times a^2 = 2\gamma a^3$

$$F_{\text{side}} = \frac{1}{2}\gamma h A = \frac{1}{2} \times \gamma \times 2a \times (a \times 2a)$$

$$= 2\gamma a^3$$

$$F_{base}/F_{\text{base}} = \frac{2\gamma a^3}{2\gamma a^3} = 1$$

Ans. (c)

309. $\frac{p_1 - p_2}{\gamma} = \left(\frac{s_2}{s_1} - 1\right)h$

$$= \left(\frac{13.6}{0.9} - 1\right) \times 0.6 = 8.47 \text{ m of oil}$$

Ans. (d)

310. Weight of aluminium block, $W = mg$

$$= 12 \times 10 = 120 \text{ N}$$

Volume of aluminium block, $V = \frac{12}{2.4} = 5 \text{m}^3$

Volume of oil displaced $= V = 5\text{m}^3$

Buoyant force, $F_B = V\rho_0 g = 5 \times 0.8 \times 10$

$= 40$ N

Tension in the wire, $T = W - F_B$

$= 120 - 40 = 80$ N **Ans.** (d)

311. $GM = BM - BG = \frac{I}{V} - BG$

$$I = \frac{bd^3}{12} = \frac{30 \times 10^3}{12} = 2500\,\text{m}^4$$

$V = bd \times \text{draft} = 30 \times 10 \times 3 = 900\ \text{m}^3$

$BG = 2.5 - 1.5 = 1.0$ m

$$GM = \frac{2500}{900} - 1 = 1.78\,\text{m}$$ **Ans.** (c)

312. Volume of vessel, $V_s = \frac{\pi}{4} d^2 h = \frac{\pi}{4} d^3$

$\tan\theta = \frac{a}{g} = \frac{g}{g} = 1$

$\tan\theta = 45°$

Volume spilled out $= \frac{\pi d^2}{4} \times \frac{d}{4}$

$= \frac{\pi d^3}{16}$

Remaining liquid$= \frac{\pi d^3}{4} - \frac{\pi d^3}{16} = \frac{3}{16}\pi d^3$

Ratio $= \frac{(3/16)}{(1/4)} = \frac{3}{4} = 0.75$ **Ans.** (d)

313. $\tau = \frac{\mu U}{b} = \frac{9.81 \times 10^{-2} \times 2}{1 \times 10^{-2}} = 19.62\,\text{N/m}^2$ **Ans.** (b)

315. A – 2, B – 3, C – 4, D – 1 **Ans.** (d)

316. $v_0 = \sqrt{2gh} = \sqrt{2 \times 9.81 \times 50 \times 10^{-3}}$

$= 0.99$ m/s **Ans.** (b)

317. $\frac{p_1}{\gamma} + \frac{v_1^2}{2g} + z_1 = \frac{p_2}{\gamma} + \frac{v_2^2}{2g} + z_2$

$32 + 4 + 0 = 0 + \frac{v_2^2}{2g} + 0$

$\frac{v_2^2}{2g} = 36$ and $\frac{v_2^2}{2g} = 4$

$\left(\frac{v_2}{v_1}\right)^2 = \frac{36}{4} = 9, \frac{v_2}{v_1} = 3$ **Ans.** (c)

318. $h_f = \frac{H}{3}$ **Ans.** (b)

319. $h_f = \frac{flv^2}{2gd} \propto \frac{v^2}{d}$

$Q = \frac{\pi}{4} d^2 v,\ v \propto \frac{Q}{d^2}$

$h_f \propto \frac{Q^2}{d^5}$

$\frac{Q_1}{Q_2} = \left(\frac{d_1}{d_2}\right)^{5/2} = \left(\frac{15}{10}\right)^{5/2} = 2.756$ **Ans.** (b)

320. For critical flow, Froude number,

$F = \frac{v}{\sqrt{gD}} = 1$

$v = \frac{Q}{A} = \frac{Q}{DT} = \frac{12}{4D} = \frac{3}{D}$

$\frac{3}{D\sqrt{9.81D}} = 1$

$D^{3/2} = 0.9578$

$D = 0.972$ m **Ans.** (c)

322. $U = U_{\max}\left(1 - \frac{r^2}{R^2}\right) = 2U_{avg}\left(1 - \frac{r^2}{R^2}\right)$

$$= 2\times5\left[1-\left(\frac{5}{10}\right)^2\right] = 7.5\text{m/s}$$ **Ans.** (a)

323. $U_{avg} = 0.817\ U_\infty$

$$\frac{U_\infty}{U_{avg}} = \frac{1}{0.817} = 1.22$$ **Ans.** (c)

324. $\frac{dp}{dx} = \frac{50}{10} = 5\,\text{kPa/m}$

$$\tau_0 = \frac{dp}{dx}\times\frac{R_0}{2} = \frac{5\times50\times10^{-3}}{2} = 0.125\,\text{kPa}$$

Ans. (b)

325. Displacement thickness,

$$\delta^* = \int_0^\delta\left(1-\frac{u}{U_\infty}\right)dy$$

$$= \int_0^\delta\left(1-\frac{y}{\delta}\right)dy = \left|y-\frac{y^2}{2\delta}\right|_0^\delta = \frac{\delta}{2}$$

Momentum thickness, $\theta = \int_0^\delta \frac{u}{U_\infty}\left(1-\frac{u}{U_\infty}\right)dy$

$$= \int_0^\delta \frac{y}{\delta}\left(1-\frac{y}{\delta}\right)dy = \int_0^\delta\left(\frac{y}{\delta}-\frac{y^2}{\delta^2}\right)dy$$

$$= \left|\frac{y^2}{2\delta}-\frac{y^3}{3\delta^2}\right|_0^\delta = \frac{\delta}{2}-\frac{\delta}{3} = \frac{\delta}{6}$$

$$\frac{\delta^*}{\theta} = \frac{\delta}{2}\times\frac{6}{\delta} = 3$$ **Ans.** (c)

326. $\text{Re}_x = \frac{vx}{v} = \frac{2\times1}{2\times10^{-5}} = 10^5$

$$\theta = \frac{0.664x}{\sqrt{\text{Re}_x}} = \frac{0.664\times1}{\sqrt{10^5}}$$

$= 2.1\times10^{-3}$ m or 2.1 mm **Ans.** (b)

328. $A-2, B-3, C-1, D-4$ **Ans.** (a)

322. $A-4, B-3, C-2, D-1$ **Ans.** (b)

330. $\psi = xy$

$$u = \frac{\partial\psi}{\partial y} = x,\ v = -\frac{\partial\psi}{\partial x} = -y$$

$$u = -\frac{\partial\phi}{\partial y} = -x,\ \phi = -\frac{x^2}{2}+c_1$$

$$v = -\frac{\partial\phi}{\partial y} = -y,\ \phi = \frac{y^2}{2}+c_2$$

$$\therefore\quad \phi = \frac{1}{2}(y^2-x^2)$$ **Ans.** (b)

331. For same Froude number

$$v_m = v_p\sqrt{l_m/l_p} = 10\sqrt{\frac{1}{25}} = 2\text{m/s}$$

Ans. (a)

332. $Q \propto H^{3/2}$

$$\frac{Q_2}{Q_1} = \left(\frac{H_2}{H_1}\right)^{3/2} = 4^{3/2} = 8$$ **Ans.** (d)

333. $\frac{1}{2}\pi r^2 z = \frac{1}{2}\pi r^2 h$

$$z = h$$

$\therefore$ Depth of liquid at the centre of cylinder is zero. Therefore, pressure shall be equal to atmospheric pressure. **Ans.** (d)

334. $(3+3+3)\,\rho hg = 3\,\rho hg$

$9\,\rho gh = 3\,\rho gh$

$H/h = 3$ **Ans.** (c)

341. $\left(\frac{vl}{\nu}\right)_p = \left(\frac{vl}{\nu}\right)_m$

$$v_m = v_p\left(\frac{l_p}{l_m}\right)\left(\frac{\nu_m}{\nu_p}\right)$$

$$= 12 \times 20 \times \frac{1}{2} = 120 \text{ m/s}$$ **Ans.** (d)

342. Force = $f(l, v, \rho, \mu, h, g)$

$$\text{Re} = \frac{vl}{\nu}, Fr = \frac{v^2}{gl}$$ **Ans.** (c)

343. $A - 2, B - 2, C - 1, D - 2$

In subsonic, adiabatic flow in pipe with friction in flow direction, temperature increases and p, ρ, v, decrease. **Ans.** (c)

345. $A - 1, B - 4, C - 3, D - 2$ **Ans.** (b)

346. Capillarly rise for clear water,

$$h \simeq \frac{30}{d} \text{mm} = \frac{30}{1} = 30 \text{ mm}$$ **Ans.** (d)

347. $\Delta p = \rho g h = 2000 \times 9.81 \times 0.1$

$= 1962 \text{ N/mm}^2$ **Ans.** (c)

348. $A - 2, B - 1, C - 4, D - 3$ **Ans.** (d)

349. $F_h = \gamma g A \bar{x} = 1000 \times 9.81 \times 20 \times 1 \times 10$

$= 1.962 \times 10^6$ N **Ans.** (c)

350. $\phi = x^2 - y^2$

$$u = -\frac{\partial \phi}{\partial x} = -2x, v = -\frac{\partial \phi}{\partial y} = 2y$$

$$\frac{\partial \psi}{\partial y} = u = -2x, \psi = -2xy + \text{const}$$

$$-\frac{\partial \psi}{\partial y} = v = 2y, \psi = -2xy + \text{const}$$

∴ Stream functin, $\psi = -2xy + \text{const}$

Ans. (a)

352. $A - 3, B - 2, C - 1, D - 4$ **Ans.** (b)

353. $A - 4, B - 1, C - 2, D - 3$ **Ans.** (d)

354. $\tau = \mu \frac{du}{dy} = 8.60(2 - 2y)$

$= 8.60 (2 - 2 \times 0.15)$

$= 14.62 \text{ N/m}^2$ **Ans.** (c)

355. $A - 3, B - 4, C - 2, D - 1$ **Ans.** (c)

358. $\frac{1}{2}\rho U^2 = p_{\text{stag}} - p_{\text{static}}$

$$U = \sqrt{\frac{380 \times 2}{1.2}} = 25.17 \text{ m/s}$$ **Ans.** (d)

367. Height of water column in U-tube

$$h = \frac{Q}{\frac{\pi}{4}d^2} = \frac{4 \times 19}{\pi \times 64 \times 10^{-2}}$$

$= 37.8$ cm of water

$$= \frac{37.8}{13.6} = 2.78 \simeq 2.8 \text{cm of } Hg$$

Difference in mercury level in two limbs = 2.8 cm

Ans. (c)

370. $$\delta^* = \int_0^\delta \left(1 - \frac{y}{\delta}\right) dy$$

$$= \left| \delta = \frac{y^2}{2\delta} \right|_0^\delta = \frac{\delta}{2}$$

$$\frac{\delta^*}{\delta} = \frac{1}{2}$$ **Ans.** (c)

376. d = 12 cm, $\frac{v^2}{2g} = 0.5$m, slope = $\frac{1}{10}$

Length of the pipe for 1: 10 slope,

$$l = \sqrt{1 + 10^2} = 10.05 \text{m}$$

$$h_f = \frac{fl}{d} \times \frac{v^2}{2g}$$

$$1 = \frac{f \times 10.05}{0.12} \times 0.5$$

$f = 0.024$ **Ans.** (c)

377. Since the EGL rises at the location of the box, therefore there is a pump at this location.

Ans. (a)

378. For pipes in series

$$\frac{l_e}{d_e^5} = \frac{l_1}{d_1^5} + \frac{l_2}{d_2^5}$$

For $d_e = d_1 = d_2 = d$ and $l_1 = l_2 = 1$

$(l_e)_s = 2l$

For pipes in parallel, $h_{f1} = h_{f2}$

$$\frac{flQ_1^2}{3d^5} = \frac{flQ_2^2}{3d^5}$$

$\therefore \quad Q_1 = Q_2$

Also $\quad Q = Q_1 + Q_2$

$\because \quad Q_1 = Q_2 = \dfrac{Q}{2}$

$$\frac{fl_eQ^2}{3d^5} = \frac{fl(Q/2)^2}{3d^5}$$

$$(l_e)_p = \frac{1}{4}$$

$$\frac{L_1}{L_2} = \frac{(l_e)_s}{(l_e)_p} = \frac{2l \times 4}{l} = 8$$ **Ans.** (b)

380. $u = Ae^x$

Continuity equation: $\dfrac{\partial u}{\partial x} + \dfrac{\partial v}{\partial y} = 0$

$$Ae^x + \frac{\partial v}{\partial y} = 0$$

$$\frac{\partial v}{\partial y} = -Ae^x$$

Integrating, we get :

$$v = \int -Ae^x dy + f(x)$$

$$= -Ae^x y + f(x)$$ **Ans.** (c)

385. $\dfrac{p}{\gamma}(h + \Delta h)s - (h_1 - \Delta h)s_1 = \dfrac{p_a}{\gamma}$

$$\frac{p - p_a}{\gamma} = (h_1 + \Delta h)s_1 - (h + \Delta h)s$$

$$= h_1 s_1 - hs + \Delta h(s_1 - s)$$ **Ans.** (a)

386. $A - 3, B - 4, C - 2, D - 1$ **Ans.** (c)

387. For irrotational flow, $\dfrac{\partial v}{\partial x} - \dfrac{\partial u}{\partial y} = 0$

(b) $\psi = 2xy, u = \dfrac{\partial \psi}{\partial y} = 2x, v = -\dfrac{\partial \psi}{\partial x} = -2y$

$$\frac{\partial v}{\partial x} = 0, \frac{\partial u}{\partial y} = 0$$ **Ans.** (b)

391. For a venturimeter,

$$Q = C_d \left[\frac{A_1 A_2}{\sqrt{A_1^2 - A_2^2}}\right]\sqrt{2gH}$$

For orifice meter, $Q = C_d A\sqrt{2gH}$

$\therefore \quad H \propto \dfrac{1}{C_d^2}$ for both.

$$\frac{H_v}{H_0} = \left(\frac{C_{d0}}{C_{dv}}\right)^2 = \left(\frac{0.61}{0.98}\right)^2$$ **Ans.** (b)

398. $A - 3, B - 1, C - 4, D - 2$ **Ans.** (c)

400. $A - 3, B - 5, C - 2, D - 1$ **Ans.** (d)

402. Dynamic loss coefficient in sudden contraction $= \left(\dfrac{1}{C_c} - 1\right)^2$

$$= \left(\frac{1}{0.62} - 1\right)^2 = 0.375$$ **Ans.** (b)

403. 1 Stoke = 10^{-4} m²/s

$$\nu = \frac{\mu}{\rho} \quad \text{or } \mu = \rho\nu$$

$\therefore$ 0.28 Stokes = 0.28 × 10–4 × 0.9 × 10^3

$= 0.0252$ N.s/m² **Ans.** (c)

404. Difference in pressure head **Ans.** (b)

$$h = h_g\left(\frac{\rho_m}{\rho_w} - 1\right) = 0.5(13.6 - 1) = 6.3\,\text{m}$$

405. $p + \rho_1 h_1 = p + \rho_2 h + \rho_3(h_1 - h)$

$$h = h_1\left[\frac{\rho_1 - \rho_3}{\rho_2 - \rho_3}\right]$$ **Ans.** (c)

406. Pressure force on the plate face,

$$F = \gamma A h_c = 9.81 \times 10^3 \times \frac{\pi}{4}[(1.2)^2 - (0.8)^2] \times 1.625$$

$= 10016$ N $\simeq 10$ kN **Ans.** (b)

407. The total force on one face of plate will remain the same. **Ans.** (b)

408. $$\frac{dp}{dz} = \gamma\left(1 + \frac{a_z}{g}\right)$$

$$dp = \gamma\left(1 + \frac{a_z}{g}\right)dz$$

$$= 0.8 \times 9.81 \times 10^3\left[1 + \frac{4.9}{9.81}\right] \times 2.5$$

$= 29.4$ kPa **Ans.** (d)

409. Let h =height of water level in the tank above the intersection of jets. Horizontal velocity of jets emanating from the orifices,

$$v_1 = \sqrt{2g(h-2)}, v_2 = \sqrt{2g(h-0.5)}$$

Horizontal distances travelled by jets in time t,

$$x_1 = v_1 t_1, x_2 = v_2 t_2$$

Vertical distance travelled by jets,

$$z_1 = \frac{1}{2}g t_1^2 = \frac{x_1^2}{4(h-2)}$$

$$z_2 = \frac{1}{2}g t_2^2 = \frac{x_2^2}{4(h-0.5)}$$

$z_1 = 2$ m

or $$\frac{x_1^2}{4(h-2)} = 2$$

or $x_1^2 = 8\ (h - 2)$

Similarly $x_2^2 = 2\ (h - 0.5)$ for $z_2 = 0.5$ m

For $x_1 = x_2$

$8(h - 2) = 2(h - 0.5)$

$h = 2.5$ m **Ans.** (b)

410. $u = t^2 + 3y, v = 3t + 3x$

$$a_x = \frac{du}{dt} + u\frac{du}{dx} + v\frac{du}{dy} = 2t + 3(3t + 3x) = 49$$

$$a_y = v + \frac{dv}{dt} + u\frac{dv}{dx} + v\frac{dv}{dy} = 3 + 3(t^2 + 3y) = 33$$

$$a = [a_x^2 + a_y^2]^{0.5} = [49^2 + 33]^{0.5} = 59\,\text{m/s}^2$$

Ans. (c)

411. $\psi = x^3 - y^3$

$$u = \frac{\partial \psi}{\partial y} = -3y^2 = -3(-1)^2 = -3$$

$$v = -\frac{\partial \psi}{\partial x} = -3x^2 = -3 \times 1^2 = -3$$

Resultant velocity,

$$V = [u^2 = v^2]^{0.5}[9 + 9]^{0.5} = 4.24\,\text{m/s}$$

Ans. (a)

413. $$h_f = \frac{flv^2}{2gd} = \frac{0.025 \times 500 \times v^2}{2 \times 9.81 \times 0.2} = 3.185\,v^2$$

$$v = \frac{Q}{A} = \frac{4Q}{\pi d^2} = \frac{4Q}{\pi \times 0.2^2} = 31.831Q$$

$h_f = 3227.5\,Q^2$

For maximum power transmission,

$$h_f = \frac{H}{3}$$

$$3227.5\,Q^2 = \frac{H}{3}, H = 9682.5\,Q^2$$

Velocity of jet,

$$v = \left[2g\left(H - \frac{H}{3}\right)\right]^{0.5} = \left[\frac{4gH}{3}\right]^{0.5}$$

or $$H = \frac{3v^2}{4g} = \left[\frac{4Q}{\pi d_n^2}\right]^2 \times \left[\frac{3}{4g}\right]$$

$$96825.5\ Q^2 = \left[\frac{16Q^2}{\pi^2 d_n^4}\right] \times \left[\frac{3}{4g}\right]$$

$$d_n^4 = \frac{16}{3227.5 \times 4 \times 9.81 \times \pi^2} = 12.8 \times 10^{-6}$$

$d_n = 59.8$ mm **Ans.** (d)

414.

(*i*) $\frac{\partial\phi}{\partial x} = 2, \frac{\partial\phi}{\partial y} = 3; \frac{\partial^2\phi}{\partial x^2} = 0 \frac{\partial^2\phi}{\partial y^2} = 0, \nabla^2\phi = 0$

(*ii*) $\frac{\partial\phi}{\partial x} = 4x, \frac{\partial\phi}{\partial y} = -6y; \frac{\partial^2\phi}{\partial x^2} = 4, \frac{\partial^2\phi}{\partial y^2} = -6, \nabla^2\phi \neq 0$

(*iii*) $\frac{\partial\phi}{\partial x} = -\sin(x-y), \frac{\partial\phi}{\partial y} = \sin(x-y)$

$$\frac{\partial^2\phi}{\partial x^2} = -\cos(x-y), \frac{\partial^2\phi}{\partial y^2} = -\cos(x-y), \nabla^2\phi \neq 0$$

(*iv*) $\frac{\partial\phi}{\partial x} = \frac{y}{(x^2+y^2)}, \frac{\partial\phi}{\partial y} = -\frac{x}{(x^2+y^2)}$

$$\frac{\partial^2\phi}{\partial x^2} = \frac{-2xy}{(x^2+y^2)^2}, \frac{\partial^2\phi}{\partial y^2} = \frac{2xy}{(x^2+y^2)^2}, \nabla^2\phi = 0$$

Therefore, 1 and 4 are velocity potentials.

Ans. (b)

415. Wall shear stress,

$$\tau_0 = \left(\frac{d}{4}\right)\left(\frac{dp}{L}\right) = \left(\frac{600 \times 10^{-3}}{4}\right)\left(\frac{70 \times 10^3}{30}\right)$$

= 350 kPa **Ans.** (b)

416. For same pressure,

$$\rho_a g h_a = \rho_w g h_w$$

$$h_a = \frac{9.81 \times 10^3 \times 75 \times 10^{-3}}{1.16 \times 9.81}$$

= 64.65 m of air

$v_a = [2gh_a]^{0.5} = [2 \times 9.81 \times 64.65]^{0.5}$

=35.6 m/s **Ans.** (c)

417. $A-4, B-3, C-2, D-1$ **Ans.** (b)

418. $A-3, B-5, C-2, D-1$ **Ans.** (c)

419. Laminar boundary layer thickness on a flat plate,

$$\delta = 5\left[\frac{\nu x}{U}\right]^{0.5}$$

$$= 5\left[\frac{16 \times 10^{-6} \times 0.4}{\frac{30 \times 10^3}{3600}}\right] = 4.38$$ **Ans.** (b)

422. $A-4, B-3, C-2, D-1$ **Ans.** (c)

409. Wall shear stress,

$$\tau_0 = \left(\frac{d}{4}\right)\left(\frac{dp}{L}\right)$$

$$\frac{dp}{L} = \frac{128\,Qu}{\pi d^4}$$

$$= \frac{128 \times 3 \times 10^3 \times 1.02 \times 10^5}{\pi \times 4^4 \times 60}$$

= 0.0812 N/m³

$$\tau_0 = \frac{4 \times 0.0812}{4} = 0.0812 \text{ dyne/cm}^3$$

Ans. (b)

425. $(Fr)_m = (Fr)_p$

$$\left(\frac{v^2}{gL}\right)_m = \left(\frac{v^2}{gL}\right)_p$$

$$\frac{v_p}{v_m} = \left(\frac{L_p}{L_m}\right)^{0.5} = \left(\frac{64}{1}\right)^{0.5} = 8$$

$v_p = 8 \times 81 = 648$ cm/s or 6.48 m/s

Ans. (b)

426. $(Fr)_m = (Fr)_p$ or $\left(\frac{v^2}{gL}\right)_m = \left(\frac{v^2}{gL}\right)_p$

$$\frac{L_m}{L_p} = \left(\frac{v_m}{v_p}\right)^2$$

Also $(Re)_m = (Re)_p$ or $\left(\frac{vL\rho}{\mu}\right)_m = \left(\frac{vL\rho}{\mu}\right)_p$

Now, $\frac{\mu_m}{\mu_p} = 0.0894$

$$\frac{L_p}{L_m} = \left(\frac{v_m}{v_p}\right)\left(\frac{\mu_p}{\mu_m}\right) = \left(\frac{\mu_p}{\mu_m}\right)^{2/3}$$

$= (11.18)^{2/3} = 5:1$ **Ans.** (c)

427. $\frac{L_m}{L_p} = \frac{1}{25}, L_p = 100\,\text{m}, v_p = 10\,\text{m/s}$

$$\frac{v_m}{v_p} = \left[\frac{L_m}{L_p}\right]^{0.5}, v_m = 10\left[\frac{1}{25}\right]^{0.5} = 2\,\text{m/s}$$

Ans. (b)

428. $A - 3, B - 4, C - 2, D - 1$ **Ans.** (b)

429. $v_x = 2x, v_y = 2y, v_z = 0$

$$a_x = v_x \frac{\partial v_x}{\partial x} + v_y \frac{\partial v_x}{\partial y}$$

$$= -2x \times (-2) + (-2) \times (0) = 4x$$

$$a_y = v_x \frac{\partial v_y}{\partial x} + v_y \frac{\partial v_y}{\partial y}$$

$= -2x + x(0) + 2y \times 2 = 4y$ **Ans.** (d)

430. Statements 1 and 2 are correct **Ans.** (b)

431. $p_B = p_{atm}$, $p_C = \gamma h(1 - a_z/g) = \gamma h(1 - g/g) = 0$

Hence, $p_B > p_C$ **Ans.** (c)

432. Since the volume V of the solid is same, hence the displaced volume of the liquid shall remain the same even after overturning so that it shall still remain in stable equilibrium.

Ans. (a)

433. Weight of water displaced in the first case **Ans.** (c)

$= 15 \times 10^{-2} \times A \times 10^3 \times g$ N

Weight of water displaced in the second case

$= 20 \times 10^{-2} \times A \times 10^3 \times s \times g$ N

Since the weight of water displaced remains the same, therefore

Specific gravity of the oil, $s = \frac{15}{20} = 0.75$

434. Thrust on the semicircular area,

$$F = \left(\frac{\pi}{8}\right) d^2 p$$

$$= \left(\frac{\pi}{8}\right) \times 1 \times 420 = 164.93\,\text{kN}$$

Moment of thrust about the straight edge = Fx

Where x = distance of the centroid of the semicircular area from the straight edge.

$$= \frac{4r}{3\pi} = \frac{4 \times 0.5}{3\pi} = 0.212\text{m}$$

Moment of thrust,

$M = 164.93 \times 0.212 = 35$ kN. m.

Ans. (a)

435. Vertical force, $F_v = \gamma\left(\frac{\pi}{4} \cdot r^2 \times \text{unit length}\right)$

Horizontal force, $F_H = \gamma(r \times \text{unit length}) \times \frac{r}{2}$

$$\frac{F_V}{F_H} = \frac{\pi}{4}\frac{r^2}{(r^2/2)} = \frac{\pi}{2}$$ **Ans.** (b)

437. Initially the level of liquid in both the legs of the manometer is same as they are exposed to atmosphere. Let x = rise of the liquid in the vertical leg due to addition of 7.5 cm³ of water in the inclined leg of the manometer, cm.

$$p_a + \gamma s_1 \times (x + x \sin 30°) - \gamma \times 1 \times 15 \times \sin 30°$$

$$= p_a$$

$$1.25 \times 1.5x = 7.5$$

$$x = 4 \text{ cm}$$ **Ans.** (a)

438. Rise due to capillary action $= \frac{4\sigma \cos\theta}{\gamma d}$

where σ = surface tension,

θ = angle of contact,

d = capillary tube dia

y= specific weight of liquid

$$2 \times 10^{-3} = \frac{4 \times 0.06 \cos °0}{15 \times 10^3 \times d} = 1.189 \text{ kg/m}^3.$$

$$d = 8 \text{ mm}$$

Ans. (b)

439. $K = \frac{dp}{(dp/p)}$

$$= (3.5 - 3) \times \frac{500}{501 - 500} = 250\text{MPa}$$

Ans. (d)

441.

$$Q \propto \sqrt{p_b} \cdot p_{b1} = h_1(s_m - 1)$$

$$= 0.2\left(\frac{13.6}{0.8} - 1\right) = 3.2\text{m}$$

$$p_{b2} = h_2\left(1 - \frac{s_{air}}{0.8}\right)$$

From $pV = mRT$, $\rho_{air} = \frac{p}{RT}$

At room temperature,

$p = 1 \text{ bar} = 1 \times 10^5\ N/m^2$

$T = 273 + 20 = 293$ K

$R = 287$ Nm/kg. K

$$\rho_{air} = \frac{1 \times 10^5}{287 \times 293} = 1.189 \text{ kg/m}^3$$

Specify gravity of air,

$$s_{air} = \frac{\text{density of air}}{\text{density of water}}$$

$$= \frac{1.189}{1000} = 1.189 \times 10^{-3}$$

$$p_{b2} = h_2\left[1 - \frac{1.189 \times 10^{-3}}{0.8}\right]$$

$$= 0.9985\, h_2$$

$$\frac{Q_2}{Q_1} = \left[\frac{p_{b2}}{p_{b1}}\right]^{0.5}$$

$$\frac{0.08}{0.16} = \left[\frac{0.9985 h_2}{3.2}\right]^{0.5}$$

$$h_2 = 0.80 \text{ m or } 80 \text{ cm}$$ **Ans.** (c)

442. $C_d = C_c C_v$ **Ans.** (c)

443. Bernouli's equation can be used only outside the boundary layer as the flow in the boundary layer is rotational. **Ans.** (b)

444.

$$\delta^* = \int_0^\delta [1 - u/U]\,dy$$

$$= \int_0^\delta [1 - (y/\delta)^{1/6}]\,dy = \delta/7$$

Ans. (d)

445. Critical depth of right angled triangular channel = $(Q^2/g)^{1/3}$ **Ans.** (a)

446. $N_s = \dfrac{NQ^{1/2}}{H^{3/4}} = \dfrac{(1/T)(L^3/T)^{1/2}}{L^{3/4}} = L^{3/4}\,T^{3/2}$

Ans. (c)

450. Let h_1, h_2 = depth of hydrometre in oil and alcohol respecitively

$$\frac{\pi}{4} \times 9 \times h_1 \times 10^{-9} \times 0.075 \times 10^3 \times 9.81 = 0.03$$

$$h_1 = 576.8 \text{ mm}$$

and $\dfrac{\pi}{4} \times 9 \times h_2 \times 10^{-9} \times 0.80 \times 10^3 \times 9.81 = 0.03$

$$h_2 = 540.8 \text{ mm}$$

$$h_1 - h_2 = 576.8 - 540.8 = 36 \text{ mm}$$

Ans. (d)

451. Let h = depth upto which the wooden block is submerged in water

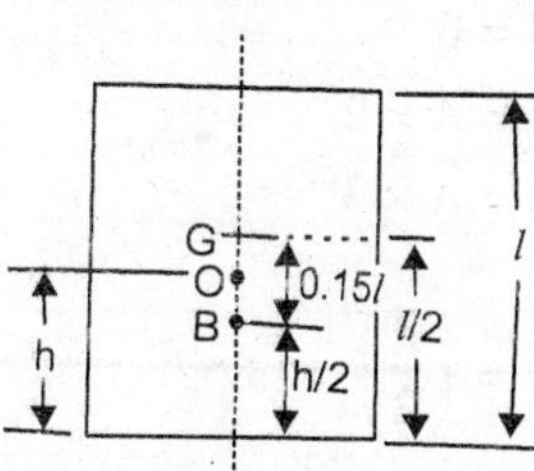

Fig. 39

A = Area of corss-section of wooden block

s = specific gravity of wooden block

Weight of block

$$= Als \times 10^3 \times 9.81 \text{ N}$$

Buoyant force

$$= Ah \times 10^3 \times 9.81 \text{ N}$$

For equilibrium of block

$$Als \times 10^3 \times 9.81 = Ah \times 10^3 \times 9.81$$

$$h = ls$$

$$\overline{BG} = \frac{l}{2} - \frac{h}{2} = \frac{l}{2} - \frac{ls}{2} = \frac{l}{2}(1-s)$$

$\therefore \quad \dfrac{l}{2}(1-s) = 0.15\,l,\ s = 0.7$ **Ans.** (c)

455. For a free cylindrical vortex, vr = const.

$$\therefore \quad v_2 r_2 = v_1 r_1$$

$$v_2 = \left(\frac{r_1}{r_2}\right) v_1$$

$$p_1 + \frac{1}{2}\rho v_1^2 = p_2 + \frac{1}{2}\rho v_2^2$$

$$p_1 - p_2 = \frac{1}{2}\rho(v_2^2 - v_1^2)$$

$$= \frac{1}{2}\rho v_1^2\left[\left(\frac{r_1}{r_2}\right)^2 - 1\right] \quad \text{...(i)}$$

For a forced cylinderical vortex, $v \propto r$

$$\therefore \quad \frac{v}{r} = \text{const,}$$

$$\frac{v_2}{r_2} = \frac{v_1}{r_1} \text{ or } v_2 = \left(\frac{r_2}{r_1}\right) v_1$$

$$\therefore \quad p_1 - p_2 = \frac{1}{2}\rho v_1^2\left[1 - \left(\frac{r_2}{r_1}\right)^2\right] \quad \text{...(ii)}$$

Equating Eqs. (i) and (ii), we get

$$1 - \left(\frac{r_1}{r_2}\right)^2 = \left[\frac{1}{2}\left(\frac{r_1}{r_2}\right)^2 - 1\right]$$

Let $\dfrac{r_2}{r_1} = m$

$$2\left(1 - \frac{1}{m^2}\right) = m^2 - 1$$

$$2 - \frac{2}{m^2} = m^2 - 1$$

$$2m^2 - 2 = m^4 - m^2$$

$$m^4 - 3m^2 + 2 = 0$$

$$m^2 = 3 \pm \frac{\sqrt{9 \pm 8}}{2} = \frac{3 \pm 1}{2} = 2$$

$$m = \sqrt{2}$$ **Ans.** (b)

457. For an irrotational flow, $\frac{\partial v}{\partial x} = \frac{\partial u}{\partial y}$

$$u = \frac{\partial \psi}{\partial y}, v = -\frac{\partial \psi}{\partial x}$$

$$\psi = y^2 - x^2$$

$$u = 2y, v = 2x$$

$$\frac{\partial v}{\partial x} = 2, \frac{\partial u}{\partial y} = 2$$

Hence $\psi = y^2 - x^2$ is irrotational **Ans.** (a)

458. $A - 2, B - 1, C - 4, D - 3$ **Ans.** (b)

461. $u_{max} = 2$ m/s, $Re = 800$

∴ Flow is laminar.

Average velocity in pipe

$$= \frac{\Delta p}{8\mu L} R_e^2$$

$$u_{max} = \frac{1}{4\mu}\left(\frac{dp}{dx}\right) R_e^2$$

$$u_{avg} = \frac{2}{2} = 1\, m/s$$

Ans. (c)

462. $A - 3, B - 4, C - 1, D - 2$ **Ans.** (d)

463. $C_d = f\left(\frac{d}{D}, \text{Re}\right)$ **Ans.** (c)

464. $v = \sqrt{2gh} = \sqrt{2 \times 9.81 \times 0.2}$

$= 1.98 = 2m/s.$

Ans. (b)

467. For equality of Froude number for model and prototype

Force ratio, $F_r = l_r^3 = 100^3 = 10^6$

Wave resistance on prototype

$= 10 \times 10^6 = 10{,}000$ kN

Ans. (c)

468. $m - n = 8 - 3 = 5$ **Ans.** (c)

472. $p = \rho gh = 10^3 \times 9.81 \times 0.1 = 981$ N/m^2

Ans. (c)

473. Total force on plate

$$F = \rho h_c A = 9.81 \times 10^3 \times (0.75 + 0.625) \times \frac{\pi}{4}(1.5)^2$$

$= 23837\text{N} \approx 24$ kN **Ans.** (c)

476. $\tau = A\left(\frac{du}{dy}\right)^n + B$

$$0 = 0 + 8$$

$$\therefore \quad B = 0$$

$$1.4 = A \times 2^n$$

$$2.6 = A \times 3^n$$

$$\frac{2.6}{1.4} = \left(\frac{3}{2}\right)^n = 1.857$$

$$n \ln 1.5 = \ln 1.857$$

$$0.405\, n = 0.619$$

$$n = 1.52 > 1$$

The fluid is dilatant. **Ans.** (d)

477. Equation of stream line is

$$\frac{dx}{u} = \frac{dy}{v} = \frac{dz}{w}$$

$$\frac{dx}{dy} = \frac{u}{v} = \frac{3xy}{\frac{3}{2}(x^2 - y^2)} = \frac{2xy}{x^2 - y^2}$$

Ans. (c)

478. $u = Ae^x$

Continuity equation:

$$\frac{\partial u}{\partial x} + \frac{\partial v}{\partial y} = 0$$

$$Ae^x + \frac{\partial v}{\partial y} = 0$$

$$\frac{\partial v}{\partial y} = -Ae^x$$

Integrating, we have

$$v = -Ae^x y$$ **Ans.** (c)

482. Displacement thickness

$$\delta^* = \int_0^\delta \left(1 - \frac{u}{U_x}\right) dy$$

$$= \int_0^\delta \left[\left(1 - \frac{y^{1/7}}{\delta}\right)\right] dy$$

$$= \left| y - \frac{7}{8} \times \frac{y^{8/7}}{\delta^{1/17}} \right|$$

$$= \delta - \frac{7}{8}\delta = \frac{\delta}{8}$$ **Ans.** (d)

484. Friction coefficient for laminar flow in pipe

$$f = \frac{64}{R_e}$$

$$R_e = \frac{vd}{\nu} = \left(\frac{4Q}{\pi d^2}\right) \times \frac{d}{\nu}$$

$$= \left(\frac{4Q}{\pi \nu}\right) \times \frac{1}{d}$$

or $\qquad \text{Re} \propto \frac{1}{d}$

Thus $f \propto d$

Frictional head loss

$$h_f = \frac{flv^2}{2gd}$$

$$= \frac{fl}{2gd}\left(\frac{4Q}{\pi d^2}\right)^2 \propto \frac{f}{d^5} \propto \frac{d}{d^5}$$

or $\qquad h_f \propto \frac{1}{d^4}$

Power loss, $\qquad P = \gamma Q h_f \propto \frac{1}{d^4}$

$\therefore \qquad n = -4$ **Ans.** (d)

485. Momentum thickness,

$$\therefore \quad \theta = \int_0^\delta \frac{u}{u_x}\left(1 - \frac{u}{U_x}\right) dy = \int_0^\delta \left[\frac{y}{\delta}\left(1 - \frac{y}{\delta}\right)\right] dy$$

$$= \int_0^\delta \left[\frac{y}{\delta} - \frac{y^2}{\delta^2}\right] dy = \left|\frac{y^2}{2\delta} - \frac{y^3}{3\delta^2}\right|_0^\delta = \frac{\delta}{6}$$

Displacement thickness,

$$\delta^* = \int_0^\delta \left(1 - \frac{u}{U_\infty}\right) dy$$

$$= \int_0^\delta \left[\left(1 - \frac{y}{\delta}\right)\right] dy = \left| y - \frac{y^2}{2\delta}\right|_0^\delta = \frac{\delta}{2}$$

$$\frac{\delta}{\delta^*} = \frac{\delta}{6} \times \frac{2}{\delta} = \frac{1}{3} \text{ or } 1:3$$ **Ans.** (d)

486. $A - 3, B - 5, C - 1, D - 2$ **Ans.** (b)

487. $A - 3, B - 1, C - 4, D - 2$ **Ans.** (d)

488. $\qquad \psi = 2xy$

$$u = \frac{\partial \psi}{\partial y} = 2x, v = \frac{\partial \psi}{\partial x} = -2y$$

At point (2, 1),

$$u = 4, v = -2$$

$$V = \sqrt{u^2 + v^2} = \sqrt{16 + 4}$$

$$= \sqrt{20} = 4.47 \text{unit}$$ **Ans.** (d)

489.

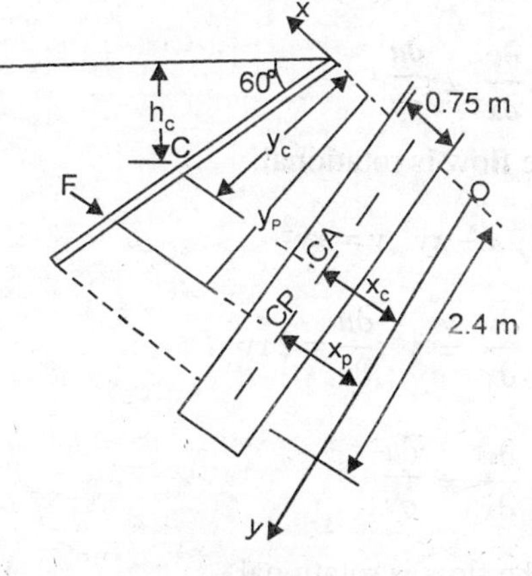

Fig. 40

$F = \gamma a y_c \sin\theta = \gamma A h_c$

$A = 0.75 \times 2.4 = 1.8 \text{ m}^2$

$y_c = \frac{2.4}{2} = 1.2 \text{ m}$

$F = 10^3 \times 9.81 \times 0.85 \times 1.8 \times 1.2 \sin 60°$
$= 15600 \text{ N or } 15.6 \text{ kN}$

490. Lowering of mercury column

$= \frac{s_k}{s_m} \times 40 = \frac{0.8}{13.6} \times 40 = 2.354 = 2.4 \text{ m}$

Ans. (a)

491. Capillary rise, $h = \frac{(4\cos\theta)\times\sigma}{\gamma d}$

or $h \propto \frac{1}{d}$ **Ans.** (c)

492. $\Delta p = \frac{8\sigma}{d} = \frac{8\times 0.1}{2\times 10^{-2}} = 40 \text{ N/m}^2$ **Ans.** (c)

495. For flow to be irrotational,

$\frac{\partial v}{\partial x} = \frac{\partial u}{\partial y}$

(A) $u = y, \quad v = \frac{-3}{2}x$

$\frac{\partial v}{\partial x} = -\frac{3}{2}, \frac{\partial u}{\partial y} = 1$

$\therefore \quad \frac{\partial v}{\partial x} \neq \frac{\partial u}{\partial y}$

The flow is rotational.

(B) $u - xy^2, v = xy^2$

$\frac{\partial v}{\partial x} = y^2, \frac{\partial u}{\partial y} = 2xy$

$\therefore \quad \frac{\partial v}{\partial x} \neq \frac{\partial u}{\partial y}$

The flow is rotational **Ans.** (a)

497. $u = 2x^2 + y^2, v = -4xy$

Convective accelation,

$a_{cx} = u\frac{\partial u}{\partial x} + v\frac{\partial u}{\partial y}$

$= (3x^2 + y^2)(4x) - 4x \times 2y$

$= 8x^3 + 4x^2 - 8xy^2 = 8x^3 - 4xy^2$

$(a_{cx})_{(1,2)} = 8 - 4 \times 1 = -8 \text{units}$ **Ans.** (c)

502. $35 \times y = 5 \times x$

$y = \frac{x}{3}$ **Ans.** (c)

503. $h = x\left(\frac{s_g - s_a}{s_a}\right) = x(s_g / s_a - 1)$ **Ans.** (d)

510. $\frac{dz}{dx} = \tan\theta = \frac{a_z}{g} = \tan 45° = 1$

$a_z = g$ **Ans.** (c)

512. $v_1 A_1 = n v_2 A_2$

$20 \times \frac{x}{4} \times (2\times 10^{-2})^2 = 100 \times \frac{\pi}{4}(2\times 10^{-3})^2 \times v_2$

$v_2 = 20 \text{ m/s}$ **Ans.** (d)

513. $Q_m = Q_p\left(\frac{l_m}{l_p}\right) = 4\left(\frac{1}{20}\right)^{5/2} = 0.0022$

Ans. (d)

515. $y_p = y_m\left(\frac{l_p}{l_m}\right) = 0.2 \times 20 = 4\text{m}$ **Ans.** (c)

517. (a) For a rotational flow, vorticity is not equal to zero, i.e., $\frac{\partial v}{\partial x} \neq \frac{\partial u}{\partial y}$.

(a) $\frac{\partial u}{\partial y} = 1, \frac{\partial v}{\partial x} = \frac{3}{2}$

Here, the flow is rotational **Ans.** (a)

SUBJECT INDEX